Edited by
Andrei K. Yudin

Catalyzed Carbon-Heteroatom Bond Formation

Further Reading

Nugent, T. C. (Ed.)

Chiral Amine Synthesis

Methods, Developments and Applications

2010

ISBN: 978-3-527-32509-2

Ricci, A. (Ed.)

Amino Group Chemistry

From Synthesis to the Life Sciences

2008

ISBN: 978-3-527-31741-7

Bäckvall, J.-E. (Ed.)

Modern Oxidation Methods

2011

ISBN: 978-3-527-32320-3

Bandini, M., Umani-Ronchi, A. (Eds.)

Catalytic Asymmetric Friedel-Crafts Alkylations

2009

ISBN: 978-3-527-32380-7

Bullock, R. M. (Ed.)

Catalysis without Precious Metals

2010

ISBN: 978-3-527-32354-8

Drauz, K., Gröger, H., May, O. (Eds.)

Enzyme Catalysis in Organic Synthesis

2011

ISBN: 978-3-527-32547-4

Royer, J., Husson, H. P. (Eds.)

Asymmetric Synthesis of Nitrogen Heterocycles

2009

ISBN: 978-3-527-32036-3

Dupont, J, Pfeffer, M. (Eds.)

Palladacycles

Synthesis, Characterization and Applications

2008

ISBN: 978-3-527-31781-3

Plietker, B. (Ed.)

Iron Catalysis in Organic Chemistry

Reactions and Applications

2008

ISBN: 978-3-527-31927-5

Edited by
Andrei K. Yudin

Catalyzed Carbon-Heteroatom Bond Formation

WILEY-VCH Verlag GmbH & Co. KGaA

The Editor

Prof. Andrei K. Yudin
University of Toronto
Department of Chemistry
St. George Street 80
Toronto, ON M5S 3H6
Canada

Library of Congress Card No.: applied for

British Library Cataloguing-in-Publication Data
A catalogue record for this book is available from the British Library.

Bibliographic information published by the Deutsche Nationalbibliothek
The Deutsche Nationalbibliothek lists this publication in the Deutsche Nationalbibliografie; detailed bibliographic data are available on the Internet at http://dnb.d-nb.de.

Composition Thomson Digital, Noida
Printing and Binding Fabulous Printers Pte. Ltd., Singapore
Cover Design Schulz Grafik-Design, Fußgönheim

Printed in Singapore
Printed on acid-free paper

ISBN: 978-3-527-32428-6

Contents

Catalyzed Carbon-Heteroatom Bond Formation. Edited by Andrei K. Yudin

ISBN: 978-3-527-32428-6

Preface

Metal catalyzed carbon-heteroatom bond forming processes constitute a vibrant area of research that continues to serve as an unmatched source of challenges. The cover of the book you hold in your hands provides a pictorial representation of a typical landscape in transition metal catalysis. The roads connecting the carbon center with heteroatoms depict catalyzed pathways. These roads are often indirect, they go via valleys and they climb over steep hills. There is almost always more than one way to connect the nodes on this map. Continuing effort in this important area is a testament to how difficult finding an optimal solution to a given bond forming reaction really is. I owe a great deal of gratitude to an outstanding cast of authors who wrote 11 outstanding chapters you will find in this book. I am grateful to these individuals for agreeing to participate in this important undertaking and for delivering superb and comprehensive chapters. I would also like to express gratitude to my students, Igor Dubovyk and Lawrence Cheung, for proof reading some of the chapters.

May 2010 *Andrei Yudin*
Toronto, Canada

Catalyzed Carbon-Heteroatom Bond Formation. Edited by Andrei K. Yudin

ISBN: 978-3-527-32428-6

List of Contributors

Valentine P. Ananikov
Russian Academy of Sciences
Zelinsky Institute of Organic Chemistry
Leninsky Prospect 47
Moscow 119991
Russia

Bruce A. Arndtsen
McGill University
Department of Chemistry
801 Sherbrooke Street
West Montreal, QC H3A 2K6
Canada

Irina P. Beletskaya
Lomonosov Moscow State University
Chemistry Department
Vorob'evy gory
Moscow 119899
Russia

Vy M. Dong
University of Toronto
Department of Chemistry
80 St. George Street
Toronto, Ontario
Canada M5S 3H6

Peter K. Dornan
University of Toronto
Department of Chemistry
80 St. George Street
Toronto, Ontario
Canada M5S 3H6

Alexander S. Dudnik
University of Illinois at Chicago
Department of Chemistry
845 West Taylor Street, 4500 SES,
M/C 111
Chicago, IL 60607-7061
USA

Matthew J. Fleming
Solvias AG
Chemical Development and
Catalysis Dept.
P.O. Box
4002
Basel
Switzerland

Valery V. Fokin
The Scripps Research Institute
Department of Chemistry
10550 North Torrey Pines Road
La Jolla, CA 92037
USA

Catalyzed Carbon-Heteroatom Bond Formation. Edited by Andrei K. Yudin

ISBN: 978-3-527-32428-6

Vladimir Gevorgyan
University of Illinois at Chicago
Department of Chemistry
845 West Taylor Street, 4500 SES,
M/C 111
Chicago, IL 60607-7061
USA

Béatrice Jacques
University of Strasbourg
Institut de Chimie, UMR 7177
4 Rue Blaise Pascal
67000 Strasbourg Cedex
France

Mark Lautens
University of Toronto
Department of Chemistry
Devenport Chemical Laboratories
80 St. George St.
Toronto, Ontario M5S 3H6
Canada

Hélène Lebel
Université de Montréal
Département de Chimie
2900 Boul Edouard Montpetit
Montreal, Quebec H3T, J4
Canada

Yingdong Lu
McGill University
Department of Chemistry
801 Sherbrooke Street
West Montreal, QC H3A 2K6
Canada

Duy H. Mai
University of Michigan
Department of Chemistry
930 N. University
Ann Arbor, MI 48109-1055
USA

Kilian Muñiz
ICIQ - Institut Català d'Investigació
Química
Avgda. Països Catalans 16
43007 Tarragona
Spain

Joshua D. Neukom
University of Michigan
Department of Chemistry
930 N. University
Ann Arbor, MI 48109-1055
USA

Feijie Song
Duke University
French Family Science Center
Durham, NC 27708-0346
USA

Ross A. Widenhoefer
Duke University
French Family Science Center
Durham, NC 27708-0346
USA

John P. Wolfe
University of Michigan
Department of Chemistry
930 N. University
Ann Arbor, MI 48109-1055
USA

Charles S. Yeung
University of Toronto
Department of Chemistry
80 St. George Street
Toronto, Ontario
Canada M5S 3H6

1
Synthesis of Saturated Five-Membered Nitrogen Heterocycles via Pd-Catalyzed C—N Bond-Forming Reactions

John P. Wolfe, Joshua D. Neukom, and Duy H. Mai

1.1
Introduction

Saturated five-membered nitrogen heterocycles, such as pyrrolidines, indolines, and isoxazolidines, appear as subunits in a broad array of biologically active and medicinally significant molecules [1]. As such, the synthesis of these compounds has been of longstanding interest. Many classical approaches to the construction of these heterocycles involve the use of C—N bond-forming reactions such as reductive amination, nucleophilic substitution, or dipolar cycloaddition for ring closure [2]. Although these methods have proven quite useful, their substrate scope and functional group tolerance is often limited.

In recent years, a number of powerful new transformations have been developed that involve the use of palladium-catalyzed C—N bond-forming reactions for construction of the heterocyclic ring [3]. These transformations frequently occur under mild conditions, tolerate a broad array of functional groups, and proceed with high stereoselectivity. In addition, the use of palladium catalysis allows for highly convergent multicomponent coupling strategies, which generate several bonds and/or stereocenters in a single process. This chapter describes recent approaches to the synthesis of saturated five-membered nitrogen heterocycles via Pd-catalyzed C—N bond forming reactions.

1.2
Pd-Catalyzed Amination of Aryl Halides

One of the most versatile and widely employed methods for the construction of aryl C—N bonds is the palladium-catalyzed cross coupling of amines with aryl halides and related electrophiles [4]. These reactions are believed to occur as shown in Scheme 1.1, with the coupling initiated by oxidative addition of the aryl halide to a Pd^0 complex. The resulting intermediate **1** is converted to a palladium(aryl)(amido) complex **2** through reaction with the amine substrate in the presence of base. Finally,

Catalyzed Carbon-Heteroatom Bond Formation. Edited by Andrei K. Yudin

ISBN: 978-3-527-32428-6

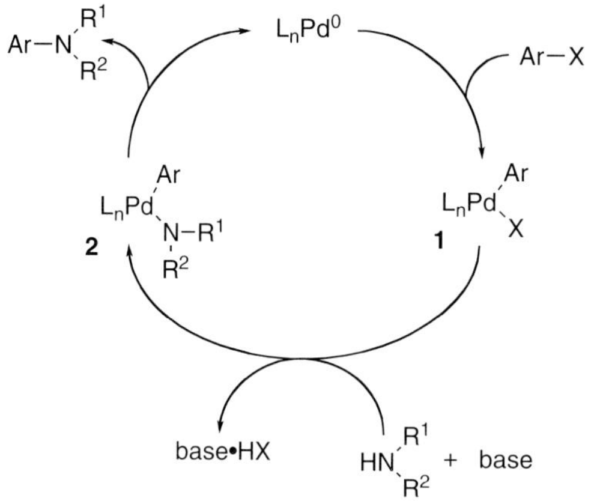

Scheme 1.1

C–N bond-forming reductive elimination affords the desired aniline derivative with concomitant regeneration of the palladium catalyst.

Although these reactions are most commonly used for intermolecular C–N bond formation, intramolecular versions of these reactions have occasionally been employed for the synthesis of saturated nitrogen heterocycles [5]. For example, Buchwald has described the synthesis of oxindoles and indolines through intramolecular reactions of aryl halides bearing pendant amines or amides (Eq. (1.1)) [6]. The conditions are amenable to the generation of indoline derivatives bearing amide, carbamate, or sulfonamide protecting groups. A two-flask sequence involving a four-component Ugi reaction followed by an intramolecular N-arylation that affords 3-amino oxindoles has also been developed (Eq. (1.2)) [7], and a number of other nitrogen heterocycles including ureas [8] and indolo[1,2-*b*]indazoles [9] have been prepared using this method.

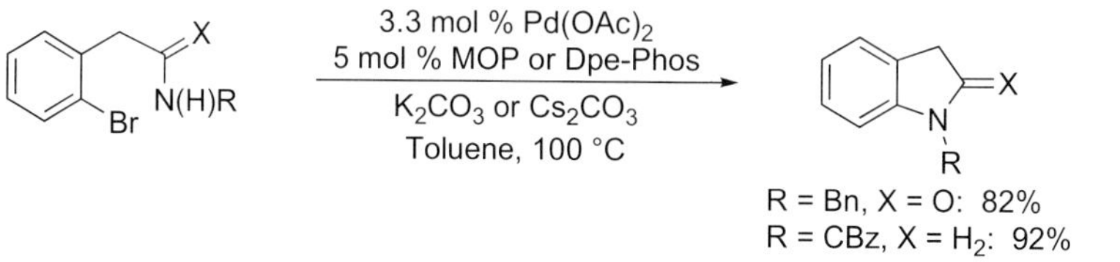

3.3 mol % $Pd(OAc)_2$, 5 mol % MOP or Dpe-Phos, K_2CO_3 or Cs_2CO_3, Toluene, 100 °C

R = Bn, X = O: 82%
R = CBz, X = H_2: 92%

(1.1)

CH_3CO_2H + *t*BuN; + $BuNH_2$

1) MeOH, rt
2) 5 mol % $Pd(dba)_2$, 5 mol % Me-Phos, K_2CO_3, Toluene, CH_3CN, μW, 100 °C

60%

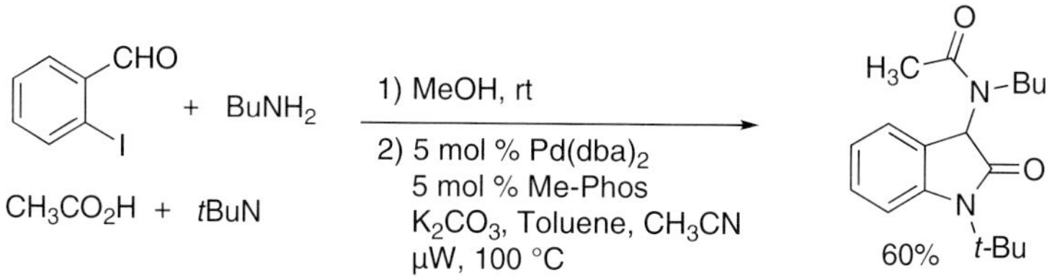

(1.2)

Intramolecular Pd-catalyzed or -mediated N-arylation reactions have been employed in the synthesis of several natural products [5]. For example, pyrroloindoline **4**, which represents the mitomycin ring skeleton was generated via the intramolecular N-arylation of **3** (Eq. (1.3)) [10]. Other targets generated using this strategy include asperlicin [11], the cryptocarya alkaloids cryptaustoline and cryptowoline [12], and the CPI subunit of CC-1065 [13].

TIPSO H N H OTf **3**
$Pd(OAc)_2$/BINAP
Cs_2CO_3, Toluene, 100 °C
44%
TIPSO H N **4**

(1.3)

A number of interesting one-pot or two-pot sequences of Pd-catalyzed reactions have been developed that involve intramolecular N-arylation processes [14]. For example, a two flask sequence of Negishi coupling followed by intramolecular C–N bond formation has been employed for the synthesis of substituted indolines (Eq. (1.4)) [14a]. Lautens has recently described an elegant one-flask sequence of intermolecular C–H bond functionalization followed by intramolecular N-arylation for the preparation of substituted indolines [14b]. As shown below (Eq. (1.5)), the Pd-catalyzed coupling of 2-iodotoluene with 2-bromopropylamine **5** in the presence of norbornene provided indoline **6** in 55% yield.

I Br + IZn Bn N H Boc
1) 3.3 mol % $Pd_2(dba)_3$
13.3 mol % P(o-tol)$_3$
DMF, rt, 39%
2) 3.3 mol % $Pd_2(dba)_3$
13.3 mol % P(o-tol)$_3$
Cs_2CO_3, Toluene, 100 °C, 74%
Bn N Boc

(1.4)

I + Br N H Ar **5**
Ar = *p*-nitrophenyl
10 mol % $Pd(OAc)_2$
22 mol % P(2-fur)$_3$
norbornene
Cs_2CO_3, DMF, 135 °C
55%
6 N Ar

(1.5)

1.3
Synthesis of Saturated Nitrogen Heterocycles via Alkene, Alkyne, or Allene Aminopalladation Reactions

A number of approaches to the synthesis of saturated five-membered nitrogen heterocycles involve alkene, alkyne, or allene aminopalladation as a key step [2b,g].

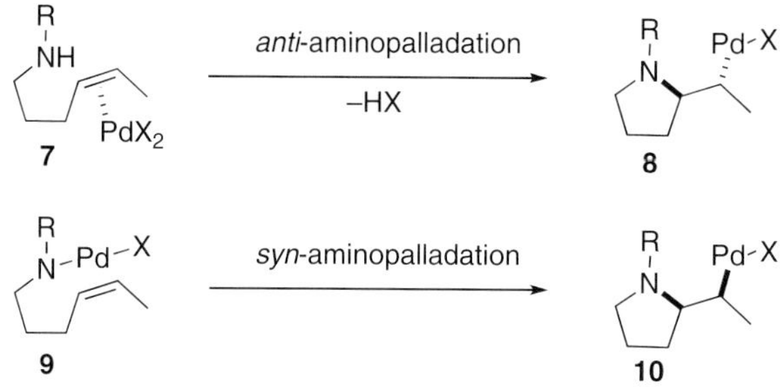

Scheme 1.2

The aminopalladation step can occur by either outer-sphere *anti*-aminopalladation or via inner-sphere *syn*-aminopalladation, and the mechanism can be dependent on substrate structure and reaction conditions. The *anti*-aminopalladation processes generally involve coordination of the unsaturated moiety to Pd^{II}, followed by external attack by a pendant nitrogen nucleophile (e.g., Scheme 1.2, **7** to **8**). In contrast, the *syn*-aminopalladations occur via formation of a palladium amido complex (e.g., **9**), which then undergoes migratory insertion of the alkene into the Pd–N bond to provide **10**. Heterocycle-forming reactions that proceed via aminopalladation of an unsaturated group can be broadly classified into four categories: (i) oxidative amination reactions of alkenes; (ii) hydroamination reactions of alkenes and alkynes; (iii) carboamination reactions of alkenes, alkynes, and allenes; and (iv) haloamination and diamination reactions of alkenes.

1.3.1
Pd^{II}-Catalyzed Oxidative Amination of Alkenes

The first examples of Pd-catalyzed oxidative amination reactions of alkenes were described by Hegedus in 1978 for the construction of indoles [15], and dihydropyrrole derivatives [16]. Although these reactions proceed in good yield with catalytic amounts of palladium, a stoichiometric amount of a co-oxidant, such as benzoquinone (BQ) or $CuCl_2$, was required to facilitate catalyst turnover. In recent years, several groups have explored the extension of this chemistry to the synthesis of saturated nitrogen heterocycles, with a focus on the use of O_2 as a mild, environmentally benign co-oxidant. Early advances in this area were reported independently by Larock and Andersson [17]. For example, treatment of **11** with a catalytic amount of $Pd(OAc)_2$ in the presence of O_2 in DMSO solvent afforded pyrrolidine **12** in 93% yield (Eq. (1.6)). The oxidative amination reactions are believed to proceed via either *syn*- or *anti*-aminopalladation to provide **13**, which then undergoes β-hydride elimination to afford the heterocyclic product. The Pd(H)X intermediate is converted to a Pd^0 complex via loss of HX, and is then subsequently re-oxidized to Pd^{II} by oxygen in the presence of DMSO. This method has also been employed for the generation of indolines and bicyclic pyrrolidines bearing sulfonyl or carbamate protecting groups (Eq. (1.7)) [17, 18].

11 N(H)Ts → 5 mol % $Pd(OAc)_2$, NaOAc, DMSO, O_2, 93% → 12 (Ts, N)

Pd^{II} → 13 (Ts, N, Pd-OAc) → 12

Pd(H)(OAc) —(–HOAc)→ Pd^0 —(O_2, DMSO)→ Pd^{II} (1.6)

P(H)N, P = Ts, CO_2Bn, n = 1, 2 → 10 mol % $Pd(OAc)_2$, DMSO, O_2, 55 °C, 90–96% → (H, H, N, P) (1.7)

In recent years, there has been a considerable focus on the development of new reaction conditions that use only molecular oxygen as the co-oxidant and do not require DMSO solvent [19]. Considerable progress has been made through the use of palladium catalysts supported by pyridine or N-heterocyclic carbenes as ligands. For example, Stahl has demonstrated that the 2-allylaniline derivative **14** is transformed to indoline **15** in 79% yield upon treatment with 5 mol% $IMesPd(TFA)_2$ and 10 mol% benzoic acid (Eq. (1.8)) [19d]. Stoltz has reported the conversion of amide **16** to lactam **17** under similar reaction conditions (Eq. (1.9)) [19b]. Through elegant mechanistic studies Stahl has shown that the stereochemistry of the aminopalladation step is dependent on reaction conditions, and both *syn*- and *anti*-aminopalladation mechanistic pathways are accessible in oxidative amination reactions [20].

14 (Cl, N(H)Ts) → 5 mol % $IMesPd(TFA)_2$, 10 mol % $PhCO_2H$, Toluene, O_2, 80 °C, 79% → 15 (Cl, N, Ts) (1.8)

16 (O, N(H)R) → 5 mol % $Pd(TFA)_2$, 20 mol % pyridine, MS3Å, Toluene, O_2, 80 °C → 17 (O, NR); R = Ts: 88%; R = OBn: 82% (1.9)

A related approach to the synthesis of nitrogen heterocycles also proceeds via Pd^{II}-catalyzed alkene aminopalladation, but involves substrates bearing allylic acetates or allylic hydroxy groups [21, 22]. In contrast to the oxidative amination reactions described above, these transformations are terminated by β-elimination of the acetate or hydroxy group (rather than β-hydride elimination). This approach alleviates the need for added oxidants, but does require the use of slightly more complex substrates. Nonetheless, this method is quite useful, and has been applied to the synthesis of several natural products [23]. In addition, a very interesting approach to the asymmetric synthesis of oxazolidinones involves treatment of tosylcarbamate **18** (generated *in situ* from the corresponding alcohol) with a catalytic amount of chiral Pd^{II} catalyst **21** (Eq. (1.10)) [24]. This reaction affords **19** in 81% yield and 91% ee by way of intermediate **20**.

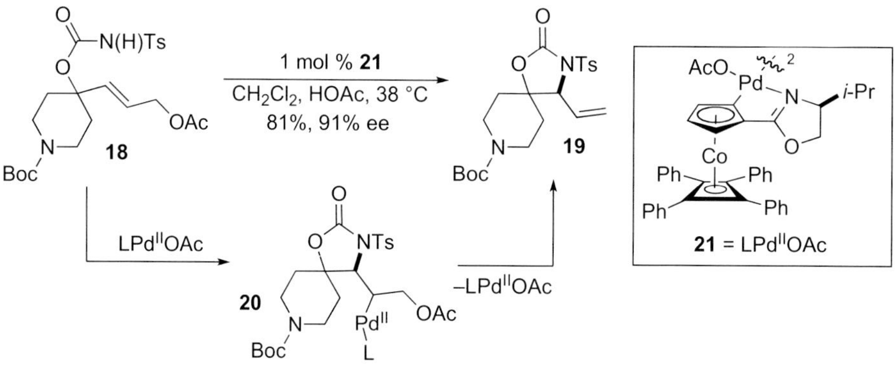

(1.10)

This strategy has also been employed for the synthesis of pyrrolidines [25]. For example, treatment of **22** with 15 mol% $PdCl_2(PhCN)_2$ afforded **23** in 77% yield as a single diastereomer (Eq. (1.11)) [25b]. The mild reaction conditions allow cyclization without epimerization of the amino ester stereocenter.

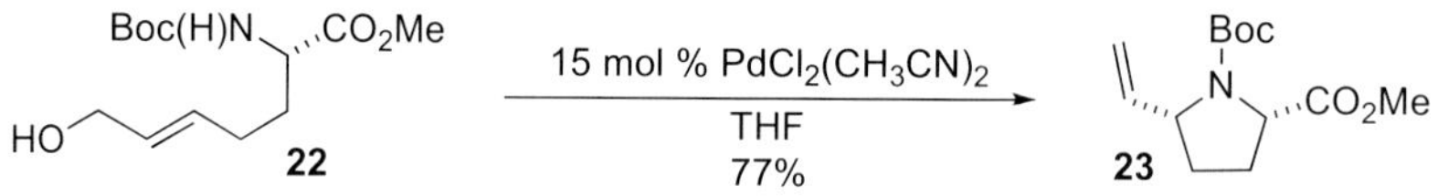

(1.11)

1.3.2
Pd-Catalyzed Hydroamination Reactions of Alkenes and Alkynes

The hydroamination of alkenes and alkynes has been of longstanding interest in organometallic chemistry [26]. Much of the early work in this area focused on early transition metal or lanthanide metal catalyst systems. However, much recent progress has been made in late-metal catalyzed hydroamination chemistry, and several interesting hydroamination reactions that afford nitrogen heterocycles have been developed using palladium catalysts.

Palladium-catalyzed intramolecular hydroamination reactions of alkynes that afford pyrrolidine derivatives were initially reported by Yamamoto in 1998 [27] and have been the subject of detailed investigation over the past ten years [28]. In a representative example, alkyne **24** was converted to **25** in 86% yield upon treatment with $Pd(PPh_3)_4$ as catalyst (Eq. (1.12)) [28c]. This transformation has been employed in the synthesis of the natural product indolizidine 209D [29], and asymmetric variants have also been developed that afford pyrrolidine products with up to 95% ee [30]. A related hydroamidation that affords lactam products has also been described [31], and hydroamination reactions of amines bearing tethered allenes are also known [32].

(1.12)

Although Pd-catalyzed intramolecular hydroamination reactions of alkynes have been known for ten years, analogous transformations of unactivated alkenes have only recently been developed [33]. Key to the success of these studies was the use of a cationic palladium complex bearing a pyridine-derived P–N–P pincer ligand (**29**). For example, treatment of **26** with catalytic amounts of **29**, $AgBF_4$, and $Cu(OTf)_2$ led to the formation of pyrrolidine **27** in 88% yield with 4 : 1 dr (Eq. (1.13)). Detailed mechanistic studies have indicated these transformations proceed via alkene coordination to the metal complex followed by outer-sphere aminopalladation to provide **28**. Protonolysis of the metal–carbon bond with acid generated *in situ* leads to formation of the product with regeneration of the active catalyst.

(1.13)

An interesting tandem intermolecular/intramolecular hydroamination reaction of cycloheptatriene with substituted anilines has been developed by Hartwig for the synthesis of tropene derivatives [34]. As shown in Eq. (1.14), the coupling of **30** with **31** provided **32** in 73% yield. The mechanism of this transformation is believed to involve acid-assisted formation of an η^3-pentadienylpalladium complex **33**, which is then captured by the aniline nucleophile to afford the allylpalladium intermediate **34**. Intramolecular attack of the aniline nitrogen on the allylpalladium moiety affords the observed heterocycle.

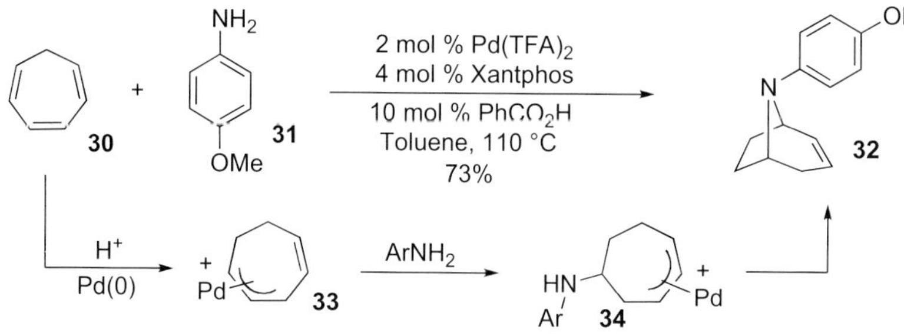

(1.14)

1.3.3
Pd^0-Catalyzed Carboamination Reactions of Alkenes

Over the past several years our group has been involved in the development of new Pd^0-catalyzed carboamination reactions between aryl or alkenyl halides and alkenes bearing a pendant nitrogen functionality [35, 36]. In a representative example, treatment of Cbz-protected amine **35** with 3-bromopyridine and a catalytic amount of Pd(OAc)$_2$/Dpe-Phos in the presence of Cs_2CO_3 afforded pyrrolidine **36** in 74% yield with >20 : 1 dr (Eq. (1.15)) [36e]. This method has been applied to a stereocontrolled synthesis of (+)-preussin [37], and is also effective with substrates bearing disubstituted alkenes (Eq. (1.16)) [36f]. The reactions appear to proceed via an unusual mechanism involving intramolecular *syn*-aminopalladation of a palladium(aryl)(amido) complex (e.g., **37**) followed by C–C bond-forming reductive elimination of the resulting intermediate **38**. Intramolecular variants of this transformation in which the aryl halide is appended to the alkene have also been described [38], and a one-flask tandem Pd-catalyzed N-arylation/carboamination reaction sequence has been developed for the conversion of primary amine substrates to *N*-aryl-2-arylmethyl indoline and pyrrolidine derivatives [39].

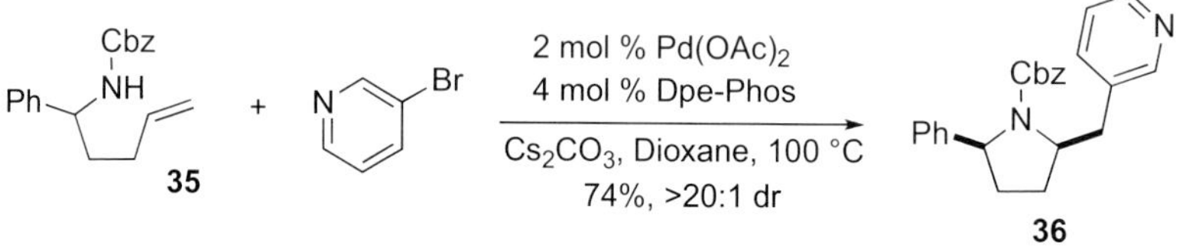

(1.15)

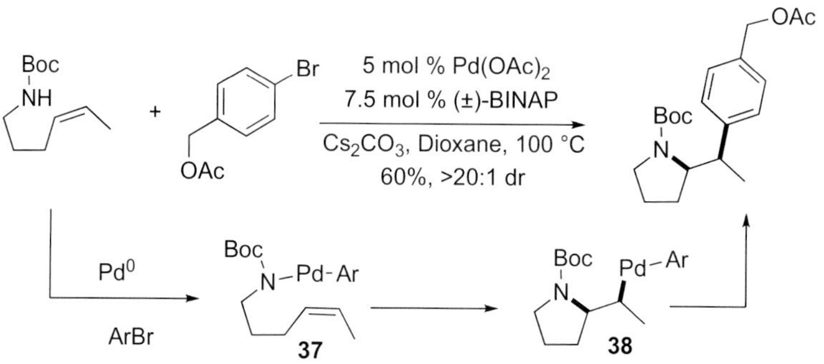

(1.16)

In addition to providing stereoselective access to substituted pyrrolidines, this method has been employed for the construction of a number of different nitrogen heterocycles including imidazolidin-2-ones (Eq. (1.17)) [40], and isoxazolidines [41]. A highly stereoselective synthesis of *cis*- and *trans*-3,5-disubstituted pyrazolidines has been developed in which the presence or absence of an N-1 protecting group controls product stereochemistry [42]. For example, treatment of **39** with 4-bromobiphenyl and a palladium catalyst in the presence of NaO*t*Bu affords the trans-disubstituted product **41** (Eq. (1.18)), whereas subjection of **40** to similar reaction conditions affords the cis-disubstituted product **42** (Eq. (1.19)).

1 mol % $Pd_2(dba)_3$
2 mol % Xantphos
NaO*t*Bu, Toluene, 110 °C
88%, 12:1 dr

(1.17)

2 mol % $Pd(OAc)_2$
2 mol %(±)-BINAP
NaO*t*Bu, Toluene, 110 °C
52%, >20:1 dr

39: R = Boc
40: R = H

2 mol % $Pd(OAc)_2$
2 mol % Dpe-Phos
NaO*t*Bu, Toluene, 70 °C
72%, 10:1 dr

(1.18, 1.19)

Balme has reported a one-pot three-component alkene carboamination between propargylic amines, alkylidene malonates, and aryl halides [43]. For example, treatment of *N*-methyl propargylamine (2 equiv), dimethyl benzylidene malonate (2 equiv) and 1,4-diiodobenzene (1 equiv) with *n*-BuLi and a palladium catalyst provided **43** as a single diastereomer (Eq. (1.20)) [43a]. The formation of the C−N bond in this process does not appear to be metal catalyzed. Instead, initial conjugate addition of the nitrogen nucleophile to the activated alkene affords a malonate anion, which undergoes carbopalladation followed by reductive elimination to afford the pyrrolidine product.

5 mol % $PdCl_2(PPh_3)_2$
*n*BuLi, DMSO, THF
54%

(1.20)

1.3.4
Pd^{II}-Catalyzed Carboamination Reactions of Alkenes

Two recent reports have described Pd^{II}-catalyzed carboamination reactions involving two alkenes that afford pyrrolidine products. Building on early work by Oshima that employed stoichiometric amounts of palladium [44], Stahl has developed an intermolecular Pd-catalyzed coupling of *N*-allylsulfonamide derivatives with enol ethers or styrene derivatives that affords substituted pyrrolidines in high yields with moderate diastereoselectivity [45]. For example, treatment of **44** with styrene in the presence of Pd^{II} and Cu^{II} co-catalysts, with methyl acrylate added for catalyst stability, provided **45** in 97% yield with 1.9: 1 dr (Eq. (1.21)). This reaction proceeds through intermolecular aminopalladation of styrene to afford **46**. Intramolecular carbopalladation then provides intermediate **47**, and subsequent β-hydride elimination yields product **45**.

5 mol % $Pd(OAc)_2$
5 mol % $Cu(TFA)_2$
0.5 equiv methyl acrylate
Toluene, air
97%, 1.9:1 dr
N(H)Ts 44 + Ph → TsN Ph 45
PdX_2 → TsN Ph PdX 46 → TsN Ph PdX 47 → −HPdX

(1.21)

Yang has reported a related tandem cyclization for the synthesis of pyrroloindoline derivatives that also proceeds though a mechanism involving alkene aminopalladation followed by carbopalladation of a second alkene [46]. As shown below, the 2-allylaniline derivative **48** was converted to **49** in 95% yield through treatment with a catalyst composed of $Pd(OAc)_2$ and pyridine (Eq. (1.22)). Use of (−)-sparteine as a ligand in this reaction provided **49** with up to 91% ee.

10 mol % $Pd(OAc)_2$
40 mol % Pyridine
Toluene, O_2
95%
O NH 48 → O N 49

(1.22)

1.3.5
Pd-Catalyzed Carboamination Reactions of Alkynes, Allenes, and Dienes

A few examples of Pd^0-catalyzed carboamination reactions between alkyne-tethered amines and aryl halides have also been reported [28d, 47]. For example, treatment of amino ester derivative **50** with PhI in the presence of K_2CO_3 using $Pd(PPh_3)_4$ as catalyst led to the formation of **51** in 80% yield with complete retention of enantiomeric purity (Eq. (1.23)) [28d]. In contrast to the Pd^0-catalyzed carboamination reactions of alkenes with aryl halides described above, the reactions of alkynes usually proceed via *anti*-aminopalladation, although products resulting from *syn*-aminopalladation have been obtained in some cases [48]. In addition to carboamination reactions that employ aryl halides as coupling partners, several transformations involving other electrophiles, such as acrylate derivatives, have been described (Eq. (1.24)) [49].

50 (>99% ee) + Ph–I → (10 mol % $Pd(PPh_3)_4$, K_2CO_3, CH_3CN, Bu_4NCl, 60 °C, 80%) → **51** (>99% ee) (1.23)

+ acrolein → (5 mol % $Pd(OAc)_2$, NaI, THF, 72%) → (1.24)

Several examples of Pd^0-catalyzed carboamination reactions between allenes and aryl or alkenyl halides have been reported [50]. For example, treatment of allene **52** with iodobenzene in the presence of K_2CO_3 and 2 mol% $Pd(PPh_3)_4$ afforded pyrrolidine **53** in 78% yield (Eq. (1.25)) [50a]. Mechanisms involving alkene aminopalladation (similar to the reactions of alkynes and alkenes noted above) have occasionally been invoked to explain these reactions. However, in many instances these transformations may involve intermediate π-allylpalladium complexes. Due to this mechanistic ambiguity, these transformations have been included in this section for comparison with the related reactions of alkenes and alkynes. Similar reactions involving allylic halides have also been described (Eq. (1.26)) [51].

52 + Ph–I → (2 mol % $Pd(PPh_3)_4$, K_2CO_3, DMF, 70 °C, 78%) → **53** (1.25)

→ (10 mol % $PdCl_2(PhCN)_2$, Et_3N, THF, 65%) → (1.26)

Cross-coupling carboamination reactions between allenes and 2-haloaniline derivatives or halogenated allylic amines have also been employed for the generation of substituted indolines, and use of an appropriate chiral catalyst for these transformations leads to formation of enantioenriched products [52]. For example, Larock has described the synthesis of indoline **56** via the Pd-catalyzed reaction of aryl iodide **54** with allene **55** (Eq. (1.27)) [52a]. The best asymmetric induction was obtained using chiral bisoxazoline ligand **57**. These reactions appear to proceed via intermediate π-allylpalladium complexes [53].

(1.27)

Ma has developed a three-component allene carboamination reaction for the stereoselective synthesis of 2,5-*cis*-disubstituted pyrrolidine derivatives [54]. A representative transformation involving allene **58**, 4-iodoanisole, and imine **59** that generates **60** in 90% yield is shown below (Eq. (1.28)). The reaction is believed to proceed through the intermediate π-allylpalladium complex **62**, which is formed by carbopalladation of the alkene to give **61** followed by addition of the malonate anion to the activated imine. Intramolecular capture of the allylpalladium moiety by the pendant nitrogen nucleophile affords the pyrrolidine product. A related asymmetric synthesis of pyrazolidines that employs azodicarboxylates as one of the electrophilic components has also been reported [55]. The pyrazolidine products are obtained with up to 84% ee when chiral bis oxazolines are employed as ligands.

(1.28)

An interesting Pd-catalyzed diene carboamination reaction that involves urea-directed C–H activation was recently reported [56]. For example, treatment of *N*-aryl urea **63** with an activated diene in the presence of 10 mol% $Pd(OAc)_2$, 50 mol% TsOH, Ac_2O, and benzoquinone provided **64** in 90% yield (Eq. (1.29)). The transformation is initiated by directed palladation of the arene by a palladium tosylate complex (formed *in situ*) to yield **65**. Carbopalladation of the diene generates allylpalladium complex **66**, which is then trapped by the urea to afford the observed product.

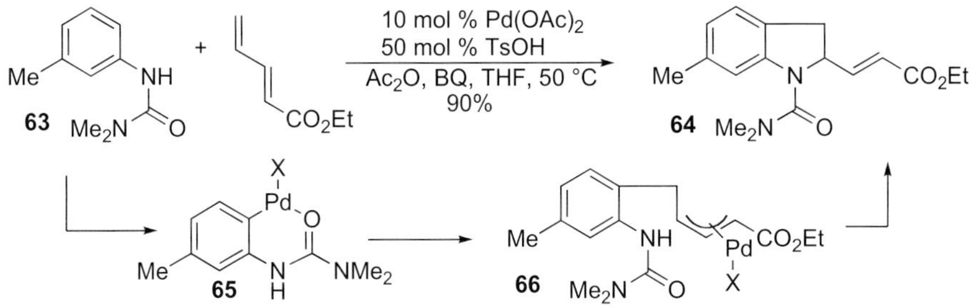

(1.29)

1.3.6
Vicinal Difunctionalization of Alkenes and Allenes

Reactions that effect addition of two heteroatoms across an alkene are very powerful methods for the generation of heterocycles, and have significant potential synthetic utility. Several important advances in this area that involve the use of palladium catalysts have recently been reported [57]. Interestingly, many of these may involve high oxidation state Pd^{IV} complexes as intermediates.

Palladium-catalyzed intramolecular aminobromination and aminochlorination reactions of alkenes have been employed for the conversion of unsaturated amides, carbamates, and sulfonamides to indolines and pyrrolidines [58]. As shown below (Eq. (1.30)), treatment of **67** with 10 mol% $Pd(TFA)_2$ in the presence of excess $CuBr_2$ or $CuCl_2$ affords **68** and **69** in moderate to good yields [58a]. In some cases superior results are obtained using $PdCl_2(CH_3CN)_2$ as the catalyst and NCS as the stoichiometric oxidant, as demonstrated in the conversion of **70** to **71** in 90% yield (Eq. (1.31)) [58b]. The alkene aminohalogenation reactions are believed to proceed via initial aminopalladation of the substrate followed by oxidative halogenation of the resulting alkylpalladium complex.

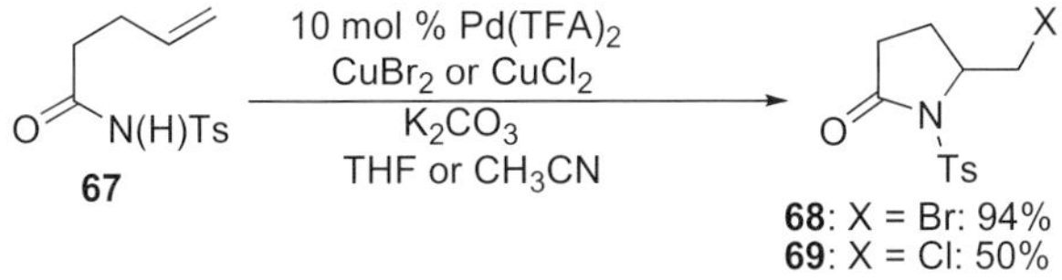

(1.30)

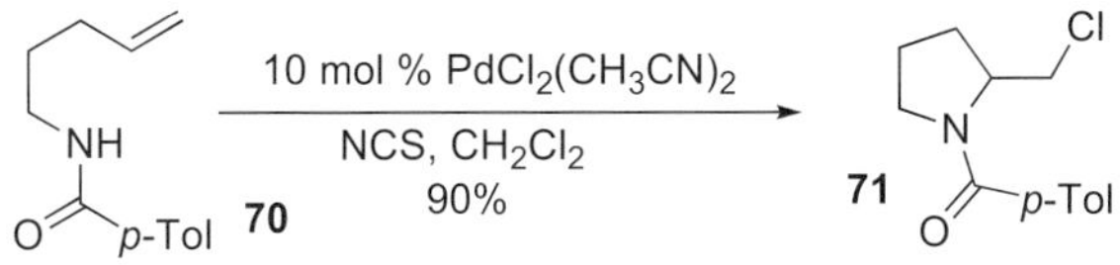

(1.31)

Related aminohalogenation reactions of allenes, such as the conversion of **72** to **73**, can be effected under similar reaction conditions (Eq. (1.32)) [59]. However, these latter transformations appear to proceed via a different mechanism involving allene bromopalladation followed by nucleophilic trapping of the resulting π-allylpalladium intermediate (e.g., **74**).

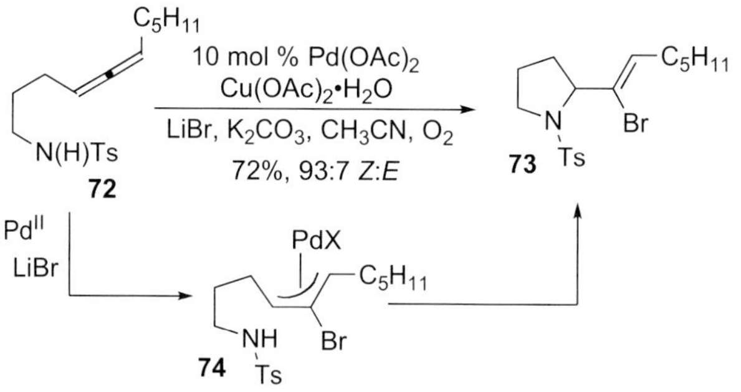

(1.32)

A very interesting Pd^{II}-catalyzed aminoacetoxylation reaction of alkenes was recently developed jointly by the Sorensen and Lee groups [60]. In a representative example, treatment of **75** with $Pd(OAc)_2$ and $PhI(OAc)_2$ provides oxazolidinone **76** in 65% yield with >20: 1 dr (Eq. (1.33)). This transformation is believed to proceed via a Pd^{II}/Pd^{IV} catalytic cycle that is initiated by *anti*-aminopalladation of the alkene to afford **77**. The intermediate Pd^{II} complex is then oxidized by $PhI(OAc)_2$ to alkyl Pd^{IV} intermediate **78**, which undergoes C–O bond-forming reductive elimination to afford **76**.

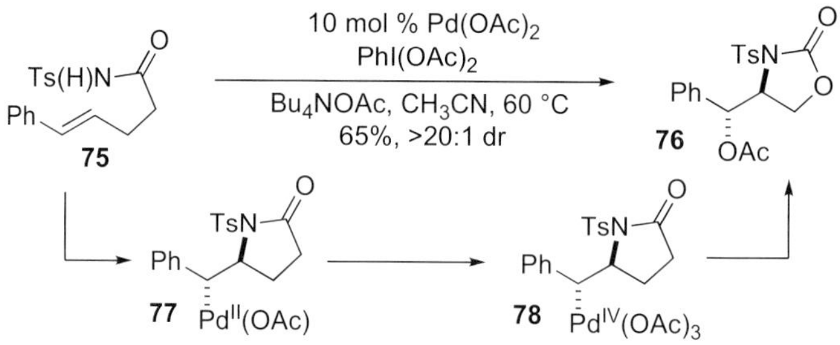

(1.33)

Three different approaches to the synthesis of five-membered cyclic ureas have recently been described that involve Pd-catalyzed alkene diamination reactions. In a series of very interesting papers, Muniz has described the conversion of alkenes bearing pendant ureas to imidizolidin-2-one derivatives using catalytic amounts of $Pd(OAc)_2$ in the presence of an oxidant such as $PhI(OAc)_2$ or $CuBr_2$ [61, 62]. For example, these conditions were used to effect the cyclization of **79** to **80** in 78% yield (Eq. (1.34)) [62a]. These reactions proceed via a mechanism similar to that shown above in Eq. (1.33), except that the heteropalladation may occur in a *syn*- rather than *anti*- fashion, and the reductive elimination occurs with intramolecular formation of a C–N bond rather than intermolecular formation of a C–O bond. The alkene diamination reactions have also been employed for the synthesis of bisindolines (Eq. (1.35)) [63] and bicyclic guanidines (Eq. (1.36)) [64].

79 → 80: 25 mol % $Pd(OAc)_2$, $PhI(OAc)_2$, CH_2Cl_2, 78% (1.34)

10 mol % $Pd(OAc)_2$, $PhI(OAc)_2$, Me_4NCl, NaOAc, DMF, 77% (1.35)

10 mol % $Pd(OAc)_2$, $CuBr_2$, K_2CO_3, DMF, 89% (1.36)

The intermolecular diamination of 1,3-dienes with acyclic ureas to provide monocyclic or bicyclic urea derivatives has been achieved by Lloyd-Jones and Booker-Milburn through use of a palladium catalyst combined with either benzoquinone or O_2 as an oxidant [65]. For example, diene **81** was converted to urea **82** in 82% yield (Eq. (1.37)). These transformations are mechanistically distinct from the reactions described above, and appear to involve intermediate π-allylpalladium complexes.

81 + EtNHC(O)NHEt → 82: 5 mol % $PdCl_2(CH_3CN)_2$, BQ, DME, 60 °C, 82% (1.37)

A much different strategy was developed by Shi for the conversion of 1,3-dienes or trienes to cyclic ureas [66]. As shown below, treatment of conjugated diene **84** with di-*t*-butyldiaziridinone **83** and in the presence of $Pd(PPh_3)_4$ as catalyst led to the formation of **85** in 94% yield with >20: 1 dr (Eq. (1.38)). This reaction is believed to occur via oxidative addition of **83** to Pd^0 to generate **86**, which undergoes aminopalladation to afford allylpalladium complex **87**. Reductive elimination from **87** affords the urea product. An asymmetric variant of this transformation that provides products with up to 95% ee has also been reported [67].

83 + 84 → 85: 10 mol % $Pd(PPh_3)_4$, C_6D_6, 65 °C, 94%, >20:1 dr; 83 → (L_nPd^0) → 86 → (diene) → 87 → 85 (1.38)

The scope of this chemistry has recently been extended to terminal alkene substrates [68]. For example, 1-hexene was transformed to **88** in 68% yield under solvent-free conditions using $Pd(PPh_3)_4$ as catalyst (Eq. (1.39)). Asymmetric induction has also been achieved in these reactions, and ees of up to 94% have been obtained with a catalyst supported by a chiral phosphoramidite ligand [68c]. The mechanism of the terminal alkene diamination reactions has not yet been fully elucidated, but it appears likely that allylic C–H activation/amination is involved.

83 + H_5C_2 (1-hexene) → (5 mol % $Pd(PPh_3)_4$, 65 °C, 68%) → **88**

(1.39)

1.4
Synthesis of Nitrogen Heterocycles via Intermediate π-Allylpalladium Complexes

The intramolecular addition of nitrogen nucleophiles to intermediate π-allylpalladium complexes is a valuable method for the synthesis of saturated nitrogen heterocycles [69]. A number of different strategies have been employed for the generation of the reactive intermediate π-allylpalladium complexes, such as oxidative addition of alkenyl epoxides, allylic acetates, allylic carbonates, and related electrophiles to Pd^0. These intermediates have also been accessed through carbopalladation or heteropalladation reactions of allenes (as described above), vinyl cyclopropanes, or 1,3-dienes, and recent approaches involving allylic C–H activation have also been developed.

1.4.1
Reactions Involving Oxidative Addition of Allylic Electrophiles

One of the most common methods employed for the generation of allylpalladium complexes involves oxidative addition of allylic electrophiles to Pd^0. This transformation has been explored by several groups, and has been the topic of recent reviews [69]. A representative example of this process was demonstrated in a recent total synthesis of (+)-Biotin [70]. The key step in the synthesis was an intramolecular amination of **89**, which provided bicyclic urea derivative **90** in 77% yield (Eq. (1.40)). In contrast to the Pd^{II}-catalyzed reactions of allylic acetates bearing pendant amines described above (Eq. (1.10)), which proceed via alkene aminopalladation, Pd^0-catalyzed reactions of these substrates occur via initial oxidative addition of the allylic acetate to provide an intermediate π-allylpalladium complex (e.g., **91**). This intermediate is then captured by the pendant nucleophile (e.g., **91** to **90**) in a formal reductive elimination process to generate the product and regenerate the Pd^0 catalyst. Both the oxidative addition and the reductive elimination steps occur with inversion of configuration when soft nucleophiles are employed, which results in overall retention of configuration at the carbonate-bearing carbon stereocenter. Related transformations of propargylic electrophiles have also been reported [71].

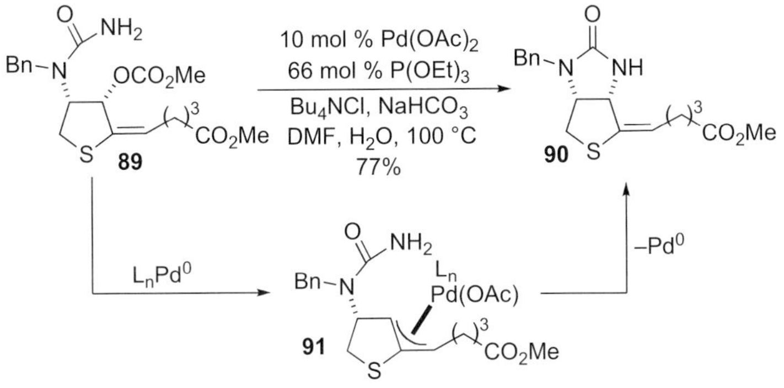

(1.40)

A number of studies have focused on the development and application of asymmetric versions of Pd-catalyzed allylic alkylation reactions [72]. Trost has developed a class of ligands, including **94** and **95**, that provide excellent yields and enantioselectivities for many of these reactions. For example, treatment of allylic acetate **92** with 1 mol% of allylpalladium chloride and 3 mol% of ligand **94** led to the formation of **93** in 97% yield and 91% ee (Eq. (1.41)) [73]. Asymmetric desymmetrization reactions of *meso*-bis-carbamates that provide heterocyclic products have also been described [74].

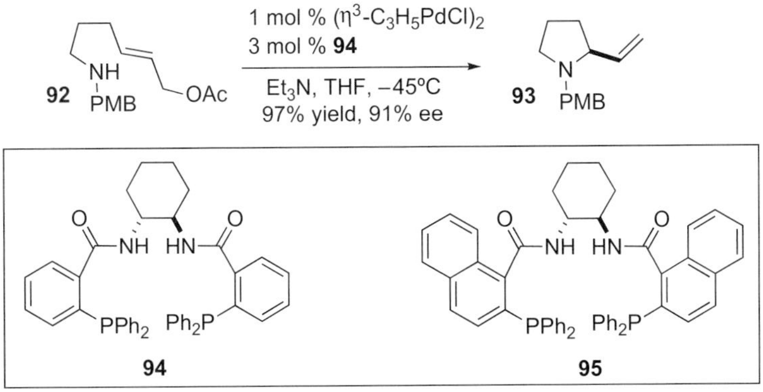

(1.41)

Recently, Trost has employed this methodology in a tandem one-pot ene–yne coupling/enantioselective allylation process [75]. This transformation was used for the construction of pyrrolidine **99**, an intermediate in a formal synthesis of kainic acid [75b]. As shown below (Eq. (1.42)), ruthenium-catalyzed coupling of **96** with **97** provided intermediate allyl ether **98**. Addition of 2 mol% of allylpalladium chloride and 6 mol% of chiral ligand **94** to the reaction vessel led to the formation of **99** in 92% yield and 94% ee.

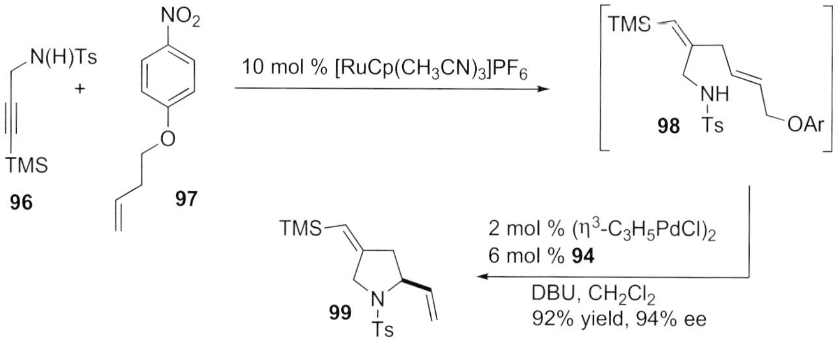

(1.42)

Alkenyl epoxides and aziridines have also been widely utilized as electrophiles in reactions that proceed via intermediate π-allylpalladium complexes [69], including reactions that form five-membered nitrogen heterocycles [76]. In a representative transformation, alkenyl epoxide **100** was coupled with isocyanate **101** in the presence of a catalyst generated *in situ* from $Pd_2(dba)_3$ and $P(Oi\text{-}Pr)_3$ to afford oxazolidin-2-one **102** in quantitative yield (Eq. (1.43)) [76b]. The reaction is initiated by oxidative addition of **100** to Pd^0 to afford **103**, which reacts with the isocyanate to yield **104**. The product is then generated by trapping the allylpalladium complex with the pendant carbamate anion. Related transformations involving the use of imines in place of isocyanates allow the construction of 1,3-oxazolidines [77], and syntheses of substituted pyrrolidines from 2-vinyloxiranes bearing tethered nitrogen nucleophiles have also been reported [78].

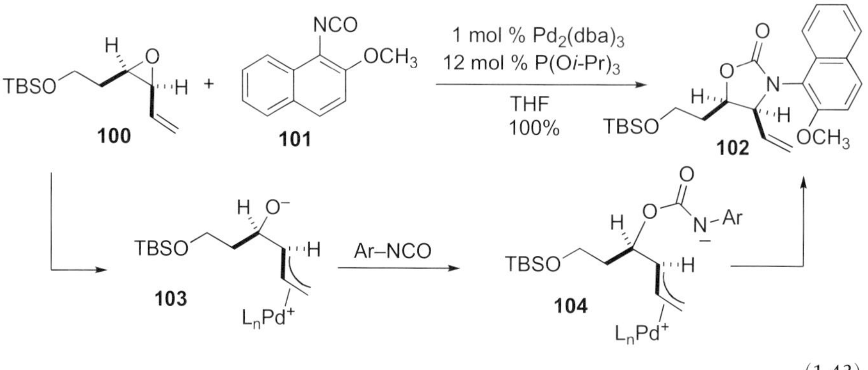

(1.43)

Related reactions of vinylaziridines [79] or activated vinylcyclopropanes [80] with isocyanates and other heterocumulenes have been developed for the construction of cyclic ureas and similar heterocycles. For example, Trost has recently described a dynamic kinetic asymmetric reaction of aziridine **105** with phenylisocyanate that affords urea **106** in 82% yield with 99% ee (Eq. (1.44)) [81]. The interconversion of stereoisomers occurs via η^3–η^1–η^3 isomerization processes after oxidative addition of the aziridine to Pd^0.

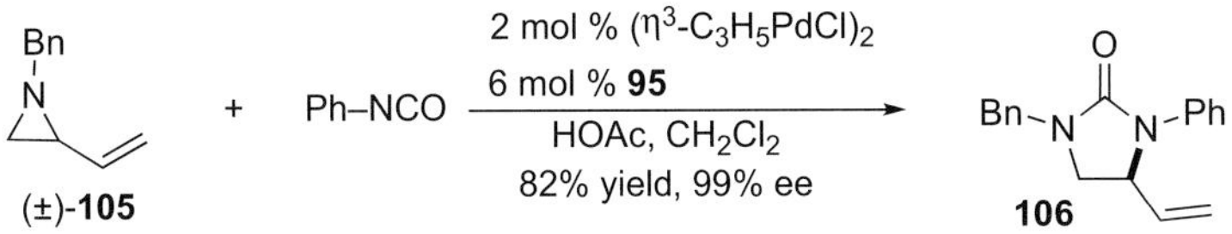

(1.44)

A related strategy has been used for the conversion of 5,5-divinyl oxazolidinones to highly substituted pyrrolidines [82]. For example, treatment of **107** with a Pd^0 catalyst in the presence of activated alkene **108** provides **109** in 95% yield (Eq. (1.45)). The reaction proceeds via oxidative addition followed by decarboxylation to afford the allylpalladium complex **110**. Intermolecular conjugate addition to give **111**, followed by intramolecular trapping of the allylpalladium moiety, affords the observed products.

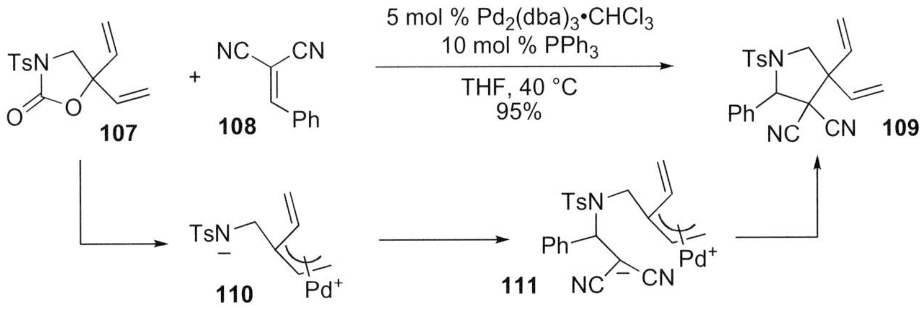

(1.45)

1.4.2
Reactions Involving π-Allylpalladium Intermediates Generated via Alkene Carbopalladation

A number of approaches to the generation of intermediate allylpalladium complexes in heterocycle synthesis involve initial Heck-type carbopalladation of an alkene with an alkenyl halide, followed by rearrangement of the resulting alkylpalladium species via reversible β-hydride elimination processes [83, 84]. For example, treatment of alkenyl sulfonamide **112** with 1-iodocyclopentene in the presence of catalytic amounts of $Pd(OAc)_2$ and $P(o\text{-tol})_3$ provided pyrrolidine **113** in 93% yield (Eq. (1.46)) [84a]. The C–N bond forming step occurs from the π-allylpalladium complex **116**, which results from β-hydride elimination of **114** followed by hydridopalladation of **115**. This strategy has also been employed for the synthesis of lactams from ω-olefinic amides [83b]. In addition, intramolecular versions of the pyrrolidine-forming reactions have been developed, such as the conversion of **117** to **118** (Eq. (1.47)) [84b,c].

112 + → 113
5 mol % $Pd(OAc)_2$
10 mol % $P(o\text{-tol})_3$
Bu_4NCl, Na_2CO_3
CH_3CN, 90 °C
93%
114 → 115 → 116

(1.46)

117 → 118
5 mol % $Pd(OAc)_2$
10 mol % $P(o\text{-tol})_3$
Bu_4NCl, Na_2CO_3
DMF, 65 °C
78%

(1.47)

Another approach to the construction of five-membered nitrogen heterocycles by way of intermediate π-allylpalladium complexes involves carbopalladation reactions of 1,3- or 1,4-dienes [85]. For example, Larock has described the coupling of *N*-tosyl-2-iodoaniline with 1,3-cyclohexadiene, which affords **119** in 87% yield (Eq. (1.48)). The allylpalladium complex **120** is a key intermediate in this transformation. Asymmetric versions of these reactions that generate pyrrolidine products have also been described [86]. Related Heck reactions that employ vinylcyclopropanes as diene surrogates have also been reported, although lengthy reaction times (3–4 days) are often required for transformations of these substrates [87].

5 mol % $Pd(dba)_2$
Na_2CO_3, Bu_4NCl
DMF, 100 °C
87%
119, 120

(1.48)

The use of Heck reactions of dienes for the construction of nitrogen heterocycles has been applied to an elegant synthesis of (−)-spirotryprostatin B [88]. As shown below (Eq. (1.49)), the intramolecular Heck reaction of **121** afforded the complex pentacycle **122**, which was converted to the natural product after cleavage of the SEM protecting group.

121 → 122

10 mol % $Pd_2(dba)_3$•$CHCl_3$
40 mol % $P(o\text{-tol})_3$
KOAc, THF, 70 °C
72%, 1:1 dr

(1.49)

1.4.3
Reactions Involving Aminopalladation of 1,3-Dienes

Bäckvall and coworkers have developed a stereoselective palladium catalyzed 1,4-addition to cyclic 1,3-dienes that produces pyrrolidine [89a] or lactam products [89b]. For example, treatment of **123** with catalytic $Pd(OAc)_2$ and excess LiCl affords **125** (Eq. (1.50)). Alternatively, treatment of **123** with catalytic $Pd(OAc)_2$ and excess LiOAc affords **127** (Eq. (1.51)). Both reactions proceed via *anti*-aminopalladation to afford an allylpalladium complex (**124** or **126**), which is then captured by an external nucleophile. Outer-sphere attack of chloride ion on **124** results in net *syn*-addition to the diene to give **125**, whereas inner-sphere attack of acetate results in *anti*-addition to provide **127**.

123 → 124 → 125: 5 mol % $Pd(OAc)_2$; BQ, LiCl, acetone, HOAc; 97%

123 → 126 → 127: 5 mol % $Pd(OAc)_2$; BQ, HOAc, acetone; 85%

(1.50, 1.51)

1.4.4
Generation of Allylpalladium Intermediates through C–H Activation

A recent example of isoxazolidine synthesis that involves generation of an allylpalladium complex via allylic C–H activation was reported by the White group [90]. As shown below (Eq. (1.52)), treatment of homoallylic carbamate **128** with $Pd(OAc)_2$ in the presence of sulfoxide ligand **130** and phenylbenzoquinone (PhBQ) provided **129** in 72% yield with 6: 1 dr. A related transformation that affords indoline products has been described by Larock [17a].

10 mol % $Pd(OAc)_2$, 10 mol % **130**, PhBQ, THF, 45 °C, 72%, 6:1 dr; **128** → **129**; Ph(O)S–S(O)Ph **130**

(1.52)

1.5
Synthesis of Nitrogen Heterocycles via Pd-Catalyzed 1,3-Dipolar Cycloaddition Reactions

Although the use of 1,3-dipolar cycloaddition reactions that form carbon–heteroatom bonds is fairly common using traditional synthetic methods [2], palladium-catalyzed dipolar cycloaddition reactions of this type are rather rare. However, a few reports have described an interesting and synthetically useful approach to the synthesis of pyrrolidines via Pd-catalyzed [3 + 2] cycloaddition reactions of trimethylenemethane with imines [91]. In very recent studies, Trost has developed an asymmetric variant of these reactions that provides access to enantioenriched pyrrolidine derivatives [92]. For example, treatment of trimethylenemethane precursor **131** with imine **132** proceeds to afford **133** in 84% yield and 91% ee when a catalyst composed of Pd $(dba)_2$ and ligand **134** is used (Eq. (1.53)).

131 (TMS, OAc) + **132** (Ph, N-Boc) → 5 mol % $Pd(dba)_2$, 10 mol % **134**, Toluene, 4 °C, 84% yield, 91% ee → **133**; ligand **134** (2-Nap, 2-Nap)

(1.53)

Although many Pd-catalyzed [3 + 2] cycloaddition reactions employ **131** as a trimethylenemethane precursor, readily available 2-methylenepropane-1,3-diols and their corresponding benzyl ethers have also been used as sources of trimethylenemethane [93]. This approach allows construction of more highly substituted pyrrolidine derivatives than can be generated from **131**. For example, treatment of **135** with **136** in the presence of diethylzinc and a palladium catalyst afforded pyrrolidine **137** in 92% yield as a single diastereomer (Eq. (1.54)).

135 (Me, OBn OBn) + **136** (Ph, N-PMP) → 10 mol % $Pd(OAc)_2$, 20 mol % Bu_3P, Et_2Zn, THF, reflux, 92%, >20:1 dr → **137** (Me, Ph, N-PMP) (1.54)

An unusual class of Pd-catalyzed [3 + 2] cycloaddition reactions between activated aziridines and heterocumulenes such as isocyanates and carbodiimides has been extensively examined by Alper and coworkers [94]. These transformations led to the preparation of ureas, carbamates, and other heterocycles in good yields. For example, treatment of **138** with phenylisocyanate afforded urea **139** in 72% yield (Eq. (1.55)) [94a]. The mechanism of these reactions presumably involves oxidative addition of the aziridine to Pd^0, followed by insertion of the isocyanate into the Pd–N bond and C–C bond-forming reductive elimination (similar to the reactions of vinylaziridines described in the section above, although allylpalladium intermediates are obviously not involved).

138 + Ph–N=C=O → (10 mol % $PdCl_2(PhCN)_2$, Toluene, 120 °C, 72%) → **139**

(1.55)

1.6
Synthesis of Nitrogen Heterocycles via Carbonylative Processes

Many of the concepts and strategies outlined above have been employed in carbonylative processes, which provide more highly functionalized heterocyclic products through incorporation of one or more units of CO [95]. Palladium-catalyzed carbonylative transformations that afford saturated five-membered nitrogen heterocycles can be broadly divided into three major categories: (i) processes involving CO insertion into a Pd–C_{Ar} or Pd–$C_{Alkenyl}$ bond, followed by intramolecular capture by a pendant nucleophile; (ii) transformations involving CO insertion into a Pd–heteroatom bond; and (3) Wacker-type processes wherein *anti*-heteropalladation of a carbon–carbon multiple bond precedes CO insertion.

1.6.1
Transformations Involving CO Insertion into Aryl or Alkenyl Pd-Carbon Bonds

Palladium-catalyzed carbonylative reactions of aryl or alkenyl bromides bearing pendant nitrogen nucleophiles have been studied for over 30 years, and have proven useful for the construction of a variety of different heterocyclic compounds [96]. For example, treatment of **140** with catalytic amounts of $Pd(OAc)_2$ and PPh_3 in the presence of Bu_3N under an atmosphere of CO afforded lactam **141** in 65% yield (Eq. (1.56)) [96a]. This transformation presumably occurs via oxidative addition of the aryl bromide to Pd^0 to provide **142**, which undergoes insertion of CO into the Pd–C bond to yield **143**. Intramolecular capture of the acylpalladium intermediate **143** by the tethered amine gives the desired heterocyclic product.

N(H)Bn Br 140 → [2 mol % $Pd(OAc)_2$, 4 mol % PPh_3; Bu_3N, CO, 100 °C, 63%] → N–Bn 141 O

142 PdBr N(H)Bn → CO → 143 N(H)Bn PdBr O

(1.56)

The insertion of CO into Pd–carbon bonds has also been employed in several tandem/cascade reactions that afford five-membered nitrogen heterocycles [97]. A representative example of this approach to the construction of heterocycles involves synthesis of isoindolinones via the Pd-catalyzed coupling of 2-bromobenzaldehyde with two equivalents of a primary amine under an atmosphere of CO [97b]. As shown below (Eq. (1.57)), this method was used for the preparation of **144** in 64% yield. The mechanism of this reaction is likely via initial, reversible condensation of 2-bromobenzaldehyde with 2 equiv of the amine to form an aminal **145**. Oxidative addition of the aryl bromide to Pd^0 followed by CO insertion provides the acylpalladium species **146**, which is then captured by the pendant aminal to afford the observed product. An alternative mechanism involving intramolecular imine insertion into the Pd–C bond of a related acylpalladium species, followed by formation of a palladium-amido complex and C–N bond-forming reductive elimination has also been proposed [97b].

CHO Br + $BnNH_2$ → [1.5 mol % $PdCl_2(PPh_3)_2$, 4 mol % PPh_3; CO, Et_3N, 100 °C, 64%] → HN–Bn N–Bn O 144

N-Bn Br ⇌ 145 HN-Bn N(H)-Bn Br → [Pd(0), CO] → 146 HN-Bn N(H)-Bn PdBr O

(1.57)

A sequence involving intermolecular Pd-catalyzed carbonylative amidation followed by intramolecular Michael addition has been employed for the construction of isoindolin-1-ones [97d]. For example, treatment of **147** with 4-methoxyaniline under standard conditions for Pd-catalyzed carbonylative coupling gave isoindolinone

148 in 79% yield (Eq. (1.58)). This transformation is effective with a wide array of primary aliphatic and aromatic amine nucleophiles.

NC, I (**147**) + NH_2, OMe → 10 mol % $Pd(OAc)_2$, 20 mol % PPh_3, CO, Cs_2CO_3, Toluene, 90 °C, 79% → CN, N–PMP, O (**148**)

(1.58)

A synthesis of phthalimides via double carbonylative coupling of *ortho*-diiodo arenes with anilines has also been reported [98]. The conversion of **149** to **150** proceeded in good yield using a Pd/PPh_3-based catalyst system (Eq. (1.59)).

MeO, MeO, I, I (**149**) + NH_2 → 3 mol % $PdCl_2(PPh_3)_2$, DBU, DMA, 115 °C, 75% → MeO, MeO, O, N–Ph, O (**150**)

(1.59)

1.6.2 Transformations Involving CO Insertion Into a Pd–Heteroatom Bond

A number of interesting methods for the synthesis of heterocycles that employ palladium carbonylation chemistry involve formal CO insertion into a Pd–heteroatom bond via either an inner-sphere or outer-sphere mechanism [99]. Several groups have employed this strategy for the generation of isoxazolidines and ureas via Pd-catalyzed carbonylation reactions of 1,2-amino alcohols or 1,2-diamines [100]. For example, Gabriele and coworkers have synthesized oxazolidin-2-ones in high yields using a PdI_2/KI/air catalyst system [100b]. Oxazolidin-2-one **152** was obtained in 96% yield from amino alcohol substrate **151** using only 0.05 mol% of PdI_2 (Eq. (1.60)). In a similar fashion, diamine **153** was transformed to 1,3-dihydrobenzoimidazol-2-one **154** in 70% yield (Eq. (1.61)) [100d]. The first carbon heteroatom bond is formed by CO insertion into the palladium amido complex **155**, followed by intramolecular trapping of the resulting acylpalladium intermediate **156** with the second heteroatom. This leads to reduction of the Pd^{II} catalyst to Pd^0, and the presence of air (oxygen) is required to regenerate the catalytically active Pd^{II} species.

HO, NH_2 (**151**) → 0.05 mol % PdI_2, 0.5 mol % KI, CO, air, DME, 100 °C, 96% → O, NH, O (**152**) (1.60)

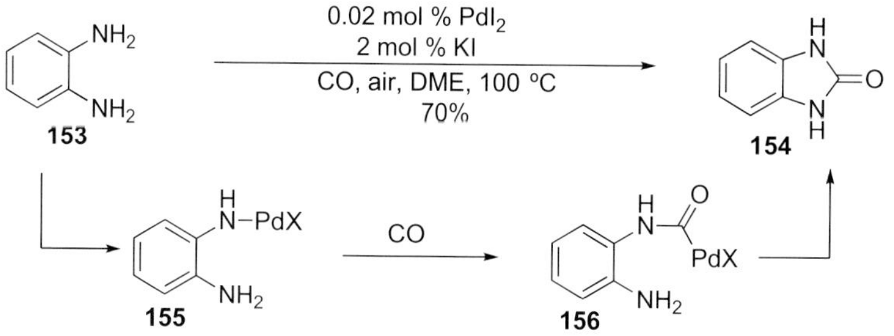

(1.61)

Several transformations involving CO insertion into a Pd–heteroatom bond have been developed that lead to incorporation of two molecules of CO into the heterocyclic product. This approach to heterocycle synthesis is exemplified by a synthesis of dihydroindolones reported by Gabriele [101]. As shown below, treatment of *ortho*-alkynyl aniline **157** with a Pd^{II} catalyst under CO in methanol afforded **158** in 50% yield (Eq. (1.62)). A similar strategy has been employed for the conversion of alkene **159** to pyrrolidinone **160** (Eq. (1.63)) [102].

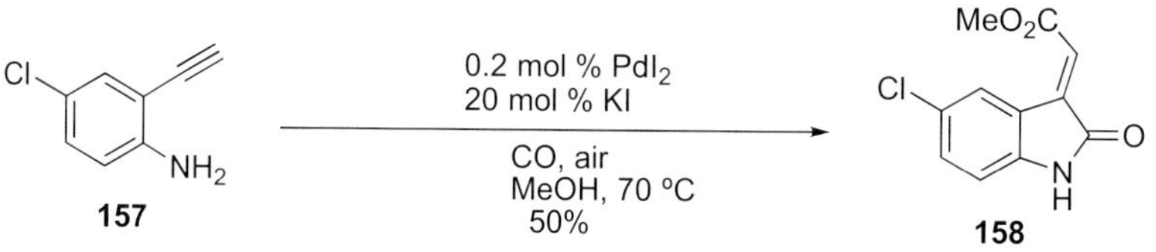

(1.62)

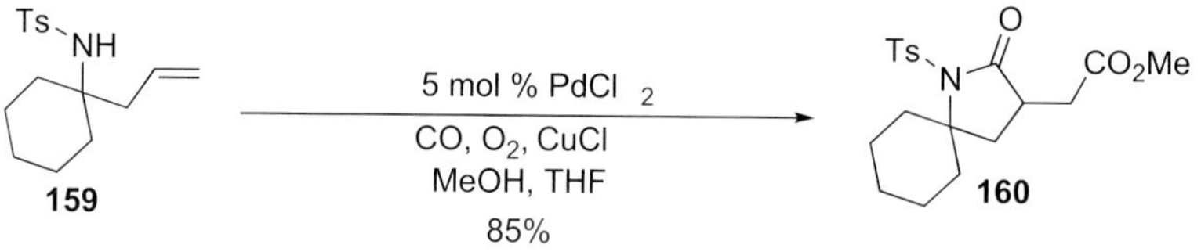

(1.63)

1.6.3
Wacker-Type Carbonylative Processes

Pd-catalyzed carbonylative processes that involve Wacker-type *anti*-aminopalladation of alkenes, alkynes or allenes have been widely employed in the construction of nitrogen heterocycles. Early studies using stoichiometric amounts of palladium to effect intramolecular alkene aminopalladation followed by carbonylation were reported by Danishefsky in 1983 [103]. A number of elegant studies by Tamaru subsequently led to the development of catalytic versions of these reactions, and extended the scope of this chemistry to allow the generation of a wide array of nitrogen heterocycles, including oxazolidin-2-ones, imidazolidin-2-ones, pyrrolidines, and isoxazolidines [104]. For example, treatment of carbamate **161** with 5 mol% $PdCl_2$ in the presence of $CuCl_2$ under CO using trimethyl orthoacetate as solvent provides oxazolidin-2-one **162** in 70% yield with >20: 1 dr (Eq. (1.64)) [104e]. The mechanism of this reaction involves *anti*-aminopalladation of the alkene to afford **163**, which undergoes CO insertion to form **164**. Capture of the acylpalladium intermediate **164** with methanol (formed *in situ* from trimethyl orthoacetate) gives the heterocyclic product.

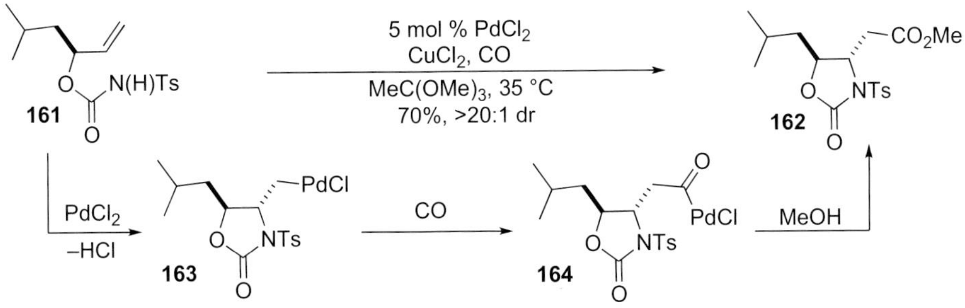

(1.64)

This strategy has been used for the construction of bridged bicyclic nitrogen heterocycles, and has been applied to a formal total synthesis of the alkaloid natural product (±)-ferruginine [105]. In addition, an asymmetric variant of this reaction has been developed by Sasai and coworkers. As shown below (Eq. (1.65)), use of a catalyst composed of $Pd(TFA)_2$ and spiro bis(isoxazoline) ligand **167** effected the conversion of sulfonamide **165** to pyrrolidine **166** in 95% yield and 60% ee [106]. Although this transformation requires high catalyst loadings and very long reaction times (7 days), it is clear that there is potential for achieving asymmetric induction in these systems, and development of new catalysts for these reactions is likely to be an area of future investigation.

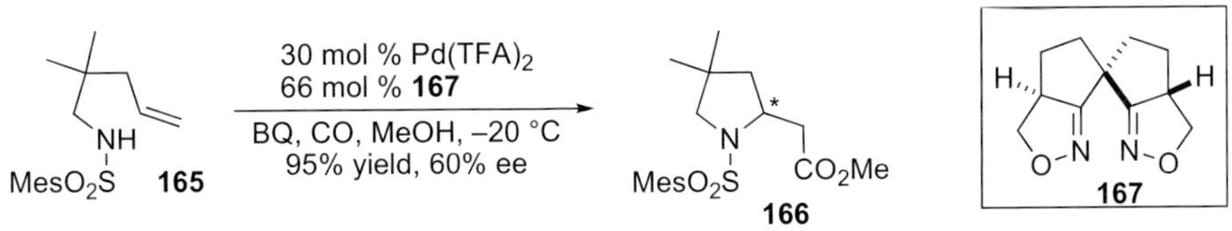

(1.65)

Alkene aminopalladation/carbonylation has also been employed in the synthesis of fused bis(heterocycles) [107]. As shown below, treatment of **168** with 10 mol% $PdCl_2$ in the presence of $CuCl_2$ under CO provided **169** in 66% yield (Eq. (1.66)) [107a]. This strategy has been used for the preparation of 1,4-iminoglycitols [107b], and has been applied to a concise synthesis of the Geissman–Waiss Lactone, which is a key intermediate in the synthesis of necine bases [107c].

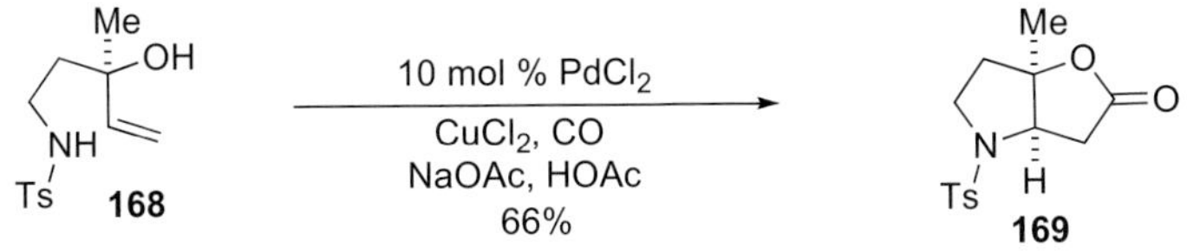

(1.66)

A number of Pd-catalyzed carbonylative processes have employed allenes as substrates for the synthesis of nitrogen heterocycles [108]. For example, subjection of substituted allene **170** to reaction conditions similar to those employed in related reactions of alkenes led to the formation of pyrrolidine **171** in 68% yield with 2: 1 dr (Eq. (1.67)) [108d]. Modest asymmetric induction has been achieved in these transformations using simple chiral auxiliaries [108b,d]. This strategy was employed in an asymmetric synthesis of pumiliotoxin 251 D, which involved the aminocarbonylation of allene **172** to pyrrolidine **173** as a key step (Eq. (1.68)) [108b].

Ph, NH, Ts **170** → 10 mol % $PdCl_2$; $CuCl_2$, CO, Et_3N, MeOH; 68%, 2:1 dr → Ph, N, Ts, CO_2Me **171** (1.67)

NH, Ph, OH **172** → 20 mol % $PdCl_2(PhCN)_2$; $CuCl_2$, CO, MeOH; 71%, 2.5:1 dr → **173** N, Ph, OH, CO_2Me (1.68)

1.7
Summary and Future Outlook

Over the past several decades, research in the field of palladium catalysis has resulted in the development of a myriad of transformations that provide access to saturated nitrogen heterocycles. Many of these transformations effect the formation of several bonds, and/or proceed with good to excellent levels of stereocontrol. Despite the many advances made in this field, discoveries of new reactivity are still being reported with great frequency, and this promises to remain a fruitful area of research for many years to come. In particular, the development of new palladium catalysts will likely lead to improvements in the scope of existing transformations, and will also open up new reaction pathways that can be applied to unsolved problems in heterocyclic chemistry.

References

1 (a) For reviews on biologically active nitrogen heterocycles, see: Kang, E.J. and Lee, E. (2005) *Chem. Rev.*, **105**, 4348–4378; (b) Saleem, M., Kim, H.J., Ali, M.S., and Lee, Y.S. (2005) *Nat. Prod. Rep.*, **22**, 696–716; (c) Bermejo, A., Figadere, B., Zafra-Polo, M.-C., Barrachina, I., Estornell, E., and Cortes, D. (2005) *Nat. Prod. Rep.*, **22**, 269–309; (d) Daly, J.W., Spande, T.F., and Garraffo, H.M. (2005) *J. Nat. Prod.*, **68**, 1556–1575; (e) Hackling, A.E. and Stark, H. (2002) *Chem. BioChem.*, **3**, 946–961; (f) Lewis, J.R. (2001) *Nat. Prod. Rep.*, **18**, 95–128.

2 (a) For reviews on the synthesis of nitrogen heterocycles, see: Bellina, F. and Rossi, R. (2006) *Tetrahedron*, **62**, 7213–7256; (b) Coldham, I. and Hufton, R. (2005) *Chem. Rev.*, **105**, 2765–2810; (c) Pyne, S.G., Davis, A.S., Gates, N.J., Hartley, J.P., Lindsay, K.B., Machan, T., and Tang, M. (2004) *Synlett*, 2670–2680; (d) Felpin, F.-X. and Lebreton, J. (2003) *Eur. J. Org. Chem.*, 3693–3712;

(e) Confalone, P.N. and Huie, E.M. (1988) *Org. React.*, **36**, 1–173; (f) Gothelf, K.V. and Jorgensen, K.A. (2000) *Chem. Commun.*, 1449–1458; (g) Kanemasa, S. (2002) *Synlett*, 1371–1387.

3 (a) For recent reviews on the use of palladium catalysis for the synthesis of heterocycles, see: D'Souza, D.M. and Müller, T.J.J. (2007) *Chem. Soc. Rev.*, **36**, 1095–1108; (b) Minatti, A. and Muniz, K. (2007) *Chem. Soc. Rev.*, **36**, 1142–1152; (c) Zeni, G. and Larock, R.C. (2006) *Chem. Rev.*, **106**, 4644–4680; (d) Zeni, G. and Larock, R.C. (2004) *Chem. Rev.*, **104**, 2285–2309; (e) Wolfe, J.P. and Thomas, J.S. (2005) *Curr. Org. Chem.*, **9**, 625–656; (f) Balme, G., Bossharth, E., and Monteiro, N. (2003) *Eur. J. Org. Chem.*, 4101–4111; (g) Balme, G., Bouyssi, D., Lomberget, T., and Monteiro, N. (2003) *Synthesis*, 2115–2134; (h) Li, J.J. and Gribble, G. (2000) *Palladium in Heterocyclic Chemistry*, Pergamon Press, New York.

4 (a) For reviews, see: Surry, D.S. and Buchwald, S.L. (2008) *Angew. Chem., Int. Ed.*, **47**, 6338–6361; (b) Schlummer, B. and Scholz, U. (2004) *Adv. Synth. Catal.*, **346**, 1599–1626; (c) Hartwig, J.F. (2002) *Modern Arene Chemistry* (ed. D. Astruc), Wiley-VCH, Weinheim, pp. 107–168; (d) Muci, A.R. and Buchwald, S.L. (2002) *Top. Curr. Chem.*, **219**, 131–209; (e) Hartwig, J.F. (1999) *Pure Appl. Chem.*, **71**, 1417–1424.

5 The first example of intramolecular Pd-mediated N-arylation was reported by Boger, who employed this reaction for construction of the aromatic core of lavendamycin methyl ester. See: Boger, D.L., Duff, S.R., Panek, J.S., and Yasuda, M. (1985) *J. Org. Chem.*, **50**, 5782–5789.

6 (a) Guram, A.S., Rennels, R.A., and Buchwald, S.L. (1995) *Angew. Chem., Int. Ed.*, **34**, 1348–1350; (b) Wolfe, J.P., Rennels, R.A., and Buchwald, S.L. (1996) *Tetrahedron*, **52**, 7525–7546; (c) Yang, B.H. and Buchwald, S.L. (1999) *Org. Lett.*, **1**, 35–37.

7 Bonnaterre, F., Bois-Choussy, M., and Zhu, J. (2006) *Org. Lett.*, **8**, 4351–4354.

8 McLaughlin, M., Palucki, M., and Davies, I.W. (2006) *Org. Lett.*, **8**, 3311–3314.

9 Zhu, Y., Kiryu, Y., and Katayama, H. (2002) *Tetrahedron Lett.*, **43**, 3577–3580.

10 Coleman, R.S. and Chen, W. (2001) *Org. Lett.*, **3**, 1141–1144.

11 He, F., Foxman, B.M., and Snider, B.B. (1998) *J. Am. Chem. Soc.*, **120**, 6417–6418.

12 Cämmerer, S.S., Viciu, M.S., Stevens, E.D., and Nolan, S.P. (2003) *Synlett*, 1871–1873.

13 Ganton, M.D. and Kerr, M.A. (2007) *J. Org. Chem.*, **72**, 574–582.

14 (a) Deboves, H.J.C., Hunter, C., and Jackson, R.F.W. (2002) *J. Chem. Soc., Perkin Trans. 1.*, 733–736; (b) Thansandote, P., Raemy, M., Rudolph, A., and Lautens, M. (2007) *Org. Lett.*, **9**, 5255–5258; (c) Omar-Amrani, R., Schneider, R., and Fort, Y. (2004) *Synthesis*, 2527–2534.

15 Hegedus, L.S., Allen, G.F., Bozell, J.J., and Waterman, E.L. (1978) *J. Am. Chem. Soc.*, **100**, 5800–5807.

16 Hegedus, L.S. and McKearin, J.M. (1982) *J. Am. Chem. Soc.*, **104**, 2444–2451.

17 (a) Larock, R.C., Hightower, T.R., Hasvold, L.A., and Peterson, K.P. (1996) *J. Org. Chem.*, **61**, 3584–3585; (b) Rönn, M., Bäckvall, J.E., and Andersson, P.G. (1995) *Tetrahedron Lett.*, **36**, 7749–7752.

18 Barreiro, J., Peçanha, E., and Fraga, C. (1998) *Heterocycles*. **48**, 2621–2629.

19 (a) Stahl, S.S. (2004) *Angew. Chem., Int. Ed.*, **43**, 3400–3420; (b) Trend, R.M., Ramtohul, Y.K., Ferreira, E.M., and Stoltz, B.M. (2003) *Angew. Chem., Int. Ed.*, **42**, 2892–2895; (c) Trend, R.M., Ramtohul, Y.K., and Stoltz, B.M. (2005) *J. Am. Chem. Soc.*, **127**, 17778–17788; (d) Rogers, M.M., Wendlandt, J.E., Guzei, I.A., and Stahl, S.S. (2006) *Org. Lett.*, **8**, 2257–2260; (e) Fix, S.R., Brice, J.L., and Stahl, S.S. (2002) *Angew. Chem., Int. Ed.*, **41**, 164–166.

20 Liu, G. and Stahl, S.S. (2007) *J. Am. Chem. Soc.*, **129**, 6328–6335.

21 (a) Lu, X. (2005) *Top. Catal.*, **35**, 73–86; (b) Lei, A., Liu, G., and Lu, X. (2002) *J. Org. Chem.*, **67**, 974–980; (c) Hirai, Y.,

Watanabe, J., Nozaki, T., Yokoyama, H., and Yamaguchi, S. (1997) *J. Org. Chem.*, **62**, 776–777.

22 (a) For related transformations of alkynes that afford five-membered nitrogen heterocycles, see: Kozawa, Y. and Mori, M. (2002) *Tetrahedron Lett.*, **43**, 1499–1502; (b) Kimura, M., Wakamiya, Y., Horino, Y., and Tamaru, Y. (1997) *Tetrahedron Lett.*, **38**, 3963–3966.

23 Yokoyama, H. and Hirai, Y. (2008) *Heterocycles*, **75**, 2133–2153.

24 (a) Overman, L.E. and Remarchuk, T.P. (2002) *J. Am. Chem. Soc.*, **124**, 12–13; (b) Kirsch, S.F. and Overman, L.E. (2005) *J. Org. Chem.*, **70**, 2859–2861.

25 (a) Banfi, L., Basso, A., Cerulli, V., Guanti, G., and Riva, R. (2008) *J. Org. Chem.*, **73**, 1608–1611; (b) Eustache, J., Van de Weghe, P., Le Nouen, D., Uyehara, H., Kabuto, C., and Yamamoto, Y. (2005) *J. Org. Chem.*, **70**, 4043–4053.

26 (a) Müller, T.E. and Beller, M. (1998) *Chem. Rev.*, **98**, 675–704; (b) Beller, M., Seayad, J., Tillack, A., and Jiao, H. (2004) *Angew. Chem., Int. Ed.*, **43**, 3368–3398; (c) Hartwig, J.F. (2004) *Pure Appl. Chem.*, **76**, 507–516.

27 Meguro, M. and Yamamoto, Y. (1998) *Tetrahedron Lett.*, **39**, 5421–5424.

28 (a) Kadota, I., Shibuya, A., Lutete, L.M., and Yamamoto, Y. (1999) *J. Org. Chem.*, **64**, 4570–4571; (b) Müller, T.E., Grosche, M., Herdtweck, E., Pleier, A.-K., Walter, E., and Yan, Y.-K. (2000) *Organometallics*, **19**, 170–183; (c) Bajracharya, G.B., Huo, Z., and Yamamoto, Y. (2005) *J. Org. Chem.*, **70**, 4883–4886; (d) Wolf, L.B., Tjen, K.C.M.F., ten Brink, H.T., Blaauw, R.H., Hiemstra, H., Schoemaker, H.E., and Rutjes, F.P.J.T. (2002) *Adv. Synth. Catal.*, **344**, 70–83.

29 Patil, N.T., Pahadi, N.K., and Yamamoto, Y. (2005) *Tetrahedron Lett.*, **46**, 2101–2103.

30 (a) Patil, N.T., Lutete, L.M., Wu, H., Pahadi, N.K., Gridnev, I.D., and Yamamoto, Y. (2006) *J. Org. Chem.*, **71**, 4270–4279; (b) Narsireddy, M. and Yamamoto, Y. (2008) *J. Org. Chem.*, **73**, 9698–9709.

31 Patil, N.T., Huo, Z., Bajracharya, G.B., and Yamamoto, Y. (2006) *J. Org. Chem.*, **71**, 3612–3614

32 Meguro, M. and Yamamoto, Y. (1998) *Tetrahedron Lett.*, **39**, 5421–5424.

33 (a) Michael, F.E. and Cochran, B.M. (2006) *J. Am. Chem. Soc.*, **128**, 4246–4247; (b) Cochran, B.M. and Michael, F.E. (2008) *J. Am. Chem. Soc.*, **130**, 2786–2792.

34 Sakai, N., Ridder, A., and Hartwig, J.F. (2006) *J. Am. Chem. Soc.*, **128**, 8134–8135.

35 (a) For reviews, see: Wolfe, J.P. (2008) *Synlett*, 2913–2937; (b) Wolfe, J.P. (2007) *Eur. J. Org. Chem.*, 571–582.

36 (a) Ney, J.E. and Wolfe, J.P. (2004) *Angew. Chem., Int. Ed.*, **43**, 3605–3608; (b) Bertrand, M.B. and Wolfe, J.P. (2005) *Tetrahedron*, **61**, 6447–6459; (c) Ney, J.E. and Wolfe, J.P. (2005) *J. Am. Chem. Soc.*, **127**, 8644–8651; (d) Ney, J.E., Hay, M.B., Yang, Q., and Wolfe, J.P. (2005) *Adv. Synth. Catal.*, **347**, 1614–1620; (e) Bertrand, M.B., Leathen, M.L., and Wolfe, J.P. (2007) *Org. Lett.*, **9**, 457–460; (f) Bertrand, M.B., Neukom, J.D., and Wolfe, J.P. (2008) *J. Org. Chem.*, **73**, 8851–8860.

37 Bertrand, M.B. and Wolfe, J.P. (2006) *Org. Lett.*, **8**, 2353–2356.

38 Nakhla, J.S., Kampf, J.W., and Wolfe, J.P. (2006) *J. Am. Chem. Soc.*, **128**, 2893–2901.

39 (a) Lira, R. and Wolfe, J.P. (2004) *J. Am. Chem. Soc.*, **126**, 13906–13907; (b) Yang, Q., Ney, J.E., and Wolfe, J.P. (2005) *Org. Lett.*, **7**, 2575–2578.

40 (a) Fritz, J.A., Nakhla, J.S., and Wolfe, J.P. (2006) *Org. Lett.*, **8**, 2531–2534; (b) Fritz, J.A. and Wolfe, J.P. (2008) *Tetrahedron*, **64**, 6838–6852.

41 (a) Lemen, G.S., Giampietro, N.C., Hay, M.B., and Wolfe, J.P. (2009) *J. Org. Chem.*, **74**, 2533–2540; (b) Peng, J., Lin, W., Yuan, S., and Chen, Y. (2007) *J. Org. Chem.*, **72**, 3145–3148; (c) Peng, J., Jiang, D., Lin, W., and Chen, Y. (2007) *Org. Biomol. Chem.*, **5**, 1391–1396; (d) Dongol, K.G. and Tay, B.Y. (2006) *Tetrahedron Lett.*, **47**, 927–930.

42 Giampietro, N.C. and Wolfe, J.P. (2008) *J. Am. Chem. Soc.*, **130**, 12907–12911.

43 (a) Azoulay, S., Monteiro, N., and Balme, G. (2002) *Tetrahedron Lett.*, **43**, 9311–9314; (b) Martinon, L., Azoulay, S., Monteiro, N., Kündig, H.E.P., and Balme, G. (2004) *J. Organomet. Chem.*, **689**, 3831–3836.

44 Fugami, K., Oshima, K., and Utimoto, K. (1987) *Tetrahedron Lett.*, **28**, 809–812.

45 Scarborough, C.C. and Stahl, S.S. (2006) *Org. Lett.*, **8**, 3251–3254.

46 Yip, K.–T., Yang, M., Law, K.–L., Zhu, N.–Y., and Yang, D. (2006) *J. Am. Chem. Soc.*, **128**, 3130–3131.

47 (a) Bouyssi, D., Cavicchioli, M., and Balme, G. (1997) *Synlett*, 944–946; (b) Jacobi, P.A. and Liu, H. (1999) *Org. Lett.*, **1**, 341–344; (c) Jacobi, P.A. and Liu, H. (2000) *J. Org. Chem.*, **65**, 7676–7681.

48 Karstens, W.F.J., Stol, M., Rutjes, F.P.J.T., Kooijman, H., Spek, A.L., and Hiemstra, H. (2001) *J. Organomet. Chem.*, **624**, 244–258.

49 (a) Lei, A. and Lu, X. (2000) *Org. Lett.*, **2**, 2699–2702; (b) Liu, G. and Lu, X. (2001) *Org. Lett.*, **3**, 3879–3882.

50 (a) Davies, I.W., Scopes, D.I.C., and Gallagher, T. (1993) *Synlett*, 85–87; (b) Kang, S.–K., Baik, T.–G., and Kulak, A.N. (1999) *Synlett*, 324–326; (c) Kang, S.–K., Baik, T.–G., and Hur, Y. (1999) *Tetrahedron*, **55**, 6863–6870; (d) Karstens, W.F.J., Stol, M., Rutjes, F.P.J.T., and Hiemstra, H. (1998) *Synlett*, 1126–1128; (e) Karstens, W.F.J., Rutjes, F.P.J.T., and Hiemstra, H. (1997) *Tetrahedron Lett.*, **38**, 6275–6278; (f) Grigg, R., Kilner, C., Mariani, E., and Sridharan, V. (2006) *Synlett*, 3021–3024.

51 (a) Kimura, M., Tanaka, S., and Tamaru, Y. (1995) *J. Org. Chem.*, **60**, 3764–3772; (b) Kimura, M., Fugami, K., Tanaka, S., and Tamaru, Y. (1992) *J. Org. Chem.*, **57**, 6377–6379; (c) Karstens, W.F.J., Klomp, D., Rutjes, F.P.J.T., and Hiemstra, H. (2001) *Tetrahedron*, **57**, 5123–5130.

52 (a) Larock, R.C. and Zenner, J.M. (1995) *J. Org. Chem.*, **60**, 482–483; (b) Larock, R.C., He, Y., Leong, W.W., Han, X., Refvik, M.D., and Zenner, J.M. (1998) *J. Org. Chem.*, **63**, 2154–2160; (c) Desarbre, E. and Merour, J.–Y. (1996) *Tetrahedron Lett.*, **37**, 43–46; (d) Grigg, R., Mariani, E., and Sridharan, V. (2001) *Tetrahedron Lett.*, **42**, 8677–8680.

53 Zenner, J.M. and Larock, R.C. (1999) *J. Org. Chem.*, **64**, 7312–7322.

54 Ma, S. and Jiao, N. (2002) *Angew. Chem., Int. Ed.*, **41**, 4737–4740.

55 (a) Ma, S., Jiao, N., Zheng, Z., Ma, Z., Lu, Z., Ye, L., Deng, Y., and Chen, G. (2004) *Org. Lett.*, **6**, 2193–2196; (b) Yang, Q., Jiang, X., and Ma, S. (2007) *Chem. Eur. J.*, **13**, 9310–9316; (c) Shu, W., Yang, Q., Jia, G., and Ma, S. (2008) *Tetrahedron*, **64**, 11159–11166.

56 Houlden, C.E., Bailey, C.D., Ford, J.G., Gagne, M.R., Lloyd-Jones, G.C., and Booker-Milburn, K.I. (2008) *J. Am. Chem. Soc.*, **130**, 10066–10067.

57 Jensen, K.H. and Sigman, M.S. (2008) *Org. Biomol. Chem.*, **6**, 4083–4088.

58 (a) Manzoni, M.R., Zabawa, T.P., Kasi, D., and Chemler, S.R. (2004) *Organometallics*, **23**, 5618–5621; (b) Michael, F.E., Sibbald, P.A., and Cochran, B.M. (2008) *Org. Lett.*, **10**, 793–796.

59 (a) Jonasson, C., Horvath, A., and Bäckvall, J.–E. (2000) *J. Am. Chem. Soc.*, **122**, 9600–9609; (b) Jonasson, C., Karstens, W.F.J., Hiemstra, H., and Bäckvall, J.–E. (2000) *Tetrahedron Lett.*, **41**, 1619–1622.

60 Alexanian, E.J., Lee, C., and Sorensen, E.J. (2005) *J. Am. Chem. Soc.*, **127**, 7690–7691.

61 Muniz, K., Hövelmann, C.H., Streuff, J., and Campoz-Gomez, E. (2008) *Pure Appl. Chem.*, **80**, 1089–1096.

62 (a) Streuff, J., Hövelmann, C.H., Nieger, M., and Muniz, K. (2005) *J. Am. Chem. Soc.*, **127**, 14586–14587; (b) Muniz, K., Hövelmann, C.H., and Streuff, J. (2008) *J. Am. Chem. Soc.*, **130**, 763–773; (c) Muniz, K., Hövelmann, C.H., Campos-Gomez, E., Barluenga, J., Gonzalez, J.M., Streuff, J., and Nieger, M. (2008) *Chem. Asian J.*, **3**, 776–788; (d) Muniz, K., Streuff, J., Chavez, P., and Hövelmann, C.H. (2008) *Chem. Asian J.*, **3**, 1248–1255.

63 Muniz, K. (2007) *J. Am. Chem. Soc.*, **129**, 14542–14543.

64 Hövelmann, C.H., Streuff, J., Brelot, L., and Muniz, K. (2008) *Chem. Commun.*, 2334–2336.
65 Bar, G.L.J., Lloyd-Jones, G.C., and Booker-Milburn, K.I. (2005) *J. Am. Chem. Soc.*, **127**, 7308–7309.
66 (a) Du, H., Zhao, B., and Shi, Y. (2007) *J. Am. Chem. Soc.*, **129**, 762–763; (b) Xu, L., Du, H., and Shi, Y. (2007) *J. Org. Chem.*, **72**, 7038–7041.
67 Du, H., Yuan, W., Zhao, B., and Shi, Y. (2007) *J. Am. Chem. Soc.*, **129**, 11688–11689.
68 (a) Du, H., Yuan, W., Zhao, B., and Shi, Y. (2007) *J. Am. Chem. Soc.*, **129**, 7496–7497; (b) Wang, B., Du, H., and Shi, Y. (2008) *Angew. Chem., Int. Ed.*, **47**, 8224–8227; (c) Du, H., Zhao, B., and Shi, Y. (2008) *J. Am. Chem. Soc.*, **130**, 8590–8591.
69 (a) Hyland, C. (2005) *Tetrahedron*, **61**, 3457–3471; (b) Patil, N.T. and Yamamoto, Y. (2006) *Top. Organomet. Chem.*, **19**, 91–113.
70 Seki, M., Mori, Y., Hatsuda, M., and Yamada, S.–i. (2002) *J. Org. Chem.*, **67**, 5527–5536.
71 (a) Ohno, H., Okano, A., Kosaka, S., Tsukamoto, K., Ohata, M., Ishihara, K., Maeda, H., Tanaka, T., and Fujii, N. (2008) *Org. Lett*, **10**, 1171–1174.
72 (a) Trost, B.M., Machacek, M.R., and Aponick, A. (2006) *Acc. Chem. Res.*, **39**, 747–760; (b) Trost, B.M. and Crawley, M.L. (2003) *Chem. Rev.*, **103**, 2921–2943; (c) Trost, B.M. (1996) *Acc. Chem. Res.*, **29**, 355–364; (d) Trost, B.M. and Van Vranken, D.L. (1996) *Chem. Rev.*, **96**, 395–422.
73 Trost, B.M., Krische, M.J., Radinov, R., and Zanoni, G. (1996) *J. Am. Chem. Soc.*, **118**, 6297–6298.
74 (a) Trost, B.M. and Patterson, D.E. (1998) *J. Org. Chem.*, **63**, 1339–1341; (b) Lee, S.-g., Lim, C.W., Song, C.E., Kwan, K.M., and Jun, C.H. (1999) *J. Org. Chem.*, **64**, 4445–4451.
75 (a) Trost, B.M. and Machacek, M.R. (2002) *Angew. Chem., Int. Ed.*, **41**, 4693–4697; (b) Trost, B.M., Machacek, M.R., and Faulk, B.D. (2006) *J. Am. Chem. Soc.*, **128**, 6745–6754.
76 (a) Trost, B.M. and Sudhakar, A.R. (1987) *J. Am. Chem. Soc.*, **109**, 3792–3794; (b) Trost, B.M. and Sudhakar, A.R. (1988) *J. Am. Chem. Soc.*, **110**, 7933–7935; (c) Trost, B.M. and Hurnaus, R. (1989) *Tetrahedron Lett.*, **30**, 3893–3896.
77 Shim, J.–G. and Yamamoto, Y. (1999) *Tetrahedron Lett.*, **40**, 1053–1056.
78 (a) Takahashi, K., Haraguchi, N., Ishihara, J., and Hatakeyama, S. (2008) *Synlett*, 671–674; (b) Noguchi, Y., Uchiro, H., Yamada, T., and Kobayashi, S. (2001) *Tetrahedron Lett.*, **42**, 5253–5256.
79 Butler, D.C.D., Inman, G.A., and Alper, H. (2000) *J. Org. Chem.*, **65**, 5887–5890.
80 Yamamoto, K., Ishida, T., and Tsuji, J. (1987) *Chem. Lett.*, 1157–1158.
81 Trost, B.M. and Fandrick, D.R. (2003) *J. Am. Chem. Soc.*, **125**, 11836–11837.
82 (a) Knight, J.G., Tchabanenko, K., Stoker, P.A., and Harwood, S.J. (2005) *Tetrahedron Lett.*, **46**, 6261–6264; (b) Knight, J.G., Stoker, P.A., Tchabanenko, K., Harwood, S.J., and Lawrie, K.W.M. (2008) *Tetrahedron*, **64**, 3744–3750.
83 (a) Larock, R.C., Pace, P., Yang, H., Russell, C.E., Cacchi, S., and Fabrizi, G. (1998) *Tetrahedron*, **54**, 9961–9980; (b) Pinho, P., Minnaard, A.J., and Feringa, B.L. (2003) *Org. Lett.*, **5**, 259–261.
84 (a) Larock, R.C., Yang, H., Weinreb, S.M., and Herr, R.J. (1994) *J. Org. Chem.*, **59**, 4172–4178; (b) Harris, G.D. Jr., Herr, R.J., and Weinreb, S.M. (1992) *J. Org. Chem.*, **57**, 2528–2530; (c) Harris, G.D. Jr., Herr, R.J., and Weinreb, S.M. (1993) *J. Org. Chem.*, **58**, 5452–5464.
85 (a) Larock, R.C., Berrios-Pena, N., and Narayanan, K. (1990) *J. Org. Chem.*, **55**, 3447–3450; (b) Iyer, S., Ramesh, C., and Kulkarni, G.M. (2001) *Synlett*, 1241–1244; (c) Larock, R.C., Berrios-Pena, N.G., Fried, C.A., Yum, E.K., Tu, C., and Leong, W. (1993) *J. Org. Chem.*, **58**, 4509–4510; (d) Back, T.G., Bethell, R.J., Parvez, M., and Taylor, J.A. (2001) *J. Org. Chem.*, **66**, 8599–8605.
86 Flubacher, D. and Helmchen, G. (1999) *Tetrahedron Lett.*, **40**, 3867–3868.
87 (a) Larock, R.C. and Yum, E.K. (1990) *Synlett*, 529–530; (b) Larock, R.C. and

Yum, E.K. (1996) *Tetrahedron*, **52**, 2743–2758.

88 Overman, L.E. and Rosen, M.D. (2000) *Angew. Chem., Int. Ed.*, **39**, 4596–4599.

89 (a) Bäckvall, J.–E. and Andersson, P.G. (1990) *J. Am. Chem. Soc.*, **112**, 3683–3685; (b) Andersson, P.G. and Bäckvall, J.-E. (1992) *J. Am. Chem. Soc.*, **114**, 8696–8698.

90 Fraunhoffer, K.J. and White, M.C. (2007) *J. Am. Chem. Soc.*, **129**, 7274–7276.

91 (a) Jones, M.D. and Kemmitt, R.D.W. (1986) *J. Chem. Soc., Chem. Commun.*, 1201–1202; (b) Trost, B.M. and Marrs, C.M. (1993) *J. Am. Chem. Soc.*, **115**, 6636–6645; (c) Oh, B.H., Nakamura, I., Saito, S., and Yamamoto, Y. (2003) *Heterocycles*, **61**, 247–257.

92 Trost, B.M., Silverman, S.M., and Stambuli, J.P. (2007) *J. Am. Chem. Soc.*, **129**, 12398–12399.

93 Kimura, M., Tamaki, T., Nakata, M., Tohyama, K., and Tamaru, Y. (2008) *Angew. Chem., Int. Ed.*, **47**, 5803–5805.

94 (a) Baeg, J.–O., Bensimon, C., and Alper, H. (1995) *J. Am. Chem. Soc.*, **117**, 4700–4701; (b) Baeg, J.–O., and Alper, H. (1994) *J. Am. Chem. Soc.*, **116**, 1220–1224.

95 (a) Tamaru, Y. and Yoshida, Z.-i. (1987) *J. Organomet. Chem.*, **334**, 213–223; (b) Negishi, E. (1992) *Pure Appl. Chem.*, **64**, 323–334; (c) Tamaru, Y. and Kimura, M. (1997) *Synlett*, 749–757; (d) El Ali, B. and Alper, H. (2000) *Synlett*, 161–171; (e) Gabriele, B., Salerno, G., and Costa, M. (2004) *Synlett*, 2468–2483; (f) Vizer, S.A., Yerzhanov, K.B., Al Quntar, A.A.A., and Dembitsky, V.M. (2004) *Tetrahedron*, **60**, 5499–5538.

96 (a) Mori, M., Chiba, K., and Ban, Y. (1978) *J. Org. Chem.*, **43**, 1684–1687; (b) Mori, M., Washioka, Y., Urayama, T., Yoshiura, K., Chiba, K., and Ban, Y. (1983) *J. Org. Chem.*, **48**, 4058–4067; (c) Grigg, R., Sridharan, V., Suganthan, S., and Bridge, A.W. (1995) *Tetrahedron*, **51**, 295–306; (d) Crisp, G.T. and Meyer, A.G. (1995) *Tetrahedron*, **51**, 5585–5596.

97 (a) Cho, C.S., Chu, D.Y., Lee, D.Y., Shim, S.C., Kim, T.J., Lim, W.T., and Heo, N.H. (1997) *Synth. Commun.*, **27**, 4141–4158; (b) Cho, C.S., Jiang, L.H., Lee, D.Y., Shim, S.C., Lee, H.S., and Cho, S.-D. (1997) *J. Heterocycl. Chem.*, **34**, 1371–1374; (c) Cho, C.S., Shim, H.S., Choi, H.-J., Kim, T.-J., and Shim, S.C. (2002) *Synth. Commun.*, **32**, 1821–1827; (d) Gai, X., Grigg, R., Khamnaen, T., Rajviroongit, S., Sridharan, V., Zhang, L., Collard, S., and Keep, A. (2003) *Tetrahedron Lett.*, **44**, 7441–7443.

98 Perry, R.J. and Turner, S.R. (1991) *J. Org. Chem.*, **56**, 6573–6579.

99 Gabriele, B., Salerno, G., Costa, M., and Chiusoli, G.P. (2003) *J. Organomet. Chem.*, **687**, 219–228.

100 (a) Chiarotto, I. and Feroci, M. (2001) *Tetrahedron Lett.*, **42**, 3451–3453; (b) Gabriele, B., Mancuso, R., Salerno, G., and Costa, M. (2003) *J. Org. Chem.*, **68**, 601–604; (c) Li, F. and Xia, C. (2004) *J. Catal.*, **227**, 542–546; (d) Gabriele, B., Salerno, G., Mancuso, R., and Costa, M. (2004) *J. Org. Chem.*, **69**, 4741–4750; (e) Gabriele, B., Plastina, P., Salerno, G., Mancuso, R., and Costa, M. (2007) *Org. Lett.*, **9**, 3319–3322.

101 Gabriele, B., Salerno, G., Veltri, L., Costa, M., and Massera, C. (2001) *Eur. J. Org. Chem.*, 4607–4613.

102 Mizutani, T., Ukaji, Y., and Inomata, K. (2003) *Bull. Chem. Soc. Jpn.*, **76**, 1251–1256.

103 Danishefsky, S. and Taniyama, E. (1983) *Tetrahedron Lett.*, **24**, 15–18.

104 (a) Tamaru, Y., Hojo, M., Higashimura, H., and Yoshida, Z. (1988) *J. Am. Chem. Soc.*, **110**, 3994–4002; (b) Tamaru, Y., Hojo, M., and Yoshida, Z. (1988) *J. Org. Chem.*, **53**, 5731–5741; (c) Tamaru, Y., Tanigawa, H., Itoh, S., Kimura, M., Tanaka, S., Fugami, K., Sekiyama, T., and Yoshida, Z.–I. (1992) *Tetrahedron Lett.*, **33**, 631–634; (d) Harayama, H., Okuno, H., Takahashi, Y., Kimura, M., Fugami, K., Tanaka, S., and Tamaru, Y. (1996) *Tetrahedron Lett.*, **37**, 7287–7290; (e) Harayama, H., Abe, A., Sakado, T., Kimura, M., Fugami, K., Tanaka, S., and Tamaru, Y. (1997) *J. Org. Chem.*, **62**, 2113–2122; (f) Bates, R.W. and Sa-Ei, K. (2002) *Org. Lett.*, **4**, 4225–4227.

105 Ham, W.–H., Jung, Y.–H., Lee, K., Oh, C.–Y., and Lee, K.–Y. (1997) *Tetrahedron Lett.*, **38**, 3247–3248.

106 Shinohara, T., Arai, M.A., Wakita, K., Arai, T., and Sasai, H. (2003) *Tetrahedron Lett.*, **44**, 711–714.

107 (a) Tamaru, Y., Kobayashi, T., Kawamura, S.I., Ochiai, H., and Yoshida, Z.-i. (1985) *Tetrahedron Lett.*, **26**, 4479–4482; (b) Hümmer, W., Dubois, E., Gracza, T., and Jäger, V. (1997) *Synthesis*, 634–642; (c) Takahata, H., Banba, Y., and Momose, T. (1991) *Tetrahedron: Asymmetry*, **2**, 445–448.

108 (a) Lathbury, D., Vernon, P., and Gallagher, T. (1986) *Tetrahedron Lett.*, **27**, 6009–6012; (b) Fox, D.N.A. and Gallagher, T. (1990) *Tetrahedron*, **46**, 4697–4710; (c) Fox, D.N.A., Lathbury, D., Mahon, M.F., Molloy, K.C., and Gallagher, T. (1991) *J. Am. Chem. Soc.*, **113**, 2652–2656; (d) Gallagher, T., Davies, I.W., Jones, S.W., Lathbury, D., Mahon, M.F., Molloy, K.C., Shaw, R.W., and Vernon, P. (1992) *J. Chem. Soc., Perkin Trans. 1*, 433–440; (e) Kimura, M., Saeki, N., Uchida, S., Harayama, H., Tanaka, S., Fugami, K., and Tamaru, Y. (1993) *Tetrahedron Lett.*, **34**, 7611–7614.

2
Transition Metal Catalyzed Approaches to Lactones Involving C—O Bond Formation

Charles S. Yeung, Peter K. Dornan, and Vy M. Dong

2.1
Introduction

Lactones found in nature, particularly those with ring sizes larger than eight (i.e., macrolactones), exhibit important biological and medicinal activities and find application as pheromones, insecticides, antibiotics, as well as cytotoxic and anti-angiogenesis agents (Figure 2.1) [1–3]. Enzymatic construction of these architectures typically involves a ring closure via intramolecular addition of an alcohol to an activated thioester [4]. In an analogous manner, traditional lactone synthesis typically involves a retrosynthetic disconnection between the C—O single bond via carboxylic acid or alcohol activation (Scheme 2.1) [1, 3].

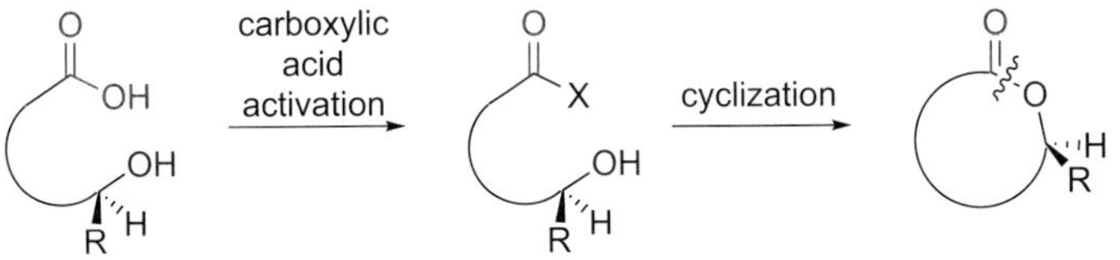

carboxylic acid activation → cyclization →

Yamaguchi: 2,4,6-benzoyl chloride, NEt_3, cat. DMAP
Corey-Nicolaou: 2,2'-dipyridyldisulfide, PPh_3
Keck: DCC, DMAP

Scheme 2.1 Traditional C—O bond formation strategies by carboxylic acid activation [1, 5].

In contrast, modern transition metal catalysis has provided many complementary synthetic approaches to lactones that have enabled a greater number of retrosynthetic connections. Using a transition metal, lactone synthesis can be achieved without the need for stoichiometric activating agents, hence addressing issues of both cost and atom economy (i.e., minimization of waste). In addition, extension to asymmetric variants allows direct synthesis of chiral lactones from achiral starting materials. Herein, we discuss lactonization strategies that involve the formation of C—O bonds made possible by employment of transition metals. Emphasis will be placed on mechanistic understanding, asymmetric transformations, and application to total

Catalyzed Carbon-Heteroatom Bond Formation. Edited by Andrei K. Yudin

ISBN: 978-3-527-32428-6

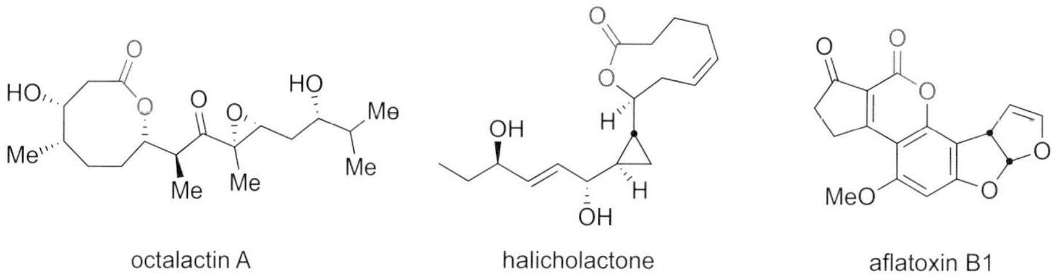

Figure 2.1 Natural products containing lactone functionalities [1, 4].

synthesis. Other methods involving C–C bond formation (e.g., olefin metathesis) are beyond the scope of this chapter [2].

2.2 Synthesis of Lactones Involving CO

2.2.1 Carbonylation of C–X Bonds

Transition metal catalyzed insertion of CO has proven to be an excellent method for the synthesis of lactones [6–8]. Access to lactones via this approach typically involves Pd(0) catalysis via: (i) oxidative addition of a C–X bond, (ii) insertion of CO, (iii) intramolecular addition of O-based nucleophiles, and (iv) reductive elimination (Scheme 2.2).

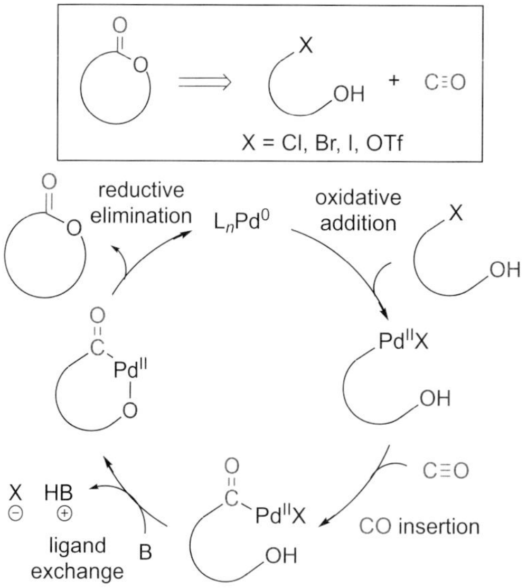

Scheme 2.2 Mechanism for Pd(0)-catalyzed carbonylation of C–X bonds terminated by OH [6].

The use of OH nucleophiles for cyclization was first reported by Ban [9] and Negishi [10]. Subsequently, Shibasaki and coworkers demonstrated that desymmetrization of alkenyl halides could yield chiral α-methylene γ- and δ-lactones with chiral bidentate phosphine ligands (Scheme 2.3). The presence of silver salts was crucial to the observed enantioselectivity (up to 53%). It is postulated that without these salts, the bidentate ligand can only bind through one arm, thus leading to poor selectivity; however, the use of silver salts generates a cationic palladium species which can offer two coordination sites to the ligand [11]. In some cases, over-oxidation of the 2° alcohol led to ketone formation.

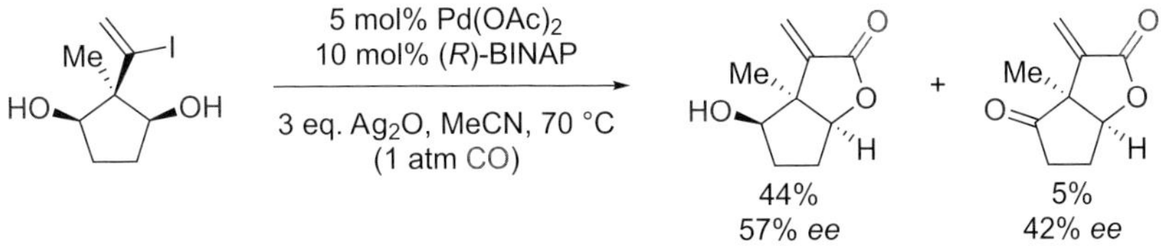

5 mol% $Pd(OAc)_2$, 10 mol% (*R*)-BINAP, 3 eq. Ag_2O, MeCN, 70 °C (1 atm CO); 44%, 57% *ee* + 5%, 42% *ee*

Scheme 2.3 Desymmetrization of alkenyl halides by carbonylative lactonization [11].

Applications of cyclocarbonylation to the preparation of natural products include the synthesis of (±)-aristolactone [12], (+)-hamabiwalactone B [13], (+)-asimicin, and (+)-bullatacin [14], all of which involve 4-hydroxybutenolide motifs (Scheme 2.4).

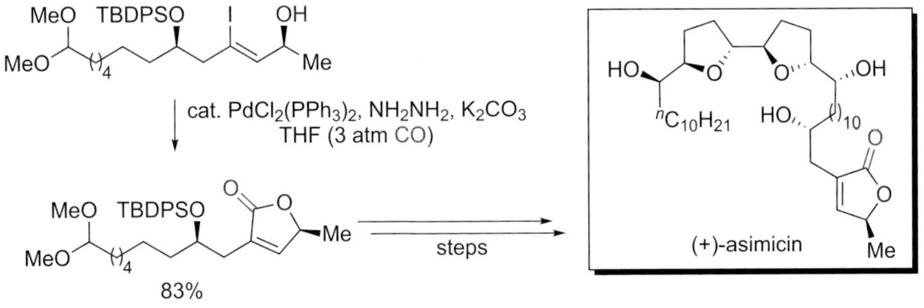

cat. $PdCl_2(PPh_3)_2$, NH_2NH_2, K_2CO_3, THF (3 atm CO); 83%; steps; (+)-asimicin

Scheme 2.4 Total synthesis of (+)-asimicin [14].

Carbonylative macrolactonization is also possible. Takahashi and coworkers successfully prepared a library of macrosphelide targets using solid phase synthesis under Pd(0) catalysis (Scheme 2.5). By using the split-and-pool technique, 122 compounds were synthesized [15].

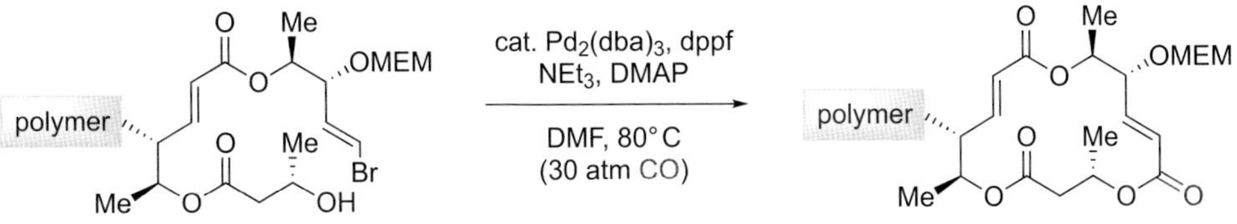

cat. $Pd_2(dba)_3$, dppf, NEt_3, DMAP, DMF, 80 °C (30 atm CO)

Scheme 2.5 Combinatorial synthesis of a macrosphelide library [15].

The nucleophilic addition of O-enolates, unlike alcohols, produces lactones with C=C bonds adjacent to the ester functionality [6, 7]. Butenolides, for instance, can be easily prepared via this strategy (Scheme 2.6). When α,β-unsaturated aryl ketones are used, it is proposed that water acts as a proton source and that CO reduces Pd(II) to Pd(0) to complete the catalytic cycle. While NEt_3 is known to reduce Pd(II), the reaction also worked with non-reducing bases, such as pyridine, but not with K_2CO_3, a dehydrating agent. Negishi and coworkers further extended their methodology to a multicomponent reaction of aryl iodides and alkynes in the presence of CO requiring 1 equiv of water [16], similar to a method reported earlier by Alper using Co(0) under phase transfer conditions and a relatively low pressure of CO (Scheme 2.7) [17]. These two examples are notable for the incorporation of 2 equivalent of CO. The analogous preparation of six-membered benzolactones has also been achieved, not only with Pd(0) but also employing Cu(I) and Ni(0) as catalysts [18, 19]. Because O-enolates can also be accessed via insertion of CO, tandem processes have also been reported for the synthesis of small-ring lactones (Scheme 2.8) [18, 20, 21].

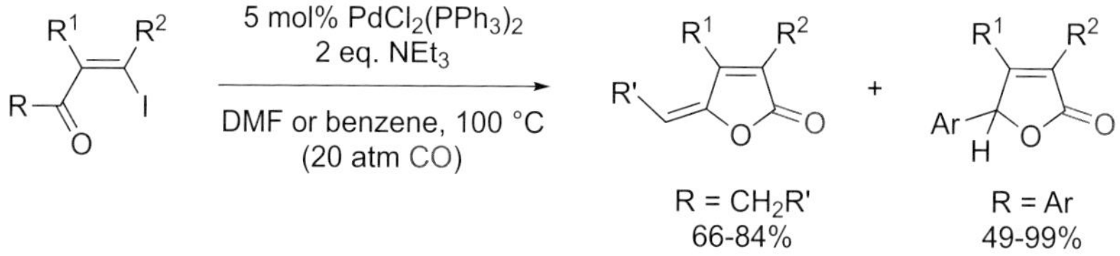

Scheme 2.6 Butenolides by carbonylation of C–I bonds terminated by O-enolates [16].

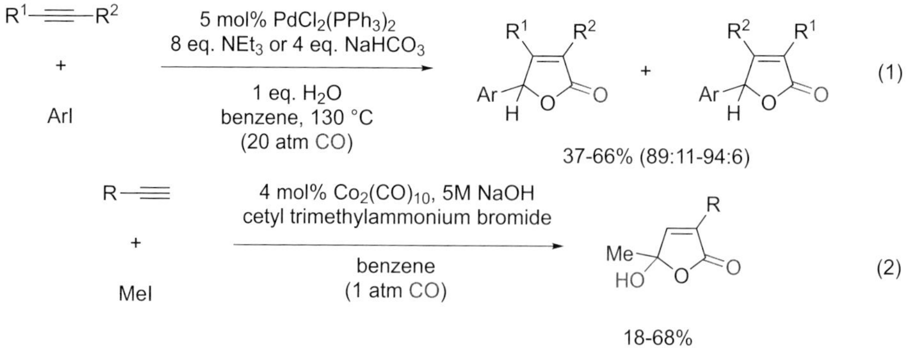

Scheme 2.7 Multicomponent approaches to lactones from alkynes and CO [16, 17].

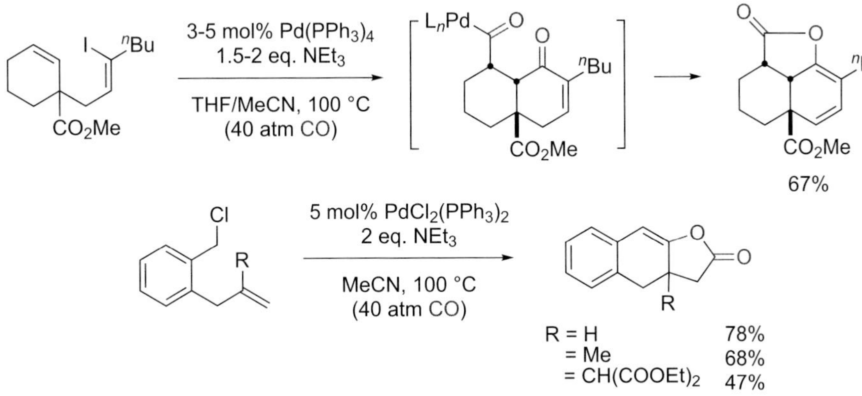

Scheme 2.8 Double carbonylation of C–X bonds terminated by O-enolates [18, 20, 21].

2.2.2
Carbonylation of C–M Bonds

Carbonylation of C–M bonds with transition metal catalysts has received significantly less attention than the corresponding insertion chemistry of C–X bonds [6]. With Pd(II) as the catalyst, transmetallation initiates the cyclocarbonylation catalytic cycle (Scheme 2.9). An oxidant is typically required to regenerate Pd(II) from Pd(0) following reductive elimination.

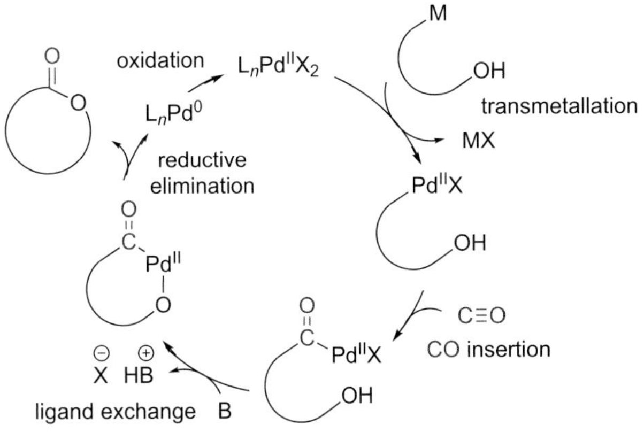

Scheme 2.9 Mechanism for Pd(II)-catalyzed carbonylation of C–M bonds terminated by OH.

Larock reported the use of stoichiometric amounts of thallium salts for direct C–H metallation and cyclocarbonylation for the synthesis of benzofuranones and benzohexanones (Scheme 2.10). Surprisingly, this reaction occurs in the absence of external oxidant because Tl(III) salts generated via transmetallation can reoxidize Pd(0) to Pd(II). Lithium chloride and magnesium oxide were added to suppress

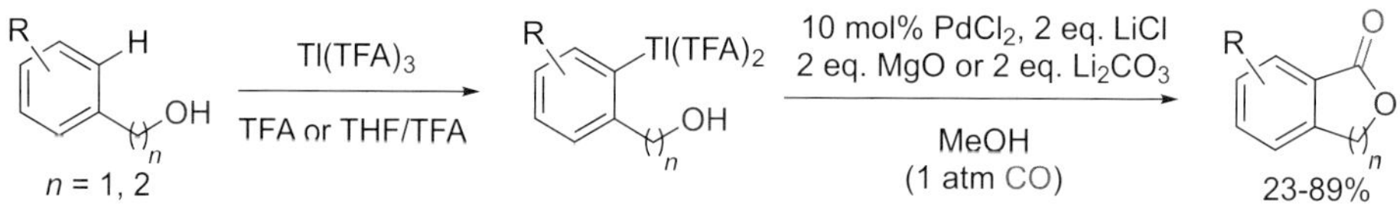

Scheme 2.10 Benzolactones by carbonylation of C–H bonds terminated by OH [22].

competing biphenyl formation [22]. This approach was applied to both 5- and 6-membered rings (phthalides and 3,4-dihydroisocoumarins, respectively) and was used in a concise total synthesis of pseudomeconin. *Meta*-substituted benzyl alcohols gave excellent regioselectivity in the thallation step, as only 5-substituted phthalides were observed, presumably because the position between substituents on the benzyl alcohol is too sterically hindered [22].

Kočovský *et al.* later demonstrated that C–Hg bonds can be used for stereospecific lactone synthesis. Both *cis*- and *trans*-fused bicyclic lactones can be prepared by ring-opening reactions of cyclopropanes followed by cyclocarbonylation of the organomercurial intermediate (Scheme 2.11). *p*-Benzoquinone was found to be superior to Cu(II) as a stoichiometric oxidant [23].

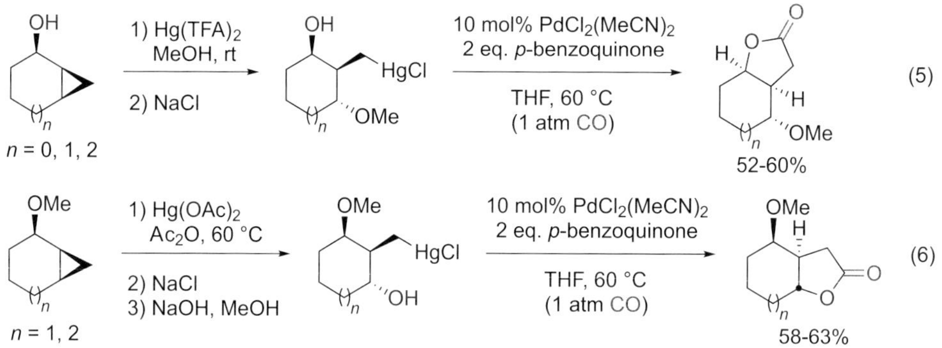

Scheme 2.11 Bicyclic lactones by carbonylation of C–Hg bonds terminated by OH [23].

2.2.3 Hydrocarbonylation of C=C and C≡C Bonds

Addition of CO to C=C and C≡C bonds provides an alternative approach to lactones distinct from that of C–X and C–M carbonylation [6, 24]. In particular, hydrocarbonylation involves the formation of new C–H and C–C bonds (Scheme 2.12). Pd-catalyzed transformations of this type proceed through either hydropalladative or carbopalladative pathways [6]. In the carbopalladative mechanism, a Pd alkoxide undergoes carbonylation, yielding an acylpalladium species. Subsequent insertion into the C=C or C≡C bond generates the desired lactone. Hydropalladation, on the other hand, is favored under reducing or acidic conditions and occurs via Pd hydride addition to the unsaturation, followed by CO insertion, and reductive elimination.

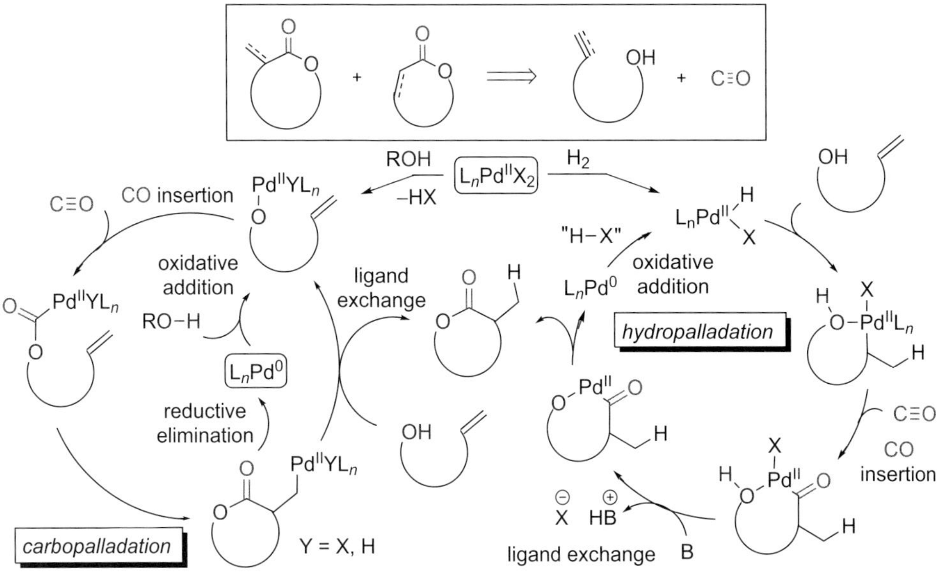

Scheme 2.12 Mechanisms for Pd-catalyzed hydrocarbonylation of C=C bonds [6, 25].

The Alper group reported the first cyclocarbonylation of alkenes [26, 27], approximately 10 years after the first account of alkyne lactonization [28]. Subsequent developments in the field include application to medium-size rings [29], and use of other metals, including $Rh_6(CO)_{16}$ [30]. Selective *endo* or *exo*-α,β-unsaturated lactone formation from homopropargyl alochols is possible [31]. When DMF is used as a solvent, the exocyclic α,β-unsaturated δ-lactone is formed, presumably through a carbopalladation mechanism. However, when MeCN is used as a solvent, the product distribution depends strongly on the choice of phosphine and counterion, and under appropriate conditions the endocyclic product can be obtained preferentially. In this case, an alternative hydropalladation mechanism is proposed, since a *syn*-carbopalladation would result in a highly strained 6-membered lactone with a *trans* double bond [31]. The corresponding reaction of allenes was only realized recently with catalytic $Ru_3(CO)_{12}$ yielding five and six membered α,β-unsaturated lactones [32].

Alper and coworkers also reported an asymmetric cyclocarbonylation of allylic alcohols using a Pd(0) system with proline-based bisphosphine (−)-BPPM (Scheme 2.13), chosen due to its structural similarity to dppb. Moderate to excellent

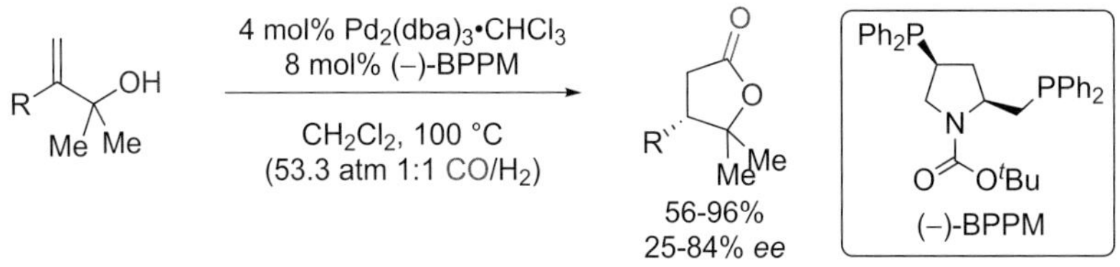

Scheme 2.13 Asymmetric hydrocarbonylation of allylic alcohols [33].

yields and enantioselectivities were observed (up to 96%, 84% *ee*) [33]. *o*-Isopropenylphenols also underwent enantioselective lactonization with (+)-DIOP as the ligand (Scheme 2.14) [34].

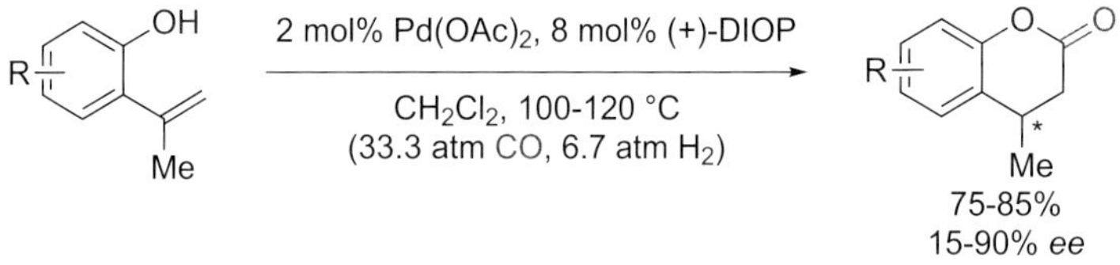

2 mol% $Pd(OAc)_2$, 8 mol% (+)-DIOP
CH_2Cl_2, 100-120 °C
(33.3 atm CO, 6.7 atm H_2)
75-85%
15-90% *ee*

Scheme 2.14 Asymmetric hydrocarbonylation of *o*-isopropenylphenols [34].

Ogawa extended hydrocarbonylation to Pt(0) by demonstrating that catalytic amounts of aromatic thiols can promote the cyclization of alkynyl alcohols (Scheme 2.15). The authors propose a Pt sulfide intermediate in the catalytic cycle generated via S—H oxidative addition (i.e., carbopalladation; see Scheme 2.12), followed by CO insertion, acylplatination and reductive elimination of the vinylic C—H bond. Cyclization of the pendant alcohol onto the thioester affords the lactone. It was observed that certain aromatic thiols could also undergo tandem conjugate addition with the lactone product [35].

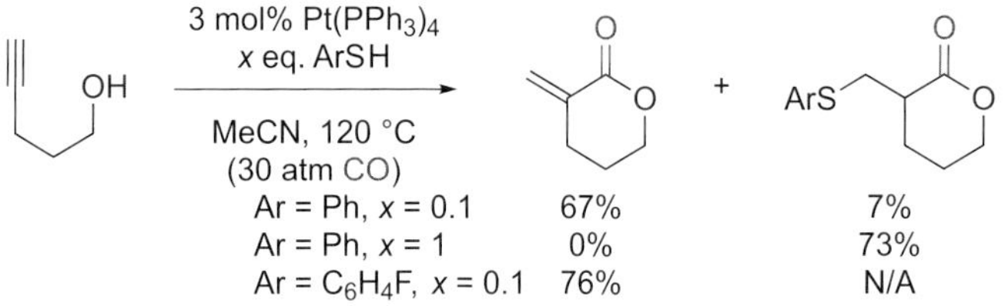

3 mol% $Pt(PPh_3)_4$
x eq. ArSH
MeCN, 120 °C
(30 atm CO)

Ar = Ph, *x* = 0.1	67%	7%
Ar = Ph, *x* = 1	0%	73%
Ar = C_6H_4F, *x* = 0.1	76%	N/A

Scheme 2.15 Pt(0)-catalyzed hydrocarbonylation of alkynes and tandem conjugate addition [35].

2.2.4 Carbocarbonylation of C=C and C≡C Bonds

Catalytic formation of two new C—C bonds from alkene and alkyne starting materials is another variant of lactone synthesis employing CO as a C_1 source. Under Pd(0) catalysis, alkenylmetallation can be terminated by carbonylative lactonization; however, this requires careful selection of reaction conditions to favor direct CO insertion into the formed Pd—C bond followed by alcoholysis of the acylpalladium intermediate over other reactions, such as β-H elimination [36]. To this end, Negishi and coworkers reported a carbocarbonylation reaction for the synthesis of polycyclic frameworks from a polyyne starting material initiated by oxidative addition into a vinyl iodide (Scheme 2.16) [37]. In a single step, five rings could be formed with high efficiency (66%). Surprisingly, MeOH was found to be the optimal solvent despite competitive intermolecular trapping. The corresponding methyl ester accounted for less than 2–3% of the crude mixture. It is also notable that 5-*exo*-carbopalladation of alkynes was found to outcompete carbonylation followed by 6-*exo*-acylpalladation. The corresponding reaction for alkenes involving O-enolates was discussed in Section 2.2.1 and is primarily limited to 5- and 6-membered lactone synthesis [36].

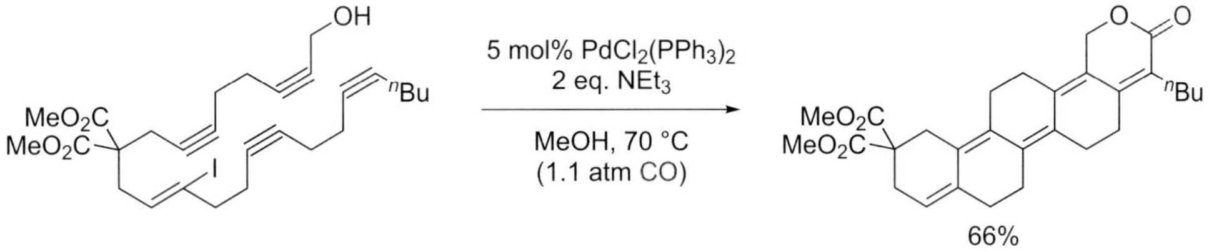
OH
^{n}Bu
MeO$_2$C
MeO$_2$C
I
5 mol% PdCl$_2$(PPh$_3$)$_2$
2 eq. NEt$_3$
MeOH, 70 °C
(1.1 atm CO)
O
O
^{n}Bu
MeO$_2$C
MeO$_2$C
66%

Scheme 2.16 Tandem alkenylcarbonylation of a polyyne [37].

Dicarbonylation can also provide access to lactones in which a 1,2-diester is formed directly from a C=C or C≡C bond. In both cases, an oxidant is required for the regeneration of Pd(II) from Pd(0) (Scheme 2.17) [38–40]. Ukaji *et al.* demonstrated the first asymmetric intramolecular alkene dicarbonylation using a chiral bisoxazoline ligand with Cu(I) and O_2 as the oxidant (Scheme 2.18) [41].

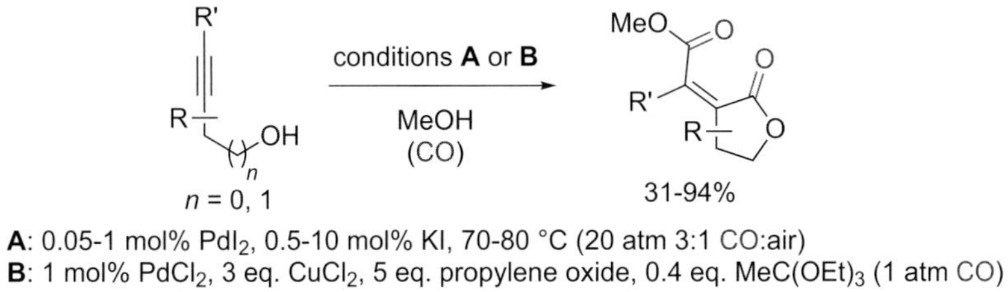
R'
R
OH
n
n = 0, 1
conditions **A** or **B**
MeOH
(CO)
MeO
O
O
R'
R
O
31-94%
A: 0.05-1 mol% PdI$_2$, 0.5-10 mol% KI, 70-80 °C (20 atm 3:1 CO:air)
B: 1 mol% PdCl$_2$, 3 eq. CuCl$_2$, 5 eq. propylene oxide, 0.4 eq. MeC(OEt)$_3$ (1 atm CO)

Scheme 2.17 Dicarbonylation of alkynes [38–40].

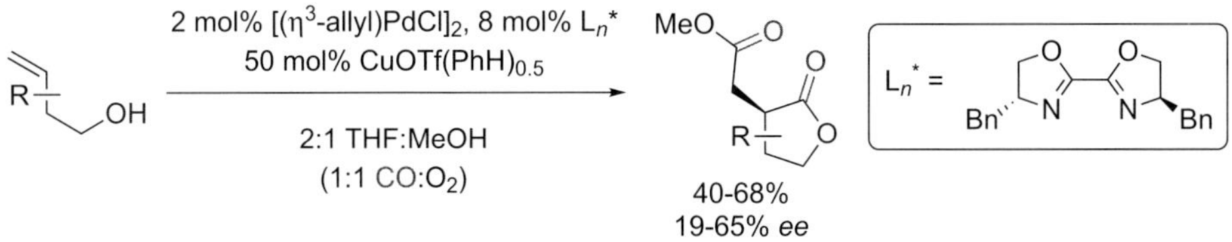
R
OH
2 mol% [(η^3-allyl)PdCl]$_2$, 8 mol% L$_n$*
50 mol% CuOTf(PhH)$_{0.5}$
2:1 THF:MeOH
(1:1 CO:O$_2$)
MeO
O
O
R
O
40-68%
19-65% *ee*
L$_n$* =
O
O
Bn
N
N
Bn

Scheme 2.18 Asymmetric dicarbonylation of alkenes [41].

2.2.5
Heterocarbonylation of C=C and C≡C Bonds

Heterocarbonylation of C=C or C≡C bonds involves the formation of new C–X and C–C bonds through heteropalladation of the C–C unsaturation followed by intramolecular trapping with a pendant alcohol [24]. This mode of reactivity is possible for intramolecular carboxy-, amino-, and oxycarbonylation of alkenes (Scheme 2.19), providing rapid access to fused bicyclic heterocycles [42–45]. An asymmetric oxycarbonylation via kinetic resolution was recently reported by the Gracza group (Scheme 2.20) [46]. In this case, *p*-benzoquinone was used as the oxidant and a moderate 31–62% *ee* was obtained using a bisoxazoline ligand.

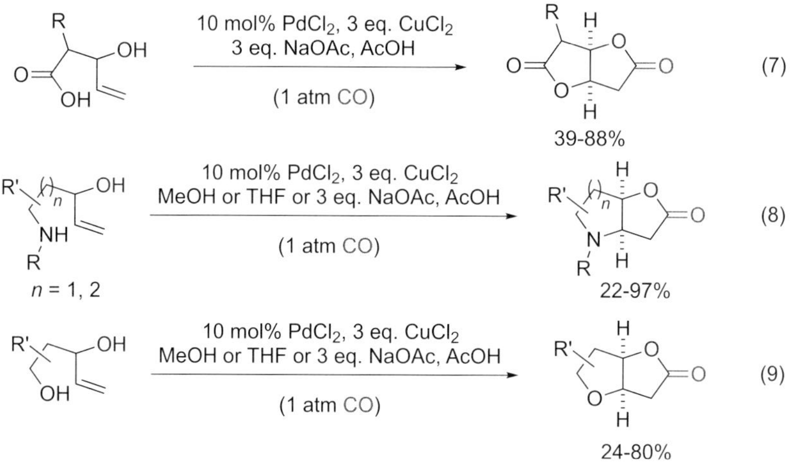

10 mol% $PdCl_2$, 3 eq. $CuCl_2$
3 eq. NaOAc, AcOH
(1 atm CO)
39-88% (7)

n = 1, 2
10 mol% $PdCl_2$, 3 eq. $CuCl_2$
MeOH or THF or 3 eq. NaOAc, AcOH
(1 atm CO)
22-97% (8)

10 mol% $PdCl_2$, 3 eq. $CuCl_2$
MeOH or THF or 3 eq. NaOAc, AcOH
(1 atm CO)
24-80% (9)

Scheme 2.19 Heterocarbonylation of alkenes by Pd(II) [42–44].

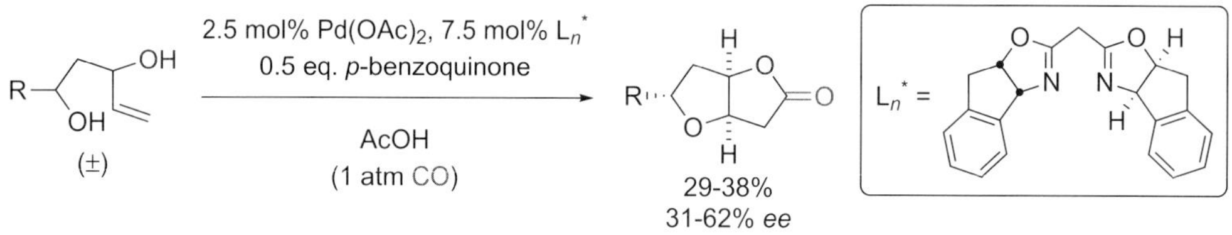

(±)
2.5 mol% $Pd(OAc)_2$, 7.5 mol% L_n^*
0.5 eq. *p*-benzoquinone
AcOH
(1 atm CO)
29-38%
31-62% *ee*
L_n^* =

Scheme 2.20 Asymmetric oxycarbonylation of alkenes by Pd(II) [46].

Matsuda *et al.* reported an innovative silylcarbonylation of alkynes for the synthesis of β-lactones from propargylic alcohols under Rh(0) catalysis (Scheme 2.21), noteworthy due to traditional challenges in preparing 4-membered rings [47]. Trapping of the acylrhodium intermediate with a second equivalent of silane to give an aldehyde was a competing side reaction; however, lactonization could be favored by the use of sterically hindered silanes (t-$BuMe_2SiH$) and strong base (DBU).

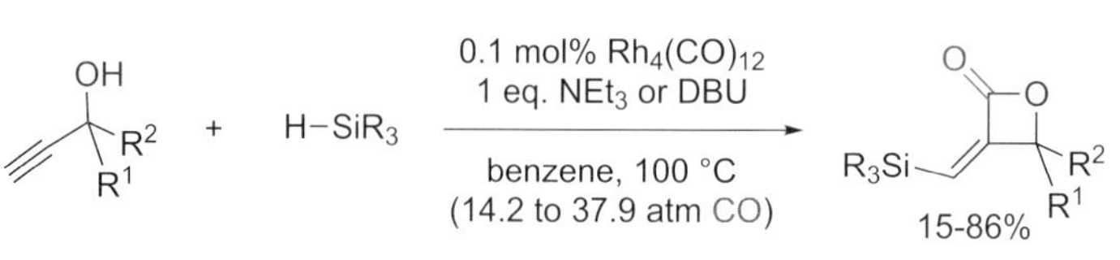

+ $H{-}SiR_3$
0.1 mol% $Rh_4(CO)_{12}$
1 eq. NEt_3 or DBU
benzene, 100 °C
(14.2 to 37.9 atm CO)
15-86%

Scheme 2.21 Silylcarbonylation of alkynes by Rh(0) [47].

2.2.6
Miscellaneous Lactone Syntheses Involving CO

One alternative approach to lactone synthesis with CO in the absence of C–X and C–M bonds involves the decarboxylation of cyclic carbonates where lactone formation occurs with concomitant loss of CO_2 (Scheme 2.22) [48–50]. Since CO_2 is the sole byproduct, this process is considerably more green than carbonylation of C–X and C–M bonds.

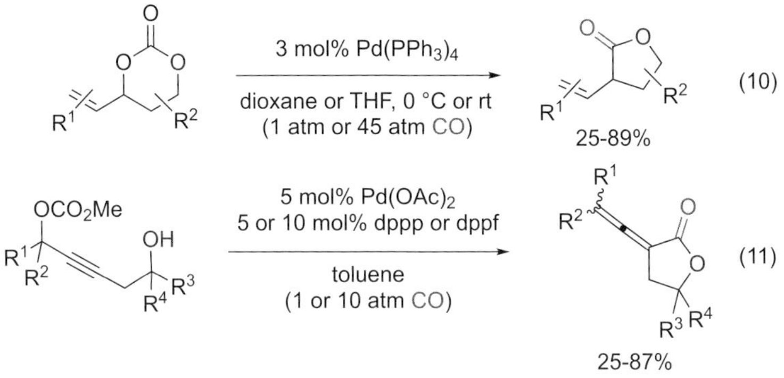

3 mol% $Pd(PPh_3)_4$
dioxane or THF, 0 °C or rt
(1 atm or 45 atm CO)
25-89% (10)

5 mol% $Pd(OAc)_2$
5 or 10 mol% dppp or dppf
toluene
(1 or 10 atm CO)
25-87% (11)

Scheme 2.22 Decarboxylative lactone formation with Pd(0) [48–50].

Transition metal catalyzed ring expansions of cyclic ethers to lactones under pressures of CO [51, 52] have been reported for tetrahydrofuran [53], oxetanes, and epoxides [54–56]. Carbonylation of epoxides is particularly important since β-lactone products are challenging synthetic targets (see Section 2.2.5). Using $Co(CO)_4^-$ in combination with a Lewis acidic Al-salen counterion, the reaction of (*R*)-propylene oxide and CO occurs with stereochemical retention (Scheme 2.23) [57]. The mechanism is believed to involve Lewis acid activation of the epoxide followed by nucleophilic ring opening with $Co(CO)_4^-$ [58].

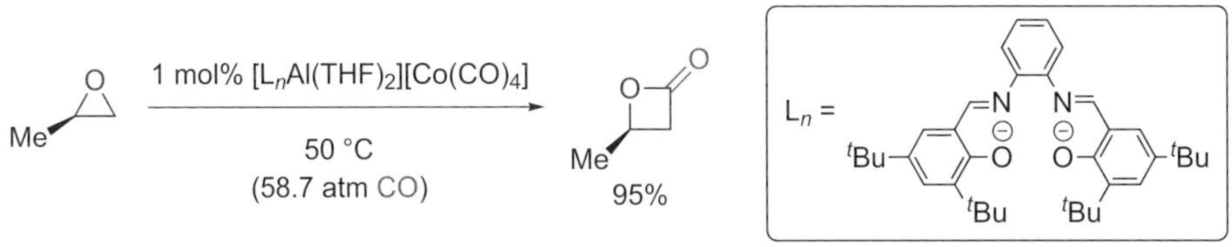

1 mol% $[L_nAl(THF)_2][Co(CO)_4]$
50 °C
(58.7 atm CO)
95%

Scheme 2.23 Carbonylative ring expansion of (*R*)-propylene oxide [57].

Heterocycloaddition with CO can also provide facile access to lactones. The intramolecular hetero-Pauson–Khand [2 + 2 + 1] cycloaddition of aldehydes or ketones generates bicyclic lactones from acyclic precursors. Catalytic reactions employing Ti [59, 60] and Ru [61, 62] via radical and ionic mechanisms, respectively, have been reported. An enantioselective variant has also been achieved with a C2-symmetric titanocene derivative (Scheme 2.24) [60].

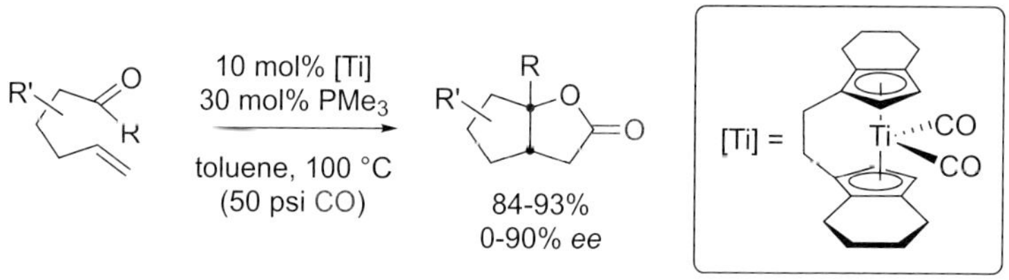

Scheme 2.24 Asymmetric intramolecular hetero-Pauson–Khand cyclocarbonylation [60].

2.3 Synthesis of Lactones via C=C and C≡C Addition

2.3.1 Hydrocarboxylation of C=C and C≡C Bonds

Hydrocarboxylation of alkenes or alkynes involves the formal addition of a carboxylic acid O—H bond across a C=C or C≡C bond (Scheme 2.25) [63]. In particular, intramolecular hydrocarboxylation provides an atom economical strategy for lactone synthesis. The aforementioned reaction is thermodynamically favorable but there is a large intrinsic kinetic barrier for this type of cyclization, thus requiring the addition of a catalyst. Catalysts for hydrocarboxylation typically facilitate addition by alkene or alkyne binding. This process increases the inherent electrophilicity of the C=C and C≡C bonds, respectively. Subsequent protonolysis (or β-H elimination under Pd catalysis) regenerates the catalytic species.

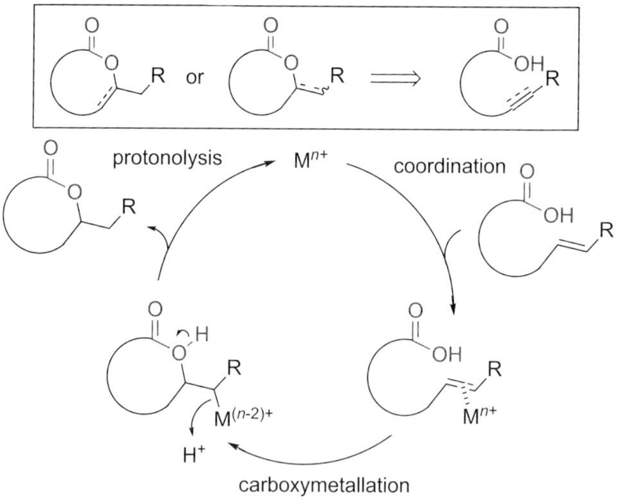

Scheme 2.25 Generic mechanism for hydrocarboxylation of alkenes [64].

A variety of highly Lewis acidic metal salts have been used to catalyze intramolecular hydrocarboxylation, including Ag(I), Pd(II), Au(I), Hg(II), Rh(I), Ni(II), and Zn(II) (Scheme 2.26) [63–68]. Depending on the metal of choice, different regioisomeric products can be formed. For instance, hydrocarboxylation of alkynoic acids with Au(I), Hg(II), and Rh(I) generates the *exo*-alkene product, while Pd(II) can yield both *endo*- and *exo*-products depending on the length of the tether [65–68]. In most cases, carboxymetallation occurs in an *anti*-fashion [64, 66].

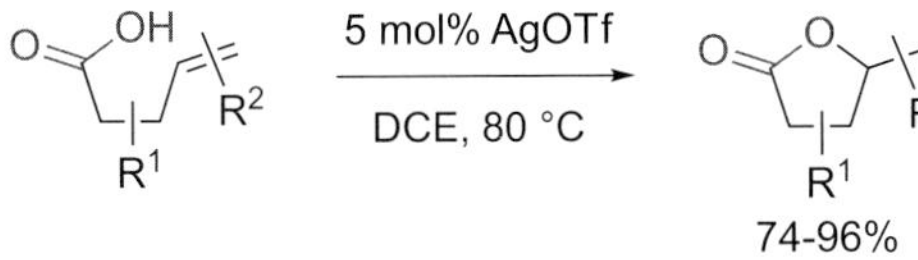

5 mol% AgOTf, DCE, 80 °C; 74-96% (12)

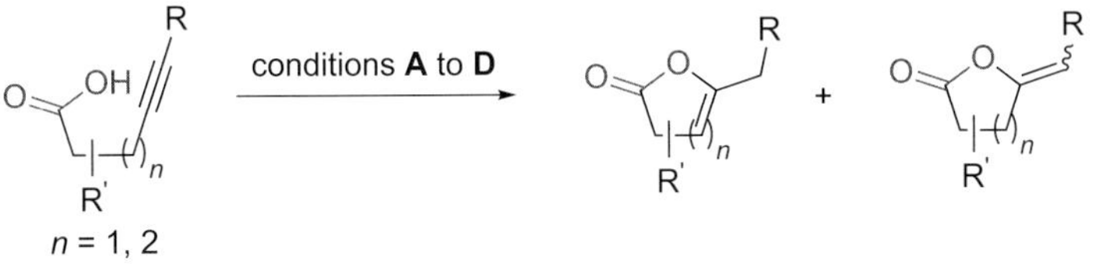

conditions **A** to **D** (13)

n = 1, 2

A: 0.1 mol% $Pd(PhCN)_2Cl_2$, 0.3 mol% NEt_3, THF, reflux (38-100% *endo* or *exo*)
B: 1-5 mol% AuCl, MeCN (72-97% *exo*)
C: 10 mol% $Hg(OAc)_2$ or $Hg(TFA)_2$, CH_2Cl_2 (49-86% *exo*)
D: 2-4.3 mol% $[Rh(Cy_2PCH_2CH_2PCy_2)Cl]_2$, CH_2Cl_2 (76-93% *exo*)

Scheme 2.26 Hydrocarboxylation of alkenes and alkynes [64–68].

Toste and coworkers reported the first asymmetric intramolecular hydrocarboxylation of an allene by combining a chiral ligand with a chiral counterion on a Au(I) catalyst (Scheme 2.27). Chiral ion pair formation was crucial to the high selectivity [69]. In another account, a related carboxyallene species was proposed to be an intermediate in a Pd-catalyzed cyclization of alkynoic acids [70].

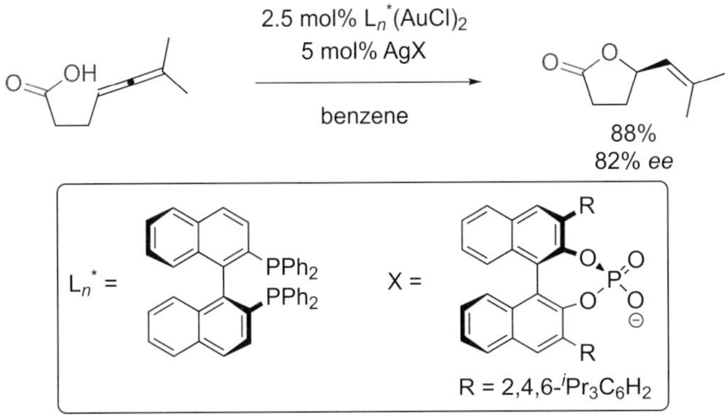

2.5 mol% $L_n^*(AuCl)_2$, 5 mol% AgX, benzene; 88%, 82% *ee*

$R = 2,4,6\text{-}^iPr_3C_6H_2$

Scheme 2.27 Asymmetric hydrocarboxylation of allenes by Au(I) [69].

Since alkenes and alkynes can be prepared via Heck and Sonogashira reactions, respectively, tandem processes are also possible for lactone synthesis [71–77]. Appropriately, *o*-iodobenzoic acids can generate both phthalide and isocoumarin derivatives under Pd catalysis, with a CuI co-catalyst favoring the former and

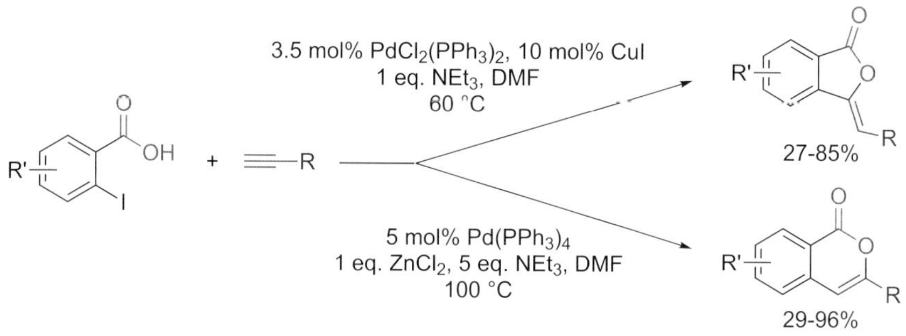

Scheme 2.28 Regioselective tandem Sonogashira/hydrocarboxylation of alkynes [71, 72].

stoichiometric $ZnCl_2$ the latter (Scheme 2.28) [71, 72]. Carboxyl-directed C—H activation is also possible. Under Rh catalysis, carborhodation can lead to isocoumarin products (Scheme 2.29) [77, 78].

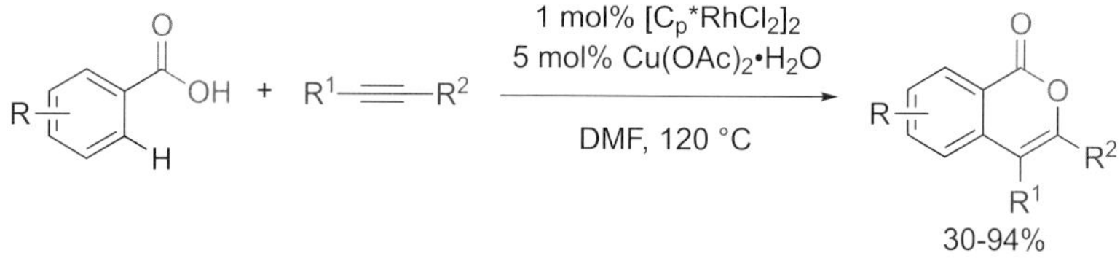

Scheme 2.29 Isocoumarins by oxidative arylcarboxylation of alkynes by Rh(I) [77, 78].

A similar type of oxidative cyclization of alkenes with both Pd and Rh is possible [75, 77, 78]. In particular, Miura showed that benzoic acids could first be oxidatively coupled with terminal alkenes, followed immediately by oxidative lactonization to give a mixture of isocoumarins and phthalides (Scheme 2.30) [75]. Because of the propensity of sp^2 C—H bonds to undergo cyclorhodation, tandem oxidative Heck couplings have also been achieved (Scheme 2.31) [77, 78].

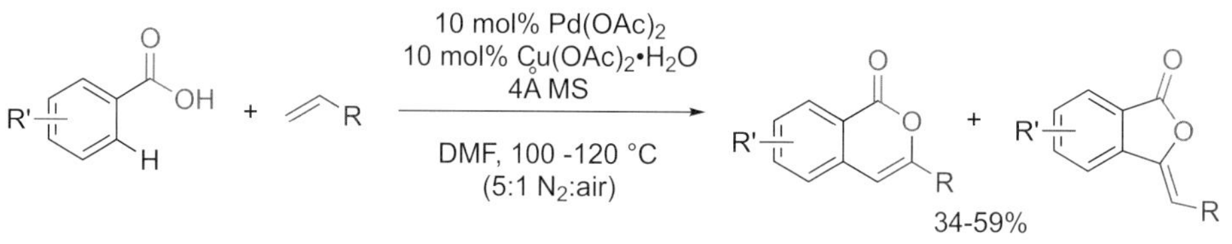

Scheme 2.30 Tandem C—H activation/hydrocarboxylation of alkenes by Pd(II) [75].

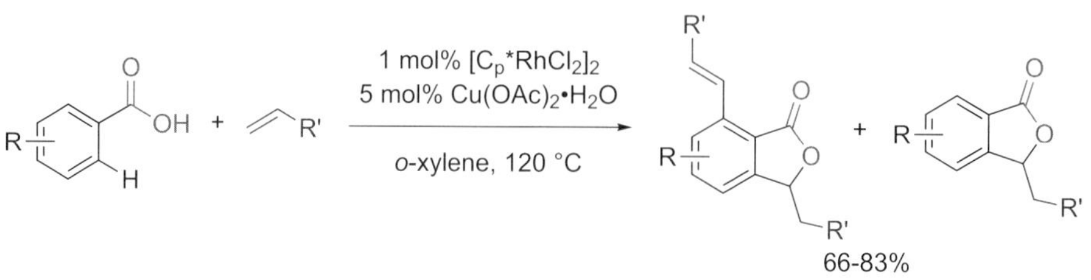

Scheme 2.31 Phthalides by C—H activation/hydrocarboxylation of alkenes by Rh(I) [77, 78].

Larock reported direct thallation as one alternative, demonstrating successful intermolecular trapping with alkenes, alkenyl halides, allenes, and vinylcyclopropanes (Scheme 2.32) [76].

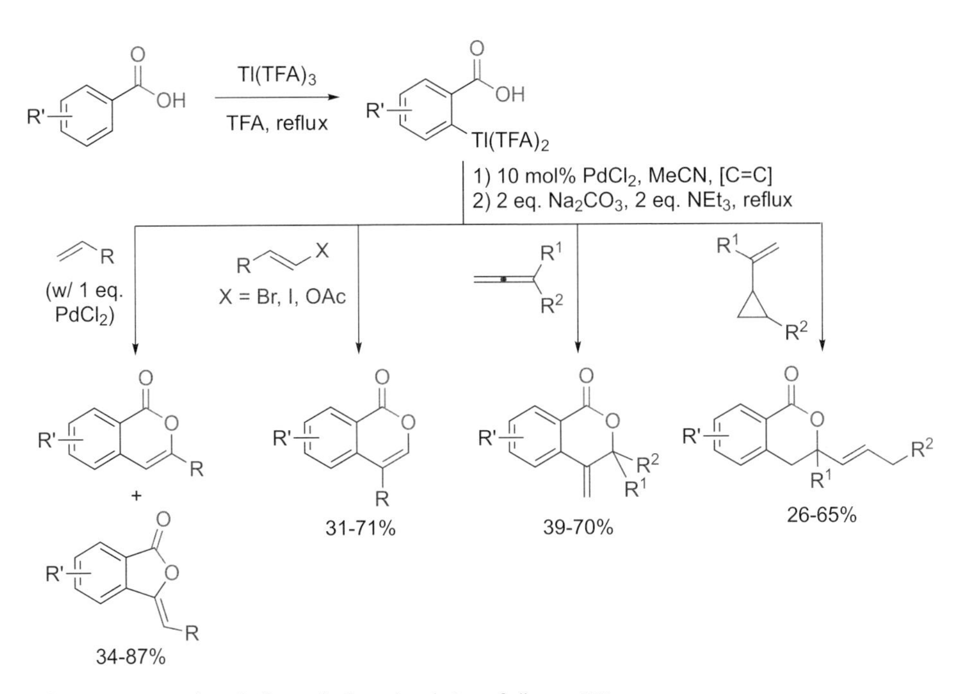

Scheme 2.32 Tandem thallation/hydrocarboxylation of alkenes [76].

Surprisingly, methyl esters are also suitable substrates whereby intramolecular cyclization occurs with concomitant loss of methyl iodide. Larock used internal alkynes as coupling partners for lactone synthesis (Scheme 2.33) [74]. The proposed mechanism involves oxidative addition of Pd(0) to the aryl iodide, followed by addition across the alkyne and cyclization of the carbonyl O of the ester to form an oxonium ion. Reductive elimination followed by loss of the methyl group then yields the product [74]. Shen and coworkers also reported a variant utilizing *o*-2,2-dibromovinylbenzoates (Scheme 2.34) [79].

OMe + R^1—≡—R^2 → (10 mol% $Pd(OAc)_2$, 1 eq. Na_2CO_3, 1 eq. LiCl; DMF or MeCN, 100 °C) 10-79%

Scheme 2.33 Isocoumarins by carbopalladation of internal alkynes terminated by esters [74].

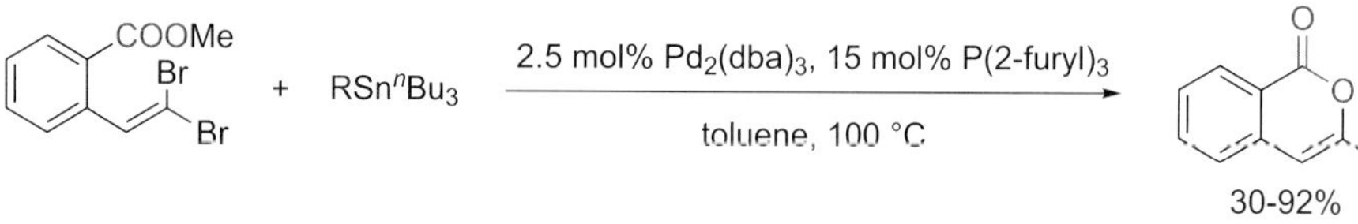

COOMe Br Br + RSn^nBu_3 → 2.5 mol% $Pd_2(dba)_3$, 15 mol% $P(2\text{-furyl})_3$; toluene, 100 °C → O O R 30-92%

Scheme 2.34 Isocoumarins by Stille coupling of methyl *o*-2,2-dibromovinyl benzoates [79].

In catalytic reactions involving Pd(II) salts, carboxypalladation yields an alkylpalladium species that can often undergo β-H elimination instead of protonolysis. Subsequently, Stoltz and coworkers demonstrated that Wacker-type processes can also afford lactones under oxidative conditions (Scheme 2.35). The proposed mechanism involves Pd(II) coordination to the alkene, followed by oxypalladation and β-H elimination. After elimination of HX to form Pd(0), aerobic oxidation is required to regenerate a Pd(II) species. The net result is olefin transposition to an adjacent position [80].

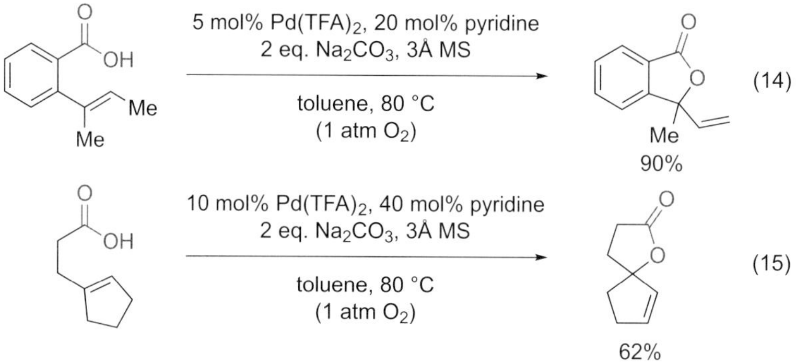

O OH Me Me → 5 mol% $Pd(TFA)_2$, 20 mol% pyridine, 2 eq. Na_2CO_3, 3Å MS; toluene, 80 °C (1 atm O_2) → O O Me 90% (14)

O OH → 10 mol% $Pd(TFA)_2$, 40 mol% pyridine, 2 eq. Na_2CO_3, 3Å MS; toluene, 80 °C (1 atm O_2) → O O 62% (15)

Scheme 2.35 Oxidative Wacker reactions of alkenylalkanoic acids [80].

2.3.2
Carbo- and Oxy-Carboxylation of C=C and C≡C Bonds

Carbocarboxylation of alkenes or alkynes involves the formation of new C—C and C—O bonds across the C=C or C≡C unsaturation. Unlike hydrocarboxylation, carbo- and oxy-carboxylation are not atom economical strategies since the mechanisms typically involve two distinct bond forming events. The most common approach involves oxidative addition to an aryl or alkenyl halide followed by addition to a C=C [81] or C≡C bond [82] (Scheme 2.36). The mechanism involves subsequent C—C bond formation instead of protonolysis (see Scheme 2.25).

Dong and coworkers recently reported a novel intramolecular carbocarboxylation of alkenes in a formal [3 + 2] cycloaddition as catalyzed by Pd (Scheme 2.37) [83]. In this reaction, C—Cl, C—C, and C—O bonds are formed in a single step. The proposed mechanism involves a *trans*-chloropalladation, followed by *syn*-oxypalladation, and

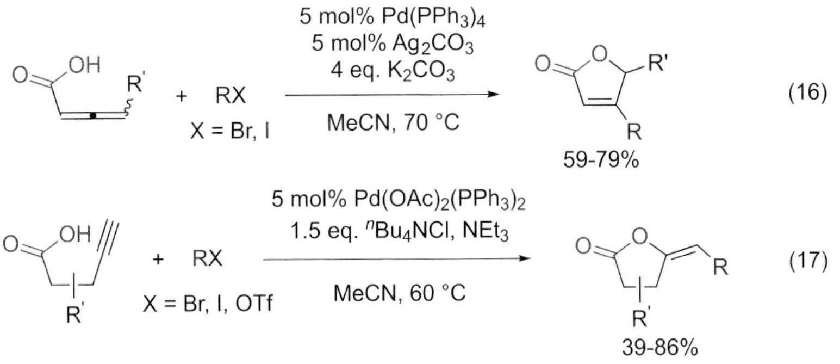

$$\text{allenoic acid} + RX \xrightarrow[\text{MeCN, 70 °C}]{\text{5 mol\% Pd(PPh}_3)_4\text{, 5 mol\% Ag}_2\text{CO}_3\text{, 4 eq. K}_2\text{CO}_3} \text{butenolide (59-79\%)} \quad X = Br, I \qquad (16)$$

$$\text{alkynoic acid} + RX \xrightarrow[\text{MeCN, 60 °C}]{\text{5 mol\% Pd(OAc)}_2\text{(PPh}_3)_2\text{, 1.5 eq. }^n\text{Bu}_4\text{NCl, NEt}_3} \text{alkylidene lactone (39-86\%)} \quad X = Br, I, OTf \qquad (17)$$

Scheme 2.36 Arylcarboxylation of allenes and alkynes [81, 82].

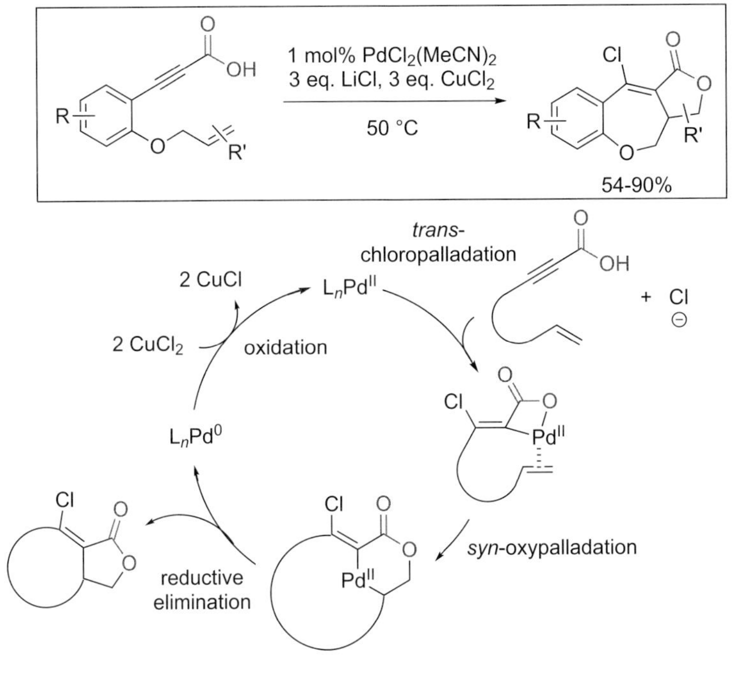

1 mol% $PdCl_2(MeCN)_2$, 3 eq. LiCl, 3 eq. $CuCl_2$, 50 °C; 54-90%

trans-chloropalladation; oxidation: 2 $CuCl_2$ → 2 CuCl; L_nPd^{II}; L_nPd^0; + $Cl^{\ominus}$; syn-oxypalladation; reductive elimination

Scheme 2.37 Intramolecular carbocarboxylation of alkenes *en route* to a 6,7,5-tricyclic core [84].

finally reductive elimination in a cascade sequence generating a 6,7,5-tricyclic core containing a lactone motif (Scheme 2.38) [83].

Oxycarboxylation of alkenes, unlike carbocarboxylation, yields two new C–O bonds in a single step. The Dong group also reported the first example of alkene dioxygenation as a route to lactone synthesis via Pd catalysis (Scheme 2.37). The proposed mechanism for this transformation involves a novel Pd(II)/Pd(IV) pathway made possible by the hypervalent iodine-mediated oxidation that occurs

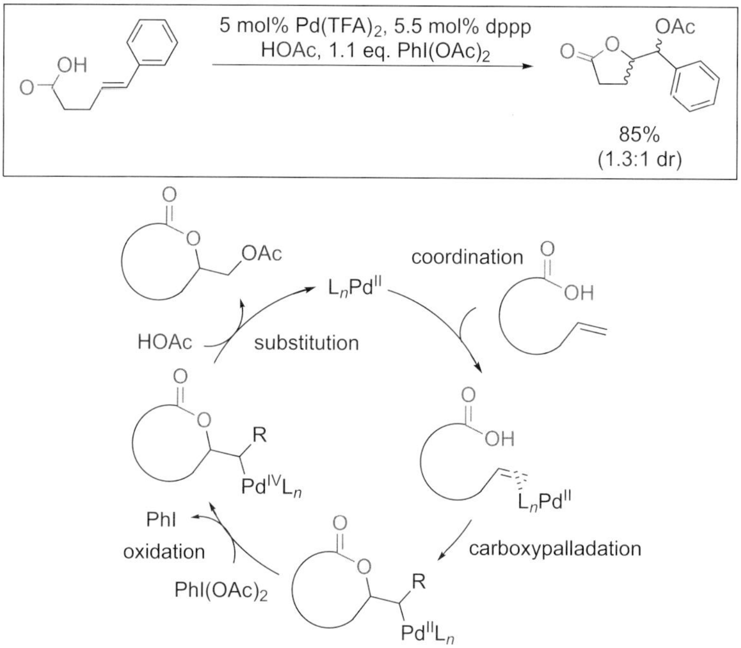

Scheme 2.38 Mechanism of oxycarboxylation of alkenes [84].

in preference to β-H elimination. S_N2-type displacement of the highly electrophilic Pd(IV) center liberates the desired lactone product [84].

2.4 Synthesis of Lactones via C=O Hydroacylation

2.4.1 Aldehyde Hydroacylation

Hydroacylation of carbonyl groups involves the formal addition of an aldehydic C–H bond across a C=O bond, a unique strategy for lactone synthesis that also exhibits complete atom economy. Aldehyde oxidation occurs with concomitant reduction of a neighboring carbonyl group, an orthogonal approach to traditional cyclization strategies since it employs only C=O bonds. Furthermore, in the case of ketone hydroacylation, C–H addition results in the formation of a new stereogenic center, providing direct access to chiral lactones without requiring the predetermined chirality of a secondary alcohol in traditional methods involving carboxylic acid activation (Scheme 2.39) [85].

Bosnich and coworkers reported the first example of aldehyde hydroacylation (i.e., Tischenko reaction) in the presence of a cationic Rh(I) catalyst (Scheme 2.40). The proposed mechanism occurs via C–H activation of the aldehyde, C=O insertion,

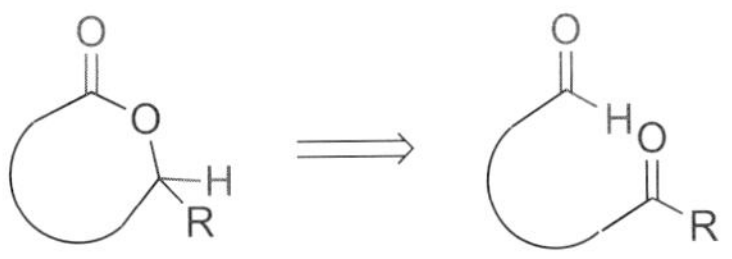

Scheme 2.39 Hydroacylation of C=O bonds [85].

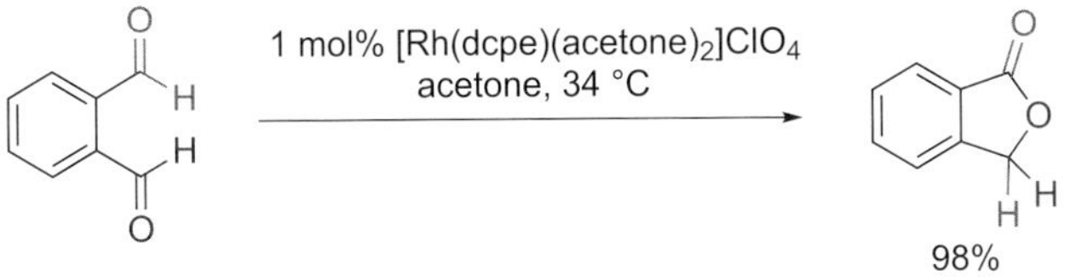

Scheme 2.40 Aldehyde hydroacylation by Rh(I) [86].

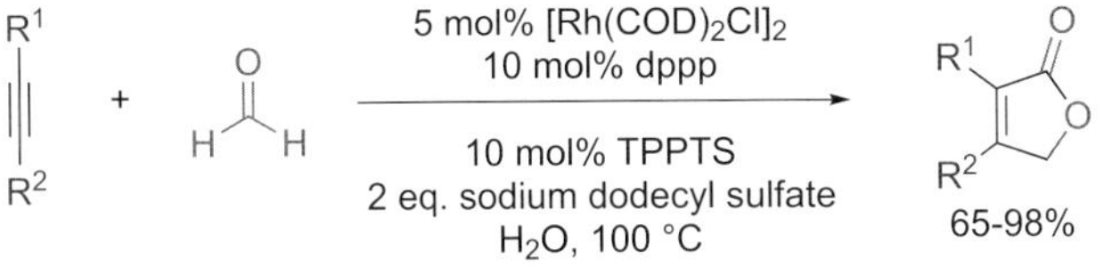

Scheme 2.41 Multicomponent approach to lactones from alkynes and formaldehyde [87].

and subsequent reductive elimination to furnish the phthalide product (see Scheme 2.44) [86]. The Morimoto group developed a multicomponent variant of this transformation by combining alkynes with formaldehyde, proposed to be a source of both CO and H_2 (Scheme 2.41). Accordingly, the authors propose the intermediacy of a 1,4-dialdehyde [87].

2.4.2
Ketone Hydroacylation

Ketone hydroacylation is more challenging than the corresponding aldehyde hydroacylation presumably due to a slower rate of ketone insertion, rendering decarbonylation kinetically competent. For instance, 4-oxo-4-phenylbutanal reacts to generate the desired lactone product (59%) in addition to the decarbonylation product (40%) (Scheme 2.42) [86].

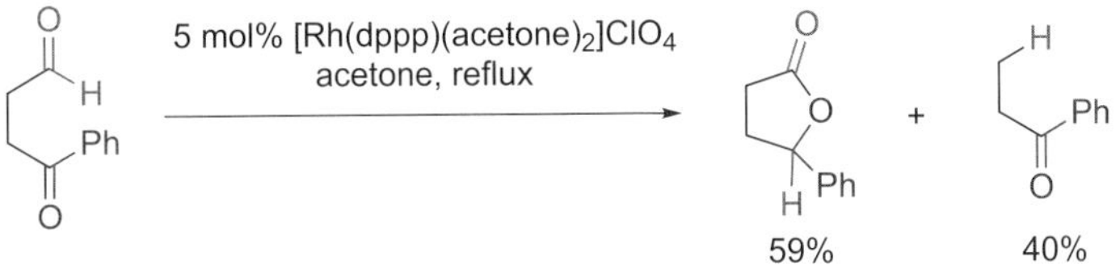

Scheme 2.42 Ketone hydroacylation by Rh(I) [86].

Interestingly, Suzuki and coworkers demonstrated that Ir(III) catalysts could promote formal ketone hydroacylation of 1,5-ketoaldehydes, yielding either isocoumarins or 3,4-dihydroisocoumarins (Scheme 2.43). The presence of pivaldehyde as an oxidant favors the formation of isocoumarins. The mechanism of this transformation involves a transfer hydrogenation process (see Section 2.5.1) [88].

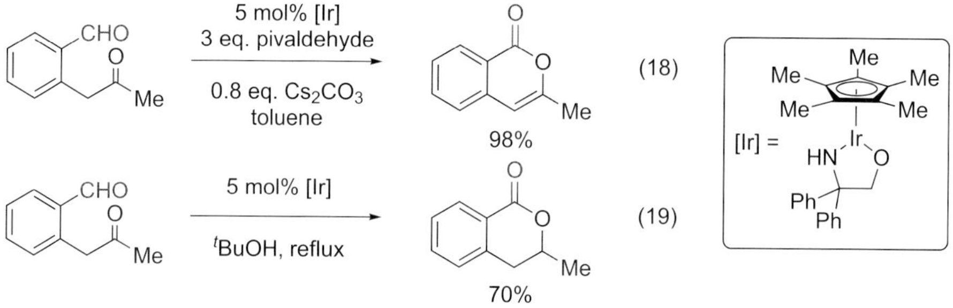

5 mol% [Ir], 3 eq. pivaldehyde, 0.8 eq. Cs_2CO_3, toluene → 98% (18)

5 mol% [Ir], ^{t}BuOH, reflux → 70% (19)

Scheme 2.43 Isocoumarin/3,4-dihydroisocoumarins by ketone hydroacylation [88].

The Dong group reported the first asymmetric ketone hydroacylation with 1,6-keto aldehydes bearing an ether linkage under Rh(I) catalysis (Scheme 2.43) [85]. Consistent with the observations of Bosnich [86], electron-rich diphosphine ligands were shown to give superior reactivity. The use of (*R*)-DTBM-SEGPHOS as ligand gave

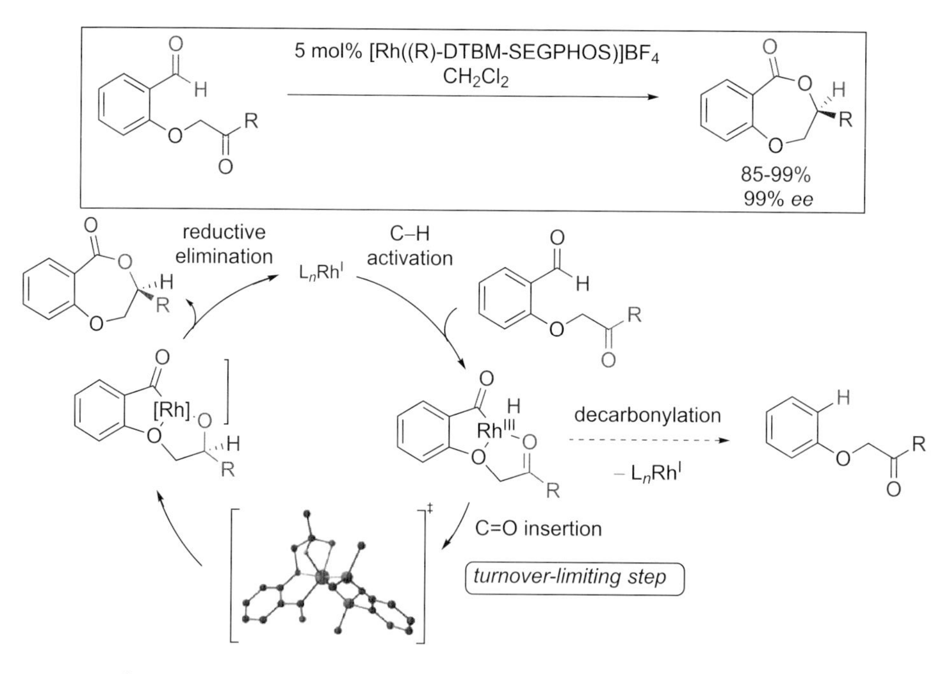

Scheme 2.44 Asymmetric ketone hydroacylation by Rh(I) [85].

excellent yields (85–99%) and enantioselectivities (≥99%). Arene formation by decarbonylation was a competitive pathway [85].

In order to gain insights into the mechanism of this unique asymmetric transformation, detailed experimental and theoretical studies were undertaken [89]. Hammett plot ($\varrho = 0.29$ with dppp) and kinetic isotope effect experiments ($k_H/k_D = 1.8$ with dppp; 3.3 with (*R*)-DTBM-SEGPHOS) in conjunction with DFT studies suggest that ketone insertion into the Rh−H is the turnover limiting step of the catalytic cycle, contrary to previous reports of olefin hydroacylation, where reductive elimination has been shown to be turnover limiting [90]. The presence of the ether O appears to play a pivotal role in suppressing competing decarbonylation by coordination to Rh throughout the catalytic cycle [89]. Hence, substitution of the O atom with an S atom had minimal effect on ketone hydroacylation, while incorporation of a methylene spacer resulted in complete shutdown of desired reactivity, even at elevated temperatures.

Recently, ketone hydroacylation has been extended to the asymmetric synthesis of phthalides (Scheme 2.45) [91]. Since the ketoaldehyde substrates lack an ether O, decarbonylation was found to be a major competing pathway when non-coordinating counteranions (e.g., BF_4^-) were used. However, a strong counterion effect was observed, with NO_3^- being optimal. This methodology was applied to a short synthesis of (*S*)-(-)-3-*n*-butylphthalide, an extract of celery [91].

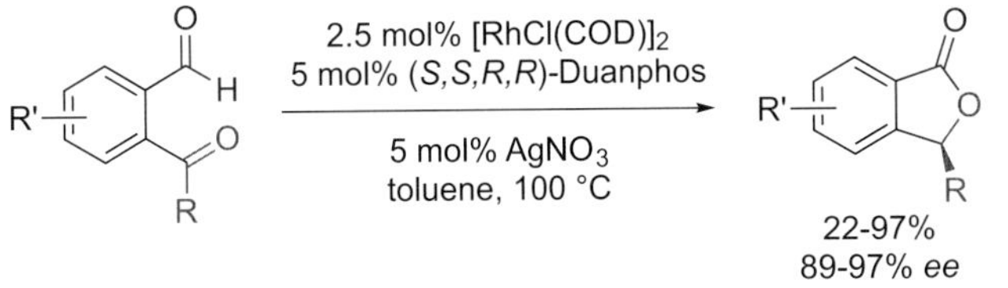

2.5 mol% $[RhCl(COD)]_2$
5 mol% (*S,S,R,R*)-Duanphos
5 mol% $AgNO_3$
toluene, 100 °C
22-97%
89-97% *ee*

Scheme 2.45 Asymmetric synthesis of phthalides via ketone hydroacylation [91].

2.4.3
[4 + 2] Annulation

Based on earlier work in Rh-catalyzed alkyne hydroacylation, the Tanaka group discovered that a rhodacycle generated upon oxidative addition of a 2-alkynylbenzaldehyde derivative underwent efficient dimerization in a formal [4 + 2] annulation (Scheme 2.46) [92]. This strategy involves formal hydroacylation across one C≡C

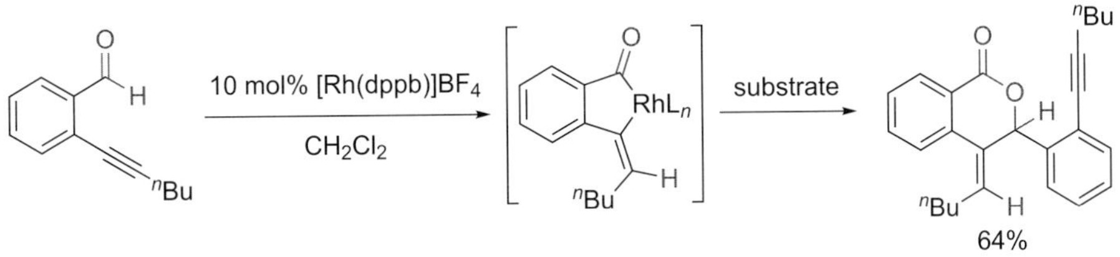

10 mol% $[Rh(dppb)]BF_4$
CH_2Cl_2
substrate
64%

Scheme 2.46 Dimerization of 2-(hex-1-ynyl)benzaldehyde by Rh(I) [92].

bond and one C=O bond, generating a valuable lactone product in high efficiency (Scheme 2.47). In this process, two new C—H bonds and two new C—O bonds are formed.

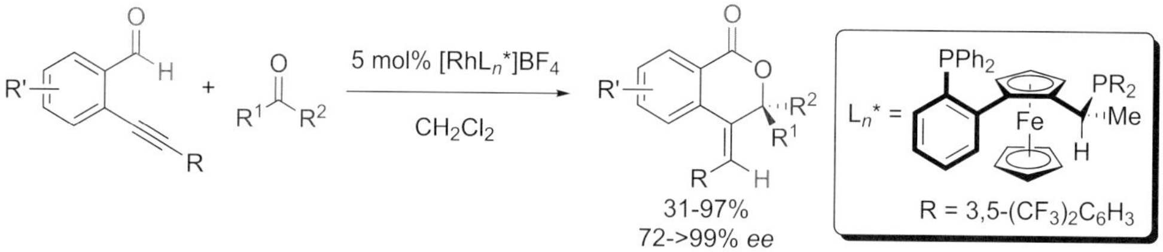

5 mol% $[RhL_n^*]BF_4$, CH_2Cl_2; 31-97%, 72->99% *ee*; L_n^* = ; R = 3,5-$(CF_3)_2C_6H_3$

Scheme 2.47 Enantioselective [4 + 2] annulation via formal alkyne/ketone hydroacylation [92].

Upon careful examination of reaction parameters, Tanaka and coworkers successfully developed a highly enantioselective lactone synthesis by reaction with both aldehydes and activated ketones as the coupling partners for the synthesis of spirocyclic benzopyranones. Cationic Rh catalysts were again found to be the most efficient. Importantly, homodimerization of the parent 2-alkynylbenzaldehydes was not observed under the optimized conditions [92].

2.5 Miscellaneous Syntheses of Lactones

2.5.1 Oxidative Lactonization of Diols

Oxidative lactonization of diols has received significant attention from the synthetic community for application in total synthesis, particularly catalytic variants involving Ir [93, 94], Ru [95–99], Cu [100], Pd [101], and Rh [102] (Scheme 2.48). The H-acceptors used include ketones [93, 94, 97], α,β-unsaturated ketones [99, 102], bromobenzene [101], and *N*-methylmorpholine oxide [96]. Asymmetric induction in this type of oxidative lactonization has been primarily limited to desymmetrization of *meso*-diols, although low enantiomeric excesses were obtained with Ru and Rh [95, 102]. Ir, however, catalyzes the desired transformation and gives enantioenriched lactone products in good selectivities (Scheme 2.49) [94] via transfer hydrogenation [88]. In this case, asymmetric induction is believed to occur at the stage of

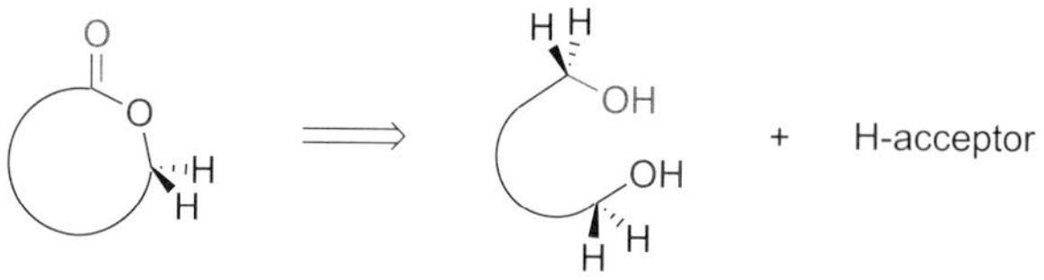

+ H-acceptor

Scheme 2.48 Lactone synthesis by oxidative lactonization of diols.

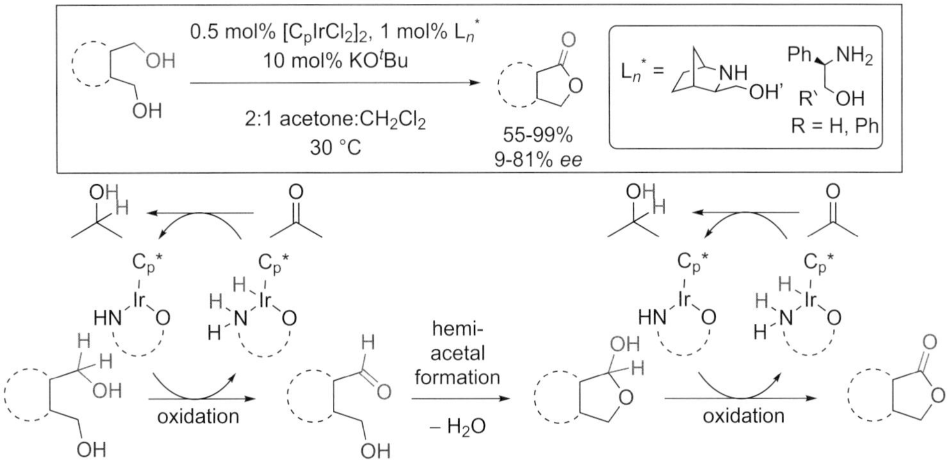

Scheme 2.49 Asymmetric oxidative lactonization of *meso* diols by Ir(I) [88, 94].

alcohol oxidation, as the lactol with the same *ee* as the final product was detected in the early stages of the reaction. This also indicates that no significant kinetic resolution of lactol intermediates takes place. However, under prolonged exposure to the reaction conditions, some erosion of enantioselectivity was observed.

2.5.2 Reductive Cyclization of Ketoacids and Ketoesters

Complementary to the oxidative diol lactonization approach, a substrate containing both ketone and ester functionalities may undergo reduction to liberate the desired lactone product following ring closure. Noyori and coworkers demonstrated that Ru-BINAP complexes were excellent catalysts for the asymmetric reduction of ketones with neighboring carboxylic acid and ester functionalities capable of undergoing the desired lactonization in high yields and selectivities (Scheme 2.50) [103, 104].

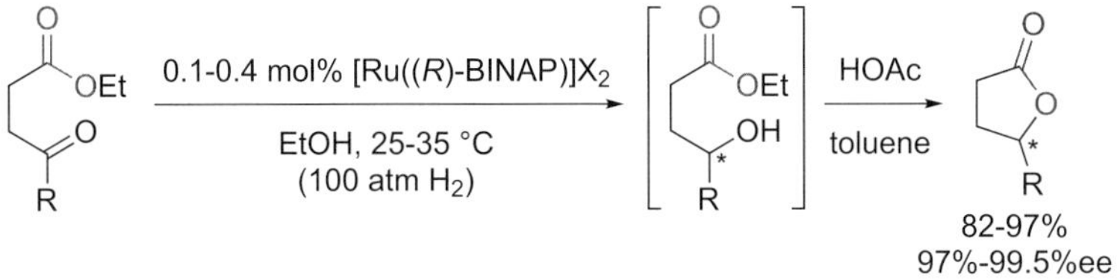

Scheme 2.50 Asymmetric reductive lactonization of ketoesters by Ru(I) [104].

Transfer hydrogenation is an alternative strategy that has also been successfully achieved, using either isopropanol (Scheme 2.51) [105] or sodium formate (Scheme 2.52) [106] as the stoichiometric reductant in place of H_2. High selectivities and efficiencies are possible.

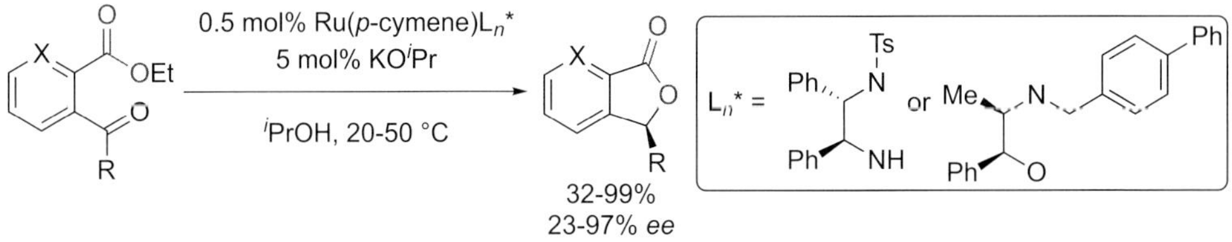

Scheme 2.51 Phthalides by transfer hydrogenation of ketoesters by Ru(I) [105].

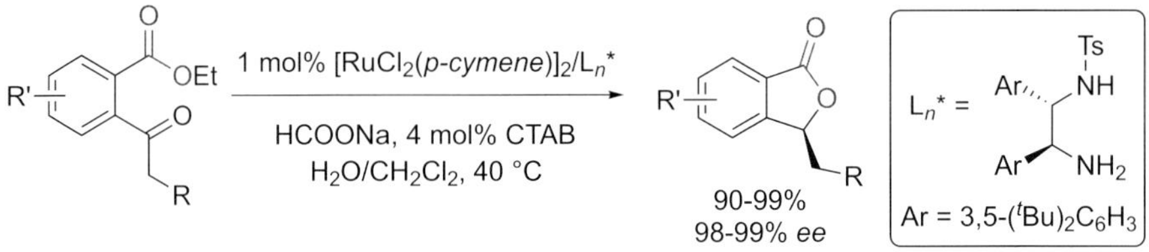

Scheme 2.52 Phthalides by transfer hydrogenation of ketoesters by Ru(I) [106].

2.5.3
C–H Oxygenation

Direct formation of C–O bonds via sp^3 C–H oxygenation is an innovative strategy for lactone synthesis. The first example reported in the literature employed catalytic amounts of Cu and Ag in a radical-type oxygenation [107]. A Pt-catalyzed variant was later developed by the Chang group (Scheme 2.53) [108]. Recently, White and coworkers have shown that allylic C–H oxygenation can be used for macrolactonization without the need for performing experiments at high dilution. Using a Pd(II)/bis-sulfoxide catalyst, 14- to 19-membered lactones could be prepared. An allyl palladium carboxylate is proposed as a templating intermediate to facilitate ring closure (Scheme 2.54) [109].

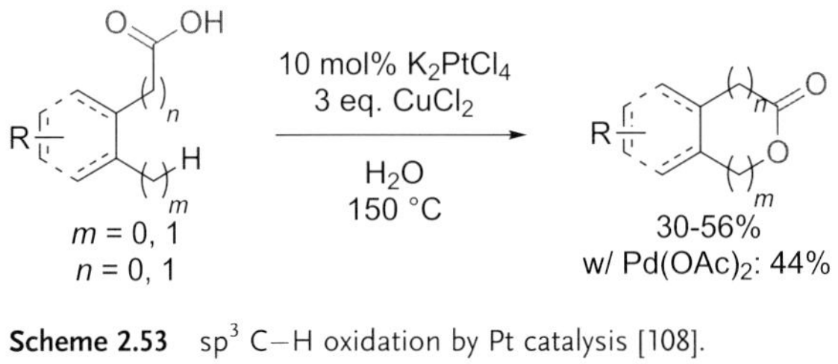

Scheme 2.53 sp^3 C–H oxidation by Pt catalysis [108].

Employing this strategy, White and coworkers successfully applied a late stage C–H oxidation for the total synthesis of 6-deoxyerythronolide B with excellent diasteroselectivity (>40: 1) (Scheme 2.55). Interestingly, upon addition of tetrabutylammonium fluoride as an additive, the opposite diastereomer was favored (1: 1.3) due to disruption of the Pd chelate intermediate [110]. The authors reasoned that the

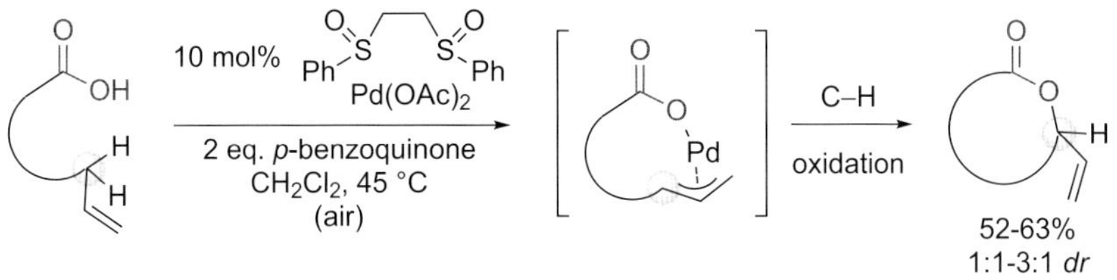

Scheme 2.54 Macrolactones via allylic C–H oxidation by Pd catalysis [109].

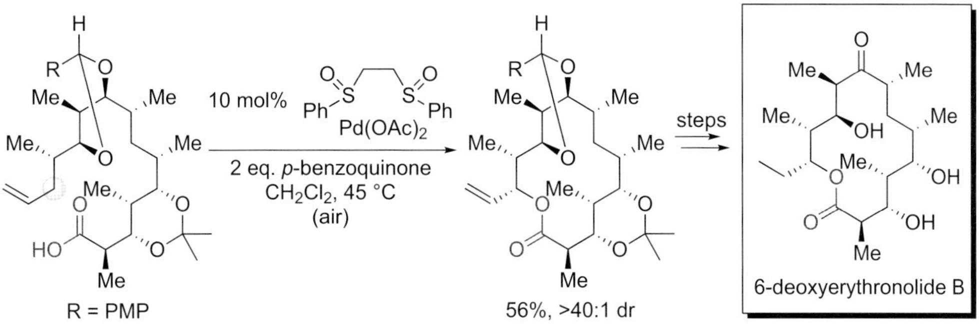

Scheme 2.55 Late stage C–H oxidation in the total synthesis of 6-deoxyerythronolide B [110].

latter reaction suffered from lower diastereoselectivity because of binding of the Pd on the face opposite the key C–O bond-forming event.

2.5.4 Ring Closure of Benzoic Acids with Dihaloalkanes

Another C–H activation strategy for lactone synthesis involves the ring closure of benzoic acids with dihaloalkanes [111]. Yu and coworkers recently developed a directed C–H functionalization/alkylation approach using Pd catalysts. Direct access to phthalides and 3,4-dihydroisocoumarins was possible using dibromomethane and 1,2-dichloroethane, respectively (Scheme 2.56).

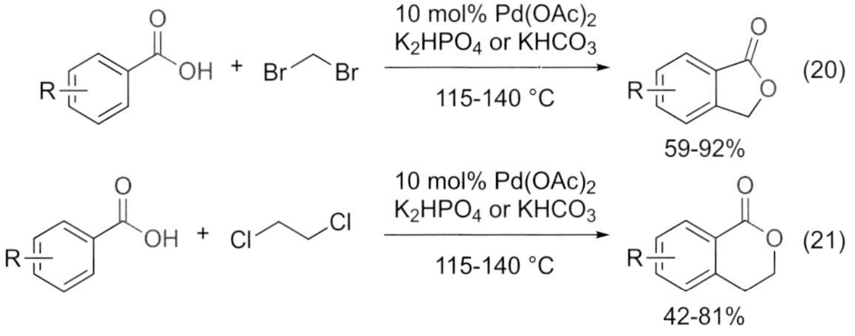

Scheme 2.56 Isocoumarin synthesis by directed C–H activation/dihaloalkylation [111].

2.5.5
Baeyer–Villiger Oxidation of Cyclic Ketones

The Baeyer–Villiger oxidation/rearrangement of cyclic ketones yields lactones directly, resulting in the formation of two new C—O bonds at the expense of one C—C bond (Scheme 2.57) [5]. The first accounts of transition metal catalysis for this transformation, however, were reported in 1994, using chiral Cu [112] and Pt [113] catalysts in a kinetic resolution strategy (Scheme 2.58). Recent developments of this methodology include a parallel kinetic resolution with Zr catalysts to generate two regioisomeric products (Scheme 2.59) [114].

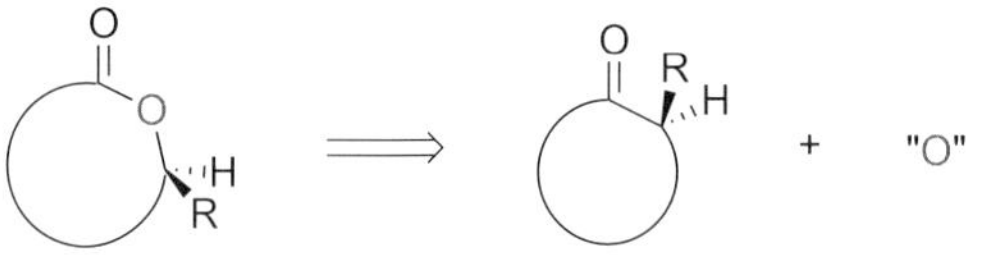

Scheme 2.57 Baeyer–Villiger oxidation of cyclic ketones [5].

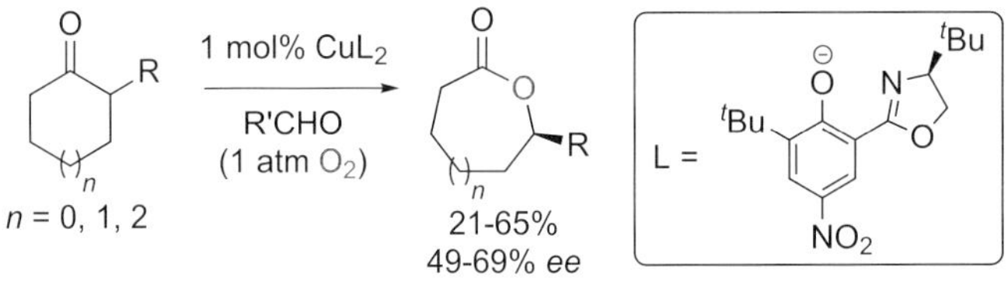

Scheme 2.58 Kinetic resolution of cyclic ketones [112].

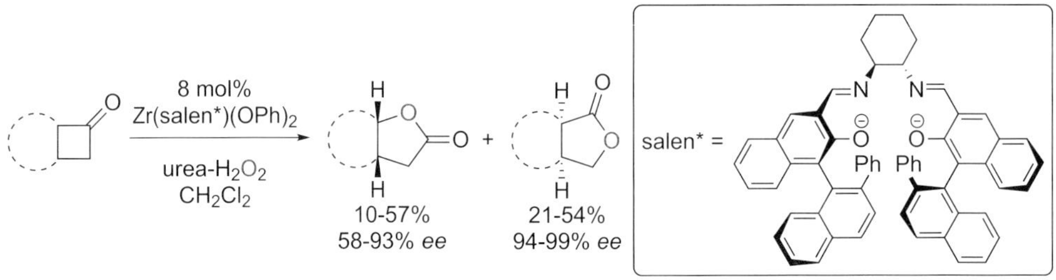

Scheme 2.59 Regiodivergent parallel kinetic resolution of cyclic ketones [114].

2.5.6
Ring Opening of Cyclopropanes with Carboxylic Acids

A unique account of lactonization via addition of a carboxylic acid to a cyclopropane ring was disclosed by Yudin and coworkers where Pd acts as a π-acid to the highly strained 3-membered ring [115]. Complementary to hydrocarboxylation of C=C and C≡C bonds (see Section 2.3.1), 5-membered and 6-membered lactones can be prepared via this strategy (Scheme 2.60). The authors postulate a mechanism involving ring opening of the cyclopropane ring to liberate an alkene followed by Wacker oxidation. However, direct carboxypalladation of the cyclopropane ring cannot be ruled out [115].

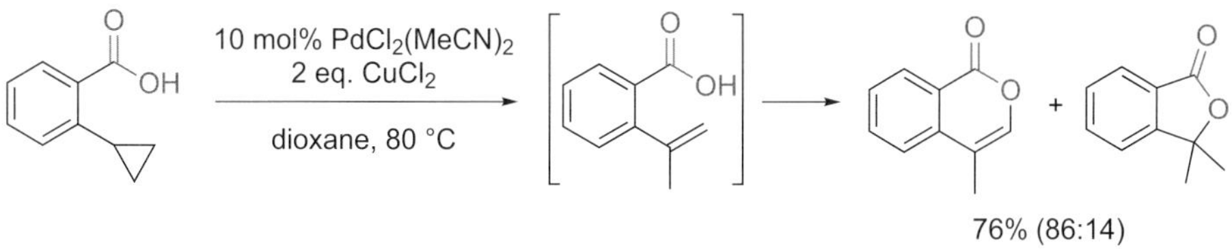

OH
10 mol% PdCl2(MeCN)2
2 eq. CuCl2
dioxane, 80 °C
OH
O
O
76% (86:14)

Scheme 2.60 Isocoumarins and phthalides by Pd-catalyzed cyclopropane carboxylation [115].

2.5.7
Ring Closure of *o*-Iodobenzoates with Aldehydes

Lactone products can also be formed via Grignard-type addition to aldehydes from *o*-iodobenzoates [116]. The Cheng group developed an enantioselective Co-catalyzed method for preparation of 5-membered phthalide products from methyl *o*-iodobenzoate and aldehydes (Scheme 2.61). While Ni and Pd catalysts were ineffective, the authors propose a Co(I)/Co(III) mechanism involving oxidative addition of the sp^2 C–I bond followed by addition and ring closure [116]. Co(III) is reduced to Co(I) by the zinc metal present in solution.

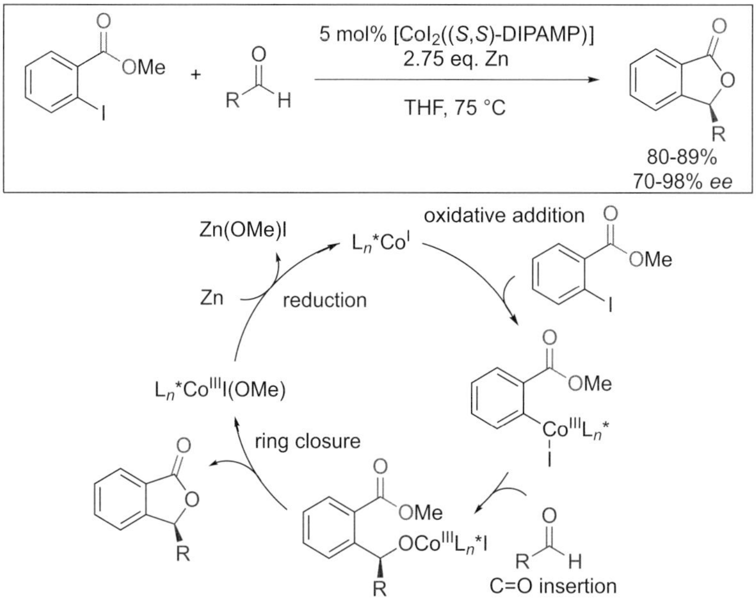

OMe
I
R H
5 mol% [CoI2((S,S)-DIPAMP)]
2.75 eq. Zn
THF, 75 °C
R
80-89%
70-98% ee
oxidative addition
Zn(OMe)I
Ln*CoI
Zn
reduction
Ln*CoIIII(OMe)
ring closure
CoIIILn*
OCoIIILn*I
C=O insertion

Scheme 2.61 Phthalides by Co-catalyzed Grignard-type addition to aldehydes [116].

2.5.8 Synthesis of Lactones Involving CO_2

Although much less prevalent than the use of CO in lactone synthesis, direct incorporation of CO_2 for the synthesis of lactones is advantageous due to lower toxicity while maintaining high atom efficiency. Louie and coworkers demonstrated that a formal [2 + 2 + 2] cycloaddition of bis-alkynes with CO_2 yields pyrone derivatives under Ni(0) catalysis (Scheme 2.62) [117]. It is proposed that an initial cyclometallation of one alkyne and CO_2 yields a Ni(II) metallacycle, followed by insertion of the other alkyne and reductive elimination. Pd(0) has also demonstrated catalytic activity in carboxylations of methoxyallene [118] and 1,3-butadiene [119].

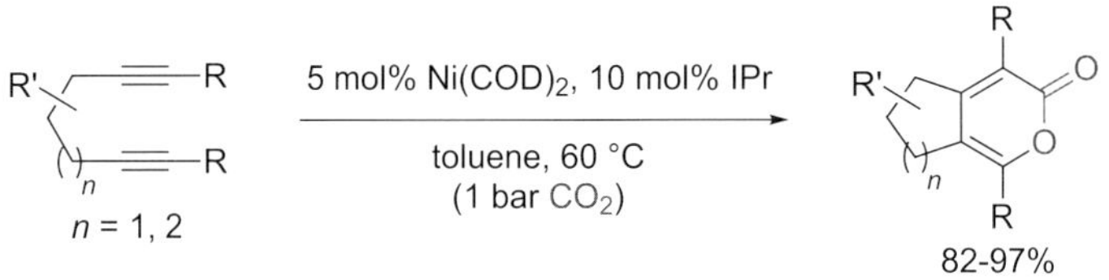

R' R R n n = 1, 2 — 5 mol% Ni(COD)$_2$, 10 mol% IPr; toluene, 60 °C (1 bar CO_2) → R' n R O O R 82-97%

Scheme 2.62 [2 + 2 + 2] cycloaddition of bis-alkynes and CO_2 by Ni(0) [117].

2.5.9 Michael Addition of α,β-Unsaturated *N*-Acylpyrrolidines

Application of transition metal catalysts as Lewis acids provides yet another mechanistic paradigm for lactone synthesis. Kanemasa used Ni(II) with a bisoxazoline ligand for the synthesis of hexahydrobenzopyranones from α,β-unsaturated *N*-acylpyrrolidines and Meldrum's acid. An asymmetric variant was later reported (Scheme 2.63) [120, 121].

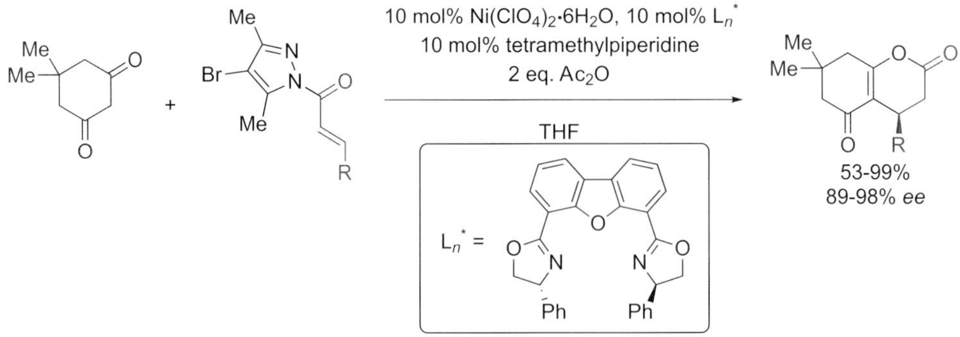

Me Me O O + Me Br N N Me O R — 10 mol% Ni(ClO$_4$)$_2$·6H$_2$O, 10 mol% L_n^*; 10 mol% tetramethylpiperidine; 2 eq. Ac$_2$O; THF → Me Me O O O R 53-99% 89-98% *ee*

L_n^* = O N Ph O N O Ph

Scheme 2.63 Michael addition of α,β-unsaturated *N*-acylpyrrolidines [120].

2.5.10
[2 + 2] Cycloaddition of Ketenes and Aldehydes

Complementary to the hetero-Pauson–Khand reaction (see Section 2.2.6), [2 + 2] cycloaddition of ketenes and aldehydes provides rapid access to β-lactones from relatively simple starting materials. Cationic $[Pd(dppb)(CH_3CN)_2]BF_4$ was selected as the optimal catalyst (Scheme 2.64) [122, 123]. A tandem allylic rearrangement/ring expansion sequence was also developed that provides elaboration of β-lactones to 3,6-dihydro-2*H*-pyran-2-ones in a one-pot process (Scheme 2.65) [123].

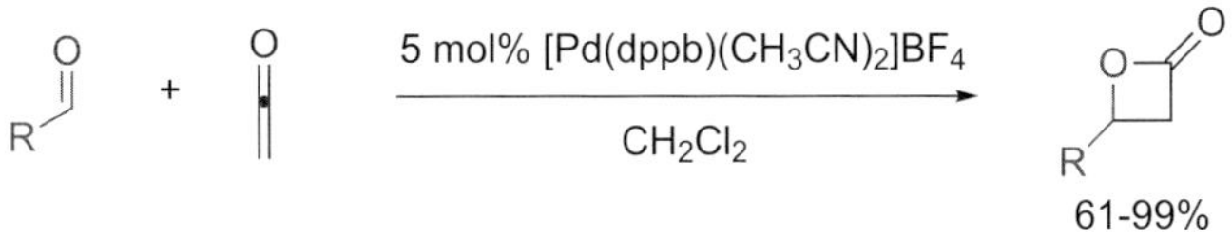

Scheme 2.64 [2 + 2] cycloaddition of ketenes and aldehydes [122, 123].

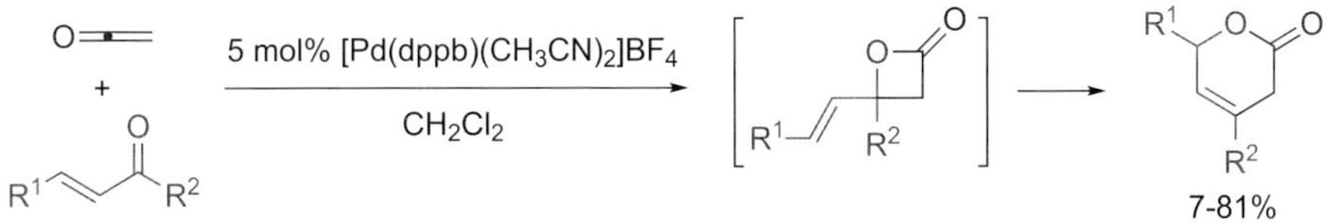

Scheme 2.65 Tandem [2 + 2] cycloaddition/allylic rearrangement to pyran-2-ones [122, 123].

2.5.11
Tandem Cross-Metathesis/Hydrogenation Route to Lactones

Ru-catalyzed cross-metathesis, hydrogenation, and subsequent ring closure has been successfully combined into a one-pot tandem process for lactone synthesis. Importantly, compatibility between the catalytic systems is crucial to success. By employing the second generation Grubbs–Hoveyda catalyst in addition to PtO_2, γ- and δ-lactones can be synthesized under an atmospheric pressure of H_2 (Scheme 2.66) [124]. Cossy and coworkers demonstrated that neither the Ru catalyst nor PtO_2 catalyzed the esterification of the two starting materials, ruling out a ring closing metathesis mechanism.

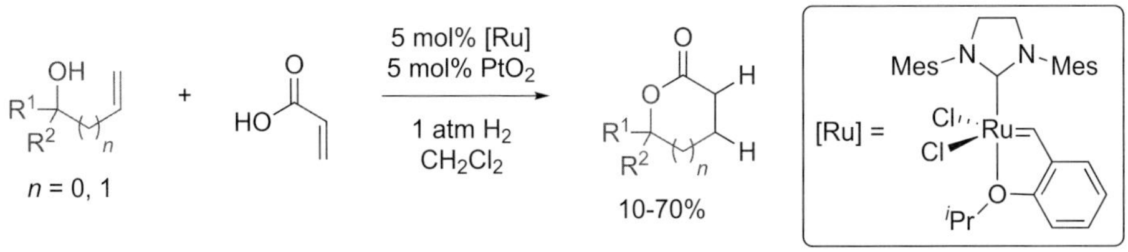

Scheme 2.66 Tandem cross-metathesis/hydrogenation route to lactones [124].

2.5.12
Modern Catalytic Variants of Classical Macrolactonizations

Extension of the traditional Yamaguchi/Corey-Nicolaou/Keck macrolactonization protocols (see Scheme 2.1) [5] to catalytic variants can address some of the drawbacks of classical methods. Mixed anhydrides can undergo cyclization with Lewis acids such as Sc, Ti, Zr, and Hf [125, 126]. Trost *et al.* reported the use of ethoxyacetylene as a stoichiometric activating agent, installed via Ru catalysis [127], later applied to the synthesis of apicularen A (Scheme 2.67) [128].

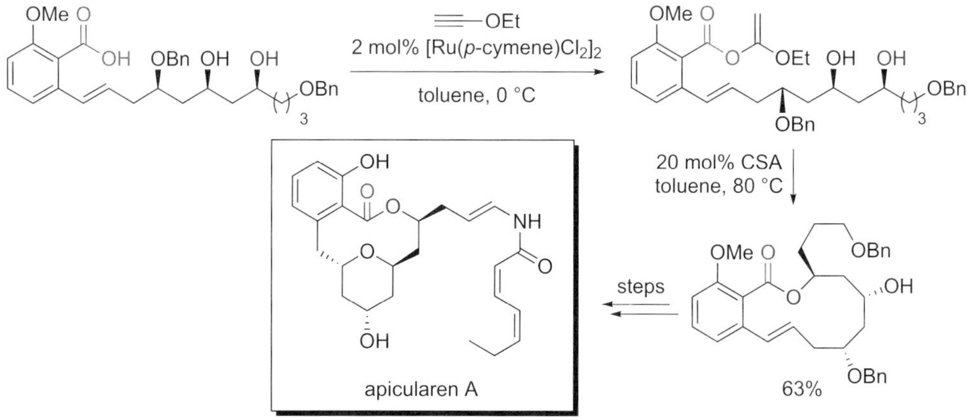

Scheme 2.67 Total synthesis of apicularen A [128].

A Rh-catalyzed ring contraction involving silyl protection of both carboxylic acid and alcohol moieties has also been reported for medium-size lactones synthesis (Scheme 2.68) [129].

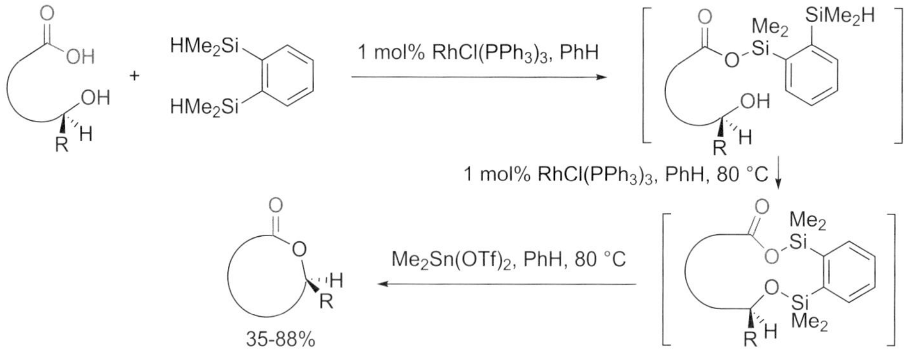

Scheme 2.68 Ring contraction of doubly silyl-protected ω-hydroxyacids [129].

2.6
Conclusions and Outlook

Methods for lactone synthesis by transition metal catalysis involving C—O formation developed over the past 50 years have demonstrated much promise. Indeed, lactones have inspired the discovery of new organometallic transformations, design of metal catalysts, and detailed understanding of reaction mechanisms. Issues of waste minimization and stereoselectivity have been addressed. Future developments for chiral lactone synthesis will likely focus on establishing efficient transformations with broad scope and application in complex molecule total synthesis, especially in regards to macrolactonization where entropic costs often plague intramolecular reactivity with undesired intermolecular reactions.

References

1 Parenty, A., Moreau, X., and Campagne, J.-M. (2006) *Chem. Rev.*, **106**, 911–939.
2 Gradillas, A. and Pérez-Castells, J. (2006) *Angew. Chem. Int. Ed.*, **45**, 6086–6101.
3 Ohba, Y., Takatsuji, M., Nakahara, K., Fujioka, H., and Kita, Y. (2009) *Chem. Eur. J.*, **15**, 3526–3537.
4 Staunton, J. and Weissman, K.J. (2001) *Nat. Prod. Rep.*, **18**, 380–416.
5 Kürti, L. and Czakó, B. (2005) *Strategic Applications of Named Reactions in Organic Synthesis*, Elsevier, Inc., Burlington.
6 Farina, V. and Eriksson, M. (2002) Chapter VI.2.1.2, in *Handbook of Organopalladium Chemistry for Organic Synthesis*, John Wiley & Sons Inc., New York.
7 Negishi, E.-i. and Makabe, H. (2002) Chapter VI.2.3, in *Handbook of Organopalladium Chemistry for Organic Synthesis*, John Wiley & Sons Inc., New York.
8 Colquhoun, H.M., Thompson, D.J., and Twigg, M.V. (1991) *Carbonylation: Direct Synthesis of Carbonyl Compounds*, Plenum, New York, NY.
9 Mori, M., Chiba, K., Inotsume, N., and Ban, Y. (1979) *Heterocycles*, **12**, 921–924.
10 Negishi, E.-i. and Tour, J.M. (1986) *Tetrahedron Lett.*, **27**, 4869–4872.
11 Suzuki, T., Uozumi, Y., and Shibasaki, M. (1991) *J. Chem. Soc. Chem. Commun.*, 1593–1595.
12 Marshall, J.A., Lebreton, J., DeHoff, B.S., and Jenson, T.M. (1987) *Tetrahedron Lett.*, **28**, 723–726.
13 Liao, B. and Negishi, E.-i. (2000) *Heterocycles*, **52**, 1241–1249.
14 Hoye, T.R. and Tan, L. (1995) *Tetrahedron Lett.*, **36**, 1981–1984.
15 Takahashi, T., Kusasaka, S.-i., Doi, T., Sunazuka, T., and Omura, S. (2003) *Angew. Chem. Int. Ed.*, **42**, 5230–5234.
16 Copéret, C., Sugihara, T., Wu, G., Shimoyama, I., and Negishi, E.-i. (1995) *J. Am. Chem. Soc.*, **117**, 3422–3431.
17 Alper, H., Currie, J.K., and Abbayes, H.d. (1978) *J. Chem. Soc. Chem. Commun.*, 311–312.
18 Negishi, E.-i., Copéret, C., Sugihara, T., Shimoyama, I., Zhang, Y., Wu, G., and Tour, J.M. (1994) *Tetrahedron*, **50**, 425–436.
19 Negishi, E.-i., Makabe, H., Shimoyama, I., Wu, G., and Zhang, Y. (1998) *Tetrahedron*, **54**, 1095–1106.
20 Shimoyama, I., Zhang, Y., Wu, G., and Negishi, E.-i. (1990) *Tetrahedron Lett.*, **31**, 2841–2844.
21 Wu, G., Shimoyama, I., and Negishi, E. (1991) *J. Org. Chem.*, **56**, 6506–6507.
22 Larock, R.C. and Fellows, C.A. (1982) *J. Am. Chem. Soc.*, **104**, 1900–1907.
23 Kočovský, P., Grech, J.M., and Mitchell, W.L. (1996) *Tetrahedron Lett.*, **37**, 1125–1128.
24 Ali, B.E. and Alper, H. (2000) *Synlett*, 161–171.

25 Godard, C., Muñiz, B.K., Ruiz, A., and Claver, C. (2008) *Dalton Trans.*, 853–860.
26 Alper, H. and Leonard, D. (1985) *Tetrahedron Lett.*, **26**, 5639–5642.
27 Alper, H. and Leonard, D. (1985) *J. Chem. Soc., Chem. Commun.*, 511–512.
28 Norton, J.R., Shenton, K.E., and Schwartz, J. (1975) *Tetrahedron Lett.*, **1**, 51–54.
29 Vasapollo, G., Mele, G., and Ali, B.E. (2003) *J. Mol. Catal. A*, **204–205**, 97–105.
30 Yoneda, E., Sugioka, T., Hirao, K., Zhang, S.-W., and Takakashi, S. (1998) *J. Chem. Soc., Perkin Trans. 1*, 477–484.
31 Tezuka, K., Ishizaki, Y., and Inoue, Y. (1998) *J. Mol. Catal. A*, **129**, 199–206.
32 Yoneda, E., Kaneko, T., Zhang, S.-W., Onitsuka, K., and Takahashi, S. (2000) *Org. Lett.*, **2**, 441–443.
33 Yu, W.-Y., Bensimon, C., and Alper, H. (1997) *Chem. Eur. J.*, **3**, 417–431.
34 Dong, C. and Alper, H. (2004) *J. Org. Chem.*, **69**, 5011–5014.
35 Ogawa, A., Kawabe, K.-i., Kawakami, J.-i., Mihara, M., Hirao, T., and Sonoda, N. (1998) *Organometallics*, **17**, 3111–3114.
36 Negishi, E.-i. and Copéret, C. (2002) Chapter IV.3.3, in *Handbook of Organopalladium Chemistry for Organic Synthesis*, John Wiley & Sons Inc., New York.
37 Sugihara, T., Copéret, C., Owczarczyk, Z., Harring, L.S., and Negishi, E.-i. (1994) *J. Am. Chem. Soc.*, **116**, 7923–7924.
38 Gabriele, B., Salerno, G., Pascali, F.D., Costa, M., and Chiusoli, G.P. (1997) *J. Chem. Soc., Perkin Trans. 1*, 147–154.
39 Gabriele, B., Costa, M., Salerno, G., and Chiusoli, G.P. (1994) *J. Chem. Soc., Chem. Commun.*, 1429–1441.
40 Tamaru, Y., Hojo, M., and Yoshida, Z. (1991) *J. Org. Chem.*, **56**, 1099–1105.
41 Ukaji, Y., Miyamoto, M., Mikuni, M., Takeuchi, S., and Inomata, K. (1996) *Bull. Chem. Soc. Jpn.*, **69**, 735–742.
42 Tamaru, Y., Higashimura, H., Naka, K., Hojo, M., and Yoshida, Z. (1985) *Angew. Chem. Int. Ed. Engl.*, **24**, 1045–1046.
43 Tamaru, Y., Hojo, M., and Yoshida, Z. (1988) *J. Org. Chem.*, **53**, 5731–5741.
44 Tamaru, Y., Kobayashi, T., Kawamura, S.-i., Ochiai, H., Hojo, M., and Yoshida, Z.-i. (1985) *Tetrahedron Lett.*, **26**, 3207–3210.
45 Paddon-Jones, G.C., McErlean, C.S.P., Hayes, P., Moore, C.J., Konig, W.A., and Kitching, W. (2001) *J. Org. Chem.*, **66**, 7487.
46 Kapitán, P. and Gracza, T. (2008) *ARKIVOC*, **8**, 8–17.
47 Matsuda, I., Ogiso, A., and Sato, S. (1990) *J. Am. Chem. Soc.*, **112**, 6120–6121.
48 Bando, T., Tanaka, S., Fugami, K., Yoshida, Z.-i., and Tamaru, Y. (1992) *Bull. Chem. Soc. Jpn.*, **65**, 97–110.
49 Tamaru, Y., Bando, T., Hojo, M., and Yoshida, Z.-i. (1987) *Tetrahedron Lett.*, **28**, 3497–3500.
50 Mandai, T., Tsujiguchi, Y., Matsuoka, S., Saito, S., and Tsuji, J. (1995) *J. Organomet. Chem.*, **488**, 127–131.
51 Khumtaveeporn, K. and Alper, H. (1995) *Acc. Chem. Res.*, **28**, 414–422.
52 Church, T.L., Getzler, Y.D.Y.L., Byrne, C.M., and Coates, G.W. (2007) *Chem. Commun.*, 657–674.
53 Reppe, W., Kroper, H., Pistor, H.J., and Weissbarth, O. (1953) *Justus Liebigs Ann. Chem.*, **582**, 87–116.
54 Kamiya, Y., Kawato, K., and Ohta, H. (1980) *Chem. Lett.*, 1549–1552.
55 Alper, H., Arzoumanian, H., Petrignani, J.-F., and Saldana-Maldonado, M. (1985) *J. Chem. Soc., Chem. Commun.*, 340–341.
56 Lee, J.T., Thomas, P.J., and Alper, H. (2001) *J. Org. Chem.*, **66**, 5424–5426.
57 Getzler, Y.D.Y.L., Mahadevan, V., Lobkovsky, E.B., and Coates, G.W. (2002) *J. Am. Chem. Soc.*, **124**, 1174–1175.
58 Molnar, F., Luinstra, G.A., Allmendinger, M., and Rieger, B. (2003) *Chem. Eur. J.*, **9**, 1273–1280.
59 Kablaoui, N.M., Hicks, F.A., and Buchwald, S.L. (1997) *J. Am. Chem. Soc.*, **119**, 4424–4431.
60 Mandal, S.K., Amin, S.R., and Crowe, W.E. (2001) *J. Am. Chem. Soc.*, **123**, 6457–6458.
61 Chatani, N., Morimoto, T., Fukumoto, Y., and Murai, S. (1998) *J. Am. Chem. Soc.*, **120**, 5335–5336.
62 Kang, S.-K., Kim, K.-J., and Hong, Y.-T. (2002) *Angew. Chem. Int. Ed.*, **41**, 1584–1586.
63 Alonso, F., Beletskaya, I.P., and Yus, M. (2004) *Chem. Rev.*, **104**, 3079–3160.

64 Yang, C.-G., Reich, N.W., Shi, Z., and He, C. (2005) *Org. Lett.*, **7**, 4553–4556.
65 Lambert, C., Utimoto, K., and Nozaki, H. (1984) *Tetrahedron Lett.*, **25**, 5323–5326.
66 Genin, E., Toullec, P.Y., Antoniotti, S., Brancour, C., Gent, J.-P., and Michelet, V. (2006) *J. Am. Chem. Soc.*, **128**, 3112–3113.
67 Krafft, G.A. and Katzenellenbogen, J.A. (1981) *J. Am. Chem. Soc.*, **103**, 5459–5466.
68 Chan, D.M.T., Marder, T.B., Milstein, D., and Taylor, N.J. (1987) *J. Am. Chem. Soc.*, **109**, 6385–6388.
69 Hamilton, G.L., Kang, E.J., Mba, M., and Toste, F.D. (2007) *Science*, **317**, 496–499.
70 Huo, Z., Patil, N.T., Jin, T., Pahadi, N.K., and Yamamoto, Y. (2007) *Adv. Synth. Catal.*, **349**, 680–684.
71 Kundu, N.G. and Pal, M. (1993) *J. Chem. Soc., Chem. Commun.*, 86–88.
72 Liao, H.-Y. and Cheng, C.-H. (1995) *J. Org. Chem.*, **60**, 3711–3716.
73 Kundu, N.G., Pal, M., and Nandi, B. (1998) *J. Chem. Soc., Perkin Trans. 1*, 561–568.
74 Larock, R.C., Doty, M.J., and Han, X. (1999) *J. Org. Chem.*, **64**, 8770–8779.
75 Miura, M., Tsuda, T., Satoh, T., Pivsa-Art, S., and Nomura, M. (1998) *J. Org. Chem.*, **63**, 5211–5215.
76 Larock, R.C., Varaprath, S., Lau, H.H., and Fellows, C.A. (1984) *J. Am. Chem. Soc.*, **106**, 5274–5284.
77 Ueura, K., Satoh, T., and Miura, M. (2007) *Org. Lett.*, **9**, 1407.
78 Ueura, K., Satoh, T., and Miura, M. (2007) *J. Org. Chem.*, **72**, 5362.
79 Wang, L. and Shen, W. (1998) *Tetrahedron Lett.*, **39**, 7625–7628.
80 Trend, R.M., Ramtohul, Y.K., Ferreira, E.M., and Stoltz, B.M. (2003) *Angew. Chem. Int. Ed.*, **42**, 2892–2895.
81 Ma, S. and Shi, Z. (1998) *J. Org. Chem.*, **63**, 6387–6389.
82 Arcadi, A., Burini, A., Cacchi, S., Delmastro, M., Marinelli, F., and Pietroni, B.R. (1992) *J. Org. Chem.*, **57**, 976–982.
83 Li, Y., Jardine, K., Tan, R., Song, D., and Dong, V.M. (2009) *Angew. Chem. Int. Ed.*, **48**, 9690–9692.
84 Li, Y., Song, D., and Dong, V.M. (2008) *J. Am. Chem. Soc.*, **130**, 2962–2964.
85 Shen, Z., Khan, H.A., and Dong, V.M. (2008) *J. Am. Chem. Soc.*, **130**, 2916–2917.
86 Bergens, S.H., Fairlie, D.P., and Bosnich, B. (1990) *Organometallics*, **9**, 566–571.
87 Fuji, K., Morimoto, T., Tsutsumi, K., and Kakiuchi, K. (2005) *Chem. Commun.*, 3295–3297.
88 Suzuki, T., Yamada, T., Watanabe, K., and Katoh, T. (2005) *Bioorg. Med. Chem. Lett.*, **15**, 2583–2585.
89 Shen, Z., Dornan, P.K., Khan, H.A., Woo, T.K., and Dong, V.M. (2009) *J. Am. Chem. Soc.*, **131**, 1077–1091.
90 Roy, A.H., Lenges, C.P., and Brookhart, M. (2007) *J. Am. Chem. Soc.*, **129**, 2082–2093.
91 Phan, D.H.T., Kim, B., and Dong, V.M. (2009) *J. Am. Chem. Soc.*, **131**, 15608–15609.
92 Hojo, D., Noguchi, K., Hirano, M., and Tanaka, K. (2008) *Angew. Chem. Int. Ed.*, **47**, 5820–5822.
93 Suzuki, T., Morita, K., Tsuchida, M., and Hiroi, K. (2002) *Org. Lett.*, **4**, 2361–2363.
94 Suzuki, T., Morita, K., Matsuo, Y., and Hiroi, K. (2003) *Tetrahedron Lett.*, **44**, 2003–2006.
95 Nozaki, K., Yoshida, M., and Takaya, H. (1994) *J. Organomet. Chem.*, **473**, 253–256.
96 Bloch, R. and Brillet, C. (1991) *Synlett*, 829–830.
97 Murahashi, S., Naota, T., Ito, K., Maeda, Y., and Taki, H. (1987) *J. Org. Chem.*, **52**, 4319–4327.
98 Shvo, Y., Blum, Y., Reshef, D., and Menzin, M. (1982) *J. Organomet. Chem.*, **226**, C21–C24.
99 Ishii, Y., Osakada, K., Ikariya, T., Saburi, M., and Yoshikawa, S. (1986) *J. Org. Chem.*, **51**, 2034–2039.
100 Kyrides, L.P. and Zienty, F.B. (1946) *J. Am. Chem. Soc.*, **68**, 1385.
101 Tamaru, Y., Yamada, Y., Inoue, K., Yamamoto, Y., and Yoshida, Z. (1983) *J. Org. Chem.*, **48**, 1286–1292.
102 Ishii, Y., Suzuki, K., Ikariya, T., Saburi, M., and Yoshikawa, S. (1986) *J. Org. Chem.*, **51**, 2822–2824.
103 Kitamura, M., Ohkuma, T., Inoue, S., Sayo, N., Kumobayashi, H., Akutagawa, S., Ohta, T., Takaya, H., and Noyori, R. (1988) *J. Am. Chem. Soc.*, **110**, 629–631.

104 Ohkuma, T., Kitamura, M., and Noyori, R. (1990) *Tetrahedron Lett.*, **31**, 5509–5512.
105 Everaere, K., Scheffler, J.-L., Montreux, A., and Carpentier, J.-F. (2001) *Tetrahedron Lett.*, **42**, 1899–1901.
106 Zhang, B., Xu, M.-H., and Lin, G.-Q. (2009) *Org. Lett.* doi: 10.1021/ol901674k.
107 Bertrand, M.P., Oumar-Mahamat, H., and Surzur, J.M. (1985) *Tetrahedron Lett.*, **26**, 1209–1212.
108 Lee, J.M. and Chang, S. (2006) *Tetrahedron Lett.*, **47**, 1375–1379.
109 Fraunhoffer, K.J., Prabagaran, N., Sirois, L.E., and White, M.C. (2006) *J. Am. Chem. Soc.*, **128**, 9032–9033.
110 Stang, E.M. and White, M.C. (2009) *Nature Chem.*, **1**, 547–551.
111 Zhang, Y.-H., Shi, B.-F., and Yu, J.-Q. (2009) *Angew. Chem. Int. Ed.*, **48**, 6097–6100.
112 Bolm, C., Schlingloff, G., and Weickhardt, K. (1994) *Angew. Chem. Int. Ed. Engl.*, **33**, 1848–1849.
113 Gusso, A., Baccin, C., Pinna, F., and Strukul, G. (1994) *Organometallics*, **13**, 3442–3451.
114 Watanabe, A., Uchida, T., Irie, R., and Katsuki, T. (2004) *Proc. Natl. Acad. Sci. USA*, **101**, 5737–5742.
115 He, Z. and Yudin, A.K. (2006) *Org. Lett.*, **8**, 5829–5832.
116 Chang, H.-T., Jeganmohan, M., and Cheng, C.-H. (2007) *Chem. Eur. J.*, **13**, 4356–4363.
117 Louie, J., Gibby, J.E., Farnworth, M.V., and Tekavec, T.N. (2002) *J. Am. Chem. Soc.*, **124**, 15188–15189.
118 Tsuda, T., Yamamoto, T., and Saegusa, T. (1992) *J. Organomet. Chem.*, **429**, C46–C48.
119 Behr, A. and Juszak, K.-D. (1983) *J. Organomet. Chem.*, **255**, 263–268.
120 Itoh, K., Hasegawa, M., Tanaka, J., and Kanemasa, S. (2005) *Org. Lett.*, **7**, 979–981.
121 Itoh, K. and Kanemasa, S. (2003) *Tetrahedron Lett.*, **44**, 1799–1802.
122 Hattori, T., Suzuki, Y., Uesugi, O., Oi, S., and Miyano, S. (2000) *Chem. Commun.*, 73–74.
123 Hattori, T., Suzuki, Y., Ito, Y., Hotta, D., and Miyano, S. (2002) *Tetrahedron*, **58**, 5215–5223.
124 Cossy, J., Bargiggia, F., and BouzBouz, S. (2003) *Org. Lett.*, **5**, 459–462.
125 Shiina, I. (2004) *Tetrahedron*, **60**, 1587–1599.
126 Mukaiyama, T., Izumi, J., Miyashita, M., and Shiina, I. (1993) *Chem. Lett.*, **22**, 907–910.
127 Trost, B.M. and Chisholm, J.D. (2002) *Org. Lett.*, **4**, 3743–3745.
128 Petri, A.F., Bayer, A., and Maier, M.E. (2004) *Angew. Chem. Int. Ed.*, **43**, 5821–5823.
129 Mukaiyama, T., Izumi, J., and Shiina, I. (1997) *Chem. Lett.*, 187–188.

3
The Formation of Csp^2–S and Csp^2–Se Bonds by Substitution and Addition Reactions Catalyzed by Transition Metal Complexes

Irina P. Beletskaya and Valentine P. Ananikov

3.1
Introduction

Transition-metal-catalyzed activation of Csp^2–X bonds (X = halogens or OTf) has become a widely recognized method for the construction of new Csp^2–C and Csp^2–heteroatom bonds. Low valent metal complexes have been used as catalyst precursors in these reactions and in many cases excellent results have been achieved with Pd(0) catalysts. The products of these reactions containing Ar–S and vinyl–S groups and the corresponding selenium derivatives are in demand in organic synthesis, pharmaceutical and material sciences [1–3]. For example, application in the pharmaceutical field includes the synthesis of selective M2 muscarinic receptor antagonists, COX-2 inhibitors (nonsteroidal anti-inflammatory drugs study), MAP kinase p38 (intracellular signal transduction cascade) inhibitors, inhibitors of human immunodeficiency virus type 1 (HIV-1) protease, antagonists of leukocyte function-associated antigen-1, hypoxia-directed bioreductive cytotoxins, inhibitors of the ATPase activity of human papillomavirus E1 helicase, inhibitors of folate-synthesizing enzymes of cell cultures of *Candida albicans*, antitumor agents and folate-dependent enzyme inhibitors in cancer cells [2].

Cross-coupling of Ar(or vinyl)–X with organometallic species, terminal alkynes, and olefins, as well as carbonylation in the presence of nucleophiles, represent well-known synthetic routes for C–C bond formation. As far as C–heteroatom bond formation is concerned, powerful synthetic procedures have been developed for selective formation of the C–P (P = R_2P, $R_2P(O)$, R(R'O)P(O), $(RO)_2P(O)$), C–SR, C–SeR, C–NR_2, and so on, bonds. All of these catalytic methods are based on substitution reactions. The catalytic cycle of a typical substitution reaction includes the following stages: (i) oxidative addition, (ii) transmetallation, and (iii) C–heteroatom reductive elimination.

Another fascinating opportunity for the formation of vinylic derivatives with Csp^2–heteroatom bond(s) involves addition reactions of Z–Z and Z–H bonds to the triple bond of alkynes (Z = heteroatom). The catalytic cycle of these addition reactions includes the following stages: (i) oxidative addition; (ii) alkyne coordination

Catalyzed Carbon-Heteroatom Bond Formation. Edited by Andrei K. Yudin

ISBN: 978-3-527-32428-6

and insertion, and (iii) C–Z reductive elimination to form a carbon–heteroatom bond (or protonolysis to form a C–H bond). Thus, compared to the cross-coupling reactions discussed above one may notice that the main difference is in the second stage of the catalytic cycle – alkyne insertion instead of transmetallation. Such mechanistic change leads to important practical advantage in terms of Green Chemistry, since addition reactions are characterized by 100% atom efficiency and do not lead to the formation of by-products.

In the present chapter we highlight recent developments in the Csp^2–S and Csp^2–Se bond formation utilizing both approaches – transition-metal-catalyzed cross-coupling and addition reactions. Comparative analysis of both approaches in terms of synthetic application and mechanism for the preparation of aryl(vinyl) sulfides and aryl(vinyl) selenides has been carried out. In the present chapter the literature was covered until November 2009 – the date of submission of the manuscript.

3.2 Catalytic Cross-Coupling Reactions

3.2.1 Pd-Catalyzed Transformations

The first successful Ar–heteroatom bond formation reaction was demonstrated in 1978 by the Japanese chemists Kosugi and Migita for a C–S bond [4]. Diaryl sulfides and arylalkyl sulfides were obtained from thiols and ArX derivatives in moderate to good yields (Scheme 3.1).

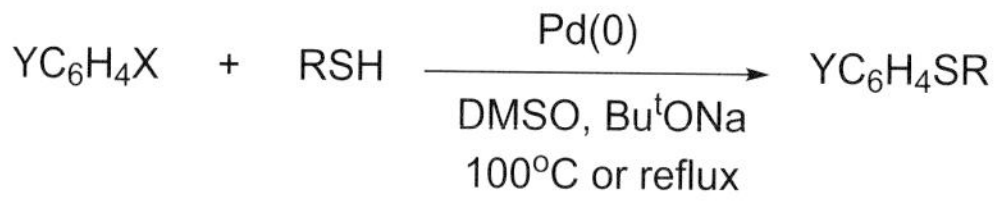

$$YC_6H_4X + RSH \xrightarrow[\text{DMSO, Bu}^t\text{ONa, 100°C or reflux}]{\text{Pd(0)}} YC_6H_4SR$$

X = I, Br
Y = H, *p*-Me, *p*-MeO, *p*-Cl
R = Ph, Et
Pd(0) = $Pd(PPh_3)_4$

Scheme 3.1

This reaction was further extended to involve vinyl halides and was found to proceed in a stereospecific manner with retention of the geometry of the C=C double bond and resulting in high product yields (Scheme 3.2) [5].

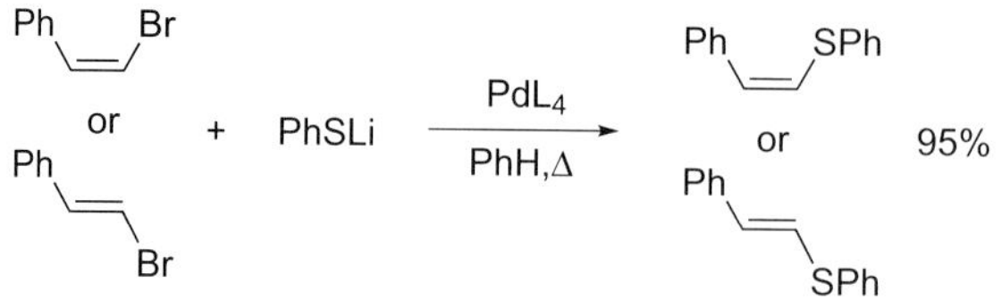

$$\text{PhCH=CHBr } (Z \text{ or } E) + PhSLi \xrightarrow[\text{PhH, }\Delta]{PdL_4} \text{PhCH=CHSPh } (Z \text{ or } E) \quad 95\%$$

Scheme 3.2

The reactions with tin derivatives Bu_3SnSR or Me_3SnSR without a base are characterized by quantitative yields and, in the case of vinylhalcogenides, high stereospecificity [6–8]. A representative example of such a reaction involving Me_3SnSR is shown on Scheme 3.3 [9].

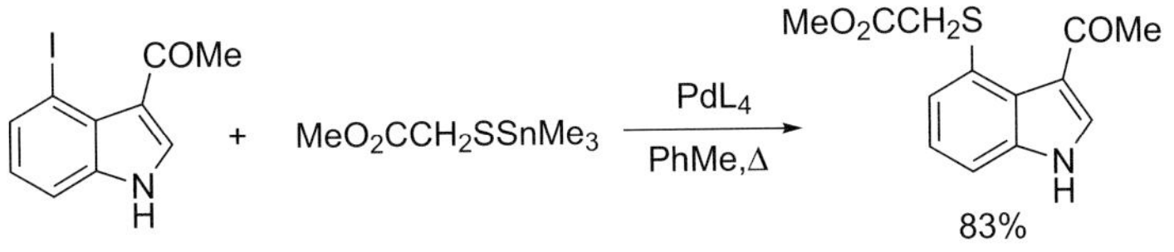

$MeO_2CCH_2SSnMe_3$ + (iodo-3-acetylindole) $\xrightarrow[PhMe, \Delta]{PdL_4}$ (MeO_2CCH_2S-3-acetylindole) 83%

Scheme 3.3

Cross-coupling of the silyl derivatives of thiols R_3SiSH led to the formation of vinyl (or aryl) silylsubstituted sulfides (Scheme 3.4) [10]. Further transformation of the products in alkylation or Pd-catalyzed alkenylation reactions was reported as a route to unsymmetric RSR' sulfides as final products. This stepwise synthetic approach allowed avoidance of the use of some scarcely available Ar(Het)SH, which otherwise would be required to prepare the same final products in a single step utilizing cross-coupling.

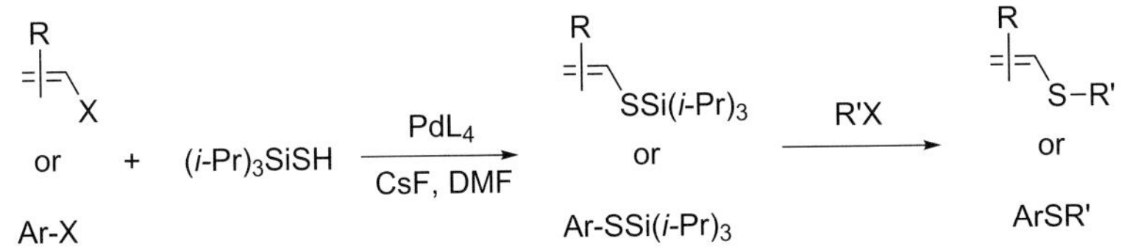

R-CH=CH-X or Ar-X + $(i\text{-}Pr)_3SiSH$ $\xrightarrow[CsF, DMF]{PdL_4}$ R-CH=CH-$SSi(i\text{-}Pr)_3$ or Ar-$SSi(i\text{-}Pr)_3$ $\xrightarrow{R'X}$ R-CH=CH-S-R' or ArSR'

Scheme 3.4

This method has been used for the preparation of bis-pyrimidine thioesters [11], for example:

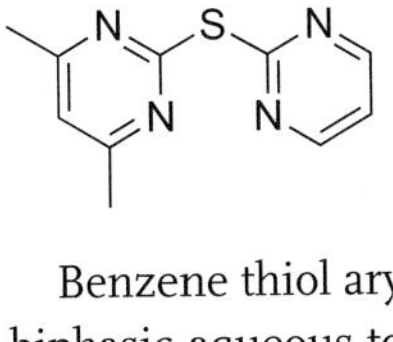

Benzene thiol arylation with bromobenzene was reported by Foa *et al.* utilizing a biphasic aqueous-toluene system (Scheme 3.5) [12].

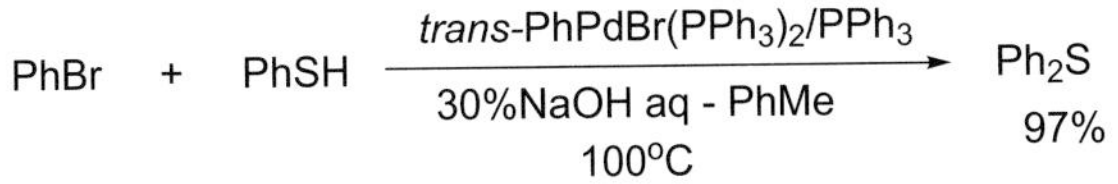

PhBr + PhSH $\xrightarrow[30\%NaOH\ aq\ -\ PhMe,\ 100^{\circ}C]{trans\text{-}PhPdBr(PPh_3)_2/PPh_3}$ Ph_2S 97%

Scheme 3.5

Arylation of amino acid derivatives with a thiol group was carried out in the Pd_2dba_3/dppf/NMP/Et_3N system [13, 14]. It should be mentioned that, even in the earlier stages of development of this catalytic methodology, nickel [15] and copper [16] complexes were considered as useful potential replacements for the Pd catalyst. However, in both cases high temperature was required to carry out the reaction.

An important point clarifying the mechanistic picture of the reaction of interest was reported in the study of Hartwig *et al.* [17] Investigation of the stoichiometric C–S reductive elimination reaction on Pd complexes with bidentate ligand indicated the need for an extra amount of phosphine ligand (Scheme 3.6). This finding clearly emphasized a noticeable difference between the cross-coupling reactions involving formation of C–heteroatom and C–C bonds. In addition, it was reported that coupling reactions involving aryl- and vinyl-thiol derivatives proceed mor easily than with their alkyl analogs.

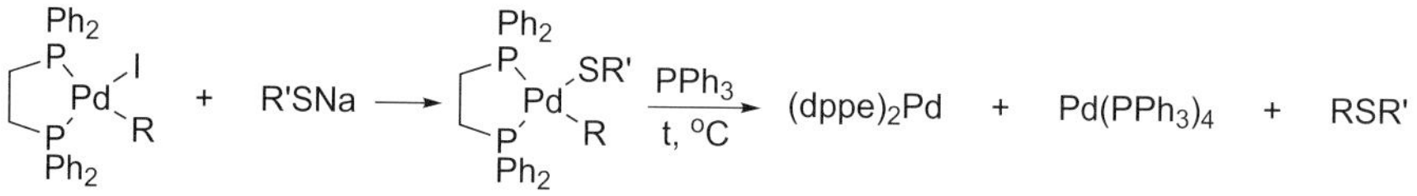

(dppe)Pd(I)(R) + R'SNa → (dppe)Pd(SR')(R) —[PPh_3, t, °C]→ $(dppe)_2Pd$ + $Pd(PPh_3)_4$ + RSR'

Scheme 3.6

Noticeably, an interest in the detailed studies of these cross-coupling reactions to form C–S and C–Se bonds was greatly facilitated after discovery of the amination reaction 17 years after the first report of Kosugi and Migita. As in the case of the amination reaction, the studies of the C–S cross-coupling catalytic procedures were devoted to the development of new catalytic systems, especially to finding new ligands, which would extend the scope of the arylation reagents to include aryl triflates, aryl tosilates, unactivated aryl bromides and aryl chlorides. Another important goal is to carry out the reaction under mild conditions.

In 1998, upon screening the available ligands Zheng *et al.* found that $Pd(OAc)_2$/BINAP/LiCl and, particularly, the $Pd(OAc)_2$/Tol/BINAP system can catalyze cross-coupling reactions of various triflates [18]. However, a relatively high amount of the catalyst was required for the reaction (Scheme 3.7). Various $ArSBu^n$ products were synthesized in the $Pd(OAc)_2$/Tol/BINAP system using NaHMDS and $NaN(SiMe_3)_2$ as bases [19].

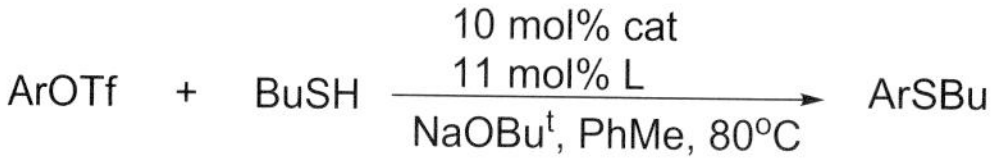

ArOTf + BuSH —[10 mol% cat, 11 mol% L; $NaOBu^t$, PhMe, 80°C]→ ArSBu

Scheme 3.7

Surprisingly, the phospine oxide t-$Bu_2P(O)H$ suggested by Li *et al.* was found to be an efficient ligand for the C–S cross-coupling reaction, which made it possible to carry out the reaction at 110 °C, even with unactivated aryl chlorides (Scheme 3.8) [20, 21].

This catalytic system has been used in the reaction of 1-chloro-1-cyclopentene (Scheme 3.9) [22].

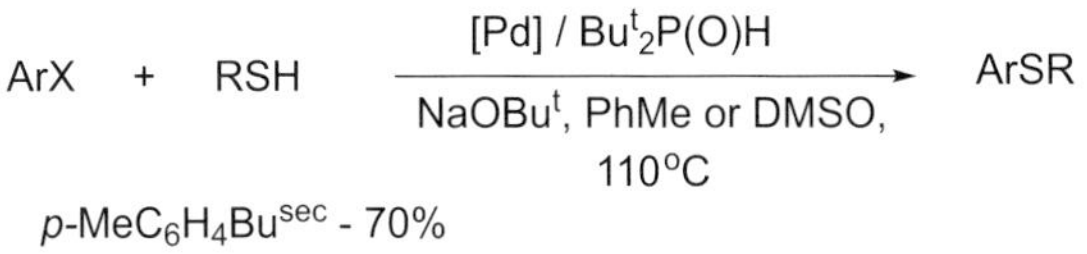

ArX + RSH —[Pd] / $Bu^t_2P(O)H$; $NaOBu^t$, PhMe or DMSO, 110°C→ ArSR

p-$MeC_6H_4Bu^{sec}$ - 70%

Scheme 3.8

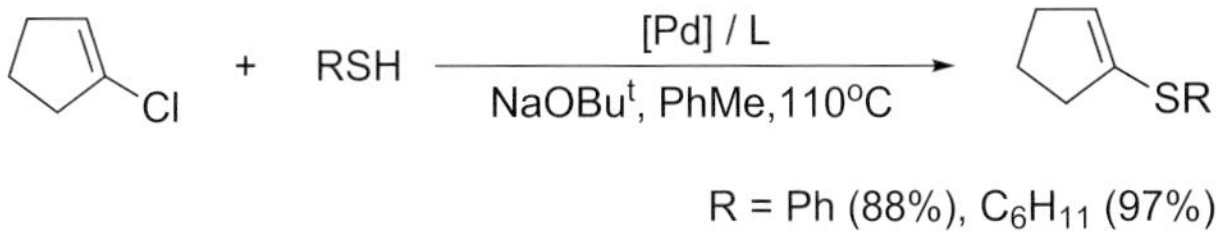

1-chlorocyclopentene + RSH —[Pd] / L; $NaOBu^t$, PhMe, 110°C→ 1-(SR)cyclopentene

R = Ph (88%), C_6H_{11} (97%)

Scheme 3.9

Unsymmetrical aromatic and heteroaromatic thioesters were obtained from Ar(Py)I and Ar(Het)SH in the Pd_2dba_3/DPE-phos/Bu^tOK system in toluene at 100 °C in high yields [23]. A series of mono- and bidentate ligands studied by Buchwald *et al.* in the reaction of p-$MeOC_6H_4Br$ led to selection of two efficient ligands for the reaction with benzene thiol, DiPPF and DPE-phos, but only one ligand was active in the reaction with Bu^tSH–DiPPF (2 mol% catalyst, Pd : P = 1 : 1.2, Bu^tONa, dioxane, 100 °C) [24]. The catalytic system with the DiPPF ligand was also used in the reaction of various alkyl thiols with ArBr (including *o*-substituted), and also with unactivated aryl chlorides:

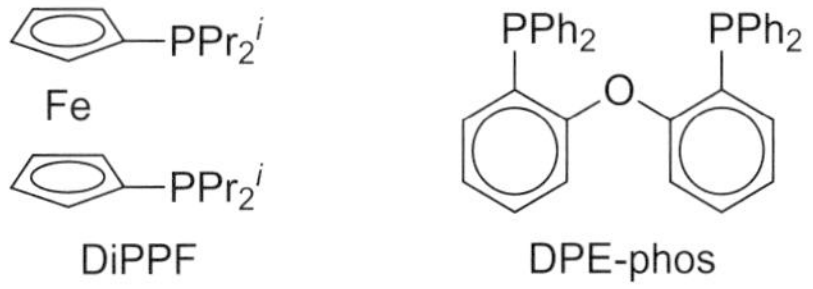

DiPPF: Fe, PPr^i_2, PPr^i_2 DPE-phos: PPh_2, O, PPh_2

Another excellent example of an efficient bis-phosphine ligand, CyPF-t-Bu, developed by Hartwig *et al.* made it possible to carry out cross-coupling of ArCl and ArOTf with thiols (Scheme 3.10) [25]. This bis-phosphine ligand possessed high ability for coordination of the metal species and was well balanced to allow oxidative

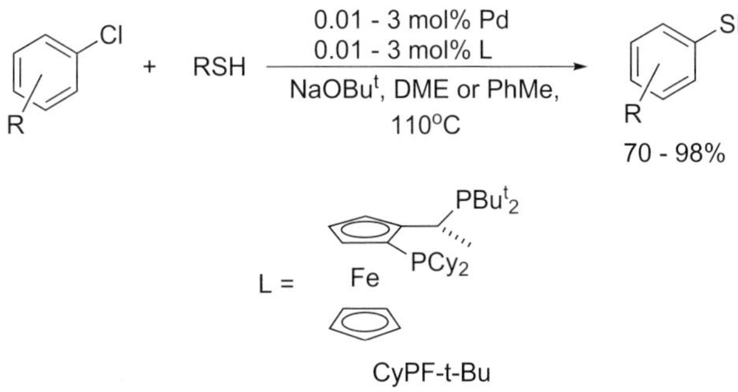

R-substituted chlorobenzene + RSH —0.01 - 3 mol% Pd; 0.01 - 3 mol% L; $NaOBu^t$, DME or PhMe, 110°C→ R-substituted aryl–SR, 70 - 98%

L = CyPF-t-Bu (Fe, PBu^t_2, PCy_2)

Scheme 3.10

addition and reductive elimination stages. This catalytic system has shown high efficiency in the reaction with p-MeC_6H_4I and p-MeC_6H_4Br, thus making it possible to utilize very low concentration of the catalyst (up to 0.001 mol%) [26].

Good efficiency was reported for the Pd_2dba_3–Xantphos catalytic system, which was successfully used to carry out cross-coupling reaction of ArBr, ArOTf and p-$NO_2C_6H_4Cl$ with alkyl- and aryl thiols [27]. Pd complexes with Xantphos and BINAP ligands were used to introduce aryl sulfide and alkyl sulfonyl groups into the meso-position of porphyrins leading to the formation of mono- and disubstituted derivatives [28]. Special reagents for application in tomography were synthesized via the reaction of 6-iodo-2(4′-*N,N*-dimethylamino) phenylimidazo[1,2-a]pyridine with $RSSnMe_3$; DiPPF and tol-BINAP were reported as ligands of choice for the reaction [29]. The key step in the synthesis of immunomudulator KRP-203 was carried out using Xantphos ligand [30]:

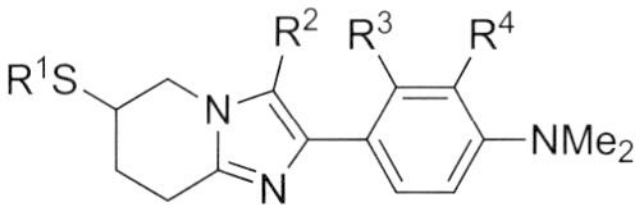
R^{1}S
R^2 R^3 R^4
N
N
NMe$_2$

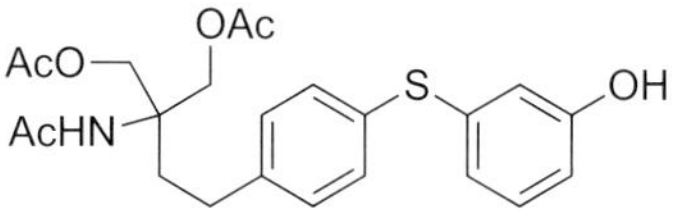
OAc
AcO
AcHN
S
OH

KRP-203

Three-component reaction of *o*-bromothiophenol and *o*-iodobromobenzene with amines led to the formation of a C−S bond and two C−N bonds in the Pd_2dba_3–dppf–$NaOBu^t$ system [31]. The reaction was carried out under microwave heating and resulted in good to high yields of product.

To carry out cross-coupling reactions of ArX and RSH, Bu^nONa or Bu^tONa were used most frequently. Krief *et al.* have shown that NaH or CsOH can be a more convenient choice, although the reaction may require a higher temperature (Scheme 3.11) [32].

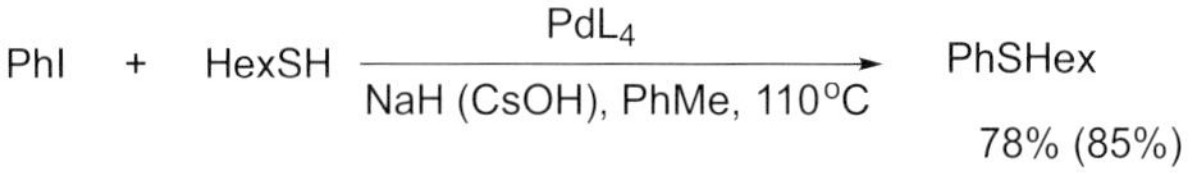
PhI + HexSH
PdL$_4$
NaH (CsOH), PhMe, 110°C
PhSHex
78% (85%)

Scheme 3.11

A combinatorial technique was employed for construction of a library of m-$(HOCH_2)C_6H_4SR$ products (Scheme 3.12) [33].

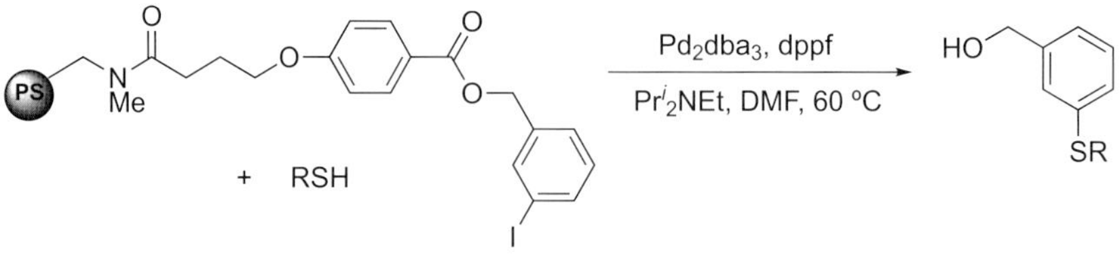
PS
N
Me
O
O
O
O
I
+ RSH
Pd$_2$dba$_3$, dppf
Pr$^i_2$NEt, DMF, 60 °C
HO
SR

Scheme 3.12

Synthesis of diaryl selenides via the cross-coupling reaction of Ar'SeH is a more difficult task due to the scarce availability of this type of reagent and their easy oxidation in air. Reaction with the tin derivative $Ar'SeSnBu_3$ (readily available from Ar'_2Se_2 and $Bu_3SnSnBu_3$) was suggested instead as a route to unsymmetric Ar'SeAr under Pd-catalyzed conditions (Scheme 3.13) [34]. However, the reaction with ArBr required a higher temperature, which may also facilitate disproportionation of $R_3SnSeAr'$ to $(R_3Sn)_2Se$ and Ar'_2Se.

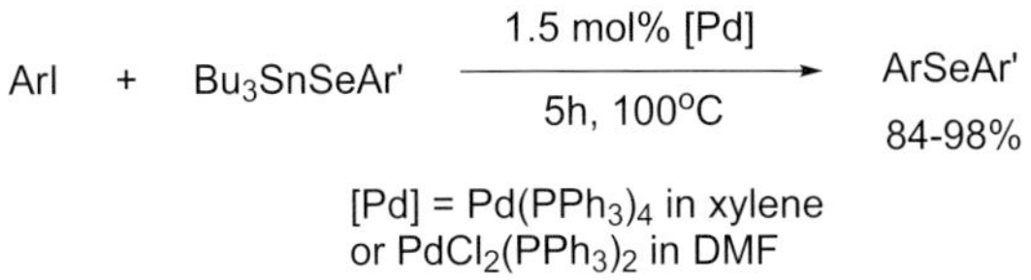

$$\mathrm{ArI} + \mathrm{Bu_3SnSeAr'} \xrightarrow[\text{5h, 100°C}]{\text{1.5 mol\% [Pd]}} \mathrm{ArSeAr'}\ \ 84\text{-}98\%$$

$[Pd] = Pd(PPh_3)_4$ in xylene
or $PdCl_2(PPh_3)_2$ in DMF

Scheme 3.13

Diaryl diselenides may serve as a source of ArSe-groups, and diaryl disulfides as a source of ArS-groups. A one-pot procedure, involving reduction of Ph_2Se_2 by Na/ NH_3 followed by further transformation into the tin derivative and cross-coupling, was developed to form PhSeAr [35]. Reduction of the PhZZPh by Zn (Z = S, Se) was used as a first step in the cross-coupling reaction with ArBr (Scheme 3.14) [36]. The ArSR disulfides were obtained in high yields, while a lower yield was observed for the ArSeR derivative. It should be mentioned, that a convenient procedure was recently reported for the quantitative reduction of diselenides to form selenols (Zn in acidic media) [37].

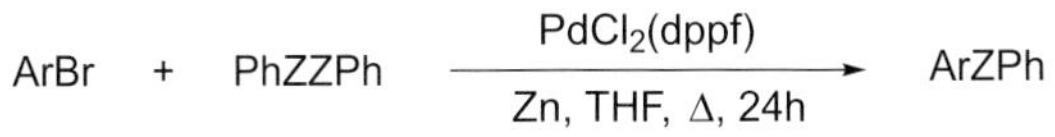

$$\mathrm{ArBr} + \mathrm{PhZZPh} \xrightarrow[\text{Zn, THF, }\Delta\text{, 24h}]{\mathrm{PdCl_2(dppf)}} \mathrm{ArZPh}$$

Scheme 3.14

In(I) was reported as a good reducing agent for the Z–Z bond. The anions formed can be easily alkylated under non-catalytic conditions [38, 39] and also arylated and vinylated under Pd-catalyzed conditions (Scheme 3.15) [40]. For the *E*-isomer the reaction is stereospecific, while for the *Z*-isomer a mixture of isomers was detected. It should be pointed out that for reduction of PhSSPh other reducing agents, such as NaH or Na_2SO_3, may also be used [41].

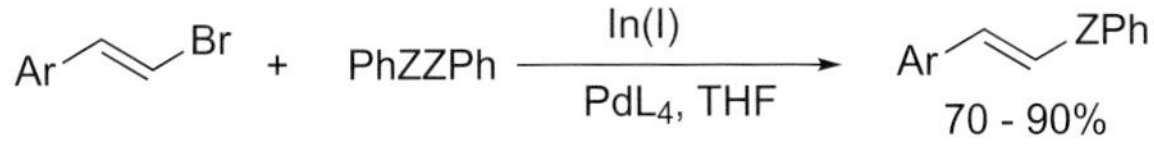

$$\mathrm{Ar{\sim}Br} + \mathrm{PhZZPh} \xrightarrow[\mathrm{PdL_4,\ THF}]{\mathrm{In(I)}} \mathrm{Ar{\sim}ZPh}\ \ 70 - 90\%$$

Scheme 3.15

In addition to the –SAr groups, the scope of cross-coupling reactions was extended to $-SO_2Ar$ and –S–COOMe groups. The reaction of sulfonium acid salts with aryl- and vinyl halides or triflates (Xantphos, Cs_2CO_3, Bu_4NCl, PhMe, 80 °C for ArI or 120 °C for ArBr and ArOTf) made it possible to synthesize various sulfones (Scheme 3.16) [42].

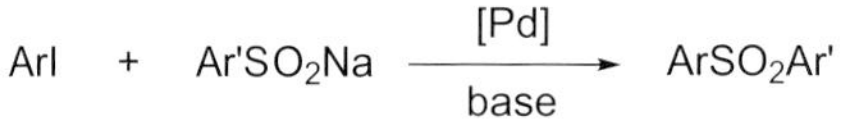

$$\text{ArI} + \text{Ar'SO}_2\text{Na} \xrightarrow[\text{base}]{[\text{Pd}]} \text{ArSO}_2\text{Ar'}$$

Scheme 3.16

Based on catalytic cross-coupling reaction between the p-$MeC_6H_4SO_2Na$ and p-BrC_6H_4I, followed by reaction with pyridine, a target compound containing a motif of β-3 adrenagic receptor agonist was prepared in 81% yield:

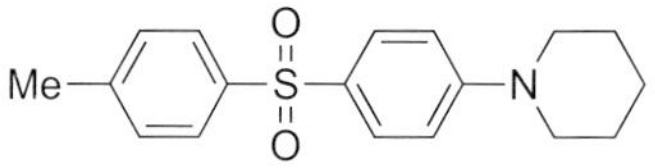

The Pd-catalyzed C–S cross-coupling reaction was carried out starting with aryl bromides and aryl triflates and potassium thioacetate (Scheme 3.17) [43]. Various sulfur compounds (such as ArSH, ArSR, ArSCl, ArSOCl) can be obtained from these products in a single step.

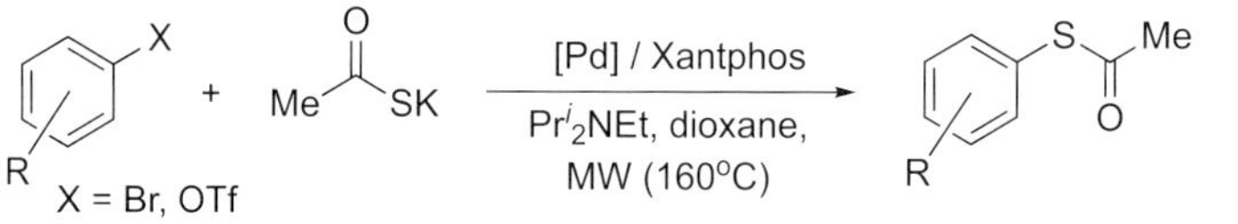

[Pd] / Xantphos
Pr^i_2NEt, dioxane, MW (160°C)
X = Br, OTf

Scheme 3.17

The method has been used for the synthesis of pyrrolidine constrained phenethylamine DPP-IV inhibitor (Scheme 3.18).

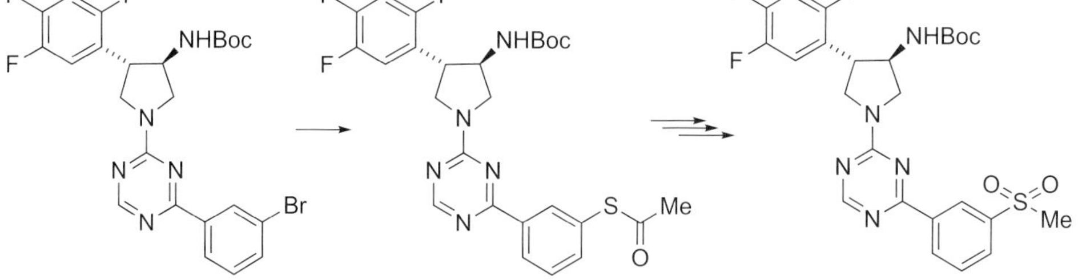

Scheme 3.18

Note, that the above discussed method is a useful alternative for the reaction between ArSM and RCOCl, which was previously employed for the formation of PhCOSePh using $PhSeSnBu_3$ as a starting material. Some authors have carried out this reaction under Pd-catalyzed conditions, as well as alkylation reaction leading to $ArCOCH_2SePh$ [44], although both reactions can easily take place under non-catalytic conditions without Pd [34, 45].

3.2.2
Ni-Catalyzed Transformations

Under carefully optimized conditions nickel complexes have shown excellent performance in the catalytic cross-coupling reactions. The possibility of Ni catalysis in Csp^2–S bond formation was first shown by Cristau *et al.* in 1981 [15]. The reaction of Ar_2S formation was carried out with $NiBr_2$ as a catalyst precursor and *o*-$(Ph_2P)_2C_6H_4$ as a ligand. The same authors have shown the possibility of the cross-coupling reaction with vinyl halides (Scheme 3.19) [46].

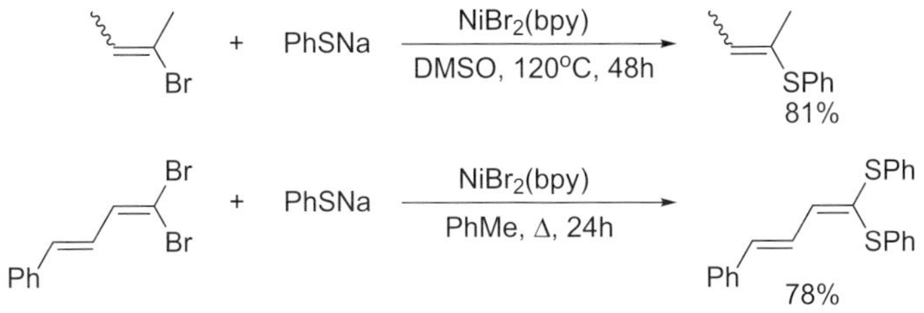

+ PhSNa → [$NiBr_2$(bpy); DMSO, 120°C, 48h] → SPh 81%

+ PhSNa → [$NiBr_2$(bpy); PhMe, Δ, 24h] → SPh, SPh 78%

Scheme 3.19

Diaryl sulfides were synthesized under mild conditions (room temperature to 60 °C) through the reaction of XC_6H_4SH (X = H, *p*-Me, *p*-MeO) and YC_6H_4X (X = I, Br; Y = H, Me, MeCO, etc.) in the catalytic system based on $NiBr_2$–Zn–dppf (Scheme 3.20) [47].

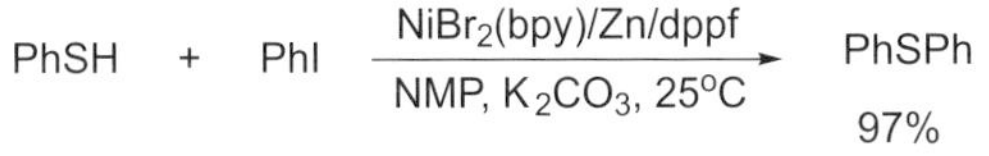

PhSH + PhI → [$NiBr_2$(bpy)/Zn/dppf; NMP, K_2CO_3, 25°C] → PhSPh 97%

Scheme 3.20

The $NiBr_2$(bpy) complex was found to be an efficient catalyst precursor, not only in the synthesis of ArSAr′ derivatives, but also in the catalytic formation of ArSeAr′ compounds (via the cross-coupling reaction of aryl selenols). Polymer-supported borohydride was used as a reducing agent and polymer-supported aryl iodide as a reagent. The reaction was carried out with high yield and resulted in high purity of the products leading to the construction of a library of supported species (Scheme 3.21) [48].

Using different reducing agents, disulfides and diselenides can be involved in the cross-coupling reaction. *In situ* reduction of diselenides with polymer-supported borohydride and the same Ni catalyst gave moderate to high yields of products in the reaction with Ar(Het)I (Scheme 3.22) [49].

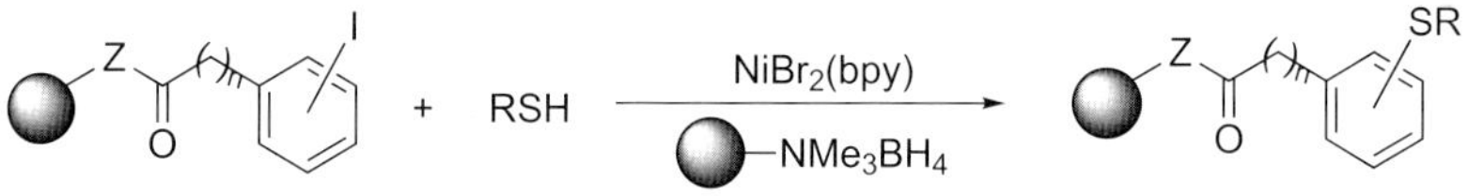

$NiBr_2(bpy)$, NMe_3BH_4; RSH; Z; O; I; SR

Scheme 3.21

R^1, R^2, R^3, R^4, $Se)_2$ + I, X, COOH — $NiBr_2(bpy)$/"H^-", EtOH, THF, 65 °C → Se, COOH

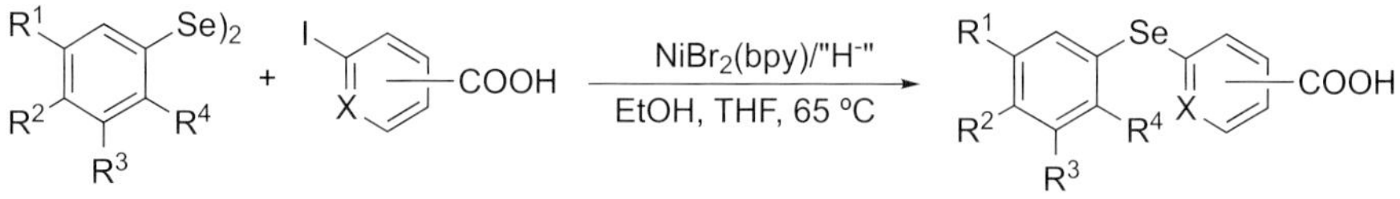

Scheme 3.22

In the reaction involving disulfides R_2S_2 the reduction step was carried out using Zn and the same Ni catalyst precursor (Scheme 3.23) [50].

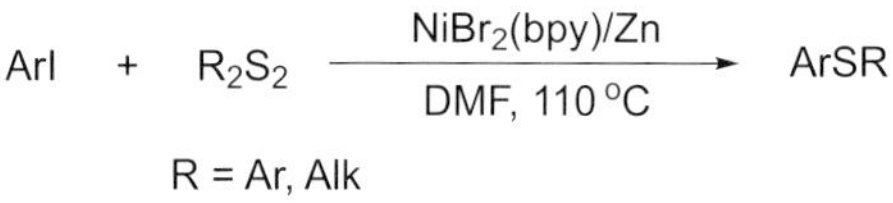

$$ArI + R_2S_2 \xrightarrow[\text{DMF, 110 °C}]{NiBr_2(bpy)/Zn} ArSR$$

R = Ar, Alk

Scheme 3.23

Pincer complexes of nickel (PCP)NiCl were reported as efficient catalysts for the C–S cross-coupling reaction (Scheme 3.24). With the exception of R = Bu^t, high product yields were observed [51]. The involvement of Ni(I) and Ni(III) complexes in the mechanism of the catalytic cycle was proposed for this particular case.

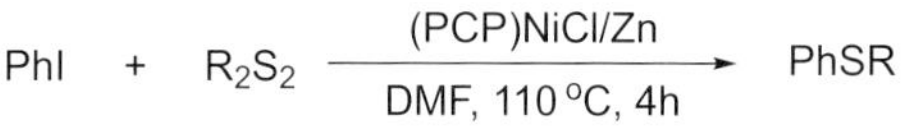

$$PhI + R_2S_2 \xrightarrow[\text{DMF, 110 °C, 4h}]{(PCP)NiCl/Zn} PhSR$$

Scheme 3.24

Nickel complexes with *NNN* nitrogen pincer ligands showed good efficiency in allyl- and arylthiolation of iodobenzene [52, 53]:

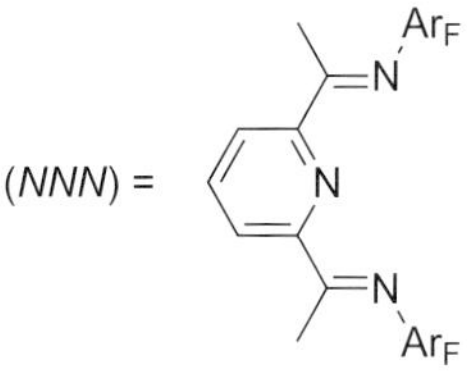

(*NNN*) = N, N, N, Ar_F, Ar_F

It was reported recently that ligand-free nickel, $NiCl_2{\bullet}6H_2O$, can also act as an efficient catalyst for the reaction of ArI and RSH in ionic liquid media and led to the PhSR product in high yields [54]. Using the model reaction of p-MeC_6H_4SH and p-$NO_2C_6H_4I$ the possibility of recycling was shown for five cycles without noticeable loss of catalytic activity.

3.2.3 Cu-Catalyzed Transformations

Recent renaissance in Ulmann chemistry has opened new opportunities for successful implementation of novel cross-coupling approaches, especially useful for carbon–heteroatom bond formation under mild conditions with copper compounds. Numerous publications have appeared dealing with thiolation and selenation of aryl- and alkenyl halides using various Cu complexes as catalysts. A high reaction temperature of 200–300 °C [55] was significantly reduced, to 100 °C and less. Therefore, a simplified reaction technique and cheaper solvents may be utilized in synthetic procedures. Lower temperature was also of much importance to avoid side reactions.

Main difficulties encountered in the field of Cu-catalyzed cross-coupling reactions have arisen from the absence of clear understanding of the mechanism, which is needed for efficient catalyst design and selection of proper ligands. This fact also hampers classification of available data on Cu-catalyzed cross-coupling reactions.

A cross-coupling reaction with tin derivatives $Ar'SeSnBu_3$ was carried out for various Ar(Het)Br in good to high yields using Cu(I) complexes (Scheme 3.25) [56]. The copper catalyst has shown much better performance than Pd and Ni systems in

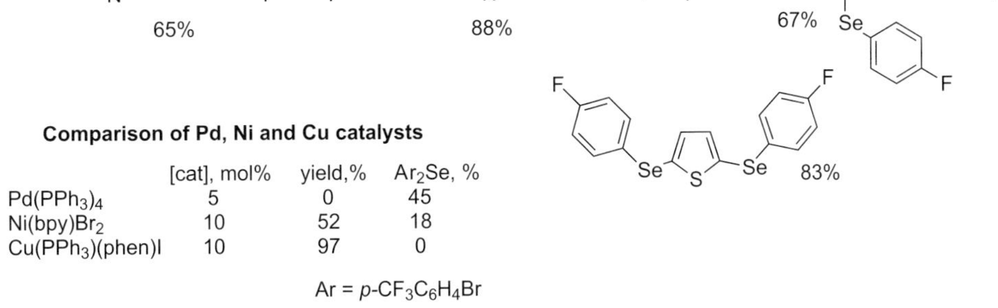

[Cu], DMF, 110 °C, 4-20h

65% 88% 67% 83%

Comparison of Pd, Ni and Cu catalysts

	[cat], mol%	yield,%	Ar_2Se, %
$Pd(PPh_3)_4$	5	0	45
$Ni(bpy)Br_2$	10	52	18
$Cu(PPh_3)(phen)I$	10	97	0

Ar = p-$CF_3C_6H_4Br$

Scheme 3.25

terms of higher product yield and the absence of disproportionation product Ar_2Se (Scheme 3.25).

For *o*-substituted aryl iodides and aryl thiols the catalytic system with $CuPF_6(MeCN)_4$ complex has been successfully applied (Scheme 3.26) [57]. An intramolecular version of the reaction leading to the formation of diaryl thioesters has also been reported.

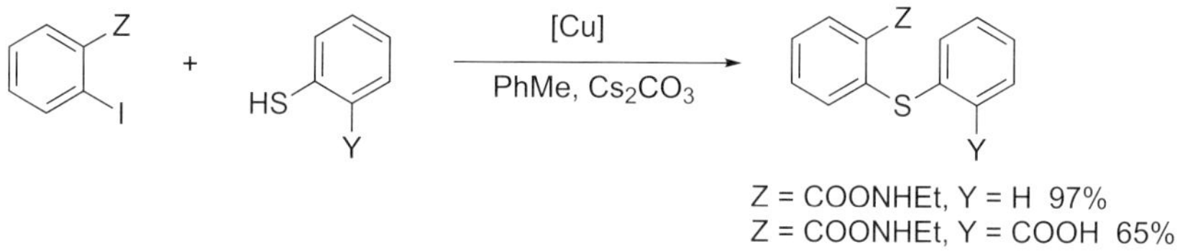

[Cu]; PhMe, Cs_2CO_3

Z = COONHEt, Y = H 97%
Z = COONHEt, Y = COOH 65%

Scheme 3.26

Diaryl sulfides in high yields were obtained in the reaction of aryl iodides and activated aryl bromides under CuBr-catalyzed conditions using Schwesinger's phosphazene base, which, most likely, was also coordinated as a ligand (Scheme 3.27) [58]. Similar results, but with longer reaction time, were observed using DBU (1,8-diazabicyclo[5.4.0]undec-7-ene).

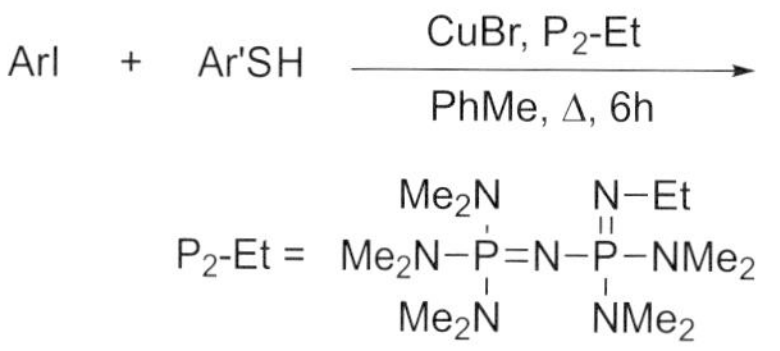

$$\text{ArI} + \text{Ar'SH} \xrightarrow[\text{PhMe, }\Delta\text{, 6h}]{\text{CuBr, P}_2\text{-Et}} \text{ArSAr'}$$

P_2-Et = $(Me_2N)_3P{=}N{-}P(NMe_2)_2{=}N{-}Et$

Scheme 3.27

Venkataraman *et al.* have suggested neocuproin as a ligand, which significantly enhanced the scope of the reaction to include aryl(hetaryl) iodides with electron-donating and electron-accepting substituents, various aryl thiols, *o,o*-$Me_2C_6H_3SH$ and alkyl thiols (Scheme 3.28) [59].

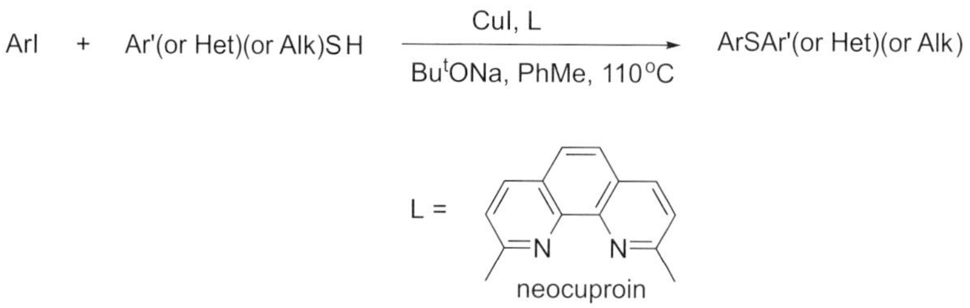

$$\text{ArI} + \text{Ar'(or Het)(or Alk)SH} \xrightarrow[\text{Bu}^t\text{ONa, PhMe, 110}^\circ\text{C}]{\text{CuI, L}} \text{ArSAr'(or Het)(or Alk)}$$

L = neocuproin

Scheme 3.28

The CuI–neocuproin (1: 1) complex was used in the arylation of 8-mercaptoadenine (Scheme 3.29). The catalytic system was used for the synthesis of a series of 8-arylsulfanyl adenine derivatives [60].

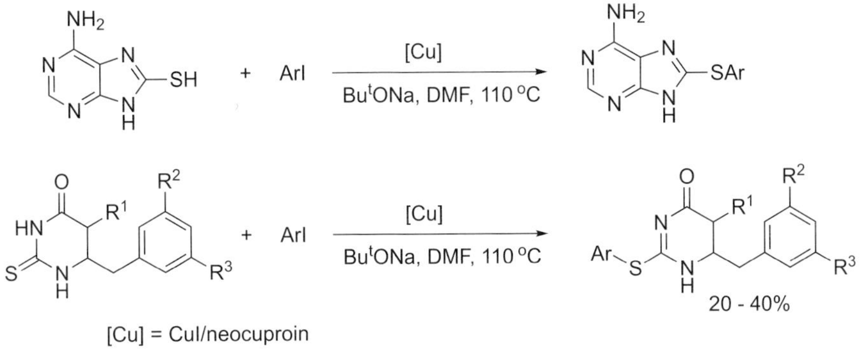

[Cu] = CuI/neocuproin

Scheme 3.29

A soluble Cu(I) complex has been successfully used by Venkataraman *et al.* in C–S cross-coupling when more reactive vinyl iodides were utilized as a reagent (Scheme 3.30) [61].

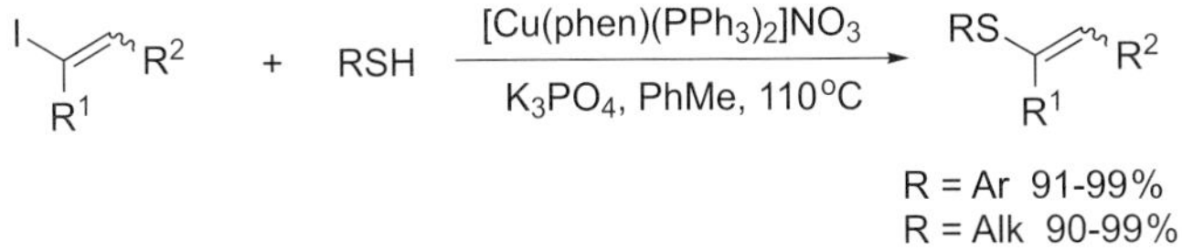

Scheme 3.30

Buchwald has suggested a simpler ligand, ethylene glycol, to carry out cross-coupling reaction with aryl thiols and alkyl thiols using a smaller amount of CuI and lower temperature (Scheme 3.31) [62].

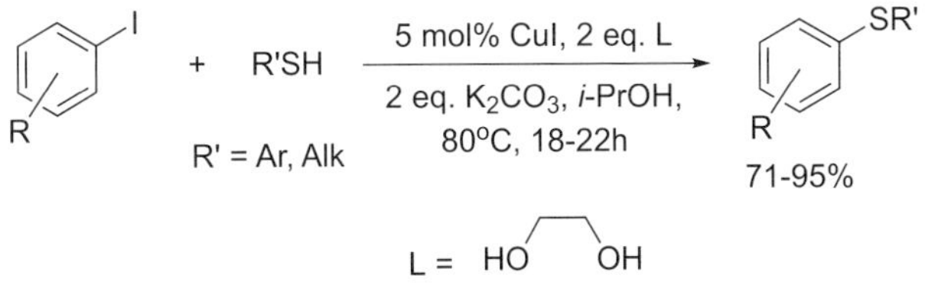

Scheme 3.31

Even better results were achieved using the tridentate oxygen ligand – 1,1,1-tris(hydroxymethyl)ethane (Scheme 3.32). Nearly quantitative yields were found in the reaction involving various aryl iodides (including *o*-MeO-substituted) with aryl thiols and alkyl thiols [63].

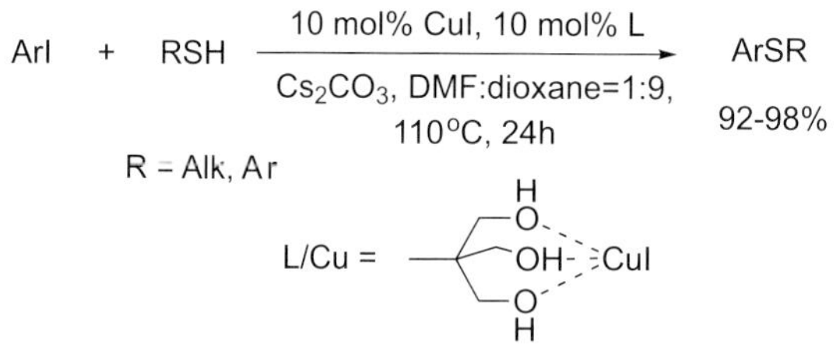

$$\mathrm{ArI} + \mathrm{RSH} \xrightarrow[\text{Cs}_2\text{CO}_3,\ \text{DMF:dioxane}=1:9,\ 110\,^\circ\text{C},\ 24\text{h}]{10\ \text{mol\% CuI, 10 mol\% L}} \mathrm{ArSR}\ \ 92\text{-}98\%$$

R = Alk, Ar

L/Cu = [tris(hydroxymethyl)ethane ligand coordinated to CuI]

Scheme 3.32

To carry out the reaction with aryl iodides, including o-HOC_6H_4I, the Cu catalyst with a new ligand bearing oxime and phosphineoxide groups was developed (Scheme 3.33) [64]. The products were obtained in good to high yields. It was shown that CuI may be replaced by readily available and air stable Cu_2O. Note, that use of Cu_2O without additives requires stoichiometric amounts and high temperatures (160 °C) [65, 66].

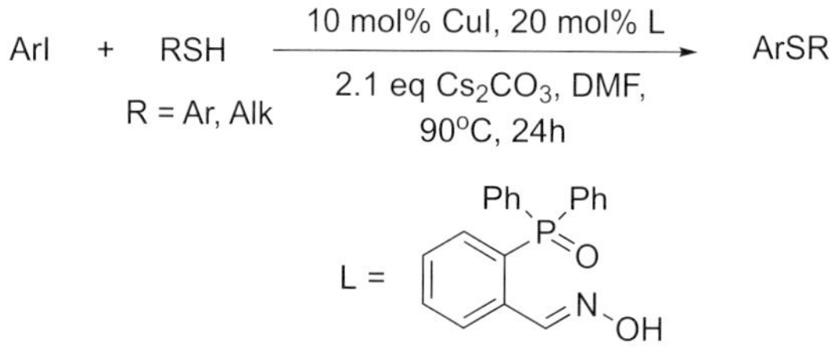

$$\mathrm{ArI} + \mathrm{RSH} \xrightarrow[2.1\ \text{eq Cs}_2\text{CO}_3,\ \text{DMF},\ 90\,^\circ\text{C},\ 24\text{h}]{10\ \text{mol\% CuI, 20 mol\% L}} \mathrm{ArSR}$$

R = Ar, Alk

L = [2-(diphenylphosphoryl)benzaldehyde oxime: Ph, Ph, P=O, N–OH]

Scheme 3.33

Significant progress was achieved using amino acids as ligands in the Cu-catalyzed reaction: not only aryl iodides, but also aryl bromides (without electron-donating substituents) were used as starting materials and led to product formation in 90–98% and 72–85% yields, respectively (Scheme 3.34) [67]. The catalytic system was found to be very sensitive to the nature of the solvent and base. A plausible mechanism was reported and the involvement of Cu(I)/Cu(III) species was proposed (Scheme 3.35).

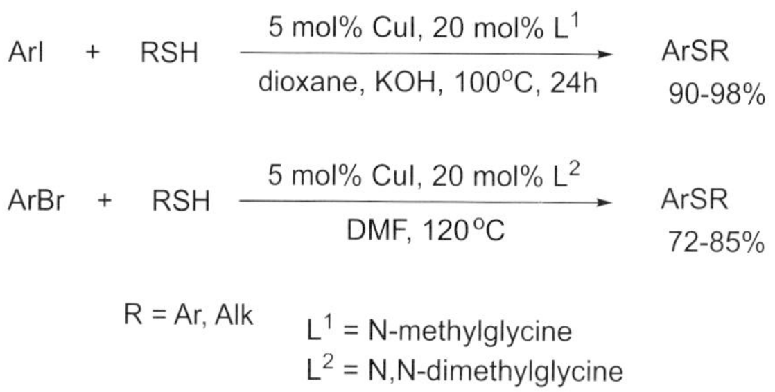

$$\mathrm{ArI} + \mathrm{RSH} \xrightarrow[\text{dioxane, KOH, }100\,^\circ\text{C, 24h}]{5\ \text{mol\% CuI, 20 mol\% L}^1} \mathrm{ArSR}\ \ 90\text{-}98\%$$

$$\mathrm{ArBr} + \mathrm{RSH} \xrightarrow[\text{DMF, }120\,^\circ\text{C}]{5\ \text{mol\% CuI, 20 mol\% L}^2} \mathrm{ArSR}\ \ 72\text{-}85\%$$

R = Ar, Alk

L^1 = N-methylglycine
L^2 = N,N-dimethylglycine

Scheme 3.34

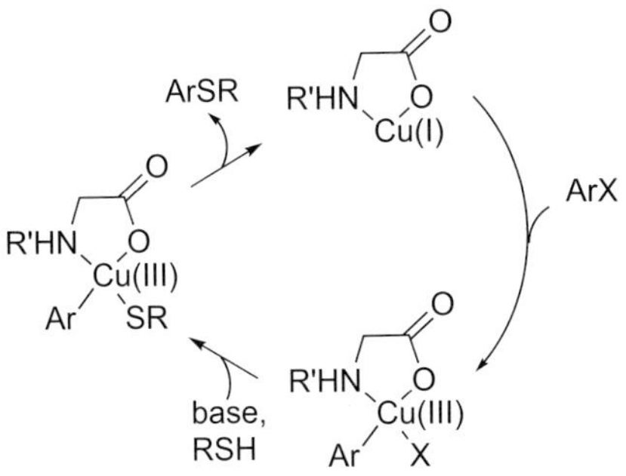

Scheme 3.35

Using L-proline as ligand it was possible to carry out the reaction under milder conditions.

A very efficient CuI–benzotriazol catalytic system was reported, by Verma *et al.*, to involve aryl bromides in the cross-coupling reaction (Scheme 3.36) [66]. For all types of aryl bromides high yields were achieved in the reaction with aryl thiols and alkyl thiols using only small amounts of catalyst and ligand (high TON values). Note, that for the reactions with alkyl thiols the temperature was decreased from 100 °C to 80 °C. For ArX bearing COOH and CHO groups Cs_2CO_3 was added instead of Bu^tOK. The mechanism of this catalytic cycle was also suggested to involve Cu(I)/Cu(III) species (Scheme 3.37).

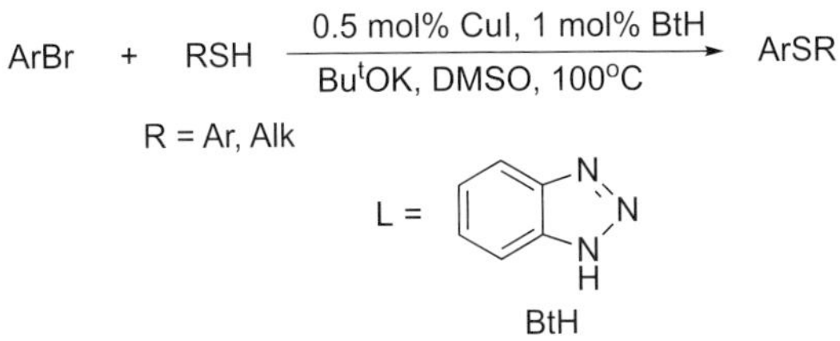

$$\text{ArBr} + \text{RSH} \xrightarrow[\text{Bu}^t\text{OK, DMSO, 100}^\circ\text{C}]{\text{0.5 mol\% CuI, 1 mol\% BtH}} \text{ArSR}$$

R = Ar, Alk

L = BtH

Scheme 3.36

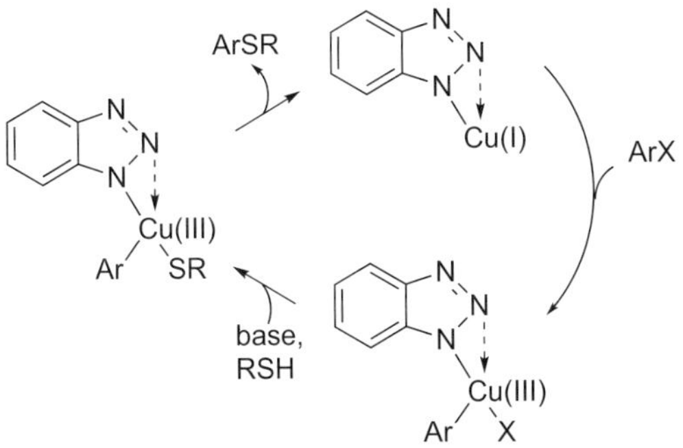

Scheme 3.37

For a particular case using *o*-bromobenzoic acid as a substrate, a binary Cu/Cu_2O catalyst was proposed with 2-ethoxyethanol as a solvent and ligand (Scheme 3.38) [68]. It was reported that in this catalytic system the formation of C–S bond was observed to be faster than formation of C–O and C–N bonds.

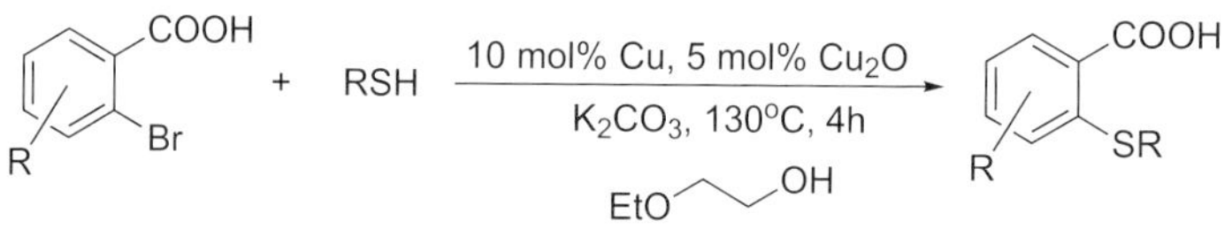

Scheme 3.38

Catalytic cross-coupling reactions were shown to be of much help in preparing several important compounds. For example, thiazene-substituted thioglycosides were readily synthesized via the reaction of the corresponding thiols with aryl iodides in the CuI/Py catalytic system in MeCN (Scheme 3.39) [69].

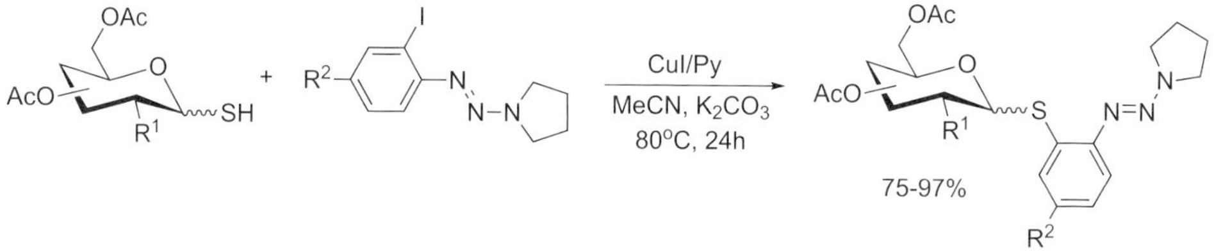

Scheme 3.39

Applying microwave (MW) heating in the reaction of ArI and ArBr with RSH in $CuI/NMP/Cs_2CO_3/195\,^\circ C$ catalytic system made it possible to shorten reaction time to 2–6 h; however, with moderate yields of 64–89% [70]. An example of the catalytic cross-coupling reaction of aryl bromides with thiols in ionic liquids was reported ($[BMIM][BF_4]/K_2CO_3/CuI$/L-proline/110–120 °C) [71]. High yields and high *E/Z*-selectivity were found, but catalyst recycling led to significant loss of activity. CuI without additional ligands was used as a catalyst in the reaction of thiols with iodochalcogeno alkenes to give product in moderate to good yields [72].

The topic of special interest is to develop cross-coupling methodology to carry out the reaction between ArX and RSH in water in the absence of organic ligands. Recently, it was reported that thiophenol reacts with phenyl halides in water in the presence of Bu_4NBr (TBAB) at 80 °C (Scheme 3.40) [73]. Product yields in the reaction

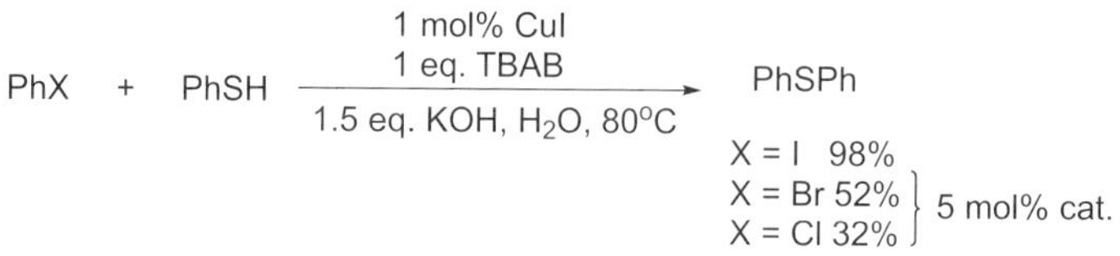

Scheme 3.40

involving ArSH and $PhCH_2SH$ were as high as 91–99%, while with BuSH the yield dropped to 15%.

A high catalytic activity of Cu nanoparticles was observed in the reaction between aryl iodides and thiols under MW heating at 120 °C (Scheme 3.41) [74]. The reaction was completed in minutes, which is much faster than the CuI-catalyzed transformation under MW conditions [70]. In the case of ArSH high product yields of 82–98% were found, while in the case of AlkSH significantly lower yields were observed due to side reaction leading to the formation of Alk_2S_2. The authors have noticed that using copper powder in the same conditions led to only 25–40% yield of products. Analogous results were obtained in the reaction with Cu-NPs supported on the hexagonal microporous silica [75].

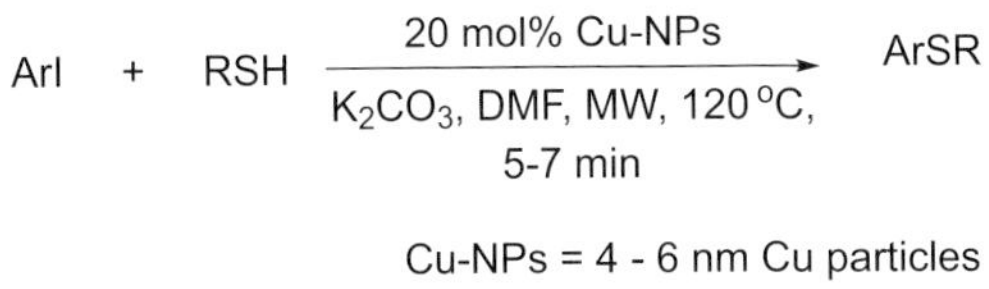

$$ArI + RSH \xrightarrow[K_2CO_3,\ DMF,\ MW,\ 120\,^{\circ}C,\ 5\text{-}7\ min]{20\ mol\%\ Cu\text{-}NPs} ArSR$$

Cu-NPs = 4 - 6 nm Cu particles

Scheme 3.41

Nano-sized CuO was also found to be an efficient catalyst for the reaction of ArI and RSH (for both ArSH and AlkSH) in DMSO at 80 °C (Scheme 3.42) [76]. The products of the reaction were obtained in nearly quantitative yields (85–99%). Catalyst recycling was demonstrated for three successive cycles, although the yield of the reaction was decreased. Introducing electron-donating groups into the ArI significantly decreased product yield. Other examples of C–S bond formation in the Cu-cat./KOH system have also been reported [77–79].

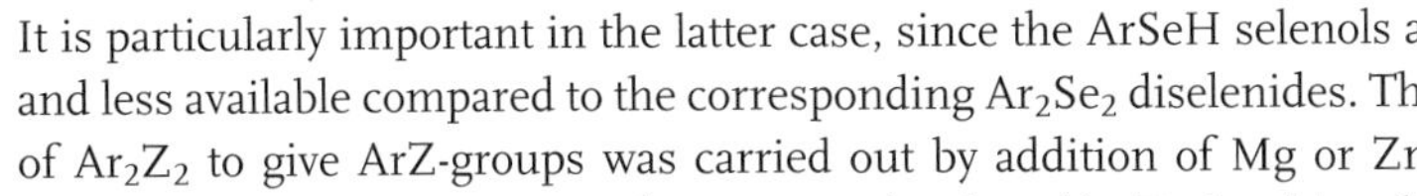

$$HS(CH_2)_6SH + PhI \xrightarrow[DMSO,\ 80^{\circ}C,\ N_2,\ 5h]{CuO,\ KOH} PhS(CH_2)_6SPh$$

90%

Scheme 3.42

As we already discussed in the Pd-catalyzed transformations, disulfides and diselenides may also be used as a source of ArZ-groups in the Cu-catalyzed reactions. It is particularly important in the latter case, since the ArSeH selenols are unstable and less available compared to the corresponding Ar_2Se_2 diselenides. The reduction of Ar_2Z_2 to give ArZ-groups was carried out by addition of Mg or Zn metal. An example of the [Cu]/bpy/Mg catalytic system developed by Taniguchi *et al.* is shown in Scheme 3.43 [80].

$$Ph_2Z_2 + Ar(or\ Het)I \xrightarrow[DMF,\ 110\,^{\circ}C]{[Cu]/bpy/Mg} Ar(or\ Het)ZPh$$

[Cu] = CuI or CuO

Z = Se 70-95%
Z = S 67-95%

Scheme 3.43

Under MW heating (160–200 °C) a wide range of chalcogenides, including tellurium derivatives, and ArBr with electron-donating groups (p-HOC_6H_4Br, p-$Me_2NC_6H_4Br$), were successfully reacted with thiols forming cross-coupling products in good to high yields [81]. Alkenyl derivatives of selenium were synthesized in ionic liquids in 65–91% yields using Zn as reducing agent (Scheme 3.44) [82].

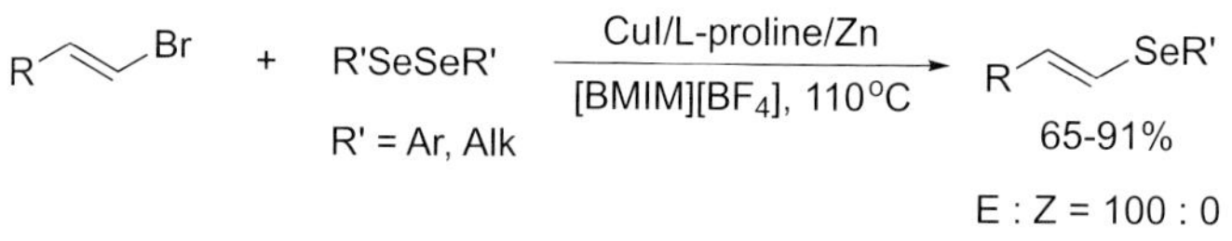

R–CH=CH–Br + R'SeSeR' (R' = Ar, Alk) → [CuI/L-proline/Zn, [BMIM][BF_4], 110°C] → R–CH=CH–SeR' 65-91%, E : Z = 100 : 0

Scheme 3.44

Cu complexes were found to be suitable catalysts for the C–S bond formation starting with ArX and sulfonic acid salts, leading to the formation of $ArSO_2R$ sulfones (Scheme 3.45) [83]. The reaction was limited to aryl iodides only, while aryl bromides did not react.

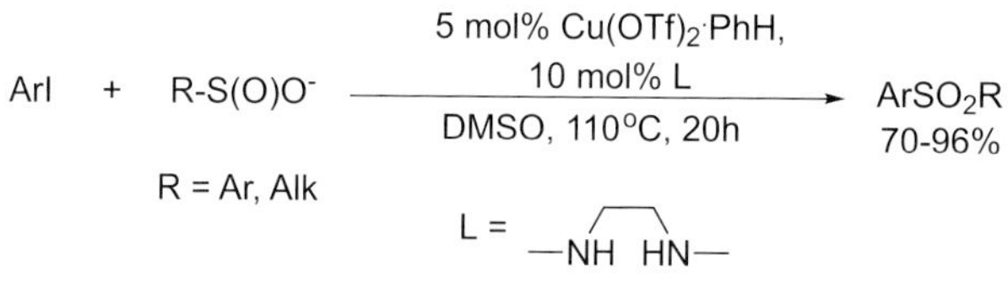

ArI + R-S(O)O$^-$ (R = Ar, Alk) → [5 mol% Cu(OTf)$_2$·PhH, 10 mol% L, DMSO, 110°C, 20h] → $ArSO_2R$ 70-96%

L = —NH HN— (ethylenediamine bridge)

Scheme 3.45

It is interesting to point out that, if CuI catalyst was used in a combination with the sodium salt of L-proline, both ArI and ArBr with electron-donating and electron-accepting substituents gave sulfones, even at lower temperature (Scheme 3.46) [84]. Both types of sulfones were formed in good to high yields.

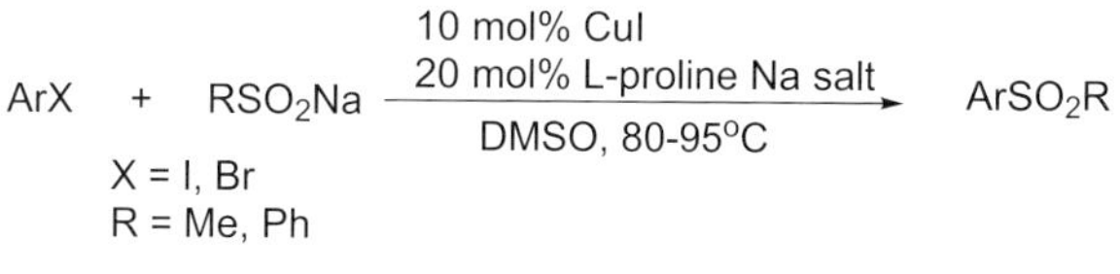

ArX + RSO_2Na (X = I, Br; R = Me, Ph) → [10 mol% CuI, 20 mol% L-proline Na salt, DMSO, 80-95°C] → $ArSO_2R$

Scheme 3.46

The reaction of RSO_2Na with ArX or vinyl bromide was carried out in a binary mixture of solvents consisting of DMSO and anion-functionalized ionic liquids (Scheme 3.47) [85].

Cu complexes with 1,10-phenatroline (phen) ligand catalyzed arylation of thiobenzoic acid in high yields (Scheme 3.48) [86]. In contrast to ArI, which readily reacted with the thiobenzoic acid, only traces of products were observed in the reaction with PhBr.

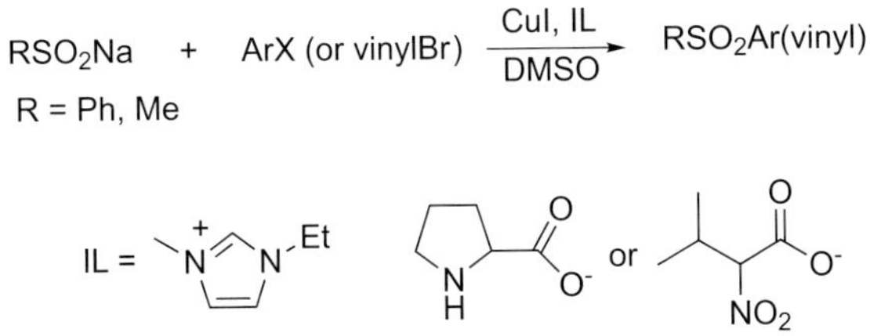

RSO_2Na + ArX (or vinylBr) —CuI, IL / DMSO→ RSO_2Ar(vinyl)
R = Ph, Me

IL = [1-methyl-3-ethylimidazolium] ; [prolinate] or [2-nitro-3-methylbutanoate, NO_2]

Scheme 3.47

ArI + HSC(O)Ph —10 mol% CuI, 20 mol% phen / Pr^i_2NEt, PhMe, 110°C→ ArSC(O)Ph 94-100%

Ar = XC_6H_4, naphtyl, Py

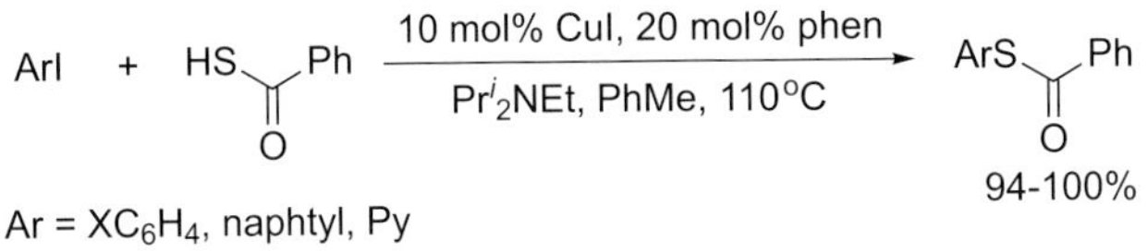

Scheme 3.48

The catalytic system based on CuI and *N,N*-dimethylglycine ligand was found to catalyze C–S cross-coupling of aryl iodides and vinyl bromides with sodium dithiocarbamates (Scheme 3.49) [87]. As in the previous case, PhBr did not react in this catalytic system.

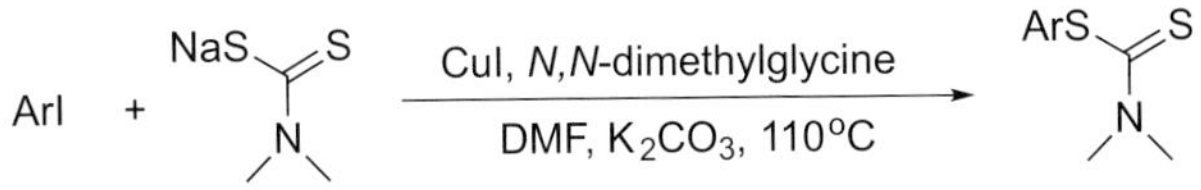

ArI + $NaSC(S)NMe_2$ —CuI, *N,N*-dimethylglycine / DMF, K_2CO_3, 110°C→ $ArSC(S)NMe_2$

Scheme 3.49

Oxidative cross-coupling of aryl boronic acids with alkylthiols (including cysteine) was reported under heating in the $Cu(OAc)_2$/Py-DMF system [88]. In the case of Cu-catalyzed reactions organic boronic derivatives were often used as the cross-coupling partner together with disulfide, diselenide and ditelluride as a donor of the ArZ-group. As an example, the Cu-catalyzed reaction without ligands is shown on Scheme 3.50 [89]. In similar conditions the reaction with the CuI/bpy catalytic system was also carried out (Scheme 3.51) [90]. However, for R = Bu the yield was much lower and for Z = Te the reaction took place only with $PhB(OH)_2$.

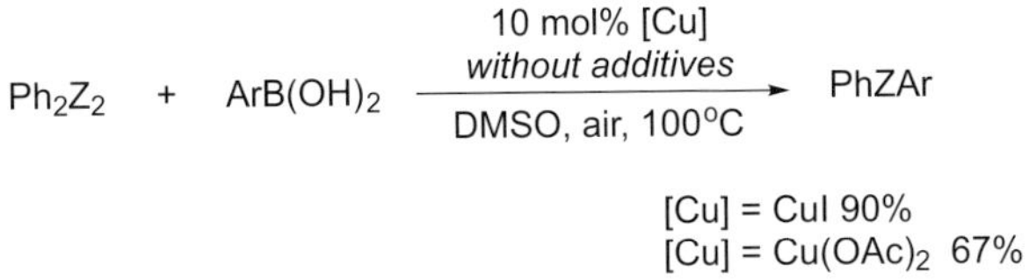

Ph_2Z_2 + $ArB(OH)_2$ —10 mol% [Cu], *without additives* / DMSO, air, 100°C→ PhZAr

[Cu] = CuI 90%
[Cu] = $Cu(OAc)_2$ 67%

Scheme 3.50

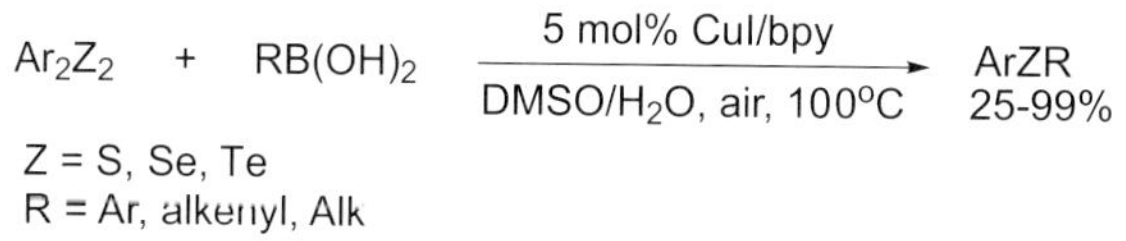

$$Ar_2Z_2 + RB(OH)_2 \xrightarrow[\text{DMSO/H}_2\text{O, air, 100°C}]{\text{5 mol\% CuI/bpy}} \text{ArZR } 25\text{-}99\%$$

Z = S, Se, Te
R = Ar, alkenyl, Alk

Scheme 3.51

Alkenyl derivatives of selenium and tellurium were synthesized in the reaction of diphenyl dichalcogenides and potassium vinyltrifluoroborates in good yields in the case of Ar_2Z_2, while only 45–53% yields were found in the case of Alk_2Se_2 (Scheme 3.52) [91]. This catalytic reaction did not require an addition of base.

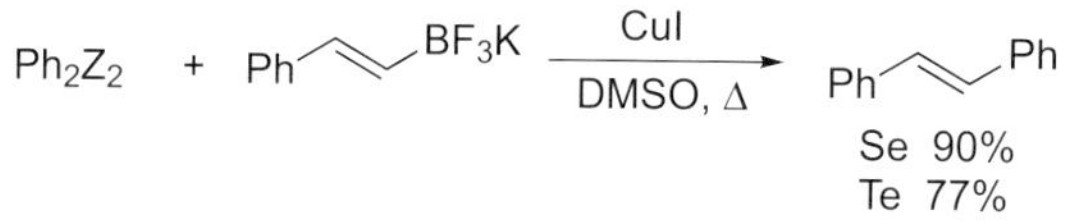

$$Ph_2Z_2 + Ph\text{CH=CH}BF_3K \xrightarrow[\text{DMSO, }\Delta]{\text{CuI}} Ph\text{CH=CH}Ph$$

Se 90%
Te 77%

Scheme 3.52

In ionic liquid it was possible to carry out oxidative coupling of aryl boronic acid with the sodium salt of sulfonic acid (Scheme 3.53) [92].

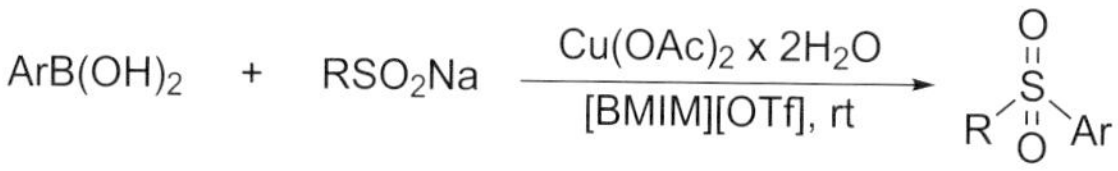

$$ArB(OH)_2 + RSO_2Na \xrightarrow[\text{[BMIM][OTf], rt}]{\text{Cu(OAc)}_2 \times 2\text{H}_2\text{O}} R\text{–S(=O)}_2\text{–Ar}$$

Scheme 3.53

3.2.4
Other Transition Metals as Catalysts

Catalytic reactions aimed at the formation of Csp^3–S and Csp^3–Se bonds are known to take place with rhodium, ruthenium and cobalt complexes. A particularly promising area with tremendous development in recent years concerns the investigation and possible application of iron catalysts.

Wilkinson complex was reported to catalyze alkylation and arylation reactions with thiols and selenols formed after reduction of R'_2Z_2 by hydrogen (Scheme 3.54) [93].

In a similar manner the reaction with various alkylation reagents took place using $RuCl_3$ as catalyst and applying diselenides reduction with Zn (Scheme 3.55) [94].

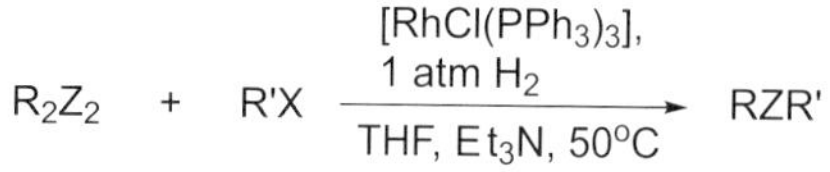

$$R_2Z_2 + R'X \xrightarrow[\text{THF, Et}_3\text{N, 50°C}]{[\text{RhCl(PPh}_3)_3],\ 1\text{ atm H}_2} RZR'$$

Scheme 3.54

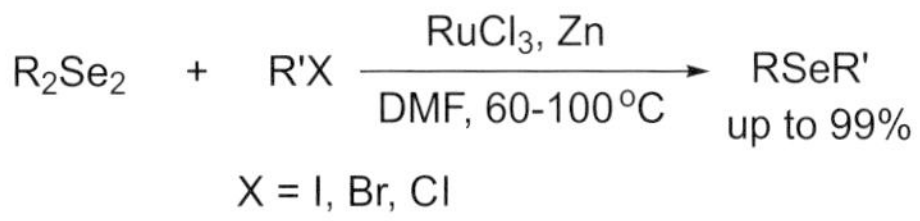

$$R_2Se_2 + R'X \xrightarrow[\text{DMF, 60-100}^{\circ}\text{C}]{\text{RuCl}_3\text{, Zn}} \underset{\text{up to 99\%}}{RSeR'}$$

X = I, Br, Cl

Scheme 3.55

Cross-coupling reaction leading to the formation of a Csp^2–S bond was possible with a cobalt catalyst. The best results were achieved using the CoI_2(dppe)/Zn catalytic system (Scheme 3.56) [95]. This new cross-coupling reaction is of much interest and complements very well the other known transition-metal-catalyzed reactions with Pd, Ni and Cu catalysts.

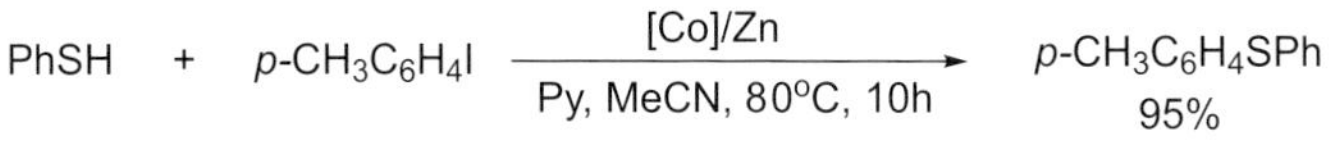

$$PhSH + p\text{-}CH_3C_6H_4I \xrightarrow[\text{Py, MeCN, 80}^{\circ}\text{C, 10h}]{[\text{Co}]/\text{Zn}} \underset{95\%}{p\text{-}CH_3C_6H_4SPh}$$

Scheme 3.56

The reaction was successfully carried out with various aryl(hetaryl) iodides and bromides involving different aryl thiols and alkyl thiols. A plausible catalytic cycle includes reduction of Co(II) complexes to Co(I), substitution of iodide ligand by SR leading to the formation of ArSCo(I), oxidative addition of ArX to this cobalt complex with formation of Co(III) derivative, followed by reductive elimination (Scheme 3.57) resulted in the product formation and regeneration of Co(I) catalyst.

$$Ar'S\text{-}Co(III)\text{-}Ar \longrightarrow Ar'SAr + Co(I)$$

Scheme 3.57

Recently, Bolm *et al.* have shown that such cross-coupling reaction is possible with iron catalyst leading to the formation of Ar(Het)SAr′ from the corresponding thiols (Scheme 3.58) [96]. The reaction was successful only with the Ar(Het)SH thiols, while the alkyl thiols AlkSH did not react. Undoubtedly, cheap and efficient iron catalysts should be further developed to carry out various C–Z bond formation reactions.

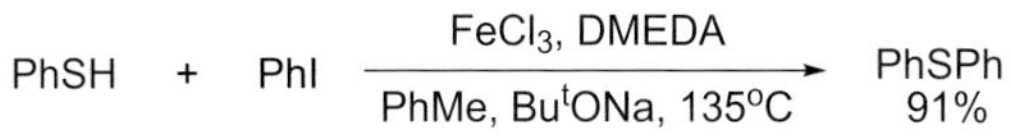

$$PhSH + PhI \xrightarrow[\text{PhMe, Bu}^t\text{ONa, 135}^{\circ}\text{C}]{\text{FeCl}_3\text{, DMEDA}} \underset{91\%}{PhSPh}$$

Scheme 3.58

In a subsequent study this reaction was further extended to include ArI and Ar(Het)SH with high selectivity and good yields (with the exception of $PhCH_2SH$) in the $FeCl_3 \bullet H_2O$/cationic 2,2′-bipyridyl (L) system in water [97]. The catalyst with the cationic ligand L in water solutions preserved the activity over several cycles:

L = $Br^- \ Me_3N^+$ ... $^+NMe_3 \ Br^-$

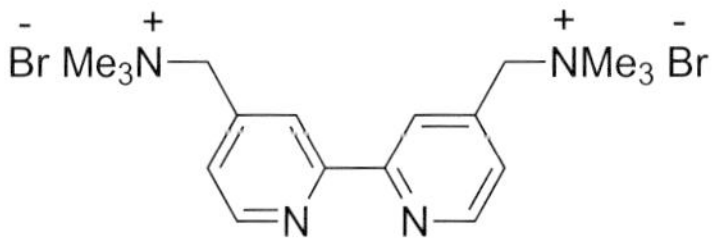

cationic 2,2'-bipyridyl ligand

Although much interest has been shown recently in the possible application of various iron catalysts, care should be taken to understand the nature of the active species in the catalytic cycle. A recent study by Buchwald and Bolm raised a question on the role of trace quantities of other metals, particularly Cu [98].

3.3
Catalytic Addition of RZ–ZR Derivatives to Alkynes (Z ≡ S, Se)

As we have already mentioned, alkenyl chalcogenides are in demand in the synthesis of various organic compounds and metal complexes (as ligands). Alkenyl chalcogenides can be prepared not only by the substitution reaction discussed in the previous section, but also by addition to the triple bond of alkynes (in contrast to aryl chalcogenides, which can be prepared only by a substitution reaction). Due to high atom efficiency and the absence of by-products, addition reactions have a good potential in terms of Green Chemistry requirements.

It is important to understand the mechanistic difference between these two approaches for C–Z bond formation. Chemical selectivity is a main concern in the design of cross-coupling reactions in order to improve the ratio of hetero-/homo-coupling products and to avoid reduction of the C–halogen bond. However, examination of the addition reactions raised other important problems – stereoselectivity in the case of the addition of dichalcogenides and regioselectivity for the addition of thiols and selenols [99].

In this section we describe the available literature on the addition reaction utilizing dichalcogenides RZZR as reagents. We do not discuss non-catalytic addition reactions carried out without transition metal catalysts, this topic has already been addressed in several publications (see [100–103] and references therein). It was shown that the non-catalytic reactions led to different products – a mixture of *E-/Z-*isomers is very often formed in the addition of RZZR to alkynes. Our goal is to concentrate on the selective formation of *Z*-isomers in the catalytic addition of RZZR to the triple bond of alkynes.

3.3.1
Pd and Ni-Catalyzed Formation of Vinyl Chalcogenides

In the pioneering study of Ogawa, Sonoda *et al.*, published in 1991, it was shown that Ph_2S_2 and Ar_2Se_2 in the $Pd(PPh_3)_4$-catalyzed conditions can be added to terminal alkynes (benzene, 80 °C) in good yields (54–98%) and with high selectivity towards formation of the *Z*-isomer (Scheme 3.59), with the exception of phenylacetylene, which readily underwent thermal side-reaction and resulted in the formation of significant amounts of the *E*-isomer [104]. The catalytic reaction was not suitable for Alk_2Z_2 (only 24% yield and $E/Z = 25/75$ selectivity with Bu_2Se_2; 6% yield with $(PhCH_2)_2S_2$), neither internal alkynes were involved in the reaction.

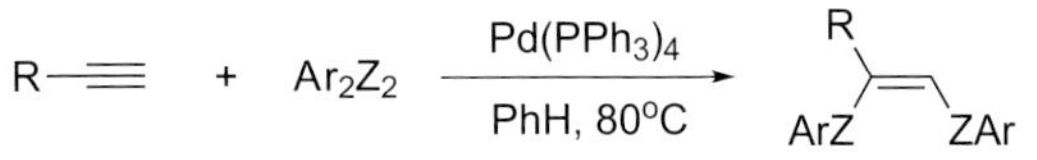

$$\text{R}\!-\!\!\equiv \; + \; \text{Ar}_2\text{Z}_2 \xrightarrow[\text{PhH, 80°C}]{\text{Pd(PPh}_3)_4} \text{(Z)-ArZ–CH=C(R)–ZAr}$$

Scheme 3.59

The catalytic cycle suggested for this reaction involves oxidative addition of Ar_2Z_2 to Pd(0) with formation of Pd disulfide or diselenide, alkyne coordination and insertion into the Pd–Z bond and reductive elimination (Scheme 3.60).

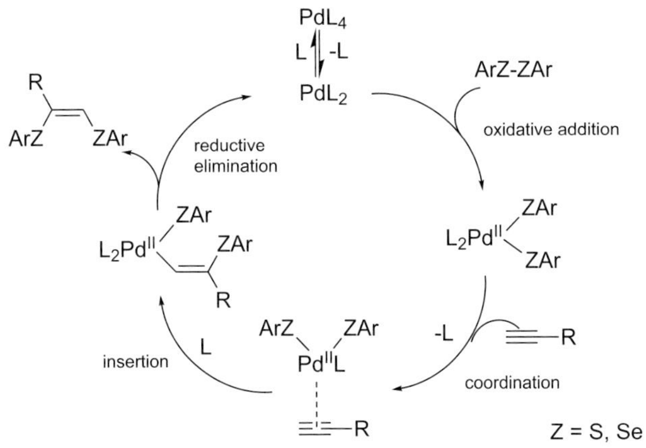

PdL4, L / -L, PdL2, ArZ-ZAr, oxidative addition, L2PdII(ZAr)2, -L, ≡–R, coordination, (ArZ)2PdIIL(≡–R), L, insertion, L2PdII–CH=C(R)(ZAr) (ZAr), reductive elimination, R / ArZ ZAr, Z = S, Se

Scheme 3.60

Important mechanistic data were provided in the study of the carbonylative addition reaction, which has proven that alkyne insertion is both stereo- and regio-selective with the Pd atom attached to the β-position and substituent R in the α-position of the alkyne unit [105]. Carbonylation of this intermediate alkenyl complex of Pd led to the formation of a single product:

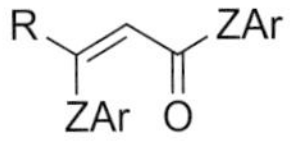

R–C(ZAr)=CH–C(=O)–ZAr

If propargylic alcohols were involved in the carbonylative addition, the reaction resulted in the formation of lactonization product in moderate to good yields (Scheme 3.61) [105].

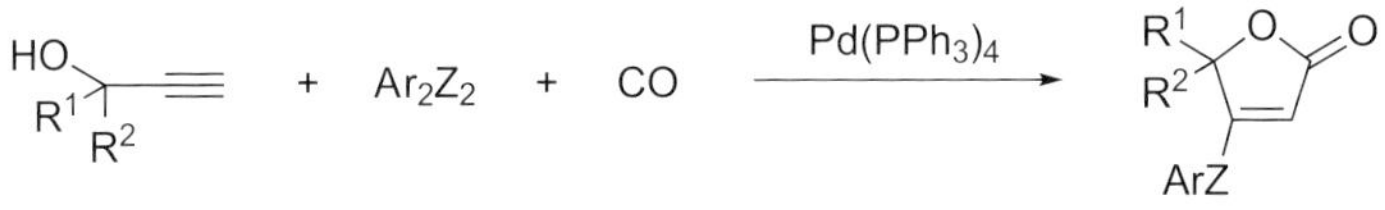

Scheme 3.61

Introducing halogen substituent into the *o*-position of the phenyl ring substantially facilitated the reaction due to coordination with the Pd atom in *syn*-fashion [106], which was also confirmed by the study of alkyne insertion into the Pt–S bond [107]:

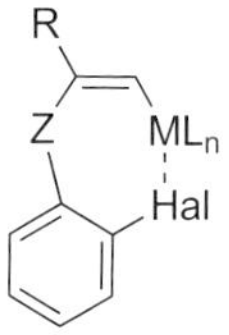

Subsequent attempts to carry out the synthetic procedure with $Pd(PPh_3)_4$ catalyst (Scheme 3.59) gave product yields of less than 65% [108, 109]. The origins of this behavior were revealed in the study of the nature of the intermediate Pd complexes involved in the reaction. It was found that dinuclear and polymeric Pd species, formed after oxidative addition or halogen substitution in the catalyst precursors, play an important role in the catalytic cycle (Scheme 3.62) [110, 111].

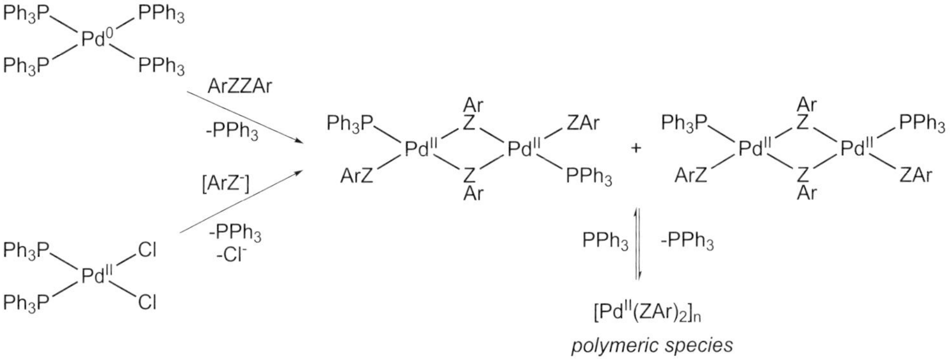

Scheme 3.62

Dinuclear complexes of trans- and cis-geometry (trans-complexes are more thermodynamically stable) have the ArZ–Pd–ZAr structural unit, which is needed to operate in the catalytic cycle. The observed rate constant of the addition reaction of Ph_2Se_2 to alkyne was linearly dependent on the concentration of dinuclear complexes $[Pd(SePh)_4(PPh_3)_2]$ [108]. Theoretical study at the density functional theory level has shown that alkyne insertion in the dinuclear complex preferentially involves the terminal ZAr group rather than the bridging ZAr group coordinated in η^2 mode [112]. A deficiency of the phosphine ligand under catalytic conditions resulted in rapid formation of polymeric species and catalyst deactivation (Scheme 3.62). This observation has suggested that the catalytic reaction should be carried out in the presence of some excess of the phosphine ligand. Indeed, an additional amount of the ligand (2–4 equiv) improved the performance of the catalytic reaction and gave reproducibly high product yields easily for various alkynes and dichalcogenides [108, 109].

The addition reaction carried out with 1,6-heptadiyne as reagent and different amounts of Ph_2S_2 has been shown to proceed through a stepwise pathway (Scheme 3.63) [109]. From the mechanistic point of view, the product structure indicates that the Pd center binds to the terminal position of the alkyne, otherwise the formation of a cyclic compound as a product would be expected (see also the carbonylative addition reaction discussed above). Independent confirmation of the regioselectivity of the insertion step was provided by carrying out the addition reaction with cyclopropylacetylene, as shown in Scheme 3.63 (i.e., the absence of homoallylic rearrangement product).

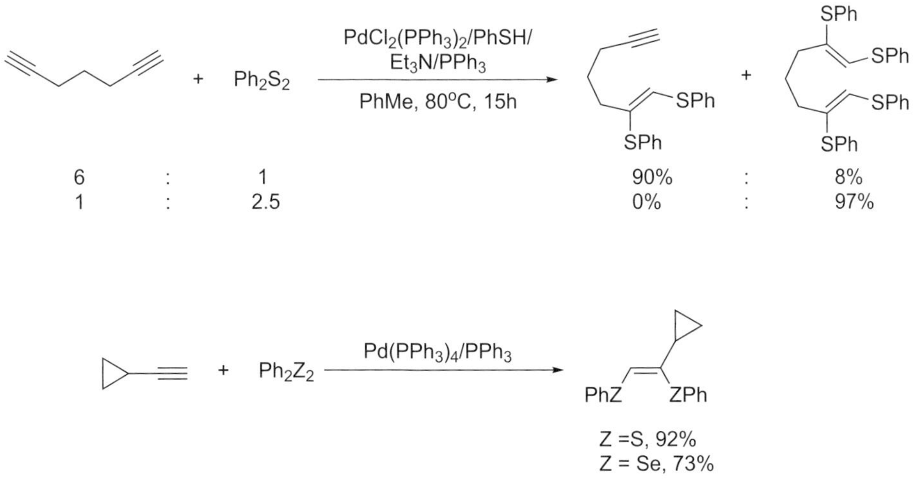

+ Ph_2S_2 — $PdCl_2(PPh_3)_2$/PhSH/Et_3N/PPh_3, PhMe, 80°C, 15h → SPh, SPh + SPh, SPh, SPh, SPh

6 : 1 — 90% : 8%
1 : 2.5 — 0% : 97%

+ Ph_2Z_2 — $Pd(PPh_3)_4$/PPh_3 → PhZ ZPh

Z =S, 92%
Z = Se, 73%

Scheme 3.63

For synthetic purposes, Pd catalyst on a polymeric support was developed and the addition reaction was carried out with excellent product yields of 94–99% and excellent stereoselectivity >99 : 1 (Scheme 3.64) [113]. After completion of the reaction the catalyst was easily isolated by filtration and a pure product was obtained after solvent evaporation (>98% purity without any additional purification procedures). Unfortunately, this synthetic approach was not applicable to Ph_2Se_2 addition to alkynes, since, at 140 °C, triphenylphosphine bound to the polymer reacted with Ph_2Se_2, resulting in selenium atom transfer to the phosphorus (Se=PR_3) and formation of Ph_2Se as a by-product. The mechanism of this side reaction was addressed in a joint experimental and theoretical study, which revealed the relationship between the C–Z and Z–Z bonds activation by Pd complexes [114].

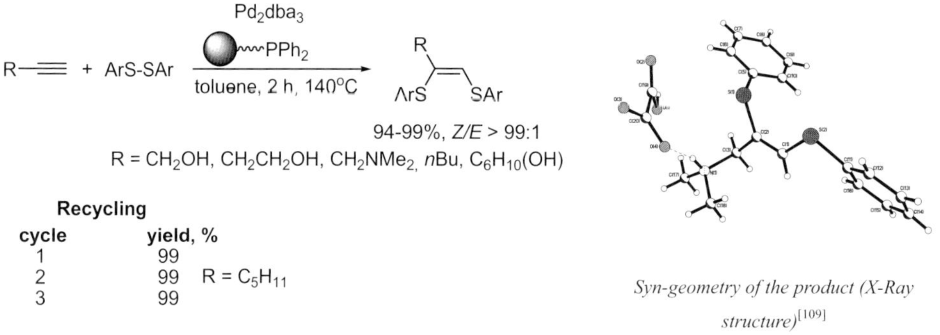

Pd_2dba_3, polymer-supported PPh_2

R–≡ + ArS-SAr → (toluene, 2 h, 140 °C) → (Z)-ArS(R)C=CH(SAr)

94-99%, Z/E > 99:1

R = CH_2OH, CH_2CH_2OH, CH_2NMe_2, nBu, $C_6H_{10}(OH)$

Recycling

cycle	yield, %
1	99
2	99
3	99

R = C_5H_{11}

Syn-geometry of the product (X-Ray structure)[109]

Scheme 3.64

These difficulties were overcome by developing a synthetic procedure under molten state conditions (solvent-free). At 80–140 °C the ligand (PPh_3) and dichalcogenide R_2Z_2 formed a melt, which readily dissolved the alkyne and the catalyst (Scheme 3.65) [115]. In such catalytic conditions the reaction with Ph_2S_2 was complete after 5 min with 1 mol% of the catalyst and at 140 °C the reaction was complete after 1 h with only 0.01 mol% of the catalyst. Both reactions involving Ar_2S_2 and Ar_2Se_2 were carried out with excellent yields and selectivity. A kinetic study has shown that solvent-free reaction in the melt is much faster than the corresponding reaction in solution. In addition, the catalyst can be easily isolated by rapid flash chromatography and re-used in the catalytic reaction without noticeable loss of catalytic activity (recycling of homogeneous catalyst from the solvent was not possible due to decomposition of the Pd species). The reaction was successfully scaled for the preparation of addition products on the gram-scale. A mechanistic study has confirmed the presence of dinuclear Pd intermediate complexes in the reaction [115].

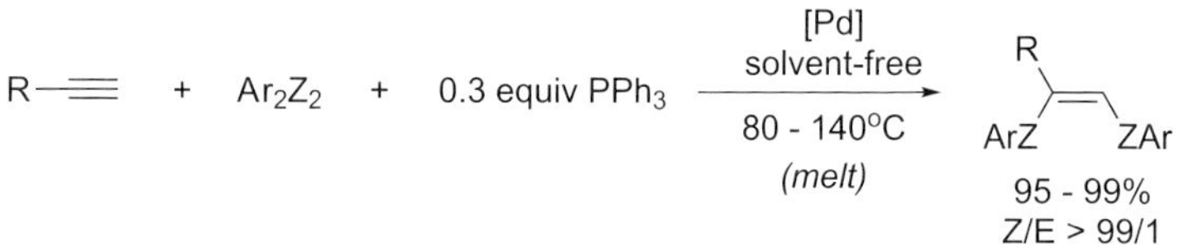

R–≡ + Ar_2Z_2 + 0.3 equiv PPh_3 → ([Pd], solvent-free, 80 - 140 °C, *(melt)*) → ArZ(R)C=CH(ZAr)

95 - 99%, Z/E > 99/1

Scheme 3.65

By using microwave irradiation the reaction time was shortened to 10 min and the catalyst amount was decreased to 0.1 mol% of Pd [114]. Various simple species like Pd(PPh_3)$_4$, Pd(OAc)$_2$, or $PdCl_2$ may, in the presence of PPh_3, be utilized as catalyst precursors. To study catalytic reactions in the molten-state systems a special NMR technique was developed [116]. Using this technique intermediate dinuclear Pd complexes with the same trans-/cis- isomers ratio were detected in solvent-free/melt systems after conventional and microwave heating [116].

$Pd(PPh_3)_4$-catalyzed addition of diaryl diselenides to the terminal alkynes was carried out in ionic liquids with high yields and selectivity at 60 °C (Scheme 3.66) [117]. The possibility of catalyst recycling was shown for these catalytic conditions.

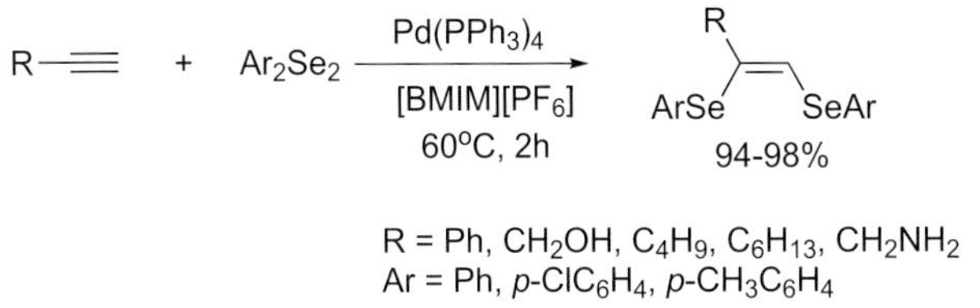

R≡ + Ar_2Se_2 —[$Pd(PPh_3)_4$, [BMIM][PF_6], 60°C, 2h]→ R(ArSe)C=CH(SeAr), 94-98%

R = Ph, CH_2OH, C_4H_9, C_6H_{13}, CH_2NH_2
Ar = Ph, p-ClC_6H_4, p-$CH_3C_6H_4$

Scheme 3.66

In contrast to Pd catalysts, Pt complexes were found to be inactive in the catalytic reaction between the Ar_2Z_2 and alkynes [108]. It was shown that $Pt(PPh_3)_4$ undergoes oxidative addition of diaryl dichalcogenides with a final formation of *trans*-[Pt$(ZAr)_2(PPh_3)_2$]. A mechanistic study at lower temperature has shown that the *cis*-[Pt$(ZAr)_2(PPh_3)_2$] complex was formed first after the oxidative addition step. At room temperature this complex underwent irreversible isomerization to the trans-derivative [108]. It is known that *cis*-$PtCl_2(PPh_3)_4$ and *cis*-PtCl(SPh)$(PPh_3)_4$ complexes also undergo isomerization to the corresponding trans-isomers, for the latter complex the isomerization took place in the presence of pyridine [118].

The *cis*-[Pt$(SePh)_2(PPh_3)_2$] complex was isolated and its structure was characterized with X-ray analysis. It was found that this complex does mediate the addition of the S–S bond to alkynes (Scheme 3.67) [108].

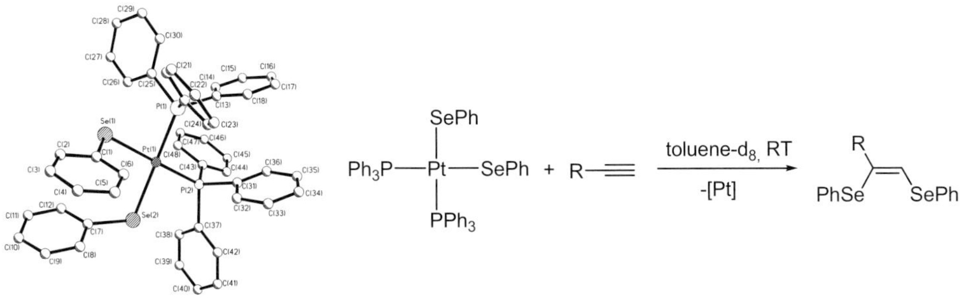

cis-[Pt$(SePh)_2(PPh_3)_2$] [108]

Ph_3P–Pt(SePh)(SePh)(PPh_3) + R≡ —[toluene-d_8, RT; -[Pt]]→ R(PhSe)C=CH(SePh)

Scheme 3.67

The question of the influence of various phosphine ligands was addressed in the study of the Pd-catalyzed reaction of Ph_2S_2 addition to alkynes (Table 3.1) [119]. Good catalytic activity was observed with Ar_3P ligands (entries 1, 2; Table 3.1). Interestingly, trialkylphosphite ligands $(AlkO)_3P$ (Alk = ^{n}Bu, ^{i}Pr) were even more active in the catalytic reaction (entries 8, 9; Table 3.1), while $P(OPh)_3$ led only to a trace amount of the product. This finding was implemented in the $Pd_2dba_3/(OiPr)_3P$ catalytic system

Table 3.1 Ligand effect in the palladium-catalyzed Ph_2S_2 addition to 3-butyn-1-ol.[a] [119].

Entry	Ligand	Yield, %
1	PPh_3	90
2	$P(p\text{-}FC_6H_4)_3$	88
3	PCy_3	15
4	PBu_3	2
5	DPPB	2
6	DPPE	0
7	$P(OPh)_3$	4
8	$P(OBu)_3$	99
9	$P(O\textit{i}Pr)_3$	99

a) Conditions: toluene, 80 °C, 15 h, 1.5 mol% of Pd_2dba_3 and 30 mol% of the ligand.

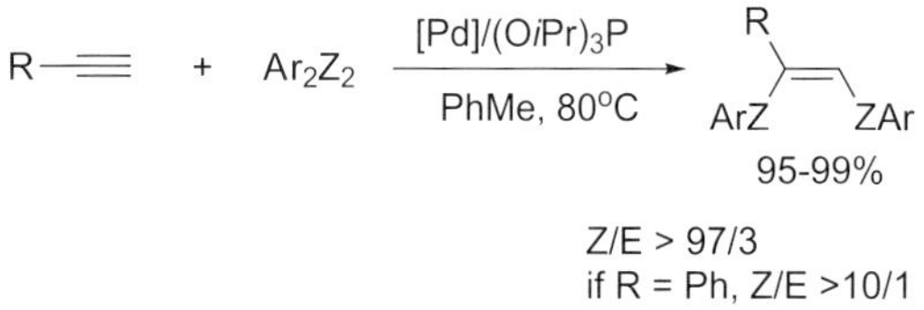

R—≡ + Ar_2Z_2 —[Pd]/(O*i*Pr)$_3$P, PhMe, 80°C→ (R)(ArZ)C=CH(ZAr) 95-99%

Z/E > 97/3
if R = Ph, Z/E >10/1

Scheme 3.68

aimed at high selectivity $Z/E > 97/3$ for typical alkynes and good selectivity $Z/E > 10/1$ for the activated alkyne (Scheme 3.68).

The $P(O\textit{i}Pr)_3$ ligand was found to be very convenient not only for synthetic purposes, but also for mechanistic studies. In the ^{1}H NMR spectra of the [Pd]/Ar_2Z_2/(O*i*Pr)$_3$P catalytic system the signals of the ligand (both free and coordinated) were clearly distinguished and did not overlap with spectral lines of the Ar_2Z_2 reagent and ArZ-Pd-Ar species [119]. This is in sharp contrast with traditional [Pd]/Ar_2Z_2/Ar'_3P catalytic systems, where signal overlap of the Ar/Ar′ groups made ^{1}H NMR spectra hardly interpretable.

Using ^{1}H and ^{31}P NMR in the model [Pd]/Ph_2S_2/(O*i*Pr)$_3$P catalytic system trans- and cis-mononuclear and dinuclear metal complexes were detected. The equilibrium constant for the dissociation of the dinuclear metal complexes was measured, $K = 0.066 \pm 0.007$ M at 30 °C [119]. Although the equilibrium is shifted towards dinuclear complexes, in excess of the (O*i*Pr)$_3$P ligand the mononuclear *trans*-[Pd(SPh)$_2$((O*i*Pr)$_3$P)$_2$] complex has been isolated and characterized by X-ray analysis (Figure 3.1).

Until recently, transition-metal-catalyzed addition of the Alk_2Z_2 to alkynes remained a challenging problem. The only available synthetic route to these products involved stepwise reaction of $R'_3SiS\text{-}SSiR'_3$ and alkynes, followed by alkylation (Scheme 3.69) [120, 121].

The nature of the organic group in the dichalcogenide plays an important role in the reactivity of these compounds with metal complexes. In fact, the S–S bond dissociation energy in Alk_2S_2 in higher than with Ph_2S_2: 74 and 55 kcal mol^{-1},

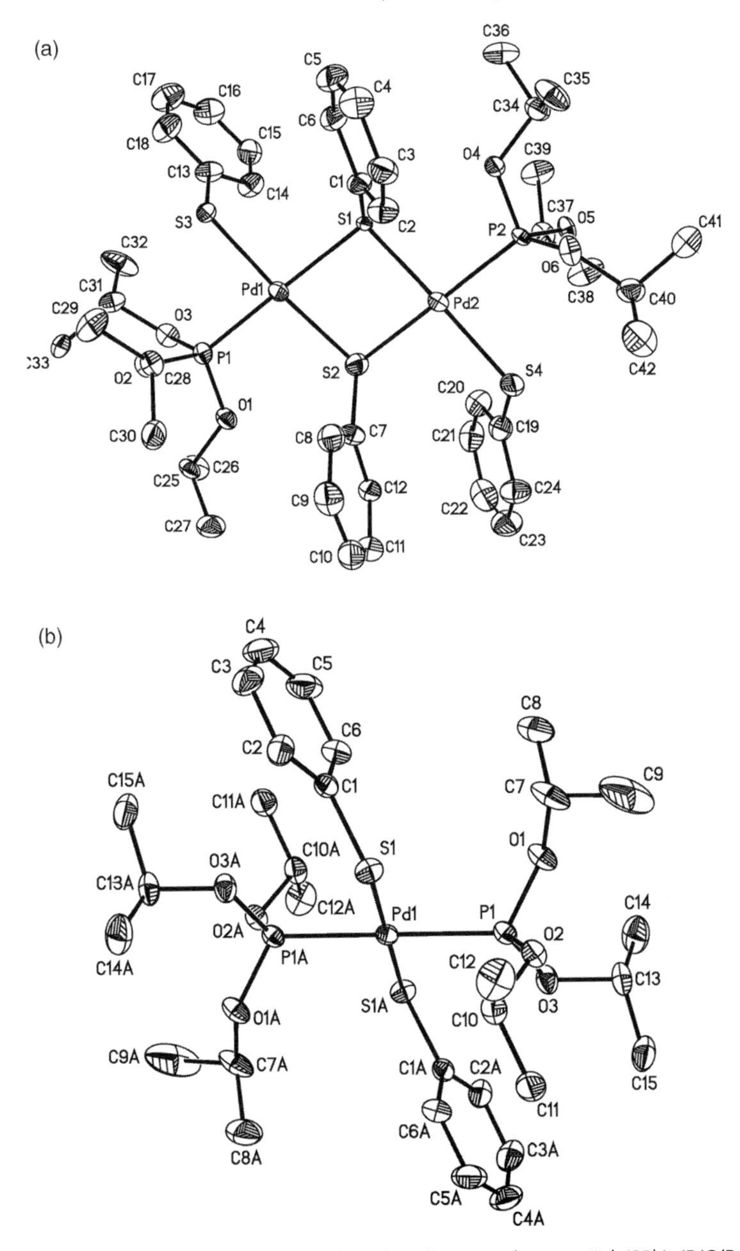

Figure 3.1 Molecular structures of the dinuclear complex *trans*-$[Pd_2(SPh)_4(P(OiPr)_3)_2]$ (a) and mononuclear complex *trans*-$[Pd(SPh)_2(P(OiPr)_3)_2]$ (b) complexes [119].

respectively [122]. This could be one of the reasons why different ligands are required to furnish the oxidative addition step involving these dichalcogenides. Indeed, it was found that Pd catalyst with Cy_2PhP ligand did catalyze stereoselective Alk_2Z_2 addition to the triple bond of alkynes (Scheme 3.70) [122]. The reaction was carried out at

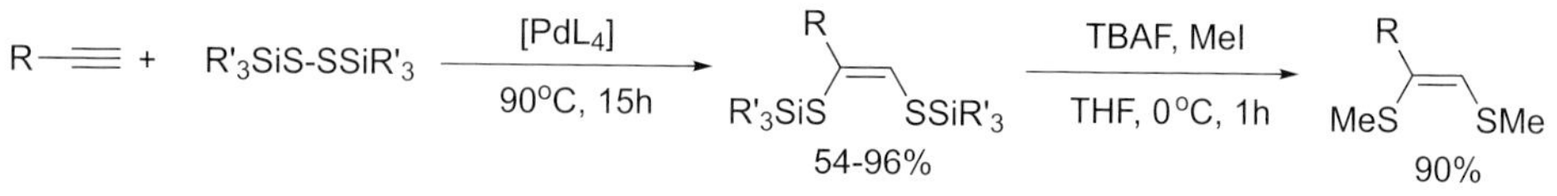

R—≡ + $R'_3SiS\text{-}SSiR'_3$ → ($[PdL_4]$, 90°C, 15h) → R, R'_3SiS, $SSiR'_3$ 54-96% → (TBAF, MeI, THF, 0°C, 1h) → R, MeS, SMe 90%

Scheme 3.69

140 °C for 12 h and gave high yields for Z = S and good yields for Z = Se. The lower yields in the case of diselenides were caused by a side-reaction of Alk_2Se_2 to Alk_2Se.

A much more efficient catalytic system was designed based on Ni complexes. Using 3 mol% of simple catalyst precursor $Ni(acac)_2$ and 30 mol% of the Me_2PhP ligand, a highly stereoselective addition of Alk_2Z_2 (Alk = Me, Bu^n, Pr^i, Cy) to alkynes was performed under solvent-free conditions (Scheme 3.70). The ligand played a dramatic role in these catalytic reactions. For both metals R_2PhP were found to be the ligands of choice: R = Cy for Pd and R = Me for Ni [122]. In spite of several efforts only terminal alkynes were involved in the reaction, while internal alkynes remained inactive.

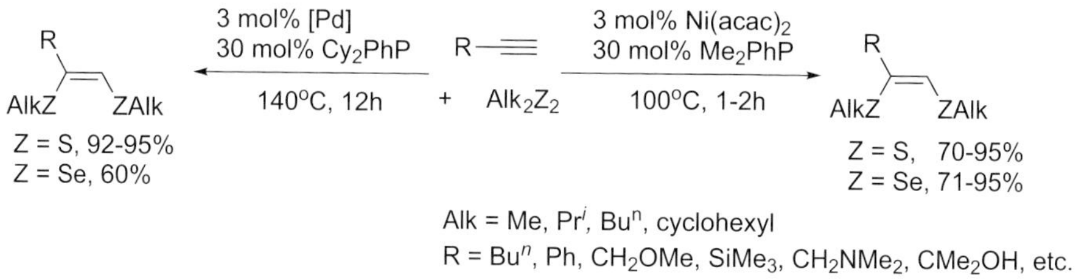

R, AlkZ, ZAlk; Z = S, 92-95%; Z = Se, 60% ← (3 mol% [Pd], 30 mol% Cy_2PhP, 140°C, 12h) ← R—≡ + Alk_2Z_2 → (3 mol% $Ni(acac)_2$, 30 mol% Me_2PhP, 100°C, 1-2h) → R, AlkZ, ZAlk; Z = S, 70-95%; Z = Se, 71-95%

Alk = Me, Pr^i, Bu^n, cyclohexyl
R = Bu^n, Ph, CH_2OMe, $SiMe_3$, CH_2NMe_2, CMe_2OH, etc.

Scheme 3.70

Cyclic vinyl chalcogenides were synthesized via the addition reaction of terminal alkynes to cyclic disulfides and diselenides [123]. In the Pd/Cy_2PhP catalytic system (L/Pd = 10/1) the reaction of 1-hexyne with methyl(±)α-lipoate led to product formation with 67% after 2 h and 99% after 8 h. As in the case of acyclic Alk_2Z_2 discussed above the ligand effect had crucial influence on the performance of the catalytic reaction. For example, using Cy_2PhP and $CyPh_2P$ resulted in 99% and 19% product yields, respectively. It was found that the ligand effect is important not only from the point of view of performance in the catalytic cycle, but also in avoiding catalyst polymerization and initiating the formation of active species via the "leaching" process (Scheme 3.71) [123]. Using NMR monitoring, stoichiometric oxidative addition (40 °C) and alkyne insertion reactions (140 °C) were studied. Mononuclear and polynuclear Pd complexes were detected with ^{31}P and 2D $^1H - ^{31}P$ HMQC NMR. The catalytic activity of the $[Pd_n(ZR)_m(Cy_2PhP)_2]$ species was significantly decreased upon increasing the value of n. Most likely, both factors – lower solubility and decreased number of terminal ZR groups (the terminal groups are more reactive towards alkyne insertion compared to η^2-coordinated bridging ZR groups) – are responsible for the observed decrease in the catalytic activity.

The Ni-based catalytic system was synthetically superior (100 °C for Ni instead of 140 °C for Pd) [123]. Five-membered and six-membered dichalcogenides

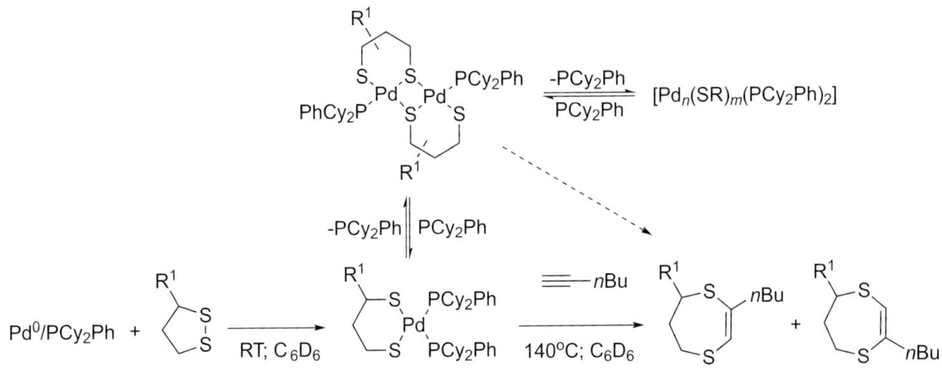

$R^1 = (CH_2)_4COOMe$
Pd/L=1/10

Scheme 3.71

reacted smoothly with alkynes and resulted in the product formation in 50–91% yields (Scheme 3.72). Ligand excess was required to maintain high activity of the catalyst (L/Ni = 10/1). The same effect was also reported for the Ni-catalyzed reaction of acyclic analogs [122]. Without excess of ligand, the formation of insoluble metal species was observed, which were isolated and studied by scanning electron microscopy (SEM). According to the SEM study the isolated solid possessed a flaky structure. The smallest structural units were of 100–500 nm in size and 40–90 nm in thickness [123]. This solid was dissolved in toluene in the presence of an excess of the ligand and showed the desired catalytic activity in the addition reaction. Without phosphine ligand the solid was inactive in the catalytic reaction.

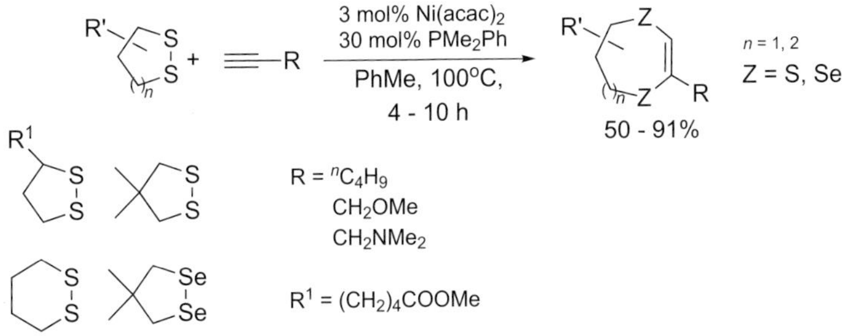

Scheme 3.72

The plausible mechanism of the catalytic cycle involves oxidative addition of the cyclic dichalcogenide to the ML_x species, alkyne coordination and insertion into the M–Z bond, and reductive elimination (Scheme 3.73) [123]. Although only

M = Pd, L = PCy_2Ph
M = Ni, L = PMe_2Ph

Scheme 3.73

mononuclear metal complexes are shown in Scheme 3.73, dinuclear and polynuclear metal species may also contribute to the product formation.

3.3.2
Ni-Catalyzed Synthesis of Dienes

A further study of the ligand effect in the Ni-catalyzed addition of diaryl dichalcogenides to terminal alkynes revealed that, in several cases, a mixture of alkenes and 1,3-dienes was formed (Scheme 3.74) [124]. The most selective reaction towards formation of dienes was observed with L = Cy_2PhP and Cy_3P. These ligands completely suppressed formation of alkenes and gave ~75 : 25 ratio in favor of the asymmetric diene.

Under optimized conditions (MeCN, 40 °C, 5 h, 3 mol% $Ni(acac)_2$, 30 mol% $PPhCy_2$) the synthetic procedure was carried out with good yields and the asymmetric dienes were isolated in pure form using flash-chromatography:

70% 75% 71% 62%
67% 74% 69%

According to solution phase NMR analysis and solid state X-ray analysis, the structures possess *s-gauche* configuration of the central C–C bond in contrast to the *s-trans* configuration reported for the majority of known 1,3-dienes. This is the first

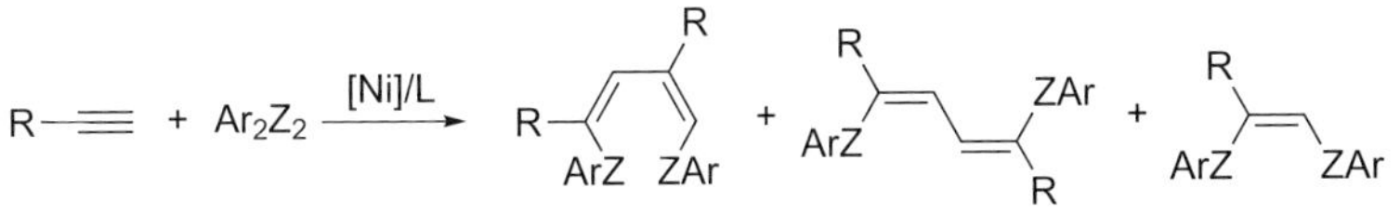

Scheme 3.74

example of this synthetic procedure to this type of dienes, which cannot be accessed by any other method with similar efficiency.

The results have suggested that the intermediate vinyl complex of Ni formed after reaction with the first molecule of alkyne may undergo different pathways (Scheme 3.75). One of the options is a very well-known pathway, which involves intramolecular C–Z reductive elimination and leads to the formation of alkene. If the intermediate vinyl complex is stable enough, it may react with the second alkyne molecule via insertion into the Ni–Z or Ni–C bonds, leading to symmetrical or unsymmetrical 1,3-dienes, respectively. According to theoretical calculations formation of unsymmetrical diene is preferred from an energetic point of view, in agreement with experimental findings [124].

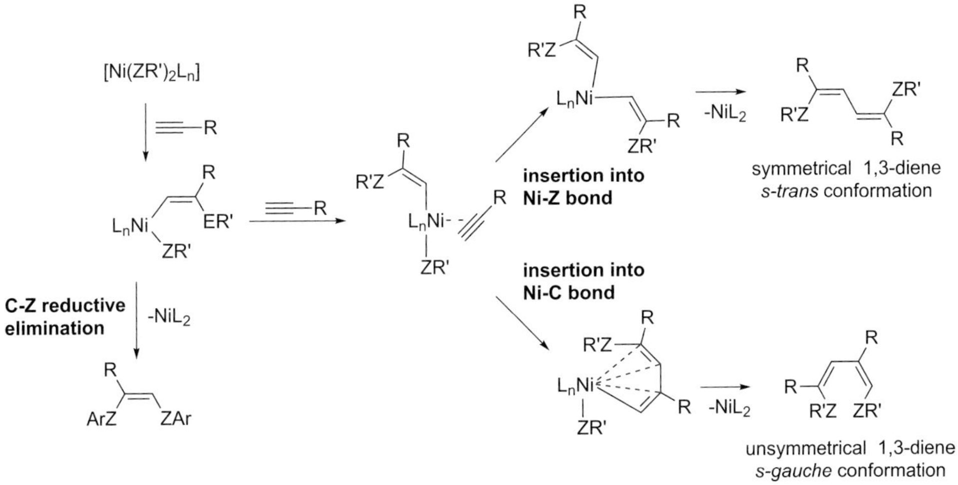

Scheme 3.75

3.3.3
Rh-Catalyzed Reactions

The reaction of terminal alkynes with disulfides in the presence of $RhH(PPh_3)_4$/dppf catalyst resulted in C–H bond cleavage and formation of a C_{SP}–S bond (Scheme 3.76) [125].

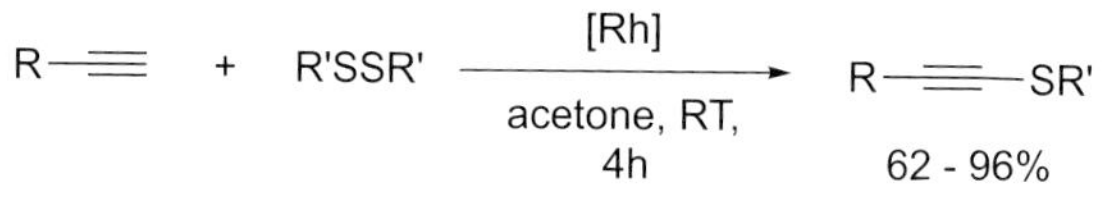

R—≡ + R'SSR' —[Rh], acetone, RT, 4h→ R—≡—SR' (62 - 96%)

Scheme 3.76

However, the same complex catalyzed stereoselective addition reaction in the mixture of alkyne with diaryl disulfide and diphenyl diselenide (Scheme 3.77) [126].

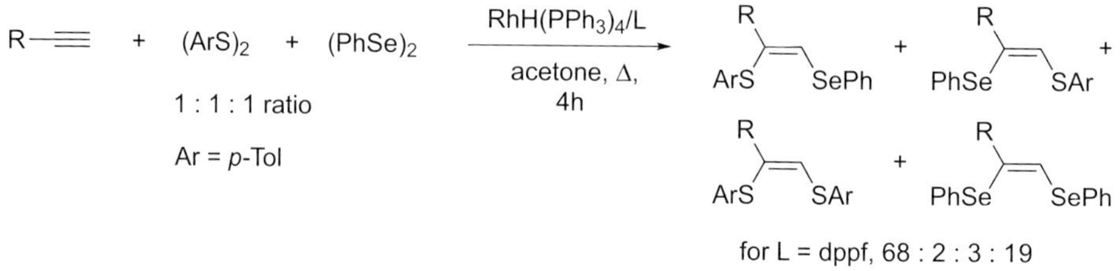

R—≡ + $(ArS)_2$ + $(PhSe)_2$ —$RhH(PPh_3)_4$/L, acetone, Δ, 4h→ R(ArS)C=CH(SePh) + R(PhSe)C=CH(SAr) + R(ArS)C=CH(SAr) + R(PhSe)C=CH(SePh)

1 : 1 : 1 ratio

Ar = *p*-Tol

for L = dppf, 68 : 2 : 3 : 19

Scheme 3.77

Syn-addition of dialkyl disulfides was catalyzed with the same Rh complex in the presence of 3 mol% of trifluoromethanesulfonic acid (Scheme 3.78) [127]. The reaction was very sensitive to the ligand effect and the best performance was observed with 3 mol% of $RhH(PPh_3)_4$ and 12 mol% of tris(p-methoxyphenyl) phosphine. In some cases the Rh complex with dppf ligand also showed good catalytic activity.

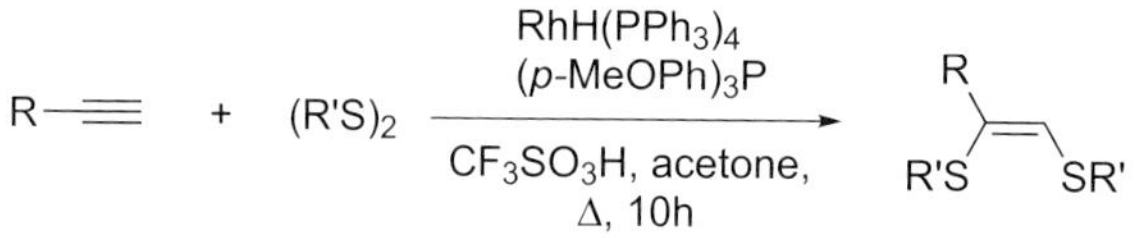

R—≡ + $(R'S)_2$ —$RhH(PPh_3)_4$, (*p*-MeOPh)$_3$P, CF_3SO_3H, acetone, Δ, 10h→ R(R'S)C=CH(SR')

Scheme 3.78

3.3.4 Catalytic Addition of S-X and Se-X Bonds to Alkynes

A new Csp^2–S or Csp^2–Se bond may be formed via the addition reaction of substrates with S–B, S(Se)–P or S–Si bonds to alkynes. $Pd(PPh_3)_4$ catalyst was found to be active in the thioboration [128], thiophosphorylation [129], thioselenation [130], thiosilylation [131] and thiogermanylation [132] reactions. In all of the studied cases the preferred regioselectivity of the addition reaction was to form the R(R'S)C=CHX products (X = B, P, Si, Ge).

The same mechanism was proposed in the cyanothiolation of alkynes with PhSCN in the $Pd(PPh_3)_4$-catalyzed reaction (120 °C, 66 h, PhH) [133]. The intermediate *trans*-$Pd(SPh)(CN)(PPh_3)_2$ complex was isolated from the corresponding stoichiometric reaction and the complex led to the same product – *Z*-(PhS)RC=CH(CN) – as compared with the catalytic reaction. The stereochemical results of the reaction with PhSCN depended on the nature of the alkyne, while the reaction with AlkSCN did not occur.

A theoretical study of the reaction mechanism has shown that alkyne insertion in the NC–Pd–SH model complex is easier for the Pd–S bond than for the Pd–C bond [134]. The study has also revealed the reasons why the product of PhSeCN addition to alkynes was not observed.

The Csp^2–Se bond was also formed in the multicomponent reaction of the sulfonylamine, diphenyldisulfide, alkyne and CO under $Pd(PPh_3)_4$-catalyzed conditions (Scheme 3.79) [135, 136]. It was proposed that the intermediate Pd complex formed after carbonylation (see Scheme 3.79) underwent nucleophile attack of the amide instead of reductive elimination. The other pathways leading to the same product are also possible [135, 136].

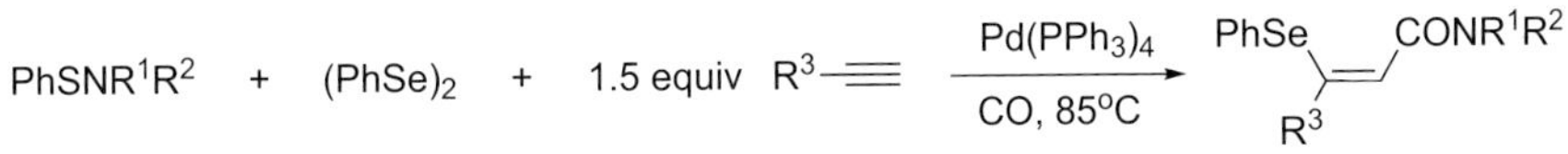

$PhSNR^1R^2$ + $(PhSe)_2$ + 1.5 equiv R^3–≡ —[$Pd(PPh_3)_4$; CO, 85°C]→ PhSe, $CONR^1R^2$, R^3

Scheme 3.79

3.3.5
Catalytic Addition to Allenes

Vinyl chalcogenides were also prepared by transition-metal-catalyzed RZZR addition to the double bond of allenes. Ogawa *et al.* have reported allenes bis-selenation with Ph_2Se_2 and $Pd(PPh_3)_4$ catalyst (Scheme 3.80) [137]. The product was obtained in good to high yields, but with poor stereoselectivity. Only with cyclohexyl allene was there a good yield combined with high selectivity, $E/Z = 93/7$. A detailed computational study of regioselective bis-selenation of allenes catalyzed by palladium complexes has been carried out by Wu *et al.* [138]

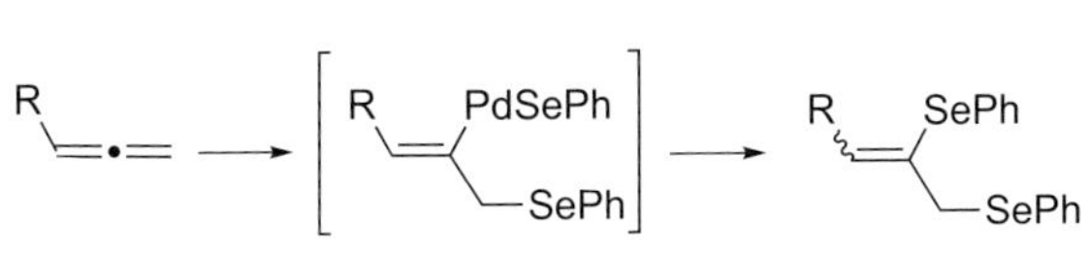

R → [R, PdSePh, SePh] → R, SePh, SePh

Scheme 3.80

3.4
Catalytic Addition of RZ-H Derivatives to Alkynes (Z ≡ S, Se)

In this section we describe the available literature on the addition reaction of thiols and selenols RZH (Z = S, Se). We do not discuss non-catalytic addition reactions carried out without transition metal catalysts as this topic has already been addressed in several publications (see [100–103, 139–142] and references therein). It was shown that the non-catalytic reactions led to a different outcome: the anti-Markovnikov products are formed in the addition of RZH to alkynes. Our goal is to concentrate on the selective formation of the scarcely available Markovnikov isomer by RZH addition to the triple bond of alkynes.

3.4.1
Pd and Ni-Catalyzed Addition of Thiols and Selenols

In 1992 Ogawa, Sonoda *et al.* carried out the first catalytic addition of aromatic thiols [143] and selenols [144] to alkynes with $Pd(OAc)_2$. Although the Markovnikov isomer was the major product of the reactions, the yields were not very high [145]. The catalytic reaction was accompanied with non-catalytic addition, leading to the anti-Markovnikov isomers (free radical or nucleophilic reactions) as well as double bond isomerization in the case of thiols (THF, 67 °C) and selenols (benzene, 80 °C) [143, 144]. The isomerization reaction was especially pronounced with $Pd(PhCN)_2Cl_2$ catalyst [146]. It is interesting to note that the intermediate metal complex taking part in the catalytic reaction was denoted as $Pd(SPh)_2L_n$ [146].

The mechanism of the catalytic hydrothiolation and hydroselenation reactions differs from the bis-thiolation and bis-selenation reactions at the product formation stage. For the Z–H bond addition to alkynes the last stage of the catalytic cycle is protonolysis or C–H reductive elimination. It was found that, independently of the catalyst precursor (either Pd^0 or Pd^{II} derivatives), the same catalytically active species $[Pd(ZR)_2]_n$ were formed (Scheme 3.81) [147, 148].

$$RZH + Pd^0 \longrightarrow [Pd(ZR)_2]_n + H_2$$

$$RZH + Pd^{II}X_2 \longrightarrow [Pd(ZR)_2]_n + 2\,HX$$

Scheme 3.81

It is interesting to compare the catalytic activity of Pd chalcogenide species in Z–Z and Z–H addition reactions [99]. The $[Pd(ZR)_2]_n$ complexes were totally inactive in the addition of disulfides and diselenides to alkynes, where C–Z reductive elimination was involved as a last product-forming step (see discussion in Section 3.3). However, in the presence of the phosphine ligands significant amounts of the bis-thiolation and bis-selenation products of the *Z*-geometry were detected with the [Pd

$(ZR)_2]_n$ catalyst, in addition to the expected Markovnikov product (Scheme 3.82) [147–149]. Therefore, in the presence of PR_3 ligands the catalyst used for the addition of Z–H bonds to alkynes demonstrated some activity and high selectivity in the Z–Z bond addition reaction as well. The mechanistic changes in these reaction conditions were initiated by catalyst leaching [123].

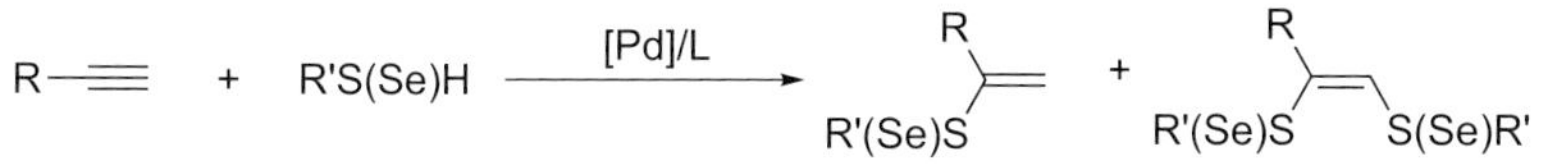

Scheme 3.82

Using pyridine as solvent for $Pd(OAc)_2$-catalyzed phenyl selenol addition to alkynes resulted in better regioselectivities and yields (although in some cases the yields were not very high) [150]. Most likely, pyridine facilitated the formation of soluble Pd complexes in this catalytic system.

Hydrothiolation of conjugated enynes with PhSH was carried out with $Pd(OAc)_2$ catalyst in THF at 50 °C and led to the formation of 1,3-diened in 41–75% yield after 14–18 h (Scheme 3.83).

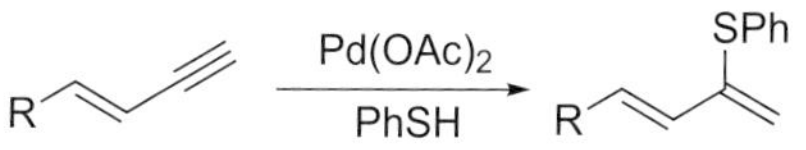

Scheme 3.83

Han *et al.* have shown that reaction between PhSH and 1-octyne can be catalyzed by the $N(PPh_2Me)_4$ complex with high selectivity in the presence of catalytic amounts of $Ph_2P(O)OH$ acid (Scheme 3.84). However, the role of the $Ph_2P(O)OH$ additive in this reaction remained unclear.

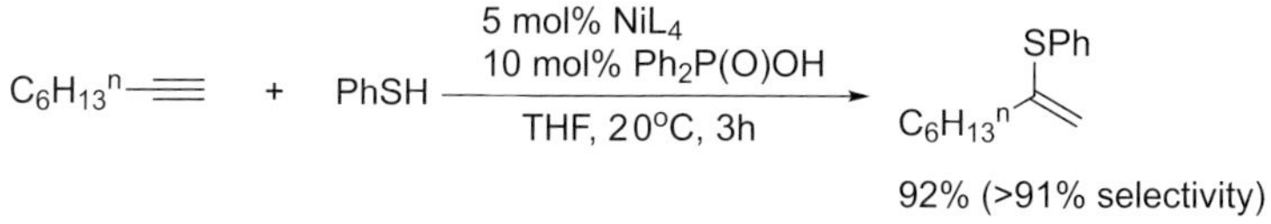

Scheme 3.84

In a simpler $NiCl_2/Et_3N$ catalytic system, high selectivity and good yields in the hydrothiolation reaction were achieved in the presence of a radical trap, which suppressed the side-reaction [149]. For the activated alkynes with R = Ph and COOMe significant amounts of anti-Markovnikov products were obtained. Regioselectivity of the reaction depended on the alkyne: RZH ratio and much better yields were achieved using $Ni(acac)_2$ as a catalyst precursor [151].

Further exploration of the higher activity of the Ni complexes compared to Pd analogs led to the discovery of a novel nano-sized catalytic system with superior performance for hydrothiolation and hydroselenation reactions of alkynes [152, 153]. Furthermore, it was found that with a simple catalyst precursor – $Ni(acac)_2$ – the reaction was carried out with excellent yields and excellent selectivity, even at room temperature. Both terminal and internal alkynes were successfully involved in the addition reaction. This catalytic system was tolerant to various functional groups in alkynes and was easily scaleable for the synthesis of 1–50 g of product (Scheme 3.85) [152, 153]. The proposed mechanism of the catalytic reaction involved: (i) catalyst self organization with nano-sized particles formation, (ii) alkyne insertion into the Ni–Z bond and (iii) protonolysis with RZH (Scheme 3.86).

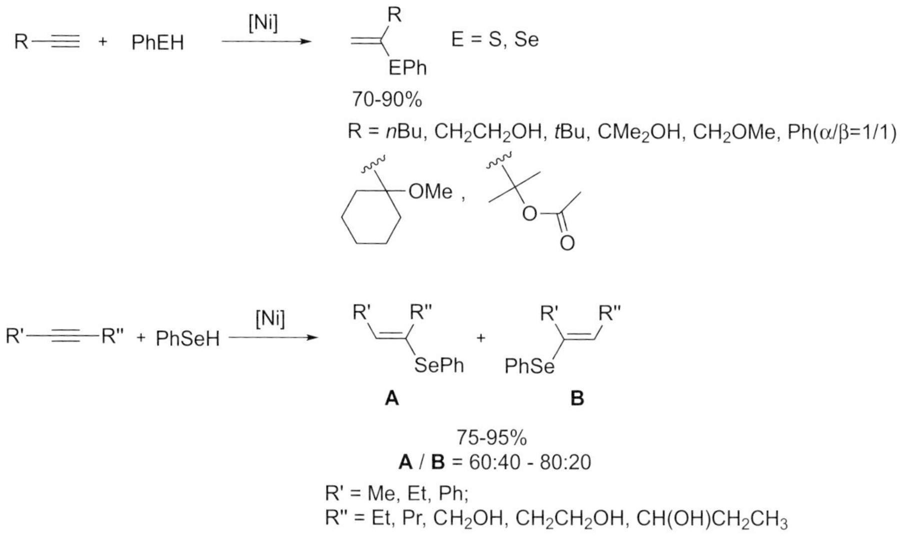

R—≡ + PhEH —[Ni]→ (R)(EPh)C=CH$_2$ E = S, Se

70-90%

R = nBu, CH_2CH_2OH, tBu, CMe_2OH, CH_2OMe, Ph(α/β=1/1)

R'—≡—R" + PhSeH —[Ni]→ A + B

75-95%

A / B = 60:40 - 80:20

R' = Me, Et, Ph;

R" = Et, Pr, CH_2OH, CH_2CH_2OH, $CH(OH)CH_2CH_3$

Scheme 3.85

According to elemental analysis the same nickel complex $[Ni(ZPh)_2]_n$ was formed when starting with various precursors under different reaction conditions. However, the catalytic activity and selectivity of the catalyst differed dramatically. An SEM study of the catalyst morphology (Figure 3.2) revealed round-shaped particles of 300 ± 90 nm size in the case of the $Ni(acac)_2$ catalyst precursor – the catalyst, which has shown the best performance in the catalytic reaction [152]. Another type of catalyst was formed in the $NiCl_2/Et_3N$ system, where the $[Et_3NH][Cl]$ salt served as a support for the $[Ni(ZPh)_2]_n$ particles (the salt was formed upon reaction of PhZH with the $NiCl_2$ in the presence of the amine) [153]. Comparative kinetic measurements have revealed a linear dependence between the catalyst activity and particle sizes in the μm range of sizes. An exponential increase in the catalytic activity was observed upon

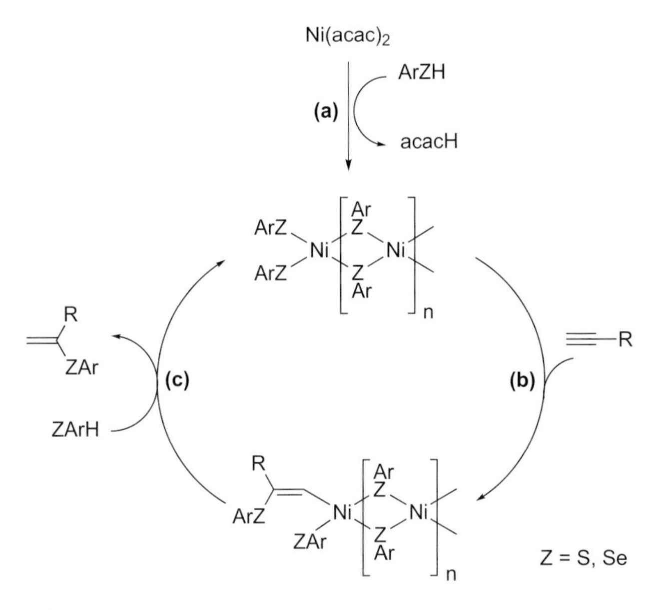

Scheme 3.86

Figure 3.2 SEM images of catalyst particles prepared from $Ni(acac)_2$ (D – 4000×, E – 10000×) and $NiCl_2/Et_3N$ system (C – 2000×, D – 9000×) [153].

decreasing the particle sizes into the nm region [152, 153]. The nano-structured Ni catalyst showed high activity and excellent syn-selectivity for a broad range of terminal and internal alkynes:

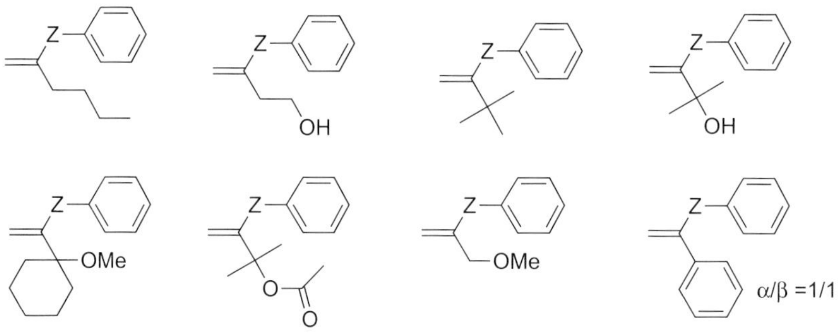

α/β =1/1

Isolated yields 70-90% (Z=S, Se)

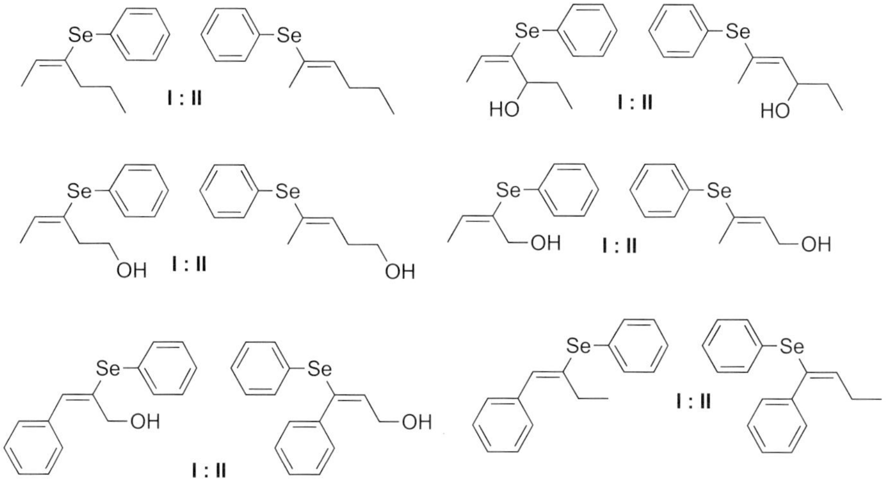

Isolated yields 75-95% (**I** : **II** = 60 : 40 - 80 : 20)

The heterogeneous hydrothiolation reaction carried out with $[Ni(ZAr)_2]_n$ catalyst particles was compared with the homogeneous reaction based on the CpNi(NHC)SAr catalyst (NHC – N-heterocyclic carbene ligand) [154]. Since only one SAr group was present in this catalyst it avoided the formation of the bis-thiolation product. This homogeneous catalyst, especially with NHC = *N*,*N*′-bis(2,4,6-trimethylphenyl)-imidazol-2-ylidene (IMes), showed good performance in the hydrothiolation reaction, resulting in product formation in 61–87% yields. However, a higher temperature of 80 °C was required compared to the heterogeneous reaction and only terminal alkynes were involved in the reaction [154].

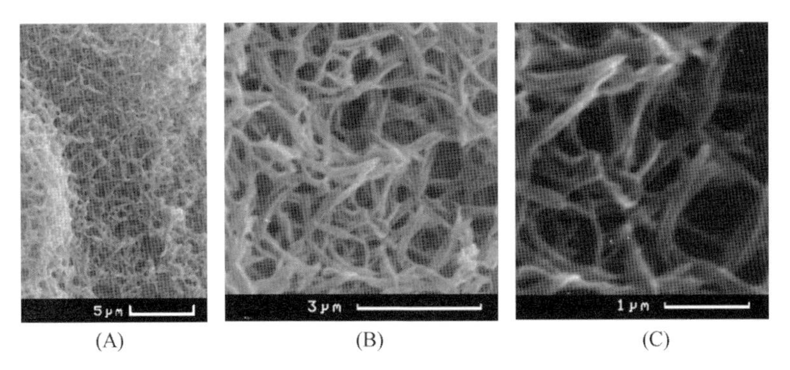

Figure 3.3 Low magnification (A – 1000×) and high magnification (B – 8000×, C – 16000×) SEM images of $[Pd(SCy)_2]_n$ [155].

The challenging problem of the regioselective addition of alkyl thiols to alkynes was solved with nano-sized $[Pd(SAlk)_2]_n$ catalyst, which was formed in a self-organized manner from the $Pd(OAc)_2$ precursor in the thiol/alkyne mixture *in situ* (Figure 3.3) [155]. Catalyst particle size and morphology were of crucial importance to achieve high activity and selectivity in the addition reaction (Table 3.2).

Hydrothiolation of 1-alkynylphosphines was catalyzed by $P(OAc)_2$ in EtOH and gave *Z*-1-phosphino-2-thio-1-alkene as product (Scheme 3.87) [156]. Surprisingly, the addition reaction took place in the anti-fashion, which probably means an external nucleophilic attack on the coordinated triple bond as a key-step instead of the alkyne insertion observed in the other examples.

3.4.2
Rh and Pt-Catalyzed Addition of Thiols to Alkynes

Somewhat contradictory data were published in 1995 [157] and later in 2003 [158] concerning the regioselectivity of the thioformylation reaction in the presence of CO (Scheme 3.88). The Markovnikov derivatives were obtained after alkyne and CO insertion steps with the $RhH(CO)(PPh_3)_3$ catalyst, however, mainly anti-Markovnikov isomer was formed with Wilkinson complex in the hydroformylation reaction (Scheme 3.89) [146]. The same stereochemistry was found in the reaction of AlkSH [159].

If $Pt(PPh_3)_4$ catalyst was utilized, the α,β-unsaturated thioester was formed as a major product (Scheme 3.90) [160]. The formation of *trans*-$[PtH(SPh)(PPh_3)_2]$ in the catalytic cycle was proposed [161].

It was reported that the pyrazolyl-borate complex of rhodium ($Tp^*Rh(PPh_3)_2$, Tp^* = hydrotris(3,5-dimethylpyrazolyl)borate) is active not only in the hydrothiolation with ArSH, but also with AlkSH in good to high yields of Markovnikov isomer at room temperature (Scheme 3.91) [162, 163]. However the reaction of 1-octyne gave a mixture of isomers in 70% yield. In addition to the double bond isomerization, a Markovnikov/anti-Markovnikov ratio of 12 : 1 was found in the case of 1-octyne.

Table 3.2 Pd-nanoparticles-catalyzed regioselective RSH addition to alkynes under microwave heating (NMR and isolated yields are given).

Entry	Alkyne	Thiol	Product	Yield	Selectivity
1	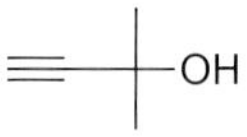 OH	CySH	CySH OH SCy 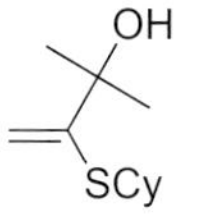	99(92)	>99 : 1
2	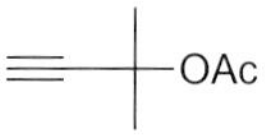 OAc	CySH	CySH OAc SCy 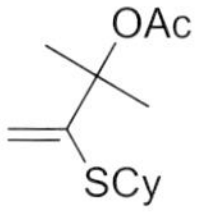	98(88)	>99 : 1
3	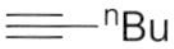 ^{n}Bu	CySH	CySH ^{n}Bu SCy 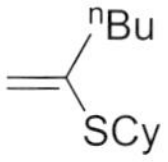	98(75)	95 : 5
4	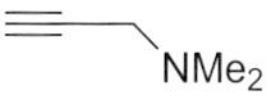 NMe_2	CySH	CySH NMe_2 SCy 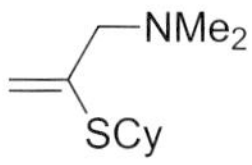	96(87)	91 : 9
5	 Ph	CySH	CySH Ph SCy 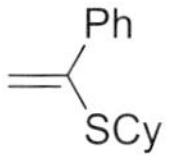	98(77)	84 : 16
6	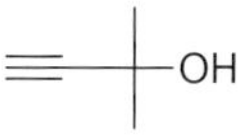 OH	BnSH	BnSH OH SBn 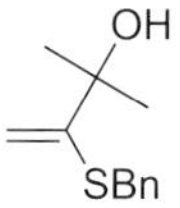	88(74)	>99 : 1
7	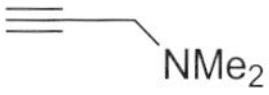 NMe_2	BnSH	BnSH NMe_2 SBn 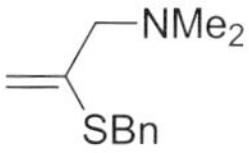	68(57)	97 : 3

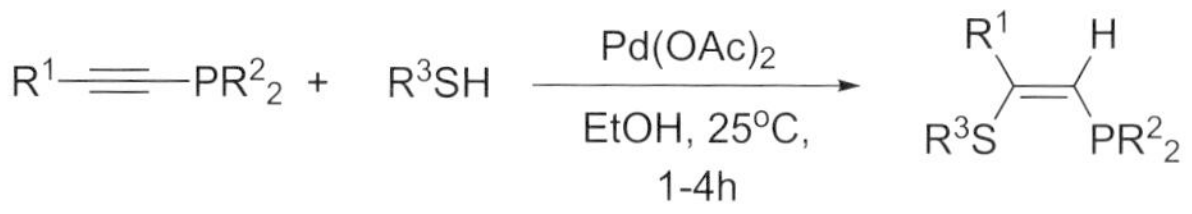

R^1—≡—PR^2_2 + R^3SH → [$Pd(OAc)_2$; EtOH, 25°C, 1-4h] → R^1, H, R^3S, PR^2_2

Scheme 3.87

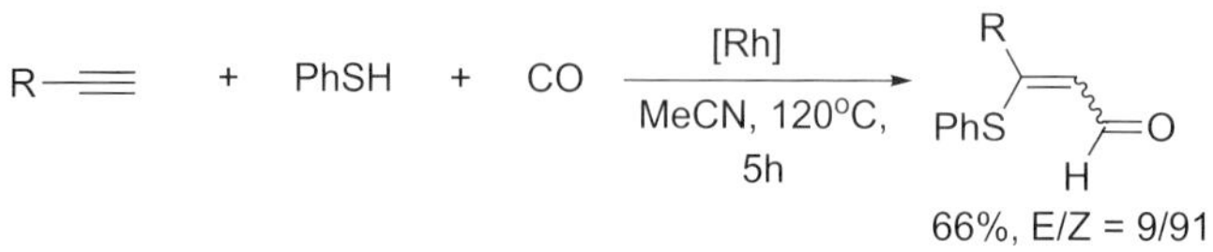

R—≡ + PhSH + CO → [[Rh]; MeCN, 120°C, 5h] → R, PhS, H, =O

66%, E/Z = 9/91

Scheme 3.88

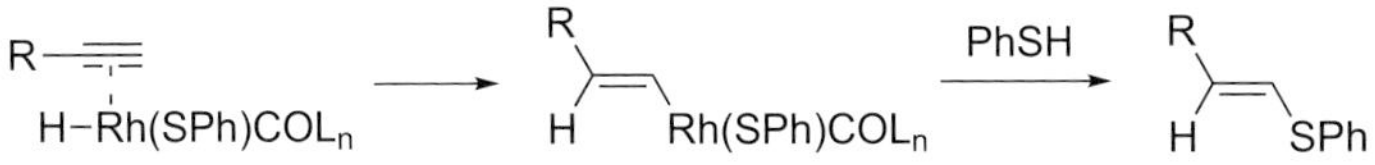

Scheme 3.89

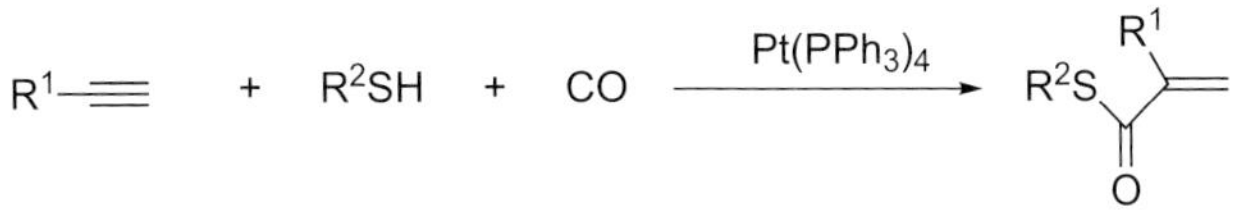

Scheme 3.90

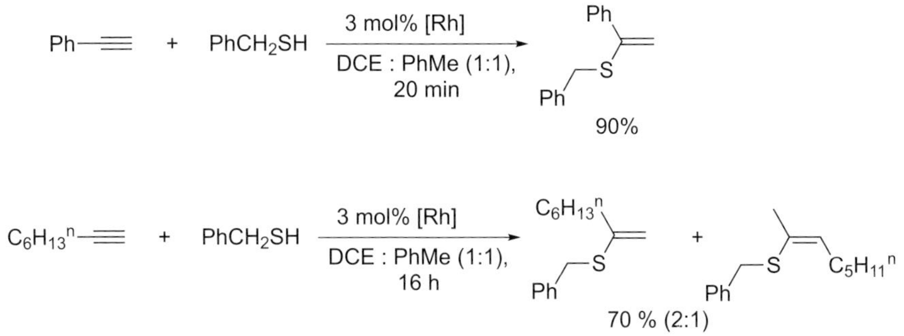

Scheme 3.91

Further development and improvement of this Rh system was recently described [164, 165].

3.4.3
Catalytic Addition of Thiols and Selenols to Allenes

In the $Pd(OAc)_2$-catalyzed reaction the products of PhSH and PhSeH addition to the internal double bond of allenes were formed in good to high yields (Scheme 3.92) [166, 167].

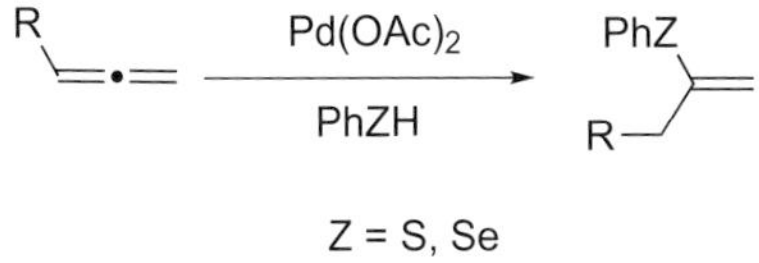

Scheme 3.92

The reaction of allenes with cyclohexylthiol in the presence of CO depends on the nature of the transition metal complex [168]. Pd complexes did not catalyze this transformation, Rh complexes resulted in copolymerization products of allene and CO. In good yields, but with low regio- and stereoselectivitites, the addition reaction was carried out with $Pt(PPh_3)_4$ catalyst (Scheme 3.93).

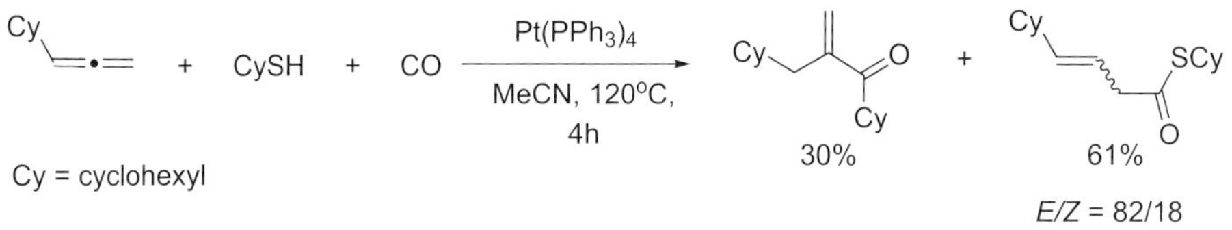

Cy = cyclohexyl

$Pt(PPh_3)_4$, MeCN, 120°C, 4h; 30%; 61%, E/Z = 82/18

Scheme 3.93

3.5 Conclusions

The following conclusions can be drawn after analysis of the Csp^2–S and Csp^2–Se bond formation in the transition-metal-catalyzed substitution and addition reactions involving RZZR and RZH species:

- Careful optimization of the Pd- and Ni-catalyzed reactions and selection of proper ligands made it possible to carry out cross-coupling reactions starting with various types of Ar(vinyl)X and RZH.
- Rapid development of the Cu-mediated reactions has led to several excellent catalytic systems based on different copper compounds as catalyst precursors. Further studies of Cu-catalyzed reactions are expected to focus on the mechanism of catalytic reactions to allow rational catalyst design.
- Rapid development and application of Fe catalysis in this field is anticipated in the next few years.
- Phosphine complexes of Pd and Ni are the catalysts of choice for the RZZR addition reactions to terminal alkynes. High yields and the excellent selectivity of the addition reaction make this catalytic methodology attractive from the point of view of Green chemical synthesis. One significant remaining problem is to effect stereoselective addition of dichalcogenides to internal alkynes.
- Morphological control of heterogeneous Pd and Ni particles has led to development of superior catalysts for regioselective Markovnikov-type addition of RZH to terminal alkynes and stereoselective addition to internal alkynes under mild conditions.

In conclusion, we want to point out that successful reports from many research groups have clearly emphasized the tremendous development of this fascinating field. Convenient and efficient synthetic procedures have been developed to prepare important classes of organic compounds. Undoubtedly, several new catalytic systems and synthetic approaches will be discovered and implemented soon.

References

1 (a) Bernardi, F., Csizmadia, I.G., and Mangini, A. (1985) Organic sulfur chemistry. Theoretical and experimental advances, in *Studies in Organic Chemistry*, **19**, Elsevier, Amsterdam; (b) Ager, D.J. (1982) *Chem. Soc. Rev.*, **11**, 493–522; (c) Block, E. (1978) *Reactions of Organosulfur Compounds*, Academic Press, New York; (d) Patai, S. and Rappoport, Z. (eds) *The Chemistry of Organic Selenium and Tellurium Compounds*, vol. 1–2, John Wiley & Sons, New York, (1986–1987); (e) Wirth, T. (ed.) (2000) *Organoselenium Chemistry: Modern Developments in Organic Synthesis*, Springer Verlag, Berlin, New York; (f) McReynolds, M.D., Dougherty, J.M., and Hanson, P.R. (2004) *Chem. Rev.*, **104**, 2239–2258; (g) Zyk, N.V., Beloglazkina, E.K., Belova, M.A., and Dubinina, N.S. (2003) *Russ. Chem. Rev.*, **72**, 769–786; (h) Sizov, A.Yu., Kovregin, A.N., and Ermolov, A.F. (2003) *Russ. Chem. Rev.*, **72**, 357–374; (i) Mangini, A. (1987) *Sulfur Rep.*, **7**, 313–334.

2 Some representative examples in the pharmaceutical field and biologically active compounds: (a) Gangjee, A., Zeng, Y., Talreja, T., McGuire, J.J., Kisliuk, R.L., and Queener, S.F. (2007) *J. Med. Chem.*, **50**, 3046–3053; (b) Faucher, A.-M., White, P.W., Brochu, C., Maitre, C.G., Rancourt, J., and Fazal, G. (2004) *J. Med. Chem.*, **47**, 18–21; (c) Otzen, T., Wempe, E.G., Kunz, B., Bartels, R., Lehwark-Yvetot, G., Hänsel, W., Schaper, K.-J., and Seydel, J.K. (2004) *J. Med. Chem.*, **47**, 240–253; (d) Liu, L., Stelmach, J.E., Natarajan, S.R., Chen, M.-H., Singh, S.B., Schwartz, C.D., Fitzgerald, C.E., O'Keefe, S.J., Zaller, D.M., Schmatz, D.M., and Doherty, J.B. (2003) *Bioorg. Med. Chem. Lett.*, **13**, 3979–3982; (e) Hu, W., Guo, Z., Chu, F., Bai, A., Yi, X., Cheng, G., and Li, J. (2003) *Bioorg. Med. Chem. Lett.*, **13**, 1153–1160; (f)Wang, Y., Chackalamannil, S., Chang, W., Greenlee, W., Ruperto, V., Duffy, R.A., McQuade, R., and Lachowicz, J.E. (2001) *Bioorg. Med. Chem. Lett.*, **11**, 891–894; (g) Liu, G., Link, J.T., Pei, Z., Reilly, E.B., Leitza, S., Nguyen, B., Marsh, K.C., Okasinski, G.F., von Geldern, T.W., Ormes, M., Fowler, K., and Gallatin, M. (2000) *J. Med. Chem.*, **43**, 4025–4040; (h) Sun, Z.Y., Botros, E., Su, A.D., Kim, Y., Wang, E., Baturay, N.Z., and Kwon, C.H. (2000) *J. Med. Chem.*, **43**, 4160–4168; (i) Artico, M., Silvestri, R., Pagnozzi, E., Bruno, B., Novellino, E., Greco, G., Massa, S., Ettorre, A., Loi, A.G., Scintu, F., and La Colla, P. (2000) *J. Med. Chem.*, **43**, 1886–1891; (j) Kaldor, S.W., Kalish, V.J., Davies, J.F., Shetty, B.V., Fritz, J.E., Appelt, K., Burgess, J.A., Campanale, K.M., Chirgadze, N.Y., Clawson, D.K., Dressman, B.A., Hatch, S.D., Khalil, D.A., Kosa, M.B., Lubbehusen, P.P., Muesing, M.A., Patick, A.K., Reich, S.H., Su, K.S., and Tatlock, J.H. (1997) *J. Med. Chem.*, **40**, 3979–3985.

3 (a) Lauterbach, C. and Fabian, J. (1999) *Eur. J. Inorg. Chem.*, 1995–2004; (b) Hope, E.G. and Levason, W. (1993) *Coord. Chem. Rev.*, **122**, 109; (c) Clemenson, P.I. (1990) *Coord. Chem. Rev.*, **106**, 171–203; (d) Yu, C.J., Chong, Y., Kayyem, J.F., and Gozin, M. (1999) *J. Org. Chem.*, **64**, 2070–2079.

4 Kosugi, M., Shimizu, T., and Migita, T. (1978) *Chem. Lett.*, 13–14.

5 Murahashi, S.-I., Yamamura, M., Yanagisawa, K.-i., Mita, N., and Kondo, K. (1979) *J. Org. Chem.*, **44**, 2408–2417.

6 (a) Bumagin, N.A., Kasatkin, A.N., and Beletskaya, I.P. (1984) *Bull. Acad. Sci. USSR: Chem. Sci. (Izv. Akad. Nauk SSSR, Ser. Khim.)*, **33**, 588–594; (b) Kashin, A.N., Bumagina, I.G., Bumagin, N.A., and Beletskaya, I.P. (1981) *Russ. J. Org. Chem. (Zhurn. Org. Khim.)*, **17**, 21–28.

7 Beletskaya, I.P. (1983) *J. Organomet. Chem.*, **250**, 551–564.

8 Kosugi, M., Ogata, T., Terada, M., Sano, H., and Migita, T. (1985) *Bull. Chem. Soc. Jpn.*, **58**, 3657–3658.

9 Dickens, M.J., Gilday, J.P., Mowlem, T.J., and Widdowson, D.A. (1991) *Tetrahedron*, **47**, 8621–8634.

10 Rane, A.M., Miranda, E.I., and Soderquist, J.A. (1994) *Tetrahedron Lett.*, **35**, 3225–3226.
11 Harr, M.S., Presley, A.L., and Thorarensen, A. (1999) *Synlett*, 1579–1581.
12 Foa, M., Santi, R., and Garavaglia, F. (1981) *J. Organomet. Chem.*, **206**, C29–C32.
13 Ciattini, P.G., Morera, E., and Ortar, G. (1995) *Tetrahedron Lett.*, **36**, 4133–4136.
14 Rajagopalan, S., Radke, G., Evans, M., and Tomich, J.M. (1996) *Synth. Commun.*, **26**, 1431–1440.
15 Cristau, H.J., Chabaud, B., Chene, A., and Christol, H. (1981) *Synthesis*, 892–894.
16 Yamamoto, T. and Sekine, Y. (1984) *Can. J. Chem.*, **62**, 1544–1547.
17 Louie, J. and Hartwig, J.F. (1995) *J. Am. Chem. Soc.*, **117**, 11598–11599.
18 Zheng, N., McWilliams, J.C., Fleitz, F.J., Armstrong, J.D., and Volante, R.P. (1998) *J. Org. Chem.*, **63**, 9606–9607.
19 McWilliams, J.C., Fleitz, F.J., Zheng, N., and Armstrong, J.D. (2004) *Organic Syntheses*, Coll. **10**, 147. Vol. 79, p.43 (2002).
20 Li, G.Y., Zheng, G., and Noonan, A.F. (2001) *J. Org. Chem.*, **66**, 8677–8681.
21 Li, G.Y. (2001) *Angew. Chem. Int. Ed.*, **40**, 1513–1516.
22 Li, G.Y. (2002) *J. Org. Chem.*, **67**, 3643–3650.
23 Schopfer, U., and Schlapbach, A. (2001) *Tetrahedron*, **57**, 3069–3073.
24 Murata, M., and Buchwald, S.L. (2004) *Tetrahedron*, **60**, 7397–7403.
25 Fernandez-Rodrıguez, M.A., Shen, Q., and Hartwig, J.F. (2006) *J. Am. Chem. Soc.*, **128**, 2180–2181.
26 Fernandez-Rodriguez, M.A., Shen, Q., and Hartwig, J.F. (2006) *Chem. Eur. J.*, **12**, 7782–7796.
27 Itoh, T., and Mase, T. (2004) *Org. Lett.*, **6**, 4587–4590.
28 Gao, G.-Y., Colvin, A.J., Chen, Y., and Zhang, X.P. (2004) *J. Org. Chem.*, **69**, 8886–8892.
29 Cai, L., Cuevas, J., Peng, Y.-Y., and Pike, V.W. (2006) *Tetrahedron Lett.*, **47**, 4449–4452.
30 Hamada, M., Kiuchi, M., and Adachi, K. (2007) *Synthesis*, 1927–1929.
31 Dahl, T., Tornøe, C.W., Bang-Andersen, B., Nielsen, P., and Jørgensen, M. (2008) *Angew. Chem. Int. Ed.*, **47**, 1726–1728.
32 Krief, A., Dumont, W., and Robert, M. (2005) *Chem. Commun.*, 2167–2168.
33 Wendeborn, S., Berteina, S., Brill, W.K.-D., and De Mesmaeker, A. (1998) *Synlett*, 671–675.
34 (a) Beletskaya, I.P., Sigeev, A.S., Peregudov, A.S., and Petrovskii, P.V. (2001) *Russ. J. Org. Chem.*, **37**, 1463–1475; (b) Beletskaya, I.P., Sigeev, A.S., Peregudov, A.S., and Petrovskii, P.V. (2001) *Russ. J. Org. Chem.*, **37**, 1703–1709.
35 Bonaterra, M., Martın, S.E., and Rossi, R.A. (2006) *Tetrahedron Lett.*, **47**, 3511–3515.
36 Fukuzawa, S.-i., Tanihara, D., and Kikuchi, S. (2006) *Synlett*, 2145–2147.
37 Santi, C., Santoro, S., Testaferri, L., and Tiecco, M. (2008) *Synlett*, 1471–1474.
38 Munbunjong, W., Lee, E.H., Chavasiric, W., and Jang, D.O. (2005) *Tetrahedron Lett.*, **46**, 8769–8771.
39 Ranu, B.C., Mandal, T., and Samanta, S. (2003) *Org. Lett.*, **5**, 1439–1441.
40 Ranu, B.C., Chattopadhyay, K., and Banerjee, S. (2006) *J. Org. Chem.*, **71**, 423–425.
41 Tang, R.-y., Zhong, P., and Lin, Q.-l. (2007) *Synthesis*, 85–91.
42 Cacchi, S., Fabrizi, G., Goggiamani, A., Parisi, L.M., and Bernini, R. (2004) *J. Org. Chem.*, **69**, 5608–5614.
43 Lai, C. and Backes, B.J. (2007) *Tetrahedron Lett.*, **48**, 3033–3037.
44 Nishiyama, Y., Kawamatsu, H., Funato, S., Tokunaga, K., and Sonoda, N. (2003) *J. Org. Chem.*, **68**, 3599–3602.
45 Beletskaya, I.P., Sigeev, A.S., Peregudov, A.S., and Petrovskii, P.V. (2005) *Chem. Lett.*, **34**, 1348–1349.
46 Cristau, H.J., Chabaud, B., Labaudiniere, R., and Christol, H. (1986) *J. Org. Chem.*, **51**, 875–878.
47 Takagi, K. (1987) *Chem. Lett.*, 2221–2224.
48 Gendre, F., Yang, M., and Diaz, P. (2005) *Org. Lett.*, **7**, 2719–2722.
49 Millois, C. and Diaz, P. (2000) *Org. Lett.*, **2**, 1705–1708.
50 Taniguchi, N. (2004) *J. Org. Chem.*, **69**, 6904–6906.

51 Gomez-Benıtez, V., Baldovino-Pantaleon, O., Herrera-Alvarez, C., Toscano, R.A., and Morales-Morales, D. (2006) *Tetrahedron Lett.*, **47**, 5059–5062.
52 Baldovino-Pantaleon, O., Hernandez-Ortega, S., and Morales-Morales, D. (2005) *Inorg. Chem. Commun.*, **8**, 955–959.
53 Baldovino-Pantaleon, O., Hernandez-Ortega, S., and Morales-Morales, D. (2006) *Adv. Synth. Catal.*, **348**, 236–242.
54 Jammi, S., Barua, P., Rout, L., Saha, P., and Punniyamurthy, T. (2008) *Tetrahedron Lett.*, **49**, 1484–1487.
55 Lindley, J. (1984) *Tetrahedron*, **40**, 1433–1456.
56 Beletskaya, I.P., Sigeev, A.S., Peregudov, A.S., and Petrovskii, P.V. (2003) *Tetrahedron Lett.*, **44**, 7039–7041.
57 Kalinin, A.V., Bower, J.F., Riebel, P., and Snieckus, V. (1999) *J. Org. Chem.*, **64**, 2986–2987.
58 Palomo, C., Oiarbide, M., López, R., and Gómez-Bengo, E. (2000) *Tetrahedron Lett.*, **41**, 1283–1286.
59 Bates, C.G., Gujadhur, R.K., and Venkataraman, D. (2002) *Org. Lett.*, **4**, 2803–2806.
60 He, H., Llauger, L., Rosen, N., and Chiosis, G. (2004) *J. Org. Chem.*, **69**, 3230–3232.
61 Bates, C.G., Saejueng, P., Doherty, M.Q., and Venkataraman, D. (2004) *Org. Lett.*, **6**, 5005–5008.
62 Kwong, F.Y. and Buchwald, S.L. (2002) *Org. Lett.*, **4**, 3517–3520.
63 Chen, Y.-J. and Chen, H.-H. (2006) *Org. Lett.*, **8**, 5609–5612.
64 Zhu, D., Xu, L., Wu, F., and Wan, B. (2006) *Tetrahedron Lett.*, **47**, 5781–5784.
65 Kondo, T. and Mitsudo, T.-a. (2000) *Chem. Rev.*, **100**, 3205–3220.
66 Verma, A.K., Singh, J., and Chaudhary, R. (2007) *Tetrahedron Lett.*, **48**, 7199–7202.
67 Deng, W., Zou, Y., Wang, Y.-F., Liu, L., and Guo, Q.-X. (2004) *Synlett*, 1254–1258.
68 Liu, S., Pestano, J.P.C., and Wolf, C. (2007) *Synthesis*, **22**, 3519–3527.
69 Nauš, P., Lešetický, L., Smrček, S., Tišlerová, I., and Štícha, M. (2003) *Synlett*, 2117–2122.
70 Wu, Y.-J. and He, H. (2003) *Synlett*, 1789–1790.
71 Zheng, Y., Du, X., and Bao, W. (2006) *Tetrahedron Lett.*, **47**, 1217–1220.
72 Manarin, F., Roehrs, J.A., Wilhelm, E.A., and Zeni, G. (2008) *Eur. J. Org. Chem.*, 4460–4465.
73 Rout, L., Saha, P., Jammi, S., and Punniyamurthy, T. (2008) *Eur. J. Org. Chem.*, 640–643.
74 Ranu, B.C., Saha, A., and Jana, R. (2007) *Adv. Synth. Catal.*, **349**, 2690–2696.
75 Gonzalez-Arellano, C., Luque, R., and Macquarrie, D.J. (2009) *Chem. Commun.*, 1410–1412.
76 Rout, L., Sen, T.K., and Punniyamurthy, T. (2007) *Angew. Chem. Int. Ed.*, **46**, 5583–5586.
77 Costantino, L., Ferrari, A.M., Gamberini, M.C., and Rastelli, G. (2002) *Bioorg. Med. Chem.*, **10**, 3923–3931.
78 Lory, P.M.J., Agarkov, A., and Gilbertson, S.R. (2006) *Synlett*, 3045–3048.
79 Bonnet, B., Soullez, D., Girault, S., Maes, L., Landry, V., Davioud-Charvet, E., and Sergheraert, C. (2000) *Bioorg. Med. Chem.*, **8**, 95–103.
80 Taniguchi, N. and Onami, T. (2004) *J. Org. Chem.*, **69**, 915–920.
81 Kumar, S. and Engman, L. (2006) *J. Org. Chem.*, **71**, 5400–5403.
82 Chang, D. and Bao, W. (2006) *Synlett*, 1786–1788.
83 Baskin, J.M. and Wang, Z. (2002) *Org. Lett.*, **4**, 4423–4425.
84 Zhu, W. and Ma, D. (2005) *J. Org. Chem.*, **70**, 2696–2700.
85 Bian, M., Xu, F., and Ma, C. (2007) *Synthesis*, **19**, 2951–2956.
86 Sawada, N., Itoha, T., and Yasuda, N. (2006) *Tetrahedron Lett.*, **47**, 6595–6597.
87 Liu, Y. and Bao, W. (2007) *Tetrahedron Lett.*, **48**, 4785–4788.
88 Herradura, P.S., Pendola, K.A., and Guy, R.K. (2000) *Org. Lett.*, **2**, 2019–2022.
89 Wang, L., Wang, M., and Huang, F. (2005) *Synlett*, 2007–2010.
90 Taniguchi, N. (2007) *J. Org. Chem.*, **72**, 1241–1245.
91 Braga, A.L., Barcellos, T., Paixao, M.W., Deobald, A.M., Godoi, M., Stefani, H.A., Cella, R., and Sharma, A. (2008) *Organometallics*, **27**, 4009–4012.

92 Kantam, M.L., Neelima, B., Sreedhar, B., and Chakravarti, R. (2008) *Synlett*, **10**, 1455–1458.
93 Ajiki, K., Hirano, M., and Tanaka, K. (2005) *Org. Lett.*, **7**, 4193–4195.
94 Zhao, X., Yu, Z., Yan, S., Wu, S., Liu, R., He, W., and Wang, L. (2005) *J. Org. Chem.*, **70**, 7338–7341.
95 Wong, Y.-C., Jayanth, T.T., and Cheng, C.-H. (2006) *Org. Lett.*, **8**, 5613–5616.
96 Correa, A., Carril, M., and Bolm, C. (2008) *Angew. Chem. Int. Ed.*, **47**, 2880–2883.
97 Wu, W.-Y., Wang, J.-C., and Tsai, F.-Y. (2009) *Green Chem.*, **11**, 326–329.
98 Buchwald, S.L. and Bolm, C. (2009) *Angew. Chem. Int. Ed.*, **48**, 5586–5587.
99 Beletskaya, I.P. and Ananikov, V.P. (2007) *Eur. J. Org. Chem.*, 3431–3444.
100 Perin, G., Lenardao, E.J., Jacob, R.G., and Panatieri, R.B. (2009) *Chem. Rev.*, **109**, 1277–1301.
101 Comasseto, J.V., Ling, L.W., Petragnani, N., and Stefani, H.A. (1997) *Synthesis*, 373–403.
102 (a) Kataeva, L.M., Kataev, E.G., and Idiyatullina, D.Ya. (1966) *J. Struct. Chem. (USSR)*, **7** (3), 380; (b) Heiba, E.-A.I. and Dessau, R.M. (1967) *J. Org. Chem.*, **32**, 3837–3840; (c) Benati, L., Montevecchi, P.C., and Spagnolo, P. (1991) *J. Chem. Soc., Perkin Trans. 1*, 2103–2109; (d) Johannsen, I., Henriksen, L., and Eggert, H. (1986) *J. Org. Chem.*, **51**, 1657–1663; (e) Back, T.G. and Krishna, M.V. (1988) *J. Org. Chem.*, **53**, 2533–2536; (f) Ogawa, A., Yokoyama, H., Yokoyama, K., Masawaki, T., Kambe, N., and Sonoda, N. (1991) *J. Org. Chem.*, **56**, 5721–5723; (g) Ogawa, A., Takami, N., Sekiguchi, M., Yokoyama, H., Kuniyasu, H., Ryu, I., and Sonoda, N. (1991) *Chem. Lett.*, 2241–2242; (h) Potapov, V.A., Amosova, S.V., Starkova, A.A., Zhnikin, A.R., Doron'kina, I.V., Beletskaya, I.P., and Hevesi, L. (2000) *Sulf. Lett.*, **23**, 229–238.
103 (a) Moro, A.V., Nogueira, C.W., Barbosa, N.B.V., Menezes, P.H., da Rocha, J.B.T., and Zeni, G. (2005) *J. Org. Chem.*, **70**, 5257–5268; (b) Zeni, G., Stracke, M.P., Nogueira, C.W., Braga, A.L., Menezes, P.H., and Stefani, H.A. (2004) *Org. Lett.*, **6**, 1135–1138; (c) Potapov, V.A., Amosova, S.V., Belozerova, O.V., Yarosh, O.G., Gendin, M.G., and Voronkov, M.G. (2003) *J. Gen. Chem.*, **73** (7), 1158; (d) Silveira, C.C., Santos, P.C.S., Mendes, S.R., and Braga, A.L. (2008) *J. Organomet. Chem.*, **693**, 3787–3790.
104 Kuniyasu, H., Ogawa, A., Miyazaki, S., Ryu, I., Kambe, N., and Sonoda, N. (1991) *J. Am. Chem. Soc.*, **113**, 9796–9803.
105 Ogawa, A., Kuniyasu, H., Sonoda, N., and Hirao, T. (1997) *J. Org. Chem.*, **62**, 8361–8365.
106 Kuniyasu, H., Yamashita, F., Terao, J., and Kambe, N. (2007) *Angew. Chem. Int. Ed.*, **46**, 5929–5933.
107 Kuniyasu, H., Takekawa, K., Yamashita, F., Miyafuji, K., Asano, S., Takai, Y., Ohtaka, A., Tanaka, A., Sugoh, K., Kurosawa, H., and Kambe, N. (2008) *Organometallics*, **27**, 4788–4802.
108 Ananikov, V.P., Beletskaya, I.P., Aleksandrov, G.G., and Eremenko, I.L. (2003) *Organometallics*, **22**, 1414–1421.
109 Ananikov, V.P., Kabeshov, M.A., Beletskaya, I.P., Aleksandrov, G.G., and Eremenko, I.L. (2003) *J. Organomet. Chem.*, **687**, 451–461.
110 Ananikov, V.P., Kabeshov, M.A., and Beletskaya, I.P. (2003) *Dokl. Chem.*, **390**, 112–114.
111 Ananikov, V.P. and Beletskaya, I.P. (2003) *Dokl. Chem.*, **389**, 81–86.
112 Beletskaya, I.P. and Ananikov, V.P. (2007) *Pure Appl. Chem.*, **79**, 1041–1056.
113 Ananikov, V.P., Kabeshov, M.A., and Beletskaya, I.P. (2005) *Synlett*, 1015–1017.
114 Ananikov, V.P., Orlov, N.V., and Beletskaya, I.P. (2005) *Russ. Chem. Bull. Int. Ed.*, **54**, 576–587.
115 Ananikov, V.P. and Beletskaya, I.P. (2004) *Org. Biomol. Chem.*, **2**, 284–287.
116 Ananikov, V.P. and Beletskaya, I.P. (2008) *Russ. Chem. Bul. Int. Ed.*, **57**, 754–760.
117 Cai, M., Wang, Y., and Hao, W. (2007) *Green Chem.*, **9**, 1180–1184.
118 Yamashita, F., Kuniyasu, H., Terao, J., and Kambe, N. (2006) *Inorg. Chem.*, **45**, 1399–1404.
119 Ananikov, V.P., Kabeshov, M.A., Beletskaya, I.P., Khrustalev, V.N., and Antipin, M.Yu. (2005) *Organometallics*, **24**, 1275–1283.

120 Gareau, Y. and Orellana, A. (1997) *Synlett*, 803–804.

121 Gareau, Y., Tremblay, M., Gauvreau, D., and Juteau, H. (2001) *Tetrahedron*, **57**, 5739–5750.

122 Ananikov, V.P., Gayduk, K.A., Beletskaya, I.P., Khrustalev, V.N., and Antipin, M.Yu. (2008) *Chem. Eur. J.*, **14**, 2420–2434.

123 Ananikov, V.P., Gayduk, K.A., Beletskaya, I.P., Khrustalev, V.N., and Antipin, M.Yu. (2009) *Eur. J. Inorg. Chem.*, 1149–1161.

124 Ananikov, V.P., Orlov, N.V., Kabeshov, M.A., Beletskaya, I.P., and Starikova, Z.A. (2008) *Organometallics*, **27**, 4056–4061.

125 Arisawa, M., Fujimoto, K., Morinaka, S., and Yamaguchi, M. (2005) *J. Am. Chem. Soc.*, **127**, 12226–12227.

126 Arisawa, M., Kozuki, Y., and Yamaguchi, M. (2003) *J. Org. Chem.*, **68**, 8964–8967.

127 Arisawa, M. and Yamaguchi, M. (2001) *Org. Lett.*, **3**, 763–764.

128 Ishiyama, T., Nishijima, K., Miyaura, N., and Suzuki, A. (1993) *J. Am. Chem. Soc.*, **115** (16), 7219–7225.

129 Han, L.-B. and Tanaka, M. (1999) *Chem. Lett.*, 863–864.

130 Han, L.-B., Choi, N., and Tanaka, M. (1996) *J. Am. Chem. Soc.*, **118**, 7000–7001.

131 Han, L.-B. and Tanaka, M. (1998) *J. Am. Chem. Soc.*, **120**, 8249–8250.

132 Ogawa, A., Kuniyasu, H., Takeba, M., Ikeda, T., Sonoda, N., and Hirao, T. (1998) *J. Organomet. Chem.*, **564**, 1–4.

133 Kamiya, I., Kawakami, J.-i., Yano, S., Nomoto, A., and Ogawa, A. (2006) *Organometallics*, **25**, 3562–3564.

134 Wang, M., Cheng, L., and Wu, Z. (2008) *Dalton Trans.*, 3879–3888.

135 Knapton, D.J. and Meyer, T.Y. (2005) *J. Org. Chem.*, **70**, 785–796.

136 Knapton, D.J. and Meyer, T.Y. (2004) *Org. Lett.*, **6**, 687–689.

137 Kamiya, I., Nishinaka, E., and Ogawa, A. (2005) *Tetrahedron Lett.*, **46**, 3649–3652.

138 Wang, M., Cheng, L., Hong, B., and Wu, Z. (2009) *Organometallics*, **28**, 1506–1513.

139 (a) Kataev, E.G. and Petrov, V.N. (1962) *J. Gen. Chem. USSR (Engl. Transl.)*, **32**, 3626;(b) Trofimov, B.A. (2002) *Curr. Org. Chem.*, **6**, 1121–1162; (c) Benati, L., Capella, L., Montevecchi, P.C., and Spagnolo, P. (1995) *J. Chem. Soc. Perkin Trans.*, 1035–1038; (d) Benati, L., Capella, L., Montevecchi, P.C., and Spagnolo, P. (1994) *J. Org. Chem.*, **59**, 2818–2823; (e) Trofimov, B.A. (1981) *Russ. Chem. Rev.*, **50**, 138–150; (f) Griesbaum, K. (1970) *Angew. Chem. Int. Ed.*, **9**, 273–287; (g) Truce, W.E. and Tichenor, G.J.W. (1972) *J. Org. Chem.*, **37**, 2391–2396; (h) Truce, W.E. and Heine, R.F. (1957) *J. Am. Chem. Soc.*, **79**, 5311–5313; (i) Potapov, V.A. and Amosova, S.V. (1996) *Russ. J. Org. Chem.*, **32**, 1099–1109; (j) Ogawa, A., Obayashi, R., Sekiguchi, M., Masawaki, T., Kambe, N., and Sonoda, N. (1992) *Tetrahedron Lett.*, **33**, 1329–1323; (k) Kataoka, T., Yoshimatsu, M., Shimizu, H., and Hori, M. (1990) *Tetrahedron Lett.*, **31**, 5927–5930; (l) Tsoi, L.A., and Patsaev, A.K., Ushanov, V.Zh., Vyaznikovtsev, L.V. (1984) *J. Org. Chem. USSR (Engl. Transl.)*, **20** (10), 1897–1902; (m) Renard, M. and Hevesi, L. (1985) *Tetrahedron*, **41**, 5939–5954; (n) Kondoh, A.K., Takami, K., Yorimitsu, H., and Oshima, K. (2005) *J. Org. Chem.*, **70**, 6468–6473.

140 (a) Silveira, C.C., Guerra, R.B., and Comasseto, J.V. (2007) *Tetrahedron Lett.*, **48**, 5121–5124; (b) Comasseto, J.V., and Brandt, C.A. (1987) *Synthesis*, 146–149; (c) Comasseto, J.V. (1983) *J. Organomet. Chem.*, **253**, 131–181.

141 (a) Yadav, J.S., Reddy, B.V.S., Raju, A., Ravindar, K., and Baishya, G. (2007) *Chem. Lett.*, **36**, 1474–1475; (b) Peppe, C., de Castro, L.B., de Azevedo Mello, M., and do Rego Barros, O.S. (2008) *Synlett*, 1165–1170; (c) Wang, Z.-L., Tang, R.-Y., Luo, P.-S., Deng, C.-L., Zhong, P., and Li, J.-H. (2008) *Tetrahedron*, **64**, 10670–10675; (d) Silva, M.S., Lara, R.G., Marczewski, J.M., Jacob, R.G., Lenardao, E.J., and Perin, G. (2008) *Tetrahedron Lett.*, **49**, 1927–1930.

142 (a) Peppe, C., Lang, E.S., Ledesma, G.N., de Castro, L.B., do Rego Barros, O.S., and de Azevedo Mello, P. (2005) *Synlett*, 3091–3095; (b) do Rego Barros, O.S., Lang, E.S., de Oliveira, C.A.F., Peppe, C.,

and Zeni, G. (2002) *Tetrahedron Lett.*, **43**, 7921–7923; (c) Dabdoub, M.J., Cassol, T.M., and Batista, A.C.F. (1996) *Tetrahedron Lett.*, **37**, 9005–9008; (d) Zeni, G., Stracke, M.P., Nogueira, C.W., Braga, A.L., Menezes, P.H., and Stefani, H.A. (2004) *Org. Lett.*, **6**, 1138–1142; (e) Silveira, C.C., Cella, R., and Vieira, A.S. (2006) *J. Organomet. Chem.*, **691**, 5861–5866; (f) Usugi, S.-I., Yorimitsu, H., Shinokubo, H., and Oshima, K. (2004) *Org. Lett.*, **6**, 601–603.

143 Kuniyasu, H., Ogawa, A., Sato, K., Ryu, I., Kambe, N., and Sonoda, N. (1992) *J. Am. Chem. Soc.*, **114**, 5902–5903.

144 Kuniyasu, H., Ogawa, A., Sato, K.-I., Ryu, I., and Sonoda, N. (1992) *Tetrahedron Lett.*, **38**, 5525–5528.

145 Ogawa, A. (2000) *J. Organomet.Chem.*, **611**, 463–474.

146 Ogawa, A., Ikeda, T., Kimura, K., and Hirao, T. (1999) *J. Am. Chem. Soc.*, **121**, 5108–5114.

147 Ananikov, V.P., Malyshev, D.A., Beletskaya, D.A., Aleksandrov, G.G., and Eremenko, I.L. (2003) *J. Organomet. Chem.*, **679**, 162–172.

148 Ananikov, V.P., Malyshev, D.A., and Beletskaya, I.P. (2002) *Russ. J. Org. Chem.*, **38**, 1475–1478.

149 Ananikov, V.P., Malyshev, D.A., Beletskaya, I.P., Aleksandrov, G.G., and Eremenko, I.L. (2005) *Adv. Synth. Catal.*, **347**, 1993–2001.

150 Kamiya, I., Nishinaka, E., and Ogawa, A. (2005) *J. Org. Chem.*, **70**, 696–698.

151 Ananikov, V.P., Zalesskiy, S.S., Orlov, N.V., and Beletskaya, I.P. (2006) *Russ. Chem. Bul. Int. Ed.*, **55**, 2109–2113.

152 Ananikov, V.P., Orlov, N.V., and Beletskaya, I.P. (2006) *Organometallics*, **25**, 1970–1977.

153 Ananikov, V.P., Orlov, N.V., and Beletskaya, I.P. (2007) *Organometallics*, **26**, 740–746.

154 Malyshev, D.A., Scott, N.M., Marion, N., Stevens, E.D., Ananikov, V.P., Beletskaya, I.P., and Nolan, S.P. (2006) *Organometallics*, **25**, 4462–4470.

155 Ananikov, V.P., Orlov, N.V., Beletskaya, I.P., Khrustalev, V.N., Antipin, M.Yu., and Timofeeva, T.V. (2007) *J. Am. Chem. Soc.*, **129**, 7252–7253.

156 Kondoh, A., Yorimitsu, H., and Oshima, K. (2007) *Org. Lett.*, **9**, 1383–1385.

157 Ogawa, A., Takeba, M., Kawakami, J.-I., Ryu, I., Kambe, N., and Sonoda, N. (1995) *J. Am. Chem. Soc.*, **117**, 7564–7565.

158 Kawakami, J.-I., Takeba, M., Kamiya, I., Sonoda, N., and Ogawa, A. (2003) *Tetrahedron*, **59**, 6559–6567.

159 Shoai, S., Bichler, P., Kang, B., Buckley, H., and Love, J.A. (2007) *Organometallics*, **26**, 5778–5781.

160 Kawakami, J.-I., Mihara, M., Kamiya, I., Takeba, M., Ogawa, A., and Sonoda, N. (2003) *Tetrahedron*, **59**, 3521–3526.

161 Ogawa, A., Kawabe, K.-I., Kawakami, J.-I., Mihara, M., Hirao, T., and Sonoda, N. (1998) *Organometallics*, **17**, 3111–3114.

162 Cao, C., Fraser, L.R., and Love, J.A. (2005) *J. Am. Chem. Soc.*, **127**, 17614–17615.

163 Misumi, Y., Seino, H., and Mizobe, Y. (2006) *J. Organomet. Chem.*, **691**, 3157–3164.

164 Yang, J., Sabarre, A., Fraser, L.R., Patrick, B.O., and Love, J.A. (2009) *J. Org. Chem.*, **74**, 182–187.

165 (a) Sabarre, A. and Love, J. (2008) *Org. Lett.*, **10**, 3941–3944. (b) Fraser, I.R., Bird, J., Wu Qu., Cao, C., Patrick, B.O., and Love, J.A. (2007) *Organometallics*, **26**, 5602–5611.

166 Ogawa, A., Kudo, A., and Hirao, T. (1998) *Tetrahedron Lett.*, **39**, 5213–5216.

167 Ogawa, A., Kawakami, J.-i., Sonoda, N., and Hirao, T. (1996) *J. Org. Chem.*, **61**, 4161–4163.

168 Kajitania, M., Kamiya, I., Nomoto, A., Kihara, N., and Ogawa, A. (2006) *Tetrahedron*, **62**, 6355–6360.

4
Palladium Catalysis for Oxidative 1,2-Difunctionalization of Alkenes

Béatrice Jacques and Kilian Muñiz

4.1
Introduction

Oxidative vicinal functionalization of alkenes represents an important reaction among alkene transformations as it allows the introduction of two functional groups within a single transformation. In the area of palladium catalysis, this type of transformation usually consists of a two-step procedure, during which the two heteroatoms are introduced into the carbon framework. Despite this simplicity at first sight, the actual reaction mechanism may be very complex and, in principle, several mechanistic pathways are possible.

In general, the initial interaction between a palladium compound, a nucleophile X and an alkene can proceed via two different pathways, which may be competing with each other (Figure 4.1)[1]. One consists of a π-complexation of the alkene to palladium [2] followed by nucleophilic attack from X via the uncomplexed face of the alkene, leading to an overall *anti*-process. Alternatively, X may be coordinated to palladium prior to the alkene coordination, resulting in an overall *syn*-addition process.

The resulting alkyl-palladium complex can undergo a variety of different reactions (Figure 4.2), presuming it can be sufficiently stabilized in order to allow incorporation of a second nucleophile. This process may proceed under reductive displacement of palladium to give a palladium(0) compound, which is then reoxidized to regenerate the palladium(II) stage under the given reaction conditions. Alternatively, the alkyl-palladium(II) intermediate may be oxidized to an alkyl-Pd(IV) complex [3, 4], which then promotes bond formation upon reductive elimination of the initial palladium(II) species. For certain reactions with copper salts as oxidants, the possibility of a transient oxidation was formulated [5]. This model consists of a concomitant oxidation of the palladium center and C–Y bond formation. Again, both *syn*- and *anti*-processes are possible for each of these replacements, opening a broad field of stereochemical diversification for these 1,2-difunctionalization reactions.

Within this chapter, we will focus on the transformation of simple alkenes and hence will not cover transformation of butadienes [6, 7]. Difunctionalization

Catalyzed Carbon-Heteroatom Bond Formation. Edited by Andrei K. Yudin

ISBN: 978-3-527-32428-6

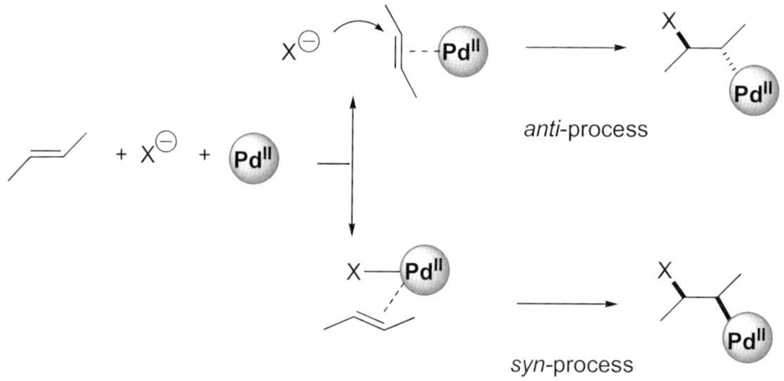

Figure 4.1 Models for palladium-promoted addition of nucleophile X to an alkene.

reactions will be limited to those reactions that lead to the incorporation of two identical or different heteroatoms. Hence, reactions which incorporate alkene functionalization followed by C–C bond formation [8] will not be considered.

4.2
Palladium-Catalyzed 1,2-Difunctionalization Reactions: Halogenation

Initially, alkene transformations that include carbon–halogen bond formation were observed as side-pathways in Wacker chemistry [9]. In particular, it was found that acetaldehyde production was a function of both the amount of free chloride and copper chloride present, as chlorohydrin appeared to be the major product at high chloride and copper chloride concentration. This aspect was extensively investigated by Henry and has been the subject of a monograph [9]. Further investigation extended this work to an enantioselective palladium(II) catalyzed chlorohydrin synthesis (Scheme 4.1) [10]. This work demonstrated that under aerobic conditions and for copper salt concentrations in the range of 3–5 M, chiral palladium complexes are capable of inducing highly regio- and enantio-selective chlorohydrin formation from simple prochiral alkenes. It was found that dimeric palladium complexes containing bridging counterion and bridging chiral BINAP ligand acted as the optimum catalyst source. It is noteworthy that the BINAP ligand is stable against oxidation under the chosen conditions, even though it is coordinated only in a bridging mode.

The underlying reactivity was then developed further by Henry. Upon replacing copper chloride by copper bromide in a water–THF solvent mixture, several alkenes could be converted into the corresponding vicinal dibromides [11]. Importantly, potentially competing by-product formation of aldehydes and ketones was low. Instead, the reaction showed reasonably high selectivity in favor of dibromination products, while the corresponding bromohydrins were not produced in this case. This was explained by the stronger nucleophilicity of bromide over chloride. Cyclopentene gave clean formation of *trans*-1,2-dibromo-cyclopentane, formation of which was explained through an *anti*-bromopalladation followed by C–Br bond

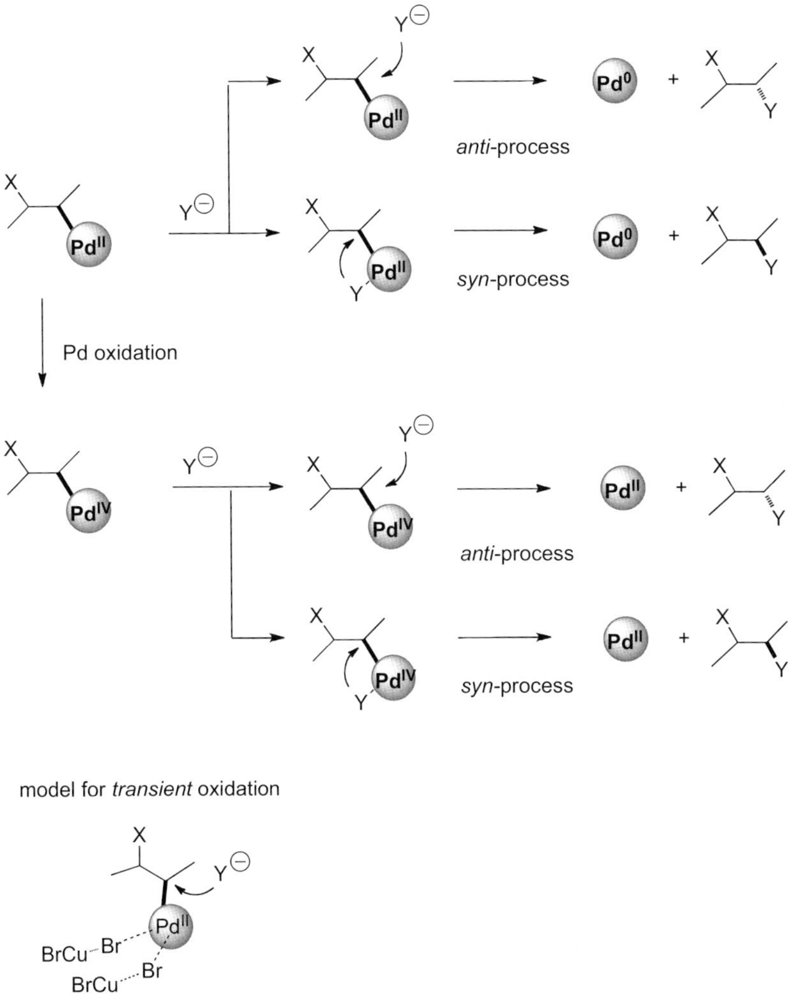

Figure 4.2 Representative pathways for nucleophilic displacement of palladium from alkyl-palladium compounds under C–Y bond formation (relative stereochemistry for the C–X bond may be *syn* or *anti*, depending on the previous pathway, see Figure 4.1.

formation from within the palladium coordination sphere, or from copper bromide species coordinated to the palladium intermediate.

4.3 Aminohalogenation Reactions

Intramolecular aminopalladations constitute a particularly versatile approach to the generation of alkyl palladium intermediates [12]. These compounds were reacted

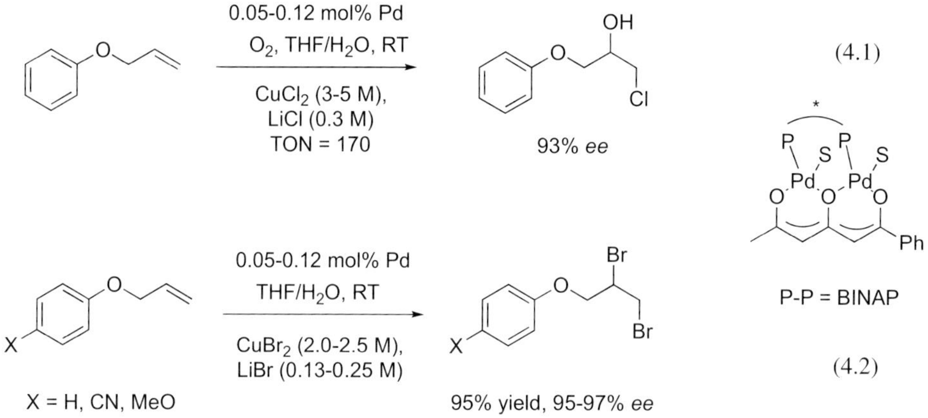

0.05-0.12 mol% Pd
O_2, THF/H_2O, RT
$CuCl_2$ (3-5 M),
LiCl (0.3 M)
TON = 170
OH
O
Cl
93% *ee*
(4.1)

P-P = BINAP
Ph

0.05-0.12 mol% Pd
THF/H_2O, RT
$CuBr_2$ (2.0-2.5 M),
LiBr (0.13-0.25 M)
Br
Br
X = H, CN, MeO
95% yield, 95-97% *ee*
(4.2)

Scheme 4.1 Palladium-catalyzed enantioselective chlorohydrin and dibromination reactions.

with various halogen sources from copper chloride and bromide to arrive at aminohalogenation reactions.

While such a process had initially been observed as an undesired side-reaction in transformations where copper salts were employed as re-oxidants [13], Chemler demonstrated that various aminohalogenation reactions proceed in THF or acetonitrile in the presence of potassium carbonate as base [14]. These reactions employ palladium trifluoroacetate or palladium dibromide as catalyst source and require a moderate excess of the copper oxidant (3–4 equiv) giving moderate to excellent yields. However, they usually suffer from rather low selectivity, either in the initial aminopalladation or via subsequent rearrangement pathways to provide mixtures of pyrrolidines and piperazines (Scheme 4.2, Eq. (4.3)). A stoichiometric control reaction in the presence of palladium bromide led only to the Wacker cyclization together with an alkene isomerization product, suggesting that the presence of copper(II) salts is crucial for the overall process. The exact role of the copper(II) salts has not yet been clarified and palladium intermediates of different oxidation states may be involved in the final stage of carbon–halogen bond formation.

Related carbamate and urea derivatives were employed by Lu in alternative cyclization processes [15]. In these reactions, lithium halogen salts were added to increase the halogen concentrations and the reactions were carried out with a catalyst from palladium acetate and in THF at room temperature (Eq. (4.4)). The reaction proceeds with complete stereospecificity, as was determined by X-ray analysis of an aminochlorination product derived from oxidation of a (*Z*)-configured starting material (Eq. (4.5)). Mechanistically, the authors favored a sequence of *trans*-aminopalladation followed by oxidative cleavage of the carbon–palladium bond with retention of configuration. In another approach, palladium-catalyzed aminochlorination was employed for the synthesis of aminoalkanols (Eq. (4.6)) [16]. Polar solvents such as glacial acetic acid were usually required in order to achieve high yield and selectivity and 70% yields were obtained for a product formation, which favored the shown L-*ido* derivative over the D-*gluco* epimer in a 19 : 1 ratio (90% *de*).

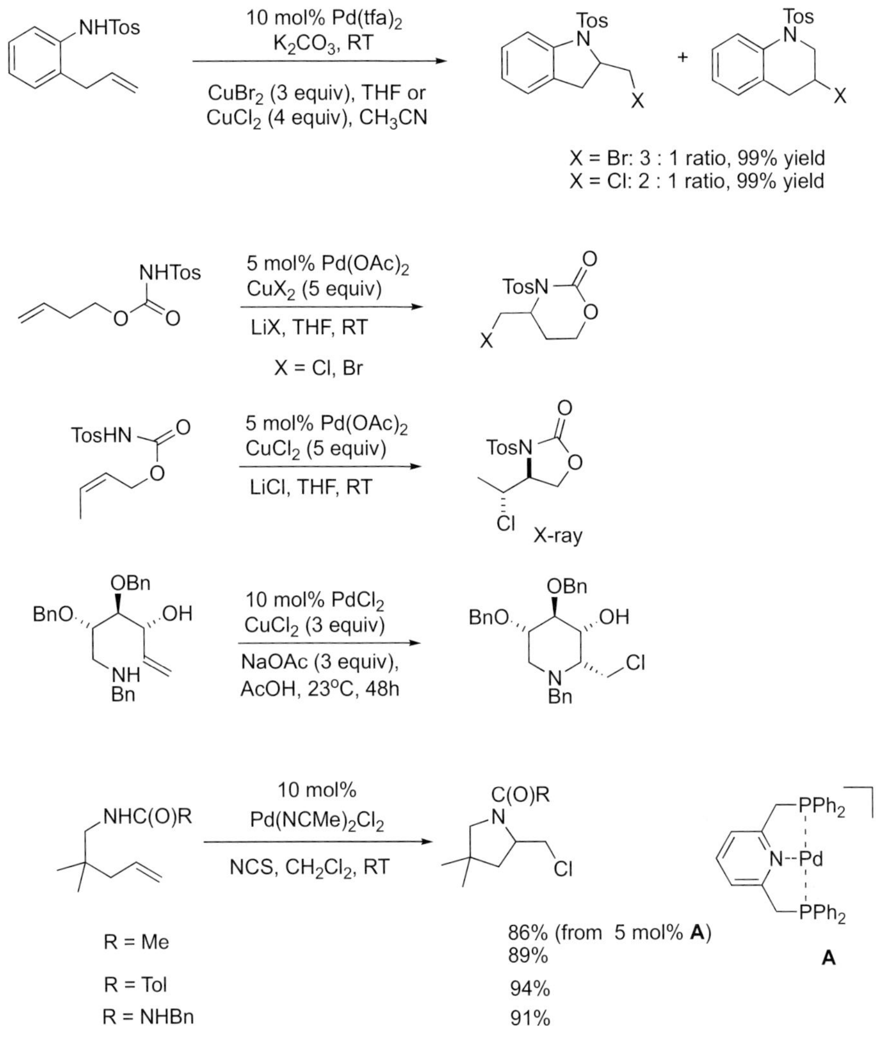

NHTos → 10 mol% $Pd(tfa)_2$, K_2CO_3, RT; $CuBr_2$ (3 equiv), THF or $CuCl_2$ (4 equiv), CH_3CN

X = Br: 3 : 1 ratio, 99% yield
X = Cl: 2 : 1 ratio, 99% yield (4.3)

5 mol% $Pd(OAc)_2$, CuX_2 (5 equiv); LiX, THF, RT
X = Cl, Br (4.4)

5 mol% $Pd(OAc)_2$, $CuCl_2$ (5 equiv); LiCl, THF, RT
X-ray (4.5)

10 mol% $PdCl_2$, $CuCl_2$ (3 equiv); NaOAc (3 equiv), AcOH, 23°C, 48h (4.6)

10 mol% $Pd(NCMe)_2Cl_2$; NCS, CH_2Cl_2, RT

R = Me	86% (from 5 mol% **A**)
	89%
R = Tol	94%
R = NHBn	91%

A: $2\,BF_4^{\ominus}$ (4.7)

Scheme 4.2 Aminohalogenation reactions employing copper salts as halogen source.

Michael recently described aminochlorination reactions from intramolecular aminopalladation followed by reaction with *N*-chlorosuccinamide (NCS) [17]. Initial reactions made use of Michael's earlier observation on stabilization of aminopalladation intermediates starting from 5 mol% ligated Pd complex **A** [18], but further work showed that simple palladium salts such as $Pd(NCMe)_2Cl_2$ could also be employed, albeit in higher catalyst loadings of 10 mol% (Eq. (4.7)). The reaction works well for a series of amides and carbamates, but not for tosylates.

Alternatively, Göttlich developed a route to 3-chloro piperidines from *N*-chloro amines (Scheme 4.3, Eq. (4.8)) [19]. The authors suggest a mechanism of

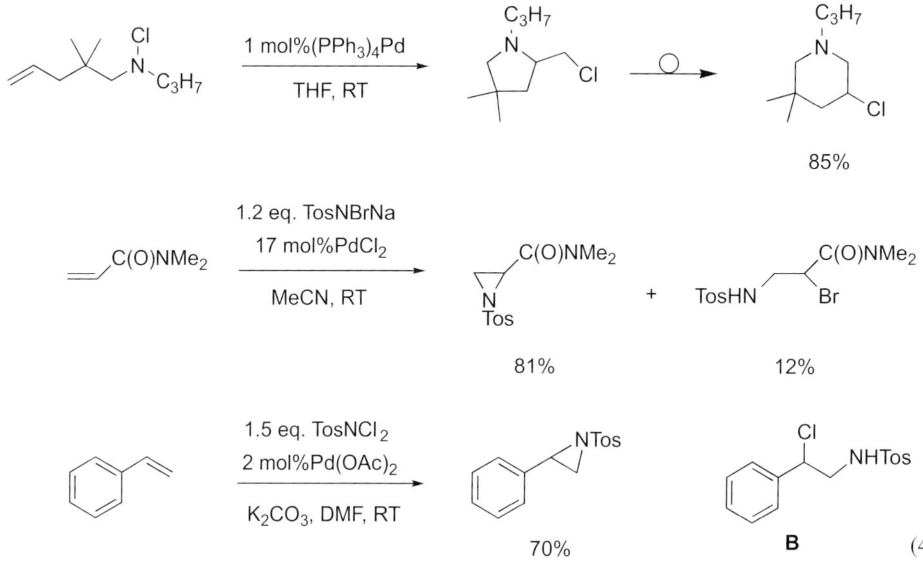

(4.8)

(4.9)

(4.10)

Scheme 4.3 Aminohalogenation reactions employing *N*-halogenated amides.

aminopalladation and palladium for chloro exchange to arrive at chloromethyl pyrrolidines, which subsequentely rearrange to the six-membered ring products. The reaction is believed to be initiated by oxidative insertion of the palladium(0) precursor into the N–Cl bond. This generates the palladium(II) oxidation state that is required for alkene functionalization. Importantly, the reaction proceeds more efficiently under a catalyst loading of 1 mol% rather than 10 mol%.

The application of a different N-halogen source was described by Branco, Prabhakar and coworkers [20]. In this case, acrylic acid derivatives underwent amination in the presence of bromamine-T to yield the corresponding aziridines as the major products. Aminobrominated side products were also observed in some cases. Scheme 4.3 shows a representative example for the oxidation of dimethyl acrylamide (Eq. (4.9)). The reactions were also investigated in the presence of stoichiometric amounts of palladium, as for example for methyl acrylate, which gave the corresponding aziridine as the only product [21]. The mechanism of this aziridination remains uncertain as both Pd(II) and Pd(IV) intermediates have been suggested, although without definite proof.

Li reported a palladium-catalyzed aziridination with *N,N*-dichlorotosylamide as nitrogen source (Eq. (4.10)) [22]. This reaction proceeds best with palladium acetate and gives the corresponding aziridines in good yields. For styrene as substrate and in the absence of potassium carbonate, the authors found that the corresponding aminochlorinated product **B** is the only product (97%). Treatment of **B** with base gave complete conversion to the corresponding aziridine, which led the authors to conclude that the reaction is actually an aminochlorination with subsequent base-mediated aziridine formation.

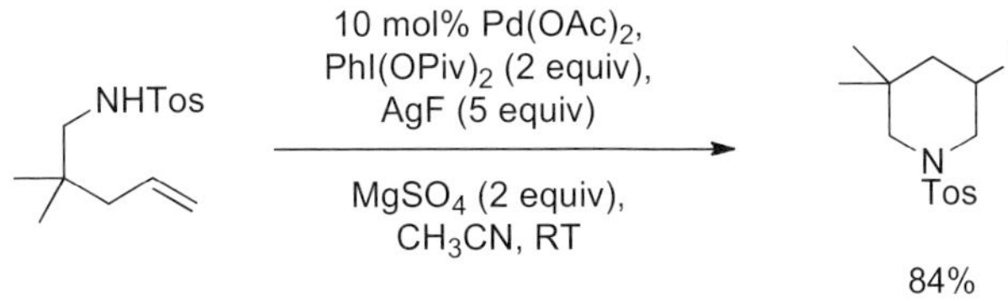

NHTos → 10 mol% $Pd(OAc)_2$, $PhI(OPiv)_2$ (2 equiv), AgF (5 equiv) / $MgSO_4$ (2 equiv), CH_3CN, RT → F, N, Tos 84% (4.11)

Scheme 4.4 Palladium-catalyzed aminofluorination.

Finally, Liu reported a novel aminofluorination procedure that makes use of the versatile intramolecular *trans*-aminopalladation [12] as an initial step, although with a noteworthy *endo*-selectivity. Formation of a palladium(IV) intermediate upon metal oxidation followed by fluorination from within the Pd coordination sphere generates the final product (Scheme 4.4) [23].

4.4 Dialkoxylation

The first recent entries in this area stem from nucleopalladation reactions of 2-vinyl phenols. These reactions were independently discovered by Le Bras, Muzart and Sigman (Scheme 4.5). The former group described a reaction employing

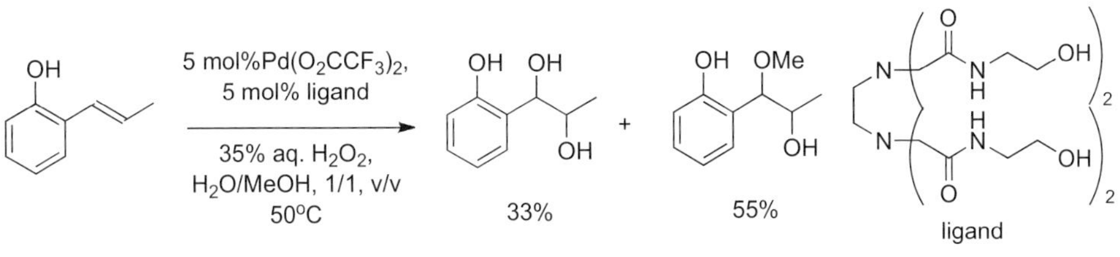

OH → 5 mol%$Pd(O_2CCF_3)_2$, 5 mol% ligand / 35% aq. H_2O_2, H_2O/MeOH, 1/1, v/v, 50°C → OH OH, OH 33% + OH OMe, OH 55%; ligand (4.12)

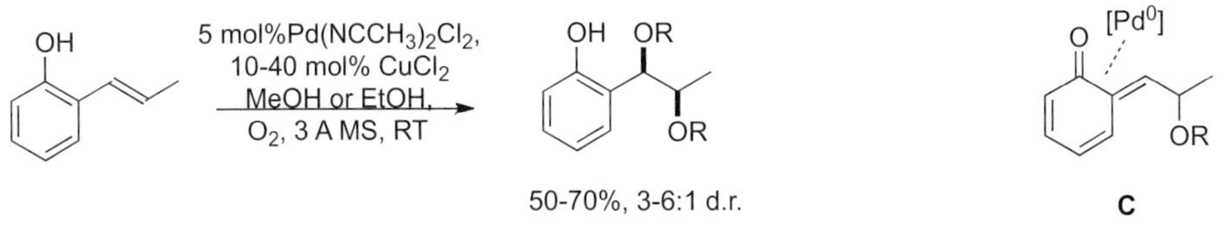

OH → 5 mol%$Pd(NCCH_3)_2Cl_2$, 10-40 mol% $CuCl_2$, MeOH or EtOH, / O_2, 3 A MS, RT → OH OR, OR 50-70%, 3-6:1 d.r.; $[Pd^0]$, O, OR **C** (4.13)

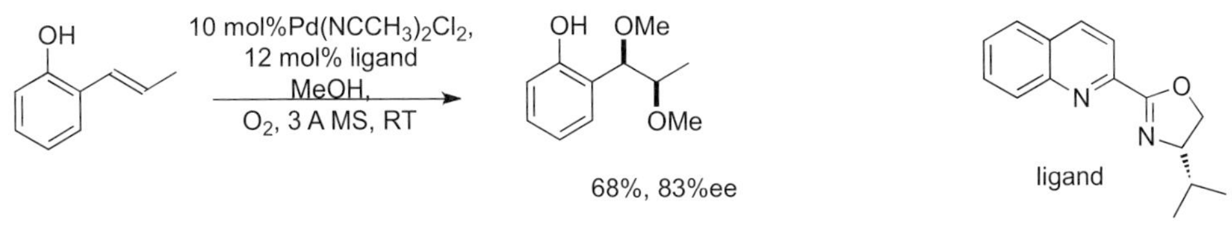

OH → 10 mol%$Pd(NCCH_3)_2Cl_2$, 12 mol% ligand, MeOH, / O_2, 3 A MS, RT → OH OMe, OMe 68%, 83%ee; ligand (4.14)

Scheme 4.5 Palladium-catalyzed dialkoxylation reactions (products from Eq. (4.11) are mixtures of diastereoisomers).

hydrogenperoxide as oxidant, which in a water/methanol solvent system gives a product mixture of the corresponding diol and monomethylated derivative (Eq. (4.12)) [24]. This reaction outcome was explained by the assumption of palladium-catalyzed epoxidation followed by nucleophilic ring opening at the benzylic position with competing nucleophiles from the solvent mixture yielding the products as mixtures of diastereomers [25].

The work by Sigman also requires the *ortho*-phenol motif and gives moderate to good diastereoselectivities for a range of different 2-vinyl phenols (Eq. (4.13)) [26]. A different mechanistic proposal was given that suggests formation of an intermediary quinone methide complex of palladium. This complex **C** would result from regioselective nucleopalladation of the alkene followed by tautomerization. Strong support for this mechanism stems from deuterium labeling experiments and from the observation that free phenol is required for the reaction.

An enantioselective version of this reaction was presented shortly thereafter, which uses the oxazoline ligand (Eq. (4.14)) [27]. The reaction results in good to excellent enantioinduction for a series of 2′-monosubstituted alkenes, while a dimethyl substitution at the terminal alkene carbon leads to 0% ee. This was interpreted by an overall mechanism for which the enantioselection is derived from the first step of alkoxypalladation followed by faster palladium loss from the corresponding quinone methide intermediate than nucleophilic attack by the second alkoxide.

A recent additional report by Song and Dong employs iodosobenzene diacetate and extends the dialkoxylation to unfunctionalized alkenes [28]. For example, stilbene is cleanly dioxygenated with reasonably high diastereoselectivity in the presence of a hydrated palladium(II) catalyst bearing a bis(phosphine) ligand (Scheme 4.6). Depending on the reaction conditions, especially the presence of acetic anhydride, either vicinal hydroxy acetates or diacetates are obtained. The reaction is also successful for substrates that incorporate a transferable oxygen group. In this case, intramolecular cyclization takes place, as for the depicted example of a tetrahydrofurane with 4-acetoxy substitution (Eq. (4.16)).

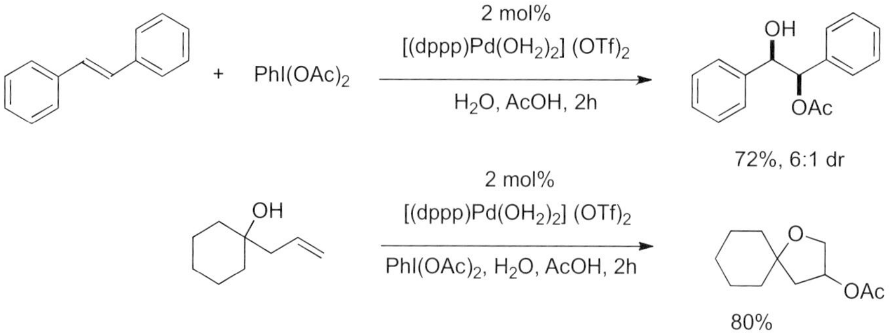

+ $PhI(OAc)_2$ → 2 mol% $[(dppp)Pd(OH_2)_2](OTf)_2$, H_2O, AcOH, 2h → OH, OAc
72%, 6:1 dr (4.15)

OH → 2 mol% $[(dppp)Pd(OH_2)_2](OTf)_2$, $PhI(OAc)_2$, H_2O, AcOH, 2h → O, OAc
80% (4.16)

Scheme 4.6 Palladium-catalyzed oxygenation reactions.

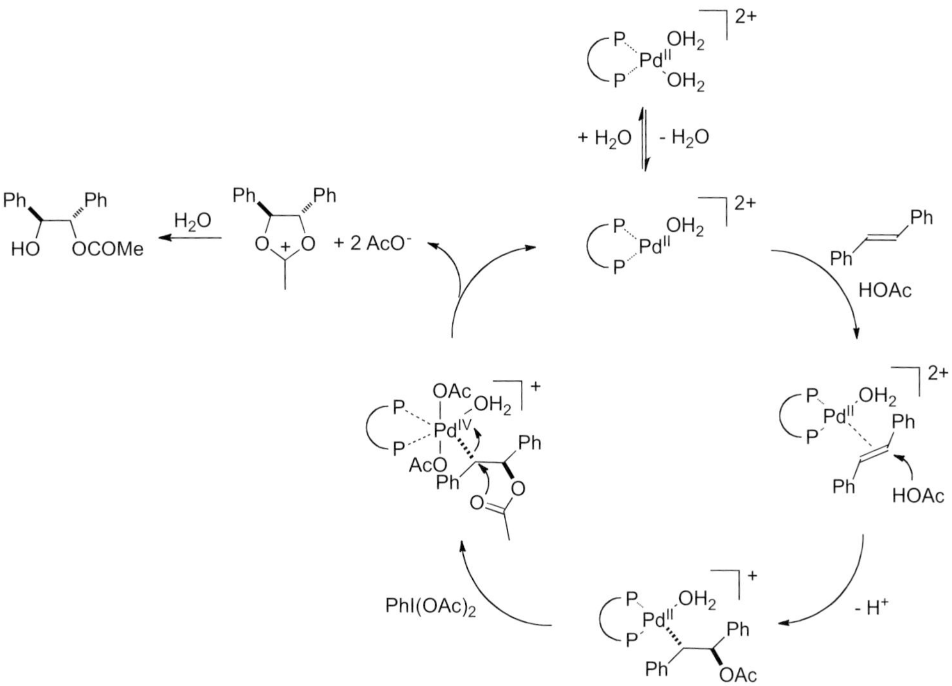

Figure 4.3 Mechanistic proposal for dioxygenation of stilbene.

The mechanism of this reaction was suggested to be as shown in Figure 4.3. Departing from hydrated cationic palladium(II) the alkene can be efficiently activated upon π-coordination. External attack of acetic acid then forms the first C–O bond and generates the alkyl palladium(II) intermediate. In the presence of the strong oxidant iodosobenzene diacetate, oxidation of palladium(II) to palladium(IV) takes place, which enables intramolecular attack from the incorporated acetate group to displace palladium in its original oxidation state (II), thereby formally closing the catalytic cycle. The resulting dioxygenated product is cleaved by water to give the observed vicinal hydroxy acetate. This mechanistic step was corroborated by the observation that a reaction in the presence of isotopically labeled water gives incorporation into the acetate carbonyl oxygen, but not into the C–O functionalities.

4.5 Aminoacetoxylation Reactions

In recent years a series of aminoacetoxylation reactions have been reported, which all appear to follow the same concept (Scheme 4.7). These approaches were initiated within an investigation on direct aminoacetoxylation of alkenes by Sorensen [29]. This protocol provides a convenient approach to various pyrrolidines, piperazines, lactams, oxazolidin-2-ones and oxazinan-2-ones containing a

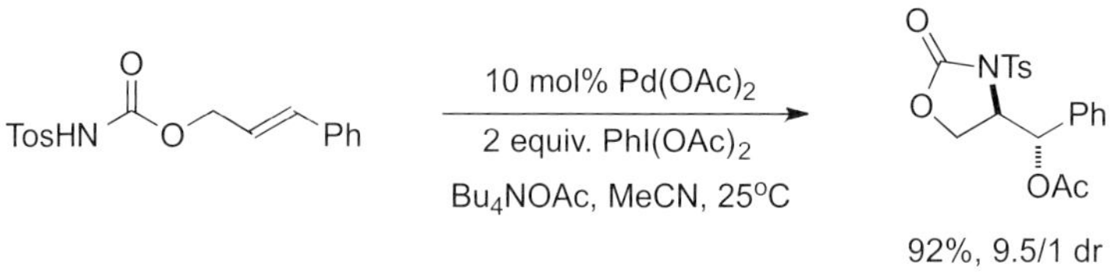
TosHN
O
O
Ph
10 mol% Pd(OAc)2
2 equiv. PhI(OAc)2
Bu4NOAc, MeCN, 25°C
NTs
Ph
OAc
92%, 9.5/1 dr

(4.17)

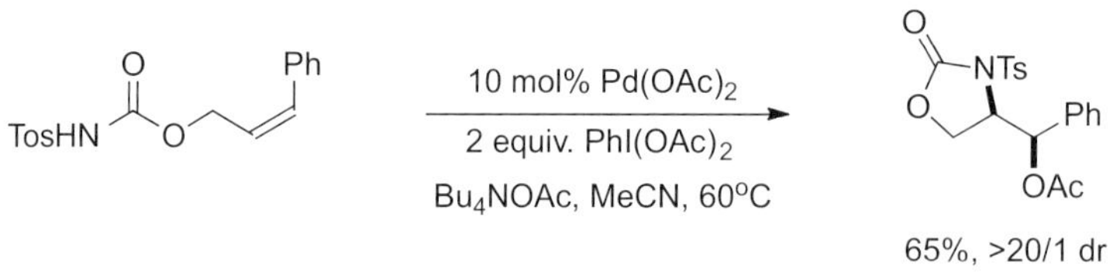
TosHN
O
O
Ph
10 mol% Pd(OAc)2
2 equiv. PhI(OAc)2
Bu4NOAc, MeCN, 60°C
NTs
Ph
OAc
65%, >20/1 dr

(4.18)

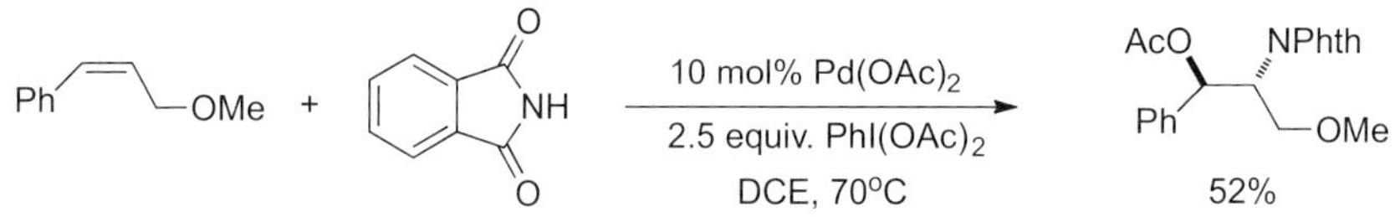
Ph
OMe
+
NH
10 mol% Pd(OAc)2
2.5 equiv. PhI(OAc)2
DCE, 70°C
AcO
NPhth
Ph
OMe
52%

(4.19)

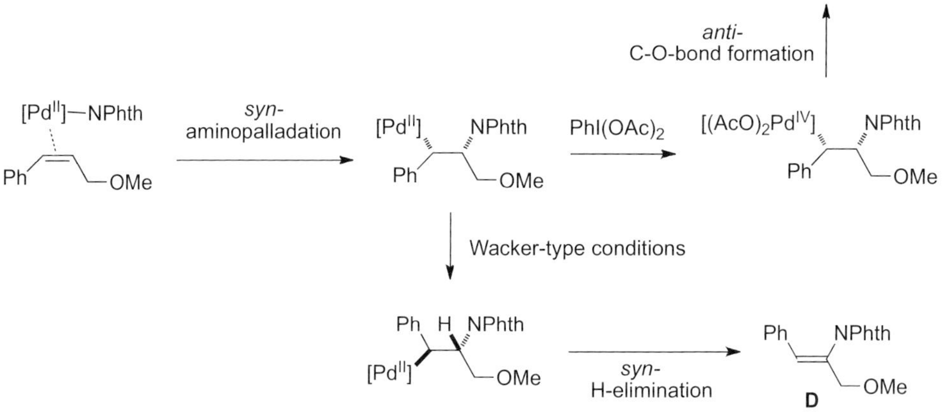
anti-
C-O-bond formation
[Pd^II]—NPhth
Ph
OMe
syn-
aminopalladation
[Pd^II]
NPhth
PhI(OAc)2
[(AcO)2Pd^IV]
Wacker-type conditions
Ph H NPhth
[Pd^II]
syn-
H-elimination
D

Scheme 4.7 Development of intra- and inter-molecular aminoacetoxylation reactions and mechanistic clarification (PhthNH = phthalimide).

vicinal acetoxy group. The reaction proceeds through initial aminopalladation and requires tosylamides or *N*-tosyl carbamates as nitrogen substituents. As in the previously discussed case of dioxygenation, the intermediary 2-aminoalkyl-palladium(II) complex is supposedly oxidized to a palladium(IV) species by iodosobenzene diacetate. Although the reaction medium has to be carefully optimized according to the substrate, it may be assumed that in all cases, the product arises from reductive acetoxylation of the palladium(IV) intermediate. Another important featureis that the reaction is not restricted to terminal alkene precursors. For example, a (*Z*)-configured cinnamyl derivative underwent clean aminoacetoxylation with complete preservation of stereoinformation, while the corresponding pure

(*E*)-isomer gave the corresponding diastereomeric product in 65% yield (Eqs. (4.17) and (4.18)).

The observed stereochemistry of this reaction outcome prompted a detailed mechanistic investigation by Stahl within a strictly intermolecular aminoacetoxylation variant, as shown for one example in Eq. (4.19) [30]. This work clarified stereochemical aspects on the basis of extensive mechanistic investigation, including the induced formation of enamide product **D** under Wacker-type conditions, the authors concluded an overall sequence of *syn*-aminopalladation and *anti*-C–O bond formation to explain the product stereochemistry. This final result agrees well for both the intra- and the inter-molecular case.

Additional aminoalkoxylation reactions were observed for certain reactions with guanidines and sulfamides (Scheme 4.8) [30, 31]. These reactions follow the earlier

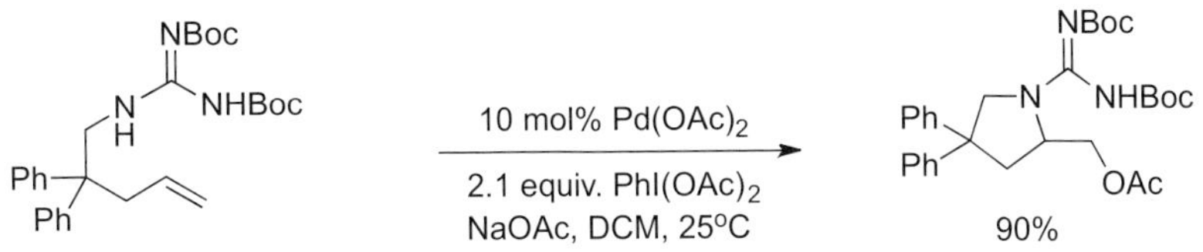

NBoc, NHBoc, N H, Ph, Ph → 10 mol% $Pd(OAc)_2$; 2.1 equiv. $PhI(OAc)_2$, NaOAc, DCM, 25°C → NBoc, N, NHBoc, Ph, Ph, OAc; 90% (4.20)

H, $N-SO_2NHCO_2tBu$ → $Pd(OAc)_2$ (10 mol%); 2.0 eq. $PhIX_2$, 1.0 eq. base, DCM, RT → O O S, N, $NHCO_2tBu$, X; 60-86% yield (4.21)

$PhIX_2$	Base	Product
$PhI(OAc)_2$	NaOAc	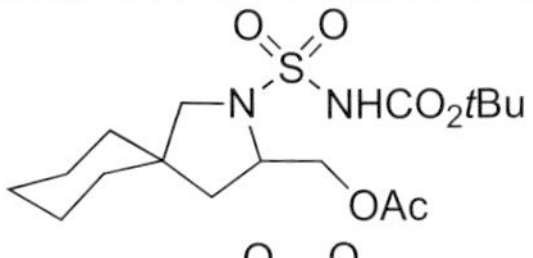 X = OAc
$PhI(OAc)_2$	NaO_2CtBu	X = OAc
$PhI(O_2CCD_3)_2$	NaOAc	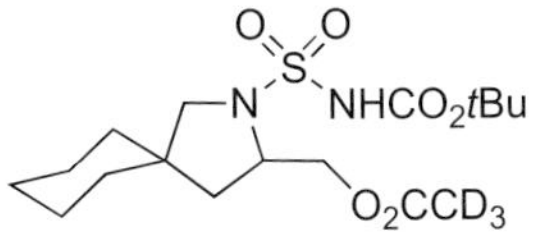 X = O_2CCD_3
$PhI(O_2CtBu)_2$	NaOAc	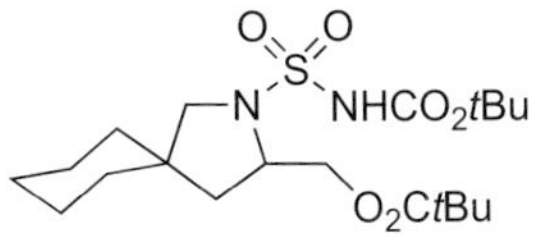 X = O_2CtBu
$PhI(O_2CtBu)_2$	NaO_2CtBu	X = O_2CtBu

Scheme 4.8 Additional aminoacetoxylation reactions employing iodosobenzene diacetate as oxidant and acetoxy source.

pathways uncovered by Sorensen and Stahl. For the latter case of sulfamates, a detailed investigation aimed to clarify the question of the exact acetate source in this kind of reaction. To this end, a series of combinations of iodosobenzene oxidants bearing different anions and anion bases convincingly demonstrated that the anions from the oxidant end up in the product. As the iodosobenzene oxidants are considered to generate a palladium(IV) intermediate, it can be concluded that the anion for the final alkyl alkoxylation stems from the coordination sphere of the palladium(IV) intermediate [32]. In addition to this observation, an experiment with deuterium-labeled alkene demonstrated that this final step proceeds with inversion of configuration following the overall mechanistic pathway from Stahl.

A somewhat different approach was pursued by Sanford, who made use of an intermolecular aminopalladation with phthalimide as nitrogen source (Scheme 4.9) [33]. Again, this reaction follows the *syn*-addition pathway of Stahl and, after coordination of the alcohol from the substrate and upon oxidation of the palladium intermediate, a six-membered ring chelate at palladium(IV) **E** was suggested. This compound undergoes direct *syn*-reductive elimination to arrive at the observed overall stereochemistry for the THF product. For chiral homoallylic substrates, the initial aminopalladation proceeds with *anti*-diastereoselectivity. Such substrate-directing behavior had already been discussed by Stahl in his aminopalladation reactions [30].

An additional reaction with transfer of nitrogen and oxygen to alkenes was also observed as a side reaction in certain diamination reactions with ureas. It is discussed within the following section.

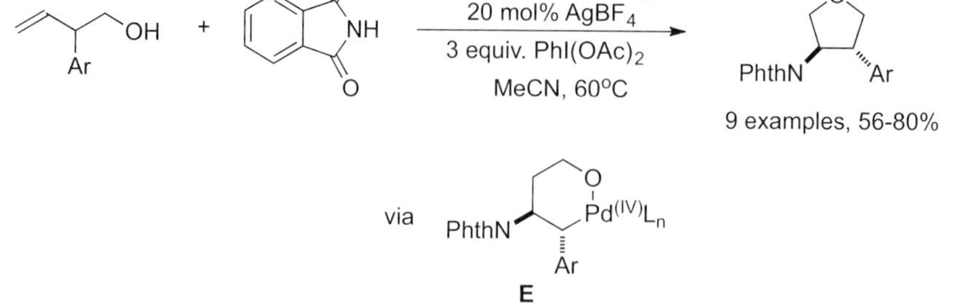

10 mol% $Pd(OAc)_2$, 20 mol% $AgBF_4$, 3 equiv. $PhI(OAc)_2$, MeCN, 60°C (4.22)

10 mol% $Pd(OAc)_2$, 20 mol% $AgBF_4$, 3 equiv. $PhI(OAc)_2$, MeCN, 60°C; 9 examples, 56-80% (4.23)

via $Pd^{(IV)}L_n$ **E**

Scheme 4.9 Aminoalkoxylation reactions to form 3-aminated tetrahydrofurane products.

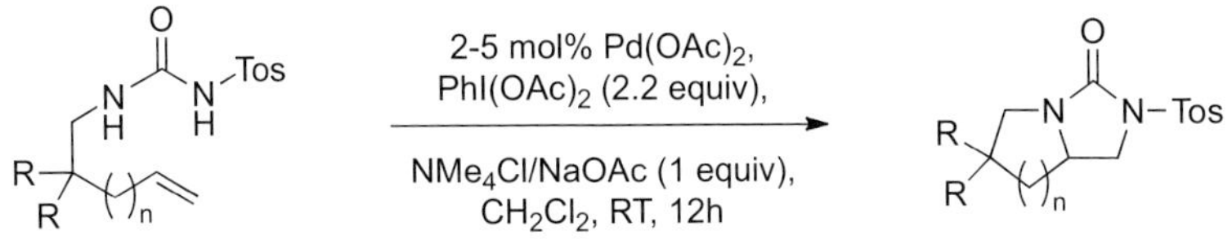

$$\text{urea substrate} \xrightarrow[\text{NMe}_4\text{Cl/NaOAc (1 equiv), CH}_2\text{Cl}_2\text{, RT, 12h}]{\text{2-5 mol\% Pd(OAc)}_2\text{, PhI(OAc)}_2\text{ (2.2 equiv),}} \text{cyclic urea (N-Tos)} \quad (4.24)$$

R = H, Me, Ph, $R_2 = (CH_2)_5$

Scheme 4.10 Cyclic ureas from intramolecular diamination of alkenes, $n = 1, 2$.

4.6
Diamination Reactions

Vicinal diamines are accessible from a consecutive transfer of two nitrogen atoms to alkenes (Scheme 4.10) [34]. This type of reaction was first investigated as an intramolecular variant employing urea groups as a tethered nitrogen source. When terminal alkene substrates were employed in such a reaction, nitrogen transfer took place exclusively giving the corresponding diamines in high yields (Eq. (4.24)). The reaction makes use of iodobenzene diacetate as oxidant and allows the general synthesis of both five- and six-membered ring annelation products.

The overall process of this transformation was the subject of a detailed study [32]. It revealed that the reaction is reminiscent of the stereochemical pathways outlined above for aminoacetoxylation reactions, hence consisting of a *syn*-aminopalladation with an *anti*-C–N bond formation as the second step. The first step was found to be rate-determining, while the final one was suggested to involve a palladium(IV) catalyst state.

This same two-step diamination process was employed in the palladium-catalyzed diamination of internal alkenes (Scheme 4.11) [35]. Here, 2,2′-diamino stilbene derivatives were converted into the corresponding diamines with complete diastereoselectivity (Eq. (4.25)). This stereochemical outcome was rationalized by the assumption of an initial amide–palladium interaction, which is unproductive regarding a possible *syn*-aminopalladation as it would generate an azetidine intermediate. Instead, *trans*-aminopalladation takes place, followed by rapid oxidation to an η^1-coordinated benzylic palladium(IV) with preservation of the relative stereochemistry. *Anti*-C-N-bond formation then displaces palladium(II) to close the catalytic cycle and to displace the C_2-symmetric diamine product.

A second set of diamination reactions was developed by changing the oxidant to copper(II) salts (Scheme 4.12) [36]. These reaction conditions proved successful for the conversion of ureas and allowed significant reduction of the catalyst loading in the case of the six-membered ring annulated products (Eq. (4.26)). Copper salt oxidation also yielded conditions for the diamination of internal alkenes (Eq. (4.27)). These reactions yield *trans*-diamines from (*E*)-alkenes and the overall pathway appears to proceed via an aminobromination where the halogen stems from the copper oxidant. The subsequent S_N2-ring closure proceeds with clean inversion of configuration, although it is not chemoselective. Hence, diamine formation is accompanied by

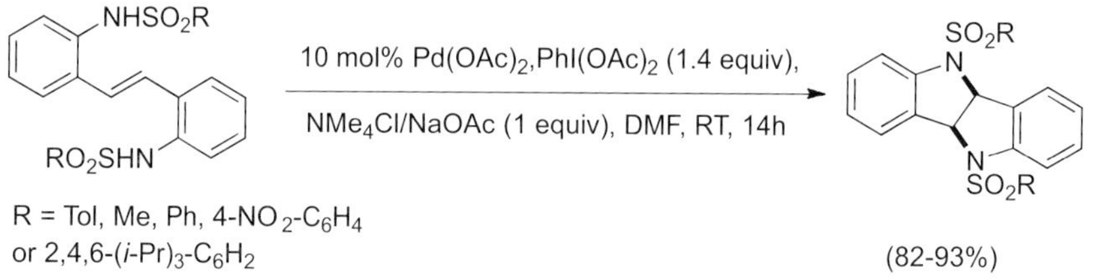

$NHSO_2R$; RO_2SHN

10 mol% $Pd(OAc)_2$, $PhI(OAc)_2$ (1.4 equiv), NMe_4Cl/NaOAc (1 equiv), DMF, RT, 14h

SO_2R N ; N SO_2R

R = Tol, Me, Ph, 4-NO_2-C_6H_4 or 2,4,6-(i-Pr)$_3$-C_6H_2

(82-93%) (4.25)

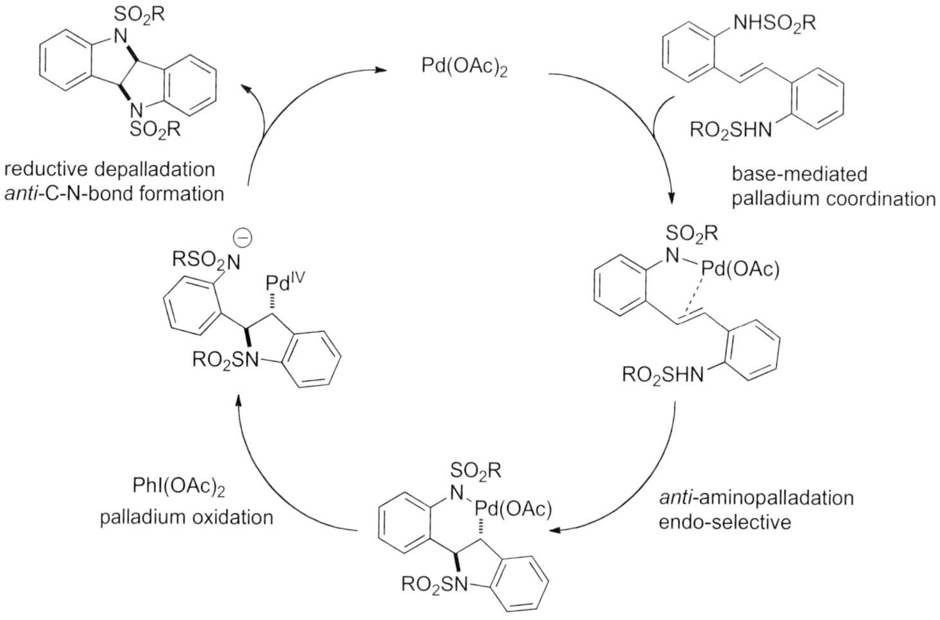

$Pd(OAc)_2$

base-mediated palladium coordination

$anti$-aminopalladation endo-selective

$PhI(OAc)_2$ palladium oxidation

reductive depalladation $anti$-C-N-bond formation

$RSO_2N^{\ominus}$; Pd^{IV} ; RO_2SN ; SO_2R N $Pd(OAc)$; RO_2SHN

Scheme 4.11 Diamination of internal alkenes.

varying amounts of isoureas, which formally constitute aminooxygenation products (Eq. (4.28)).

The intramolecular diamination of alkenes could be further extended to the synthesis of cyclic guanidines (Eq. (4.29)) [31]. In this case, copper chloride represents the optimum oxidant and the reaction proceeds equally well for both five and six-membered annelation products.

Finally, the combination of copper bromide as oxidant and ureas as nitrogen source enabled an intramolecular palladium catalyzed diamination of acrylates (Scheme 4.13) [37]. For methyl or ethyl esters, this process proceeds with excellent diastereoselectivity in favor of the *syn*-configured cyclic urea product

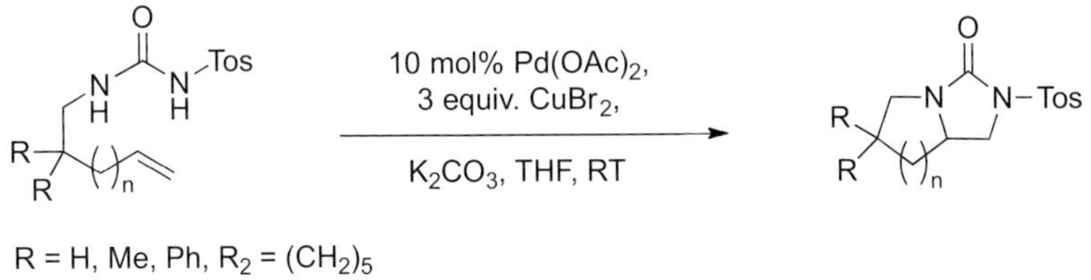

10 mol% $Pd(OAc)_2$, 3 equiv. $CuBr_2$, K_2CO_3, THF, RT (4.26)

R = H, Me, Ph, $R_2 = (CH_2)_5$

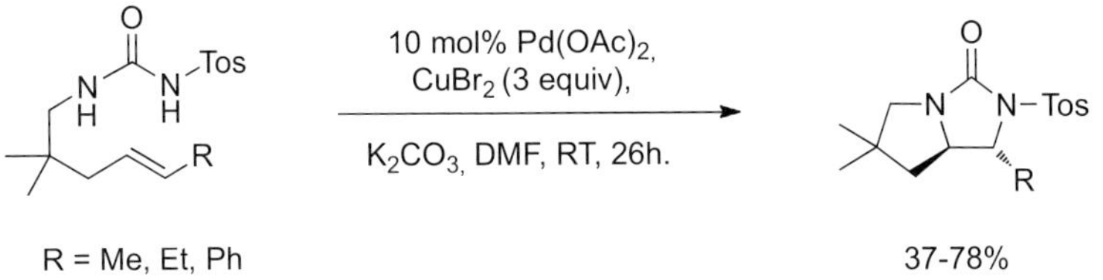

10 mol% $Pd(OAc)_2$, $CuBr_2$ (3 equiv), K_2CO_3, DMF, RT, 26h. (4.27)

R = Me, Et, Ph

37-78%

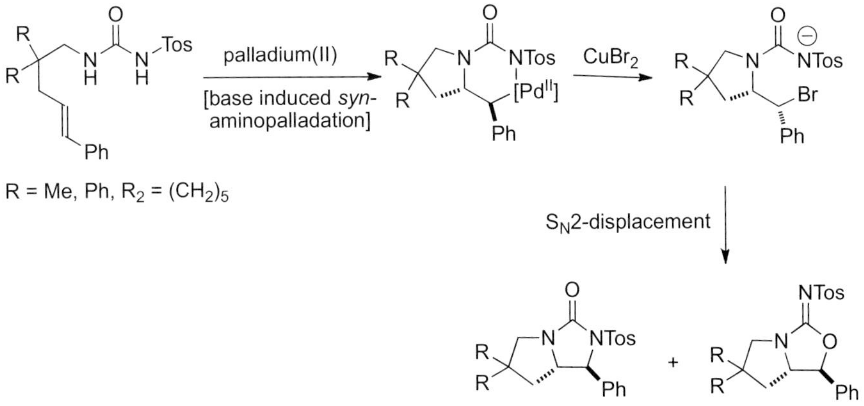

palladium(II) [base induced *syn*-aminopalladation]; $CuBr_2$; S_N2-displacement (4.28)

R = Me, Ph, $R_2 = (CH_2)_5$

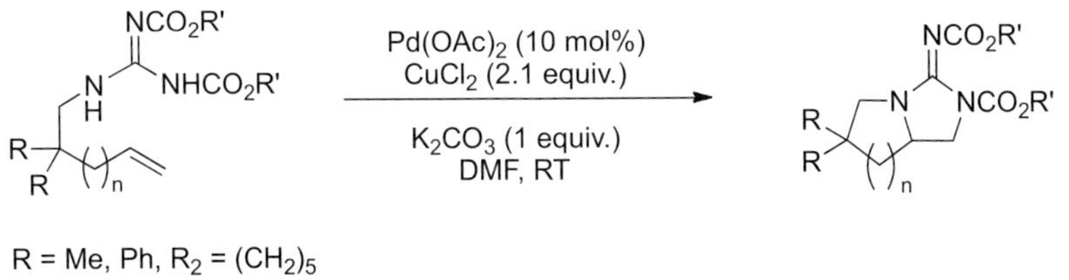

$Pd(OAc)_2$ (10 mol%), $CuCl_2$ (2.1 equiv.), K_2CO_3 (1 equiv.), DMF, RT (4.29)

R = Me, Ph, $R_2 = (CH_2)_5$

Scheme 4.12 Palladium-catalyzed diamination reactions employing copper oxidants, $n = 1, 2$.

(Eq. (4.30)). The reaction appears to follow a sequence of *syn*-aminopalladation and direct *anti*-C–N-bond closure, opposite to that in Eq. (4.28). The versatility of these diamination products as building blocks was briefly demonstrated upon conversion of one of the products into the natural alkaloid absouline within a series of steps.

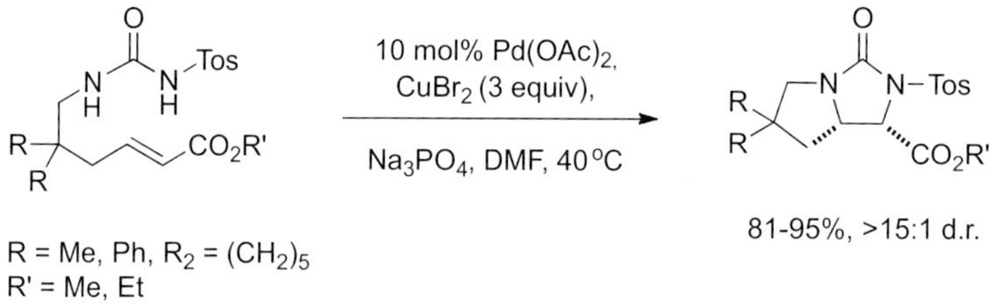

10 mol% $Pd(OAc)_2$, $CuBr_2$ (3 equiv), Na_3PO_4, DMF, 40 °C (4.30)

R = Me, Ph, $R_2 = (CH_2)_5$
R' = Me, Et

81-95%, >15:1 d.r.

(+/-)-absouline

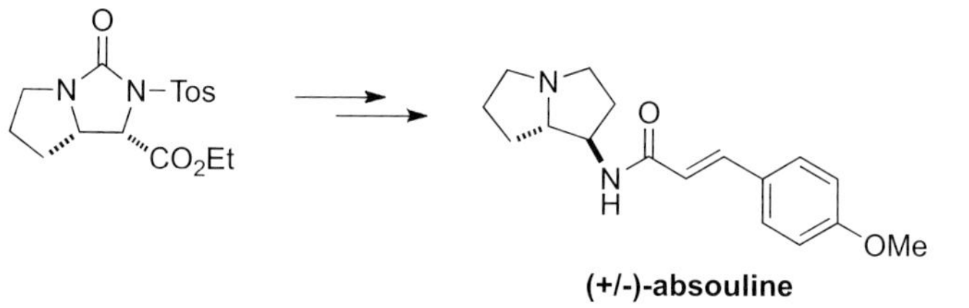

Scheme 4.13 Palladium-catalyzed diamination of acrylates.

4.7 Conclusion

Although a rather new field of research, palladium-catalyzed 1,2-difunctionalization reactions have recently gained momentum. Major advances in this area have introduced suitable reactivity to incorporate chloride, bromide, oxygen and nitrogen-based nucleophiles into carbon frameworks. The common key step of all these transformations relies on the inter- or intra-molecular functionalization of intermediary alkyl–palladium bonds, for which several methods have been developed. The potential of all these approaches for organic synthesis is obvious, and the future looks bright for further developments in this field.

References

1 (a) Tsuji, J. (2004) *Palladium Reagents and Catalysts*, Wiley, New York; (b)Bäckvall, J.-E. (1983) *Acc. Chem. Res.*, **16**, 335.

2 Coleman, J.L., Hegedus, L.S., Noton, J.R., and Finke, R.G. (1987) *Principles and Applications of Transition-Metal Chemistry*, University Science Books, Mill Valley.

3 For the concept of palladium(IV) chemistry: Canty, A.J. (1992) *Acc. Chem. Res.*, **25**, 83.

4 For a review on Pd(IV) catalysis: Muñiz, K. (2009) *Angew. Chem. Int. Ed.*, **48**, 9412.

5 For the conceptual introduction of transient palladium oxidation with $CuCl_2$: (a) Stangl, H. and Jira, R. (1970) *Tetrahedron Lett.*, **11**, 3589; (b) Hamed, O. and Henry, P.M. (1998) *Organometallics*, **17**, 5184.

6 For leading recent work on 1,2-difunctionalization of butadienes: (a) Bar, G.L.J., Lloyd-Jones, G.C., and Booker-Milburn, K.I. (2005) *J. Am. Chem. Soc.*, **127**, 7308; (b) Du, H., Zhao, B., and Shi, Y. (2007) *J. Am. Chem. Soc.*, **129**, 762; (c) Du, H., Yuan, W., Zhao, B., and Shi, Y. (2007) *J. Am. Chem. Soc.*, **129**, 7496; (d) Du, H., Yuan, W., Zhao, B., and Shi, Y. (2007) *J. Am. Chem. Soc.*, **129**, 11688; (e) Yuan, W., Du, H., Zhao, B., and Shi, Y. (2007) *Org. Lett.*, **9**, 2589; (f) Xu, L., Du, H., and Shi, Y. (2007) *J. Org. Chem.*, **72**, 7038;

(g) Xu, L. and Shi, Y. (2008) *J. Org. Chem.*, **73**, 749.

7 For leading recent work on 1,4-difunctionalization of butadienes: Bäckvall, J.E. (2004) in *Metal-Catalyzed Cross-Coupling Reactions* (eds A. de Meijere and F. Diederich), Wiley-VCH, Weinheim, pp. 479–529.

8 For leading references, see: (a) Yip, K.-T., Yang, M., Law, K.-L., Zhu, N.-Y., and Yang, D. (2006) *J. Am. Chem. Soc.*, **128**, 3130; (b) Welbes, L.L., Lyons, T.W., Cychosz, K.A., and Sanford, M.S. (2007) *J. Am. Chem. Soc.*, **129**, 5836; (c) Tong, X., Beller, M., and Tse, M.K. (2007) *J. Am. Chem. Soc.*, **129**, 4906; (d) Scarborough, C.C. and Stahl, S.S. (2006) *Org. Lett.*, **8**, 3251.

9 Henry, P.M. (ed.) (1980) *Palladium Catalyzed Oxidation of Hydrocarbons*, Reidel, Dordrecht.

10 (a) El-Qisairi, A., Hamed, O., and Henry, P.M. (1998) *J. Org. Chem.*, **63**, 2790; (b) El-Qisairi, A. and Henry, P.M. (2000) *J. Organomet. Chem.*, **603**, 50; (c) El-Qisairi, A., Qaseer, H., and Henry, P.M. (2002) *J. Organomet. Chem.*, **656**, 167.

11 El-Qisairi, A., Qaseer, H., Katsigras, G., Lorenzi, P., Trivedi, U., Tracz, S., Hartman, A., Miller, J.A., and Henry, P.M. (2003) *Org. Lett.*, **5**, 439.

12 Minatti, A. and Muñiz, K. (2007) *Chem. Soc. Rev.*, **36**, 1142.

13 Tamaru, Y., Hojo, M., Higashimura, H., and Yoshida, Z.-i. (1988) *J. Am. Chem. Soc.*, **110**, 3994.

14 Manzoni, M.R., Zabawa, T.P., Kasi, D., and Chemler, S.R. (2004) *Organometallics*, **23**, 5618.

15 Lei, A., Lu, X., and Liu, G. (2004) *Tetrahedron Lett.*, **45**, 1785.

16 Szolcsányi, P. and Gracza, T. (2006) *Tetrahedron*, **62**, 8498.

17 Michael, F.E., Sibbald, P.A., and Cochran, B.M. (2008) *Org. Lett.*, **10**, 793.

18 Michael, F.E. and Cochran, B.M. (2006) *J. Am. Chem. Soc.*, **128**, 4246.

19 Helaja, J. and Göttlich, R. (2002) *Chem. Commun.*, 720.

20 Antunes, A.M.M., Marto, S.J.L., Branco, P.S., Prabhakar, S., and Lobo, A.M. (2001) *Chem. Commun.*, 405.

21 (a) Antunes, A.M.M., Bonifacio, V.D.B., Nascimento, S.C.C., Lobo, A.M., Branco, P.S., and Prabhakar, S. (2007) *Tetrahedron*, **30**, 7009; (b) For a strictly stoichiometric palladium-assisted aziridination: Bäckvall, J.E. (1977) *J. Chem. Soc., Chem. Commun.*, 413.

22 Han, J., Li, Y., Zhi, S., Pan, Y., Timmons, C., and Li, G. (2006) *Tetrahedron Lett.*, **47**, 7225.

23 Wu, T., Yin, G., and Liu, G. (2009) *J. Am. Chem. Soc.*, **131**, 16354.

24 Chevrin, C., Le Bras, J., Hénin, F., and Muzart, J. (2005) *Synthesis*, 2615.

25 Thiery, E., Chevrin, C., Le Bras, J., Harakat, D., and Muzart, J. (2007) *J. Org. Chem.*, **72**, 1859.

26 Schultz, M.J. and Sigman, M.S. (2006) *J. Am. Chem. Soc.*, **128**, 1460.

27 Zhang, Y. and Sigman, M.S. (2007) *J. Am. Chem. Soc.*, **129**, 3076.

28 Li, Y., Song, D., and Dong, V.M. (2008) *J. Am. Chem. Soc.*, **130**, 2962.

29 Alexanian, E.J., Lee, C., and Sorensen, E.J. (2005) *J. Am. Chem. Soc.*, **127**, 7690.

30 Liu, G. and Stahl, S.S. (2006) *J. Am. Chem. Soc.*, **128**, 7179.

31 Hövelmann, C.H., Streuff, J., Brelot, L., and Muñiz, K. (2008) *Chem. Commun.*, 2334.

32 Muñiz, K., Hövelmann, C.H., and Streuff, J. (2008) *J. Am. Chem. Soc.*, **130**, 763.

33 Desai, L.V. and Sanford, M.S. (2007) *Angew. Chem. Int. Ed.*, **46**, 5737.

34 Streuff, J., Hövelmann, C.H., Nieger, M., and Muñiz, K. (2005) *J. Am. Chem. Soc.*, **127**, 14587.

35 Muñiz, K. (2007) *J. Am. Chem. Soc.*, **129**, 14542.

36 Muñiz, K., Hövelmann, C.H., Campos-Gómez, E., Barluenga, J., González, J.M., Streuff, J., and Nieger, M. (2008) *Chem. Asian J.*, **3**, 776.

37 Muñiz, K., Streuff, J., Chávez, P., and Hövelmann, C.H. (2008) *Chem. Asian J.*, **3**, 1245.

5
Rhodium-Catalyzed C–H Aminations

Hélène Lebel

The formation of C–N bonds is an important transformation in organic synthesis, as the amine functionality is found in numerous natural products and plays a key role in many biologically active compounds [1]. Standard catalytic methods to produce C–N bonds involve functional group manipulations, such as reductive amination of carbonyl compounds [2], addition of nucleophiles to imines [3], hydrogenation of enamides [4–8], hydroamination of olefins [9] or a C–N coupling reaction [10, 11]. Recently, the direct and selective introduction of a nitrogen atom into a C–H bond via a metal nitrene intermediate has appeared as an attractive alternative approach for the formation of C–N bonds [12–24].

Copper-catalyzed decomposition of benzenesulfonyl azide in the presence of cyclohexene was the first reported evidence of a metal-catalyzed nitrene insertion reaction [25]. This seminal discovery was then followed by the pioneering work of Breslow and Gellman who introduced the use of iminoiodinanes as metal nitrene precursors as well as rhodium dimer complexes as catalysts [26, 27]. They showed the formation of the corresponding benzosultam in 86% yield in the presence of rhodium (II) acetate dimer ($Rh_2(OAc)_4$) via an intramolecular metal nitrene C–H bond insertion reaction (Eq. (5.1)).

$Rh_2(OAc)_4$ (5 mol%)
MeCN, rt
86%
(5.1)

Müller then intensively studied rhodium dimer complexes as catalysts for the intermolecular amination of alkanes using iminoiodinanes, typically PhI=NTs and PhI=NNs [28–30]. Good yields were obtained for amination of benzylic, allylic and tertiary C–H bonds, which are the most reactive C–H bonds towards metal nitrenes. For instance the amination of indane provided the corresponding benzylic amine in

Catalyzed Carbon-Heteroatom Bond Formation. Edited by Andrei K. Yudin

ISBN: 978-3-527-32428-6

69% yield (Eq. (5.2)). However, an excess of the alkane substrates is required to minimize the competitive degradation of the iminoiodinane reagent.

(20 equiv) $Rh_2(OAc)_4$ (2 mol%), PhI=NNs (1 equiv), MS 4Å/CH_2Cl_2 → NHNs 69% (5.2)

High yields are obtained with iminoiodinane reagents, however, they cannot be stored and must be used immediately [15]. Moreover, they are tedious to purify, as they form polymeric structures and are insoluble in most organic solvents. Besides tosyl- and nosyl-substituted iminoiodinanes, others are almost impossible to isolate in pure form. It is to overcome these problems that one-pot processes in which the iminoiodinane species is prepared *in situ* rather than being isolated, have been developed. Che first demonstrated such a reaction for an intermolecular C–H insertion (Eq. (5.3)) [31, 32]. The reactive metal nitrene species is generated from a mixture of the requisite amine with diacetoxyiodobenzene, in the presence of manganese porphyrin catalyst.

1 equiv. $TsNH_2$ (1.5 equiv.), $PhI(OAc)_2$ (1.25 equiv.), Mn^{III}(TPFPP)Cl (1 mol%), DCM, 40 °C → NHTs 79% (5.3)

There are numerous advantages to this approach in addition to avoiding complex purification procedures. For instance, it is now possible to further diversify the nitrogen substituent of the reagent, which has a profound impact on the practicality of C–H amination. In 2001, Du Bois and Espino first reported the intramolecular C–H insertion reaction with carbamates and diacetoxyiodobenzene in the presence of rhodium (II) acetate or rhodium (II) triphenylacetate dimer and magnesium oxide [33]. The corresponding oxazolidinone was isolated in high yields (Eq. (5.4)).

O, O, NH_2 — $PhI(OAc)_2$ (1.4 equiv), $Rh_2(O_2CR)_4$ (5 mol%), MgO (2.3 equiv), DCM, 40 °C → O, HN, O; $Rh_2(OAc)_4$, 77%; $Rh_2(tpa)_4$, 79% (5.4)

This discovery has tremendously expanded the field of amination reactions, not only to perform intramolecular, but also intermolecular C–H insertion reactions. Today this approach is the most commonly used to prepare metal nitrenes, especially those derived from rhodium catalysts. *N*-Tosyloxycarbamates are alternative rhodium nitrene precursors, which do not require the use of a hypervalent iodine reagent [34]. Azides [17, 35] and haloamines [36–38] have also been reported, but usually with other catalysts than rhodium complexes.

Rhodium dimer complexes are the most widely used catalysts to perform C–H aminations. Other metal complexes that catalyze metal nitrene C–H insertion reactions, include metalloporphyrins [12, 39], silver [21, 40], copper [22], palladium [41] and gold [42] complexes.

A number of reviews have appeared recently on C–H insertion of metal nitrenes [18–20], some of them specifically dedicated to rhodium catalysts [16, 24, 43]. This chapter will focus only on the most recent aspect of rhodium nitrene chemistry to perform C–H amination of alkanes, which has not been covered in previous reviews. The literature has been surveyed from 2001 to September 2008.

5.1 Metal Nitrenes from Iminoiodinanes

Since the first report by Du Bois and Espino on the use of diacetoxyiodobenzene to generate iminoiodinanes *in situ* [33], the field of C–H amination has flourished, expanding the type of products that can be synthesized using this approach. Not only is intramolecular C–H insertion possible, furnishing oxazolidinones, imidazolidin-2-ones, 2-aminoimidazolines and oxathiazinane [44–46], but new reagents have also been developed to perform efficient intermolecular C–H insertions [47]. There are only a few nitrogen substituents that lead to stable metal nitrenes capable of performing C–H amination reactions. Most of the substituents are electron withdrawing and are designed to minimize side reactions. Indeed, alkyl or acyl nitrene precursors are rarely used as they are prone to rearrangement by migration of a group from the adjacent carbon to the electron deficient nitrogen atom to form imines or isocyanates, respectively [48]. Conversely, sulfonyl-, sulfonimidoyl-, sulfonate- and carbamoyl-substituted nitrene precursors favor the formation of the metal nitrenes over degradation.

In situ generation of the iminoiodinane also prevents the use of a large excess of the alkane substrate. Indeed, the main decomposition pathway for metal nitrene is through radical abstraction of hydrogen (via its triplet state), leading back to the formation of the starting amine (Scheme 5.1). Thus, even if the C–H insertion step is slow, it is possible, by using an excess of the hypervalent iodine reagent, to react stoichiometric amounts of alkane and amine substrates.

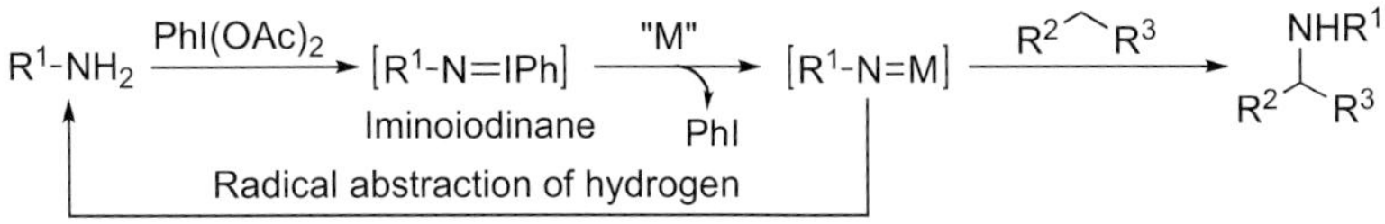

R^1-NH_2 —$PhI(OAc)_2$→ [R^1-N=IPh] (Iminoiodinane) —"M", −PhI→ [R^1-N=M] —R^2⌒R^3→ R^2CH(NHR^1)R^3

Radical abstraction of hydrogen ([R^1-N=M] → R^1-NH_2)

Scheme 5.1

5.1.1
Intramolecular C–H amination

The intramolecular reaction of a carbamate in the presence of diacetoxyiodobenzene, rhodium (II) acetate dimer and magnesium oxide produces exclusively the corresponding 5-membered ring [33]. The oxazolidinone is also formed stereospecifically (Eq. (5.5)).

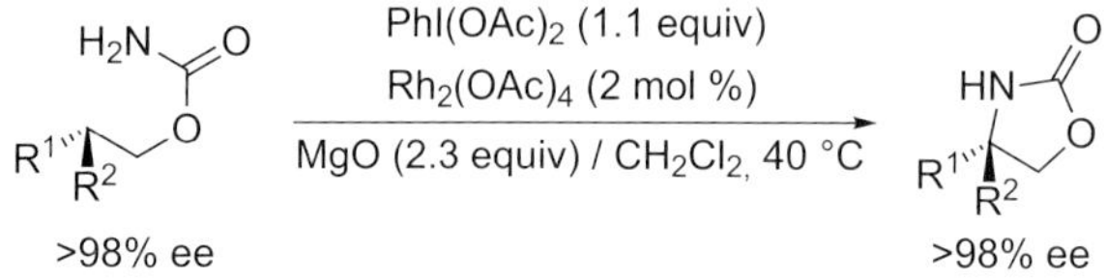

H2N
O
O
R1
R2
>98% ee
PhI(OAc)2 (1.1 equiv)
Rh2(OAc)4 (2 mol %)
MgO (2.3 equiv) / CH2Cl2, 40 °C
HN
O
O
R1
R2
>98% ee

(5.5)

The chemoselectivity of the reaction favors insertion in the most electron-rich C–H bond available to produce an oxazolidinone (Eqs. (5.6) and (5.7)) [33, 49, 50]. When a bicyclic system is formed, the C–H amination reaction leans toward the formation of the cis-isomer. The scope of the reaction is large and many examples of carbamates derived from primary and tertiary alcohols are known. The use of carbamates derived from secondary alcohols is, however, more limited to substrates having conformational bias and/or very reactive C–H bonds (benzylic and allylic) [51, 52]. Otherwise the formation of the corresponding ketone is observed, probably via a hydride shift process, similar to what is observed in C–H bond insertion with metal carbene [53, 54].

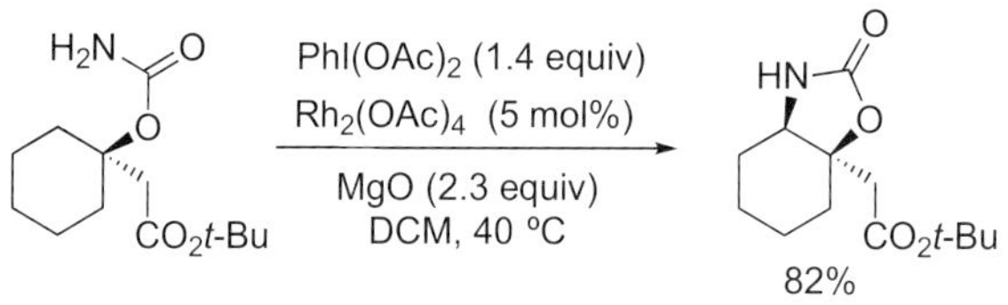

H2N
O
O
CO2t-Bu
PhI(OAc)2 (1.4 equiv)
Rh2(OAc)4 (5 mol%)
MgO (2.3 equiv)
DCM, 40 °C
HN
O
O
CO2t-Bu
82%

(5.6)

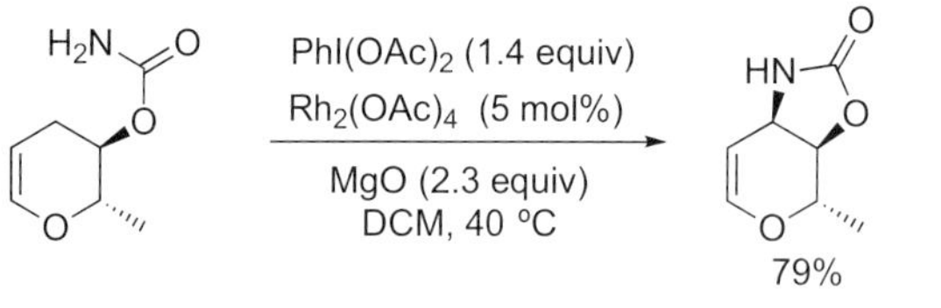

H2N
O
O
O
PhI(OAc)2 (1.4 equiv)
Rh2(OAc)4 (5 mol%)
MgO (2.3 equiv)
DCM, 40 °C
HN
O
O
O
79%

(5.7)

Du Bois and coworkers have extended such a strategy for intramolecular C–H aminations to the synthesis of other 5-membered rings. Whereas amides would have led to rearrangement to form the corresponding isocyanate, urea and guanidine derivatives produce imidazolidin-2-ones and 2-aminoimidazolines, respectively (Eqs. (5.8) and (5.9)) [46]. $Rh_2(esp)_2$, a tethered dicarboxylate-derived rhodium complex is used, as this catalyst displays better reactivity and selectivity [45].

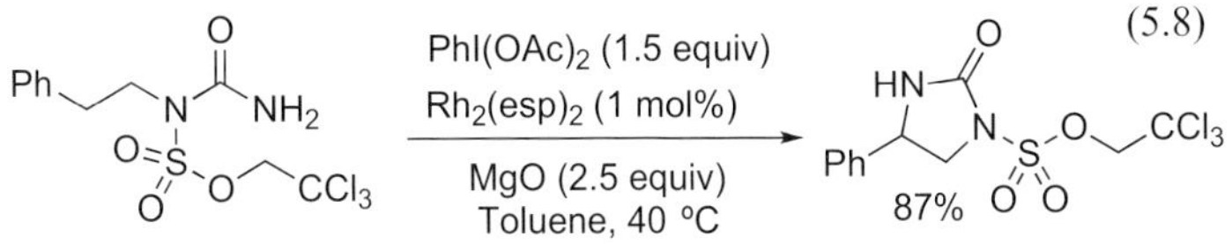

(5.8)

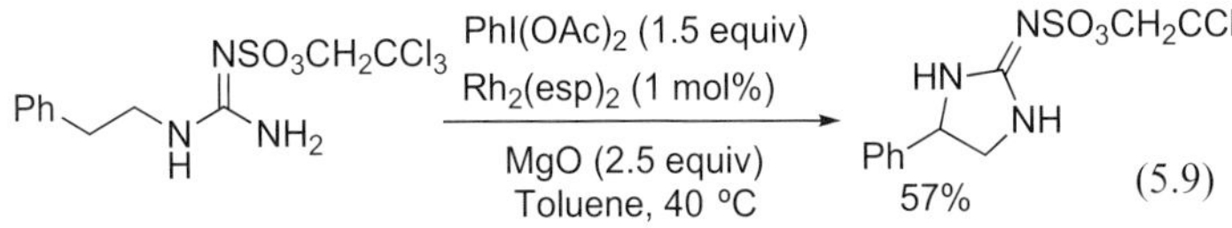

(5.9)

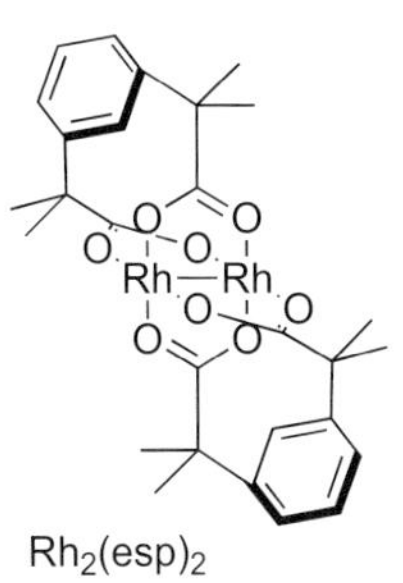

Besides the catalyst, this reaction is also strongly dependent on the nitrogen substituent. For the C–H amination to proceed, an electron-withdrawn 2,2,2-trichloroethoxysulfonyl protecting group appears to be mandatory. The required starting material is easily synthesized from either the 2,2,2-trichloroethoxysulfonyl-protected urea or the corresponding isothiourea or imidochloride (Schemes 5.2 and 5.3).

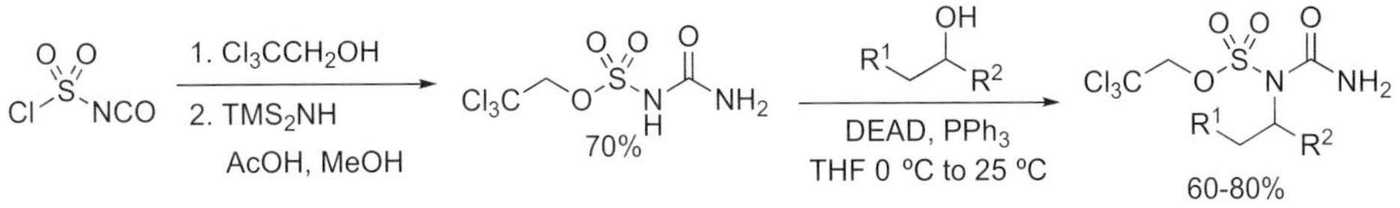

Scheme 5.2

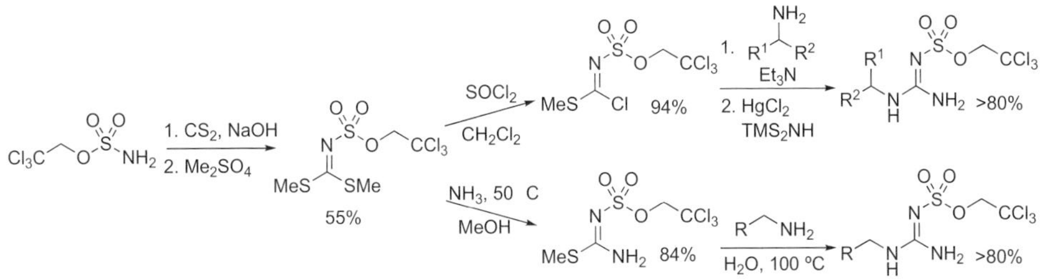

Scheme 5.3

The synthesis of oxazolidinones using a C–H insertion reaction of carbamates has been used in the total synthesis of natural products. For instance, the oxazolidinone of callipeltoside A was recently prepared according to this procedure (Scheme 5.4) [55, 56]. The use of the C–H amination has provided the shortest synthesis of the sugar moiety of this natural product. A few other amino sugars have been prepared using such an approach [49, 50]. A few other oxazolidinone-containing biologically active products, such as (–)-cytoxazone, were prepared

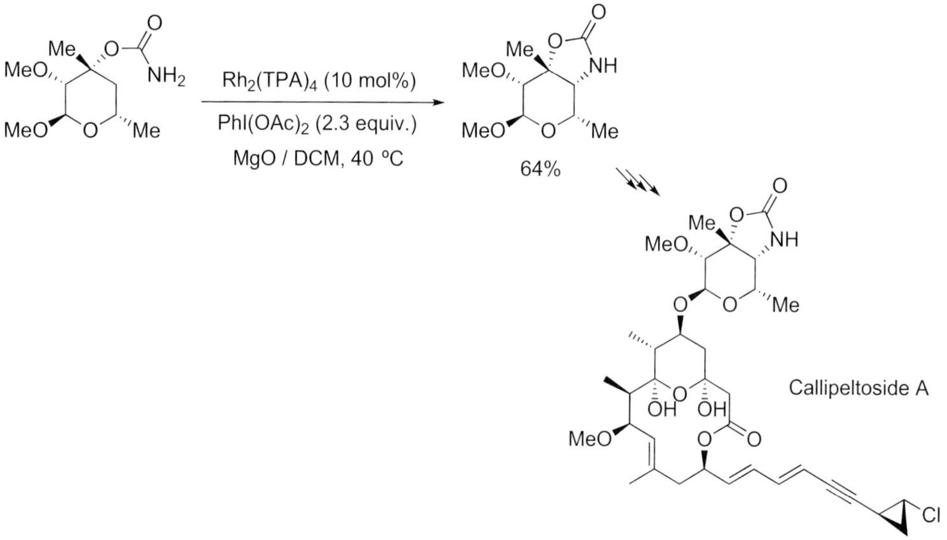

$Rh_2(TPA)_4$ (10 mol%)
$PhI(OAc)_2$ (2.3 equiv.)
MgO / DCM, 40 °C
64%
Callipeltoside A

Scheme 5.4

using C–H amination reactions [57, 58]. Furthermore, it can also be used for the preparation of chiral 1,2-amino alcohols found in other biologically active molecules [59–63]. In all cases, the reaction was successful, as long as the carbamate was derived from a primary or a tertiary alcohol.

Intramolecular C–H amination can also be used to install nitrogen functionality stereospecifically in a very complex substrate, such as a late precursor of (–)-tetrodotoxin (Scheme 5.5) [64]. This example illustrates the power of this method, in which the tethering effect allows the functionalization of a specific C–H bond.

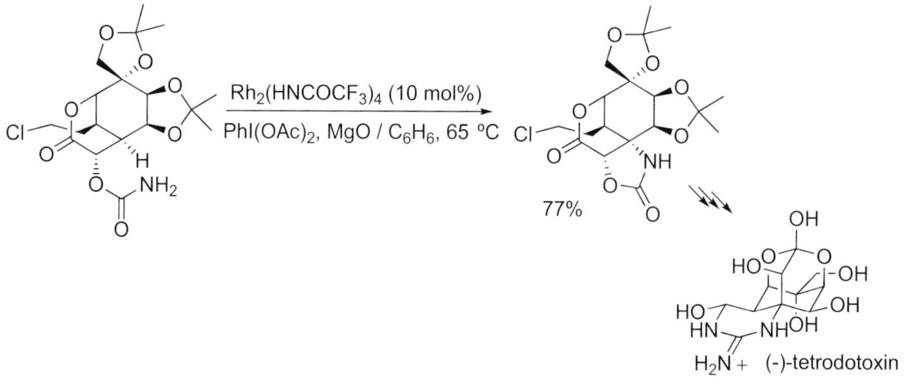

$Rh_2(HNCOCF_3)_4$ (10 mol%)
$PhI(OAc)_2$, MgO / C_6H_6, 65 °C
77%
(-)-tetrodotoxin

Scheme 5.5

C–H amination reactions via oxidation with a hypervalent iodine reagent are also possible with sulfamates [44]. The formation of the 5-membered sulfamidate is only observed if no alternative cyclization pathway is available (Eq. (5.10)).

$PhI(OAc)_2$ (1.1 equiv), $Rh_2(OAc)_4$ (2 mol%), MgO (2.3 equiv), DCM, 40 °C; 60% (5.10)

Otherwise, the corresponding 6-membered ring compound, the oxathiazinane, is favored via a selective γ-C–H insertion reaction favoring the most electron-rich C–H bond (Eq. (5.11)). This strong bias for 6-membered ring formation is presumably due to the elongated S–O and S–N bonds (1.58 Å) and the obtuse N–S–O angle (103°) of the sulfamate. The C–H insertion proceeds in good yields with sulfamates derived from primary, secondary and tertiary alcohols. The reaction is stereospecific and C–H insertion at a chiral center proceeds without racemization (Eq. (5.12)).

$PhI(OAc)_2$ (1.1 equiv), $Rh_2(OAc)_4$ (2 mol%), MgO (2.3 equiv), DCM, 40 °C; 90% (5.11)

OSO_2NH_2, >99% ee; $PhI(OAc)_2$ (1.1 equiv), $Rh_2(OAC)_4$ (2 mol %), MgO (2.3 equiv) / CH_2Cl_2, 40 °C; 91%, >99% ee (5.12)

Good to excellent diastereoselectivities, favoring the *syn*-product are obtained for the rhodium-catalyzed C–H amination with α- and β-substituted sulfamates (Figure 5.1) [65].

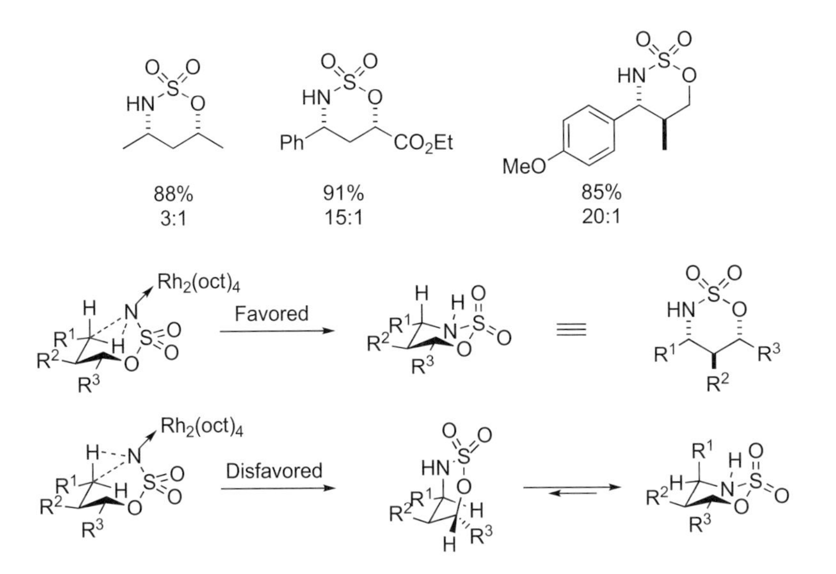

Figure 5.1 Diastereoselective amination of α- and β-substituted sulfamates.

The *syn/anti*-selectivity is influenced by the steric hindrance of the group at the α-position. On the other hand, β-substitution leads to uniformly high stereocontrol.

Chair-like transition states minimizing gauche interactions between the substituents with metal nitrene insertion into the equatorial C–H bond are used to explain the stereochemical outcome. Such an intramolecular C–H amination with sulfamates was used in a few syntheses, including the total synthesis of (+)-saxitoxin [66, 67].

Many chiral rhodium complexes have been studied to perform enantioselective intramolecular benzylic C–H amination with sulfamates. Up to now, it is $Rh_2[(S)\text{-nap}]_4$, recently disclosed by Du Bois and coworkers, that provides the highest enantioselectivities for benzylic C–H amination of sulfamates in the presence of iodosobenzene (Eq. (5.13)) [68]. This catalyst appears to be particularly resistant to the oxidizing reaction conditions. Allylic C–H bond insertions are also possible, although only in moderate yields and ee >80% for *Z*-alkenes.

PhI=O (1.2 equiv)
μS 3Å / DCM
$Rh_2[(S)\text{-nap}]_4$ (2 mol%)
85%
92% ee
(5.13)

5.1.2
Intermolecular C–H Aminations

There have been substantial developments in the design of novel reagents and catalysts for performing intermolecular C–H aminations using hypervalent iodine reagents. The research has been directed towards finding more reactive reagents, which would lead to amines containing a protecting group that is easy to cleave. Trichloroethylsulfamate has been developed for this purpose. In the presence of di-*tert*-butylacetoxyiodobenzene and 2 mol% of $Rh_2(esp)_2$, such a reagent undergoes intermolecular amination of alkanes in good yields (Eq. (5.14)) [47].

$Cl_3CCH_2O_3SNH_2$ (1 equiv)
$PhI(O_2Ct\text{-}Bu)_2$ (2 equiv)
$Rh_2(esp)_2$ (2 mol %)
C_6H_6, 23 °C
70%
(5.14)

The amination of benzylic C–H bonds proceeds in good yields using 1 equiv of amine and alkane in the presence of 2 equiv of the hypervalent iodine reagent. However, insertion in tertiary and secondary C–H bonds requires 5 equiv of substrate to undergo complete conversion (Eq. (5.15)).

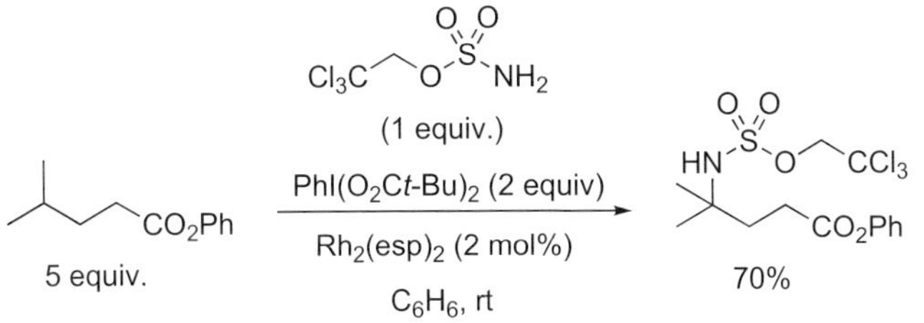

CO_2Ph (5 equiv.) → $Cl_3CCH_2OSO_2NH_2$ (1 equiv.), $PhI(O_2Ct\text{-}Bu)_2$ (2 equiv), $Rh_2(esp)_2$ (2 mol%), C_6H_6, rt → $HNSO_2OCH_2CCl_3$ / CO_2Ph, 70% (5.15)

Not only are higher yields obtained when trichloroethylsulfamate is used (compared to arylsufonylamine), but it also leads to substituted amines that are easier to deprotect, using a mixture of Zn(Cu) and HCl. Such a method has been used in the preparation of ^{15}N-labeled amines (Scheme 5.6).

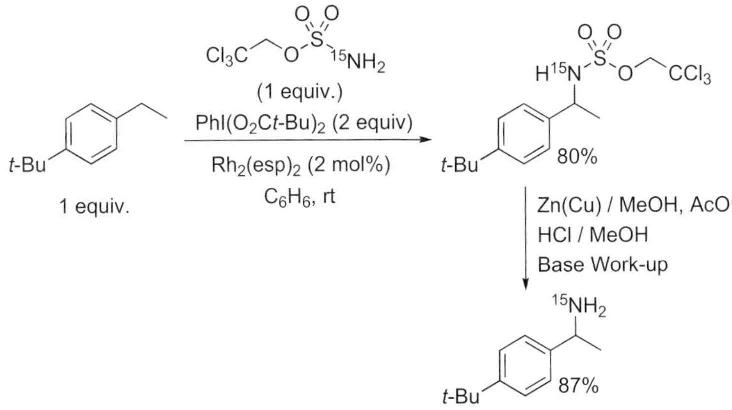

t-Bu (1 equiv.) → $Cl_3CCH_2OSO_2{}^{15}NH_2$ (1 equiv.), $PhI(O_2Ct\text{-}Bu)_2$ (2 equiv), $Rh_2(esp)_2$ (2 mol%), C_6H_6, rt → $H^{15}NSO_2OCH_2CCl_3$, t-Bu, 80%

Zn(Cu) / MeOH, AcOH; HCl / MeOH; Base Work-up → $^{15}NH_2$, t-Bu, 87%

Scheme 5.6

The most general and selective method to perform stereoselective intermolecular C–H amination is based on the use of a chiral sulfonimidamide reagent with $Rh_2[(S)\text{-nta}]_4$ as catalyst [69, 70]. Muller, Dodd, Dauban and coworkers have developed such a reagent. Equimolar amounts of alkane and nitrogen precursor are used in the presence of 1.4 equiv of di-*tert*-butylacetoxyiodobenzene and 3 mol % of the rhodium catalyst. The reaction is typically run at −35 °C for 3 days. Benzylic amines are obtained in 14 : 1 to 200 : 1 dr and allylic amines are formed in 3 : 1 to 19 : 1 dr; both products are produced in good to excellent yields (Eqs. (5.16) and 5.17).

O NTs S NH$_2$ (1.2 equiv.)
PhI(O$_2$C*t*-Bu)$_2$ (1.4 equiv.)
Rh$_2$[(*S*)-nta]$_4$ (3 mol%)
DCE/MeOH, 3/1, -35 °C
1 equiv.
TsN O S HN *p*-Tol
86%
99:1 dr

(5.16)

O N Me O O O Rh—Rh 4
Rh$_2$[(*S*)-nta]$_4$

O NTs S NH$_2$ (1.2 equiv.)
PhI(O$_2$C*t*-Bu)$_2$ (1.4 equiv.)
Rh$_2$[(*S*)-nta]$_4$ (3 mol%)
DCE/MeOH, 3/1, -35 °C
1 equiv.
TsN O S HN *p*-Tol
90%
19:1 dr

(5.17)

With a stoichiometric amount of the cycloalkane, the corresponding C–H insertion products are isolated in 48–83% yield, while the use of a fivefold excess of substrate leads to yields from 85–96%. It is possible to cleave the sulfonimidoyl moiety without epimerization using sodium naphthalenide. More conveniently, one can perform a Boc protection followed by sulfonimidoyl removal with magnesium in methanol under sonication.

It is not necessary to use the enantiopure chiral sulfonimidamide, as a kinetic resolution of racemic sulfonimidamides (2 equiv) in the presence of only 1 equiv of PhI(OCO*t*-Bu)$_2$ is operative and provides the desired product with a yield greater than 50% (Eq. (5.18)).

O NTs S NH$_2$ (2.2 equiv.)
PhI(O$_2$C*t*-Bu)$_2$ (1.1 equiv.)
Rh$_2$[(*S*)-nta]$_4$ (3 mol%)
DCE/MeOH, 3/1, -35 °C
MeO 1 equiv.
TsN O S HN p-Tol
MeO
80%
99:1 dr; 99% ee

(5.18)

Such a process is possible because of the reversibility during the *in situ* formation of the iminoiodinanes, which is indeed the current hypothesis for the mechanism of these one-pot processes (Scheme 5.7).

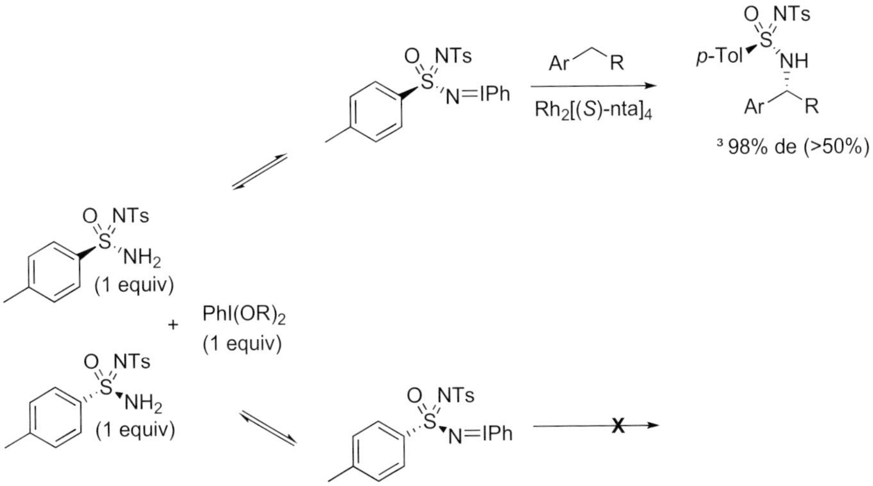

Scheme 5.7

5.1.3
Mechanism of C–H Amination using Hypervalent Iodine Reagents

The formation of the iminoiodinane from various amines and a hypervalent iodine reagent has been extensively studied by Du Bois and coworkers [47, 71]. However, it was not always possible to observe such a species. It is possible to detect a small amount (<10%) of $Cl_3CCH_2O_3SN{=}IPh$ by NMR upon mixing the sulfamate and PhI$(OCO\textit{t}\text{-Bu})_2$ (Eq. (5.19)). Trapping experiments with thioanisole to give the corresponding sulfilimine also support the intermediacy of an iminoiodinane in the catalytic cycle. An equilibrium favoring the starting material is invoked to explain why the iminoiodinane is not always observed by NMR. Kinetic data for rhodium-catalyzed intramolecular reaction with sulfamate and PhI$(OCO\textit{t}\text{-Bu})_2$ shows zero order in catalyst and first-order dependence on oxidant and sulfamate, which is in agreement with the proposed reversible formation of an iminoiodinane intermediate.

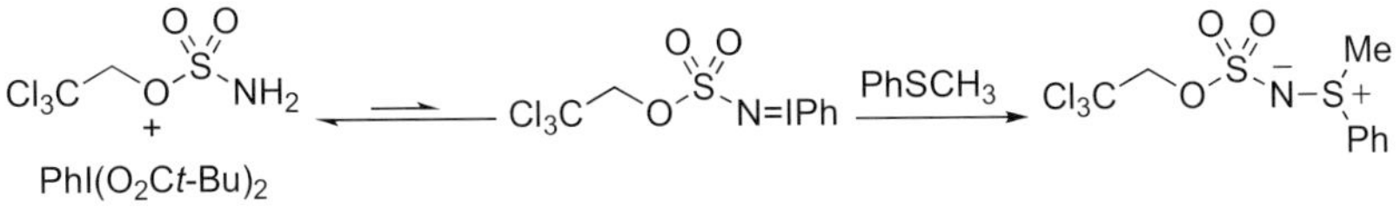

(5.19)

By analogy with the rhodium carbene intermediate proposed in the C–H insertion reaction with diazo compounds, C–H amination is believed to proceed via a rhodium nitrene species, although such an intermediate has never been characterized. However, as chiral dimeric rhodium complexes lead to the formation of enantioenriched amination products, it suggests that the metal center is closely associated with the reactive nitrogen during the C–H insertion step. Both a rhodium nitrene or rhodium phenyliminoiodinane species may be involved (Figure 5.2).

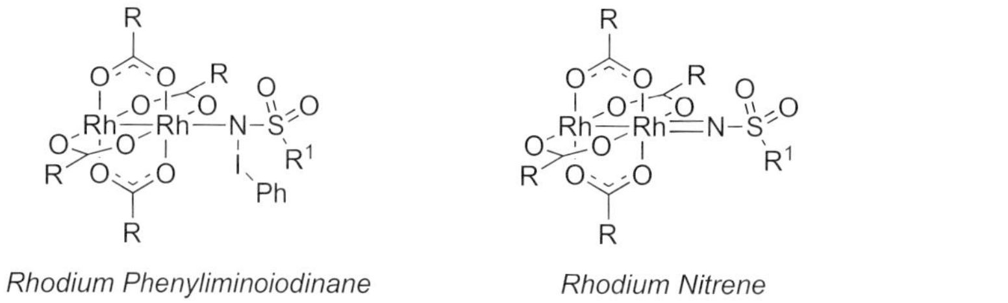

Rhodium Phenyliminoiodinane

Rhodium Nitrene

Figure 5.2 Proposed intermediates from rhodium carboxylate dimer and iminoiodinane.

There is no experimental evidence that rules out the intermediacy of rhodium phenyliminoiodinane species for C–H amination with iminoiodinane. However, according to a DFT study by Che, Phillips, and coworkers, the formation of a rhodium nitrene from an iminoiodinane reagent is energetically favored over a metalphenyliodinane species [72]. For the sake of the following argumentation, the hypothesis of a rhodium nitrene as the reactive species will be considered, according to the mechanistic pathway shown in Scheme 5.8. The reversible formation of the iminoiodinane is followed by reaction with the rhodium dimer to form the corresponding rhodium nitrene, which undergoes the C–H insertion reaction.

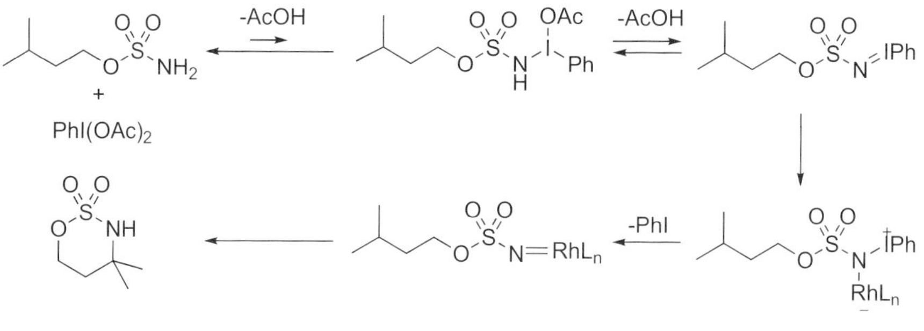

Scheme 5.8

The same DFT study suggests that the HOMO orbitals of rhodium nitrenes are not degenerate, thus the singlet and triplet states of the rhodium nitrene complex have similar energies [72]. The computational analysis also shows that a concerted ring-product formation from the singlet rhodium nitrene has a lower activation barrier than a stepwise triplet diradical recombination pathway. This is in agreement with experimental results obtained by Du Bois and coworkers with chiral substrates. The products are obtained with retention of configuration at the reactive carbon atom. For instance, when (*S*)-2-phenylpropyl acetate is treated with trichloroethylsulfamate and di-*tert*-butylacetoxyiodobenzene in the presence of a catalytic amount of a rhodium dimer complex, the desired benzylic amine is recovered with >99% ee (Eq. (5.20)) [47].

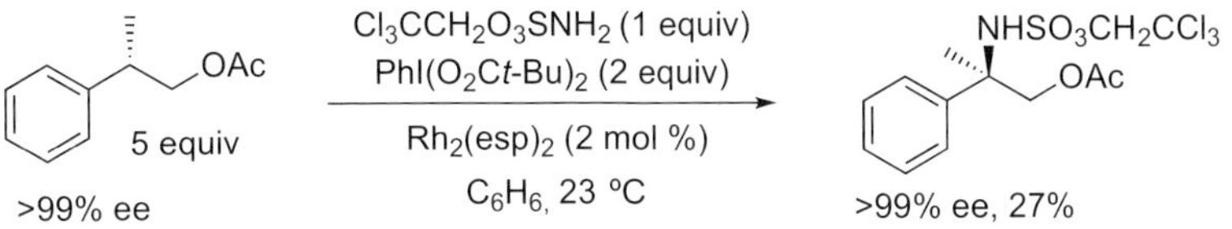

OAc 5 equiv >99% ee — $Cl_3CCH_2O_3SNH_2$ (1 equiv), $PhI(O_2Ct\text{-}Bu)_2$ (2 equiv), $Rh_2(esp)_2$ (2 mol %), C_6H_6, 23 °C → $NHSO_3CH_2CCl_3$ OAc >99% ee, 27%

(5.20)

Further evidence to exclude the triplet radical pathway includes the use of cyclopropyl substrates, which serve as a radical clock. In all cases, the reaction proceeds with no indication of ring fragmentation. The nature of the transition state of the C–H insertion step has been analyzed, via a Hammett study of the intermolecular C–H amination with *p*-substituted benzenes. A negative ρ value of -0.73 is obtained for the intermolecular reaction with trichloroethylsulfamate [71]. Such data indicate that there is a small, but significant, preference for electron-rich substrates, thus the resonance does contribute to the stabilization of a partial positive charge at the insertion carbon in the transition state. A kinetic isotope value of 1.9 is observed for competitive intramolecular C–H amination with a deuterated substrate (Eq. (5.21)).

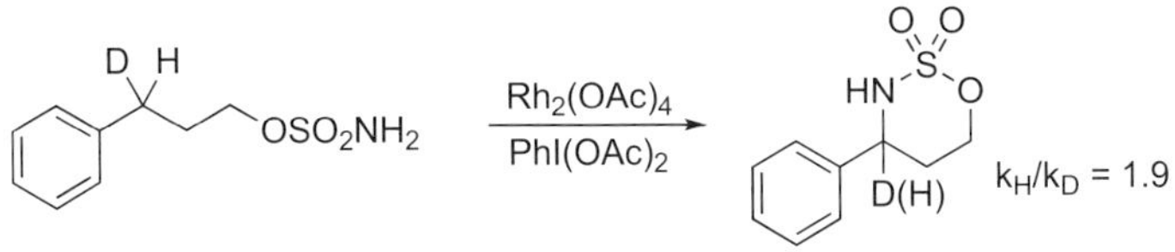

D H OSO_2NH_2 — $Rh_2(OAc)_4$, $PhI(OAc)_2$ → HN S O O O D(H) $k_H/k_D = 1.9$

(5.21)

This is in agreement with a transition state in which the C–H bond is at least partially broken. The current transition state hypothesis for the C–H insertion with a rhodium nitrene species involves an asynchronous-concerted pathway, in which the C–H–N angle is smaller than 180° (Figure 5.3).

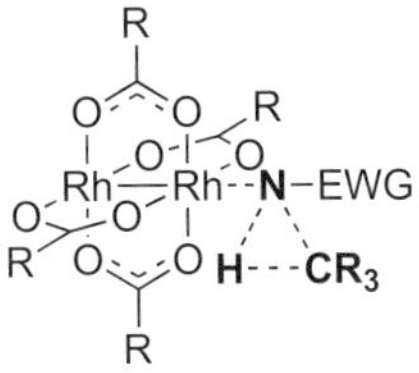

R O O R O O Rh Rh N–EWG O O R O O H CR3 R

Figure 5.3 Proposed transition state for rhodium-catalyzed C–H insertion.

5.2
Metal Nitrenes from *N*-Tosyloxycarbamates

The development of *in situ* procedures to prepare iminoiodinanes from amine precursors and hypervalent iodine reagents has had a significant impact on the progress made towards C–H amination reactions. However, there are some drawbacks associated with the use of hypervalent iodine reagents, in particular, their cost and also the generation of a stoichiometric amount (and even more) of iodobenzene as a by-product. While seeking for alternative rhodium nitrene precursors, Lebel and coworkers have disclosed the use of *N*-tosyloxycarbamates to perform C–H amination reactions [73]. Such a reaction proceeds in the presence of potassium carbonate as the base and the generated by-product is a tosylate salt, which can be easily removed by filtration or aqueous work-up. Intramolecular reactions with *N*-tosyloxycarbamates lead to oxazolidinones in high yields in the presence of potassium carbonate and rhodium (II) triphenylacetate dimer ($Rh_2(tpa)_4$) in dichloromethane at room temperature (Eq. (5.22)). *N*-Tosyloxycarbamates are stable compounds and are easy to prepare from the corresponding alcohol; it is thus possible to run the reaction on a 50-mmol scale [74]. In this case, only 1.1 equiv of potassium carbonate and 1 mol % of the rhodium catalyst are required for the reaction to proceed to completion. Furthermore, anhydrous reaction conditions are not compatible and a small amount of water is required to solubilize the base.

K_2CO_3 (1.1 equiv), $Rh_2(tpa)_4$ (1 mol%), DCM, H_2O, 0 °C to rt; 79% (5.22)

The cyclization of *N*-tosyloxycarbamates is also stereospecific and enantiopure substrates produce the corresponding oxazolidinone without racemization (Eq. (5.23)).

>98% ee; $Rh_2(tpa)_4$ (6 mol %), K_2CO_3 (3 equiv) / CH_2Cl_2, 25 °C; 73%, >98% ee (5.23)

Davies has recently developed a chiral rhodium dimer derived from adamantane glycine, $Rh_2[(S)\text{-tcptad}]_4$ [75]. He demonstrated that this was the most efficient catalyst for the enantioselective synthesis of oxazolidinones via C–H amination, using *N*-tosyloxycarbamates as nitrene precursors (Eq. (5.24)). Enantioselectivities up to 82% are observed in this case.

K_2CO_3 (3 equiv) / DCM, rt; $Rh_2[(S)\text{-tcptad}]_4$ (2 mol%); 72%, 82% ee (5.24)

Trichloroethyl-*N*-tosyloxycarbamate has been designed by Lebel and Huard specifically to achieve intermolecular C–H amination reactions and leads to Troc-protected amines which are easy to cleave [76]. The C–H amination of cyclic alkanes proceeds in good yields with 2 equiv of hydrocarbon substrate and 1 equiv of the amination reagent in the presence of 6 mol% of $Rh_2(tpa)_4$ (Eq. (5.25)). However, 5 equiv of alkane are usually required to achieve good yields with other substrates, such as aromatic compounds (Eq. (5.26)). In this case, the reaction is run without solvent.

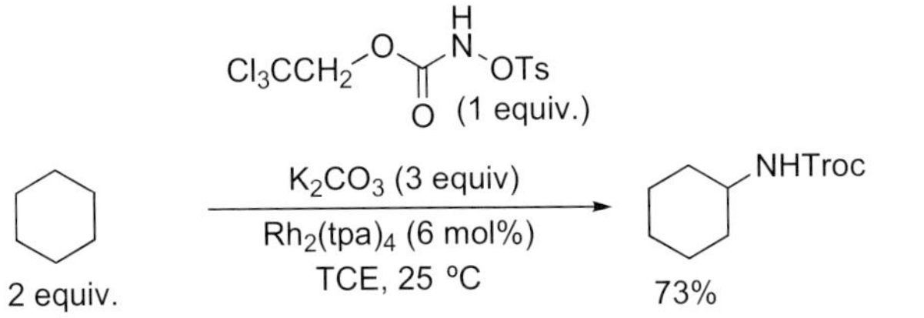

$Cl_3CCH_2OC(O)NHOTs$ (1 equiv.), K_2CO_3 (3 equiv), $Rh_2(tpa)_4$ (6 mol%), TCE, 25 °C; cyclohexane (2 equiv.) → NHTroc cyclohexane 73% (5.25)

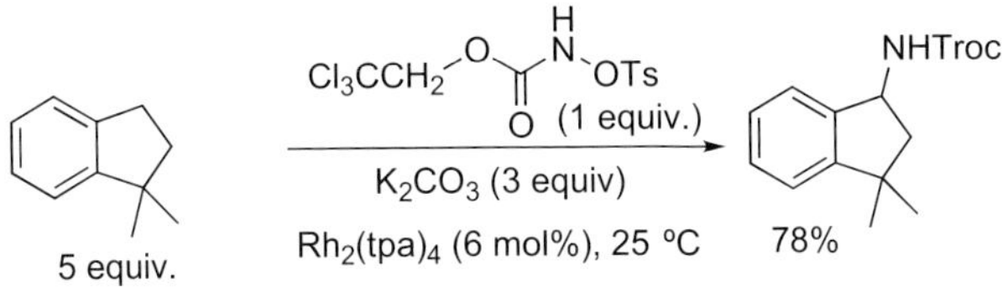

$Cl_3CCH_2OC(O)NHOTs$ (1 equiv.), K_2CO_3 (3 equiv), $Rh_2(tpa)_4$ (6 mol%), 25 °C; 1,1-dimethylindane (5 equiv.) → NHTroc 78% (5.26)

For a hydrocarbon substrate which does contain tertiary and secondary C–H bonds, a better chemoselectivity is observed with 5 mol% of $Rh_2[(S)\text{-nttl}]_4$. For instance, a 12 : 1 ratio is obtained for the C–H amination of adamantane leading to 58% of Troc-protected 1-adamantanamine, compared to 3 : 1 if $Rh_2(tpa)_4$ is used (Scheme 5.9). The protecting group is easy to cleave using Zn in acetic acid and amantadine hydrochloride is then isolated in quantitative yields [77].

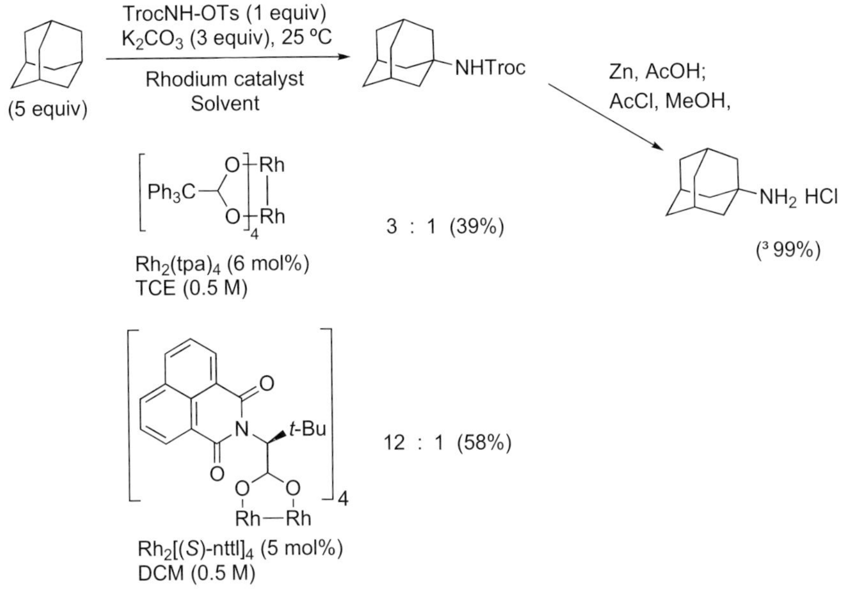

Adamantane (5 equiv) → TrocNH-OTs (1 equiv), K_2CO_3 (3 equiv), 25 °C, Rhodium catalyst, Solvent → NHTroc; Zn, AcOH; AcCl, MeOH, → NH_2 HCl (≥ 99%)

$Rh_2(tpa)_4$ (6 mol%), TCE (0.5 M): 3 : 1 (39%)

$Rh_2[(S)\text{-nttl}]_4$ (5 mol%), DCM (0.5 M): 12 : 1 (58%)

Scheme 5.9

A few mechanistic studies have been performed by Lebel and Huard to establish the catalytic cycle of this C–H amination reactions [77]. It has been shown that the deprotonation of the *N*-tosyloxycarbamate does not occur in the absence of the transition metal complex. Furthermore, the nature of the leaving group does not affect the amination pathway. For instance, the regioselectivity (tertiary vs. secondary C–H bonds) for the amination of adamantane is conserved, independent of the nature of the sulfonyloxy group (Eq. (5.27)).

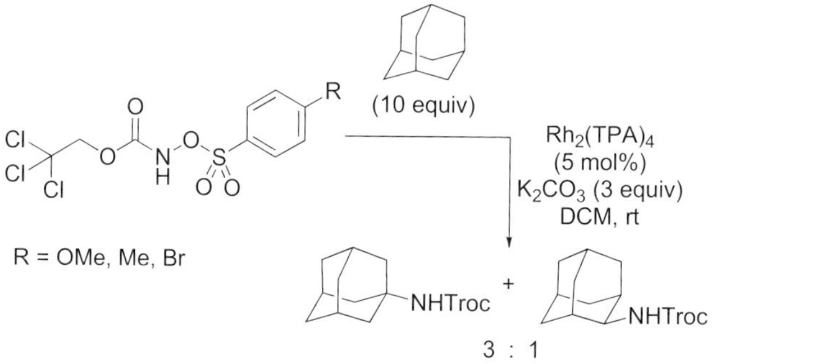

(10 equiv)

$Rh_2(TPA)_4$ (5 mol%)
K_2CO_3 (3 equiv)
DCM, rt

R = OMe, Me, Br

NHTroc + NHTroc

3 : 1

(5.27)

This observation indicates a common reactive species for all sulfonyloxycarbamate precursors, presumably a rhodium nitrene. The proposed catalytic cycle for the rhodium-catalyzed C–H amination with *N*-tosyloxycarbamates is shown in Scheme 5.10.

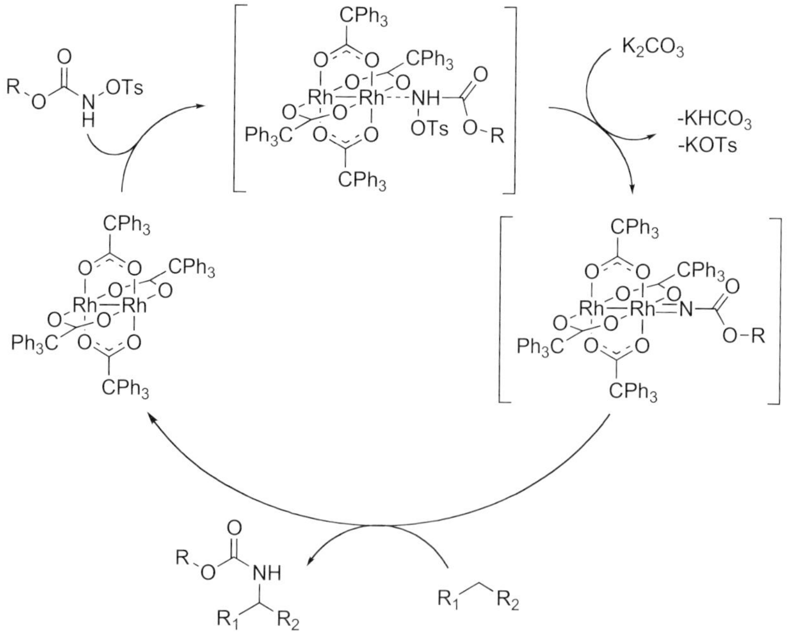

K_2CO_3

$-KHCO_3$
$-KOTs$

Scheme 5.10

The coordination of the *N*-tosyloxycarbamate to the rhodium catalyst is required for the deprotonation to occur, as no reaction is observed between potassium carbonate and the *N*-tosyloxycarbamate alone. The formation of the rhodium nitrene may proceed through the formation of a sulfonyloxy metal species which rapidly liberates the tosylate group, or the deprotonation may be concerted with the departure of the leaving group. It is postulated that the formation of the rhodium nitrene is the rate-determining step, since the rate of C–H insertion is the same for a deuterated and undeuterated substrate (Eqs. (5.28) and (5.29)).

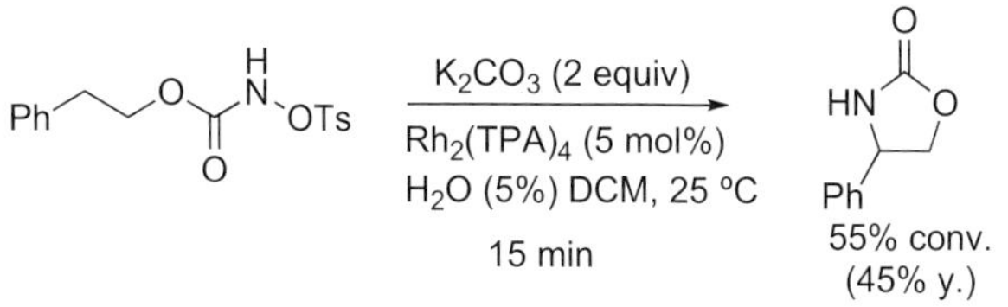

Ph O N(H) OTs, O → K_2CO_3 (2 equiv), $Rh_2(TPA)_4$ (5 mol%), H_2O (5%) DCM, 25 °C, 15 min → HN O O Ph, 55% conv. (45% y.) (5.28)

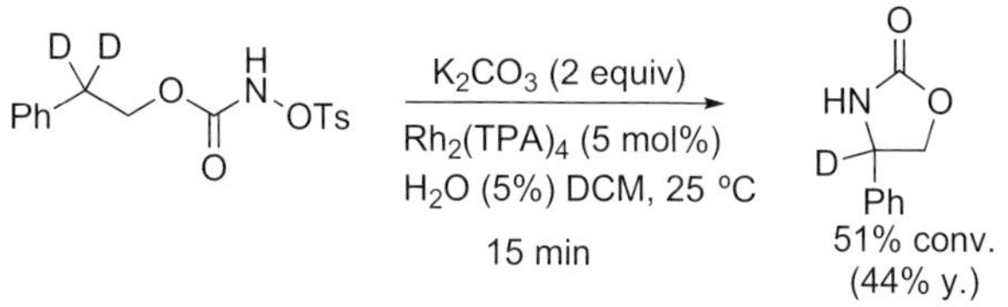

D D Ph O N(H) OTs, O → K_2CO_3 (2 equiv), $Rh_2(TPA)_4$ (5 mol%), H_2O (5%) DCM, 25 °C, 15 min → HN O O D Ph, 51% conv. (44% y.) (5.29)

Furthermore, the C–H insertion with *N*-tosyloxycarbamates appears also to involve a singlet rhodium nitrene species, as the reaction is sterospecific, and no ring opening product is observed with a radical clock substrate (Scheme 5.11).

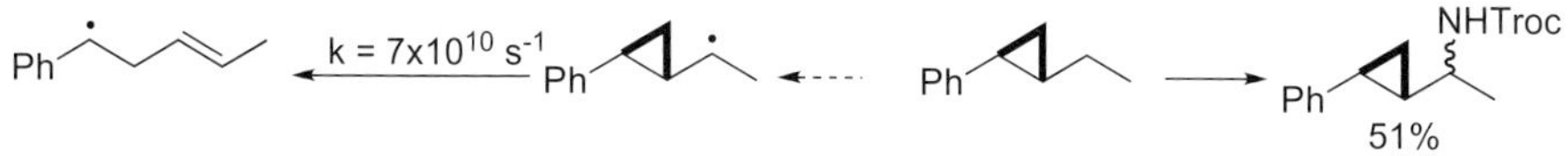

Ph ← k = 7x10^{10} s^{-1} — Ph ← - - - Ph → Ph NHTroc 51%

Scheme 5.11

A negative ϱ value of −0.47 is obtained from a Hammett study of the intermolecular C–H amination with *p*-substituted benzenes and trichloroethyl-*N*-tosyloxycarbamate. A kinetic isotope value of 5 is observed for a competitive C–H amination between a deuterated and undeuterated substrate. These data are consistent with a transition state for the C–H amination similar to the one shown previously (Figure 5.3).

In conclusion, the recent developments towards more practical reagents and catalysts to perform C–H amination reactions have provided a variety of new synthetic tools to prepare various nitrogen derivatives. Furthermore, a number of stereoselective methods have been recently disclosed. All this progress should promote the use of C–H amination reactions in total synthesis of natural products and biologically active compounds.

References

1 Ricci, A. (2000) *Modern Amination Methods*, Wiley-VCH, Weinheim.
2 Nugent, T.C., Ghosh, A.K., Wakchaure, V.N., and Mohanty, R.R. (2006) *Adv. Synth. Cat.*, **348**, 1289.
3 Friestad, G.K. and Mathies, A.K. (2007) *Tetrahedron*, **63**, 2541.
4 Bunlaksananusorn, T. and Rampf, F. (2005) *Synlett*, 2682.
5 Dai, Q., Yang, W.R., and Zhang, X.M. (2005) *Org. Lett.*, **7**, 5343.
6 Clausen, A.M., Dziadul, B., Cappuccio, K.L., Kaba, M., Starbuck, C., Hsiao, Y., and Dowling, T.M. (2006) *Org. Process Res. Dev.*, **10**, 723.
7 Kubryk, M. and Hansen, K.B. (2006) *Tetrahedron: Asymmetry*, **17**, 205.
8 Enthaler, S., Hagemann, B., Junge, K., Erre, G., and Beller, M. (2006) *Eur. J. Org. Chem.*, 2912.
9 Hultzsch, K.C. (2005) *Adv. Synth. Cat.*, **347**, 367.
10 Muci, A.R. and Buchwald, S.L. (2002) *Top. Curr. Chem.*, **219**, 131.
11 Hartwig, J.F. (2006) *Synlett*, 1283.
12 Mansuy, D. and Mahy, J.P. (1994) *Catalysis by Metal Complexes*, vol. 17 (eds F. Montanari and L. Casella), Springer, New York, p. 175.
13 Müller, P. (1997) *Advances in Catalytic Processes*, vol. 2 (ed. M.P. Doyle), Jai Press Inc., Greenwich, p. 113.
14 Müller, P. and Fruit, C. (2003) *Chem. Rev.*, **103**, 2905.
15 Dauban, P. and Dodd, R.H. (2003) *Synlett*, 1571.
16 DuBois, J. (2005) *Chemtracts*, **18**, 1.
17 Katsuki, T. (2005) *Chem. Lett.*, **34**, 1304.
18 Halfen, J.A. (2005) *Curr. Org. Chem.*, **9**, 657.
19 Davies, H.M.L. and Long, M.S. (2005) *Angew. Chem. Int. Ed.*, **44**, 3518.
20 Davies, H.M.L. (2006) *Angew. Chem. Int. Ed.*, **45**, 6422.
21 Li, Z.G. and He, C. (2006) *Eur. J. Org. Chem.*, 4313.
22 Diaz-Requejo, M.M. and Perez, P.J. (2008) *Chem. Rev.*, **108**, 3379.
23 Davies, H.M.L. and Manning, J.R. (2008) *Nature*, **451**, 417.
24 Espino, C.G. and DuBois, J. (2005) *Modern Rhodium-Catalyzed Organic Reactions* (ed. P.A. Evans), Wiley-VCH Verlag GmbH & Co. KGaA, Weinheim, p. 379.
25 Kwart, H. and Khan, A.A. (1967) *J. Am. Chem. Soc.*, **89**, 1951.
26 Breslow, R. and Gellman, S.H. (1982) *J. Chem. Soc., Chem. Commun.*, 1400.
27 Breslow, R. and Gellman, S.H. (1983) *J. Am. Chem. Soc.*, **105**, 6728.
28 Müller, P., Baud, C., Jacquier, Y., Moran, M., and Nageli, I. (1996) *J. Phys. Org. Chem.*, **9**, 341.
29 Nageli, I., Baud, C., Bernardinelli, G., Jacquier, Y., Moran, M., and Müller, P. (1997) *Helv. Chim. Acta*, **80**, 1087.
30 Mueller, P., Baud, C., and Naegeli, I. (1998) *J. Phys. Org. Chem.*, **11**, 597.
31 Au, S.M., Huang, J.S., Che, C.M., and Yu, W.Y. (2000) *J. Org. Chem.*, **65**, 7858.
32 Yu, X.Q., Huang, J.S., Zhou, X.G., and Che, C.M. (2000) *Org. Lett.*, **2**, 2233.
33 Espino, C.G. and DuBois, J. (2001) *Angew. Chem. Int. Ed.*, **40**, 598.
34 Lebel, H., Leogane, O., Huard, K., and Lectard, S. (2006) *Pure Appl. Chem.*, **78**, 363.
35 Cenini, S., Gallo, E., Caselli, A., Ragaini, F., Fantauzzi, S., and Piangiolino, C. (2006) *Coord. Chem. Rev.*, **250**, 1234.
36 Albone, D.P., Challenger, S., Derrick, A.M., Fillery, S.M., Irwin, J.L., Parsons, C.M., Takada, H., Taylor, P.C., and Wilson, D.J. (2005) *Org. Biomol. Chem.*, **3**, 107.
37 Bhuyan, R. and Nicholas, K.M. (2007) *Org. Lett.*, **9**, 3957.
38 Chanda, B.M., Vyas, R., and Bedekar, A.V. (2001) *J. Org. Chem.*, **66**, 30.
39 Yan, S.Y., Wang, Y., Shu, Y.J., Liu, H.H., and Zhou, X.G. (2006) *J. Mol. Catal. A: Chem.*, **248**, 148.
40 Li, Z.G., Capretto, D.A., Rahaman, R., and He, C.A. (2007) *Angew. Chem. Int. Ed.*, **46**, 5184.
41 Thu, H.Y., Yu, W.Y., and Che, C.M. (2006) *J. Am. Chem. Soc.*, **128**, 9048.
42 Li, Z., Capretto, D.A., Rahaman, R.O., and He, C. (2007) *J. Am. Chem. Soc.*, **129**, 12058.

43 Hansen, J. and Davies, H.M.L. (2008) *Coord. Chem. Rev.*, **252**, 545.

44 Espino, C.G., Wehn, P.M., Chow, J., and DuBois, J. (2001) *J. Am. Chem. Soc.*, **123**, 6935.

45 Espino, C.G., Fiori, K.W., Kim, M., and DuBois, J. (2004) *J. Am. Chem. Soc.*, **126**, 15378.

46 Kim, M., Mulcahy, J.V., Espino, C.G., and DuBois, J. (2006) *Org. Lett.*, **8**, 1073.

47 Fiori, K.W. and DuBois, J. (2007) *J. Am. Chem. Soc.*, **129**, 562.

48 Moody, C.J. (1991) *Comprehensive Organic Synthesis*, vol. 7 (eds B.M. Trost and I. Fleming), Pergamon Press, Oxford, p. 21.

49 Parker, K.A. and Chang, W. (2003) *Org. Lett.*, **5**, 3891.

50 Parker, K.A. and Chang, W. (2005) *Org. Lett.*, **7**, 1785.

51 Espino, C.G. and DuBois, J. (2005) *Modern Rhodium-Catalyzed Organic Reactions* (ed. P.A. Evans), Wiley-VCH Verlag GmbH & Co. KGaA, Weinheim, p. 389.

52 Yakura, T., Sato, S., and Yoshimoto, Y. (2007) *Chem. Pharm. Bull.*, **55**, 1284.

53 Doyle, M.P., Dyatkin, A.B., and Autry, C.I. (1995) *J. Chem. Soc., Perkin Trans. 1*, 619.

54 Clark, J.S., Dossetter, A.G., Wong, Y.S., Townsend, R.J., Whittingham, W.G., and Russell, C.A. (2004) *J. Org. Chem.*, **69**, 3886.

55 Trost, B.M., Gunzner, J.L., Dirat, O., and Rhee, Y.H. (2002) *J. Am. Chem. Soc.*, **124**, 10396.

56 Huang, H. and Panek, J.S. (2003) *Org. Lett.*, **5**, 1991.

57 Smith, A.B., Charnley, A.K., Harada, H., Beiger, J.J., Cantin, L.D., Kenesky, C.S., Hirschmann, R., Munshi, S., Olsen, D.B., Stahlhut, M.W., Schleif, W.A., and Kuo, L.C. (2006) *Bioorg. Med. Chem. Lett.*, **16**, 859.

58 Narina, S.V., Kumar, T.S., George, S., and Sudalai, A. (2007) *Tetrahedron Lett.*, **48**, 65.

59 Zhu, R., Snyder, A.H., Kharel, Y., Schaffter, L., Sun, Q., Kennedy, P.C., Lynch, K.R., and Macdonald, T.L. (2007) *J. Med. Chem.*, **50**, 6428.

60 Yakura, T., Yoshimoto, Y., Ishida, C., and Mabuchi, S. (2006) *Synlett*, 930.

61 Yakura, T., Yoshimoto, Y., and Ishida, C. (2007) *Chem. Pharm. Bull.*, **55**, 1385.

62 Yakura, T., Yoshimoto, Y., Ishida, C., and Mabuchi, S. (2007) *Tetrahedron*, **63**, 4429.

63 Rohde, J.J., Pliushchev, M.A., Sorensen, B.K., Wodka, D., Shuai, Q., Wang, J.H., Fung, S., Monzon, K.M., Chiou, W.J., Pan, L.P., Deng, X.Q., Chovan, L.E., Ramaiya, A., Mullally, M., Henry, R.F., Stolarik, D.F., Imade, H.M., Marsh, K.C., Beno, W.A., Fey, T.A., Droz, B.A., Brune, M.E., Camp, H.S., Sham, H.L., Frevert, E.U., Jacobson, P.B., and Link, J.T. (2007) *J. Med. Chem.*, **50**, 149.

64 Hinman, A. and DuBois, J. (2003) *J. Am. Chem. Soc.*, **125**, 11510.

65 Wehn, P.M., Lee, J.H., and DuBois, J. (2003) *Org. Lett.*, **5**, 4823.

66 Fleming, J.J. and DuBois, J. (2006) *J. Am. Chem. Soc.*, **128**, 3926.

67 Fleming, J.J., McReynolds, M.D., and DuBois, J. (2007) *J. Am. Chem. Soc.*, **129**, 9964.

68 Zalatan, D.N. and DuBois, J. (2008) *J. Am. Chem. Soc.*, **130**, 9220.

69 Liang, C.G., Robert-Pedlard, F., Fruit, C., Müller, P., Dodd, R.H., and Dauban, P. (2006) *Angew. Chem. Int. Ed.*, **45**, 4641.

70 Liang, C., Collet, F., Robert-Peillard, F., Müller, P., Dodd, R.H., and Dauban, P. (2008) *J. Am. Chem. Soc.*, **130**, 343.

71 Fiori, K.W., Espino, C.G., Brodsky, B.H., and DuBois, J. (2009) *Tetrahedron*, **65**, 3042.

72 Lin, X.F., Zhao, C.Y., Che, C.M., Ke, Z.F., and Phillips, D.L. (2007) *Chem. Asian J.*, **2**, 1101.

73 Lebel, H., Huard, K., and Lectard, S. (2005) *J. Am. Chem. Soc.*, **127**, 14198.

74 Huard, K. and Lebel, H. (2009) *Org. Synth.*, **86**, 59.

75 Reddy, R.P. and Davies, H.M.L. (2006) *Org. Lett.*, **8**, 5013.

76 Lebel, H. and Huard, K. (2007) *Org. Lett.*, **9**, 639.

77 Huard, K. and Lebel, H. (2008) *Chem. Eur. J.*, **14**, 6222.

6
The Palladium-Catalyzed Synthesis of Aromatic Heterocycles

Yingdong Lu and Bruce A. Arndtsen

6.1
Introduction

Palladium-catalyzed bond forming reactions have had a major impact on how organic chemists approach the synthesis of products. While this is often considered with regard to cross coupling reactions, Heck couplings, allylations or carbon–heteroatom bond forming reactions, similarly important advances have also been made in the use of palladium catalysis in heterocycle synthesis. Interest in the assembly of aromatic heterocycles is often tied to the importance of these compounds in biologically relevant products, including natural products, pharmaceuticals, and other biologically important structures. In addition, aromatic heterocycles are key components in materials science, molecular electronics, metal-coordinating ligands, and polymer chemistry.

There is a wide range of methods available to construct aromatic heterocycles. Nevertheless, many of these rely upon relatively involved multistep processes, especially for the assembly of highly substituted products. In addition, these can sometimes suffer from harsh reaction conditions, or display limited product diversity. As such, there has been intense recent interest in the design of new and more efficient routes to these products. In this regard, metal catalysis, and in particular palladium catalysis, has played a central role.

Palladium catalysis has become an important tool in synthesis due in large part to its ability to mediate a number of different fundamental transformations with low reaction barriers. These include oxidative addition and reductive elimination reactions, insertion and de-insertion (often β-hydride elimination), nucleophilic ligand attack, or cycloaddition (Scheme 6.1).

Notably, palladium complexes not only mediate these reactions, but often do so reversibly. This has provided fertile ground for the development of new catalytic routes to heterocycles. These exploit different combinations of the basic reactions in Scheme 6.1. For example, palladium(II) complexes are well established π-Lewis acids, with the ability to coordinate to multiple bonds and activate these towards nucleophilic attack. When this nucleophilic attack is intramolecular, this provides a

Catalyzed Carbon-Heteroatom Bond Formation. Edited by Andrei K. Yudin

ISBN: 978-3-527-32428-6

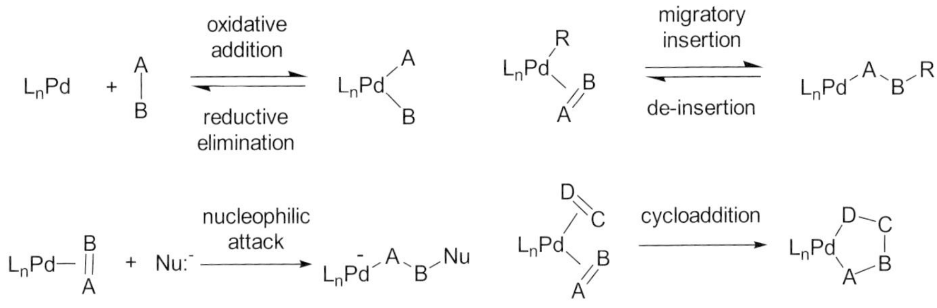

Scheme 6.1

viable method to generate heterocyclic products (Scheme 6.2a). Alternatively, oxidative addition/reductive elimination cycles have become a mainstay of palladium-catalyzed bond forming processes, including the generation of carbon–heteroatom bonds. The latter has provided a valuable alternative method to assemble heterocycles, either via direct cyclization (Scheme 6.2b), or with the concomitant insertion of substrates into the intermediate palladium–carbon bond (Scheme 6.2c). Related chemistry involves Heck-type cyclizations with olefins (Scheme 6.2d), wherein oxidative addition to generate either a palladium–carbon or palladium–heteroatom bond is followed by insertion and β-hydride elimination. In the field of palladium catalysis, the activation of typically unreactive C−H bonds (i.e., C−H bond activation) is rapidly emerging as a synthetic method for bond formation, which has also seen application in heterocycle synthesis. This includes those involving carbon–heteroatom bond formation (Scheme 6.2e). Palladium complexes can also catalyze cycloadditions, many of which have been adapted to multicomponent protocols to

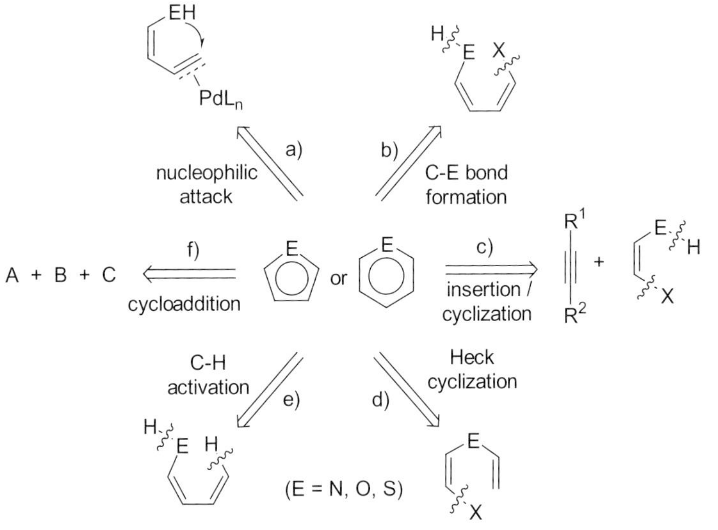

Scheme 6.2

rapidly assemble aromatic heterocycles from three or more components (Scheme 6.2f).

The use of palladium catalysis to synthesize aromatic heterocycles has been reviewed in a number of recent publications [1]. As such, rather than a complete discussion of this large field, the thrust of this review will be to highlight general examples of how palladium catalysis has become used in heterocycle synthesis. This will focus on routes that directly assemble the aromatic heterocyclic core, rather than their subsequent functionalization or the use of palladium to assemble precursors for traditional cyclocondensations, and on processes that involve generation of carbon–heteroatom bonds.

6.2 Palladium π-Lewis Acidity: Intramolecular Nucleophilic Attack on Unsaturated Bonds

6.2.1 Addition to Alkynes

Perhaps the most straightforward way in which palladium complexes can mediate heterocycle formation involves their ability to coordinate to multiple bonds, and, through accepting π-electron density, activate these substrates towards intramolecular cyclization (Scheme 6.3). This general approach can allow the assembly of a range of heterocycles, varying from simple furans, pyrroles and thiophenes to polycyclic products, such as indoles, benzofurans and others. In order to activate the multiple bond toward attack by the neighboring nitrogen, oxygen or sulfur nucleophile, the catalyst role is to withdraw electron density from the π-bond. Palladium(II) complexes are prime candidates for mediating these transformations, since they can both coordinate to π-bonds and, due to their higher oxidation state than palladium(0), have only weak backbonding to the unsaturated fragment.

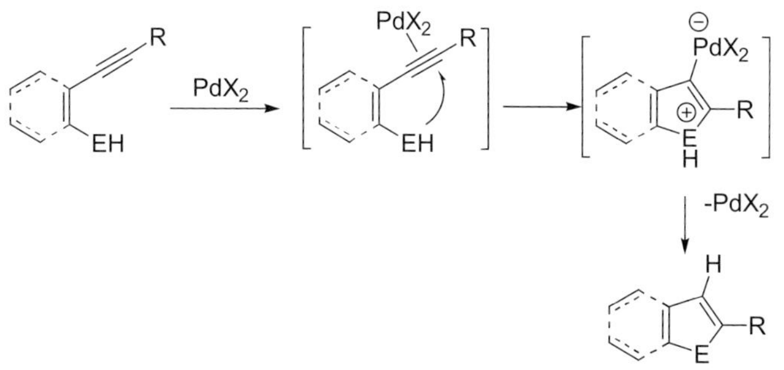

Scheme 6.3

This reaction manifold has been heavily exploited in the assembly of indoles and benzofurans [2]. For example, Taylor and McKillop reported in 1985 that *ortho*-alkynyl substituted anilines **1** can undergo cyclization in the presence of $PdCl_2$ catalysts to

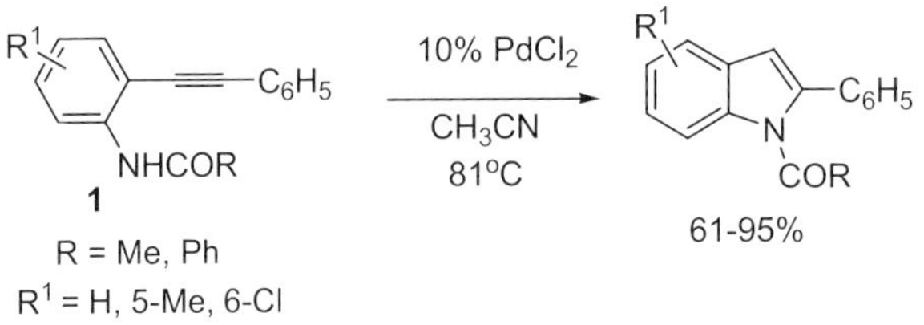

1 → 61-95% ($10\%\ PdCl_2$, CH_3CN, 81°C)

R = Me, Ph

R^1 = H, 5-Me, 6-Cl

Scheme 6.4

afford indoles (Scheme 6.4) [3]. This reaction presumably proceeds via initial coordination of $PdCl_2$ to the alkyne to allow cyclization, followed by subsequent protonation of the palladium–carbon bond to generate the product and re-form the active palladium(II) catalyst, as illustrated in Scheme 6.3.

The cyclization of 2-alkynylaniline derivatives has become a general method for the assembly of 2-substituted indoles [2, 4]. While a range of other transition metals and Lewis acids can mediate similar intramolecular cyclizations, a useful feature of palladium catalysis is that it can also be employed to assemble the cyclization precursors. The latter is often accomplished by coupling more traditional palladium-catalyzed coupling chemistry with cyclization, and provides an avenue to readily expand the diversity of indole products available. As illustrated with the examples in Scheme 6.5, this can be done via the Sonogashira coupling of **2** with R–X [5], the Stille

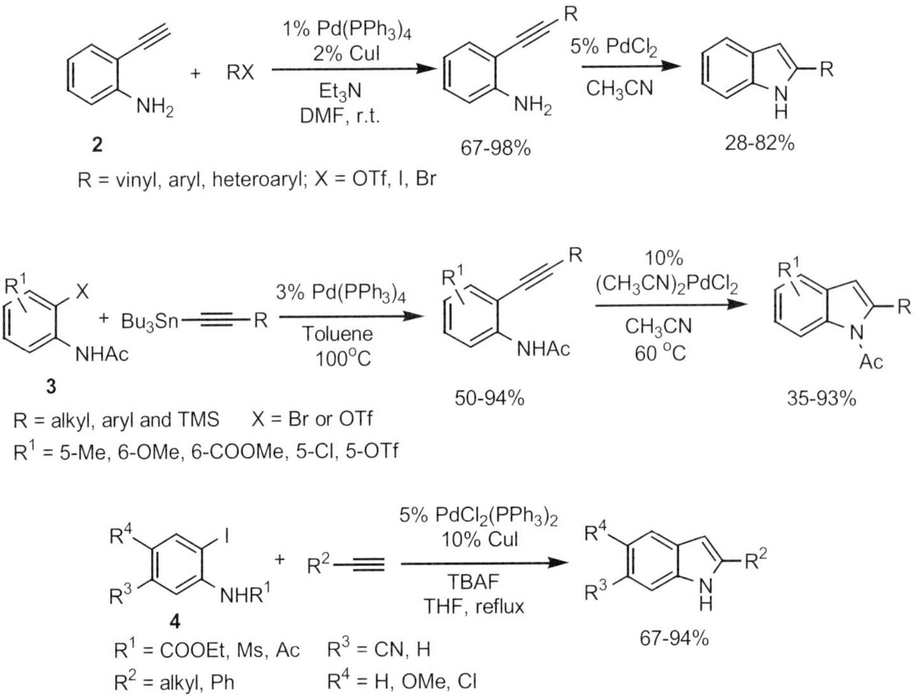

2 + RX ($1\%\ Pd(PPh_3)_4$, 2% CuI, Et_3N, DMF, r.t.) → 67-98% ($5\%\ PdCl_2$, CH_3CN) → 28-82%

R = vinyl, aryl, heteroaryl; X = OTf, I, Br

3 + Bu_3Sn–≡–R ($3\%\ Pd(PPh_3)_4$, Toluene, 100°C) → 50-94% (10% $(CH_3CN)_2PdCl_2$, CH_3CN, 60 °C) → 35-93%

R = alkyl, aryl and TMS X = Br or OTf

R^1 = 5-Me, 6-OMe, 6-COOMe, 5-Cl, 5-OTf

4 + R^2–≡ ($5\%\ PdCl_2(PPh_3)_2$, 10% CuI, TBAF, THF, reflux) → 67-94%

R^1 = COOEt, Ms, Ac R^3 = CN, H

R^2 = alkyl, Ph R^4 = H, OMe, Cl

Scheme 6.5

coupling of **3** with alkynylstannanes [6], or the Sonogashira coupling of **4** with terminal alkynes [7]. Each of these generate the analogous *ortho*-alkynylaniline derivatives for cyclization. Notably, the latter route provides a method to assemble indoles from simple haloanilines and terminal alkynes, and has been exploited by a number of research groups for indole synthesis [2].

With the development of Buchwald–Hartwig amination reactions, the amine component of these indoles can also be introduced into these precursors via palladium catalysis [8]. As shown by Ackermann, this can be coupled with aryl halide alkynylation and cyclization to provide a one-pot, three-component synthesis of substituted indoles (Scheme 6.6) [9]. In this case, simple *ortho*-dihaloarene derivatives **5** were employed as starting materials, with Sonogashira coupling occurring at the more activated aryl–iodide bond, followed by selective coupling of various alkyl or arylamines. Alternatively, Zhao has recently demonstrated that amination can be performed on both bromoalkyne **6**, followed by the aryl–bromide bond, to provide a route to 2-amidoindoles (Scheme 6.7) [10].

Scheme 6.6

Scheme 6.7

This palladium-catalyzed activation of alkynes toward nucleophilic attack can also be adapted to oxygen nucleophiles, providing a route to benzofurans. Cacchi has demonstrated this in the $Pd(OAc)_2$/CuI-catalyzed Sonogashira coupling/cyclization of 2-halophenols with terminal alkynes (Scheme 6.8) [11]. This protocol has since

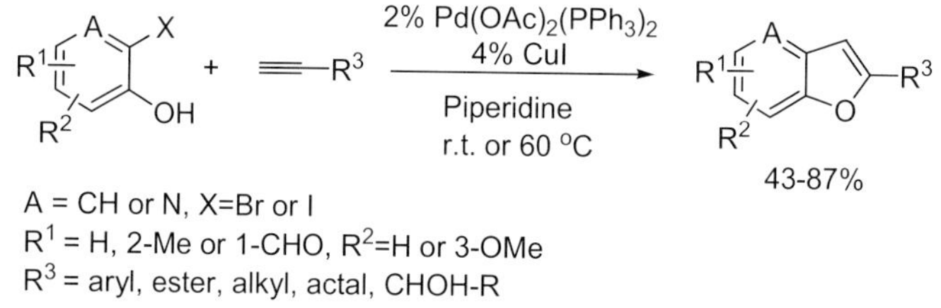

Scheme 6.8

been expanded to access a range of substituted benzofurans and related products, wherein the phenol unit is replaced with heterocycles, benzoqunione derivatives, and other polycyclic aromatic units [12].

The application of this approach to monocyclic heterocycles has also been explored, though not to the degree of indoles and benzofurans. In these systems, the aryl spacer between the amine and alkyne can be replaced by either a double bond, or a saturated fragment capable of elimination. In an early report, Utimoto showed that 1-amino-3-alkyn-2-ols can undergo cyclization to generate various 2,4-substituted pyrroles in the presence of palladium(II) catalysts (Scheme 6.9) [13]. This reaction is postulated to proceed via a similar palladium-catalyzed nucleophilic attack on the terminal alkyne carbon, followed by proton migration and subsequent alcohol elimination from **7**, thus allowing aromatization. More recently, Gabriele has prepared substituted pyrroles from (*Z*)-(2-en-4-ynyl)amines, which can undergo cycloisomerization either spontaneously (with terminal or aryl-substituted alkynes), or more commonly under palladium-catalyzed conditions (Scheme 6.10) [14]. Rearrangement to the aromatic

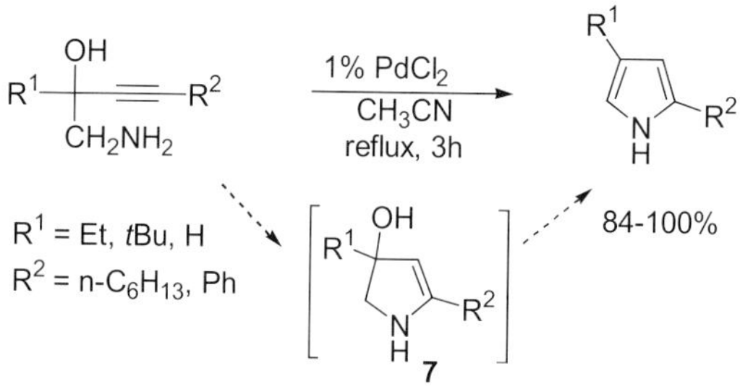

Scheme 6.9

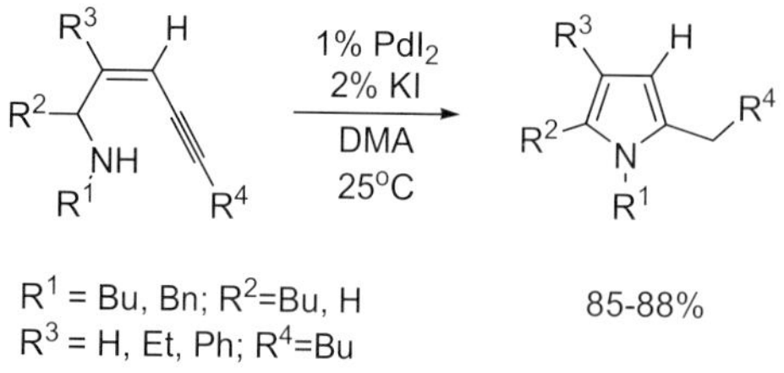

Scheme 6.10

pyrrole occurs spontaneously via proton migration, providing access to up to tetrasubstituted pyrroles in good yields.

The above approaches can also be applied to a range of other aromatic heterocycles, including furans, thiophenes and oxazoles [2, 15]. While these protocols typically require the initial assembly of the precursor for cyclization, Trost has demonstrated that palladium catalysis can also be employed to assemble the en-yn-ol **8** for cyclization to furans. (Scheme 6.11) [16]. This reaction, that proceeds via the palladium-catalyzed regioselective addition of terminal alkynes to internal propargyl alcohols, provides a biomolecular method to build up furans from these two alkyne building blocks.

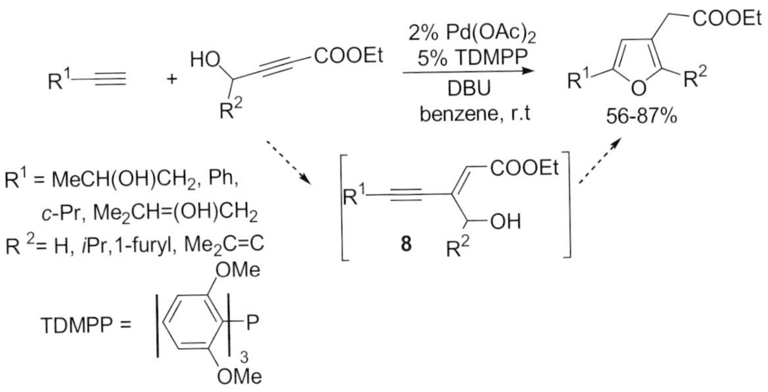

R^1 = MeCH(OH)CH_2, Ph, c-Pr, Me_2CH=(OH)CH_2
R^2 = H, iPr, 1-furyl, Me_2C=C

TDMPP = tris(2,6-dimethoxyphenyl)phosphine

2% $Pd(OAc)_2$, 5% TDMPP, DBU, benzene, r.t; 56-87%

Scheme 6.11

As an alternative to five membered aromatic heterocyclic rings, Gabriele has noted that the alkynylation of 2-ketoaniline derivatives followed by palladium-catalyzed cyclization can provide a route to quinolines (Scheme 6.12) [17]. This cyclization presumably proceeds in a similar fashion to those above, yet undergoes spontaneous dehydration upon ring closure.

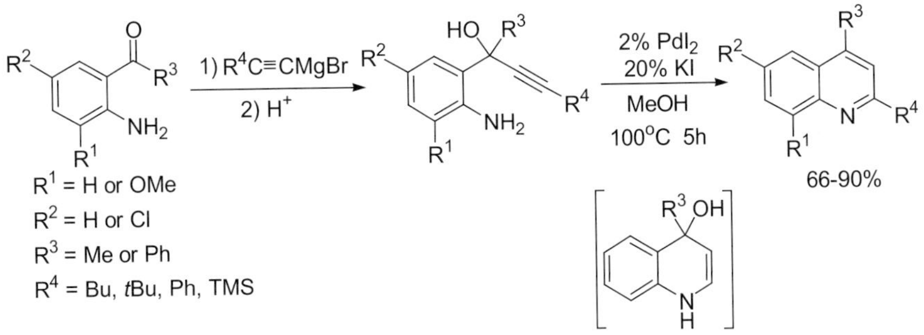

1) $R^4C{\equiv}CMgBr$ 2) H^+; 2% PdI_2, 20% KI, MeOH, 100°C 5h; 66-90%

R^1 = H or OMe
R^2 = H or Cl
R^3 = Me or Ph
R^4 = Bu, tBu, Ph, TMS

Scheme 6.12

6.2.2
Heteroatom Addition to Alkynes with Functionalization

The intramolecular addition of nucleophiles to palladium-activated triple bonds generates the palladated intermediate **9** (Scheme 6.13). While the latter can be protonated to liberate product, the reactivity of palladium can also be employed to intercept this intermediate with other bond forming reactions. These have provided methods to derivatize the heterocycle at the same time as it is generated. A range of substrates have been coupled with cyclization via this approach, including aryl or vinyl halides, allylic and related R–X substrates, amines, halides, as well as carbon monoxide and olefins.

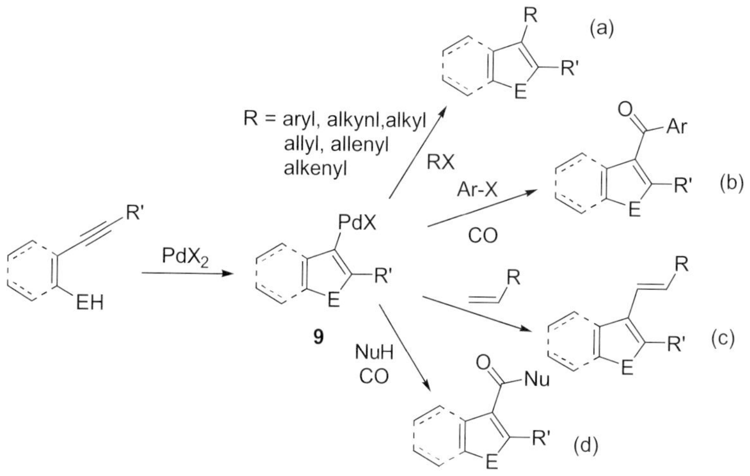

R = aryl, alkynl,alkyl allyl, allenyl alkenyl; RX; Ar-X, CO; NuH, CO; PdX_2; PdX; R'; EH; E; 9; (a); (b); (c); (d); Ar; Nu; R

Scheme 6.13

One of the most common examples of this approach involves the coupling of aryl or vinyl halides with palladium-catalyzed cyclization. This chemistry typically employs palladium(0) catalysts, which are postulated to undergo an initial oxidative addition of aryl or vinyl halides to form a palladium(II) complex to mediate cyclization. This is followed by reductive elimination of the heterocycle–aryl or heterocycle–vinyl bond. This approach has been employed to construct a range of polysubstituted indoles. For example, Cacchi has shown that the $Pd(PPh_3)_4$ catalyzed coupling of trifluoroacetanilides with aryl halides or vinyl halides/triflates proceeds to substituted indoles in good yield (Scheme 6.14) [18].

Lu has demonstrated that this aryl halide-mediated cyclization can be coupled with the generation of the alkynylaniline derivative via Sonogashira coupling, providing a multicomponent route to build up indoles from anilides, terminal alkynes and aryl halides (Scheme 6.15) [19]. Similar approaches have been applied to the construction of benzofurans [20] and substituted furans [21].

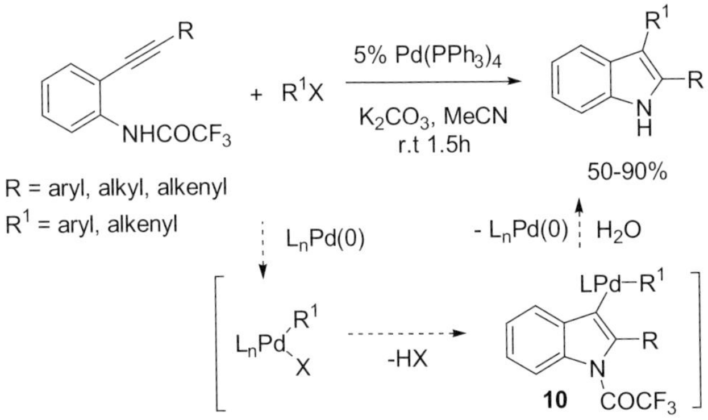

R = aryl, alkyl, alkenyl
R^1 = aryl, alkenyl

Scheme 6.14

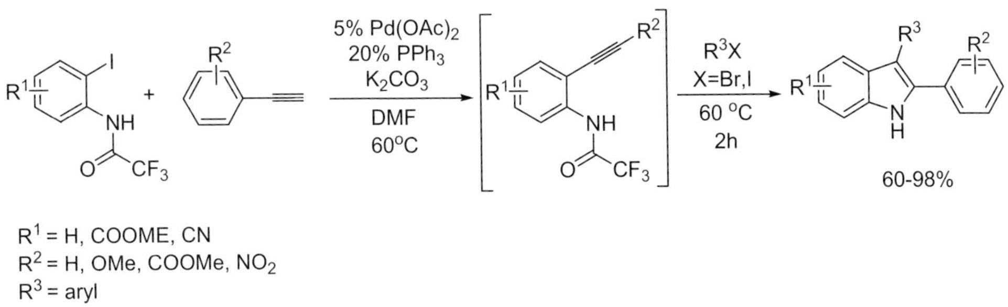

R^1 = H, COOME, CN
R^2 = H, OMe, COOMe, NO_2
R^3 = aryl

Scheme 6.15

As illustrated in Scheme 6.13, a range of other R–X substrates can be coupled with heterocycle formation, as can carbon monoxide insertion to form keto-substituted products (Scheme 6.13b) [22]. These are typically postulated to proceed in an analogous fashion to the reactions above. In addition to these, several mechanistic variants have also been developed. For example, Zhang has reported that allyl-substituted furans can be synthesized in a multicomponent reaction of ketone **11**, allylic halides and various nucleophiles (Scheme 6.16) [23]. This reaction utilizes the

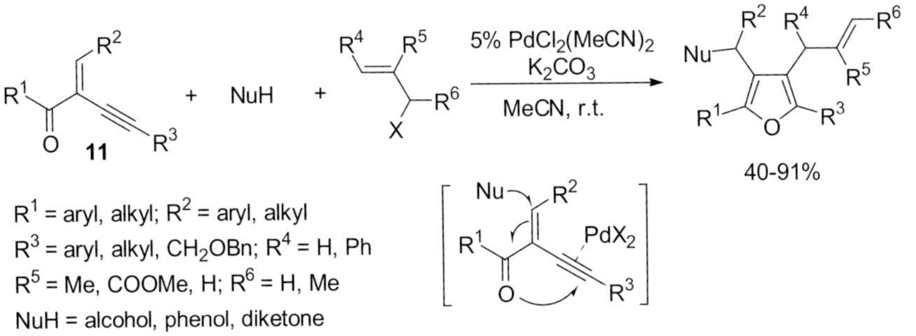

R^1 = aryl, alkyl; R^2 = aryl, alkyl
R^3 = aryl, alkyl, CH_2OBn; R^4 = H, Ph
R^5 = Me, COOMe, H; R^6 = H, Me
NuH = alcohol, phenol, diketone

Scheme 6.16

Michael addition of the nucleophile to initiate the palladium-catalyzed cyclization similar to that noted above, followed by allyl coupling.

As an alternative to external allylic substrates, Balme has demonstrated that the O-allylated precursor **12** can participate in cyclization, providing access to 3-allyl indoles (Scheme 6.17) [24]. In this case, $Pd(PPh_3)_4$ is postulated to activate the C–O bond to create a palladium-allyl complex, which then mediates cyclization and subsequent reductive elimination of the functionalized product. Other substrates can also participate in this intramolecular rearrangement, including propargylic units, which give rise to either allenyl- or propargyl-substituted benzofurans [25].

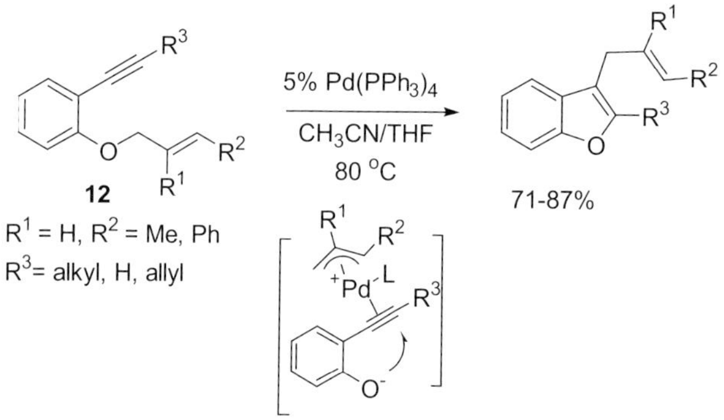

12

R^1 = H, R^2 = Me, Ph

R^3= alkyl, H, allyl

5% $Pd(PPh_3)_4$

CH_3CN/THF

80 °C

71-87%

Scheme 6.17

As a replacement for alkynylanilines, Yamamoto has reported that indoles can be generated via the $Pd(PPh_3)_4$/CuCl-catalyzed coupling of 2-alkynylarylisocyanates with allylcarbonates (Scheme 6.18) [26]. In this case, fragmentation of the carbonate anion to an alkoxide upon oxidative addition to palladium allows conversion of the isocyanate into a carbamate for subsequent cyclization. A number of substituted alkynes can participate in this reaction, and it can be performed with alcohols instead of allylcarbonates to form 3-unsubstituted indoles. A variant of this reaction involved the use of isocyanides in concert TMS-azide, providing a route to substituted *N*-cyanoindoles [27].

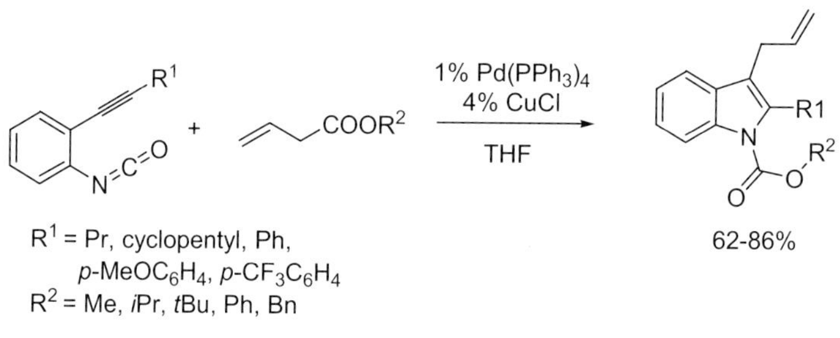

1% $Pd(PPh_3)_4$

4% CuCl

THF

62-86%

R^1 = Pr, cyclopentyl, Ph, *p*-$MeOC_6H_4$, *p*-$CF_3C_6H_4$

R^2 = Me, *i*Pr, *t*Bu, Ph, Bn

Scheme 6.18

A mechanistic alternative to the oxidative addition of R–X substrates to palladium to initiate cyclization is to rely upon palladium(II) catalysts to mediate heterocycle formation, followed by subsequent alkene insertion into the palladium–carbon bond generated (**9**, Scheme 6.13c). This can be followed by β-hydride elimination to provide a Heck-type post functionalization of these heterocycles (Scheme 6.19).

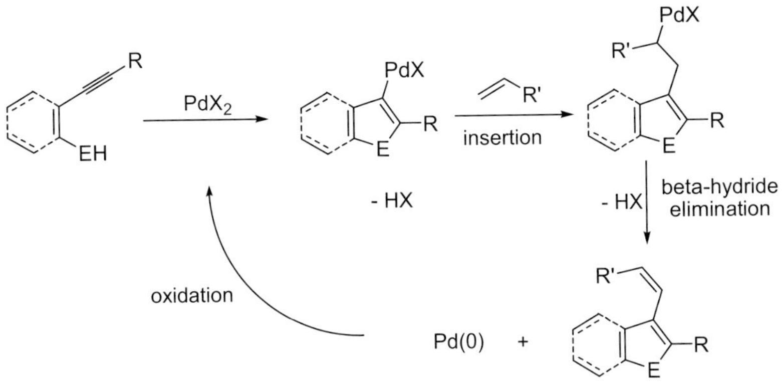

Scheme 6.19

This approach has been used to efficiently assemble 3-vinyl indoles **13** with alkenes (Scheme 6.20) [28]. It was observed in this case that the nature of the nitrogen substituent influences the ability to trap the 3-palladated indole, with carbamates providing the highest yields. The elimination of HX from the palladium after β-hydride elimination creates a Pd(0) complex that is unable to mediate subsequent cyclizations. As such, co-oxidants, such as stoichiometric copper(II) salts, are used in this reaction to regenerate the palladium(II) catalyst. However, by modifying reaction conditions, Lu has found that the addition of excess LiBr can inhibit β-hydride elimination, and instead allow the formation of the reduced product (Scheme 6.21) [29]. This not only allows access to 3-alkyl substituted indoles, but also eliminates the need for stoichiometric oxidants.

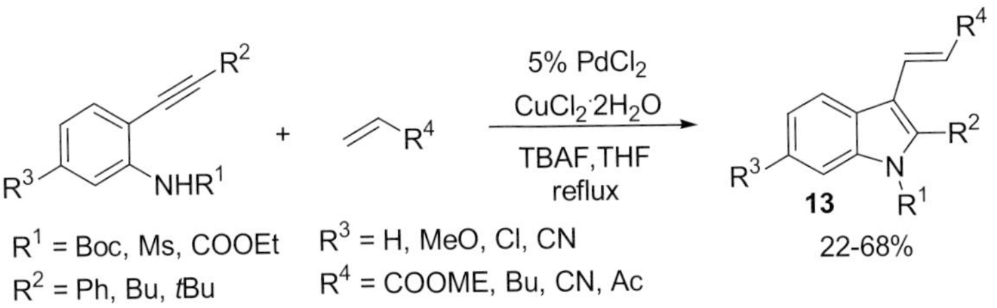

Scheme 6.20

These palladium intermediates can also be intercepted by carbonylative nucleophilic trapping (Scheme 6.13d). This approach has been applied to a range of heterocyclic cyclizations, often in concert with alcohols to generate ester-substituted

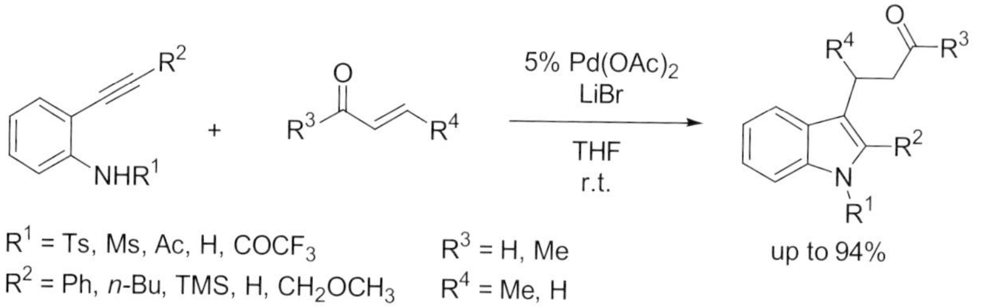

R^1 = Ts, Ms, Ac, H, $COCF_3$ R^3 = H, Me
R^2 = Ph, n-Bu, TMS, H, CH_2OCH_3 R^4 = Me, H

5% $Pd(OAc)_2$, LiBr, THF, r.t.; up to 94%

Scheme 6.21

products. In many cases, the oxidant employed can be simple oxygen. An example of this approach to pyrrole synthesis is shown in Scheme 6.22 [30].

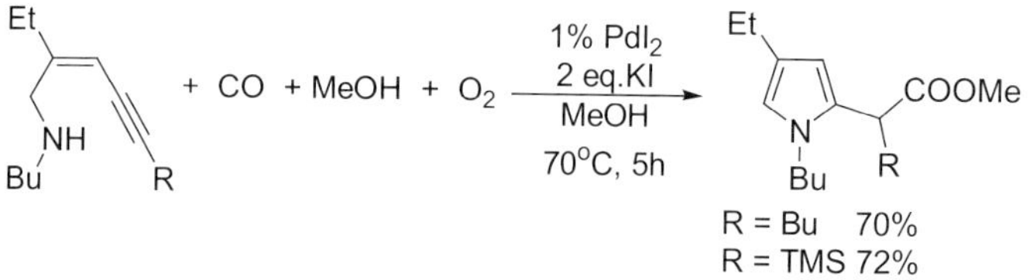

+ CO + MeOH + O_2; 1% PdI_2, 2 eq.KI, MeOH, 70°C, 5h

R = Bu 70%
R = TMS 72%

Scheme 6.22

6.2.3
Heteroatom Addition to Allenes

In much the same way that palladium complexes can activate alkynes towards intramolecular cyclization upon π-complexation, a similar reaction manifold has also been employed using allenes. This chemistry has been commonly employed in the assembly of substituted furans. For example, Huang reported that conjugated ynones can undergo isomerization under palladium-catalyzed conditions to form furans (Scheme 6.23) [31]. This reaction was postulated to proceed via a palladium-

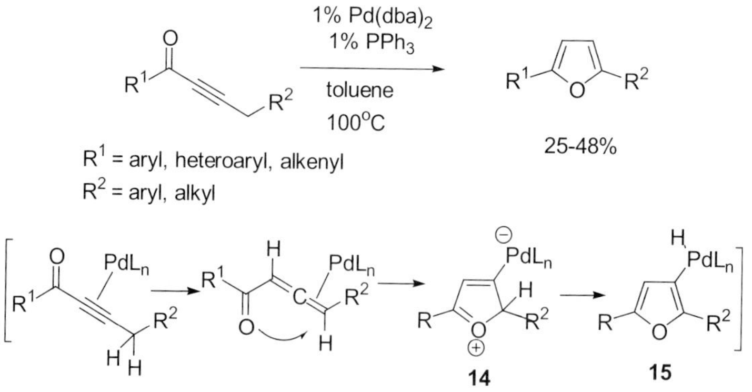

1% $Pd(dba)_2$, 1% PPh_3, toluene, 100°C; 25-48%

R^1 = aryl, heteroaryl, alkenyl
R^2 = aryl, alkyl

14 **15**

Scheme 6.23

catalyzed isomerization to an allene, which undergoes subsequent palladium-catalyzed cyclization. This process has the advantage of employing easily accessible ketones as substrates, and Ling has more recently demonstrated it can be used to generate a diverse variety of 2,5-substituted furans, or, depending upon the catalyst, bifurans [32].

As an alternative to keto-alkyne isomerization, allenylketones themselves can undergo cycloisomerization, providing access to furans and dimeric products, depending upon the palladium catalyst and allene employed (Scheme 6.24) [33]. The mechanism of this transformation has explored by Hashmi, and postulated to proceed via a Pd(II)/Pd(IV) cycle, wherein palladium mediates cyclization to form **14** (Scheme 6.23), followed by either proton migration and reductive elimination (forming simple furans) or allene coupling via either a carbene or insertion. This approach can be employed to form exclusively monomeric cycloisomerization products with more sterically encumbered substituted allenes [34].

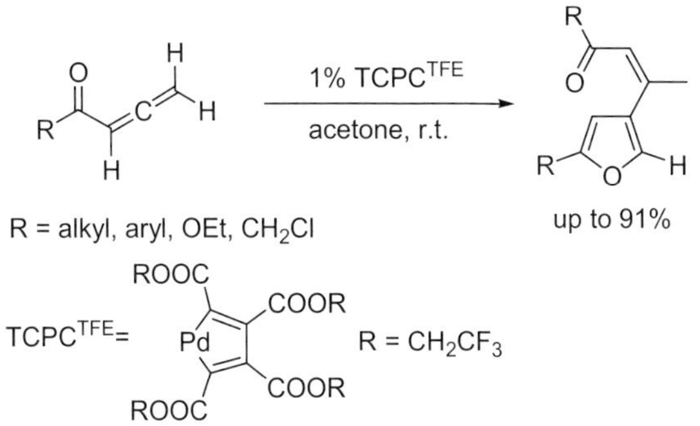

1% $TCPC^{TFE}$, acetone, r.t.

R = alkyl, aryl, OEt, CH_2Cl

up to 91%

$TCPC^{TFE}$= ROOC, COOR, Pd, COOR, ROOC; R = CH_2CF_3

Scheme 6.24

As in the above chemistry with alkynes, the palladium-bound products of this cycloisomerization (e.g., **15**, Scheme 6.23) can also be trapped with the addition of external reagents prior to the elimination step. This typically involves the oxidative addition of R−X substrates (e.g., aryl or vinyl halides, allylic substrates, etc.) to a palladium(0) catalyst to create the palladium(II) complex needed for cycloisomerization, followed by reductive elimination of the substituted furan product. As illustrated with the example in Scheme 6.25, this provides a route to selectively install substituents into the 3-furan position; a derivatization difficult via more traditional electrophilic aromatic substitution routes [35].

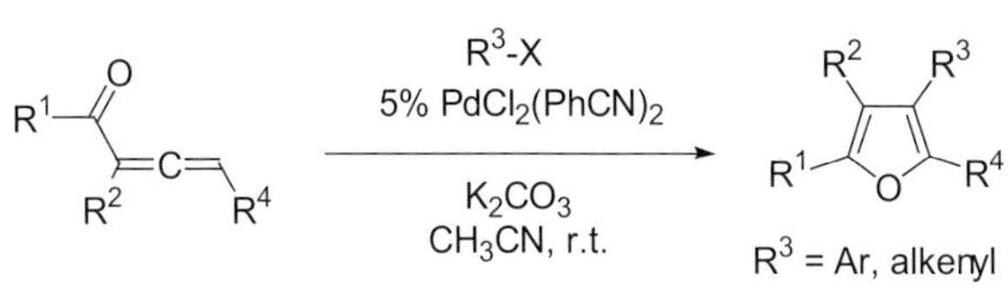

R^3-X, 5% $PdCl_2(PhCN)_2$, K_2CO_3, CH_3CN, r.t.

R^3 = Ar, alkenyl

Scheme 6.25

Alcohol tethered cumulenes can also undergo cycloisomerization to generate furan derivatives under palladium catalysis [36]. This transformation proceeds in an analogous fashion to those with alkynes, where, in this case a palladium(II) catalyst can activate the addition of the alcohol moiety across the remote double bond of **16**, followed by proton migration. These building blocks are accessible via the SmI_2-mediated ring opening of alkynyl-substituted epoxides, a process that can occur in a single reaction pot with subsequent palladium-catalyzed cyclization. As shown in Scheme 6.26, this reaction can be mediated by palladium(II) complexes to generate the simple isomerization products, coupled with the initial oxidative addition of aryl or allyl halides to generate 3-substituted furans, or even combined with Heck chemistry to create 3-vinyl furans.

Scheme 6.26

While furans have been the main focus of palladium-catalyzed allene cycloisomerization, pyrroles can also be generated via reaction of allenyl-substituted amines. A number of metal catalysts have been reported to mediate the cyclization of these substrates to pyrrolines, however, the use of palladium catalysis can allow the concomitant incorporation of aryl functionality into the 3-position, as shown in Scheme 6.27. At elevated temperatures, oxidation of the pyrroline occurs to afford pyrroles [37].

Scheme 6.27

6.2.4
Heteroatom Additions to Alkenes

This same palladium π-acidity can be employed to activate alkenes towards intramolecular cyclization with heteroatom–hydrogen bonds. In contrast to alkynes or allenes, the addition of N−H, O−H and other heteroatom–hydrogen bonds across the alkene would formally create a saturated carbon–carbon bond, rather than the unsaturation necessary for aromatic heterocycles. As such, subsequent β-hydride elimination is often required to direct these reactions towards aromatic products, in a Wacker-type alkene oxidation (Scheme 6.28).

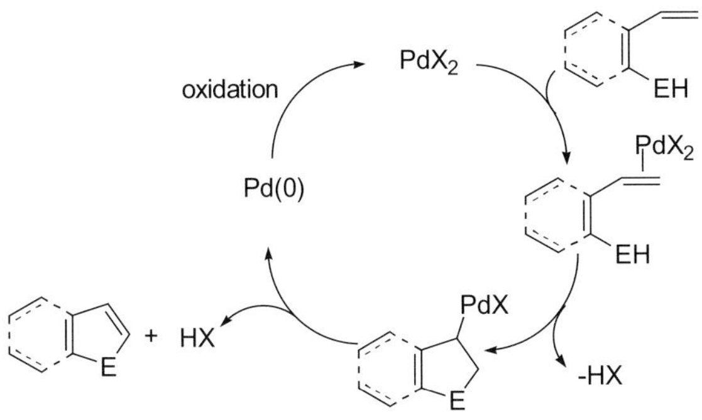

Scheme 6.28

The use of stoichiometric palladium reagents to mediate this reaction has been known for several decades, with cyclization often occurring with reduction to palladium(0), which is inactive for further reaction [38]. Closely following upon these examples, it was found that the addition of stoichiometric oxidants, such as copper(II) salts, can allow catalytic cyclization. For example, Hosokawa noted that the addition of $Cu(OAc)_2$ in the presence of oxygen with 5 mol% $Pd(OAc)_2$ can allow the catalytic cyclization of 2-allylphenols to benzofurans (Scheme 6.29) [39]. This reaction was found to proceed even at ambient temperature in DMF/water solvent, and with good diversity.

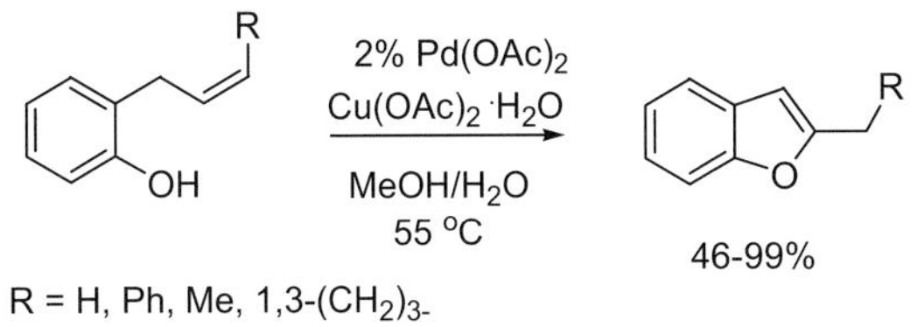

Scheme 6.29

This oxidative cyclization to alkenes has been applied to a range of aromatic heterocycles. For example, Hegedus demonstrated that both allylic and vinylic anilines can undergo palladium-catalyzed cyclization as a general route to indoles (Scheme 6.30) [40]. The formation of these 2-vinyl-*N*-tosyl-aniline substrates can also be accomplished via palladium catalysis, via a Stille coupling of 2-haloaniline derivatives with vinylstannanes, followed by subsequent conversion to the indole via oxidative cyclization [41].

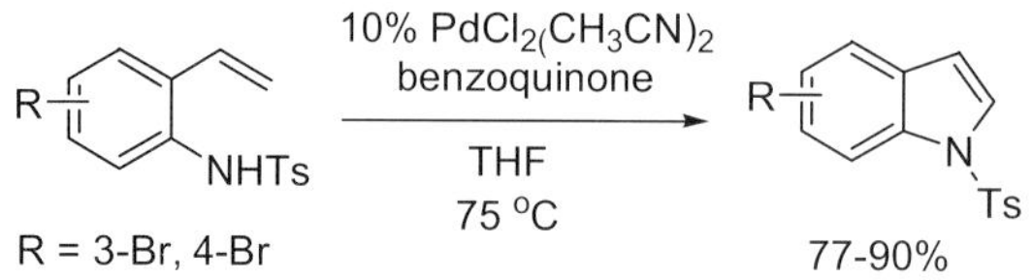

10% $PdCl_2(CH_3CN)_2$
benzoquinone
THF
75 °C
NHTs
R = 3-Br, 4-Br
N
Ts
77-90%

Scheme 6.30

This same synthetic approach can also be applied to keto-substituted alkenes, as a route to generate furans (Scheme 6.31), as well as in pyrrole synthesis (Scheme 6.32) [42]. In the first case, the enol forms of these ketones can be considered to undergo initial cyclization on the double bond, followed by β-hydride elimination (and product rearrangement) and subsequent catalyst reoxidation by $CuCl_2$. Interestingly, good selectivity was observed for the cyclization of alkyl-substituted ketone into the furan core over aryl-ketones, allowing the generation of diversely R^1/R^2-substituted furans.

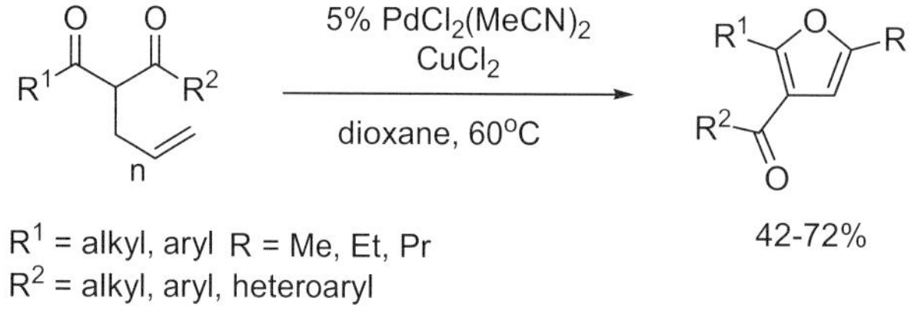

5% $PdCl_2(MeCN)_2$
$CuCl_2$
dioxane, 60°C
R^1 = alkyl, aryl R = Me, Et, Pr
R^2 = alkyl, aryl, heteroaryl
42-72%

Scheme 6.31

10% $PdCl_2(MeCN)_2$
THF, r.t.
16
R^1 = H, Me; R^2=H, Me
R^3 = H, Me; R^4=H, vinyl
up to 83%

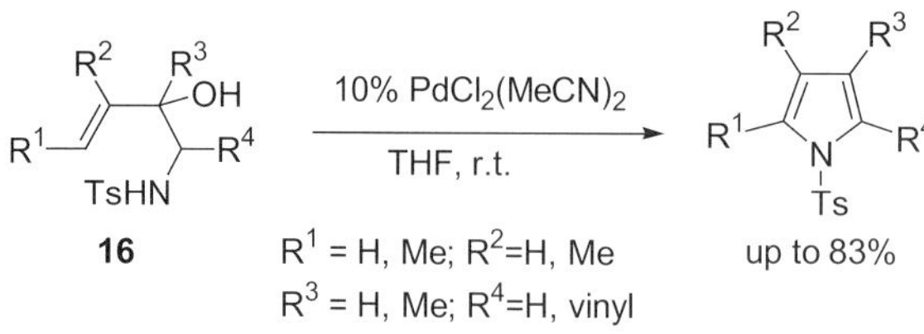

Scheme 6.32

As described above, the addition of amines to alkenes typically requires an external oxidant to regenerate the palladium(II) catalyst, due to the rapid elimination of HX from the palladium upon β-hydride elimination. An alternative, however, is to allow aromatization via β-X elimination, which does not undergo reductive elimination, and allows the palladium to remain in the Pd(II) state upon aromatization. An example of this is in the recent work of Venturello, where it was observed that the palladium-catalyzed intramolecular addition of *N*-tosyl aniline derivatives to dienes allows the generation of quinolines (Scheme 6.33) [43]. This latter reaction is believed to proceed in a similar fashion to that above, wherein coordination of the palladium catalyst to the diene initiates nucleophilic attack. However, β-hydride elimination is followed by re-insertion and finally β-tosyl elimination to regenerate the palladium (II) catalyst for further reaction. This approach both eliminates the need for external oxidants, and allows access to the six-membered quinoline ring system.

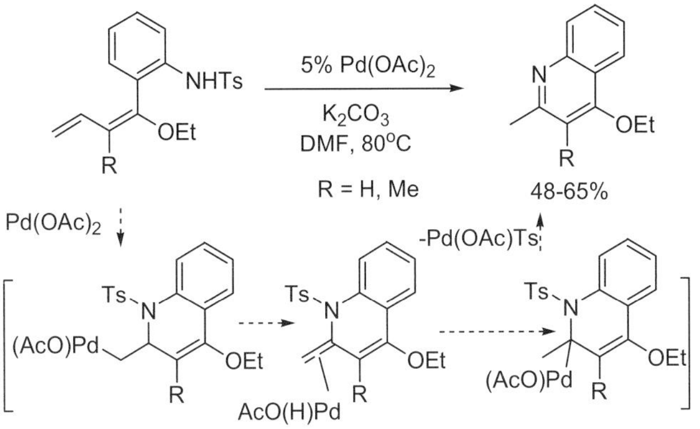

Scheme 6.33

The precursors used above in generating the aniline derivatives for indole synthesis are often nitroarenes, which can be easily derivatized with the alkene unit. As such, an interesting finding in this cyclization strategy was that these nitroarenes can themselves be directly converted into indole derivatives. This was noted under palladium catalysis in the work of Watanabe, as illustrated in Scheme 6.34 [44]. While the specific mechanism for this reaction is still unclear, it is postulated that the carbon monoxide serves to reduce the nitro unit, either to a nitrene or nitroso fragment, which undergoes cyclization with the pendant unsaturated group. Research into this transformation has found that, with the correct catalyst, $SnCl_2$ is not required, and as little as 1 atm CO can be employed, providing access to indoles with good product diversity [44b]. A useful feature of this protocol is that halo-substituted nitroarenes can be readily functionalized via palladium-catalyzed carbon–carbon or carbon–heteroatom bond forming reactions. Södenberg has demonstrated how this approach can be used to develop efficient effective routes to a range of heterocycles [45].

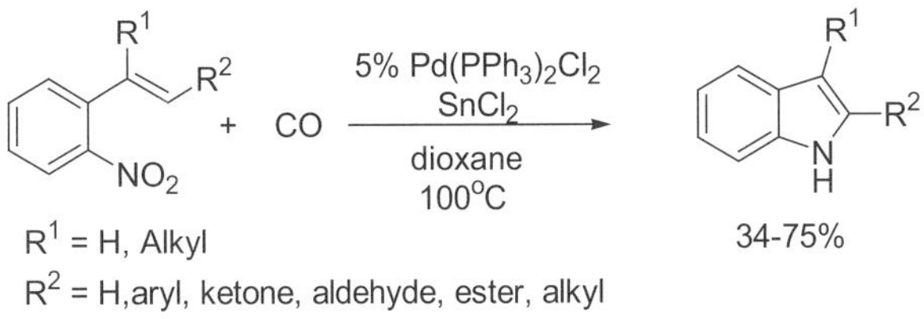

Scheme 6.34

6.3
Palladium-Catalyzed Carbon–Heteroatom Bond Forming Reactions

Perhaps the most important reaction manifold offered by palladium catalysis is its ability to undergo oxidative addition and reductive elimination reactions. When coupled with other transformations, this can provide a powerful method to construct σ-bonds within substrates. For example, the oxidative addition of aryl or vinyl halides (or analogues) to palladium(0), followed by transmetallation and with an organometallic partner and reductive elimination provides a mild route to generate carbon–carbon bonds (e.g., Suzuki, Stille, Negishi, or Sonogashira couplings). More recently, the discovery that palladium can mediate similar transformations with heteroatom–hydrogen bonds replacing these organometallic reagents has been a major advance in carbon–heteroatom bond formation. This reactivity has also provided a useful approach to the construction of heterocycles (Scheme 6.35) [1b].

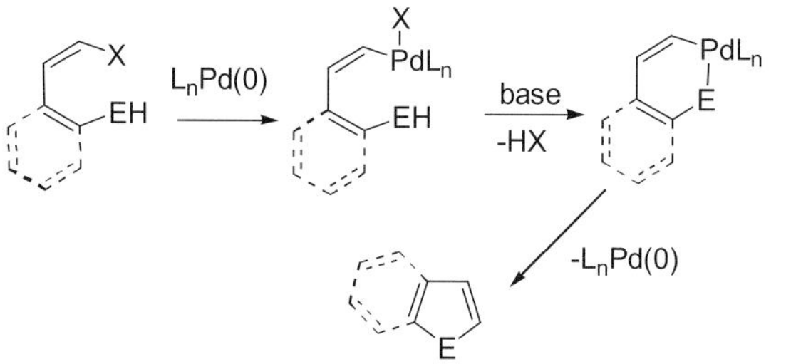

X; EH; $L_nPd(0)$; X; PdL_n; EH; base; -HX; PdL_n; E; $-L_nPd(0)$; E

Scheme 6.35

6.3.1
Palladium-Catalyzed Carbon–Nitrogen Bond Formation

Intramolecular carbon–nitrogen bond formation can be used to assemble a range of nitrogen-containing aromatic heterocycles (e.g., indoles, benzimidazoles, pyrrozoles, quinolines, etc.). An early example of the application of this chemistry to indoles was reported by Watanabe, in the intramolecular cyclization of hydrazones with *ortho*-chloroarenes (Scheme 6.36) [46]. This reaction can be considered to

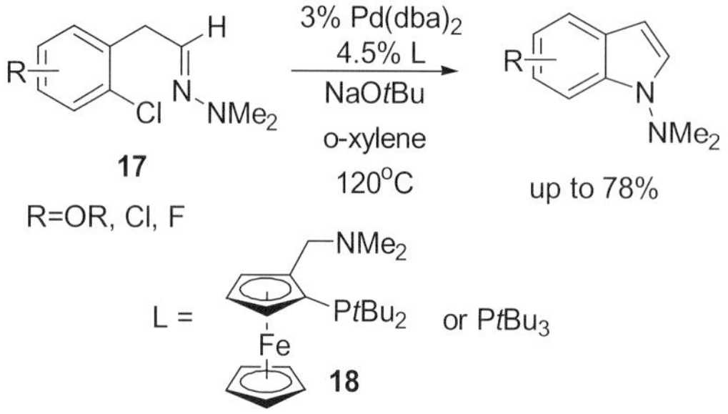

R; H; N; Cl; NMe_2; **17**; R=OR, Cl, F; 3% $Pd(dba)_2$; 4.5% L; NaO*t*Bu; o-xylene; 120°C; R; N; NMe_2; up to 78%; L = ; NMe_2; $PtBu_2$; Fe; **18**; or $PtBu_3$

Scheme 6.36

proceed via an intramolecular C–N bond formation from the enamine tautomer of **17**. In the presence of bulky, electron-rich phosphine ligands, such as $^tBu_3P$ or **18**, this proceeds in good yields. In addition to hydrazones, pre-synthesized enamines, as well as simple imines, can also undergo similar cyclizations to form indoles [47].

An analogous reaction sequence can be used to generate benzimidazoles and aminobenzimidazoles, involving the palladium-catalyzed intramolecular cyclization of aryl bromide-substituted amidines and guanidines, respectively (e.g., Scheme 6.37) [48]. With aryl bromides, simple $Pd(PPh_3)_4$ or Pd_2dba_3/PPh_3 catalysts are sufficient to mediate cyclization. This same approach is equally applicable to indazoles and polycyclic benzimidazoles [49].

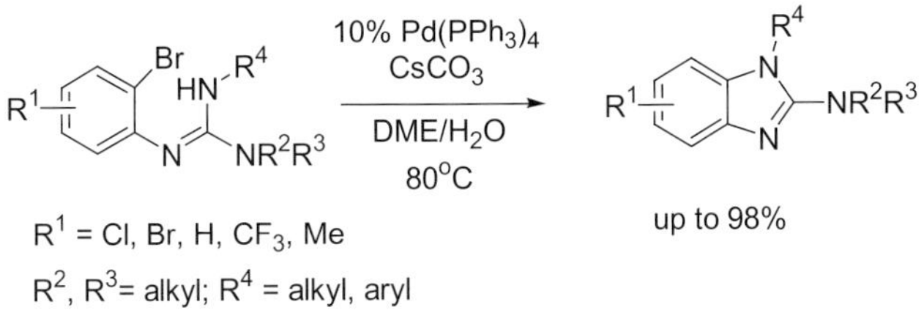

10% $Pd(PPh_3)_4$, $CsCO_3$, DME/H_2O, 80 °C; up to 98%

R^1 = Cl, Br, H, CF_3, Me
R^2, R^3 = alkyl; R^4 = alkyl, aryl

Scheme 6.37

In addition to simply mediating the formation of carbon–nitrogen bonds, palladium catalysis can be used to construct these precursors in concert with cyclization. One example of this was demonstrated by Barluenga, where palladium catalysis is employed to form two separate bonds; first the carbon–carbon bond via the arylation of the azaallylic anion of imine **19**, followed by catalytic carbon–nitrogen bond formation (Scheme 6.38) [50].

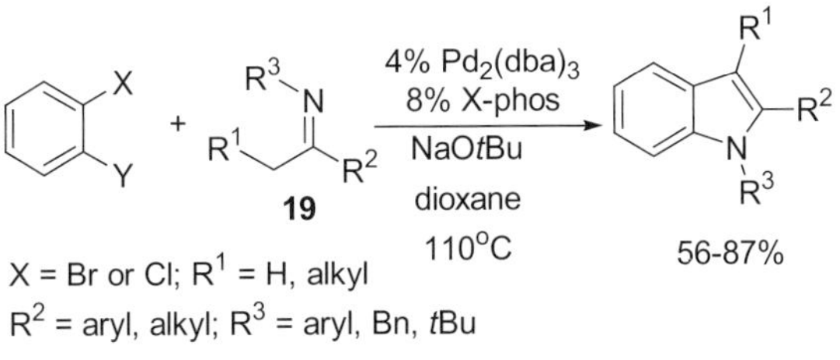

19; 4% $Pd_2(dba)_3$, 8% X-phos, NaO*t*Bu, dioxane, 110 °C; 56-87%

X = Br or Cl; R^1 = H, alkyl
R^2 = aryl, alkyl; R^3 = aryl, Bn, *t*Bu

Scheme 6.38

Palladium catalysis can also provide an avenue to derivatize the products of cyclization, often in concert with the cyclization itself. Lautens, Bisseret and Alper have all effectively demonstrated the power of this approach in the construction of indole derivatives with a high degree of modularity [51]. These tandem palladium-catalyzed methods have typically employed 1,1-dibromoalkenes in indole cyclization, wherein one vinyl bromide bond is available for palladium-catalyzed indole formation, and the second for subsequent functionalization. As shown in Scheme 6.39,

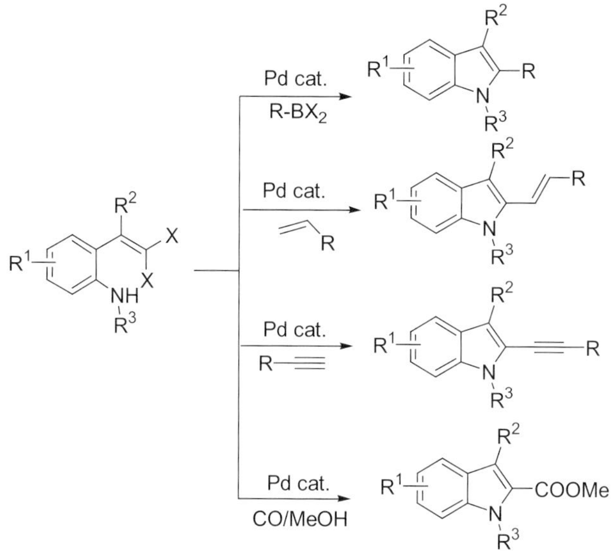

Scheme 6.39

these can be coupled with a range of external reagents, providing a useful method to build up molecular complexity, all within a single palladium-catalyzed operation.

Consecutive palladium-catalyzed carbon–nitrogen bond forming reactions can also be employed to access aromatic heterocycles. Nozaki has demonstrated this approach in the double *N*-arylation of **20** as a route to efficiently generate carbazoles (Scheme 6.40) [52]. This reaction can proceed with a range of alkyl and aryl amines, replacement of the aryl halides with sulfonates, and can even be performed consecutively to construct extended heteroacenes.

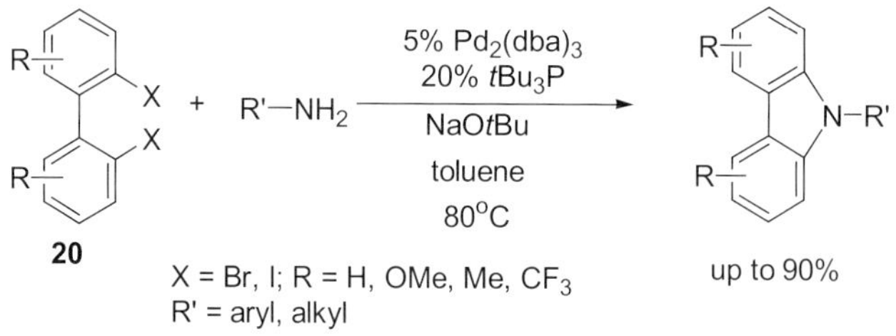

Scheme 6.40

Willis has similarly noted that indoles can be prepared via the cascade coupling of the aryl–halide and vinyl–triflate bonds in styrene **21** with primary amines (Scheme 6.41). Despite the fact that two different types of C–X bonds must be activated, this process proceeds with a single palladium catalyst (Pd_2dba_3/Xantphos), and a range of primary amines, carbamates and amides. More recently, even vinyl chlorides were found to participate in this coupling reaction [53].

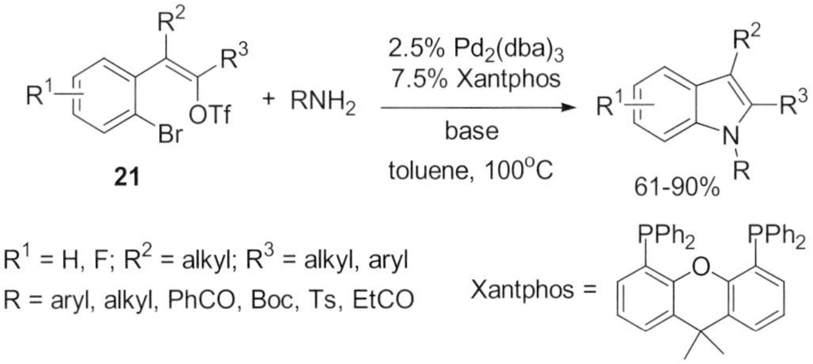

Scheme 6.41

As an alternative to primary amines in these double carbon–nitrogen bond forming reactions, Maes has demonstrated that **22** can undergo consecutive palladium-catalyzed carbon–nitrogen bond formation with *ortho*-dihalopyridine derivatives (Scheme 6.42). This reaction proceeds in good yield with a number of amidine variants, and provides a route to the construction of polycyclic imidazoles [54].

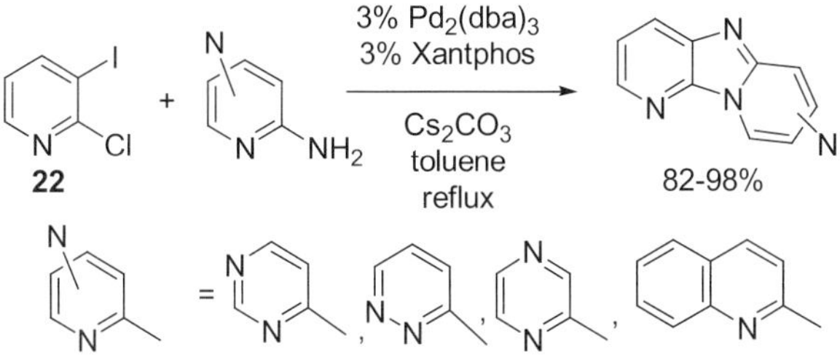

Scheme 6.42

6.3.2
Palladium-Catalyzed Carbon–Oxygen Bond Formation

Mirroring the development of palladium-catalyzed methods to construct aromatic carbon–nitrogen bonds, aromatic carbon–oxygen formation has also been developed and applied to the construction of oxygen-containing aromatic heterocycles. An early example of this is in the cyclization of 2-aryl substituted ketones to benzofurans (Scheme 6.43) [55]. This chemistry can be considered to proceed in a fashion analogous to the indole syntheses above, wherein an *in situ* generated enolate can undergo cyclization upon palladium-catalyzed oxidative addition of the aryl–bromide or aryl–chloride bond. Notably, this transformation can be equally applied to thioketones to generate benzothiophenes.

In an interesting extension, Oh has recently shown that palladium complexes can be used to catalyze the cascade cyclization of **23** to furan derivatives [56]. This reaction presumably involves the initial generation of a palladium–hydride bond under the reaction conditions, which can mediate sequential insertion of the alkyne units to

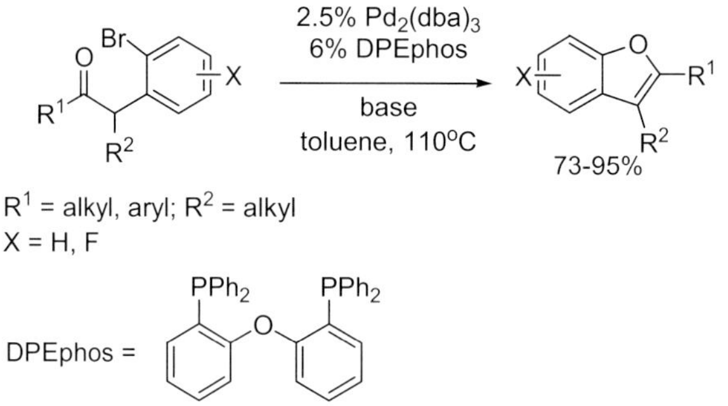

R^1 = alkyl, aryl; R^2 = alkyl
X = H, F

DPEphos =

2.5% $Pd_2(dba)_3$
6% DPEphos
base
toluene, 110 °C
73-95%

Scheme 6.43

form **23** (Scheme 6.44). In the presence of formate, a formal hydride addition can lead to the generation of an enolate anion for subsequent cyclization to create the furan product. Performing this transformation in the presence of primary amines can also allow the generation of pyrroles [57].

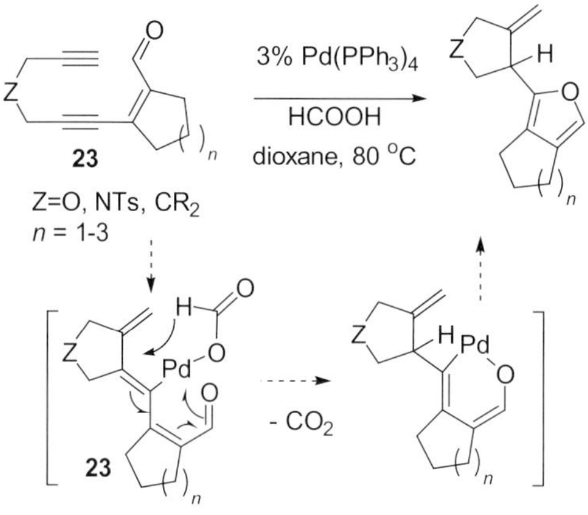

3% $Pd(PPh_3)_4$
HCOOH
dioxane, 80 °C
Z=O, NTs, CR_2
n = 1-3
- CO_2

Scheme 6.44

6.4
Palladium-Catalyzed Carbon–Heteroatom Bond Formation with Alkynes

Predating much of the work on intramolecular carbon–heteroatom bond formation for aromatic heterocycle synthesis was research of Larock and others involving the construction of indoles derivatives [1]. This reaction involves the coupling of *ortho*-halo-substituted anilines and internal alkynes. While appearing similar to the palladium-catalyzed synthesis of indoles via the generation of *ortho*-alkynylanilines (Section 2.1.1), this transformation employs internal alkynes, which cannot generate these intermediates. Instead, this reaction proceeds in a fashion similar to more recently developed carbon–nitrogen bond forming reactions (Scheme 6.45). The

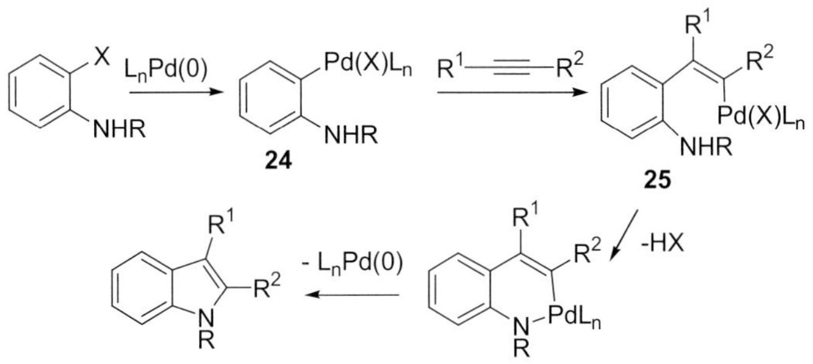

Scheme 6.45

initial step in this chemistry involves the oxidative addition of the aryl halide to generate **24**. While this intermediate cannot undergo cyclization with the pendant *ortho*-amino group, alkyne insertion can occur to form **25**, which can now undergo carbon–nitrogen bond formation via palladation of the NH bond, followed by reductive elimination.

Larock reported his initial results in this area in 1991, in the reaction of *ortho*-iodoanilines with internal alkynes [58]. This method provides an efficient route to assembly of the 2,3-disubstituted indole core from anilines and alkynes, both of which are easily generalized. This initial report employed N-protected anilines, however, subsequent research demonstrated that the parent aniline moiety can also participate in this chemistry [59].

One issue that can arise in this reaction is that of alkyne regiochemistry upon insertion into **24**. In the original reports, reasonably good selectivity has been noted to place the bulkier alkyne substituent (e.g., larger alkyl or aryl) at the 2-position, presumably due to steric influences during this alkyne insertion step. The development of further methods to control regioselectivity in these cyclizations has attracted the interest of several research groups. Larock has noted that coordinating alcohol groups in the alkyne can accentuate regioselectivity, presumably by coordination to palladium during insertion, to favor ultimately ending up in the 2-position of the indole [58, 59]. This strategy has since been shown to provide, for example, unusually high ratios of 2-(2-pyridyl)-substituted indoles relative to when the non-chelating 3- or 4-pyridyl units are employed (Scheme 6.46) [60]. A potentially more general method

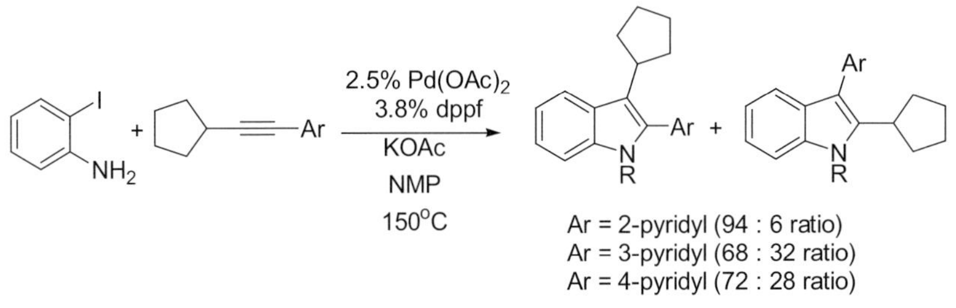

Scheme 6.46

to influence regioselectivity is to employ catalyst influences on the alkyne insertion step. This has been found by Konno, where PPh_3 favored the formation of 2-CF_3-substituted indoles, while the bulkier P(*o*Tol)$_3$ ligand instead led to the generation of the 3-CF_3-substituted products (Scheme 6.47) [61].

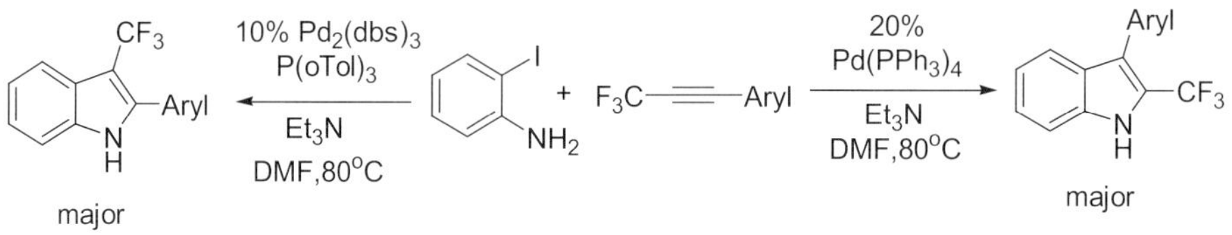

Scheme 6.47

This cyclization strategy has been employed by multiple groups as an efficient method to build up 2,3-substituted indoles from available precursors. This includes work on solid supports [62], the use of heteroaryl spacers [63], the construction of polycyclic products [64], and, more recently, moving from aryl iodides to less expensive aryl bromides and aryl chlorides [65]. The latter was demonstrated by Senanayke and coworkers with the use of the now common bulky electron-rich phosphines to activate the aryl–chloride bond for cyclization (Scheme 6.48).

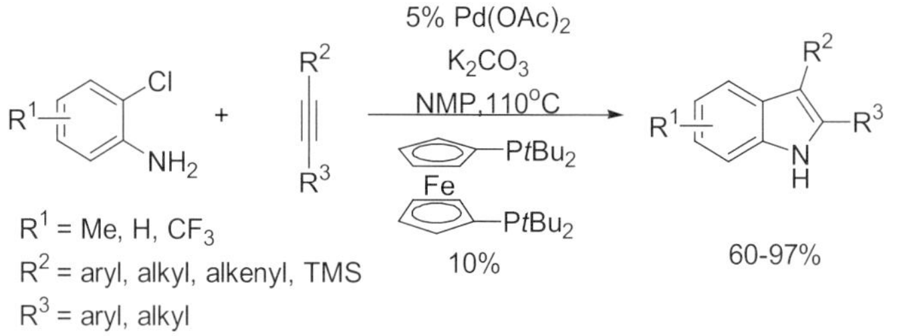

Scheme 6.48

As an alternative to the use of aniline derivatives in this chemistry, Lebel has recently reported that carboxylic acids can be coupled with sodium azide and alkynes to form indoles (Scheme 6.49) [66]. In this chemistry, a Curtius rearrangement first occurs to generate *in situ* an *ortho*-iodoarylisocyanate, which in the presence of nucleophiles can convert to the NH unit required for palladium-catalyzed indole

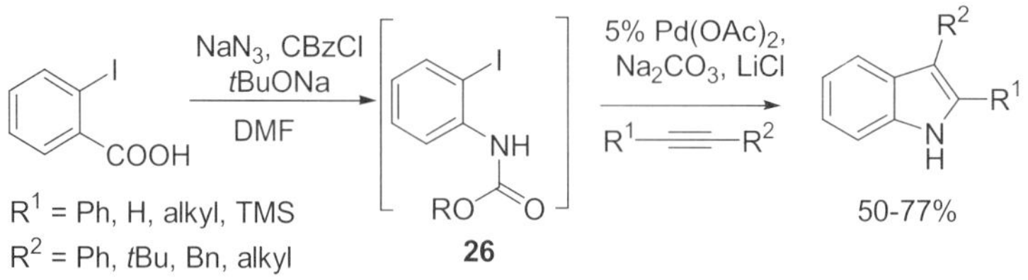

Scheme 6.49

synthesis. In addition to providing access to **26** for cyclization, this process can also be coupled with amines to generate urea variants of **26**, providing a synthesis of *N*-substituted indoles, four separate units in one pot.

These annulations can be extended beyond indole syntheses. For example, the analogous reaction with *ortho*-iodophenols can provide an effective method to build up benzofurans (Scheme 6.50) [67]. Presumably due to the lower nucleophilicity of the phenolic oxygen, these reactions generally require higher reaction temperatures. Nevertheless, similar regioselectivity was found here as with the indole synthesis, and, in particular, 2-silyl-substituted benzofuran is often obtained with high selectivity. This same approach can be applied to the construction of pyrroles and furans [68].

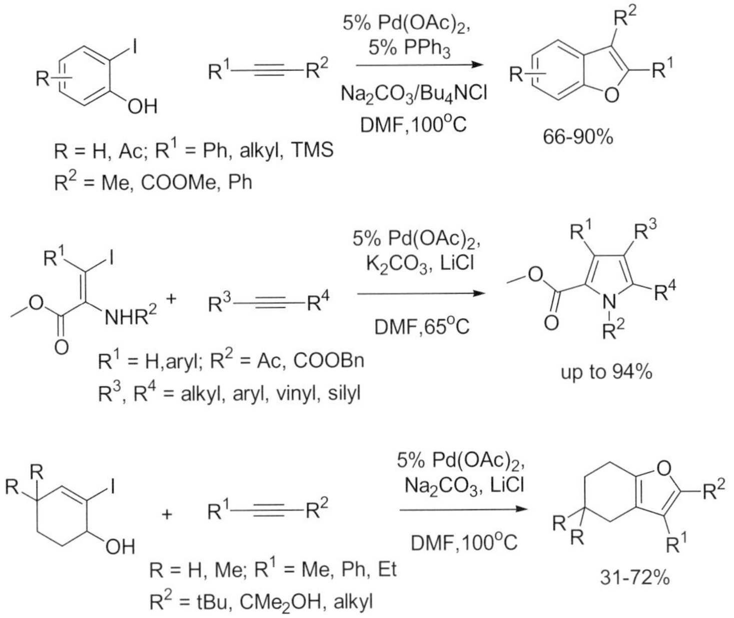

5% $Pd(OAc)_2$, 5% PPh_3, Na_2CO_3/Bu_4NCl, DMF, 100 °C; 66-90%

R = H, Ac; R^1 = Ph, alkyl, TMS
R^2 = Me, COOMe, Ph

5% $Pd(OAc)_2$, K_2CO_3, LiCl, DMF, 65 °C; up to 94%

R^1 = H,aryl; R^2 = Ac, COOBn
R^3, R^4 = alkyl, aryl, vinyl, silyl

5% $Pd(OAc)_2$, Na_2CO_3, LiCl, DMF, 100 °C; 31-72%

R = H, Me; R^1 = Me, Ph, Et
R^2 = tBu, CMe_2OH, alkyl

Scheme 6.50

This approach can also be used to generate six-membered ring heterocycles. In order to expand the five-membered heterocyclic ring of indoles, an extra carbon atom is required. Larock has found that *N-tert*-butyl-substituted imines are viable in this role, with the palladium-catalyzed coupling of imine **27** with internal alkynes leading to the formation of isoiquinolines (Scheme 6.51) [69]. This reaction presumably proceeds in a similar fashion to those for indoles above, in this case leading to an iminium salt, which can undergo loss of the stabilized formal *tert*-butyl cation and generation of isobutene. Notably, *N*-methyl, -ethyl, -isopropyl, and -benzyl imines were all found to be ineffective in this reaction, while the *N-tert*-butyl imines led to the cyclization products in high yields. This same approach can be applied to pyridine synthesis, as well as employing various aromatic spacers between the carbon–halide and imine unit [70].

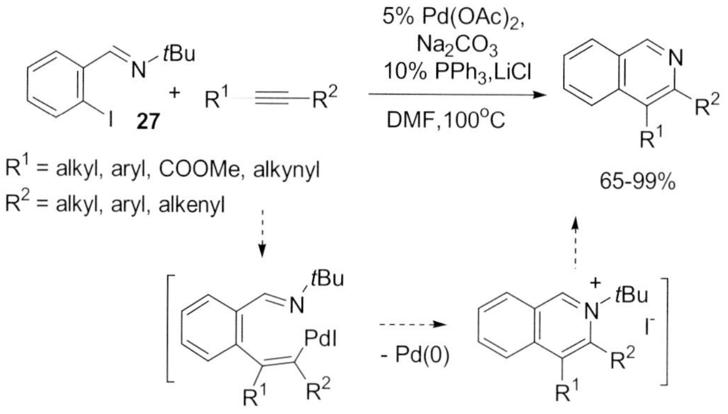

5% $Pd(OAc)_2$, Na_2CO_3, 10% PPh_3, LiCl, DMF, 100 °C

27

R^1 = alkyl, aryl, COOMe, alkynyl
R^2 = alkyl, aryl, alkenyl

PdI

- Pd(0)

65-99%

Scheme 6.51

The *N-tert*-butyl imine cyclization strategy can also be applied to the synthesis of isoquinolines via non-oxidative addition/reductive elimination processes. For example, it has been found that the *ortho*-alkynyl substituted imine **28** can undergo palladium-catalyzed cyclization in an analogous fashion to the *ortho*-alkynylanilines noted in Section 6.2.1. In this case, the extra carbon unit provides a synthesis of 2-substituted isoquinolines (Scheme 6.52) [71]. This cyclization can be coupled with the formation of **28** from imines and either TMS-substituted or terminal alkynes, providing access to a diverse range of substituted products with perfect regiocontrol.

While this cyclization method does not allow the incorporation of substituents into the 3-position, as is available with internal alkynes, in analogous chemistry to that in Section 6.2.2, the palladium bound intermediate of cyclization can be intercepted with a range of external reagents to functionalize this site. These include aryl or vinyl halides, allylic substrates, carbonylation, and Heck coupling with alkenes [72].

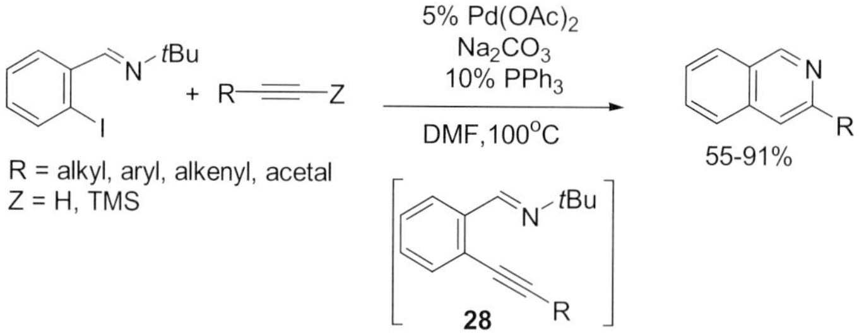

5% $Pd(OAc)_2$, Na_2CO_3, 10% PPh_3, DMF, 100 °C

R = alkyl, aryl, alkenyl, acetal
Z = H, TMS

28

55-91%

Scheme 6.52

6.5 Heck Cyclizations

The palladium Heck reaction of aryl halides (and related R–X bonds) with alkenes provides a valuable route to the assembly of substituted alkenes [73]. This reaction proceeds via a mechanism similar to cross coupling chemistry, though rather than a transmetallation step, alkene insertion followed by β-hydride elimination is the mechanism by which carbon–carbon bonds are generated (Scheme 6.53). As in the other chemistry above, this palladium-catalyzed reaction manifold can also be directly applied to heterocycle synthesis. Examples of these transformations with heteroatom bridged aryl halides and alkenes date back several decades [1].

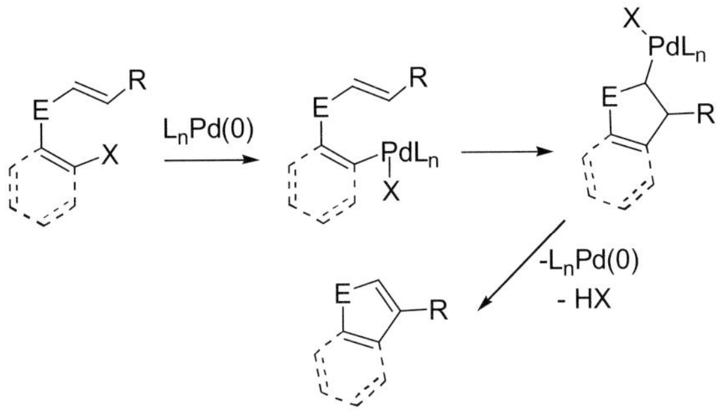

Scheme 6.53

In addition to these methods, carbon–heteroatom bond formation can also be coupled with these Heck cyclizations. One of the more straightforward methods to couple carbon–heteroatom bond formation with Heck cyclization is via enamine generation. For example, the reaction of enolizable aldehydes or ketones with 2-haloanilines provides a route to the construction of enamines for subsequent Heck cyclization. A recent example of this reaction was reported by Nazare, who demonstrated that *ortho*-chloroanilines can react with enolizable ketones to generate indoles (Scheme 6.54) [74]. This chemistry proceeds in good yields with a diverse variety of symmetrical and unsymmetrical ketones, provided there is only one enamine isomer, and can be applied to a range of substituted indoles. Zhu has demonstrated that enolizable aldehydes can be used in the cyclization [75]. This latter approach can provide a very effective method to selectively incorporate a range of alkyl, aryl or functionalized units into the 3-indole position, many of which are not easily accessible via other routes.

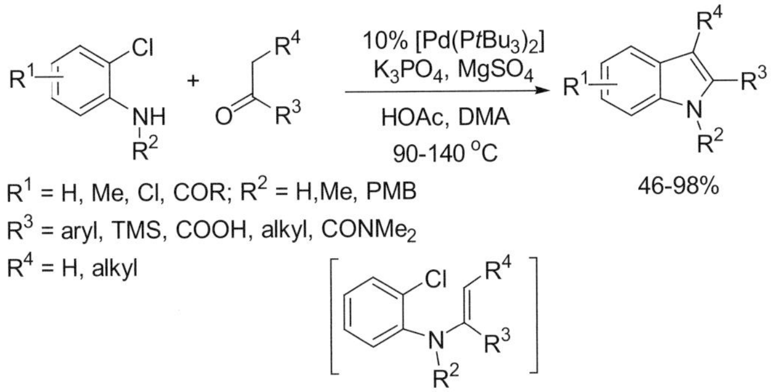

Scheme 6.54

A variant of this reaction involves the formation of the enamine precursors for Heck cyclization via the hydroamination of alkynes [76]. Examples of this transformation are shown in Scheme 6.55. Notably, the regiochemistry of these reactions can be controlled in the hydroamination step. As such this provides an interesting alternative to Larock chemistry for generating 2,3-substituted indoles (Section 6.4).

Palladium-catalyzed carbon–nitrogen bond forming reactions also provide an efficient route to generation of these precursors for Heck cyclization. Jorgensen

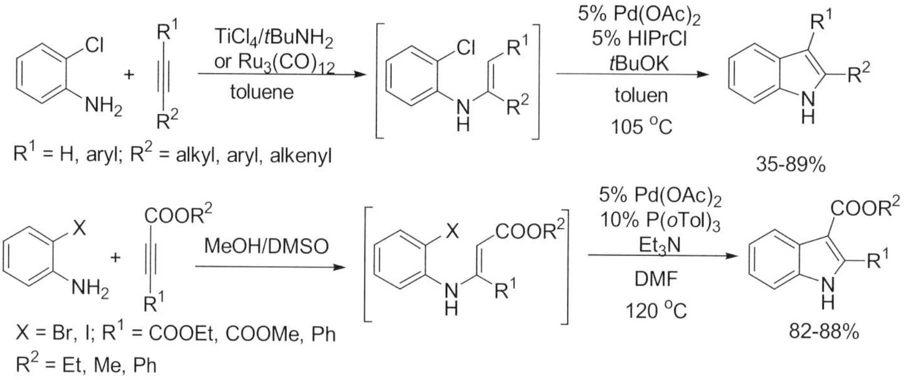

Scheme 6.55

has demonstrated how the palladium-catalyzed tandem carbon–nitrogen bond formation (with the aryl iodide)/Heck cyclization cascade between **29** and allyl amines can provide one-pot access to indoles (Scheme 6.56) [77]. Through screening, the dppf ligand in concert with Pd_2dba_3 was determined to be the optimal catalyst for these two steps, providing a route to generation of a number of 3-substituted indoles.

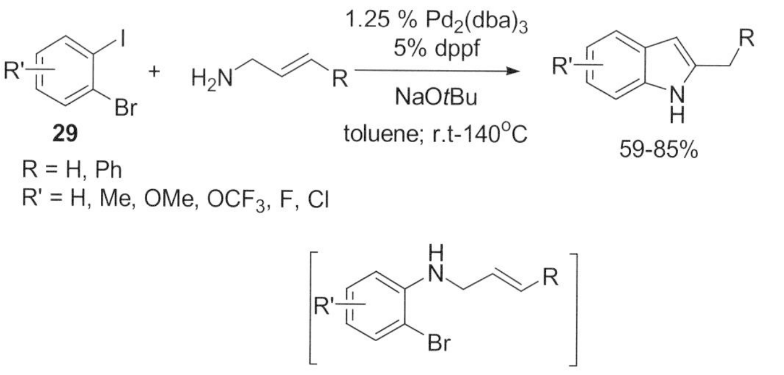

Scheme 6.56

An alternative disconnection involves the coupling of 2-haloanilines with vinyl bromides. Barluenga has shown that this approach can lead to either simple imines, or indoles, depending upon the ligand employed, with bulky phosphines allowing exclusive indole synthesis (Scheme 6.57) [78].

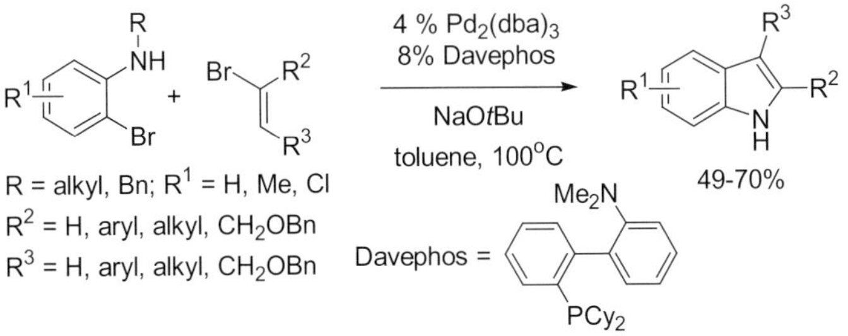

4 % $Pd_2(dba)_3$, 8% Davephos, NaO*t*Bu, toluene, 100°C; 49-70%

R = alkyl, Bn; R^1 = H, Me, Cl
R^2 = H, aryl, alkyl, CH_2OBn
R^3 = H, aryl, alkyl, CH_2OBn

Davephos = (Me_2N, PCy_2)

Scheme 6.57

Interestingly, this Heck-type palladium-catalyzed oxidative addition/insertion manifold can also be applied to the actual formation of the carbon–heteroatom bond. This was illustrated by Narasaka in the reaction of olefin-tethered oxime derivatives. This chemistry can be considered to arise from oxidative addition of the N–O bond to palladium (**30**) followed by the more classical olefin insertion and β-hydride elimination, ultimately allowing the assembly of pyrroles (Scheme 6.58) [79]. The nature of the OR unit was found to be critical in pyrrole formation, with the pentafluorobenzoylimine leading to selective cyclization and rearrangement to the aromatic product. An analogous approach has also been applied to pyridines and imidazoles [80].

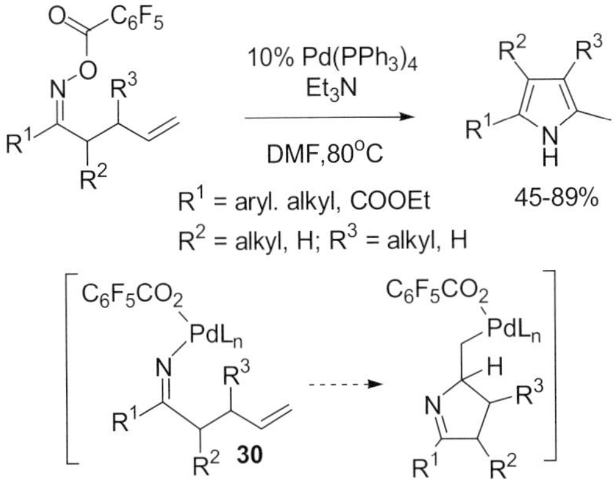

10% $Pd(PPh_3)_4$, Et_3N, DMF, 80°C; 45-89%

R^1 = aryl. alkyl, COOEt
R^2 = alkyl, H; R^3 = alkyl, H

Scheme 6.58

6.6 Palladium Catalyzed C–H Bond Activation

The last decade has seen an explosive growth in the development of palladium catalyzed C–H bond activation reactions [81]. These have provided intriguing methods to couple the C–H bond of simple arenes with activated aryl halides, and even to couple two different aromatic C–H bonded substrates. While the mechanism of this chemistry is still under investigation, it can be considered to arise from the sequential addition of two different bonds to palladium, with C–H bond activation proceeding via a palladium(II) complex (**31**, Scheme 6.59).

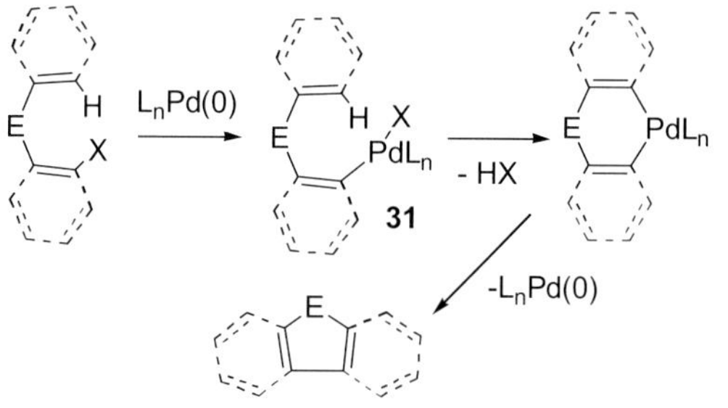

Scheme 6.59

The application of palladium-catalyzed C–H bond activation to aromatic heterocycles is growing as a viable synthetic strategy. Early examples involved substrates that incorporated the heteroatom between two aromatic units, where C–H bond activation was employed to generate the carbon–carbon bond for cyclization [82]. More recently, the generation of the carbon–heteroatom bond has been performed, either in concert with this carbon–carbon bond formation via C–H bond activation, or even using C–H activation as a route to generate the carbon–heteroatom bond in the heterocycle. In an example of the former, $Pd(OAc)_2/P^tBu_3$-catalyzed carbon–nitrogen bond formation was coupled with C–H bond activation as a route to the construction of carbazole derivatives. This was noted by Bedford in the reaction of 2-chloroanilines with simple aryl bromides (Scheme 6.60) [83]. In this reaction, it is postulated that initial coupling of the more reactive aryl bromide with the aniline N–H bond creates diarylamine **32** for a subsequent palladium-catalyzed operation, aryl–aryl coupling via C–H bond activation. This reaction proceeded in good yields with a number of aromatic precursors, and can be extended to C–O bond formation [84].

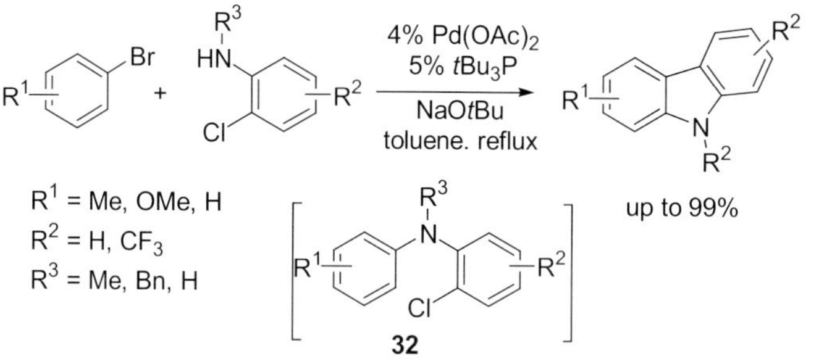

Scheme 6.60

An alternative strategy to the synthesis of nitrogen-containing heterocycles has been shown by Ackermann, with the coupling of simple anilines with **33**

(Scheme 6.61) [85]. This reaction has been found to proceed even with *ortho*-dichlorobenzene, a typically less reactive substrate towards carbon–heteroatom bond formation, and applicable to both carbazole and indole synthesis. Both 2- and 4-substituted anilines can be employed in this chemistry, as can unsymmetrically-substituted dihaloarenes.

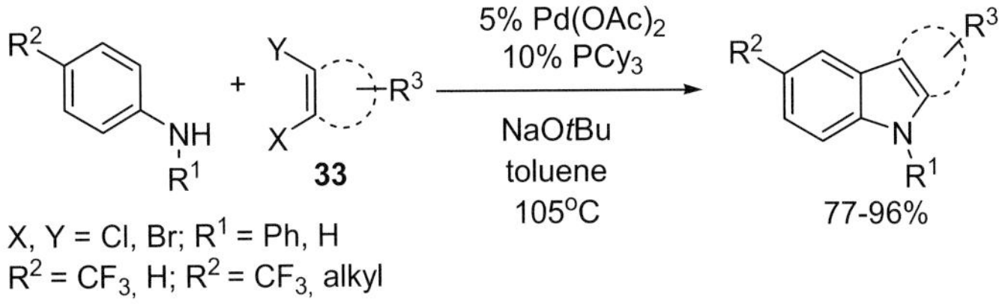

X, Y = Cl, Br; R^1 = Ph, H
R^2 = CF_3, H; R^2 = CF_3, alkyl

5% $Pd(OAc)_2$, 10% PCy_3, NaO*t*Bu, toluene, 105°C; **33**; 77-96%

Scheme 6.61

The coupling of palladium-catalyzed carbon–heteroatom formation with the direct, oxidative coupling of two C–H bonds has also been reported very recently. This reaction was noted by Fujii and Ohno in the assembly of carbazole derivatives from aryl triflates and aniline derivatives (Scheme 6.62) [86]. This consecutive C–N and C–C bond formation reaction proceeds with a single palladium catalyst, generated from $Pd(OAc)_2$/**34**, where carbon–nitrogen coupling was performed under an intert atmosphere, followed by the introduction of oxygen and acetic acid to mediate the subsequent C–H activation/cyclization. Notably, this reaction can tolerate a number of substituents, with product selectivity favoring activation of the sterically less encumbered C–H bond. A related approach has been reported by Ryu *et al.* for the effective synthesis of benzothiophenes [87].

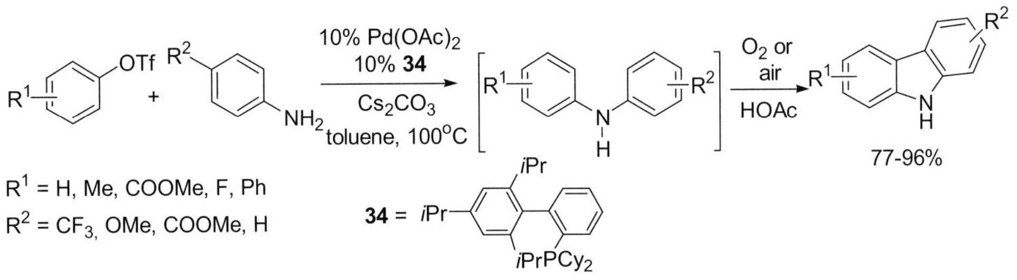

10% $Pd(OAc)_2$, 10% **34**, Cs_2CO_3, toluene, 100°C; O_2 or air, HOAc; 77-96%

R^1 = H, Me, COOMe, F, Ph
R^2 = CF_3, OMe, COOMe, H

34 = *i*Pr, *i*Pr, *i*PrPCy$_2$

Scheme 6.62

In the above chemistry, the carbon–heteroatom bonds are generally formed via established palladium-catalyzed coupling, and C–H activation is subsequently responsible for cyclization. In principle, carbon–heteroatom formation itself can also be performed via this oxidative coupling strategy. One version of this transformation was reported by Smitrovich in the reductive cyclization of 2-aryl-substituted nitroarenes (Scheme 6.63) [88]. As previously noted, carbon monoxide is postulated to reduce the nitro unit to a nitrene, which can undergo insertion into the adjacent C–H bond.

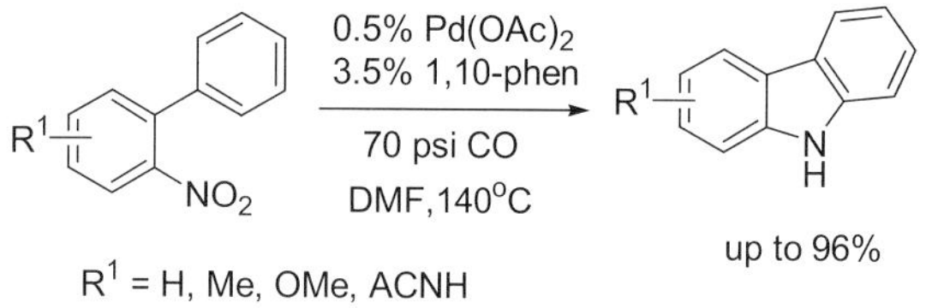

0.5% $Pd(OAc)_2$
3.5% 1,10-phen
70 psi CO
DMF, 140°C

R^1 = H, Me, OMe, ACNH

up to 96%

Scheme 6.63

More recently, Buchwald has reported that the oxidative coupling strategy for C–H bonds can be extended to the construction of aromatic heterocycles via simple C–H/N–H coupling. Building upon work on C–N coupling reactions, the authors noted that simple $Pd(OAc)_2$ can catalyze the formation of a range of carbazole to construct a range of carbazole derivatives (Scheme 6.64) [89]. Queiroz demonstrated that a similar strategy can be employed with enamides to construct indoles [90]. Both of these reactions proceed at elevated temperatures and employ copper salts in concert with air as the oxidant. However, recent work by Gaunt has shown that these cyclizations can be performed under mild conditions by modulating the oxidant employed (Scheme 6.65) [91]. The authors postulate that the role of the oxidant is in

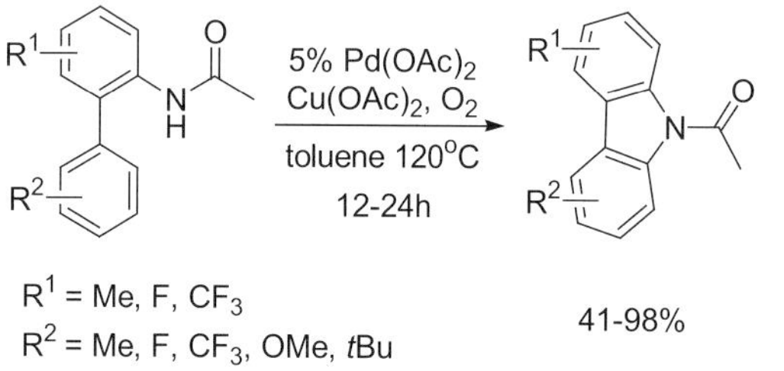

5% $Pd(OAc)_2$
$Cu(OAc)_2$, O_2
toluene 120°C
12-24h

R^1 = Me, F, CF_3
R^2 = Me, F, CF_3, OMe, *t*Bu

41-98%

Scheme 6.64

5% $Pd(OAc)_2$
$PhI(OAc)_2$
toluene , r.t.

R = Bn, alkyl, allyl

56-96%

R^1 or R^2 = CF_3, H, Me, OMe, F, COOMe, etc.

PdL_n **35** [O]

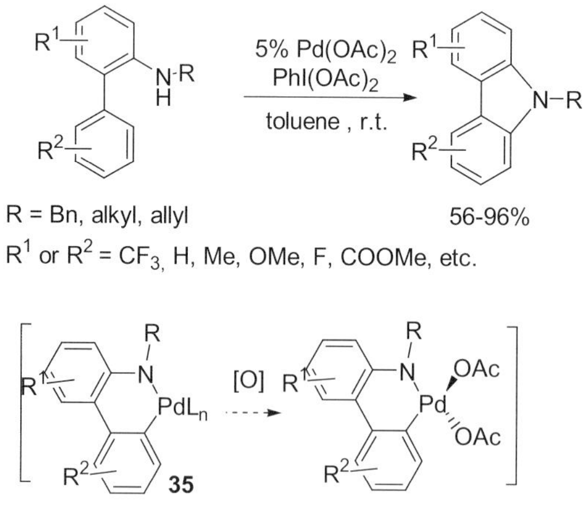

Scheme 6.65

facilitating reductive elimination from **35**, via its oxidation to a Pd(IV) intermediate, which can eliminate product more rapidly than the Pd(II) intermediate. This can allow ambient temperature coupling with $PhI(OAc)_2$.

While research into the use of C–H bond activation to assemble heterocycles is still developing, especially with regard to product diversity and selectivity with highly substituted targets, fundamental research into more efficient methods to employ this reaction manifold to assemble aromatic heterocycles is rapidly developing. As an example, Shi and coworkers have demonstrated that simple acetanilides and arenes can undergo selective coupling via consecutive bond activation reactions to form carbazoles (Scheme 6.66) [92]. This transformation involves a chelation assisted C–H activation of the acetanilide, which can undergo subsequent catalytic cyclization. While these reactions require an excess of arene to provide for selective cross coupling between the two units, under optimized conditions they can provide a straightforward route to carbazoles.

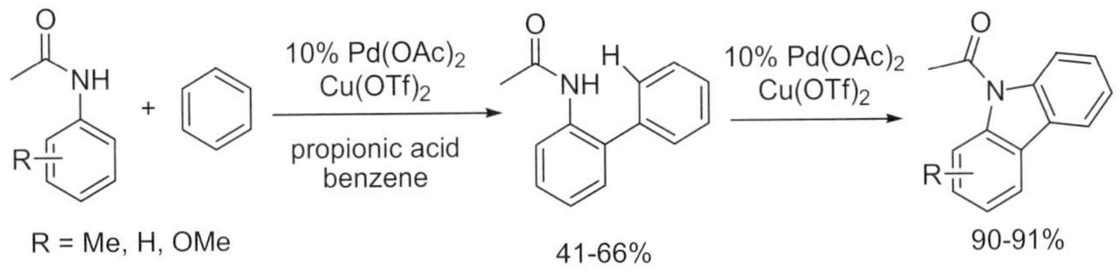

10% $Pd(OAc)_2$
$Cu(OTf)_2$
propionic acid
benzene
R = Me, H, OMe
41-66%
10% $Pd(OAc)_2$
$Cu(OTf)_2$
90-91%

Scheme 6.66

6.7 Multicomponent Coupling Reactions

The biological relevance of aromatic heterocycles has made the design of methods to generate these products both easily and with access to product diversity issues of significance. While many of the methods above show good product diversity at certain positions, in the construction of families of these products, the flexibility offered by multicomponent reactions can be appealing. In addition to their potential of streamlining syntheses by allowing multiple bond forming at once, the ability to simultaneously vary three or more building blocks within a single reaction can provide an attractive scaffold to tune heterocycle structures.

The past decade has seen a significant growth in the application of palladium catalysis to designing multicomponent reactions [93]. In the case of aromatic heterocycles, we have already noted several examples above. In addition, many multicomponent approaches fall outside these general synthetic methods, and instead rely upon alternative mechanisms. While by no means an exhaustive list, highlights of some of these general methods are given below, with an emphasis on those providing routes to product diversity.

One general reaction platform for multicomponent reactions was developed by Yamamoto, involving azides in [3 + 2] cycloaddition. For example, the reaction of internal alkynes, TMS-azide and allylic carbonates was found to provide a route to 1,2,3-substituted triazoles (Scheme 6.67) [94]. This chemistry, which is reminiscent of copper-catalyzed alkyne/substituted azide coupling, is postulated to proceed via the palladium-catalyzed oxidative addition of the allylic substrate, which allows subsequent generation of a palladium-azide for cycloaddition. This latter product can reductively eliminate to regenerate the palladium catalyst.

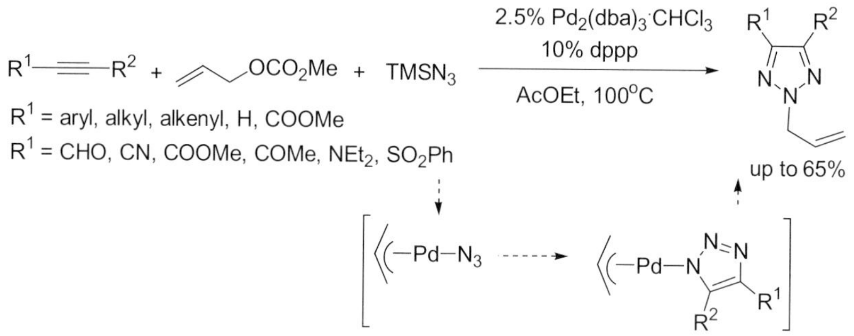

R^1—≡—R^2 + allyl-OCO_2Me + $TMSN_3$ → (2.5% $Pd_2(dba)_3 \cdot CHCl_3$, 10% dppp, AcOEt, 100°C) → N-allyl triazole, up to 65%

R^1 = aryl, alkyl, alkenyl, H, COOMe
R^1 = CHO, CN, COOMe, COMe, NEt_2, SO_2Ph

[(π-allyl)–Pd–N_3 → (π-allyl)–Pd–N(triazole)(R^1, R^2)]

Scheme 6.67

Yamamoto has since shown this same reaction protocol can provide multicomponent access to tetrazoles, via the replacement of alkyne cycloaddition with that of substituted nitriles (Scheme 6.68) [95]. This transformation proceeds with good product diversity. More recently, Barluenga has developed a bimolecular version for this reaction employing vinylbromides and sodium azide, providing access to *N*-unsubstituted triazoles [96].

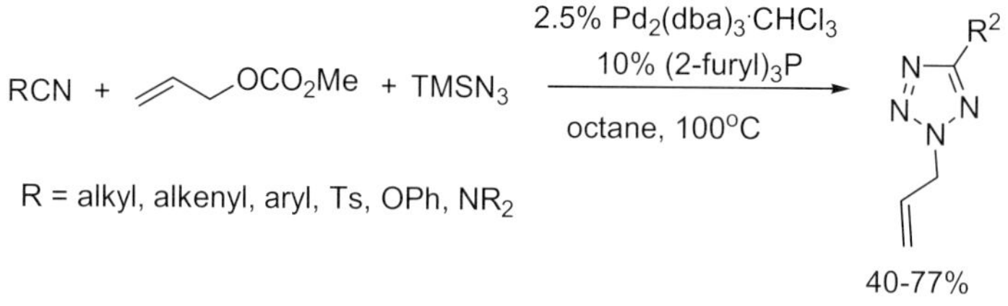

RCN + allyl-OCO_2Me + $TMSN_3$ → (2.5% $Pd_2(dba)_3 \cdot CHCl_3$, 10% $(2\text{-furyl})_3P$, octane, 100°C) → N-allyl tetrazole (R^2), 40-77%

R = alkyl, alkenyl, aryl, Ts, OPh, NR_2

Scheme 6.68

An alternative approach to multicomponent heterocycle synthesis involves the use of palladium catalysis to construct keto-alkynes for cycloaddition reactions. Müller has demonstrated the power of this approach in the construction of a range of aromatic heterocycles. For example, the palladium-catalyzed coupling of acid chlorides with terminal alkynes provides a method to assemble **36**. The trapping of this substrate can provide routes to aromatic heterocycles. As an example, the addition of amidines provides a multicomponent synthesis of pyrimidines (Scheme 6.69) [97]. This same substrate **36** is available via the carbonylative coupling of aryl halides with terminal alkynes, providing a four-component synthesis of pyrimidines [98]. **36** can also be employed in 1,3-dipolar cycloaddition reactions. For example, cycloaddition with hydroximinoyl chlorides provides a concise synthesis of isoxazaoles (Scheme 6.69), while addition of 1,3-dipolar 1-(2-oxyethyl)pyridinium salts allows the formation of indolizines [99].

2% $Pd(PPh_3)_2Cl_2$
4% CuI
Et_3N
THF, r.t.

36

up to 84%

up to 77%

R^1 = aryl, heteroaryl, *t*Bu, alkenyl
R^2 = alkyl, TMS
R^3 = aryl

Scheme 6.69

Mori (Scheme 6.70) and Stonehouse have each reported related approaches to employ *in situ* generated **36** to build up pyrazines [100]. In this case, the formation of **36** under carbonylative coupling of terminal alkynes with aryl or vinyl halides can be coupled with hydrazine addition.

+ CO + R^3HNNH_2

1% $Pd(PPh_3)_2Cl_2$
Et_3N
THF, r.t.

59-93%

R^1 = Me, OMe, H, 2-thienyl
R^2 = alkyl, aryl; R^3 = H, alkyl

Scheme 6.70

In an alternative use of this reactivity, Müller has demonstrated that the palladium-catalyzed arylation of propargyl alcohol derivatives can provide a method to assemble α,β-unsaturated ketones. This can be performed with subsequent Stetter 1,4-addition and cyclization as a route to the assembly of pyrroles (Scheme 6.71) [101]. While involving several steps, these processes can occur in a single pot, and provide an approach to assembling these heterocycles with high modularity. Alternatively, furan derivatives are available from the coupling of acid chlorides with terminal alkynes, followed by halide-initiated cyclization [102]. The unsaturated ketone products can also be induced to undergo subsequent hetero-Diels–Alder reactions, which upon rearrangement provide a multicomponent synthesis of pyridines [103].

As part of her work into multicomponent coupling reactions, Balme has reported that the palladium-catalyzed cyclization of *in situ* generated **37** with aryl or vinyl halides can provide a route to tetrahydrofuran derivatives, which can undergo

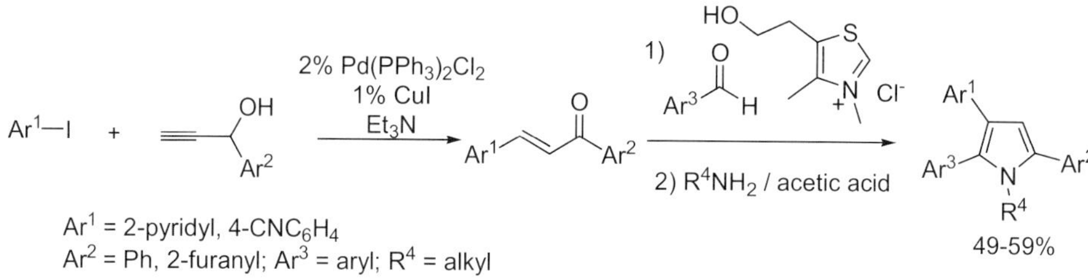

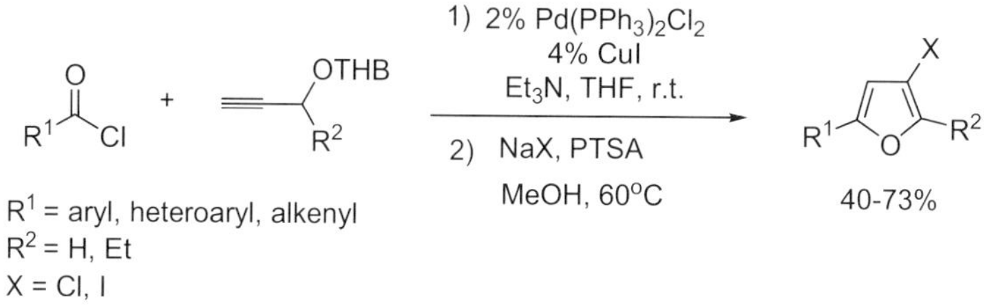

Ar^1 = 2-pyridyl, 4-CNC_6H_4
Ar^2 = Ph, 2-furanyl; Ar^3 = aryl; R^4 = alkyl

R^1 = aryl, heteroaryl, alkenyl
R^2 = H, Et
X = Cl, I

Scheme 6.71

subsequent isomerization to furans (Scheme 6.72) [104]. A similar approach with propargylamines can provide access to pyrroles [105].

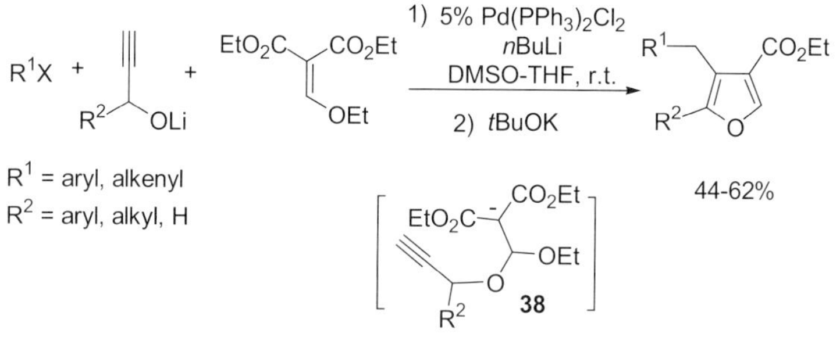

R^1 = aryl, alkenyl
R^2 = aryl, alkyl, H

Scheme 6.72

Another general multicomponent platform for heterocycle synthesis has been reported by our laboratory [106]. This chemistry employs palladium catalysis to mediate the generation of 1,3-dipolar cycloaddition reagents **39**, commonly referred to as Münchnones (Scheme 6.73) [107]. This involves the coupling of acid chlorides and imines into *N*-acyliminium salts which, upon oxidative addition to palladium, can undergo carbonylation and cyclization to **39**. In addition to forming **39**, the coupling of its generation with subsequent dipolar cycloaddition can provide routes to several heterocyclic cores. For example, by performing this reaction in the presence of alkynes, a range of polysubstituted pyrroles can be prepared (Scheme 6.74) [108]. These arise from the cycloaddition of alkyne to **39** followed by CO_2 loss, and provide, overall, a one-step route to the assembly of pyrroles from imines, acid chlorides and

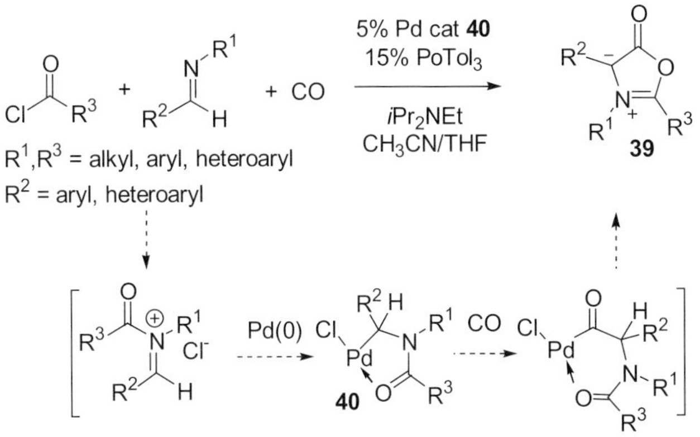

5% Pd cat **40**
15% PoTol$_3$
*i*Pr$_2$NEt
CH$_3$CN/THF

R^1,R^3 = alkyl, aryl, heteroaryl
R^2 = aryl, heteroaryl

39

40

Scheme 6.73

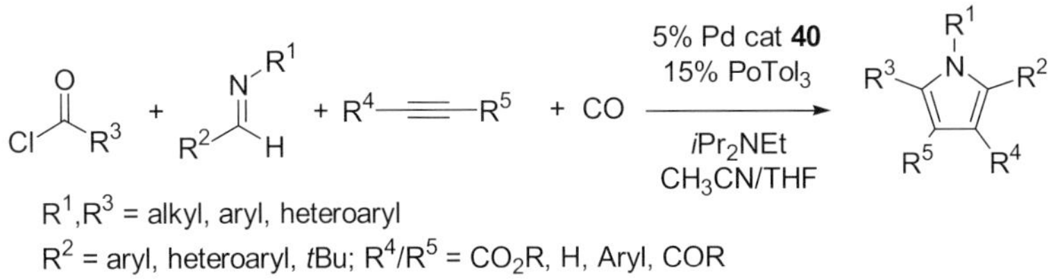

5% Pd cat **40**
15% PoTol$_3$
*i*Pr$_2$NEt
CH$_3$CN/THF

R^1,R^3 = alkyl, aryl, heteroaryl
R^2 = aryl, heteroaryl, *t*Bu; R^4/R^5 = CO_2R, H, Aryl, COR

Scheme 6.74

alkynes. This same reaction manifold can be accessed from α-aryloxyamides as iminium salt precursors [109].

This chemistry can also be used to access imidazoles. The latter can be generated via the formation of **39** in the presence of *N*-tosylimines, which undergo cycloaddition followed by subsequent aromatization to imidazoles upon CO_2 and HTs elimination (Scheme 6.75) [110]. In this case, two distinct imines are simultaneously coupled into the imidazole core. The latter arises from the inability of *N*-tosylimines to form iminium salts, but instead undergo more rapid 1,3-dipolar addition. As such, imidazoles can be formed with high regiocontrol, and with selective variation of every substituent about the heterocyclic core by modulating each building block.

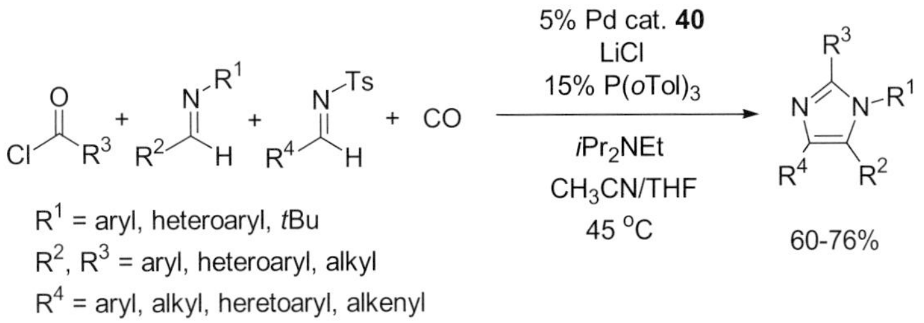

5% Pd cat. **40**
LiCl
15% P(*o*Tol)$_3$
*i*Pr$_2$NEt
CH$_3$CN/THF
45 °C

60-76%

R^1 = aryl, heteroaryl, *t*Bu
R^2, R^3 = aryl, heteroaryl, alkyl
R^4 = aryl, alkyl, heretoaryl, alkenyl

Scheme 6.75

6.8
Summary and Outlook

The use of palladium catalysis to synthesize aromatic heterocycles has undergone significant change over the past several decades. This includes the application of transformations that were not considered viable or efficient only a few years ago, such as palladium-catalyzed carbon–heteroatom bond forming reactions, or even C–H bond activation to initiate heterocycle formation. While these efforts have reflected, in part, the general development of palladium-catalyzed bond forming reactions, palladium reactivity has also been exploited to discover fundamentally new ways to assemble heterocycles. Taken together, these provide synthetic routes that display a number of promising features, ranging from modularity, atom economy, structural diversity, and/or access to increased complexity. There is certainly room for further improvement in many of these methods, particularly regarding their generality and overall synthetic efficiency. These may see partial solutions through the development of more active catalysts. In addition, the diverse chemistry available to palladium complexes (Scheme 6.1) suggest it will continue to serve as rich ground for reaction discovery, and the of design new and interesting approaches to the assembly of these products.

References

1 (a) Rubin, M., Sromek, A.W., and Gevorgyan, V. (2003) *Synlett.*, 2265–2291; (b) Zeni, G. and Larock, R.C. (2006) *Chem. Rev.*, **106**, 4644–4680; (c) Conreaux, D., Bouyssi, D., Monteiro, N., and Balme, C. (2006) *Curr. Org. Chem.*, **10**, 1325–1340; (d) Wolfe, J.P. and Thomas, J.S. (2005) *Curr. Org. Chem.*, **9**, 625–655; (e) Kirsch, G., Hesse, S., and Comel, A. (2004) *Curr. Org. Syn.*, **1**, 47–63; (f) D'Souza, D.M. and Müller, T.J.J. (2007) *Chem. Soc. Rev.*, **36**, 1095–1108.

2 Alonso, F., Beletskaya, I.P., and Yus, M. (2004) *Chem. Rev.*, **104**, 3079–3160.

3 Taylor, E.C., Katz, H.A., and Salgado-Zamara, H. (1985) *Tetrahedron Lett.*, **26**, 5963–5966.

4 For recent examples: (a) Dooleweerdt, K., Ruhland, T., and Skrydstrup, T. (2009) *Org. Lett.*, **11**, 221–224; (b) Majumdar, K.C. and Mondal, S. (2007) *Tetrahedron Lett.*, **48**, 6951–6953; (c) Sakai, H., Tsutsumi, K., Morimoto, T., and Kakiuchia, K. (2008) *Adv. Synth. Catal.*, **350**, 2498–2502.

5 Arcadi, A., Cacchi, S., and Marinelli, F. (1989) *Tetrahedron Lett.*, **30**, 2581–2584.

6 Rudisill, D.E. and Stille, J.K. (1989) *J. Org. Chem.*, **54**, 5856–5866.

7 Suzuki, N., Yasski, S., Yasuhara, A., and Sakamoto, T. (2003) *Chem. Pharm. Bull.*, **51**, 1170–1173.

8 Tang, Z.-Y. and Hu, Q.-S. (2006) *Adv. Synth. Catal.*, **348**, 846–850.

9 (a) Ackermann, L. (2005) *Org. Lett.*, **7**, 439–442; (b) Kaspar, L.T. and Aakermann, L. (2005) *Tetrahedron*, **61**, 11311–11316.

10 Yao, P.Y., Zhang, Y., Hsung, P.R., and Zhao, K. (2008) *Org. Lett.*, **10**, 4275–4278.

11 Arcadi, A., Marinelli, F., and Cacchi, S. (1986) *Synthesis*, 749–751.

12 For examples: (a) Aquila, B.M. (1997) *Tetrahedron. Lett.*, **38**, 2795–2798; (b) Arcadi, A., Cacchi, S., Di Giuseppe, S., Fabrizi, G., and Marinelli, F. (2002) *Synlett.*, 453–457; (c) Kobiyashi, K., Uneda, T., Kawakita, M., Morikawa, O., and Kobishi, H. (1997) *Tetrahedron Lett.*, **38**, 837–840; (d) Kundu, N.G., Pal, M.,

Mahanty, J.S., and De, M. (1997) *J. Chem. Soc., Perkin Trans. 1*, 2815–2820.

13 Utimoto, K., Miwa, H., and Nozaki, H. (1981) *Tetrahedron Lett.*, **22**, 4277–4278.

14 (a) Gabriele, B., Salerno, G., Fazio, A., and Bossio, M.R. (2001) *Tetrahedron Lett.*, **42**, 1339–1341; (b) Gabriele, B., Salerno, G., and Fazio, A. (2003) *J. Org. Chem.*, **68**, 7853–7861.

15 For examples: (a) Gabriele, B., Salerno, G., and Lauria, E. (1999) *J. Org. Chem*, **64**, 7687–7692; (b) Qing, F.-L., Gao, W.-Z., and Ying, J. (2000) *J. Org. Chem.*, **65**, 2003–2006; (c) Gabriele, B., Salerno, G., and Fazio, A. (2000) *Org. Lett.*, **2**, 351–352; (d) Beccalli, E.M., Borsini, E., Broggini, G., Palmisano, G., and Sottocornola, S. (2008) *J. Org. Chem.*, **73**, 4746–4749.

16 Trost, B.M. and McIntosh, M.C. (1995) *J. Am. Chem. Soc.*, **117**, 7255–7256.

17 Gabriele, B., Mancuso, R., Salerno, G., Ruffolo, G., and Plastina, P. (2007) *J. Org. Chem.*, **72**, 6873–6877.

18 For example: (a) Arcadi, A., Cacchi, S., and Fabrizi, G. (1992) *Tetrahedron Lett.*, **33**, 3915–3918; (b) Cacchi, S., Fabrizi, G., and Goggiamani, A. (2006) *Adv. Synth. Catal.*, **348**, 1301–1305; (c) Cacchi, S., Fabrizi, G., Marinelli, F., Moro, L., and Pace, P. (1997) *Synlett*, 1363–1366; (d) Cacchi, S., Fabrizi, G., Lamba, D., Marinelli, F., and Parisi, L.M. (2003) *Synthesis*, 728–734.

19 Lu, Z.B., Zhao, W., Wei, H.-X., Dufour, M., Farina, V., and Senanayake, C.H. (2006) *Org. Lett.*, **8**, 3271–3274.

20 For example: (a) Hu, Y., Nawoschik, K.J., Liao, Y., Ma, J., Fathi, R., and Yang, Z. (2004) *J. Org. Chem.*, **69**, 2235–2239; (b) Arcadi, A., Cacchi, S., Del Rosario, M., Fabrizi, G., and Marinelli, F. (1996) *J. Org. Chem.*, **61**, 9280–9288; (c) Flynn, B.L., Hamel, E., and Jung, M.K. (2002) *J. Med. Chem.*, **45**, 2670–2673.

21 (a) Bossharth, E., Desbordes, P., Monteiro, N., and Balme, G. (2003) *Org. Lett.*, **5**, 2441–2444; (b) Cacchi, S., Fabrizi, G., and Moro, L. (1997) *J. Org. Chem.*, **62**, 5327–5332; (c) Arcadi, A., Cacchi, S., Fabrizi, G., Marinelli, F., and Parisi, L.M. (2003) *Tetrahedron*, **59**, 4661–4671.

22 Examples: (a) Arcadi, A., Cacchi, S., Fabrizi, G., Marinelli, F., and Parisi, L.M. (2005) *J. Org. Chem.*, **70**, 6213–6217; (b) Arcadi, A., Cacchi, S., Fabrizi, G., and Marinelli, F. (2000) *Synlett*, 394–396; (c) Nakamura, I., Mizushima, Y., Yamagishi, U., and Yamamoto, Y. (2007) *Tetrahedron*, **63**, 8670–8676; (d) Cacchi, S., Fabrizi, G., and Pace, P. (1998) *J. Org. Chem.*, **63**, 1001–1011.

23 Xiao, Y. and Zhang, J. (2008) *Angew. Chem. Int. Ed.*, **47**, 1903–1906.

24 Monteiro, N. and Balme, G. (1998) *Synlett*, 746–747.

25 Cacchi, S., Fabrizi, G., and Moro, L. (1998) *Tetrahedron Lett.*, **39**, 5101–5104.

26 (a) Kamijo, S. and Yamamoto, Y. (2002) *Angew. Chem. Int. Ed.*, **41**, 3230–3233; (b) Kamijo, S. and Yamamoto, Y. (2003) *J. Org. Chem.*, **68**, 4764–4771.

27 Kamijo, S. and Yamamoto, Y. (2002) *J. Am. Chem. Soc.*, **124**, 11940–11945.

28 (a) Cacchi, S., Fabrizi, G., and Parisi, L.M. (2004) *Synthesis*, 1894–1889; (b) Yasuhara, A., Takeda, Y., Suzuki, N., and Sakamoto, T. (2002) *Chem. Pharm. Bull.*, **50**, 235–238.

29 Shen, Z. and Lu, X. (2006) *Tetrahedron*, **62**, 10896–10899.

30 (a) Gabriele, B., Salerno, G., De Pascali, F., Sciano, G.T., Costa, M., and Chiusoli, G.P. (1997) *Tetrahedron Lett.*, **38**, 6877–6880; (b) Gabriele, B., Salerno, G., Fazio, A., and Campana, F.B. (2002) *Chem. Commun.*, 1408–1409; (c) Gabriele, B., Salerno, G., De Pascali, F., Costa, M., and Chiusoli, G.P. (1999) *J. Org. Chem.*, **64**, 7693–7699; (d) Gabriele, B., Salerno, G., Fazio, A., and Veltri, L. (2006) *Adv. Synth. Catal.*, **348**, 2212–2222.

31 Sheng, H., Lin, S., and Huang, Y. (1986) *Tetrahedron Lett.*, **27**, 4893–4894.

32 Jeevanandam, A., Narkunan, K., and Ling, Y.-C. (2001) *J. Org. Chem.*, **66**, 6014–6020.

33 (a) Hashmi, S.K.A. (1995) *Angew. Chem. Int. Ed.*, **34**, 1582–1583; (b) Hashmi, S.K.A., Ruppert, T.L., Knfel, T., and Bats, J.W. (1997) *J. Org. Chem.*, **62**, 7295–7304.

34 Alcaide, B., Almendros, P., and Campo, T.M. (2007) *Eur. J. Org. Chem.*, 2844–2849.

35 (a) Ma, S., Zhang, J., and Lu, L. (2003) *Chem. Eur. J.*, **9**, 2447–2456; (b) Ma, S., Gu, Z., and Yu, Z. (2005) *J. Org. Chem.*, **70**, 6291–6294. Ma, S. and Li, L. (2000) *Org. Lett.*, **2**, 941–944.

36 (a) Aurrecoechea, J.M. and Pérez, E. (2001) *Tetrahedron Lett.*, **42**, 3839–3841; (b) Aurrecoechea, J.M. and Pérez, E. (2003) *Tetrahedron Lett.*, **44**, 3263–3266; (c) Aurrecoechea, J.M. and Pérez, E. (2004) *Tetrahedron*, **60**, 4139–4149; (d) Aurrecoechea, J.M., Durana, A., and Pérez, E. (2008) *J. Org. Chem.*, **73**, 3650–3653.

37 Dieter, K.R. and Yu, H. (2001) *Org. Lett.*, **3**, 3855–3858.

38 For example: (a) Hosokawa, T., Maeda, K., Koga, K., and Moritani, I. (1973) *Tetrahedron Lett.*, **14**, 739–740; (b) Hosokawa, T., Shimo, N., Maeda, K., Sonoda, A., and Murahashi, S.-I. (1976) *Tetrahedron Lett.*, **17**, 383–386; (c) Hegedus, L.S., Allen, G.F., and Waterman, E.L. (1976) *J. Am. Chem. Soc.*, **98**, 2674–2676.

39 Hosokawa, T., Ohkata, H., and Moritani, I. (1975) *Bull. Chem. Soc. Jpn.*, **48**, 1533–1536.

40 Harrington, P.J. and Hegedus, L.S. (1984) *J. Org. Chem.*, **49**, 2657–2662.

41 Krolski, M.E., Renaldo, A.F., Rudisill, D.E., and Stille, J.K. (1988) *J. Org. Chem.*, **53**, 1170–1176.

42 (a) Han, X. and Widenhoefer, R.A. (2004) *J. Org. Chem.*, **69**, 1738–1740; (b) Kimura, M., Harayama, H., Tanaka, S., and Tamaru, Y. (1994) *J. Chem. Soc., Chem. Commun.*, 2531–2533; (c) Zhang, Z., Zhang, J., Tan, J., and Wang, Z. (2008) *J. Org. Chem.*, **73**, 5180–5182.

43 Deagostino, A., Farina, V., Prandi, C., Zavattaro, C., and Venturello, P. (2006) *Eur. J. Org. Chem.*, 3451–3456.

44 (a) Akazome, M., Kondo, T., and Watanabe, Y. (1994) *J. Org. Chem.*, **59**, 3375–3380; (b) Davies, I.W., Smitrovich, J.H., Sidler, R., Qu, C., Gresham, V., and Bazaral, C. (2005) *Tetrahedron*, **61**, 6425–6437.

45 Examples: (a) Söderberg, B.C.G. and Shriver, J.A. (1997) *J. Org. Chem.*, **62**, 5838–5845; (b) Söderberg, B.C.G., Chisnell, A.C., O'Nei, S.N., and Shriver, J.A. (1999) *J. Org. Chem.*, **64**, 9731–9734; (c) Söderberg, B.C.G., Banini, S.R., Turner, M.R., Minter, A.R., and Arrington, A.K. (2008) *Synthesis*, 903–912.

46 Watanabe, M., Yamamoto, T., and Nishiyama, M. (2000) *Angew. Chem. Int. Ed.*, **39**, 2501–2504.

47 (a) Abouabdellah, A. and Dodd, R.H. (1998) *Tetrahedron Lett.*, **39**, 2119–2122; (b) Brown, J.A. (2000) *Tetrahedron Lett.*, **41**, 1623–1626.

48 (a) Brain, C.T. and Steer, T.J. (2003) *J. Org. Chem.*, **68**, 6814–6816; (b) Evindar, G. and Batey, R.A. (2003) *Org. Lett.*, **5**, 133–136.

49 (a) Song, J.J. and Yee, K.N. (2001) *Tetrahedron Lett.*, **42**, 2937–2940; (b) Venkatesh, C., Sundaram, S.G.M., Ila, H., and Junjappa, H. (2006) *J. Org. Chem.*, **71**, 1280–1283.

50 Barluenga, J., Jimenez-Aquino, A., Valdes, C., and Aznar, F. (2007) *Angew. Chem. Int. Ed.*, **46**, 1529–1532.

51 For examples: (a) Fayol, A., Fang, Y.-Q., and Lautens, M. (2006) *Org. Lett.*, **8**, 4203–4206; (b) Fang, Y.-Q., Karisch, R., and Lautens, M. (2007) *J. Org. Chem.*, **72**, 1341–1346; (c) Fang, Y.-Q. and Lautens, M. (2008) *J. Org. Chem.*, **73**, 538–549; (d) Bryan, S.C. and Lautens, M. (2008) *Org. Lett.*, **10**, 4633–4636; (e) Vieira, T.O., Meaney, L.A., Shi, Y.-L., and Alper, H. (2008) *Org. Lett.*, **10**, 4899–4901; (f) Thielges, S., Meddah, E., Bisseret, P., and Eustache, J. (2004) *Tetrahedron Lett.*, **45**, 907–910.

52 (a) Nozaki, K., Takahashi, K., Nakano, K., Hiyama, T., Tang, H.-Z., Fujiki, M., Yamaguchi, S., and Tamao, K. (2003) *Angew. Chem. Int. Ed.*, **42**, 2051–2053; (b) Kawaguchi, K., Nakano, K., and Nozaki, K. (2007) *J. Org. Chem.*, **72**, 5119–5128; (c) Nakano, K., Hidehira, Y., Takahashi, K., Hiyama, T., and Nozaki, K. (2005) *Angew. Chem. Int. Ed.*, **44**, 7136–7138; (d) Odom, S.A., Lancaster, K., Beverina, L., Lefler, K.M., Thompson, J.N., Coropceanu, V., Bredas, J.-L., Marder, S.R., and Barlow, S. (2007) *Chem. Eur. J.*, **13**, 9637–9646.

53 (a) Willis, M.C., Brace, G.N., and Holmes, I.P. (2005) *Angew. Chem. Int. Ed.*, **44**, 403–406; (b) Willis, M.C., Brace, G.N., Findlay, T.J.K., and Holmes, I.P. (2006) *Adv. Synth. Catal.*, **348**, 851–856.

54 Loones, K.T.J., Maes, B.U.W., Dommisse, R.A., and Lemiere, G.L.F. (2004) *Chem. Commun.*, 2466–2467.

55 Willis, M.C., Taylor, D., and Gillmore, T.A. (2004) *Org. Lett.*, **6**, 4755–4757.

56 Oh, C.H., Park, H.M., and Park, D.I. (2007) *Org. Lett.*, **9**, 1191–1193.

57 Oh, C.H., Park, W., and Kim, M. (2007) *Synlett.*, 1411–1415.

58 Larock, R.C. and Yum, E.K. (1991) *J. Am. Chem. Soc.*, **113**, 6689–6690.

59 Larock, R.C., Yum, E.K., and Refvik, D.M. (1998) *J. Org. Chem.*, **63**, 7652–7662.

60 Roschangar, F., Liu, J., Estanove, E., Dufour, M., Rodriguez, S., Farina, V., Hickey, E., Hossain, A., Jones, P.-J., Lee, H., Lu, B.Z., Varsolona, R., Schroder, J., Beaulieu, P., Gillard, J., and Senanayake, H.C. (2008) *Tetrahedron Lett.*, **49**, 363–366.

61 Konno, T., Chae, J., Ishihara, T., and Yamanaka, H. (2004) *J. Org. Chem.*, **69**, 8258–8265.

62 Fnaru, A., Berthault, A., Besson, T., Guillaumet, C., and Berteina-Raboin, S. (2002) *Org. Lett.*, **4**, 2613–2615.

63 For examples: (a) Wensbo, D., Erlksson, A., Jeschke, T., Annby, K.U., and Gronowitz, S. (1993) *Tetrahedron Lett.*, **34**, 2823–2826. Park, S.S., Choi, J.-K., and Yum, E.K. (1998) *Tetrahedron Lett.*, **39**, 627–630; (b) Ujjainwalla, F. and Warner, D. (2008) *Tetrahedron Lett.*, **39**, 5355–5358.

64 Gee, M.B., Lee, W.J., and Yum, E.K. (2003) *Bull. Korean Chem. Soc.*, **24**, 1193–1196.

65 (a) Shen, M., Li, C., Lu, B.Z., Hossain, A., Roschangar, F., Farina, V., and Senanayake, H.C. (2004) *Org. Lett.*, **6**, 4129–4132; (b) Cui, X., Li, J., Fu, Y., Liu, L., and Guo, Q.-X. (2008) *Tetrahedron Lett.*, **49**, 3458–3462; (c) Liu, J., Shen, M., Zhang, Y., Li, G., Khodabocus, A., Rodriguez, S., Qu, B., Farina, V., Senanayake, H.C., and Lu, B.Z. (2006) *Org. Lett.*, **8**, 3573–3575.

66 Leogane, O. and Lebel, H. (2008) *Angew. Chem. Int. Ed.*, **47**, 350–352.

67 (a) Larock, R.C., Yum, E.K., Doty, M.J., and Sham, K.C.K. (1995) *J. Org. Chem.*, **60**, 3270–3271; (b) Bishop, B.C., Cottrell, I.F., and Hands, D. (1997) *Synthesis*, 1315–1320; (c) Konno, T., Chae, J., Ishihara, T., and Yamanaka, H. (2004) *Tetrahedron*, **60**, 11695–11700.

68 (a) Crawley, M.L., Goljer, I., Jenkins, D.J., Mehlmann, J.F., Nogle, L., Dooley, R., and Mahaney, P.E. (2006) *Org. Lett.*, **8**, 5837–5840; (b) Larock, R.C., Doty, M.J., and Han, X. (1998) *Tetrahedron Lett.*, **39**, 5143–5146.

69 (a) Roesch, K.R. and Larock, R.C. (1998) *J. Org. Chem.*, **63**, 5306–5307; (b) Konno, T., Chae, J., Miyabe, T., and Ishihara, T. (2005) *J. Org. Chem.*, **70**, 10172–10174.

70 (a) Roesch, R., Zhang, H., and Larock, R.C. (2001) *J. Org. Chem.*, **66**, 8042–8051; (b) Zhang, H. and Larock, R.C. (2003) *J. Org. Chem.*, **68**, 5132–5138. Zhang, H. and Larock, R.C. (2002) *Org. Lett.*, **4**, 3035–3038.

71 (a) Roesch, K.R. and Larock, R.C. (1999) *Org. Lett.*, **1**, 553–556; (b) Roesch, K.R. and Larock, R.C. (2002) *J. Org. Chem.*, **67**, 86–94.

72 (a) Dai, G. and Larock, R.C. (2002) *Org. Lett.*, **4**, 193–196; (b) Dai, G. and Larock, R.C. (2002) *J. Org. Chem.*, **67**, 7042–7047; (c) Dai, G. and Larock, R.C. (2001) *Org. Lett.*, **3**, 4035–4038; (d) Huang, Q. and Larock, R.C. (2003) *J. Org. Chem.*, **68**, 980–988.

73 de Meijere, A. and Meyer, F.E. (1994) *Angew. Chem. Int. Ed.*, **33**, 2379–2411.

74 Nazare, M., Schneider, C., Lindenschmidt, A., and Will, D.W. (2004) *Angew. Chem. Int. Ed.*, **43**, 4526–4528.

75 Jia, J. and Zhu, J. (2006) *J. Org. Chem.*, **71**, 7826–7834.

76 For examples: (a) Ackermann, L. and Althammer, A. (2006) *Synlett.*, 3125–3129; (b) Ackermann, L., Kaspar, L.T., and Gschrei, C.J. (2004) *Chem. Commun.*, 2824–2825; (c) Kasahara, A., Izumi, T., Murakami, S., Yanai, H., and Takatori, M. (1986) *Bull, Chem. Soc. Jpn.*, **59**, 927–928; (d) Grigg, R. and Savic, V. (2000) *Chem. Commun.*, 873–874.

77 (a) Jensen, T., Pedersen, H., Bang-Andersen, B., Madsen, R., and Jørgensen, M. (2008) *Angew. Chem. Int. Ed.*, **47**, 888–890; (b) Edmondson, S.D., Mastracchio, A., and Parmee, E.R. (2000) *Org. Lett.*, **2**, 1109–1112.

78 Barluenga, J., Fernandez, A.M., Aznar, F., and Valdes, C. (2005) *Chem. Eur. J.*, **11**, 2276–2283.

79 Tsutsui, H., Kitamura, M., and Narasaka, K. (2002) *Bull. Chem. Soc. Jpn.*, **75**, 1451–1460.

80 (a) Kitamura, M., Kudo, D., and Narasaka, K. (2006) *ARKIVOC*, 148–162; (b) Zaman, S., Mitsuru, K., and Abell, A.D. (2005) *Org. Lett.*, **7**, 609–611.

81 Alberico, D., Scott, M.E., and Lautens, M. (2007) *Chem. Rev.*, **107**, 174–238.

82 For example: Ames, E.D. and Opalko, A. (1983) *Synthesis*, 234–235.

83 Bedford, R.B. and Cazin, C.S.J. (2002) *Chem. Commun.*, 2310–2311; Bedford, R.B. and Betham, M. (2006) *J. Org. Chem.*, **71**, 9403–9410.

84 Xu, H. and Fan, L.-L. (2008) *Chem. Pharm. Bull.*, **56**, 1496–1498.

85 Ackermann, L. and Althammer, A. (2007) *Angew. Chem. Int. Ed.*, **46**, 1627–1629.

86 Watanabe, T., Ueda, S., Inuki, S., Oishi, S., Fujii, N., and Ohno, H. (2007) *Chem. Commun.*, 4516–4518.

87 Ryu, C.-K., Choi, I.H., and Park, R.E. (2006) *Syn. Commun.*, **36**, 3319–3328.

88 Smitrovich, J.H. and Davies, I.W. (2004) *Org. Lett.*, **6**, 533–535.

89 (a) Tsang, W.C.P., Zheng, N., and Buchwald, S.L. (2005) *J. Am. Chem. Soc.*, **127**, 14560–14561; (b) Tsang, W.C.P., Munday, R.H., Brasche, G., Zheng, N., and Buchwald, S.L. (2008) *J. Org. Chem.*, **73**, 7603–7610.

90 Abreu, S.A., Ferreira, P.M.T., Queiroz, M.-J.R.P., Ferreira, I.C.F.R., Calhelha, R.C., and Estevinho, L.M. (2005) *Eur. J. Org. Chem.*, 2951–2957.

91 Jordan-Hore, J.J.A., Johansson, C.C.C., Gulias, M., Beck, E.M., and Gaunt, M.J. (2008) *J. Am. Chem. Soc.*, **130**, 16184–16186.

92 Li, B.-J., Tian, S.-L., Fang, Z., and Shi, Z.-J. (2008) *Angew. Chem. Int. Ed.*, **47**, 1115–1118.

93 For reviews, see ref. 1f and Balme, G., Bossharth, E., and Montiero, N. (2003) *Eur. J. Org. Chem.*, 4101–4111.

94 (a) Kamijo, S., Jin, T., Huo, Z., and Yamamoto, Y. (2002) *Tetrahedron Lett.*, **43**, 9707–9710; (b) Kamijo, S., Jin, T., Huo, Z., and Yamamoto, Y. (2003) *J. Am. Chem. Soc.*, **125**, 7786–7787; (c) Kamijo, S., Jin, T., and Yamamoto, Y. (2004) *Tetrahedron Lett.*, **45**, 689–691.

95 Kamijo, S., Jin, T., and Yamamoto, Y. (2002) *J. Org. Chem.*, **67**, 7413–7417.

96 Barluenga, J., Valdes, C., Beltran, G., Escribano, M., and Aznar, F. (2006) *Angew. Chem. Int. Ed.*, **45**, 6893–6896.

97 Karpov, A.S. and Müller, T.J.J. (2003) *Org. Lett.*, **5**, 3451–3454.

98 Karpov, A.S., Merkul, E., Rominger, F., and Muller, T.J.J. (2005) *Angew. Chem. Inter. Ed.*, **44**, 6951–6956.

99 Willy, B., Rominger, F., and Müller, T.J.J. (2008) *Synthesis*, 293–303.

100 (a) Ahmed, M.S.M., Kobayashi, K., and Mori, A. (2005) *Org. Lett.*, **7**, 4487–4489; (b) Stonehouse, J.P., Chekmarev, D.S., Ivanova, N.V., Lang, S., Pairaudeau, G., Smith, N., Stocks, N.J., Sviridov, S.I., and Utkina, L.U. (2008) *Synlett.*, 100–104; (c) Xie, F., Cheng, G., and Hu, Y. (2006) *J. Comb. Chem.*, **8**, 286–288.

101 Braun, R.U., Zeitler, K., and Mu1ller, T.J.J. (2001) *Org. Lett.*, **3**, 3297–3300.

102 (a) Karpov, A.S., Merkul, E., Oeser, T., and Müller, T.J.J. (2005) *Chem. Commun.*, 2581–2583; (b) Karpov, A.S., Merkul, E., Oeser, T., and Müller, T.J.J. (2006) *Eur. J. Org. Chem.*, 2991–3000.

103 Schramm, O.J., Dediu, N., Oeser, T., and Müller, T.J.J. (2006) *J. Org. Chem.*, **71**, 3494–3500.

104 Garcon, S., Vassiliou, S., Cavicchioli, M., Hartmann, B., Monteiro, N., and Balme, G. (2001) *J. Org. Chem.*, **66**, 4069–4073.

105 Clique, B., Anselme, C., Otto, D., Monteiro, N., and Balme, G. (2004) *Tetrahedron Lett.*, **45**, 1195–1197.

106 Arndtsen, B.A. (2009) *Chem. Eur. J.*, **15**, 302–313.

107 Dhawan, R., Dghaym, R.D., and Arndtsen, B.A. (2003) *J. Am. Chem. Soc.*, **125**, 1474–1475.

108 Dhawan, R. and Arndtsen, B.A. (2004) *J. Am. Chem. Soc.*, **126**, 468–469.

109 Lu, Y. and Arndtsen, B.A. (2008) *Angew. Chem. Int. Ed.*, **47**, 5430–5433.

110 Siamaki, A.R. and Arndtsen, B.A. (2006) *J. Am. Chem. Soc.*, **128**, 6050–6051.

7
New Reactions of Copper Acetylides: Catalytic Dipolar Cycloadditions and Beyond

Valery V. Fokin

7.1
Introduction

Among the most energetic hydrocarbons, acetylenes have built-in reactivity that can be utilized for their transformation into useful intermediates and products. Transition metals enable selective and controlled manipulation of the triple bond, opening the door to the wealth of reliable transformations. η^2-Coordination to π-acidic transition metals activates the acetylenic triple bond towards nucleophilic additions, and electrophilic additions and, especially, cyclizations have been widely utilized in the synthesis of hetero- and carbo-cycles [1]. The chemistry of terminal η^1-complexes of terminal alkynes (σ-acetylides) is dominated by the reactions with electrophiles at the α-carbon of the metal-bound alkyne. The nucleophilic activation of acetylenes by π-basic metals, which results in the formation of synthetically useful vinylidene complexes that are β-nucleophilic, is also known and is best represented by the complexes of Group 6 elements (Cr, Mo, W) and ruthenium. In contrast, vinylidene complexes of Group 11 metals, in particular copper(I), have neither been isolated nor implicated in catalysis. Yet, this first period electron-rich d^{10} metal appears to be an ideal candidate for revealing β-nucleophilic properties of its acetylide ligands. This chapter discusses such reactivity and its most prominent example to date, the copper-catalyzed azide–alkyne cycloaddition (CuAAC) reaction. Appreciation of the dynamics of exceedingly rapid ligand exchange of copper(I) acetylides in protic solutions and developing methods for generation of reactive forms of these well-known organometallic species *in the presence of organic azide* were key to the success of the CuAAC reaction. Here, we shall focus on the fundamental aspects of the CuAAC process and on its mechanism, with an emphasis on the properties of copper that enable this unique mode of reactivity.

Organic azides are unique reactive partners for copper(I) acetylides. They can be easily introduced into organic molecules and can usually be handled and stored as stable reagents. They are small and relatively non-polar functional groups that are devoid of acid–base reactivity (with the exception of strong acids and organolithium/ organomagnesium reagents). They remain quite "invisible" to and unreactive with most other chemical functionalities in nature and the laboratory. The product

Catalyzed Carbon-Heteroatom Bond Formation. Edited by Andrei K. Yudin

ISBN: 978-3-527-32428-6

1,2,3-triazole is a remarkably stable aromatic heterocycle that can serve, among other roles, as a structural analogue of a peptide linkage. Azides can, thereby, be installed on structures that one wishes to link together, and kept in place through many operations of synthesis and elaboration. They are ideal connectors for such a modular approach to synthesis, *if* effective methods to catalyze their cycloaddition are available. The complementary reactivity of copper(I) acetylides has allowed this general scheme to be put in operation for a wide array of applications and settings.

It is therefore not surprising that the CuAAC reaction has gained recognition as the premier example of click chemistry, a term coined in 2001 by Sharpless and colleagues to describe a set of "near-perfect" bond-forming reactions useful for rapid assembly of molecules with desired function [2]. Click transformations are easy to perform, give rise to their intended products in very high yields with little or no byproducts, work well under many conditions (usually especially well in water), and are unaffected by the nature of the groups being connected to each other. The potential of organic azides as highly energetic, yet very selective, functional groups in organic synthesis was highlighted by Sharpless *et al.* [2], and their dipolar cycloadditions with olefins and alkynes were placed among the reactions fulfilling the click criteria. However, the inherently low reaction rates of the azide-alkyne cycloaddition did not make it very useful in the click context, and its potential was not revealed until the discovery of its catalysis by copper under the broadly applicable solution conditions [3].

The copper-catalyzed reaction was reported simultaneously and independently by the groups of Meldal in Denmark [4] and Fokin and Sharpless in the US [3]. It transforms organic azides and terminal alkynes exclusively into the corresponding 1,4-disubstituted 1,2,3-triazoles, in contrast to the uncatalyzed reaction, which requires much higher temperatures and provides mixtures of 1,4- and 1,5-disubstituted triazole isomers. Meldal used the transformation for the synthesis of peptidotriazoles in organic solvents, starting from alkynylated amino acids attached to solid supports, whereas the Scripps group immediately turned to aqueous systems and devised a straightforward and practical procedure for the covalent "stitching" of virtually any fragments containing an azide and an alkyne functionality, noting the broad utility and versatility of the novel process "for those organic synthesis endeavors which depend on the creation of covalent links between diverse building blocks" [3]. It has since been widely employed in synthesis, medicinal chemistry, molecular biology, and materials science.

Numerous applications of the CuAAC reaction reported during the last several years have been regularly reviewed [5–12], and are continously enriched by investigators in many fields.

7.2 Azide–Alkyne Cycloaddition: Basics

The thermal reaction of terminal or internal alkynes with organic azides (Scheme 7.1A) has been known for more than a century: the first 1,2,3-triazole was

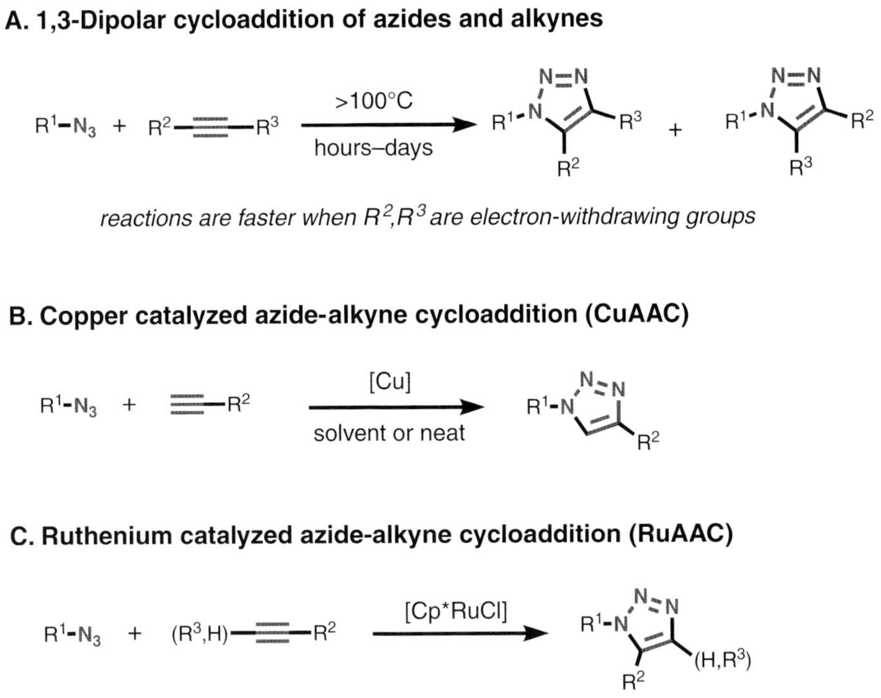

A. 1,3-Dipolar cycloaddition of azides and alkynes

$R^1{-}N_3 + R^2{-}{\equiv}{-}R^3$ —(>100°C, hours–days)→ 1,4-regioisomer (R^1, R^2, R^3) + 1,5-regioisomer (R^1, R^3, R^2)

reactions are faster when R^2,R^3 are electron-withdrawing groups

B. Copper catalyzed azide-alkyne cycloaddition (CuAAC)

$R^1{-}N_3 + {\equiv}{-}R^2$ —([Cu], solvent or neat)→ 1-R^1-4-R^2-1,2,3-triazole

C. Ruthenium catalyzed azide-alkyne cycloaddition (RuAAC)

$R^1{-}N_3 + (R^3,H){-}{\equiv}{-}R^2$ —([Cp*RuCl])→ 1-R^1-5-R^2-4-(H,R^3)-1,2,3-triazole

Scheme 7.1 Thermal cycloaddition of azides and alkynes usually requires prolonged heating and results in mixtures of both 1,4- and 1,5-regioisomers (A), whereas CuAAC produces only 1,4-disubstituted 1,2,3-triazoles at room temperature in excellent yields (B). The RuAAC reaction proceeds with both terminal and internal alkynes and gives 1,5-disubstituted and fully, 1,4,5-trisubstituted-1,2,3-triazoles.

synthesized by A. Michael from phenyl azide and diethyl acetylenedicarboxylate in 1893 [13]. The reaction has been most thoroughly investigated by Huisgen and coworkers in the1950s–1970s in the course of their studies of the larger family of 1,3-dipolar cycloaddition reactions [14–20].

Although the reaction is highly exothermic (ΔG^0 between -50 and -65 kcal mol^{-1}), its high activation barrier (approximately 25 kcal mol^{-1} for methyl azide and propyne [21]) results in exceedingly low reaction rates for unactivated reactants, even at elevated temperature. Furthermore, since the differences in HOMO–LUMO energy levels for both azides and alkynes are of similar magnitude, both dipole-HOMO- and dipole-LUMO-controlled pathways operate in these cycloadditions. As a result, a mixture of regioisomeric 1,2,3-triazole products is usually formed when an alkyne is unsymmetrically substituted. These mechanistic features can be altered, and the reaction dramatically accelerated, by making the alkyne electron-deficient, such as in propiolate or acetylene dicarboxylate esters, at the cost of opening up competitive side reactions such as conjugate addition.

Copper catalysts (Scheme 7.1B) dramatically change the mechanism and the result of the reaction, converting it to a sequence of discreet steps which culminates in the

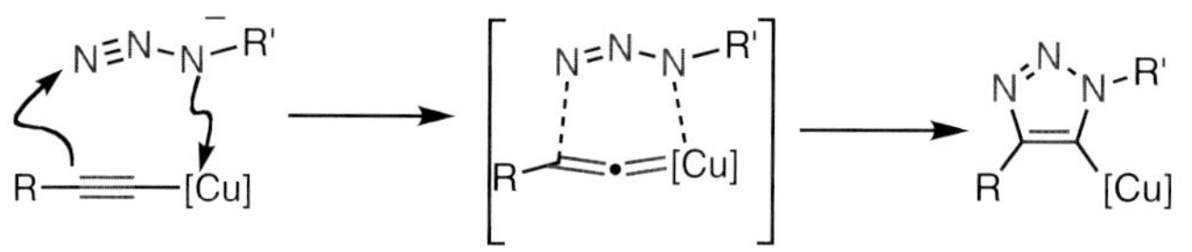

Scheme 7.2 Simplified representation of the proposed C–N bond-making steps in the reaction of copper(I) acetylides with organic azides. [Cu] denotes either a single-metal center CuL_x or a di-/oligonuclear cluster Cu_xL_y.

formation of 5-triazolyl copper intermediate (Scheme 7.2). The key C–N bond-forming event takes place between the nucleophilic, vinylidene-like β-carbon of copper (I) acetylide and the electrophilic terminal nitrogen of the coordinated organic azide.

The rate of the catalytic reaction is increased by a factor of 10^7 relative to the thermal, concerted version [21], making it conveniently fast at and below room temperature. The reaction is not significantly affected by the steric and electronic properties of the groups attached to the azide and alkyne centers. For example, primary, secondary, and even tertiary, electron-deficient and electron-rich, aliphatic, aromatic, and heteroaromatic azides usually react well with variously substituted terminal alkynes. The reaction proceeds in most protic and aprotic solvents, including water, and is unaffected by most organic and inorganic functional groups, therefore all but eliminating the need for protecting group chemistry. The 1,2,3-triazole heterocycle has the advantageous properties of high chemical stability (generally inert to severe hydrolytic, oxidizing, and reducing conditions, even at high temperature), strong dipole moment (4.8–5.6 D), aromatic character, and hydrogen bond accepting ability [22, 23]. Thus it can interact productively in several ways with biological molecules and can serve as a hydrolytically-stable replacement for the amide bond [24–26]. Compatibility of the CuAAC reaction with a broad range of functional groups and reaction conditions [27–31] has made it broadly useful across the chemical disciplines, and its applications include the synthesis of biologically active compounds, the preparation of conjugates to proteins and polynucleotides, the synthesis of dyes, the elaboration of known polymers and the synthesis of new ones, the creation of responsive materials, and the covalent attachment of desired structures to surfaces [5, 32–36].

Cu(I) remains the only catalytically active species identified so far for the conversion of azides and terminal alkynes to 1,4-disubstituted triazoles. Surveys in our laboratories of complexes of all of the first-row transition elements as well as complexes of Ag(I), Pd(0/II), Pt(II), Au(I/III), and Hg(II), among others, have all failed to produce triazoles, although interesting and complex reactivity has occasionally been seen. The unique activity of Cu(I) stems from the fortuitous combination of its ability to engage terminal alkynes in both σ- and π-interactions, and the ability to rapidly exchange these and other ligands in its coordination sphere. When organic azide is a ligand, the synergistic nucleophilic activation of the alkyne and electrophilic activation of the azide drives the formation of the first carbon–nitrogen bond.

In 2005, ruthenium cyclopentadienyl complexes were found to catalyze the formation of the complementary 1,5-disubstituted triazole from azides and terminal alkynes, and also to engage internal alkynes in the cycloaddition [37]. As one would imagine from these differences, this sister process, designated RuAAC (ruthenium-catalyzed azide–alkyne cycloaddition), is mechanistically quite distinct from its cuprous cousin. Although the scope and functional group compatibility of RuAAC are excellent [38–40], the reaction is more sensitive to the solvents and the steric demands of the azide substituents than CuAAC. Applications of RuAAC are beginning to appear [26, 41, 42].

7.3 Copper-Catalyzed Cycloadditions

7.3.1 Catalysts and Ligands

A very wide range of experimental conditions for the CuAAC have been employed since its discovery highlights the robustness of the process and its compatibility with most functional groups, solvents, and additives, regardless of the source of the catalyst. The choice of the catalyst is mostly dictated by the particular requirements of the experiment, and usually many combinations will produce desired results. The most commonly used protocols and their advantages and limitations are discussed below.

Different copper(I) sources can be utilized in the reaction, as recently summarized in necessarily partial fashion by Meldal and Tornøe [11]. Copper(I) salts (iodide, bromide, chloride, acetate) and coordination complexes such as $[Cu(CH_3CN)_4]PF_6$ and $[Cu(CH_3CN)_4]OTf$ [43, 44] have been commonly employed [45]. In general, cuprous iodide is not recommended, because of the ability of iodide anion to act as ligand for the metal. Control of Cu(I) coordination chemistry is paramount in the performance of the catalysts, and iodide tends to promote formation of polynuclear aggregates which appear to be less reactive. Similarly, high concentrations of chloride ion in water (0.5 M or above) can be deleterious. For aqueous-rich reactions, cuprous bromide and acetate are favored, as is the sulfate from *in situ* reduction of $CuSO_4$; for organic reactions, the acetate salt is generally a good choice.

Copper(II) salts and coordination complexes are not competent catalysts, and reports describing Cu(II)-catalyzed cycloadditions [46–48] are not accurate. Copper (II) is a well-known oxidizing agent for organic compounds [49]. Alcohols, amines, aldehydes, thiols, phenols, and carboxylic acids may be oxidized by the cupric ion, reducing it to the catalytically active copper(I) species in the process. Especially relevant is the family of oxidative acetylenic couplings catalyzed by the cupric species [50], with the venerable Glaser coupling [51, 52] being the most studied example. Since terminal acetylenes are necessarily present in the CuAAC reaction, their oxidation is an inevitable side process which, in turn, produces the needed catalytically active copper(I) species.

The Cu(I) oxidation state is the least thermodynamically stable form of copper, and many copper(I) complexes can be readily oxidized to catalytically inactive copper(II) species, or can disproportionate to a mixture of Cu(II) and Cu(0). The standard potential of the Cu^{2+}/Cu^{+} couple is 159 mV, but can vary widely depending on the solvent and the ligands coordinated to the metal, and is especially complex in water [53]. When present in significant amounts, the ability of Cu(II) to mediate the aforementioned Glaser-type alkyne coupling processes can result in the formation of undesired byproducts while impairing triazole formation. When the cycloaddition is performed in organic solvents using copper(I) halides as catalysts, the conditions originally reported by the group of Meldal [4], the reaction is plagued by the formation of oxidative coupling byproducts **4a–d** unless the alkyne is bound to a solid support (**5**, Scheme 7.3A and B). If the alkyne is present in solution and the azide is immobilized on the resin, only traces of the desired triazole product are formed. Therefore, when copper(I) catalyst is used directly, whether by itself or in conjunction with amine ligands, exclusion of oxygen may be required. On the other

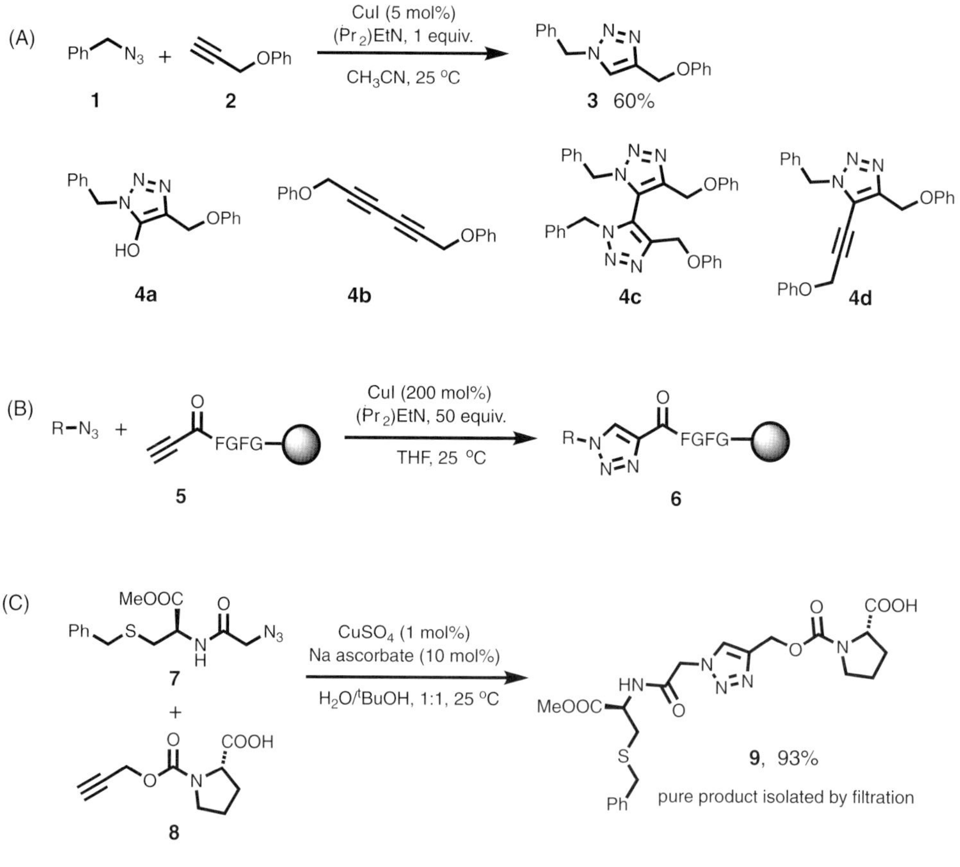

Scheme 7.3 (**A**) Oxidative coupling byproducts in the CuAAC reactions catalyzed by copper(I) salts; (**B**) CuAAC with immobilized alkyne avoids the formation of the oxidative byproducts but requires large excess of the catalyst; reactions with immobilized azide fail; (**C**) solution-phase CuAAC in the presence of sodium ascorbate.

hand, 5-alkynyl, 5-hydroxy, or bis-triazoles can be prepared in synthetically useful yields when the reaction is performed in the presence of chelating ligands and an oxidant [54, 55].

The use of a reducing agent, most commonly sodium ascorbate, introduced by Fokin and coworkers [3], is a convenient and practical alternative to oxygen-free conditions. Its combination with a copper(II) salt, such as the readily available and stable copper(II) sulfate pentahydrate or copper(II) acetate, has become the method of choice for preparative synthesis of 1,2,3-triazoles. Water or aqueous-rich solvent systems are ideal for supporting copper(I) acetylides in their reactive state, especially when they are formed *in situ*. Indeed, our examination of the reactivity of *in situ*-generated, and hence less aggregated, copper(I) acetylides with electrophilic reagents led to the development of the "aqueous ascorbate procedure" in the first place. The "aqueous ascorbate" procedure often furnishes triazole products in nearly quantitative yield and greater than 90% purity, without the need for ligands or protection of the reaction mixture from oxygen (Scheme 7.3C). Of course, copper(I) salts can also be used in combination with ascorbate, wherein it converts any oxidized copper(II) species back to the catalytically active +1 oxidation state.

The reaction can also be catalyzed by Cu(I) ions supplied by elemental copper, thus further simplifying the experimental procedure – a small piece of copper metal (wire or turning) is all that is added to the reaction mixture, followed by shaking or stirring for 12–48 h [3, 21, 56]. Aqueous alcohols (methanol, ethanol, *tert*-butanol), tetrahydrofuran, and dimethylsulfoxide can be used as solvents in this procedure. Cu(II) sulfate may be added to accelerate the reaction; however, this is not necessary in most cases, as copper oxides and carbonates, the patina on the metal surface, are sufficient to initiate the catalytic cycle. Although the procedure based on copper metal requires longer reaction times when performed at ambient temperature, it usually provides access to very pure triazole products with low levels of copper contamination. Alternatively, the reaction can be performed under microwave irradiation at elevated temperature, reducing the reaction time to 10–30 min [56–62].

The copper metal procedure is also very simple experimentally and is particularly convenient for high-throughput synthesis of compound libraries for biological screening. The reaction is very selective, and triazole products are generally isolated in >85–90% yields, and can often be submitted for screening directly. When required, trace quantities of copper remaining in the reaction mixture can be removed with an ion-exchange resin or by using solid-phase extraction techniques. Other heterogeneous copper(0) and copper(I) catalysts, such as copper nanoclusters [63], copper/cuprous oxide nanoparticles [64], and copper nanoparticles adsorbed onto charcoal [65] have also shown good catalytic activity.

Many other copper complexes involving ligands have been reported as catalysts or mediators of the CuAAC reaction. It is probably fair to say that finding ligands that kill the catalysis is more difficult than identifying those that do not. Given these considerations, quantitative comparison between various reported ligands and additives is difficult, since they are employed under widely differing conditions [11]. Instead, we offer a general survey of the published conditions categorizing the ligands into "soft" and "hard" classes by virtue of the properties of their donor centers [66]. Cu(I) is "borderline soft" in its acceptor character [67], making for a wide variety of potentially effective ligands.

The "soft" ligand class is exemplified by phosphine-containing CuAAC-active species such as the simple coordination complexes $Cu(P(OMe)_3)_3Br$ [68] and $Cu(PPh_3)_3Br$ [69, 70]. These species are favored for reactions in organic solvents, in which cuprous salts have limited solubility. A very recent report describes the bis (phosphine) complex $Cu(PPh_3)_2OAc$ as an excellent catalyst for the CuAAC reaction in toluene and dichloromethane [71]. Monodentate phosphoramidite and related donors have also been evaluated [72]. Chelating complexes involving phosphines have not found favor, except for bidentate combinations of phosphine with the relatively weakly-binding triazole unit [73]. Thiols are a potent poison of the CuAAC reaction in water, but thioethers are an underexplored member of the "soft" ligand class [74].

Several Cu(I) complexes with *N*-heterocyclic carbene ligands have been described as CuAAC catalysts at elevated temperature in organic solvents, under heterogeneous aqueous conditions (when both reactants are not soluble in water), and under neat conditions [75]. These catalyst show high activity under the solvent-free conditions, achieving turnover numbers as high as 20 000. However, their activity in solution-phase reactions is significantly lower than that of other catalytic systems (for example, a *stoichiometric* reaction of the isolated copper(I) acetylide/NHC complex with benzhydryl azide required 12 h to obtain 65% yield of the product [76], whereas under standard solution conditions even a catalytic reaction would proceed to completion within 1 h).

The category of "hard" donor ligands for CuAAC is dominated by amines. In many cases, amines are labeled as "additives" rather than "ligands," since it is often the intention to aid in the deprotonation of the terminal alkyne rather than to coordinate to the metal center. However, this assumption is not accurate, as formation of copper (I) acetylides is so facile that it occurs even in strongly acidic media (up to 20–25% H_2SO_4) [77]. Instead, the primary roles of amine ligands can be (i) to prevent the formation of unreactive polynuclear copper(I) acetylides; (ii) to facilitate the coordination of the azide to copper center at the ligand exchange step (*vide infra*); and (iii) to increase the solubility of the copper complex to deliver high solution concentrations of the necessary Cu(I)-species. In several cases of amine-based chelates, it is probable that metal binding is at least part of the productive role of the polydentate ligand. An example is the use of a hydrophobic Tren ligand for CuAAC catalysis in organic solvent at elevated temperature [78].

As befits the nature of the Cu(I) ion, by far the largest and most successful class of ligands is those of intermediate character between "hard" and "soft," particularly those containing heterocyclic donors. With rare exceptions [79], these also contain a central tertiary amine center, which can serve as a coordinating donor or base. The need for these ligands was particularly evident for reactions involving biological molecules that are handled in water in low concentrations and are not stable to heating. The appeal of azide–alkyne cycloaddition for "bioorthogonal" connectivity in aqueous media, well established by the utility of the Staudinger ligation involving organic azides and phosphine esters [80], prompted investigations of copper catalysis in water early on. Chemical transformations used in the synthesis of bioconjugates impose additional demands on the efficiency and selectivity. They must be exquisitely chemoselective, biocompatible, and fast. Despite the experimental simplicity and efficiency of the "ascorbate" procedure, the CuAAC reaction in the forms described above is simply not fast enough when the concentrations of the reactants are low, particularly in aqueous media.

The first ligand to be useful in bioconjugations was tris(benzyltriazolyl)methyl amine**10** (TBTA, Scheme 7.4), prepared using the CuAAC reaction and introduced soon after its discovery [81]. This ligand was shown to significantly accelerate the reaction and stabilize the Cu(I) oxidation state in water-containing mixtures. After its utility in bioconjugation was demonstrated by the efficient attachment of 60 alkyne-containing fluorescent dye molecules to the azide-labeled cowpea mosaic virus [28], it was widely adopted for use with such biological entities as nucleic acids [82–84], proteins [27, 85], *E. coli* bacteria [29, 86, 87], and mammalian cells [88, 89]. A resin-immobilized version of TBTA [90] has also been shown to be very useful in combinatorial and related experiments. The tris(*tert*-butyl) analog, **10b**, often shows superior activity to TBTA in organic solvents.

The poor solubility of TBTA in water prompted the development of more polar analogues such as **11a–c** (Scheme 7.4) [91–93]. At the same time, a combinatorial search for alternatives led to the identification of the commercially-available sulfonated bathophenanthroline **12** as the ligand component of the fastest water-soluble CuAAC catalyst under dilute aqueous conditions [94]. The high catalytic activity of this system made it very useful for demanding bioconjugation tasks [44, 95–100]. However, Cu•**12** complexes are strongly electron-rich and are therefore highly susceptible to oxidation in air. Ascorbate can be used to keep the metal in the +1 oxidation state, but when exposed to air, reduction of O_2 is very fast and can easily use up all of the available reducing agent. Thus, **12** must be used under inert atmosphere, which can be inconvenient, particularly when small amounts of biomolecule samples are used. A procedural solution was found in the use of an

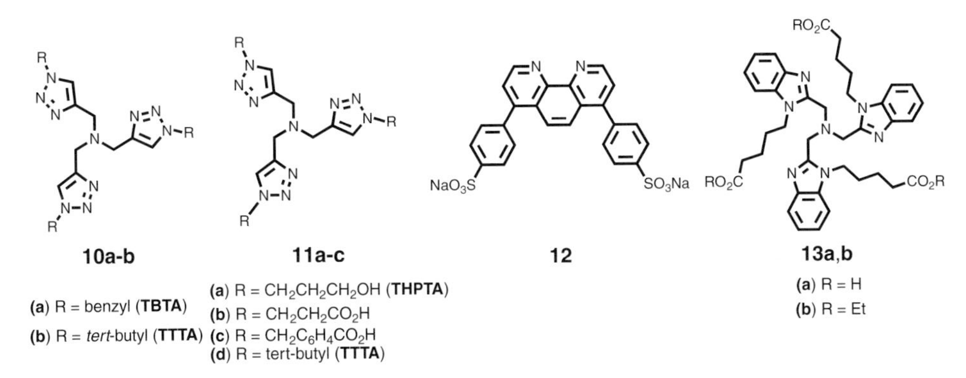

Scheme 7.4 CuAAC-accelerating ligands of choice: tris(1,2,3-triazolyl)methyl amine (TBTA), water-soluble analogs **11**, sulfonated bathophenanthroline **12**, tris(benzimidazole)methyl amine (TBIA) **13**, and "hybrid" ligand **14**.

electrochemical cell to provide the reducing equivalents to scrub O_2 out of such reactions and maintain Cu•**12** in the cuprous oxidation state, but this was again less than optimal due to the need for extra equipment and electrolyte salts [92].

The polydentate trimethylamine theme has been extended to benzimidazole, benzothiazole, oxazoline, and pyridine substituents [101, 102]. Several have provided significantly faster catalysis when quantitative rates are measured, particularly the pendant ester and water-soluble acid derivatives of the tris(benzimidazole) motif, **13a,b**.

7.3.2
CuAAC with *In Situ* Generated Azides

Although organic azides which contain six or more carbon or other heteroatoms (excluding nitrogen and oxygen) are generally stable and safe compounds, those of low molecular weight can spontaneously decompose and, therefore, could be difficult or dangerous to handle. This is especially true for small molecules with several azide functionalities that would be of much interest for the generation of polyfunctionalized structures. Indeed, small-molecule azides should never be isolated away from solvent, for example by distillation, precipitation, or recrystallization. Fortunately, the CuAAC reaction is highly tolerant of inorganic azide, even in large excess. The process can therefore be performed in a one-pot two-step sequence, whereby an *in situ* generated organic azide is immediately consumed in a reaction with a copper acetylide (Scheme 7.5). This type of process has been implemented many times in our laboratories and others from alkyl halides or arylsulfonates by S_N2 reaction with sodium azide (Scheme 7.5A). In a recent example, Pfizer chemists developed a continuous flow process wherein a library of 1,4-disubstituted 1,2,3-triazoles was synthesized from alkyl halides, sodium azide, and terminal acetylenes, with the copper catalyst required for cycloaddition being supplied from the walls of the heated copper tubing through which the reaction solution was passed (Scheme 7.5B) [103].

Aryl and vinyl azides can also be accessed in one step from the corresponding halides or triflates via a copper-catalyzed reaction with sodium azide in the presence of a catalytic amount of L-proline (Scheme 7.5C) [104]. In this fashion, a range of 1,4-disubstituted 1,2,3-triazoles can be prepared in excellent yields [105–107]. Anilines can also be converted to aryl azides by the reaction with *tert*-butyl nitrite and azidotrimethylsilane [108]. The resulting azides can be submitted to the CuAAC conditions without isolation, furnishing triazole products in excellent yields. Microwave heating further improves both reactions, significantly reducing reaction time [56, 62].

7.3.3
Mechanistic aspects of the CuAAC Reaction

Before considering possible mechanistic possibilities for the CuAAC reaction, let us highlight the fundamental reactivity of the reactants: organic azides and copper(I) acetylides. With the exception of the thermal and photochemical decomposition, the reactivity of organic azides is dominated by reactions with nucleophiles at the

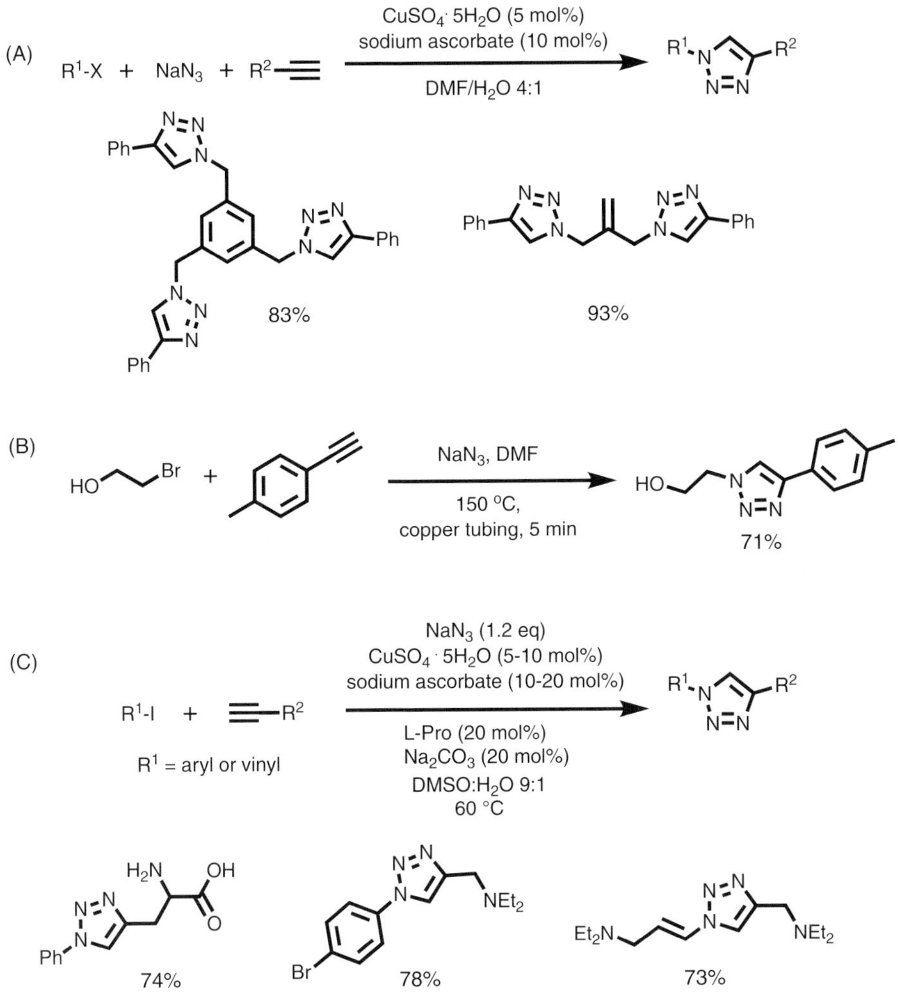

(A) R^1-X + NaN_3 + R^2—≡ → ($CuSO_4 \cdot 5H_2O$ (5 mol%), sodium ascorbate (10 mol%), DMF/H_2O 4:1)

83% 93%

(B) 71% (NaN_3, DMF, 150 °C, copper tubing, 5 min)

(C) R^1-I + ≡—R^2 (R^1 = aryl or vinyl) → (NaN_3 (1.2 eq), $CuSO_4 \cdot 5H_2O$ (5-10 mol%), sodium ascorbate (10-20 mol%), L-Pro (20 mol%), Na_2CO_3 (20 mol%), DMSO:H_2O 9:1, 60 °C)

74% 78% 73%

Scheme 7.5 One-pot syntheses of triazoles from halides at (A,B) sp^3 and (C) sp^2 carbon centers. Reaction B was performed in a flow reactor in 0.75 mm diameter Cu tubing with no added copper catalyst.

terminal N3 atom. Examples of reactions with electrophiles reacting at the proximal N1 (Scheme 7.6) are also known, although these are less common [109, 110]. These reactivity patterns are in agreement with theoretical studies of the electronic structure of the azido group. Coordination chemistry of organic azides follows the same trend, and the azide usually behaves as an L-type σ-donor via its N1 nitrogen atom, with a few exceptions when electron-rich π-donor metals engage the terminal N3 atom in back donation [111].

While the history of copper(I) acetylides dates back as far as Glaser's discovery in 1869 of the oxidative dimerization of Cu-phenylacetylide, the precise nature of the

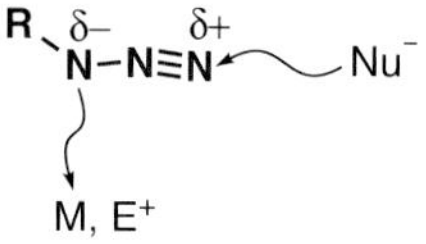

R $\delta-$ $\delta+$
N–N≡N Nu^-
M, E^+

Scheme 7.6 Common reactivity patterns of organic azides.

reactive alkynyl copper species in CuAAC is not well understood. The chief complications are (i) the tendency of copper species to form polynuclear compounds [77, 112] and (ii) the great facility of the ligand exchange at the copper center. As a result, mixtures of Cu(I), terminal alkynes, and other ligands (including solvents) usually contain multiple organocopper species in rapid equilibrium with each other. While this may make elucidation of the exact mechanism difficult, the dynamic nature of copper acetylides is a major contributor to the remarkable adaptability of the reaction to widely different conditions. Whatever the details of the interactions of Cu with alkyne during the CuAAC reaction, it is clear that Cu-acetylide species are easily formed and are productive components of the reaction mechanism.

The initial computational treatment of CuAAC focused on the possible reaction pathways available to mononuclear copper(I) acetylides and organic azides; propyne and methyl azide were chosen for simplicity (Scheme 7.7). Formation of copper(I) acetylide **15** (step A) was calculated to be exothermic by 11.7 kcal mol^{-1}, consistent with the well-known facility of this step which probably occurs through a π-alkyne copper complex intermediate. π-Coordination of alkyne to copper significantly acidifies the terminal hydrogen of the alkyne, bringing it into the proper range to be deprotonated in an aqueous medium and resulting in the formation of a σ-acetylide. The azide is then activated by coordination to copper (step B), forming intermediate **16**. This ligand exchange step is very facile (nearly thermoneutral computationally: 2.0 kcal mol^{-1} uphill when L is water). This coordination is synergistic for both reactive partners: coordination of the azide reveals the β-nucleophilic, vinylidene-like properties of the acetylide, while at the same time the azide's terminus becomes even more electrophilic. In the next step (C), the first C–N bond-forming event takes place, and a strained copper metallacycle **17** forms. This step is endothermic by 12.6 kcal mol^{-1} with a calculated barrier of 18.7 kcal mol^{-1}, which is considerably lower than the barrier for the uncatalyzed reaction (approximately 26.0 kcal mol^{-1}). This drop in the activation energy corresponds roughly to the observed rate increase, thus accounting for the observed rate acceleration accomplished by Cu(I). The formation of copper triazolide is very facile and energetically favorable. When protected by steric bulk, Cu-triazolyl complexes can be isolated from CuAAC reactions [76], and, in rare cases of low catalyst loading and high catalytic rate, step E can be turnover-limiting [102]. Alternative pathways, including the concerted cycloaddition of azide and copper acetylide, were ruled out based on even higher activation barriers than for the concerted cycloaddition. The coordination of the azide to copper via the terminal nitrogen, proposed in a recent review article by Meldal and Tornøe [11], was also examined and was found to be

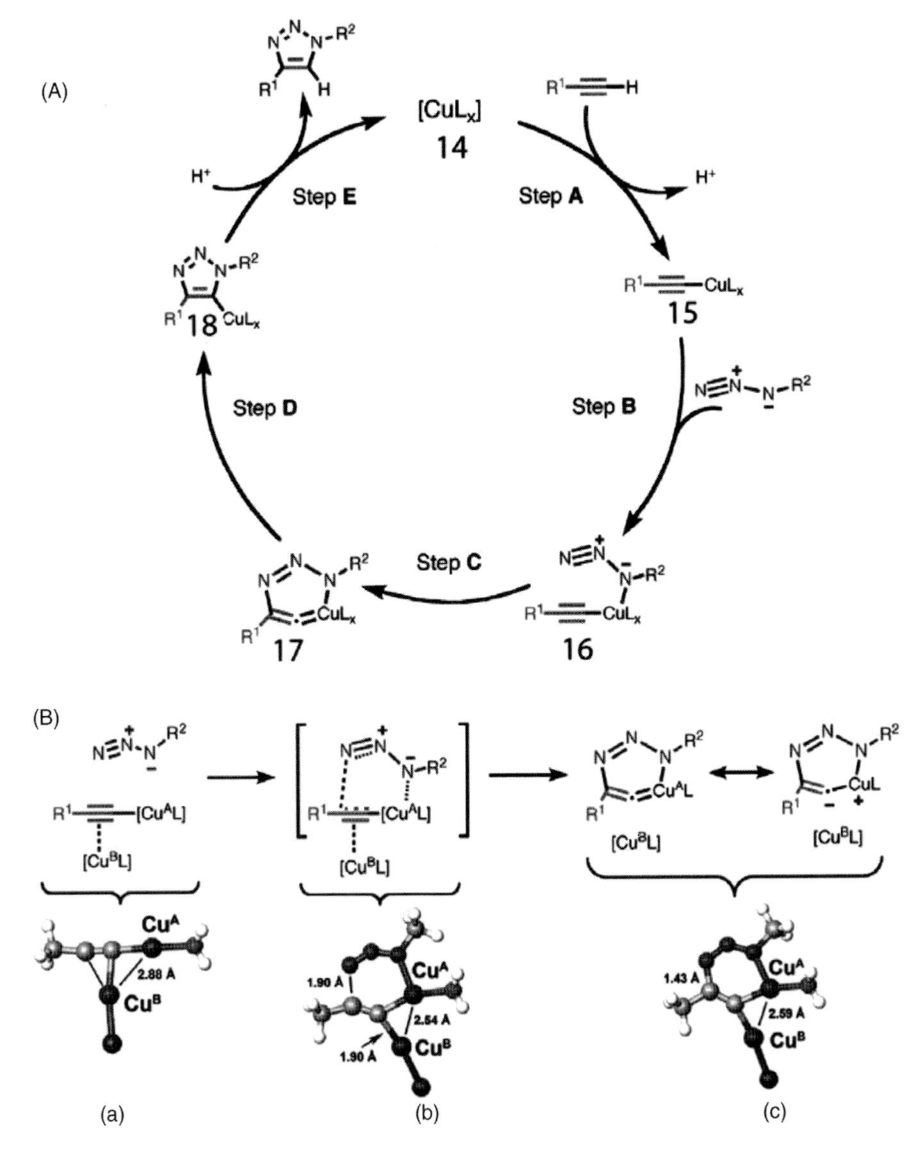

Scheme 7.7 (A) Early proposed catalytic cycle for the CuAAC reaction based on DFT calculations. (B) Introduction of a second copper(I) atom favorably influences the energetic profile of the reaction (L = H_2O in DFT calculations). At the bottom is shown the optimized structures for dinuclear Cu forms of the starting acetylide (a, corresponding to **15**), transition state for the key C–N bond-forming step (b), and the metallacycle **17** (c). The calculated structures are essentially identical when acetylide instead of chloride is used as the ancillary ligand on the second copper center (Cu^B).

energetically unfavorable (no feasible intermediates could be identified). No transition states which could lead to the 1,5-regioisomer could be located either.

The DFT investigation described above was soon followed by a study of the kinetics of the copper-mediated reaction between benzyl azide and phenylacetylene in DMSO under pseudo-first-order conditions. With low catalyst loading, saturating alkyne concentrations, and in the presence of the triazole product, the initial rate law appeared to be second order in the catalyst [113]. Second-order initial rate law was also observed for CuAAC reactions accelerated by tris(benzimidazolyl)methyl amine ligands, such as **13** in neutral Tris buffer solutions [76, 102]. The second order rate law in the catalyst during the initial stages of the reaction does not, of course, necessarily indicate the involvement of two copper atoms in the bond-forming events in the catalytic cycle. Indeed, we have learned since then that the mechanism of the reaction is more complex and the reaction often exhibits discontinuous kinetic behavior (i.e., orders in the catalyst and reagents change as the reaction progresses). However, the implication that multinuclear catalysts could be involved in catalysis prompted further examination of such possibilities both computationally and experimentally.

When the CuAAC pathway was investigated by DFT, taking into account the possibility of the involvement of dinuclear copper(I) acetylides, a further drop in the activation barrier (by approximately 3–6 kcal mol^{-1}) was revealed (Scheme 7.7B) [114, 115]. In the transition states a second copper(I) atom, Cu^B interacts strongly with the proximal acetylide carbon (C^1), as indicated by the short Cu–C distances of 1.93 and 1.90 Å.

Additional experiments further expanded our mechanistic understanding of this process and helped in formulating the experimental guidelines that follow below. Since isolated copper(I) acetylides normally exist in highly aggregated form, preparation of the reactive copper(I) acetylides *in situ*, that is, in the presence of the azide, is crucial for the success of the reaction. Although organic azides are weak ligands for copper, their interaction with the metal center is sufficient to prevent the formation of the unproductive polymeric acetylides. This is easily demonstrated by a simple experiment which can be followed visually: the addition of 5 mol% of copper(II) sulfate to the 0.1 M solution of phenyl acetylene in *tert*-butanol/water (2 : 1) results, within minutes, in the formation of dark brown phenyl copper acetylide. Removal of the precipitate after 15 min leaves supernatant virtually copper free, and the isolated phenyl copper acetylide does not react with azides unless an amine ligand is added, and even then only sluggishly. In fact, once the brown acetylide precipitate is formed, the addition of the azide alone does not revive the catalysis. When the same experiment is performed in the presence of 1 equiv of benzyl azide, the light yellow color that develops upon the addition of $CuSO_4$ disappears within about 20 min, but the reaction is far from completion at that point (about 40% conversion in the case at hand). Evidently the catalyst is going through reorganization while the catalytic cycle is turning over during the initial 20 min after the start of the reaction. Before the reaction enters the steady state regime, a number of turnovers occur, some product is formed, and both the identity and the concentration of the catalyst undergo significant changes. Such behavior, although not uncommon for catalytic reactions, clearly complicates interpretation of the initial rates results [116]. The multimodal kinetic profile is confirmed by following the heat output of the reaction using heat flow calorimetry [117].

As already mentioned, the nature of the copper counter ion also has a dramatic effect on the rate and efficiency of the reaction. For example, the cuprous iodide-catalyzed reaction takes nearly 40 min to reach the maximum rate and over 100 min to reach full conversion (0.1 M in [azide] = [alkyne] = 0.1 M, [CuI/TTTA] = 0.005 M), whereas the replacement of the iodide with much weaker coordinating tetrafluoroborate (by treating the reaction solution with 0.005 M of silver tetrafluoroborate salt) propels the reaction to completion within minutes (with v_{max} at least 10 times higher than for the CuI system).

These mechanistic insights underscore the importance of taking into consideration every event affecting the fate of the on-cycle intermediates when analyzing complex catalytic processes involving multiple equilibria, multiple competing pathways, and off-cycle dead ends. The elementary steps for the key bond-making and bond-breaking events that were previously proposed and supported by DFT calculations [21, 114, 115] still stand, we now have a better quantitative measure of catalyst activation and deactivation pathways and their effect on the overall process. Scheme 7.8 offers our current mechanistic proposal that takes into account earlier studies as well as recent results. The key reversible events that affect the catalytic

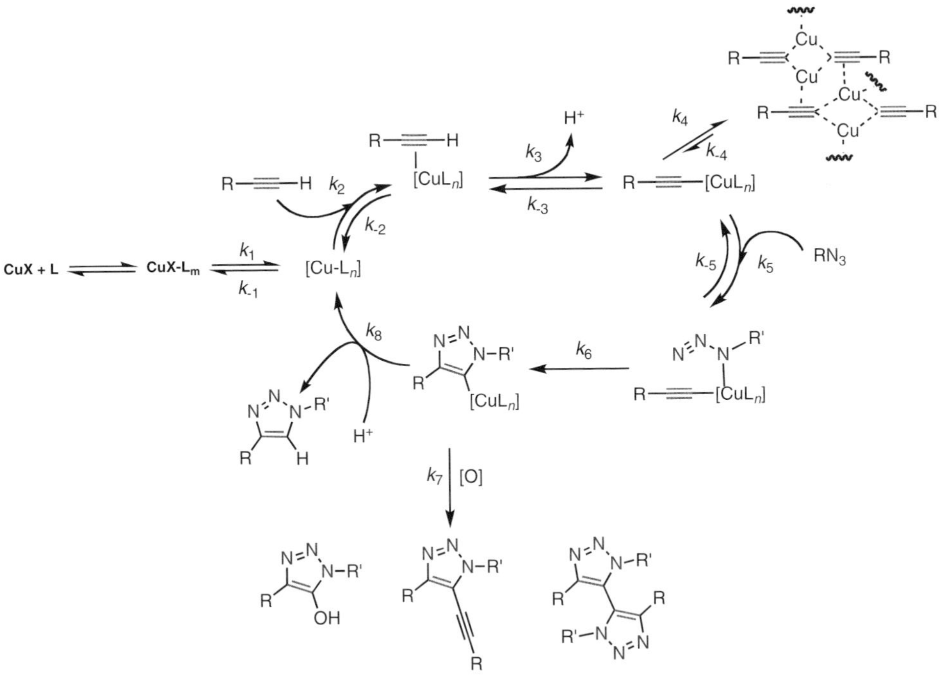

Scheme 7.8 The involvement of multiple equilibria and irreversible off-cycle pathways that affect the productive catalytic CuAAC cycle.

cycle are related to the aggregation state of the copper species (both pre-catalyst, k_1/k_{-1}, and copper acetylide complexes, k_4/k_{-4}). The stability of these complexes determines the rate with which the active catalyst is formed as well as the steady-state concentration of the active catalytic species. In those cases when formation of higher order *stable* polynuclear copper acetylide complexes is facile (as is the case for the chloride and iodide, which readily engage copper acetylides in μ-interactions), such aggregates could very well be the off-cycle catalyst reservoirs, limiting the concentration of the active catalytic species. The irreversible oxidative steps, grouped under the k_7 rate constant, result in the formation of byproducts which may themselves act as ligands for copper and alter the aggregation state of the catalyst.

Thus, the key observations and hypotheses are as follows. First, the formation of higher order polynuclear copper acetylides is detrimental to the rate and the outcome of the reaction. Therefore, solvents that promote ligand exchange (e.g., water and alcohols) are preferred over apolar, organic solvents which promote aggregation of copper species. Ill-defined catalysts perform better in the CuAAC precisely for this reason.

Second, dinuclear complexes may exhibit enhanced reactivity. However, conclusions about the molecularity of the elementary steps based on the observed rate law are tenuous at best, and are likely wrong. Indeed, to date, we have not seen a CuAAC reaction that exhibits uniform second order in the catalyst as it progresses. It is possible that nuclearity of the catalytic species is maintained throughout the catalytic cycle and, as a consequence, all elementary steps are effectively bimolecular, exhibiting the commonly observed first order in the catalyst, even though the reaction is catalyzed by a dinuclear catalyst.

Third, successful CuAAC ligands need to balance the competing requirements of binding Cu(I) strongly enough to prevent the formation of unreactive polymeric complexes, yet allowing azide to access the coordination sphere of the σ-acetylide Cu center. Too potent a binder would tie up the necessary Cu coordination sites and too weak a ligand would not prevent the formation of higher order aggregates. Tris-triazolyl amine ligands, such as TBTA **10a** and TTTA **10b**, and related tripodal ligands, successfully meet these challenges by combining weak donor ligands on the "arms" of the structure with the stronger central nitrogen donor. A possible composition of a reactive complex is exemplified by structure **19**. These ligands also accomplish the delicate task of providing sufficient electron density to the metal to promote the catalysis while not destabilizing the Cu(I) oxidation state.

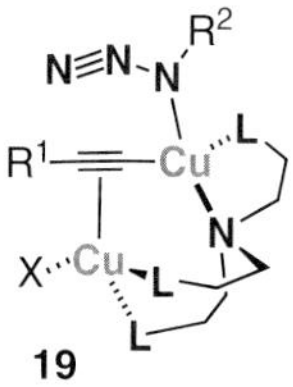

19

7.3.4
Reactions of Sulfonyl Azides

Sulfonyl azides participate in unique CuAAC reactions with terminal alkynes. Depending on the conditions and reagents, products other than the expected triazole **20** [118] can be obtained, as shown in Scheme 7.9. For example, *N*-sulfonyl azides are converted to *N*-sulfonyl amidines **21** when the reaction is conducted in the presence of amines [119]. In the aqueous conditions, *N*-acyl sulfonamides **22** are the major products [120, 121]. In addition to amines and water, the latter can be trapped with imines, furnishing *N*-sulfonyl azetidinimines **23** [122].

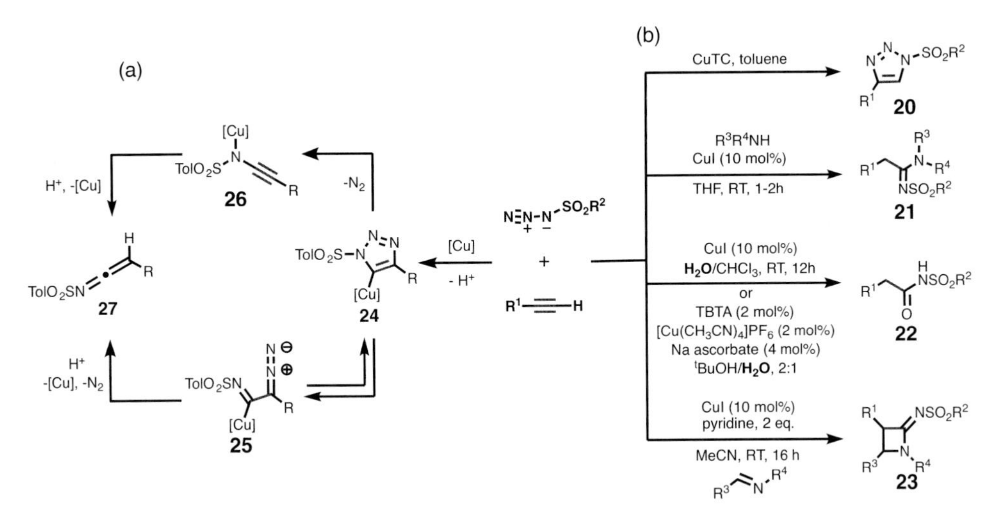

Scheme 7.9 Products of CuAAC reactions with sulfonyl azides (b). Possible pathways leading to ketenimine intermediates (a).

These products are thought to derive from the cuprated triazole intermediate **24**, which is destabilized by the strong electron-withdrawing character of the *N*-sulfonyl substituent. Ring–chain isomerization can occur to form the cuprated diazoimine **25** which, upon the loss of a molecule of dinitrogen, furnishes the *N*-sulfonyl ketenimine **27** [122]. Alternatively, copper(I) alkynamide **26** can be generated with a concomitant elimination of N_2 and, after protonation, would again generate the reactive ketenimine species **27**.

7.3.5
Sulfonyl Triazoles as Stable Carbene Precursors

As already mentioned, 1,2,3-triazoles are very stable heterocycles. However, those that bear a strong electron-withdrawing group at N1 are known to undergo ring–chain tautomerization. The ring–chain isomerism of 1-sulfonyl 1,2,3-triazoles can be exploited in synthesis. As reported recently, 1-sulfonyl triazoles could serve as

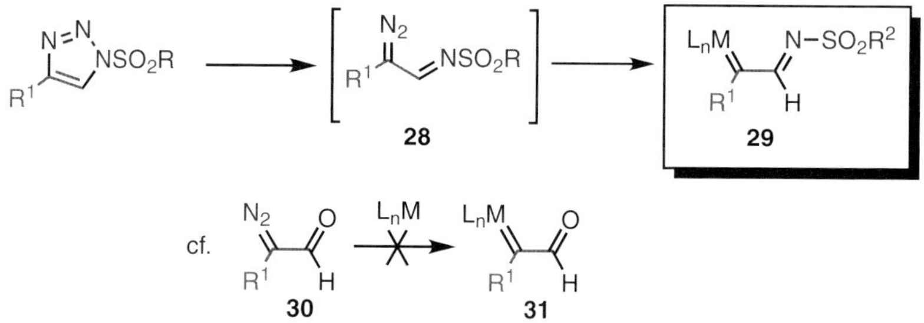

Scheme 7.10 Azavinyl carbenes from diazoimines.

precursors to the diazoimine species 15 that, in turn, could be converted to metal carbene complexes **29** (Scheme 7.10). Rhodium(II)-stabilized carbene complexes exhibit a wealth of reactivity, and this method of generating their diazo progenitors is particularly attractive considering that sulfonyl triazoles effectively become the synthetic equivalents of α-diazo aldehydes **31**, which are very unstable and cannot be converted to the corresponding metal carbenes **30**.

Thus, in the presence of a dirhodium(II) tetraoctanoate catalyst, 1-sulfonyl triazoles react with nitriles, forming imidazoles [123]. The reaction proceeds at 60–80 °C with conventional heating or can be performed in a microwave reactor, and generally provides imidazoles in good to excellent yields (Scheme 7.11A). The sufonyl group can be readily removed, revealing the parent NH-imidazole. Alternatively, sulfonyl imidazoles can be converted to 1,2,5-trisubstituted imidazoles by simple alkylation (Scheme 7.11B).

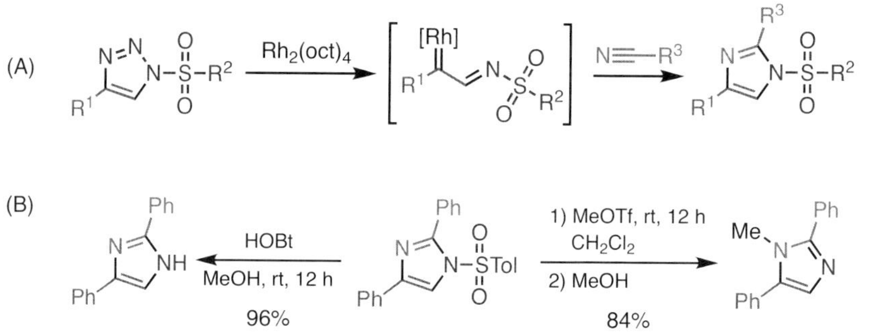

Scheme 7.11 Synthesis of imidazoles via rhodium-catalyzed transannulation of 1-sulfonyl triazoles.

The copper(I)-catalyzed synthesis of 1-sulfonyl triazoles and their subsequent transannulation with nitriles can be combined into a one-pot two-step synthesis, thus further simplifying the experimental procedure (Scheme 7.12). The catalytic amount of copper remaining in the reaction mixture after the first step evidently does not interfere with the formation or reactivity of the carbene complex.

Another example of the exquisite reactivity of rhodium(II) azavinyl carbenes is illustrated by their addition to olefins, which is a very facile process and proceeds

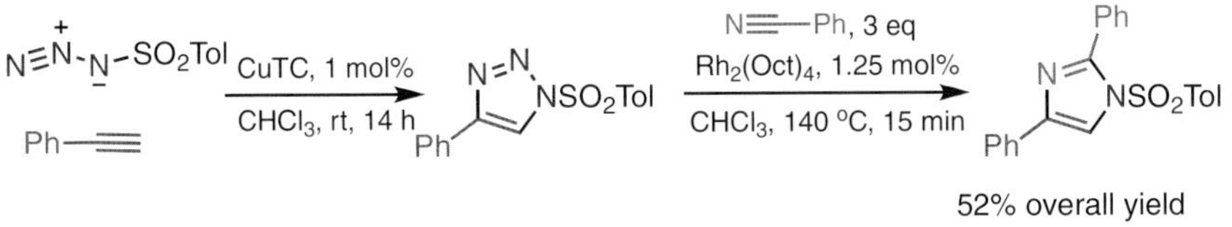

Scheme 7.12 One-pot two-step synthesis which converts sulfonyl azides, terminal alkynes, and nitriles into imidazoles. The only byproduct of the reaction is a molecule of dinitrogen.

under mild conditions with high diastereo- and enantio-selectivity. Chiral rhodium (II) carboxylate complexes, such as $Rh_2(S\text{-NTTL})_4$ and $Rh_2(S\text{-NTV})_4$ are used in the reactions, which are usually performed at elevated temperature (Scheme 7.13A) when isolated *N*-sulfonyl triazoles are used as starting materials [124]. The *N*-sulfonyl group can be easily removed, revealing virtually enantiopure cyclopropane carboxaldehydes **32**. These useful synthetic intermediates are difficult to obtain using other methods.

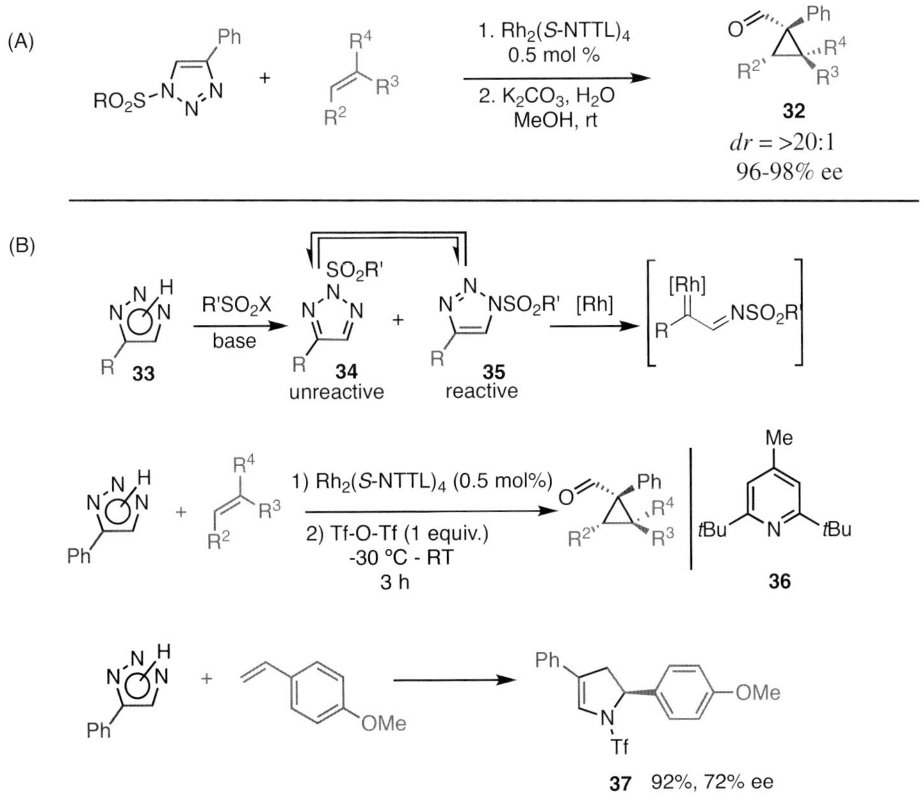

Scheme 7.13 Cyclopropanations of olefins using 1-sulfonyl 1,2,3-triazoles.

In an alternative approach, very electron-deficient and highly reactive *N*-triflyl azavinyl carbenes can be prepared by *in situ* sulfonylation of NH-triazoles **33** with triflic anhydride in the presence of a hindered pyridine base **36** (Scheme 7.13B) [125]. Although the sulfonylation step is not selective and results in the formation of both N1- and N2-sulfonylated triazoles (**34** and **35**, respectively), only the N1-isomer can undergo ring–chain isomerization and, therefore, form a reactive carbene intermediate. The resulting *N*-triflyl azavinyl carbenes exhibit exceptional reactivity towards olefins, resulting in the formation of cyclopropanes and 1,2-dihydropyrroles (**37**, when electron-rich olefins are used; Scheme 7.13B) with excellent enantio- and diastereoselectivity. The ability to introduce extremely electron-withdrawing groups into the azavinyl carbene, thereby controlling its electrophilicity, is a valuable feature of this approach.

7.3.6
Reactions of 1-Iodoalkynes

1-Iodoalkynes are stable and readily accessible internal acetylenes. As was disclosed recently, they are exceptionally reactive partners with organic azides under copper(I) catalysis (Scheme 7.14) [126]. In fact, their reactivity appears to surpass that of terminal alkynes. As an added benefit, the products of the reaction, 5-iodo-1,2,3-triazoles, are versatile synthetic intermediates amenable for further functionalization. The reaction is general, chemo- and regio-selective, and the experimental procedure is operationally simple. The 5-iodo-1,4,5-trisubstituted-1,2,3-triazoles are obtained in high yields from variously substituted organic azides and iodoalkynes. The catalysis is effected by copper(I) iodide in the presence of an amine ligand. In contrast to the CuAAC reaction, the observed rate and chemoselectivity of the reaction are strongly dependent on the nature of the ligand, and catalysis does not proceed in its absence. Both TBTA (**10a**) and its *tert*-butyl analog, TTTA (**10b**) give 5-iodotriazoles **40** as exclusive products in excellent yield.

Ph—≡—I **38** + **39** (N_3, CF_3) —CuI (5 mol%), Ligand, THF, r.t.→ **40**: R = -I; **41**: R = -H; **42**: R = -CCPh

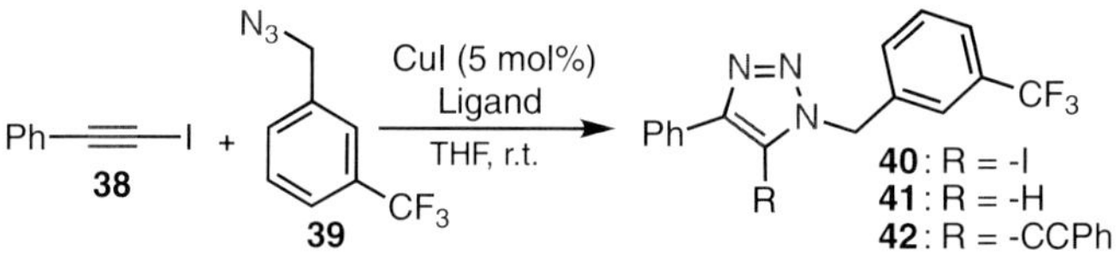

Scheme 7.14 The iodoalkyne version of the CuAAC reaction (^{i}CuAAC).

The competing pathways, which arise from dehalogenation of the iodoalkyne and the triazole product and lead to 5-proto **41** and 5-alkynyl triazoles **42**, are far too slow in the presence of the ligands.

The CuI–TTTA catalyst system exhibits excellent functional group compatibility, and 5-iodotriazoles are obtained as exclusive products in excellent yield from structurally and functionally diverse azides and 1-iodoalkynes. Due to the mild reaction conditions, high chemoselectivity, and low copper catalyst loading, reaction work-up is usually as simple as trituration followed by filtration. The scale-up is easy, and a number of 5-iodotriazoles have been prepared in multigram quantities.

1-iodoalkynes are readily obtained from terminal acetylenes by treating them with *N*-iodomorpholine **43** in the presence of CuI, giving the corresponding 1-iodoakynes within 30–60 min. The products can be isolated by passing the reaction mixture through a pad of silica gel or alumina, yielding the desired 1-iodoalkyne in high yield. It can be submitted to the reaction with an azide in a one-pot, two-step procedure (Scheme 7.15). The 1-iodoalkyne is partially purified via filtration through neutral alumina prior to the introduction of the azide component. This method gives 5-iodotriazoles with an efficiency comparable to that observed with the isolated 1-iodoalkynes.

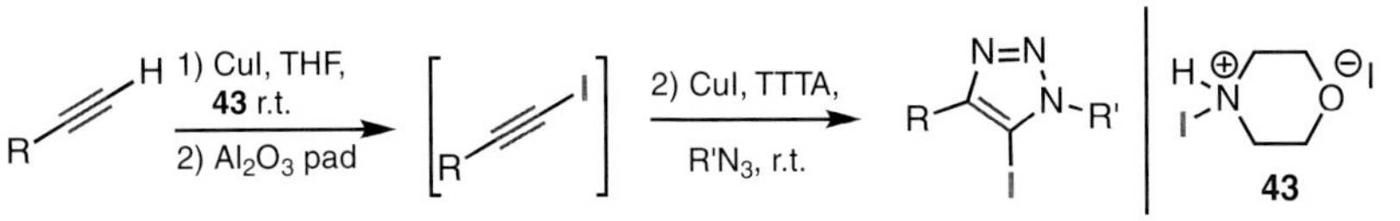

Scheme 7.15 One-pot, two-step synthesis of 5-iodo-1,2,3-triazoles.

This sequence could be further extended to the synthesis of 1,4,5-triaryl-1,2,3-triazoles **44–46** (Scheme 7.16) by assembling the 5-iodotriazole and immediately employing Pd(0)-catalyzed cross-coupling with an appropriate arylboronic acid. This simple stepwise construction obviates purification of any intermediates and simultaneously provides complete control over the placement of substituents around the 1,2,3-triazole core, allowing facile access to all regioisomeric permutations of trisubstituted triazoles **44–46**. Similar regiocontrolled synthesis would not be possible via thermal or ruthenium-catalyzed reaction due to the steric and electronic similarity of the aryl substituents (phenyl, tolyl, and *p*-methoxyphenyl).

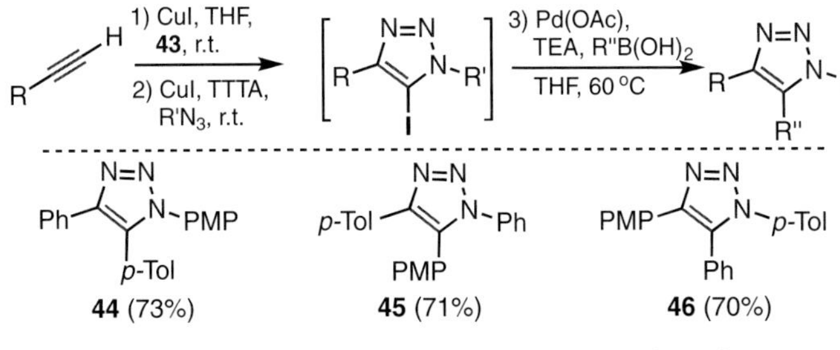

Scheme 7.16 One-pot, three-step synthesis of 1,4,5-triaryltriazoles. PMP = p-methoxyphenyl, p-Tol = p-methylphenyl.

In addition to the immediate synthetic utility of the iodo-CuAAC reactions, examination of the mechanism will provide a better understanding of both the iodo and the parent CuAAC processes. Although both reactions clearly share some common features, the modes of activation of iodo- and terminal alkynes by copper are likely distinctly different. Our current mechanistic proposals are outlined in Scheme 7.17. One possible pathway is similar to that proposed for the CuAAC and

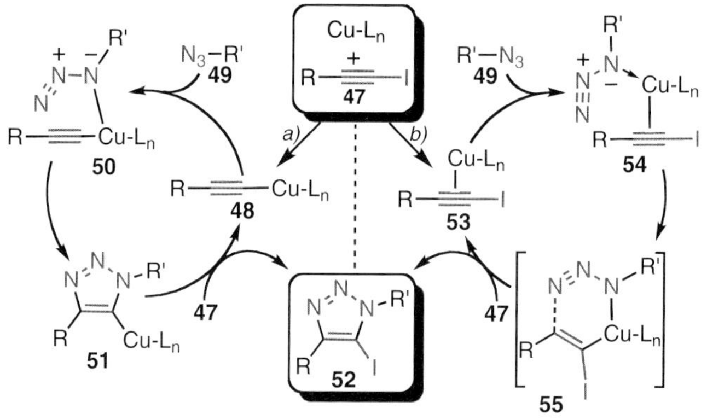

Scheme 7.17 Proposed mechanistic pathways for the Cu(I)-catalyzed azide-iodoalkyne cycloaddition.

involves the formation of the σ-acetylide complex **48** as the first key intermediate (Scheme 7.17*a*. Cu-I exchange via σ-bond metathesis with iodoalkyne **47** completes the cycle, liberating iodotriazole **52** and regenerating acetylide **48**.

Alternatively, copper may activate the iodoalkyne via the formation of a π-complex intermediate (Scheme 7.13*b*, which then engages the azide, producing complex **54**. Cyclization then proceeds via a now familiar vinylidene-like intermediate **55**, to give iodotriazole **52**. A similar transition state has been proposed to explain the involvement of di-copper intermediates in the CuAAC reaction [114, 115]. The distinctive feature of this pathway is that the C–I bond is never severed during the catalysis.

Pathway b has gained more experimental support at this time. The main argument in support of this hypothesis is the exclusive formation of the 5-iodotriazole, even when the reaction is performed in protic solvents or with the substrates containing acidic protons. If pathway a were operational, the cuprated triazole intermediate **51** could be trapped with other electrophiles, including a proton, thereby producing a mixture of the 5-iodo and 5-prototriazoles. Whatever the interactions of iodoalkynes with copper(I), they appear to be stronger than for terminal acetylenes: when the reaction is performed on a mixture of a terminal and 1-iodoalkyne, the formation of the 5-prototriazole does not begin until all iodoalkyne is consumed. This behavior may indicate the complete catalyst monopoly by the iodo cycle and different catalyst resting states for the two competing processes.

7.3.7
Reactions of Copper Acetylides with Other Dipoles

In the mechanism of the CuAAC reaction described above, the metal catalyst activates terminal alkyne for reaction with a Cu-coordinated azide. This mode of reactivity also operates with other dipolar reagents. In fact, the first example of a copper-catalzyed 1,3-dipolar cycloaddition reaction of alkynes was reported for nitriones by Kinugasa in 1972 [127]. An asymmetric version of the Kinugasa reaction was developed by Fu *et al.* in 2002 [128, 129].

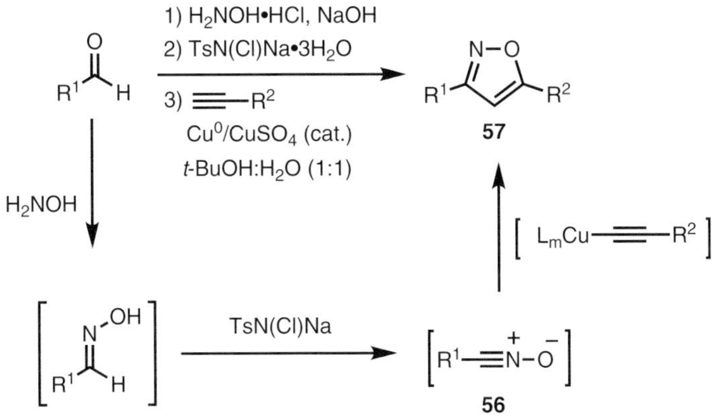

Scheme 7.18 Synthesis of 3,5-disubstituted isoxazoles from aldehydes, hydroxylamine, and terminal alkynes.

For example, Scheme 7.18 shows the case of nitrile oxides. Similarly to azides, the uncatalyzed 1,3-dipolar cycloaddition of nitrile oxides and acetylenes has long been known, but its applications to the synthesis of the corresponding heterocycle (isoxazoles) are scarce. Yields of isoxazole products are often quite low, side reactions are common, and both regioisomers may be formed (although the selectivity of nitrile oxide cycloadditions is usually higher than in reactions of azides, favoring the 3,5-isomer) [130]. Furthermore, nitrile oxides are not very stable and readily dimerize.

In contrast, Cu(I) catalysis makes possible the efficient synthesis of 3,5-disubstituted isoxazoles **57** from aromatic or aliphatic aldehydes and alkynes. Stable nitrile oxides can be isolated and subsequently submitted to the reaction [21] in isolated form and submitted to the reaction in one-pot, three-step process [131]. Here, nitrile oxide intermediates **56** are generated *in situ* via the corresponding aldoxime and halogenation/deprotonation by Chloramine-T [132]. Capture of the intermediate nitrile oxide by copper(I) acetylides occurs presumably before dimerization. In this case, the Cu catalyst was obtained from copper metal and copper(II) sulfate, and the products were isolated by simple filtration or aqueous work-up. Trace amounts of toluenesulfonamide and unreacted acetylene are easily removed by recrystallization or by passing the product through a short plug of silica gel.

Copper catalyzed reactions of azomethine imines with alkynes yielding fused nitrogen heterocycles have also been reported [133].

References

1 Larock, R.C. (2005) *Acetylene Chemistry: Chemistry, Biology, and Material Science* (eds F. Diederich, P.J., Stang, and R.R. Tykwinski), Wiley-VCH, Weinheim, pp. 51–100.

2 Kolb, H.C., Finn, M.G., and Sharpless, K.B. (2001) *Angew. Chem. Int. Ed.*, **40**, 2004–2021.

3 Rostovtsev, V.V., Green, L.G., Fokin, V.V., and Sharpless, K.B. (2002) *Angew. Chem. Int. Ed.*, **41**, 2596–2599.

4 Tornøe, C.W., Christensen, C., and Meldal, M. (2002) *J. Org. Chem.*, **67**, 3057–3062.
5 Bock, V.D., Hiemstra, H., and van Maarseveen, J.H. (2006) *Eur. J. Org. Chem.*, 51–68.
6 Fokin, V.V. (2007) *ACS Chem. Biol.*, **2**, 775–778.
7 Moses, J.E. and Moorhouse, A.D. (2007) *Chem. Soc. Rev.*, **36**, 1249–1262.
8 Wu, P. and Fokin, V.V. (2007) *Aldrichim. Acta*, **40**, 7–17.
9 Johnson, J.A., Koberstein, J.T., Finn, M.G., and Turro, N.J. (2008) *Macromol. Rapid Commun.*, **29**, 1052–1072.
10 Lutz, J.F. and Schlaad, H. (2008) *Polymer*, **49**, 817–824.
11 Meldal, M. and Tornoe, C.W. (2008) *Chem. Rev.*, **108**, 2952–3015.
12 Tron, G.C., Pirali, T., Billington, R.A., Canonico, P.L., Sorba, G., and Genazzani, A.A. (2008) *Med. Res. Rev.*, **28**, 278–308.
13 Michael, A. (1893) *J. Prakt. Chem.*, **48**, 94.
14 Huisgen, R., Knorr, R., Moebius, L., and Szeimies, G. (1965) *Chem. Ber.*, **98**, 4014–4021.
15 Huisgen, R., Moebius, L., Mueller, G., Stangl, H., Szeimies, G., and Vernon, J.M. (1965) *Chem. Ber.*, **98**, 3992–4013.
16 Huisgen, R., Moebius, L., and Szeimies, G. (1965) *Chem. Ber.*, **98**, 1138–1152.
17 Huisgen, R., Szeimies, G., and Moebius, L. (1966) *Chem. Ber.*, **99**, 475–490.
18 Huisgen, R., Szeimies, G., and Moebius, L. (1967) *Chem. Ber.*, **100**, 2494–2507.
19 Huisgen, R. (1984) *1,3-Dipolar Cycloaddition Chemistry*, vol. 1 (ed. A. Padwa), Wiley, New York, pp. 1–176.
20 Huisgen, R. (1989) *Pure Appl. Chem.*, **61**, 613–628.
21 Himo, F., Lovell, T., Hilgraf, R., Rostovtsev, V.V., Noodleman, L., Sharpless, K.B., and Fokin, V.V. (2005) *J. Am. Chem. Soc.*, **127**, 210–216.
22 Tomé, A.C. (2004) Science of synthesis: Houben-Weyl methods of molecular transformations, in *Five-Membered Hetarenes with Three or More Heteroatoms*, vol. 13 (eds R.C. Storr and T.L. Gilchrist), Thieme, Stuttgart, pp. 415–601.
23 Krivopalov, V.P. and Shkurko, O.P. (2005) *Russ. Chem. Rev.*, **74**, 339–379.
24 Brik, A., Muldoon, J., Lin, Y.-c., Elder, J.H., Goodsell, D.S., Olson, A.J., Fokin, V.V., Sharpless, K.B., and Wong, C.-h. (2003) *ChemBioChem*, **4**, 1246–1248.
25 Brik, A., Alexandratos, J., Lin, Y.-C., Elder, J.H., Olson, A.J., Wlodawer, A., Goodsell, D.S., and Wong, C.-H. (2005) *ChemBioChem*, **6**, 1167–1169.
26 Tam, A., Arnold, U., Soellner, M.B., and Raines, R.T. (2007) *J. Am. Chem. Soc.*, **129**, 12670–12671.
27 Speers, A.E., Adam, G.C., and Cravatt, B.F. (2003) *J. Am. Chem. Soc.*, **125**, 4686–4687.
28 Wang, Q., Chan, T.R., Hilgraf, R., Fokin, V.V., Sharpless, K.B., and Finn, M.G. (2003) *J. Am. Chem. Soc.*, **125**, 3192–3193.
29 Link, A.J., Vink, M.K.S., and Tirrell, D.A. (2004) *J. Am. Chem. Soc.*, **126**, 10598–10602.
30 Binder, W.H. and Kluger, C. (2006) *Curr. Org. Chem.*, **10**, 1791–1815.
31 Nebhani, L. and Barner-Kowollik, C. (2009) *Adv. Mater.*, **21**, 3442–3468.
32 Kolb, H.C. and Sharpless, K.B. (2003) *Drug Disc. Today*, **8**, 1128–1137.
33 Gierlich, J., Burley, G.A., Gramlich, P.M.E., Hammond, D.M., and Carell, T. (2006) *Org. Lett.*, **8**, 3639–3642.
34 Goodall, G.W. and Hayes, W. (2006) *Chem. Soc. Rev.*, **35**, 280–312.
35 Nandivada, H., Chen, H.-Y., Bondarenko, L., and Lahann, J. (2006) *Angew. Chem. Int. Ed.*, **45**, 3360–3363.
36 Spruell, J.M., Sheriff, B.A., Rozkiewicz, D.I., Dichtel, W.R., Rohde, R.D., Reinhoudt, D.N., Stoddart, J.F., and Heath, J.R. (2008) *Angew. Chem., Int. Ed.*, **47**, 9927–9932.
37 Zhang, L., Chen, X., Xue, P., Sun, H.H.Y., Williams, I.D., Sharpless, K.B., Fokin, V.V., and Jia, G. (2005) *J. Am. Chem. Soc.*, **127**, 15998–15999.
38 Majireck, M.M. and Weinreb, S.M. (2006) *J. Org. Chem.*, **71**, 8680–8683.
39 Rasmussen, L.K., Boren, B.C., and Fokin, V.V. (2007) *Org. Lett.*, **9**, 5337–5339.
40 Boren, B.C., Narayan, S., Rasmussen, L.K., Zhang, L., Zhao, H., Lin, Z., Jia, G., and Fokin, V.V. (2008) *J. Am. Chem. Soc.*, **130**, 8923–8930.
41 Oppilliart, S., Mousseau, G., Zhang, L., Jia, G., Thuery, P., Rousseau, B., and

Cintrat, J.-C. (2007) *Tetrahedron*, **63**, 8094–8098.
42 Horne, W.S., Olsen, C.A., Beierle, J.M., Montero, A., and Ghadiri, M.R. (2009) *Angew. Chem., Int. Ed.*, **48**, 4718–4724, S4718/4711-L S4718/4711.
43 Kubas, G.J. (1979) *Inorg. Synth.*, **19**, 90–92.
44 Gupta, S.S., Kuzelka, J., Singh, P., Lewis, W.G., Manchester, M., and Finn, M.G. (2005) *Bioconjugate Chem.*, **16**, 1572–1579.
45 Rostovtsev, V.V., Green, L., Sharpless, K.B., and Fokin, V.V. (2002) Abstracts of Papers, 224th ACS National Meeting, Boston, MA, United States, August 18–22 2002, ORGN-458.
46 Reddy, K.R., Rajgopal, K., and Kantam, M.L. (2006) *Synlett*, 957–959.
47 Fukuzawa, S., Shimizu, E., and Kikuchi, S. (2007) *Synlett*, 2436–2438.
48 Reddy, K.R., Rajgopal, K., and Kantam, M.L. (2007) *Catal. Lett.*, **114**, 36–40.
49 Nigh, W.G. (1973) *Oxidation in Organic Chemistry*, **Part B** (ed. W.S. Trahanovsky), Academic Press, New York, pp. 1–95.
50 Siemsen, P., Livingston, R.C., and Diederich, F. (2000) *Angew. Chem. Int. Ed.*, **39**, 2632–2657.
51 Glaser, C. (1869) *Chem. Ber.*, **2**, 422.
52 Glaser, C. (1870) *Ann. Chem. Pharm.*, **154**, 159.
53 Fahrni, C.J. (2007) *Curr. Opin. Chem. Biol.*, **11**, 121–127.
54 Gerard, B., Ryan, J., Beeler, A.B., and Porco, J.A.Jr. (2006) *Tetrahedron*, **62**, 6405–6411.
55 Angell, Y. and Burgess, K. (2007) *Angew. Chem. Int. Ed.*, **46**, 3649–3651.
56 Appukkuttan, P., Dehaen, W., Fokin, V.V., and Van der Eycken, E. (2004) *Org. Lett.*, **6**, 4223–4225.
57 Ermolat'ev, D., Dehaen, W., and Van der Eycken, E. (2004) *QSAR Comb. Sci,*, **23**, 915–918.
58 Khanetskyy, B., Dallinger, D., and Kappe, C.O. (2004) *J. Comb. Chem.*, **6**, 884–892.
59 Bouillon, C., Meyer, A., Vidal, S., Jochum, A., Chevolot, Y., Cloarec, J.-P., Praly, J.-P., Vasseur, J.-J., and Morvan, F. (2006) *J. Org. Chem.*, **71**, 4700–4702.
60 Appukkuttan, P. and Van der Eyeken, E. (2008) *Eur. J. Org. Chem.*, 1133–1155.
61 Lucas, R., Neto, V., Bouazza, A.H., Zerrouki, R., Granet, R., Krausz, P., and Champavier, Y. (2008) *Tetrahedron Lett.*, **49**, 1004–1007.
62 Moorhouse, A.D. and Moses, J.E. (2008) *Synlett*, 2089–2092.
63 Pachon, L.D., van Maarseveen, J.H., and Rothenberg, G. (2005) *Adv. Synth. Catal.*, **347**, 811–815.
64 Molteni, G., Bianchi, C.L., Marinoni, G., Santo, N., and Ponti, A. (2006) *New J. Chem.*, **30**, 1137–1139.
65 Lipshutz, B.H. and Taft, B.R. (2006) *Angew. Chem. Int. Ed.*, **45**, 8235–8238.
66 Pearson, R.G. (1988) *J. Am. Chem. Soc.*, **110**, 7684–7690.
67 Sivasankar, C., Sadhukhan, N., Bera, J.K., and Samuelson, A.G. (2007) *New J. Chem.*, **31**, 385–393.
68 Perez-Balderas, F., Ortega-Munoz, M., Morales-Sanfrutos, J., Hernandez-Mateo, F., Calvo-Flores, F.G., Calvo-Asin, J.A., Isac-Garcia, J., and Santoyo-Gonzalez, F. (2003) *Org. Lett.*, **5**, 1951–1954.
69 Wu, P., Feldman, A.K., Nugent, A.K., Hawker, C.J., Scheel, A., Voit, B., Pyun, J., Frechet, J.M.J., Sharpless, K.B., and Fokin, V.V. (2004) *Angew. Chem. Int. Ed.*, **43**, 3928–3932.
70 Malkoch, M., Schleicher, K., Drockenmuller, E., Hawker, C.J., Russell, T.P., Wu, P., and Fokin, V.V. (2005) *Macromolecules*, **38**, 3663–3678.
71 Gonda, Z. and Novák, Z. (2010) *Dalton Trans.*, **39**, 726–729.
72 Campbell-Verduyn, L.S., Mirfeizi, L., Dierckx, R.A., Elsinga, P.H., and Feringa, B.L. (2009) *Chem. Commun.*, 2139–2141.
73 Detz, R.J., Arevalo Heras, S., De Gelder, R., Van Leeuwen, P.W.N.M., Hiemstra, H., Reek, J.N.H., and Van Maarseveen, J.H. (2006) *Org. Lett*, **8** 3227–3230.
74 Bai, S.-Q., Koh, L.L., and Hor, T.S.A. (2009) *Inorg. Chem. (Washington, DC, U.S.)*, **48**, 1207–1213.
75 Díez-González, S. and Nolan, S.P. (2008) *Angew. Chem. Int. Ed.*, **46**, 9013–9016.
76 Nolte, C., Mayer, P., and Straub, B.F. (2007) *Angew. Chem. Int. Ed.*, **46**, 2101–2103.
77 Mykhalichko, B.M., Temkin, O.N., and Mys'kiv, M.G. (2001) *Russ. Chem. Rev.*, **69**, 957–984.

78 Candelon, N., Lastecoueres, D., Diallo, A.K., Aranzaes, J.R., Astruc, D., and Vincent, J.M. (2008) *Chem. Comm.*, 741–743.

79 Özcubukcu, S., Ozkal, E., Jimeno, C., and Pericás, M.A. (2009) *Org. Lett.*, **11**, 4680–4683.

80 Saxon, E. and Bertozzi, C.R. (2000) *Science*, **287**, 2007–2010.

81 Chan, T.R., Hilgraf, R., Sharpless, K.B., and Fokin, V.V. (2004) *Org. Lett.*, **6**, 2853–2855.

82 Weller, R.L. and Rajski, S.R. (2005) *Org. Lett.*, **7**, 2141–2144.

83 Burley, G.A., Gierlich, J., Hammond, D.M., Gramlich, Phillipp M.E., and Carell, T. (2006) *AIP Conf. Proc.*, **859**, 45–53.

84 Burley, G.A., Gierlich, J., Mofid, M.R., Nir, H., Tal, S., Eichen, Y., and Carell, T. (2006) *J. Am. Chem. Soc.*, **128**, 1398–1399.

85 Speers, A.E. and Cravatt, B.F. (2004) *Chem. Biol.*, **11**, 535–546.

86 Link, A.J. and Tirrell, D.A. (2003) *J. Am. Chem. Soc.*, **125**, 11164–11165.

87 Beatty, K.E., Xie, F., Wang, Q., and Tirrell, D.A. (2005) *J. Am. Chem. Soc.*, **127**, 14150–14151.

88 Dieterich, D.C., Link, A.J., Graumann, J., Tirrell, D.A., and Schuman, E.M. (2006) *Proc. Natl. Acad. Sci. USA*, **103**, 9482–9487.

89 Sawa, M., Hsu, T.-L., Itoh, T., Sugiyama, M., Hanson Sarah, R., Vogt Peter, K., and Wong, C.-H. (2006) *Proc. Natl. Acad. Sci. USA*, **103**, 12371–12376.

90 Chan, T.R. and Fokin, V.V. (2007) *QSAR Comb. Sci.*, **26**, 1274–1279.

91 Chan, T.R., Fokin, V.V., and Sharpless, K.B. (2004) Abstracts of Papers, 227th ACS National Meeting, Anaheim, CA, United States, March 28-April 1 2004, ORGN-041.

92 Hong, V., Udit, A.K., Evans, R.A., and Finn, M.G. (2008) *ChemBioChem*, **9**, 1481–1486.

93 Hong, V., Presolski, S.I., Ma, C., and Finn, M.G. (2009) *Angew. Chem., Int. Ed.*, **48**, 9879–9883.

94 Lewis, W.G., Magallon, F.G., Fokin, V.V., and Finn, M.G. (2004) *J. Am. Chem. Soc.*, **126**, 9152–9153.

95 Prasuhn, D.E., Yeh, R.M., Obenaus, A., Manchester, M., and Finn, M.G. (2007) *Chem. Comm.*, 1269–1271.

96 Zeng, Q., Li, T., Cash, B., Li, S., Xie, F., and Wang, Q. (2007) *Chem. Comm.*, 1453–1455.

97 Megiatto, J.D. Jr. and Schuster, D.I. (2008) *J. Am. Chem. Soc.*, **130**, 12872–12873.

98 Prasuhn, D.E. Jr., Singh, P., Strable, E., Brown, S., Manchester, M., and Finn, M.G. (2008) *J. Am. Chem. Soc.*, **130**, 1328–1334.

99 Schoffelen, S., Lambermon, M.H.L., van Eldijk, M.B., and van Hest, J.C.M. (2008) *Bioconjugate Chem.*, **19**, 1127–1131.

100 Strable, E., Prasuhn, D.E. Jr., Udit, A.K., Brown, S., Link, A.J., Ngo, J.T., Lander, G., Quispe, J., Potter, C.S., Carragher, B., Tirrell, D.A., and Finn, M.G. (2008) *Bioconjugate Chem.*, **19**, 866–875.

101 Rodionov, V.O., Presolski, S.I., Diaz, D.D., Fokin, V.V., and Finn, M.G. (2007) *J. Am. Chem. Soc.*, **129**, 12705–12712.

102 Rodionov, V.O., Presolski, S.I., Gardinier, S., Lim, Y.-H., and Finn, M.G. (2007) *J. Am. Chem. Soc.*, **129**, 12696–12704.

103 Bogdan, A.R. and Sach, N.W. (2009) *Adv. Synth. Cat.*, **351**, 849–854.

104 Zhu, W. and Ma, D. (2004) *Chem. Comm.*, 888–889.

105 Feldman, A.K., Colasson, B., and Fokin, V.V. (2004) *Org. Lett.*, **6**, 3897–3899.

106 Chittaboina, S., Xie, F., and Wang, Q. (2005) *Tetrahedron Lett.*, **46**, 2331–2336.

107 Kacprzak, K. (2005) *Synlett*, 943–946.

108 Barral, K., Moorhouse, A.D., and Moses, J.E. (2007) *Org. Lett.*, **9**, 1809–1811.

109 Sheradsky, T. (1971) *Chemistry of Functional Groups* (ed. S. Patai), John Wiley & Sons, New York, pp. 332–389.

110 Braese, S., Gil, C., Knepper, K., and Zimmermann, V. (2005) *Angew. Chem. Int. Ed.*, **44**, 5188–5240.

111 Cenini, S., Gallo, E., Caselli, A., Ragaini, F., Fantauzzi, S., and Piangiolino, C. (2006) *Coord. Chem. Rev.*, **250**, 1234–1253.

112 Vrieze, K. and Koten van, G. (1987) *Comprehensive Coordination Chemistry*, vol. 2, Pergamon, Oxford, pp. 189–245.

113 Rodionov, V.O., Fokin, V.V., and Finn, M.G. (2005) *Angew. Chem. Int. Ed.*, **44**, 2210–2215.

114 Ahlquist, M. and Fokin, V.V. (2007) *Organometallics*, **26**, 4389–4391.

115 Straub, B.F. (2007) *Chem. Comm.*, 3868–3870.

116 Rosner, T., Pfaltz, A., and Blackmond, D.G. (2001) *J. Am. Chem. Soc.*, **123**, 4621–4622.

117 Hein, J.E. and Fokin, V.V. (2010)

118 Yoo, E.J., Ahlquist, M., Kim, S.H., Bae, I., Fokin, V.V., Sharpless, K.B., and Chang, S. (2007) *Angew. Chem. Int. Ed.*, **46**, 1730–1733.

119 Bae, I., Han, H., and Chang, S. (2005) *J. Am. Chem. Soc.*, **127**, 2038–2039.

120 Cho, S.H., Yoo, E.J., Bae, I., and Chang, S. (2005) *J. Am. Chem. Soc.*, **127**, 16046–16047.

121 Cassidy, M.P., Raushel, J., and Fokin, V.V. (2006) *Angew. Chem. Int. Ed.*, **45**, 3154–3157.

122 Whiting, M. and Fokin, V.V. (2006) *Angew. Chem. Int. Ed.*, **45**, 3157–3161.

123 Horneff, T., Chuprakov, S., Chernyak, N., Gevorgyan, V., and Fokin, V.V. (2008) *J. Am. Chem. Soc.*, **130**, 14972–14974.

124 Chuprakov, S., Kwok, S.W., Zhang, L., Lercher, L., and Fokin, V.V. (2009) *J. Am. Chem. Soc.*, **131**, 18034–18035.

125 Grimster, N.P., Zhang, L., and Fokin, V.V. (2010) *J. Am. Chem. Soc.*, **132**, in press.

126 Hein, J.E., Tripp, J.C., Krasnova, L.B., Sharpless, K.B., and Fokin, V.V. (2009) *Angew. Chem., Int. Ed.*, **48**, 8018–8021, S8018/8011–S8018/8064.

127 Kinugasa, M. and Shizunobu, H. (1972) *J. Chem. Soc., Chem. Comm.*, 466–467.

128 Lo, M.M.C. and Fu, G.C. (2002) *J. Am. Chem. Soc.*, **124**, 4572–4573.

129 Shintani, R. and Fu, G.C. (2003) *Angew. Chem. Int. Ed.*, **42**, 4082–4085.

130 Lang, S.A. and Lin, Y.I. (1984) *Comprehensive Heterocyclic Chemistry*, vol. VI/4B (eds A.R. Katritzky and C.W. Rees), Pergamon, Oxford, pp. 1–130.

131 Hansen, T.V., Wu, P., and Fokin, V.V. (2005) *J. Org. Chem.*, **70**, 7761–7764.

132 Hassner, A. and Rai, K.M. (1989) *Synthesis*, **1**, 57–59.

133 Shintani, R. and Fu, G.C. (2003) *J. Am. Chem. Soc.*, **125**, 10778–10779.

8
Transition Metal-Catalyzed Synthesis of Monocyclic Five-Membered Aromatic Heterocycles

Alexander S. Dudnik and Vladimir Gevorgyan

8.1
Introduction

Aromatic heterocycles, particularly furans and pyrroles, are structural motifs found in a vast number of biologically active natural and artificial compounds [1–3]. Other examples of practical applications of furans and pyrroles include dyes, polymers, and electronic materials [4]. Moreover, these heterocycles are employed as important intermediates in organic synthesis [5–9]. The successful application of furans and pyrroles in these and many other ways and their significance in applied and fundamental areas have placed them at the forefront of contemporary organic chemistry. A variety of methodologies and different protocols for their synthesis have been reported [1–3, 10–23] and become well established throughout decades. Among the variety of novel approaches for the synthesis of furans and pyrroles, transition metal-catalyzed transformations are arguably the most attractive methodologies [24–51]. Several excellent reviews on the transition metal-catalyzed synthesis of monocyclic five-membered heterocycles have been published in the literature. Many of them were categorized by either the metal or the type of transformation. This chapter covers transition metal-catalyzed syntheses of furans and pyrroles. The main organization of this chapter is based on a heterocycle, wherein syntheses of a particular core are structured by the type of transformation and substrates engaged. Herein, we have tried to discuss equally the synthetic applicability of a method and mechanistic aspects and concepts implicated in the described transformations. A discussion of the mechanisms is given when needed to provide an idea about possible reaction pathways and the nature of the elementary processes involved in the catalytic transformation. It should be noted that the most essential and general catalytic reactions, as well as recent and conceptionally interesting transformations, are discussed in more detail in this chapter. Syntheses of furans and pyrroles via functionalization of the preexisting heterocyclic cores [37, 52–60] are not covered and, therefore, only reactions in which assembly of a heterocyclic ring occurs are described. In addition, the Pd-catalyzed synthesis of heterocycles is not covered herein, as this topic is covered in Chapter 6.

Catalyzed Carbon-Heteroatom Bond Formation. Edited by Andrei K. Yudin

ISBN: 978-3-527-32428-6

8.2
Monocyclic Five-Membered Heterocycles

8.2.1
Furans

Furans are highly important five-membered heterocycles, which are incorporated as structural units into a variety of many natural and synthetic biologically-active compounds and materials [61–65]. Moreover, a furan heterocycle itself is of significant interest as a synthon towards an array of valuable intermediates [66–73]. Thus, it is not surprising that the development of selective and general methods for the facile assembly of the furan core has attracted tremendous interest over the past decades [74–79]. In particular, development of catalytic approaches for furan syntheses under mild reaction conditions from simple starting materials still remains an active task for the synthetic community. Consequently, numerous attempts at systematization of the vast array of existing synthetic methods toward the preparation of furans via modification of the preexisting heterocyclic core [6, 80, 81] and assembly of the ring from acyclic precursors [82, 83] have been made. Among the two, the latter has a greater potential for rapid construction of diversely substituted heterocycles. Within this group of transformations catalyzed by transition metals, cycloisomerization and formal [3 + 2] cycloaddition [84] approaches arguably represent the most versatile methods and provide valuable entry into furan compounds. Accordingly, a detailed discussion of recent contributions to the area of furan synthesis via these two routes is presented below.

8.2.1.1 Synthesis of Furans via Cycloisomerization Reactions

CYCLOISOMERIZATION APPROACHES

Transition metal-catalyzed cycloisomerization of allenyl ketones into furan products was first introduced by Marshall. In the first report from this group, it was demonstrated that various alkyl-substituted furans **2** could be prepared in high yields via the Ag-[50] or the Rh(I)-catalyzed [85, 86] cycloisomerization of allenyl ketones **1** (Scheme 8.1) [87]. This type of transformation of allenyl ketones [88–90] proved to be highly efficient for the synthesis of up-to-trisubstituted furans. Furthermore, Marshall applied this strategy as the key step in the construction of the 2,5-disubstituted furanocycle **2a** (Scheme 8.2) [87].

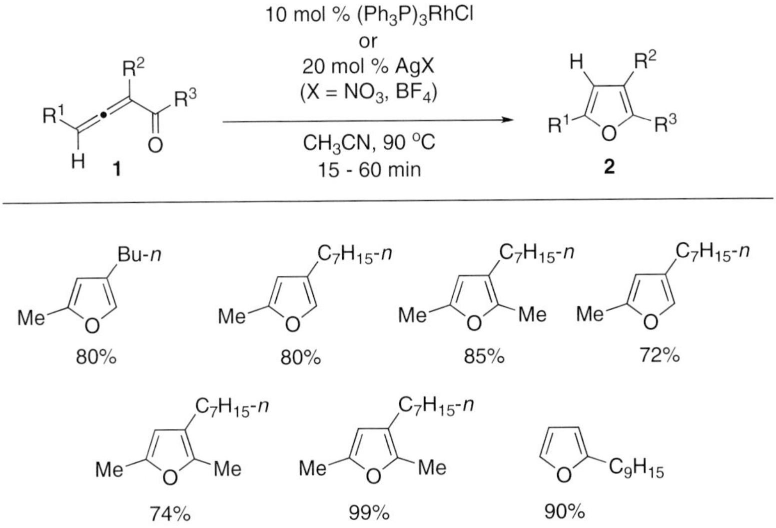

Scheme 8.1

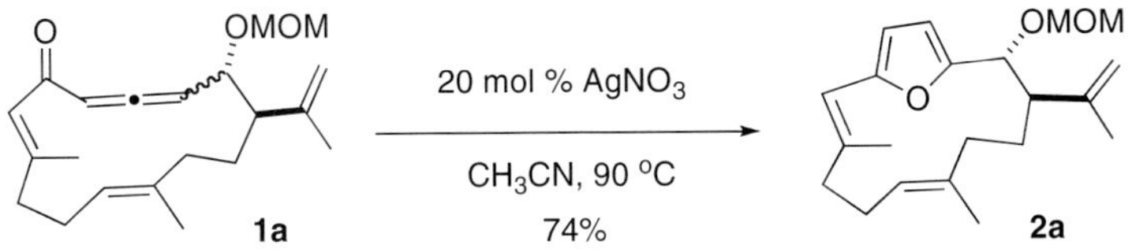

Scheme 8.2

The same group later found milder Ag-catalyzed reaction conditions and investigated the scope of this transformation for the syntheses of di- and tri-substituted functionalized furans **4**. Reactions were performed at room temperature and generally excellent yields of various sensitive furans were achieved (Scheme 8.3) [91]. Syntheses of some furanocycles [92, 93] were accomplished via employment of this protocol. Later, it was found that the cycloisomerization reaction of **3** leading to furans **4** proceeded more efficiently in the absence of $CaCO_3$ [94].

The Ag-catalyzed assembly of the furan core via the cycloisomerization approach was featured as a key step in the synthesis of several naturally-occurring furanocycles, such as (±)-kallolide B [95], (−)-kallolide B [96], rubifolide [97], kallolide A [98], and unnatural polyhydroxylated piperidine [99]. The authors proposed a mechanism for the cycloisomerization of allenylketones [94]. First, Ag-catalyst coordinates to the distal double bond of the allene moiety **7** and activates it toward the intramolecular nucleophilic attack of the carbonyl oxygen atom to produce oxonium intermediate **8**. Deprotonation of **8** gives the organosilver intermediate **9**, which undergoes a protonation–E1-type silver elimination sequence through **10** to give **6**. Alternatively **9** is directly converted into furan **6** via an S_E2 type process (Scheme 8.4).

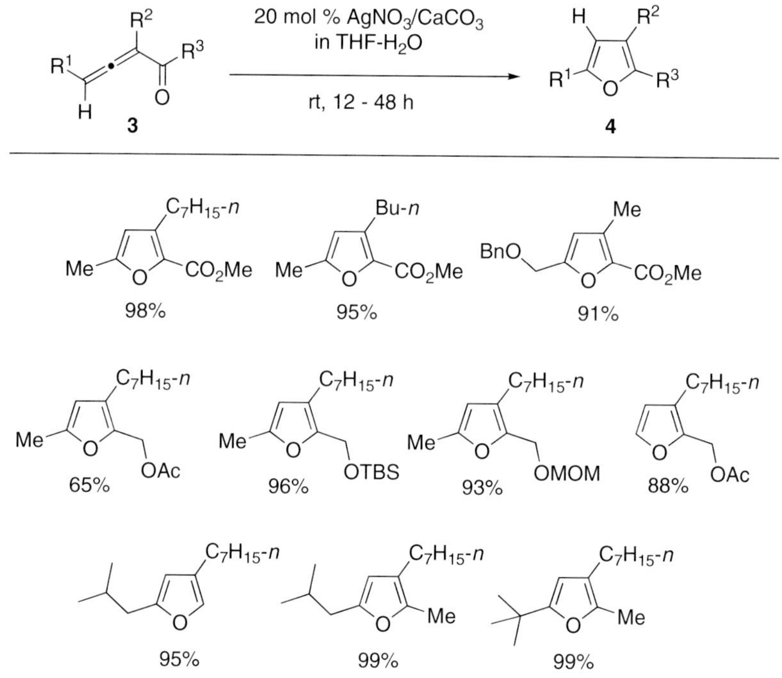

20 mol % $AgNO_3/CaCO_3$ in THF-H_2O, rt, 12 - 48 h

98%, 95%, 91%

65%, 96%, 93%, 88%

95%, 99%, 99%

Scheme 8.3

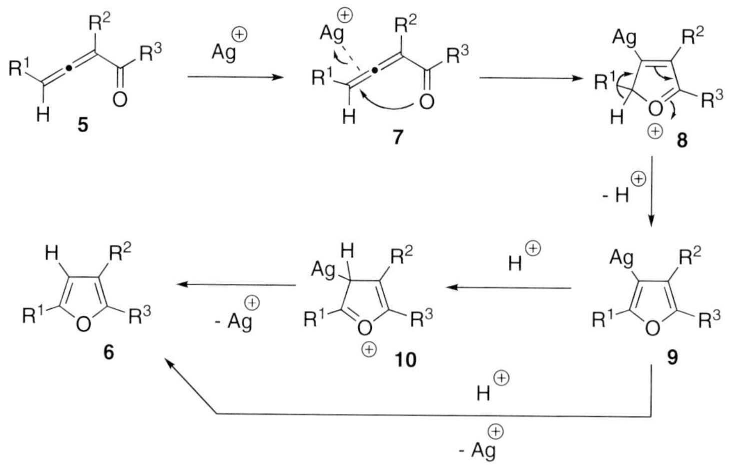

Scheme 8.4

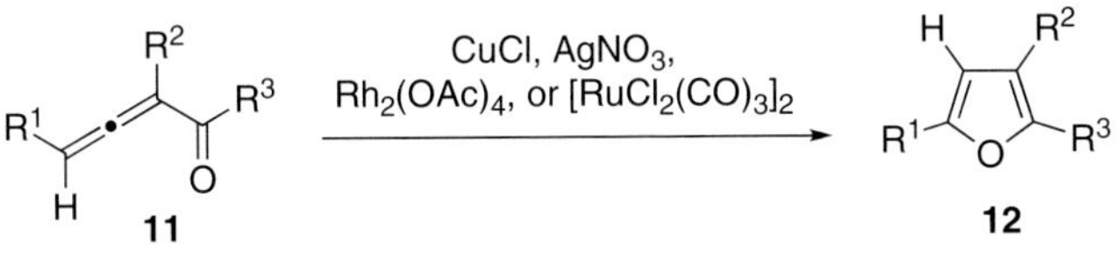

Scheme 8.5

Cycloisomerization of allenyl ketones **11** into furan products **12** was further elaborated by Hashmi (Scheme 8.5). It was found that this transformation could be effectively catalyzed by other transition metals, such as Cu(I), Ag, Rh(II), and Ru(II) [100, 101]. Several aryl-containing furans were synthesized via the silver-catalyzed protocol in high yields (Scheme 8.6) [102].

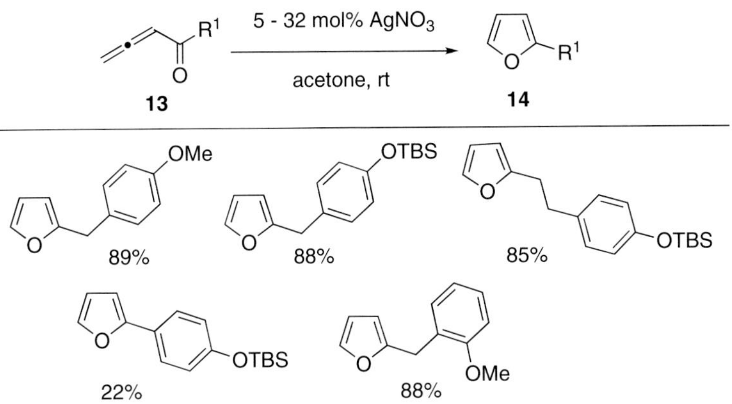

Scheme 8.6

The same researchers also reported a very facile cycloisomerization reaction of allenyl ketones **15** into furans **16** in the presence of gold(III) catalyst (Scheme 8.7) [46, 103–112]. Furthermore, the authors extended this reaction to the cycloisomerization–addition cascade process of allenyl ketones **15** with enones **17** to produce 2,5-disubstituted furans **18** (Scheme 8.8) [111]. Formation of the latter products was rationalized via two proposed pathways. According to path **A**, furan intermediate **16** undergoes an auration with Au(III)-catalyst to produce the furyl-gold species, which, upon subsequent 1,4-addition to the Michael acceptor **17**, generates intermediate **19**

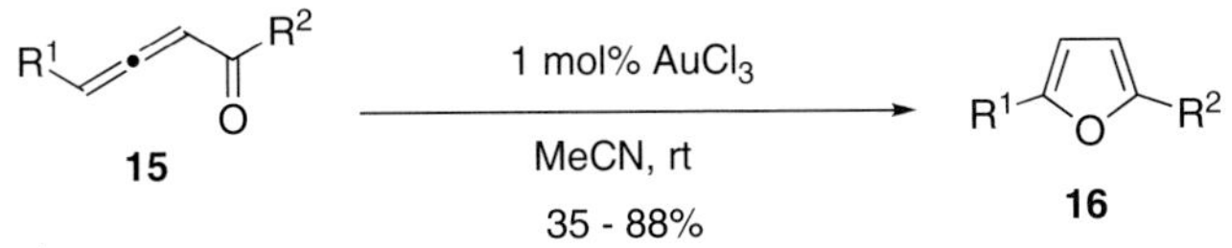

Scheme 8.7

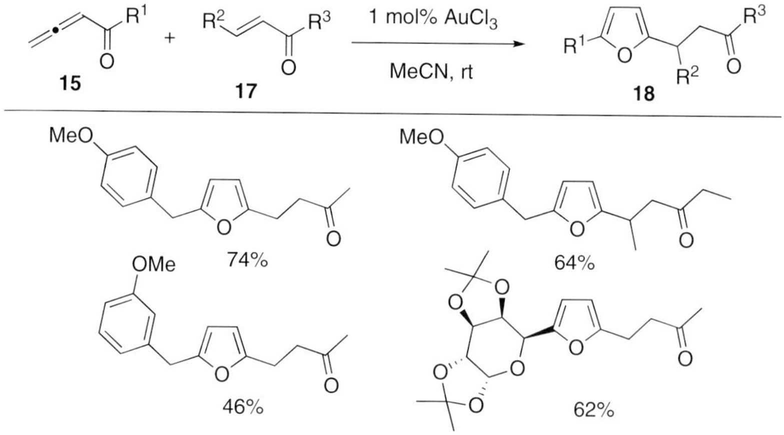

15 + 17 → (1 mol% $AuCl_3$, MeCN, rt) → 18

74% 64% 46% 62%

Scheme 8.8

(Scheme 8.9). Alternatively, $AuCl_3$ activates enone (**17′**) toward a direct electrophilic aromatic substitution at the 5-position of the furan **16** to provide **19** (Path **B**) (Scheme 8.9). Protiodeauration of **19** affords furan **18** and regenerates the Au-catalyst.

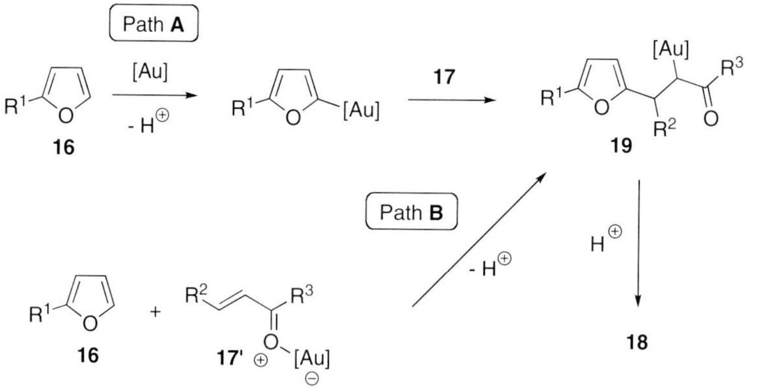

Path **A**: 16 → ([Au], - $H^{\oplus}$) → 17 → 19 → ($H^{\oplus}$) → 18

Path **B**: 16 + 17′ → (- $H^{\oplus}$) → 19

Scheme 8.9

The mercury(II)-catalyzed cycloisomerization of allenyl ketones **20** into furans **21** was reported by Tius (Scheme 8.10) [113]. A similar Hg(II) triflate–$(TMU)_2$ complex catalyst was employed for the synthesis of furans by Gosselin [114]. More recently, Che *et al.* described the cycloisomerization of allenones into furan products under the Au(III)-porphyrin complex catalysis [115]. This catalyst is recyclable and TON of 8300 can be attained.

Gevorgyan has shown that Cu(I)-catalysis can be successfully employed for the transformation of allenyl ketones **22** into the corresponding furans **23** (Scheme 8.11) [116]. Most importantly, cycloisomerization of 4-thio-substituted

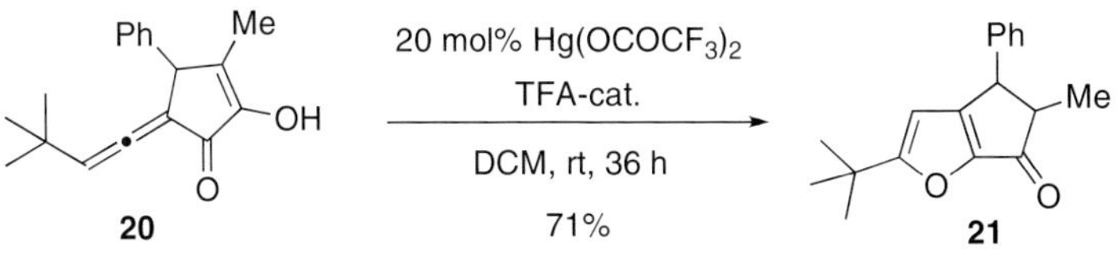

Scheme 8.10

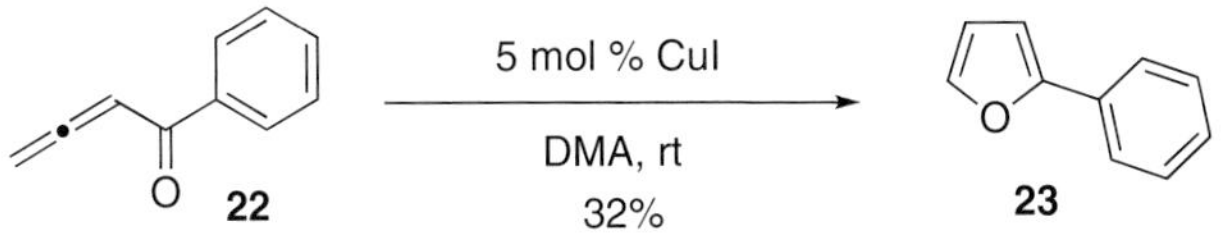

Scheme 8.11

allenones **24** proceeded very efficiently with a 1,2-migration of the phenylsulfanyl group [117], providing the 3-thio-substituted furan **25** (Scheme 8.12) [118].

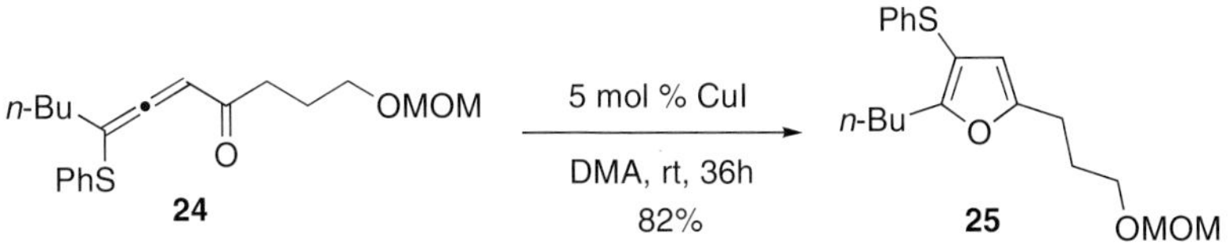

Scheme 8.12

Based on these results, the same group aimed at the development of novel protocols for the furan synthesis that will overcome the limitation of Marshall's protocol on the introduction of various groups at the C-3 position of the furan ring. Accordingly, Gevorgyan reported that a variety of multisubstituted 3-halofurans **27** could be accessed via the Au(III)-catalyzed cycloisomerization of haloallenyl ketones **26** proceeding with a 1,2-migration of iodine, bromine, or chlorine atoms. This chemistry represents a very efficient, selective and mild approach for the synthesis of up-to-fully-substituted 3-halofurans (Scheme 8.13) [119]. Iodo- and bromo-substituted substrates were shown to be more efficient in this cycloisomerization than the corresponding chloro-substituted analogs.

In addition, in the case of the cycloisomerization of ambident C-4 monosubstituted haloallenones **26**, the authors demonstrated that simply switching solvent from toluene to THF provided a regiodivergent formation of 2-halofurans **28** [119]. It was also shown that employment of gold complexes with counter-anions capable of assisting 1,2-H migration [120], such as Et_3PAuCl [119] and $Ph_3PAuOTf$ [121], led to the formation of 2-halofurans **28** with high degrees of regioselectively (Scheme 8.14).

Thorough mechanistic studies, including a high-level DFT calculations, indicated that both Au(I) and Au(III) catalysts activate the distal double bond of the allene **29** to produce cyclic zwitterionic carbenoid-like intermediates **30**, which undergo a kinetically favored 1,2-halogen migration to give the 3-halofuran **27**. However, with Au $(PR_3)L$ (L = Cl, OTf) catalysts, the stepwise counterion- or ligand-assisted H-shift (**31**)

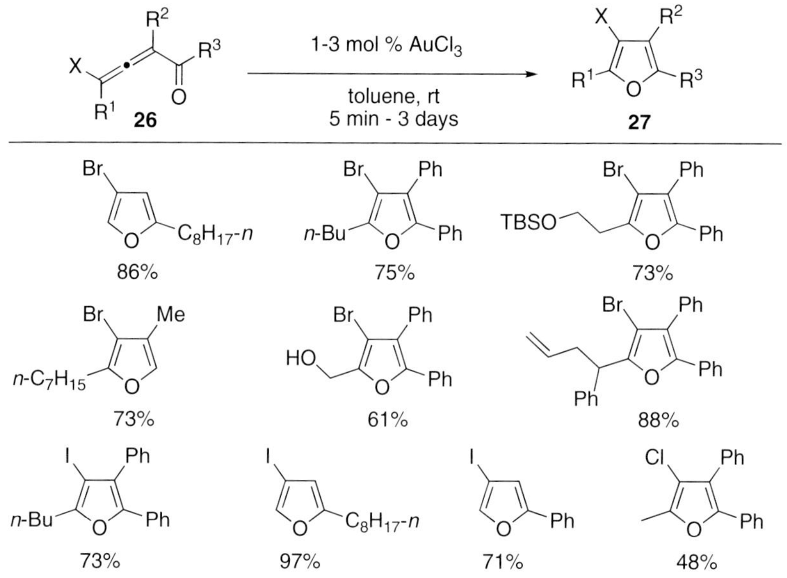

Scheme 8.13

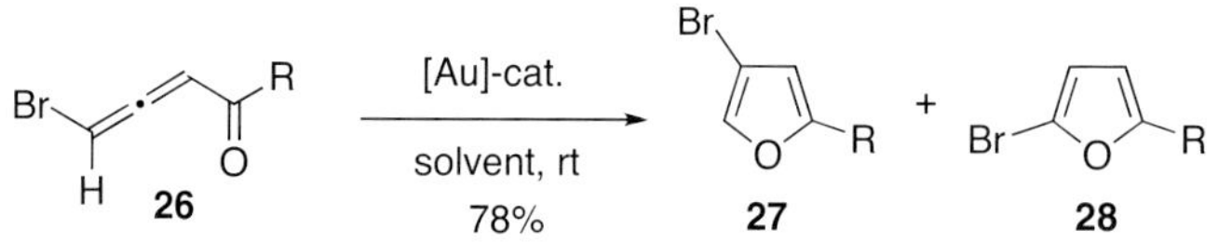

R	[Au]	Solvent	27 : 28
$n\text{-}C_8H_{17}$	$AuCl_3$	THF	5 : 95
$n\text{-}C_8H_{17}$	Et_3PAuCl	toluene	<1 : 99
Me	$Ph_3PAuOTf$	toluene	0 : 100

Scheme 8.14

was demonstrated to be the major process, leading to the 2-halofuran **28**. This observation indicates that the regioselectivity of the Au-catalyzed 1,2-H versus 1,2-Br migration processes is ligand dependent. (Scheme 8.15) [119, 121].

Aiming at incorporation of various 1,2-migrating groups into the cascade [122] cycloisomerization of allenones, Gevorgyan *et al.* recently disclosed an efficient protocol for the synthesis of furans proceeding via 1,2-migration of alkyl or aryl groups (Scheme 8.16) [123]. With transition metal catalysts, such as cationic Au(I)-, Ag-, Cu(I)-, and Cu(II)-complexes, the authors proposed this cycloisomerization to occur via a mechanism analogous to that suggested for the transformation of allenones with the 1,2-halogen migration shown in Scheme 8.15 [123, 124].

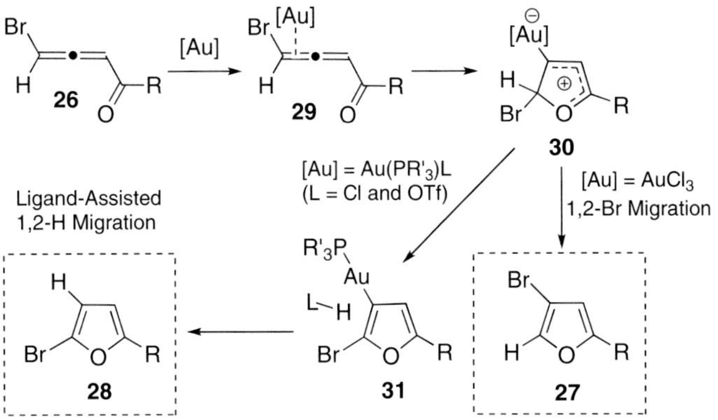

Scheme 8.15

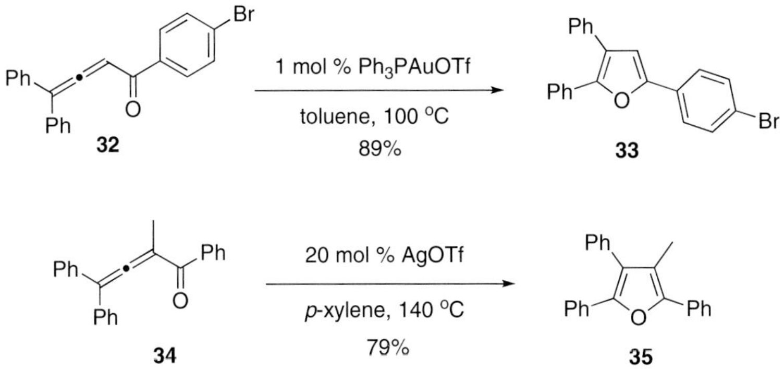

Scheme 8.16

Kirsch *et al.* reported that vinyl propargyl ethers **36** could be converted into the densely substituted furans **38** via the Au(I)-catalyzed cycloisomerization reaction (Scheme 8.17) [125] A variety of substituted furans **38** (Table 8.1) could be obtained under very mild reaction conditions at only 2 mol% catalyst loading. It is believed that this cascade process begins with the Au(I)-catalyzed Claisen-type rearrangement of **36** leading to the formation of skipped allenyl ketone **37**, which, upon the Au(I)-catalyzed 5-*exo-dig*-cyclization, provides furan **38**.

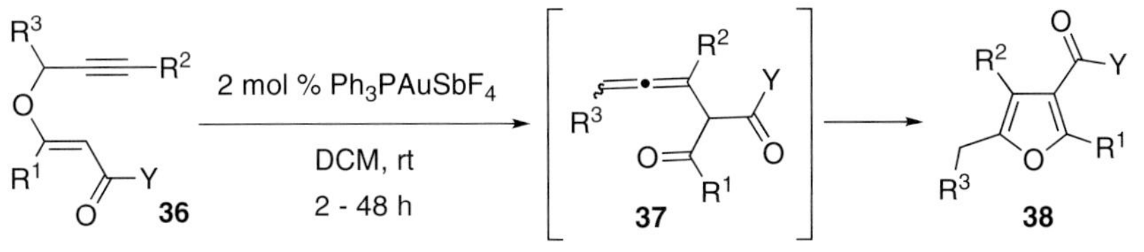

Scheme 8.17

Table 8.1 Au(i)-catalysed synthesis of furans **38**.

Entry	36				Yield[a),b)] 38, %
	R^1	Y	R^2	R^3	
1	Me	OEt	Me	H	95
2	n-C_5H_{11}	OMe	Me	H	97
3	Ph	OEt	Me	H	90
4[c)]	H	OEt	Me	H	75
5	Ph	Ph	Me	H	72
6	Me	OEt	H	H	82
7	n-C_5H_{11}	OMe	H	H	77
8	Me	OEt	Ph	H	87
9	Ph	OEt	Ph	H	90
10	H	OEt	Ph	H	84
11	Me	OEt	2-MeO-C_6H_4	H	99
12	Ph	OEt	3-thienyl	H	89
13	Ph	OEt	2-pyridyl	H	82
14	Ph	OEt	CH_2Hex-c	H	73
15	Ph	OEt	CH_2CH_2OTBS	H	72
16	Me	OEt	TBS	H	83
17[d)]	Ph	OEt	Me	Me	45

a) 0.2 mmol scale, 2 mol% $Ph_3PAuCl/AgBF_4$, rt, DCM, 0.2 M.
b) Yield of pure product after column chromatography.
c) Solvent: benzene.
d) 38 °C.

A highly efficient Ru(II)- [85, 126–132] and Pt(II)-catalyzed cycloisomerization of skipped allenyl ketones, proceeding via a formal 1,4-migration of phenylsulfanyl- or phenylselenyl groups to produce densely functionalized furans **40**, was reported by Wang and coworkers (Scheme 8.18) [133]. Since the allenyl sulfides **39** are easily accessible via the Cu(I)- or Rh(II)-catalyzed reactions of α-diazocarbonyl compounds **41** with propargyl sulfide **42**, the authors extended this chemistry to a one-pot reaction with two different catalysts (Table 8.2) and a one-pot sequential protocol with one catalyst (Table 8.3). The latter cascade approach provided the corresponding furans **40** in moderate yields. A mechanism for this cascade cycloisomerization involves an initial attack of the phenylsulfanyl group at the terminal carbon atom of the activated allene moiety **43** to give the cyclic thiolanium intermediate **44**. Upon fragmentation, the latter undergoes a 1,4-migration of the thio group to generate metal carbene **45a** or its resonance structure, metal stabilized allylic cation **45b**. Subsequently, the former cyclizes into the carbonyl ylide **46** which, upon elimination of the catalyst, produces furan **40** (Scheme 8.19).

Gevorgyan *et al.* demonstrated that somewhat unstable and not simply accessible allenones could be substituted with far more easily available conjugated alkynyl ketones [83, 134, 135] in furan syntheses. Thus, it was shown that cycloisomerization of alkynyl ketones **46** proceeded smoothly in the presence of Cu(I)-catalyst providing furans **47** in high yields (Scheme 8.20) [116]. This protocol allowed a highly efficient

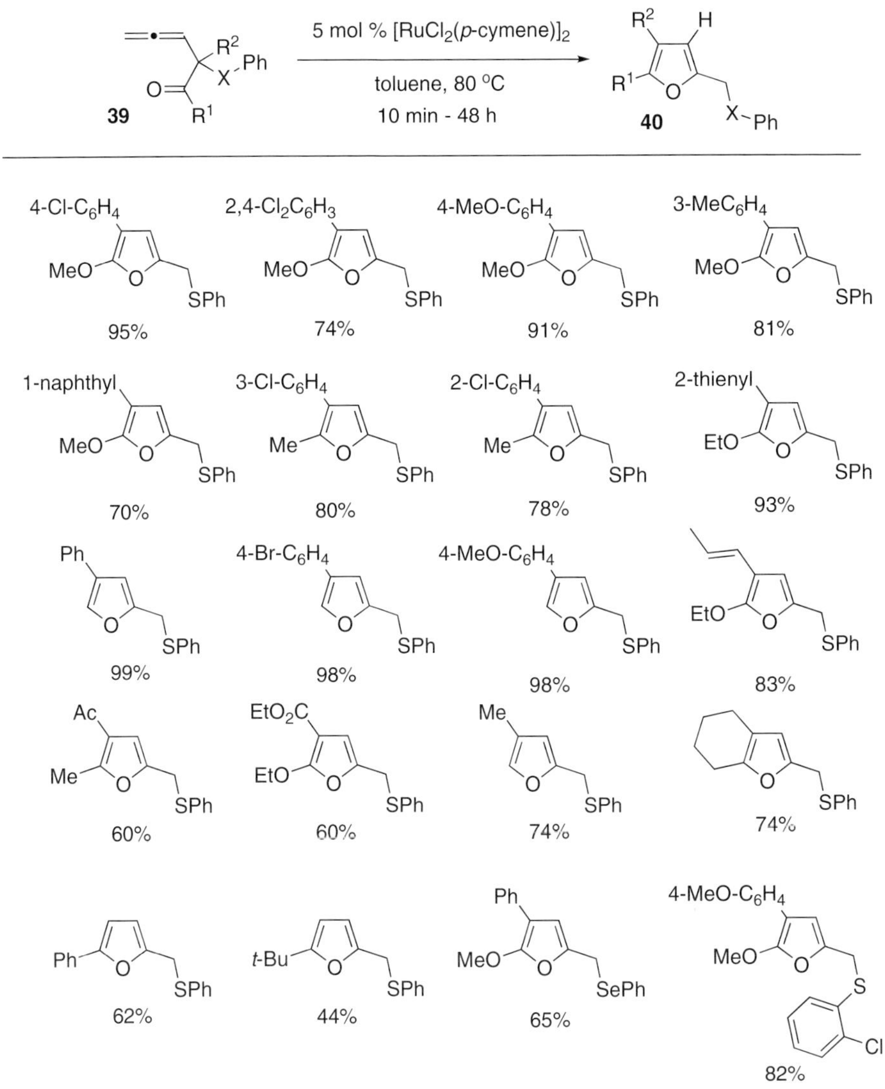

Scheme 8.18

preparation of 2-monosubstituted and 2,5-disubstituted furans possessing various labile groups. A proposed mechanism involves an initial Cu(I)/base-assisted alkynyl-allenyl isomerization of **46** into **48**. Subsequent allene activation by the Cu(I)-catalyst is followed by the 5-*endo-trig* cyclization to give the organocopper intermediate **50**. The latter, upon proton transfer, isomerizes into **51**, which then transforms into furan **47** (Scheme 8.21) [116].

This discovery guided the same group toward the development of a general concept of a transition metal-catalyzed cascade cycloisomerization involving a formal 1,2-

Table 8.2 One-pot two-catalyst synthesis of furans **40**.

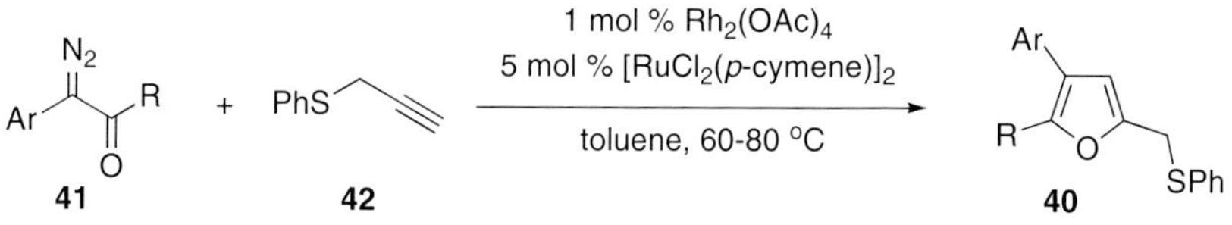

Entry	Ar	R	Yield **40**, %[a]
1	Ph	OMe	81
2	4-Cl-C_6H_4	OMe	90
3	2,4-$Cl_2C_6H_3$	OMe	72
4	4-MeO-C_6H_4	OMe	83
5	3-Me-C_6H_4	OMe	84
6	1-naphthyl	OMe	50
7	3-Cl-C_6H_4	Me	73
8	2-Cl-C_6H_4	Me	62
9	1-thienyl	OEt	82

a) Yield of isolated product after column chromatography.

Table 8.3 One-pot sequential Ru(ii)-catalyzed synthesis of furans **40**.

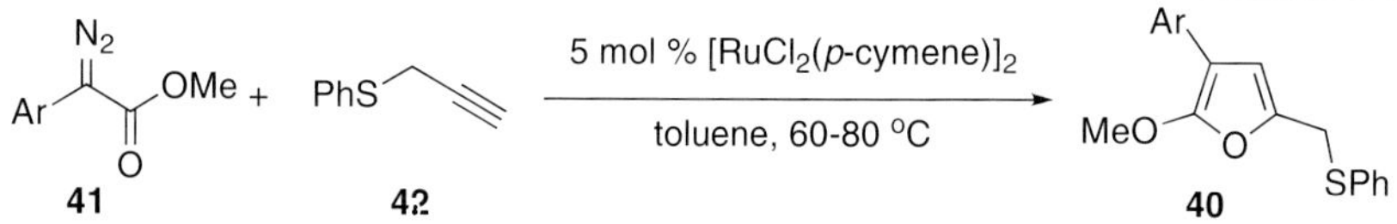

Entry	Ar	Yield **40**, %[a]
1	Ph	52
2	4-Cl-C_6H_4	67
3	2,4-$Cl_2C_6H_3$	0
4	4-MeO-C_6H_4	0
5	3-MeO-C_6H_4	68
6	4-Me-C_6H_4	45
7	3-Me-C_6H_4	61
8	4-Br-C_6H_4	54

a) Yield of isolated product after column chromatography.

migration of different functional groups in alkynyl and allenyl systems as the key step in a rapid and very facile assembly of a densely functionalized furan core. Thus, Gevorgyan and coworkers disclosed that 3-sulfanyl-substituted furans **53** could be efficiently accessed via the Cu(I)-catalyzed migratory cycloisomerization of the corresponding propargyl sulfides **52** (Scheme 8.22) [118]. It is believed that, mechanistically, this transformation occurs via an initial Cu/base-assisted prototropic rearrangement of **52** into allenyl sulfides **54** followed by the intramolecular attack

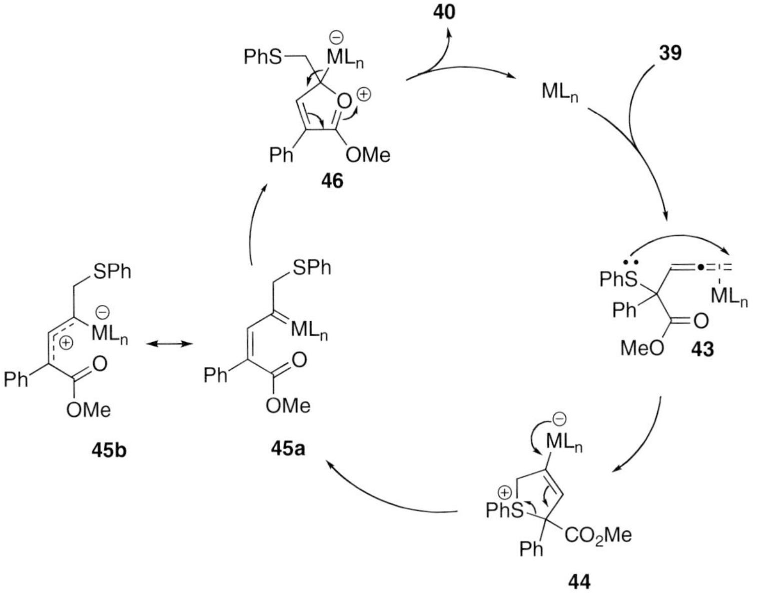

40 39 ML_n PhS ML_n Ph OMe 46 SPh ML_n Ph O OMe 45b SPh ML_n Ph O OMe 45a PhS Ph ML_n O MeO 43 ML_n PhS Ph CO_2Me 44

Scheme 8.19

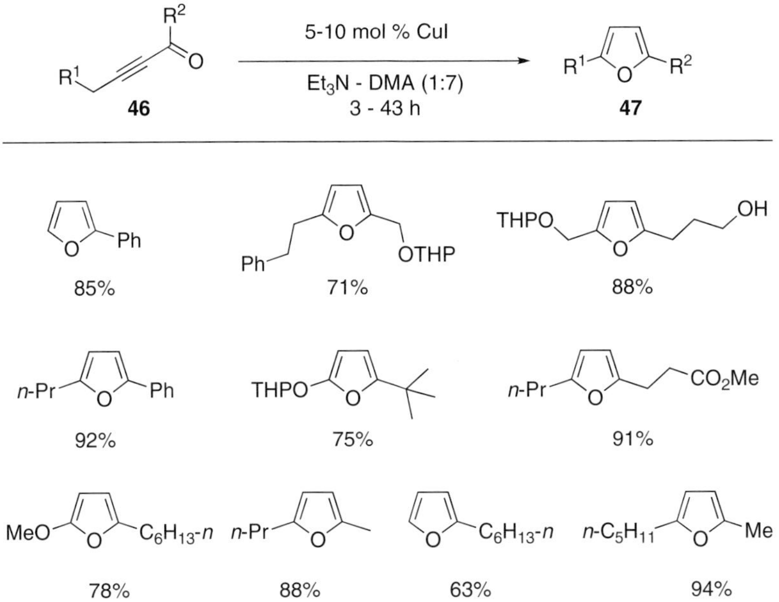

R^2 O R^1 46

5-10 mol % CuI

Et_3N - DMA (1:7)
3 - 43 h

R^1 O R^2 47

Ph 85% | Ph OTHP 71% | THPO OH 88%

n-Pr Ph 92% | THPO 75% | n-Pr CO_2Me 91%

MeO C_6H_{13}-n 78% | n-Pr 88% | C_6H_{13}-n 63% | n-C_5H_{11} Me 94%

Scheme 8.20

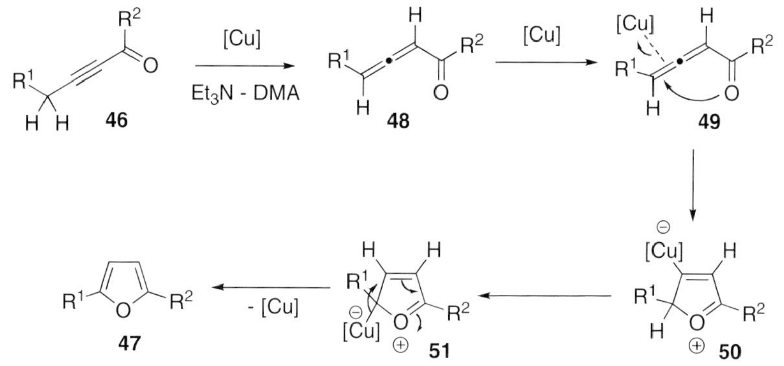

Scheme 8.21

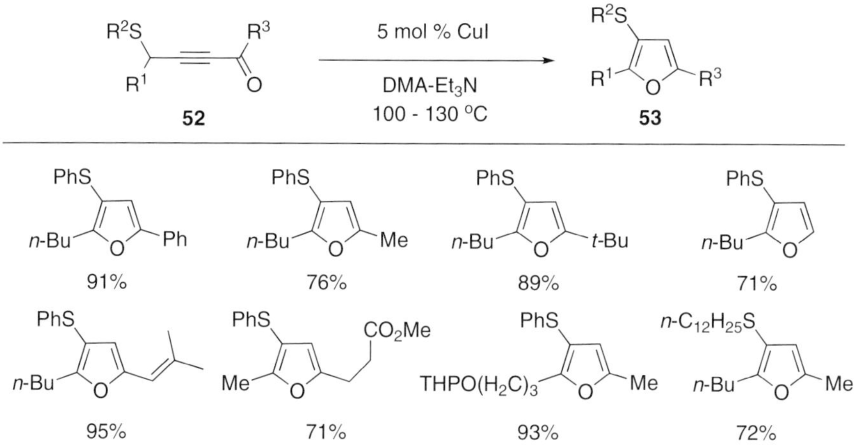

Scheme 8.22

of a sulfanyl group at the activated enone moiety **55** to produce a thiirenium intermediate **56** [118, 124]. The latter undergoes cycloisomerization via either addition–elimination (Ad_N–E) or directly via a S_N2–*vin* substitution process to give furan products (Scheme 8.23). Notably, this methodology represents the first example of 1,2-migration of the thio-group [117] which occurs from an olefinic sp^2 carbon to an sp center.

Recently, the same group extended this protocol to the Cu(I)-catalyzed synthesis of 3-selenyl-substituted furans **60** proceeding via a 1,2-migration of arylselenyl groups in propargyl selenides **58** (Scheme 8.24) [124]. Remarkably, the 1,2-migration of the seleno group was easier than that of the thio groups, allowing performance of the cycloisomerizations under significantly milder reaction conditions.

Taking into consideration that highly reactive allenones could be accessed via a formal 1,3-migration in propargyl acetates, phosphates, and sulfonates, Gevorgyan and coworkers established a series of highly efficient, practical and general methodologies toward the assembly of tri- and tetra-substituted furan cores. Thus, it was

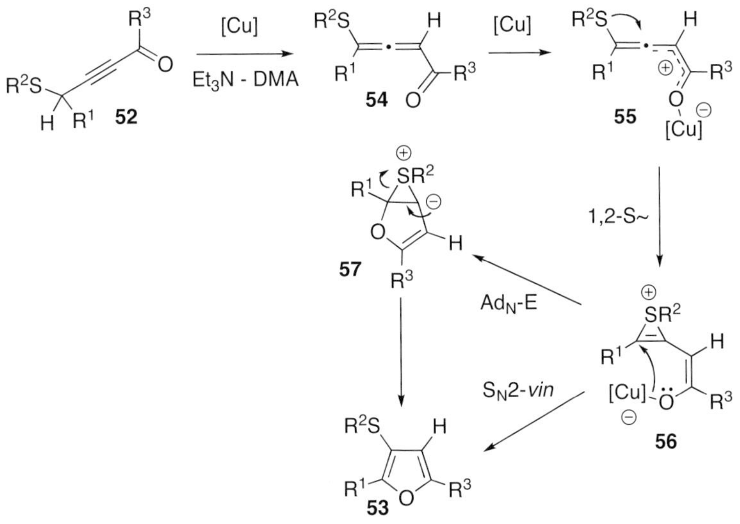

Scheme 8.23

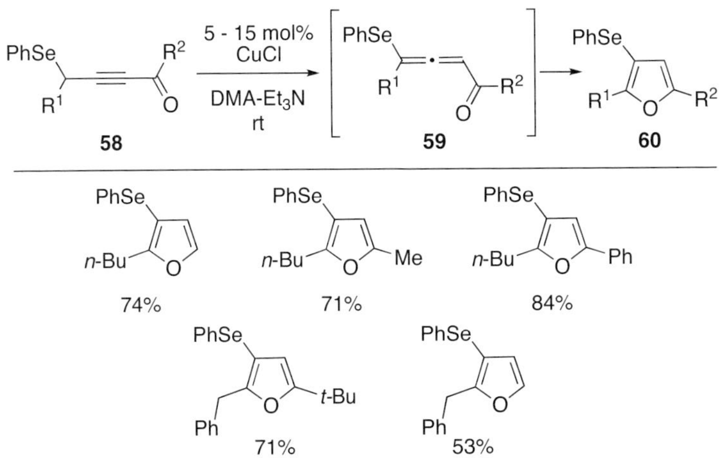

Scheme 8.24

demonstrated that a vast array of densely functionalized furans could be synthesized via formal 1,2-migration/cycloisomerization protocols from various propargyl and allenyl phosphates, acetates, and tosylates. Accordingly, conjugated propargyl acetates [136–138] **61** (Scheme 8.25) [139] and phosphates **63** (Scheme 8.26) [140] underwent the Cu(I)-catalyzed cycloisomerization into regioisomeric 4- and 3-oxyfurans **62** and **65**. Further studies on the transformation of ^{17}O-labeled phosphatyloxy-substituted substrates supported the mechanistic rationale involving a propargyl–allenyl isomerization proceeding via a formal sigmatropic

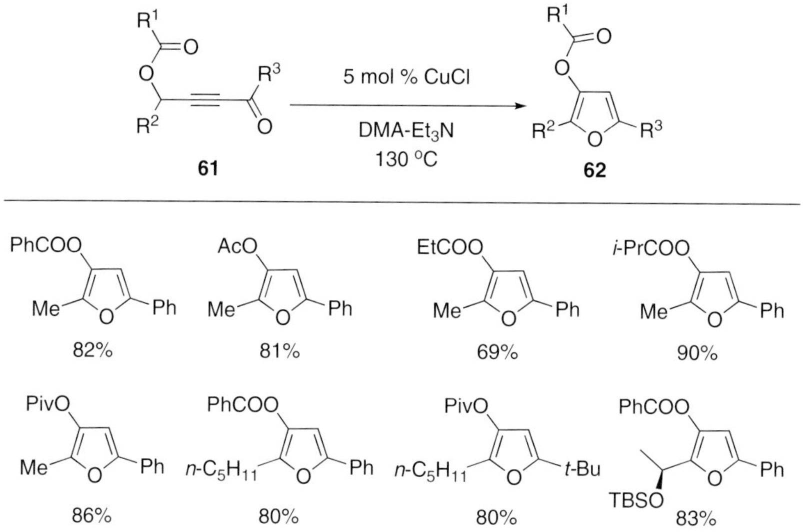

Scheme 8.25

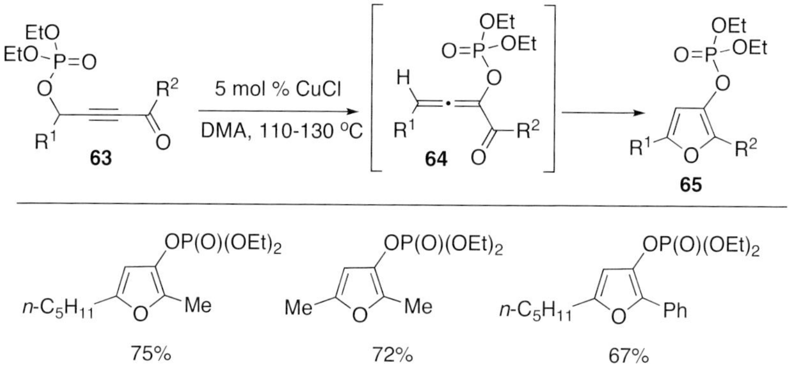

Scheme 8.26

[3,3]-phosphatyloxy shift [140]. In the case of conjugated acyloxy-substituted ketones, the prerequisite of base for the selective formation of the observed regioisomer **62** supported possible involvement of an allene intermediate **66** via a prototropic rearrangement (Scheme 8.27). According to the mechanism proposed by the authors, this mode of cycloisomerization most likely occurred with the involvement of a dioxolenylium intermediate **67** [140].

Next, Gevorgyan *et al.* disclosed an alternative route to tetrasubstituted and even fused furans via a transition metal-catalyzed migratory cycloisomerization of skipped propargylic systems. Thus, syntheses of 3-acyloxyfurans **69** from alkynyl acetates **68** occurred with a formal 1,2-migration in the presence of 5 mol% of Ag-catalyst at room temperature (Scheme 8.28) [139, 140]. Several other transition metals such as Cu(II),

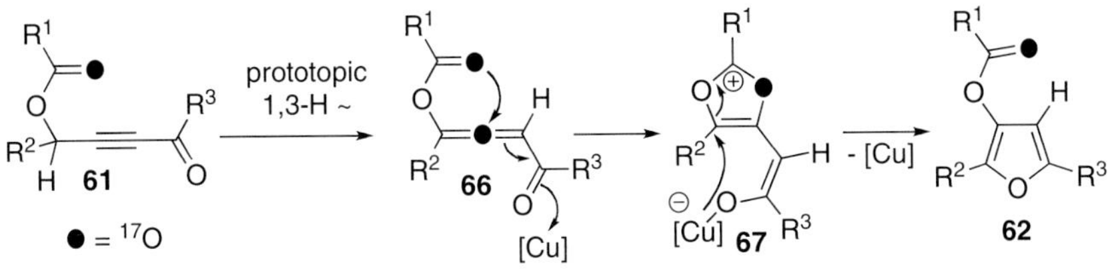

Scheme 8.27

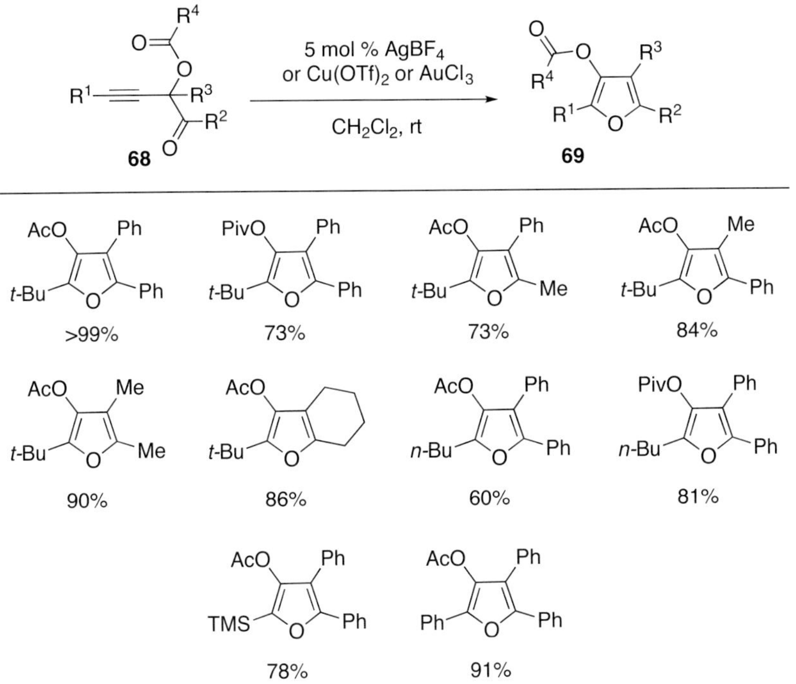

Scheme 8.28

Pd(II), Pt(II), and Au(III) were also found to catalyze this transformation. Furthermore, phosphatyloxy- (Scheme 8.29) and tosyloxy groups (Scheme 8.30) were shown to undergo an analogous 1,2-migration from propargyl or/and allenyl substrates **70**, **71** and **74**, providing highly valuable 3-phosphatyloxy- and 3-tosyloxy furans, respectively. It was suggested by the authors, that the cycloisomerization of skipped phosphatyloxy alkynyl ketones (**76**, Y = $P(OEt)_2$) proceeds through two consecutive 1,2-migrations with the formation of allene **77**, resulting in an apparent 1,3-shift. It is then followed by a subsequent 1,2-migration through competitive oxirenium and dioxolenylium pathways (Scheme 8.31). The mechanism for the cycloisomerization of skipped alkynyl ketones containing an acyloxy group (**76**, Y = CR^4) was found to follow either a Rautenstrauch-type 1,2-migration of the acyloxy group followed by

5 mol % $AgBF_4$
DCE, 60°C
65%

OEt
O=P–OEt
O
t-Bu–≡–C(Ph)–C(=O)Ph
70

5 mol % $AgBF_4$
CH_2Cl_2, rt
90%

EtO
EtO–P=O
O
t-Bu, Ph, Ph
71

5 mol % $AgBF_4$
DCE, 60°C
77%

OEt
EtO–P–O
O
t-Bu, Ph, Ph
72

Scheme 8.29

Ar
O=S=O
O
^{t}Bu–≡–C(Ph)–C(=O)Ph
73

Ar
O=S=O
O
^{t}Bu, Ph, Ph
74

1 mol %
$AgBF_4$
DCE, 60°C
82%

O
Ar–S–O
O
^{t}Bu, Ph, Ph
75

Ar = p-CH_3-C_6H_4

Scheme 8.30

cycloisomerization to the furan **78** directly or, alternatively, two 1,2-migrations of the acyloxy group to form the postulated allenyl intermediate **77**, which cycloisomerizes to the furan via another 1,2-acyloxy shift (Scheme 8.31) [140].

Y=O
O
R^2
R^1 **77** R^3
O

1,2 ~ via
dioxolenylium

1,2 ~ via
oxirenium

Y = CR^4, $P(OR^4)_2$, or $S(O)R^4$

1,2 ~

O=Y
O
R^1–≡–C(R^2)–C(=O)R^3
76

1,2 ~

O
Y–O R^2
R^1 R^3
[M] O
75

O
Y–O R^2
R^1 O R^3
78

Scheme 8.31

A very mild and facile Au(III)-catalyzed cycloisomerization of skipped alkynyl ketones **79** into 2,5-disubstituted furans **80** was first reported by Hashmi (Scheme 8.32) [110, 111]. Interestingly, this catalyst was shown to be completely inefficient for an analogous transformation of conjugated substrates [111]. In addition, an interesting cascade transformation of the propargyl ketone **81**, occurring

Scheme 8.32

via the proposed intermediate **82**, to produce spirofuran **83** was also reported (Scheme 8.32) [111].

An analogous Au(I)-catalyzed synthesis of furan **86** was recently described by Toste and coworkers [141]. Thus, sulfoxide **84** undergoes an Au(I)-catalyzed rearrangement into the homopropargylic ketone **85**, which subsequently cycloisomerizes into the furan product (Scheme 8.33).

Scheme 8.33

Nishizava reported that 2-methylfurans **88** could be synthesized via cyclization of homopropargylic ketones **87** in the presence of as little as 1 mol% of $Hg(OTf)_2$ under very mild reaction conditions and with high turnover numbers of up to 100 (Scheme 8.34) [142]. The authors believe that the reaction is initiated by the π-activation of an alkynyl group with the Hg(II)-catalyst toward an intramolecular attack of a carbonyl function. It should be noted that employment of internal alkynes in this reaction led to complex reaction mixtures and provided only poor or modest yields of the target products. An analogous transformation of γ-ketoalkynes into furans was observed by Hidai and Uemura in the presence of $PtCl_2$ catalyst [143].

Recently, Cadierno and Gimeno reported several examples of a similar Ru(II)-catalyzed reaction of terminal, as well as internal, γ-ketoalkynes, leading to the furan products [144].

An alternative to the existing protocols for furan syntheses, Zn(II)-catalyzed transformation of skipped alkynyl ketones was recently elaborated by Dembinski *et al.* Accordingly, 5-*endo*-*dig*-cycloisomerization of alkynyl ketones **89**, in the presence of the Zn(II)-catalyst, at room temperature provided 2,5-di- and 2,3,5-trisubstituted furans **90** in high yields (Scheme 8.35) [145, 146]. This route tolerates sensitive

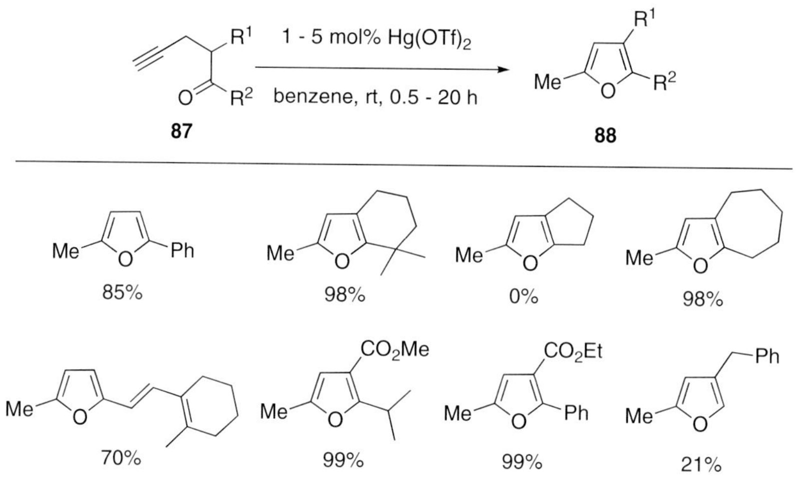

Scheme 8.34

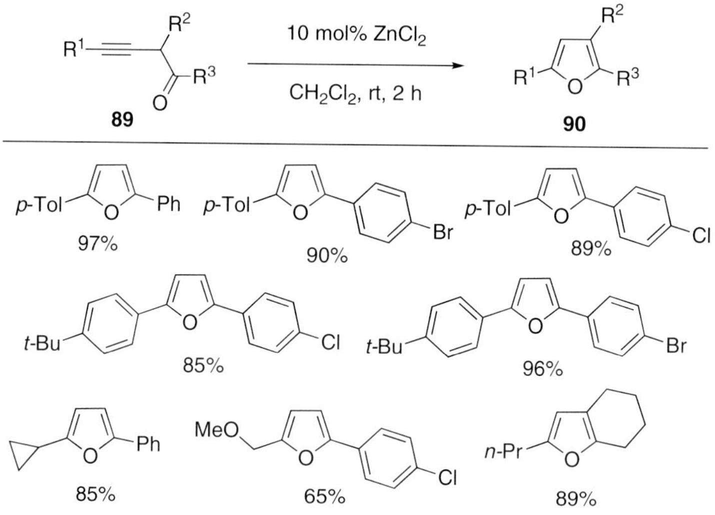

Scheme 8.35

functional groups, as illustrated with the labile propargyl ether- and cyclopropyl-containing substrates.

Furthermore, the same researchers applied the above-described chemistry for an efficient synthesis of furopyrimidine nucleosides **92** under Zn(II)- [146] or Cu(I)-catalysis (Scheme 8.36) [147]. Agrofoglio *et al.* [148, 149] and later Hudson [150] reported that several furo[2,3-*d*]pyrimidine-containing compounds with valuable properties could be efficiently assembled via the Ag-catalyzed version of this transformation.

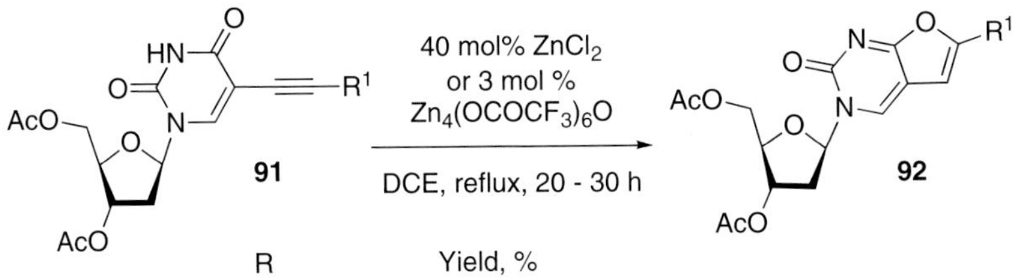

R	Yield, %
p-Me-C_6H_4	94
$CH_3(CH_2)_2$	99

Scheme 8.36

Recently, a novel approach for the synthesis of highly substituted furans was developed by Larock and coworkers (Scheme 8.37). It was reported that 2-alkynylenones [151–153] **93** underwent a facile transition metal-catalyzed cycloisomerization affording highly substituted furans **94** in the presence of the external O- or electron

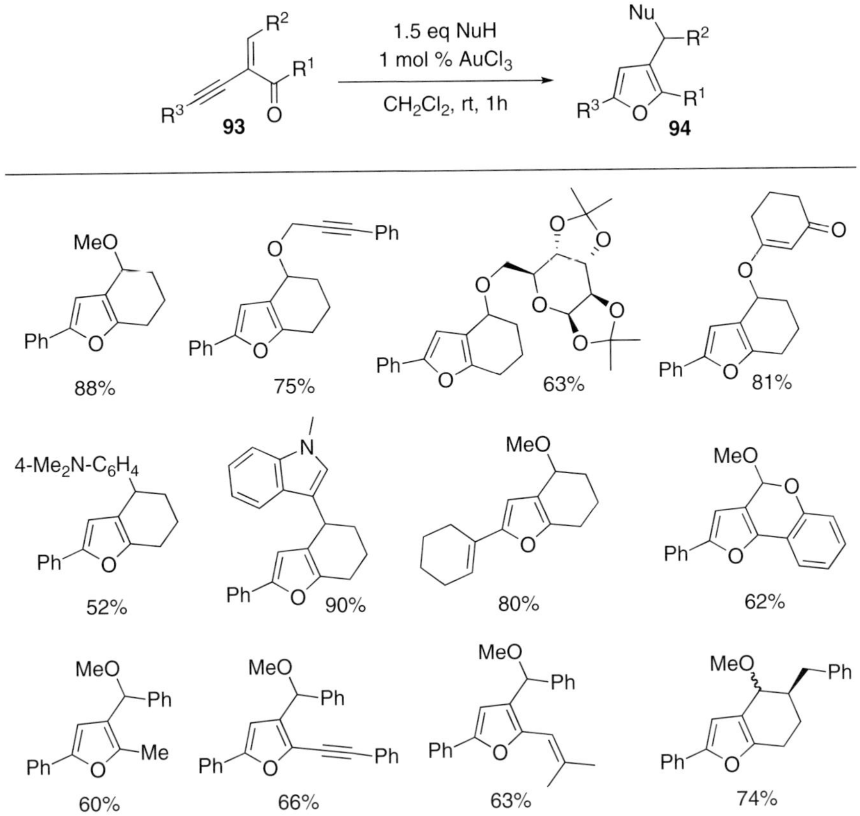

Scheme 8.37

rich C-nucleophiles [154, 155]. Several transition metal catalysts, such as AgO_2CCF_3, $Cu(OTf)_2$, and $Hg(OTf)_2$, provided good yields of the furan products, though $AuCl_3$ catalyst was proven to be superior in terms of reaction times. This method was shown to be limited to aryl- or vinylalkynes only, as cyclization of terminal- or alkyl- and silyl-substituted substrates failed to produce furans **94**.

According to the proposed mechanism, the Au-catalyst activates a carbon–carbon triple bond of **93** via **95**, which is then followed by a subsequent nucleophilic attack of the carbonyl function to produce a cyclic oxonium intermediate **96**. An intermolecular nucleophilic addition of an external nucleophile to the activated enone in **96** leads to the furyl-gold species **97**. Protiodeauration of the latter produces furan **94** and regenerates the catalyst (Scheme 8.38).

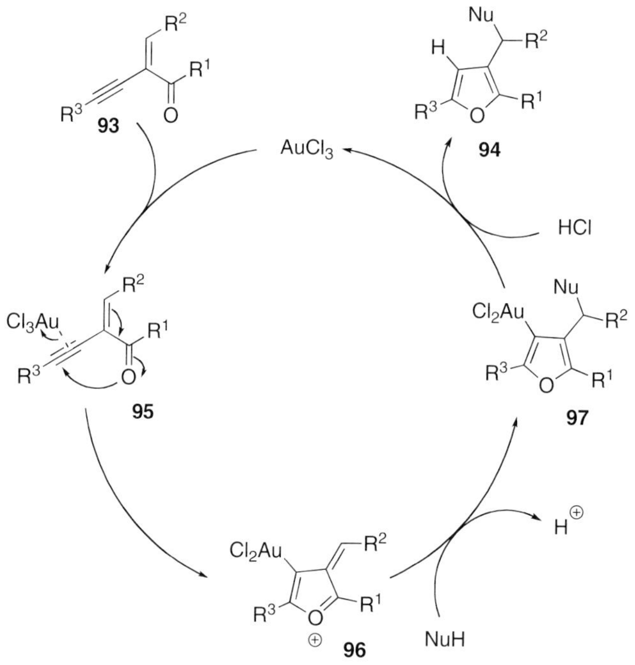

Scheme 8.38

Later, Yamamoto reported that inexpensive and air-stable Cu(I) catalyst in DMF could also catalyze this transformation (Scheme 8.39) [156]. Recently, Liang discovered that a robust air-stable and recyclable $Bu_4N[AuCl_4]$-catalyst in ionic liquids [157] could be employed for analogous cyclization of alkynyl alkenones **93** into furans **94** in high yields [158].

Finally, Oh revealed that the Pt(II)-catalyzed cycloisomerization of alkynyl alkenones **100** into furans **101** could be extended to a variety of other O-, N-, and even C-nucleophiles, which were previously shown to be unreactive in this transformation (Scheme 8.40) [159]. Furthermore, a major limitation of this chemistry to aryl- or alkenyl-substituted alkyne substrates could be alleviated under the Pt(II) catalysis.

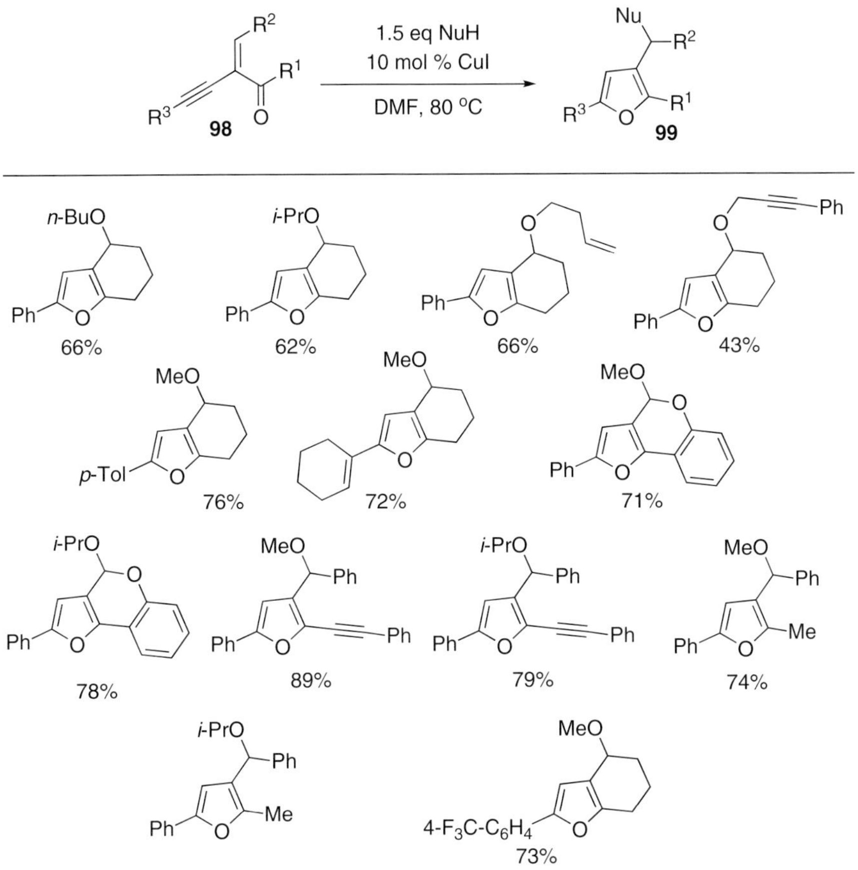

Scheme 8.39

Thus, terminal, silyl-, or alkyl-substituted alkynes were shown to provide high yields of furans **101**(Scheme 8.40). In addition, commonly reactive arylalkynyl alkenones, such as **93**, were cycloisomerized into the target compounds in the presence of Pt(II) catalyst with improved yields (72–91%). Apparently, these new conditions appeared to be far more efficient than those for the previously reported catalytic systems.

Schmalz and coworkers recently reported an interesting and highly efficient Au(I)-catalyzed cascade cycloisomerization of geminal acyl-alkynylcyclopropanes **102** into the densely functionalized furans **103**(Scheme 8.41) [160]. This reaction proceeded under very mild reaction conditions and a variety of nucleophiles, such as alcohols, including *tert*-butanol, phenols, acetic acid, 2-pyrrolidone, and indole could be employed. In addition, this transformation was shown to be catalyzed by Cu(II)- and Ag-triflates, albeit with somewhat lower efficiency. Two mechanisms, including concerted and stepwise formation of a furan ring, were proposed by the authors for this cascade transformation (Scheme 8.42).

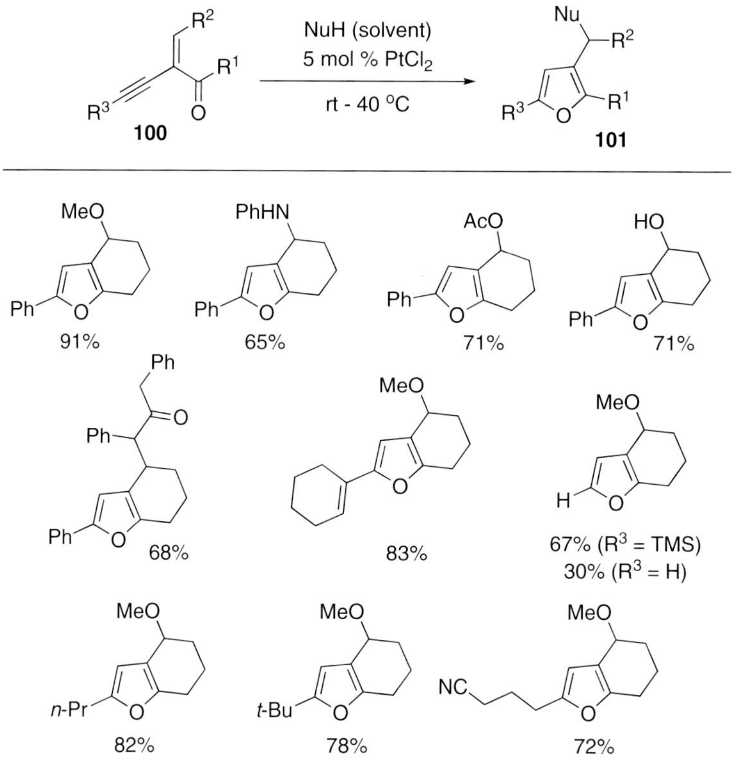

Scheme 8.40

In their recent study, Zhang and coworkers took advantage of the facile ring-opening of the cyclopropane **104** to generate a [1,4]-Au-containing dipole **105**, which could undergo a stepwise [4 + 2] cycloaddition reaction (Scheme 8.43). Thus, annulated furans **112** were obtained in good yields via the Au(I)-catalyzed cycloisomerization–annulation cascade of acyl-alkynylcyclopropanes **110** with a variety of dipolarophiles **111**, such as indoles, aldehydes, ketones, imines, and silyl enol ethers (Scheme 8.44) [161].

Dixneuf *et al.* first reported that various 2,3,5-trisubstituted furans **114** could be synthesized via a cycloisomerization of (*Z*)-pent-2-en-4-yn-1-ols **113** in the presence of Ru(II)-catalyst (Scheme 8.45) [162–164]. The corresponding furans were obtained generally in good to high yields, though this reaction was specific to terminal alkynes only. The authors proposed a mechanism based on the electrophilic activation of the alkyne moiety followed by intramolecular addition of the hydroxy function at the internal carbon atom of alkyne. A subsequent protiodemetalation–isomerization sequence furnished the furan **114** (Scheme 8.46).

Later, Marshall observed that this transformation for internal alkynes occurred in the presence of 10 mol% $AgNO_3$-catalyst supported on a silica gel, providing furan products in good yields [165]. More recently, Hashmi demonstrated that the cycloi-

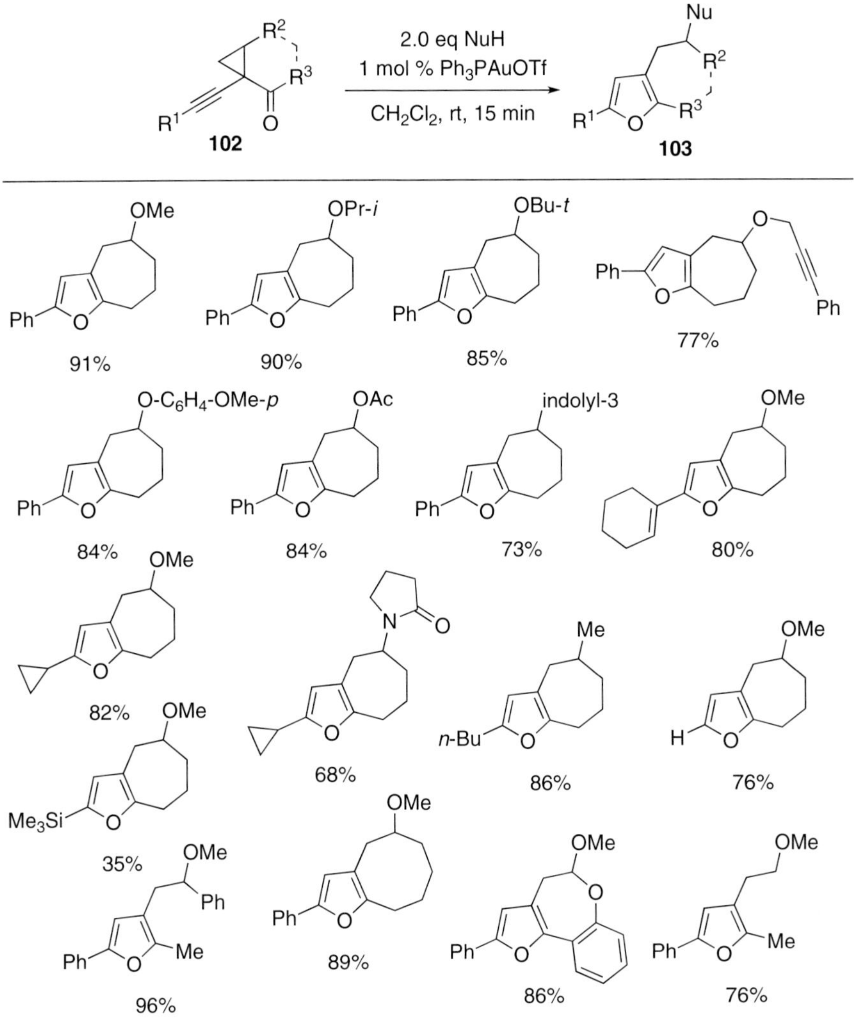

Scheme 8.41

somerization of (*Z*)-pent-2-en-4-yn-1-ols could be achieved under catalysis by $AuCl_3$ at a remarkably low 0.1 mol% loading of the catalyst at room temperature within 1 h [111]. This reaction could also be performed with high efficiency in aqueous solution in the presence of 1 mol% of various water-soluble Ru-, Rh-, and Ir-catalysts [166]. The scope of the Au-catalyzed version was further investigated by Liu. It was shown that fully substituted furans **119** possessing various functional groups could be efficiently synthesized from enynols **118** (Scheme 8.47) [167].

The Ag-catalyzed cycloisomerization of isomeric-to-**118**, β-alkynyl allylic alcohols **120**, was first investigated by Marshall. While various Ag salts provided high yields of trisubstituted furans **121**, performing cycloisomerization reaction with the catalytic

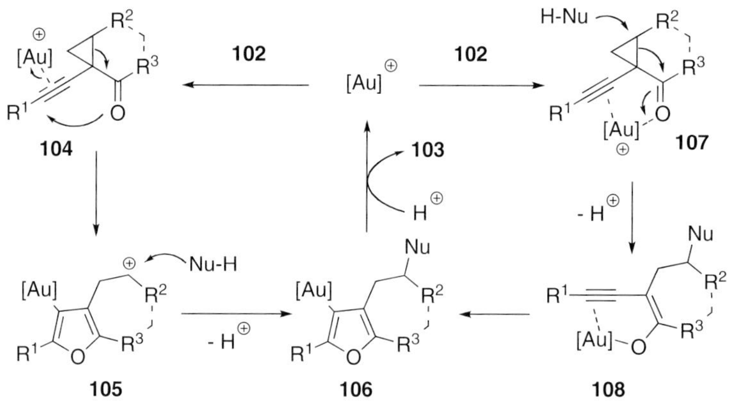

Scheme 8.42

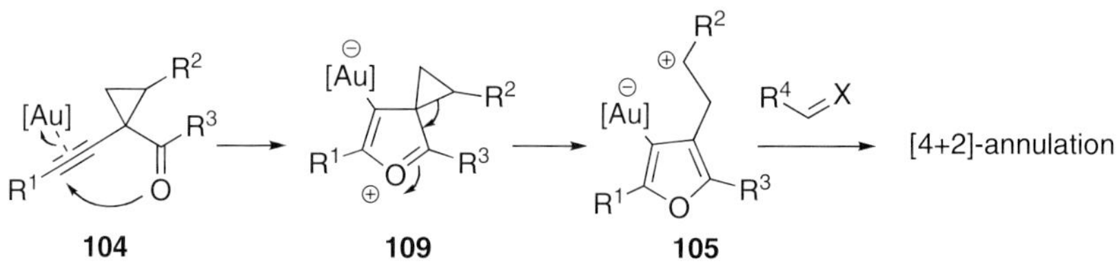

Scheme 8.43

system involving 10 mol% of $AgNO_3$ supported on silica gel and in nonpolar hexane solvent provided the best yields for the target compounds in shorter reaction times (Scheme 8.48) [165].

Recently, Lee *et al.* adopted this methodology for the cycloisomerization of allenyne-1,6-diols **122**. Accordingly, these ambident substrates underwent cyclization smoothly into various 3-vinylfurans **123** in the presence of AgOTf catalyst (Scheme 8.49) [168].

The first report of the use of alkynylepoxides as surrogates for furans came from Miller's group. It was reported that, in the presence of Hg(II) catalyst under acidic conditions, alkynylepoxides **124** underwent cycloisomerization into the corresponding 3-monosubstituted furans **125** (Scheme 8.50) [169].

Later, McDonald and coworkers disclosed that furans could be efficiently synthesized from differently functionalized terminal alkynylepoxides **126** (Scheme 8.51) [170, 171]. Accordingly, 2-mono- and 2,3-disubstituted furans **127** were obtained in high yields under mild reaction conditions upon cycloisomerization of **126** using an *in situ* generated $Et_3N{:}Mo(CO)_5$ catalyst. The authors proposed a mechanism involving an initial alkyne–vinylidene isomerization [172–174] of a terminal acetylene **126** into a reactive epoxyvinylidene carbenoid **128**. The latter undergoes a concerted rearrangement into a cyclic alkenyloxacarbenoid **129**, which, following deprotonation with the triethylamine ligand, gives a furanylmolybdenum

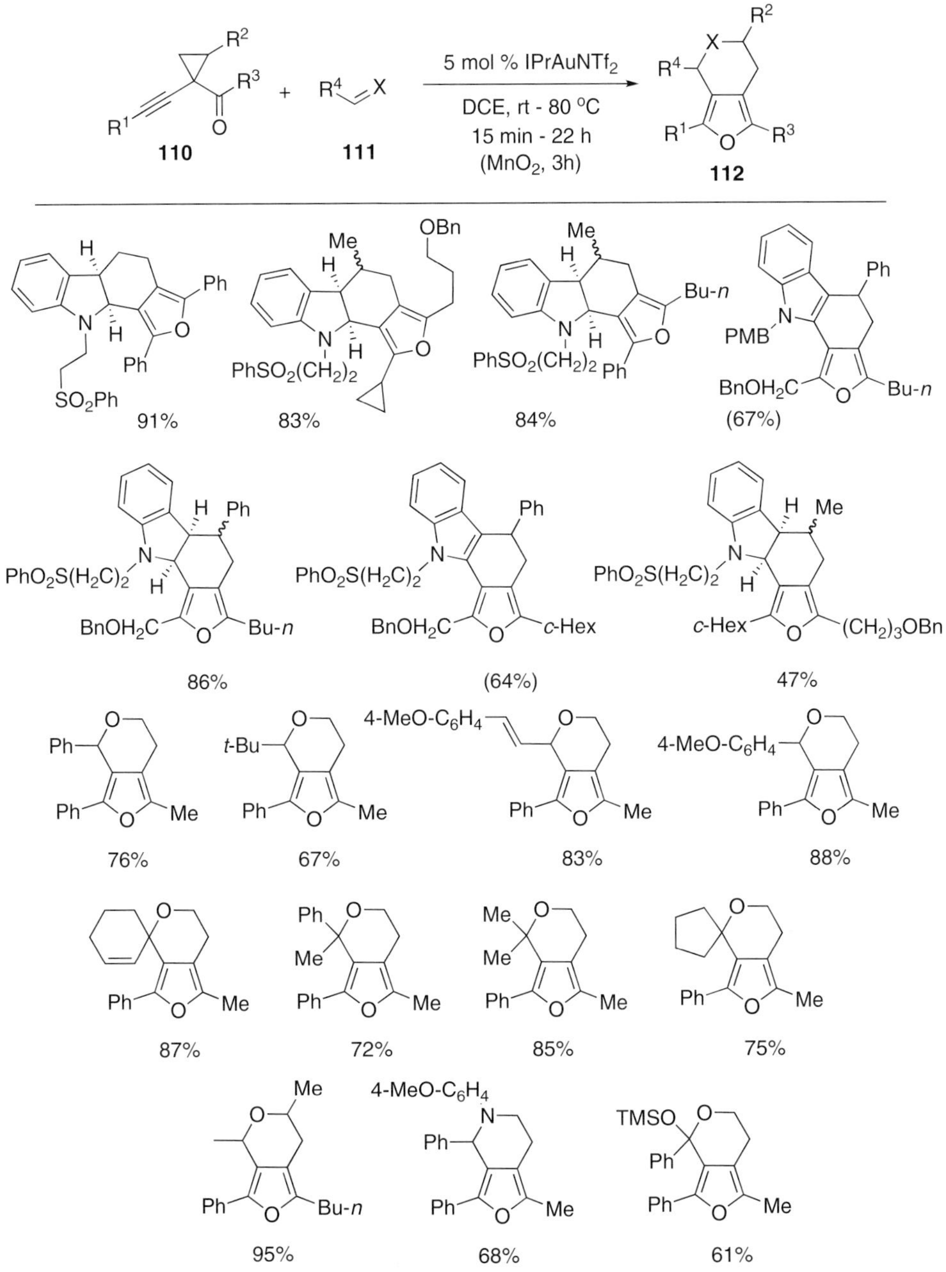

5 mol % IPrAuNTf$_2$

DCE, rt - 80 °C

15 min - 22 h

(MnO_2, 3h)

110 + **111** → **112**

91% 83% 84% (67%)

86% (64%) 47%

76% 67% 83% 88%

87% 72% 85% 75%

95% 68% 61%

Scheme 8.44

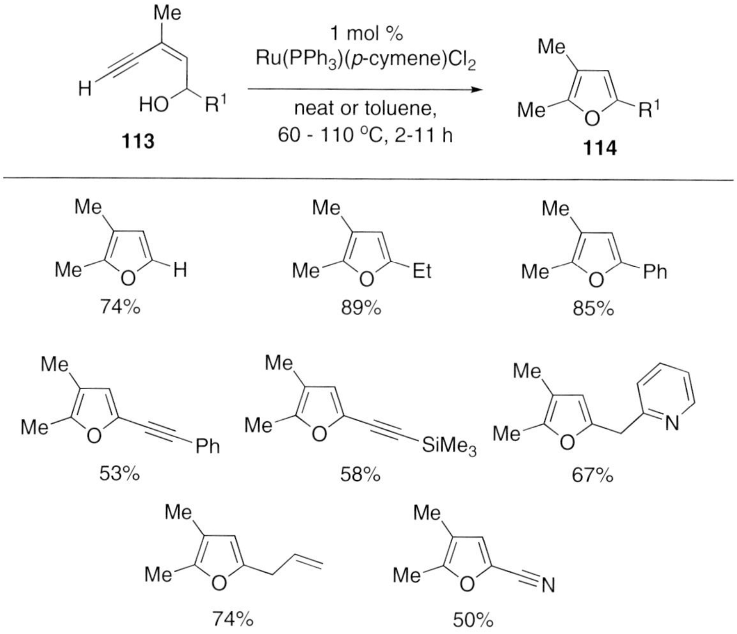

Scheme 8.45

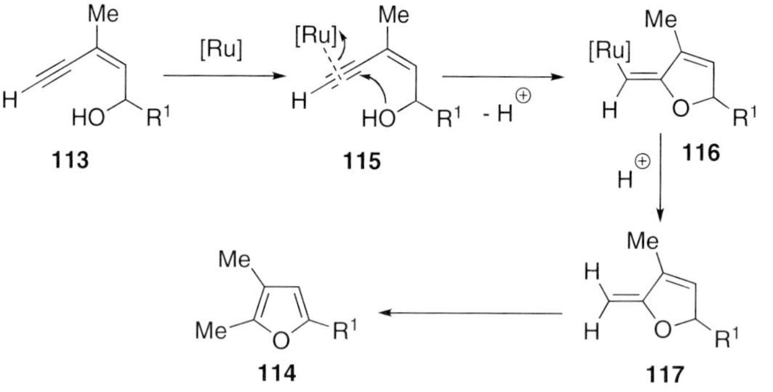

Scheme 8.46

zwitterionic intermediate **129**. A subsequent protiodemetalation furnishes the expected furan **127** and regenerates the Mo(0) catalyst (Scheme 8.52). Apparently, involvement of the requisite alkyne–vinylidene isomerization limits the applicability of this method to terminal alkynes only.

Liu's group further elaborated the scope of this reaction with the Ru(II) catalyst (Scheme 8.53) [175] A variety of 2- or 3-monosubstituted and 2,3-disubstituted furans **132**, possessing sensitive functional groups, were synthesized in good to high yields from the corresponding terminal epoxyalkynes **131** using 1–2 mol% catalyst loadings. Mechanistic studies of this Ru(II)-catalyzed transformation using labeling

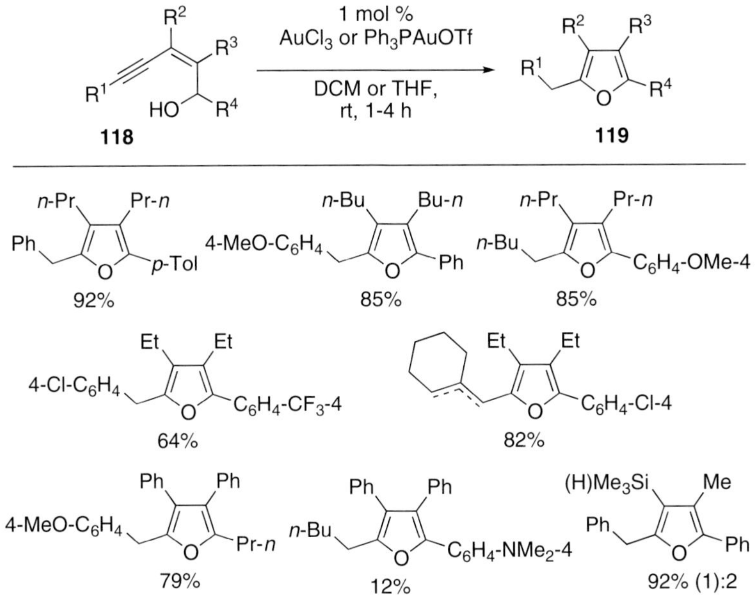

Scheme 8.47

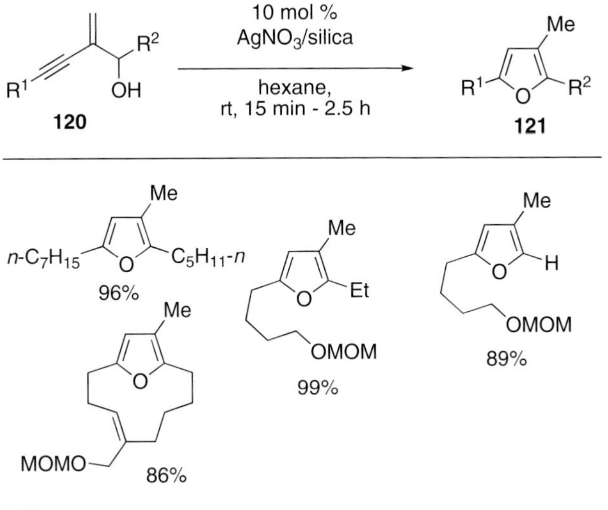

Scheme 8.48

experiments supported the involvement of the mechanism analogous to that proposed by McDonald for the Mo(0) system (Scheme 8.52).

Hashmi *et al.* reported that a major limitation of McDonald's chemistry to terminal alkynes could be alleviated if a different mode of substrate activation was engaged. Accordingly, Au(III) catalyst was shown to be quite efficient in the cycloisomerization

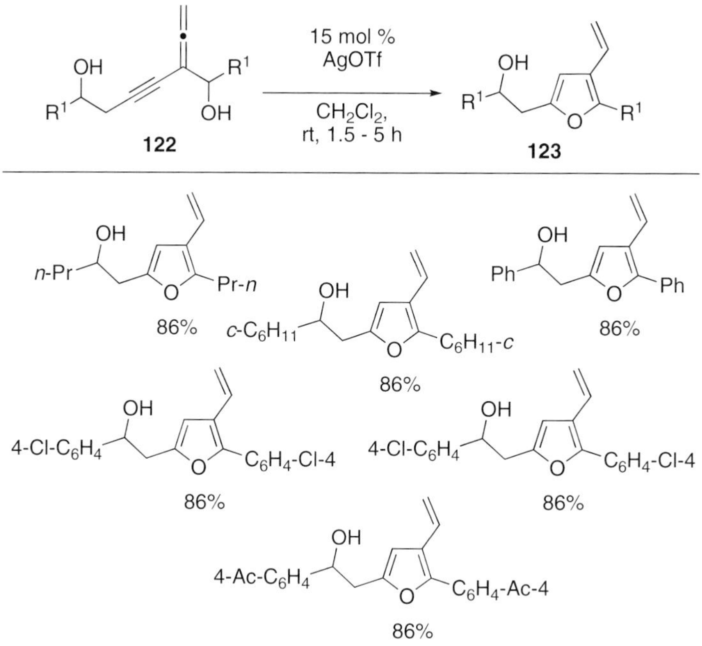

15 mol % AgOTf; CH_2Cl_2, rt, 1.5 - 5 h; **122** → **123**

86%, 86%, 86%, 86%, 86%, 86%

Scheme 8.49

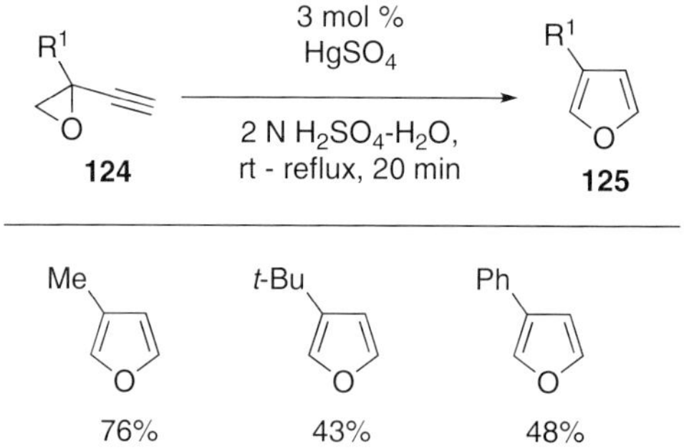

3 mol % $HgSO_4$; 2 N H_2SO_4-H_2O, rt - reflux, 20 min; **124** → **125**

Me 76%, t-Bu 43%, Ph 48%

Scheme 8.50

of epoxyacetylenes **133** possessing an internal alkyne into the corresponding 2,4-disubstituted furans **134** (Scheme 8.54) [176]. Although, a variety of functional groups were tolerated under the Au(III)-catalyzed reaction conditions, the scope of this transformation was somewhat limited to substrates bearing nucleophilic hydroxyalkyl substituents. The proposed mechanism involves initial activation of a carbon–carbon triple bond of the alkyne moiety by a π–philic Au(III) catalyst **135**. A subsequent cycloisomerization of the latter produces the furylgold intermediate **136**, which, upon proton elimination–protiodeauration sequence, affords furan **134** (Scheme 8.55).

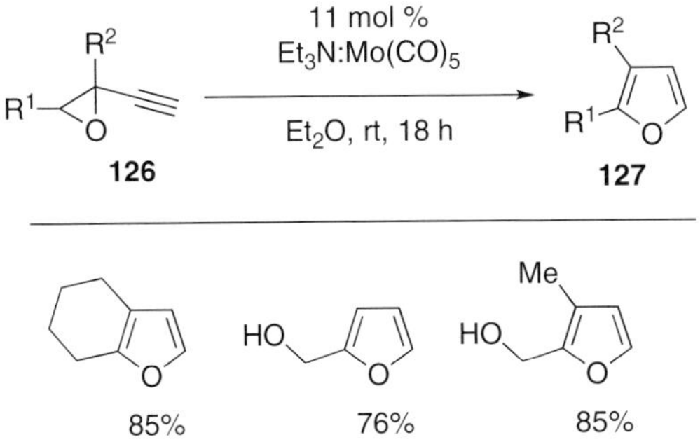

Scheme 8.51

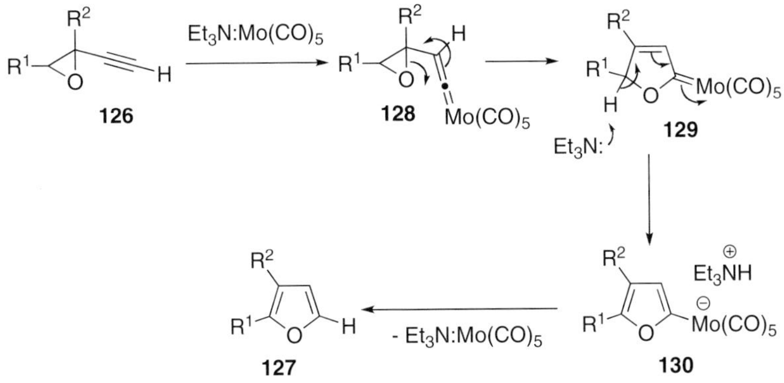

Scheme 8.52

In a recent study, Yoshida demonstrated that cycloisomerization of alkynylepoxides **138** into furans **139** could be achieved under Pt(II) catalysis (Scheme 8.56) [177]. It should be noted that the scope of this transformation was significantly improved by the employment of $PtCl_2$ catalyst in a dioxane–water system. High yields of 2,4-di- and fused 2,3,5-trisubstituted furans **139** were achieved in 10 min while using these new conditions. In addition, it was demonstrated that the reactive furylplatinum intermediate similar to **137** could be intercepted with electrophiles such as NIS, affording tetrasubstituted 3-iodofuran **141** (Scheme 8.57).

Cycloisomerization of homologous-to-**138** propargyl oxiranes **142** in the presence of transition metals was first studied by Marson (Scheme 8.58) [178, 179]. It was demonstrated that propargyl oxiranes **142**, possessing a hydroxy group at the propargylic position, were cycloisomerized into 2,5-di- and 2,3,5-trisubstituted furans upon catalysis with Hg(II) salts. The authors proposed that the initially formed common intermediate 2,3-dihydrofuran **143**, depending on the substitution pattern, underwent further different types of fragmentation processes to give functionalized furans **144** and **145**. In the case where $R^3 = H$, elimination of a water molecule from **143** afforded hydroxymethyl-substituted furan **144**, whereas for a

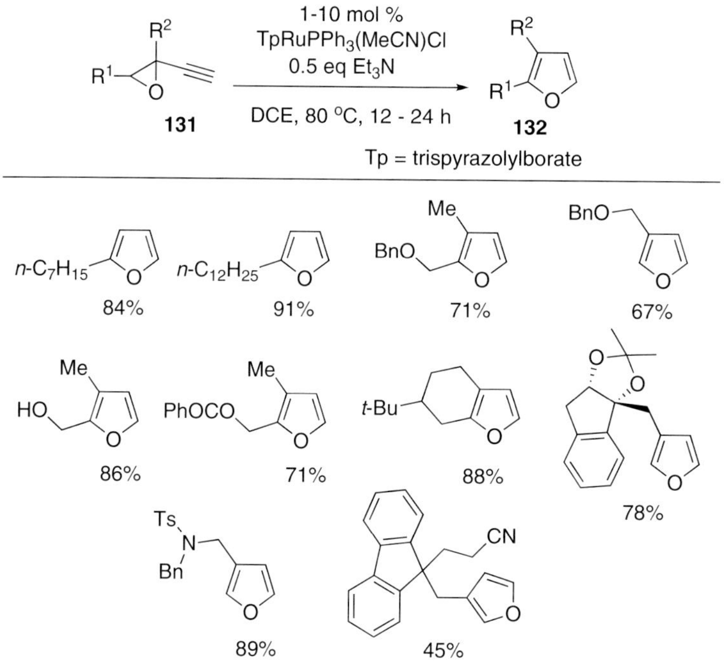

Scheme 8.53

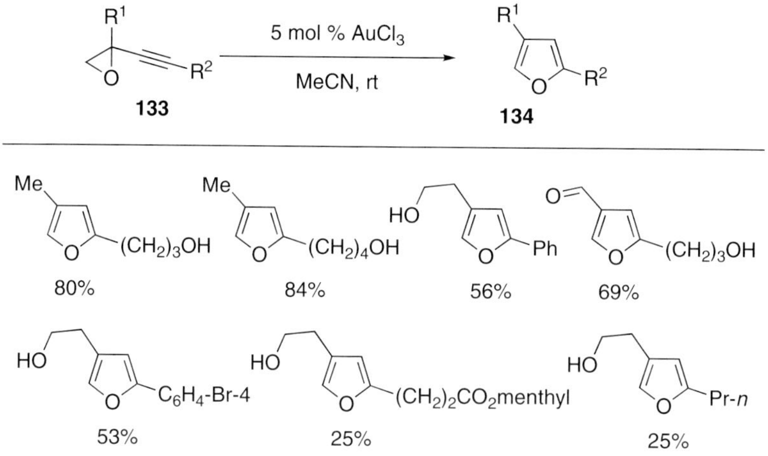

Scheme 8.54

spirodihydrofuran **143** a Grob-type fragmentation occurred, producing furans **145** with a tethered carbonyl function.

Recently, Liang reported the Au(III)-catalyzed cycloisomerization of a similar propargylic oxirane system into furans. Thus, propargylic acetates **146**, possessing

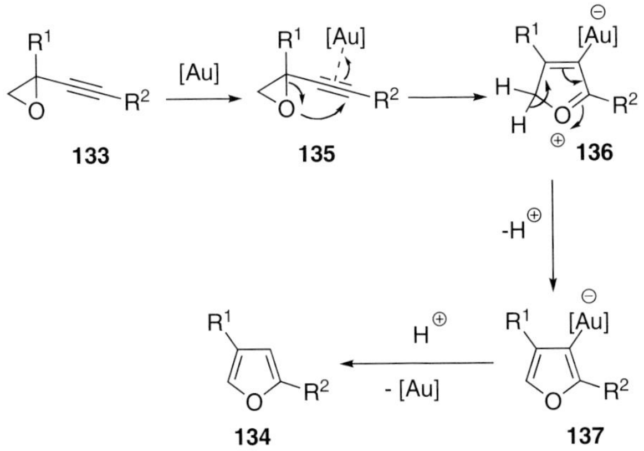

Scheme 8.55

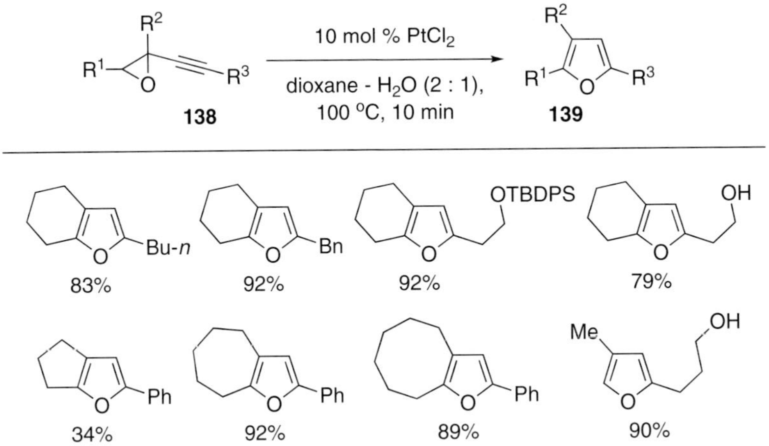

Scheme 8.56

an epoxide moiety, provided moderate to excellent yields of hydroxymethyl furan derivatives **147** in the presence of the $AuCl_3$ catalyst and an external nucleophile (Scheme 8.59) [180]. The cycloisomerization reaction occurred under mild reaction conditions and with only 2 mol% of catalyst. In addition, a variety of functionalized epoxides **146**, as well as an array of alcohols, including sterically congested ones,

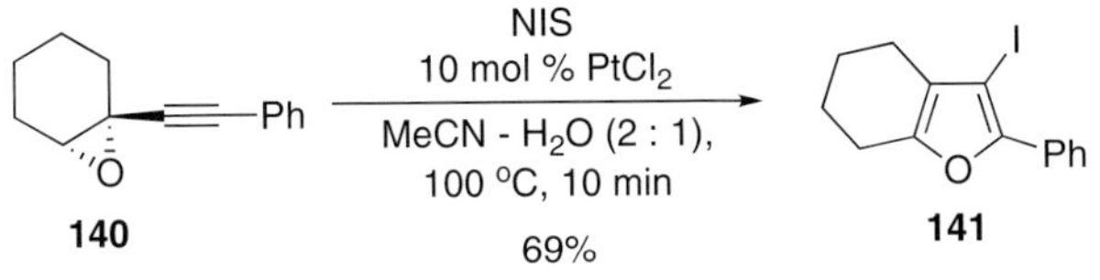

Scheme 8.57

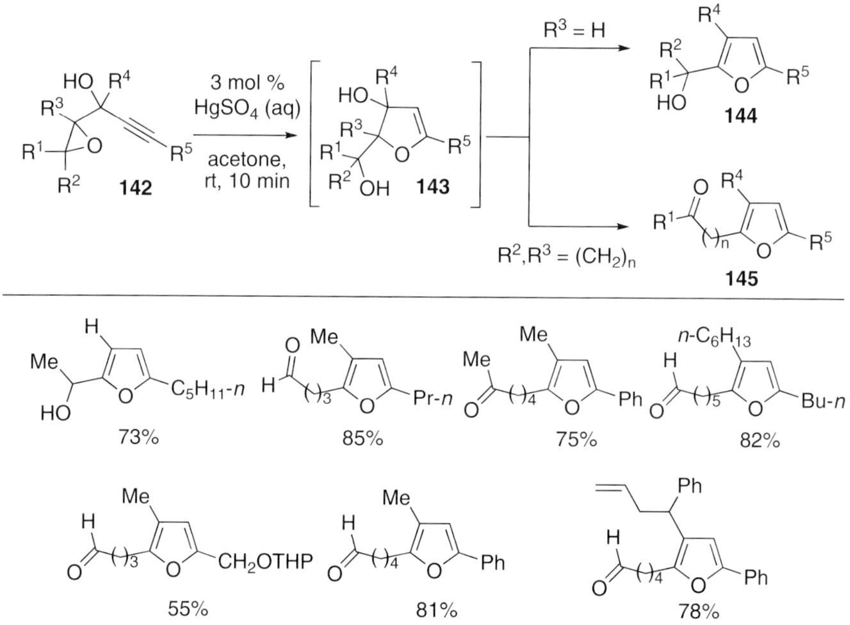

Scheme 8.58

serving as O-nucleophiles, could be employed in this transformation. According to the proposed mechanism, (Scheme 8.60) activation of the alkyne by Au(III)-catalyst in **148** triggers a subsequent domino nucleophilic attack/*anti-endo-dig* cyclization affording an organogold intermediate **149**. The latter, upon protiodemetalation, is transformed into 2,3-dihydrofuran **150**, which is followed by aromatization into the corresponding furan **147**. Alternatively, the reaction may proceed in a stepwise fashion with involvement of the cyclic oxonium ion **151** (Scheme 8.60). The epoxide ring opening in the latter, by an attack of an external O-nucleophile, generates a common organogold intermediate **149**.

Similar transformations of epoxides bearing an alkyne unit were reported to occur in the presence of $HAuCl_4$ [181], Ph_3PAuBF_4 [182], and Cu(II) triflate–iodine [183] catalytic systems. The first two protocols offer an attractive route to the synthesis of differently substituted difurylmethanes, whereas the latter provides an easy and modular access to up-to-tetrasubstituted 3-iodofurans.

The first example of a transition metal-catalyzed cycloisomerization of cyclopropenes [184, 185] into furans was demonstrated by Nefedov (Scheme 8.61) [186]. It was proposed that this rearrangement proceeds via a carbenoid intermediate [187]. Formation of the furan products via the cycloisomerization of potentially involved cyclopropene intermediates in the Rh(II)-catalyzed cyclopropenation reaction of alkynes was later reported by several research groups, including Liebeskind, Davies, and Müller [188–192].

Later, Padwa reported a regioselective Rh(II)-catalyzed room temperature cycloisomerization of trisubstituted cyclopropenyl ketones to give 2,3,4-trisubstituted

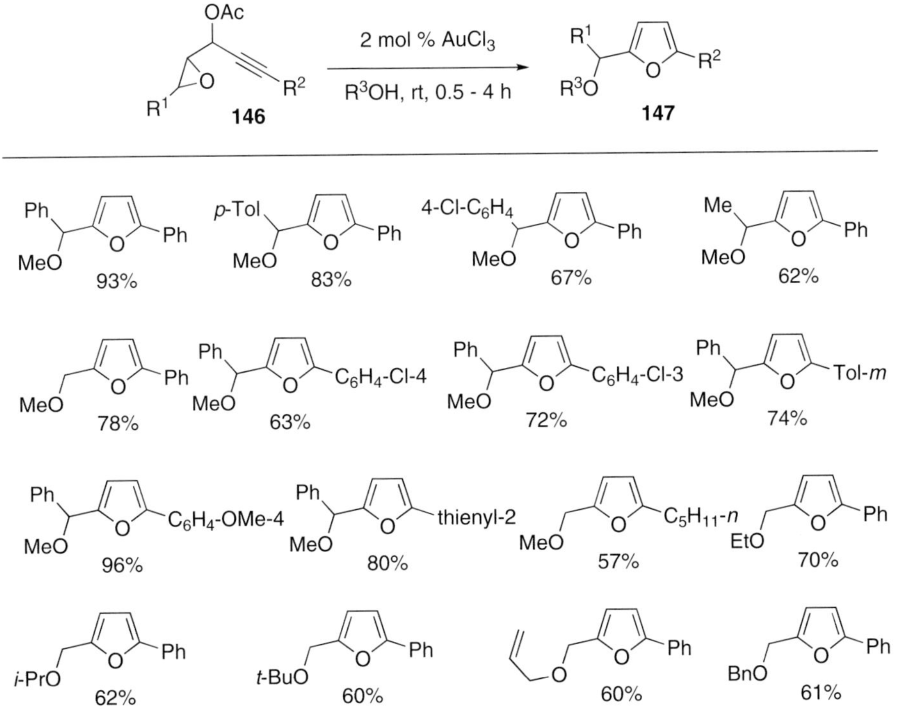

OAc
R[1] O
R[2]
146
2 mol % $AuCl_3$
R^3OH, rt, 0.5 - 4 h
R[1]
R^3O
O
R[2]
147

Ph MeO O Ph 93%
p-Tol MeO O Ph 83%
4-Cl-C_6H_4 MeO O Ph 67%
Me MeO O Ph 62%

MeO O Ph 78%
Ph MeO O C_6H_4-Cl-4 63%
Ph MeO O C_6H_4-Cl-3 72%
Ph MeO O Tol-m 74%

Ph MeO O C_6H_4-OMe-4 96%
Ph MeO O thienyl-2 80%
MeO O C_5H_{11}-n 57%
EtO O Ph 70%

i-PrO O Ph 62%
t-BuO O Ph 60%
O O Ph 60%
BnO O Ph 61%

Scheme 8.59

OAc
R[1] O
R[2]
146
[Au]
NuH OAc [Au] ⊖
R[1] O R[2]
148
-H ⊕
AcO [Au] ⊖
R[1]
Nu O R[2]
149
AcO [Au] ⊖
NuH
R[1] O ⊕ R[2]
151
-H ⊕
H ⊕ - [Au]
AcO
R[1]
R^3O O R[2]
150
- AcOH
R[1]
R^3O O R[2]
147

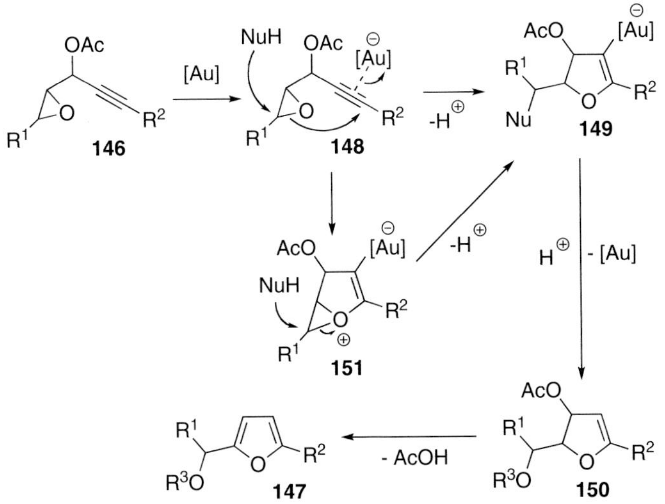

Scheme 8.60

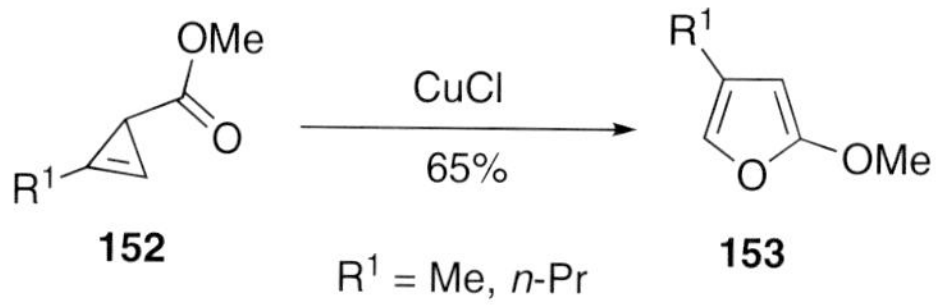

Scheme 8.61

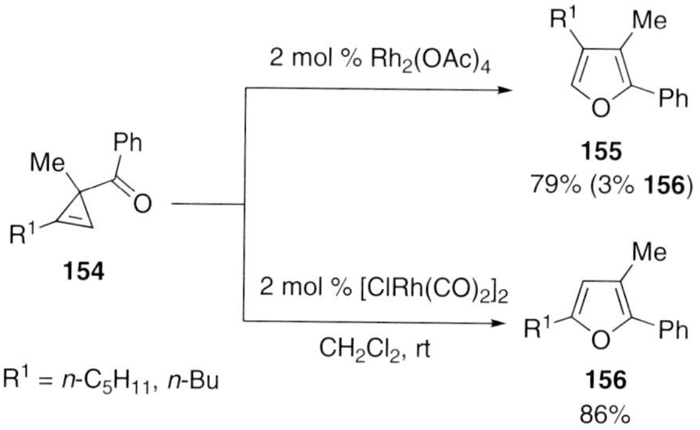

Scheme 8.62

furans **155** (Scheme 8.62) [193, 194]. Formation of the regioisomeric furans **156** as the minor products was observed under these reaction conditions. The latter complementary regioselectivity in the cycloisomerization of cyclopropenes **155** was exclusively achieved with the employment of Rh(I) catalyst. Thus, 2,3,5-trisubstituted furans **156** were obtained in high yields and as single regioisomers (Scheme 8.62). It was proposed that, in the case of the Rh(II) catalyst, a preferential electrophilic attack by the bulky Rh(II) catalyst occurred at the cyclopropene double bond to produce the best stabilized tertiary cyclopropyl carbocation **158** (Path A). A subsequent ring opening of the latter produced a key Rh-carbenoid intermediate **159**. A following cyclization led to the formation of furylrhodium zwitterionic species **160**, where elimination of the Rh(II) catalyst afforded normal product **155**. Alternatively, 6π-electrocyclization in **159** generates rhodacycle **161**, which, upon reductive elimination, produces furan **155** (Scheme 8.63). To account for the opposite regiochemistry in the case of the Rh(I) catalyst, formation of the regioisomeric to **165** metallocyclobutene species **166**, available either via a direct oxidative addition of rhodium into the C1–C3 bond of cyclopropene or through the cycloreversion–cycloaddition pathway, was suggested (Path B). Following cyclization steps similar to that involved in the Rh(II)-catalyzed reaction furnished the formation of furan **156** (Scheme 8.64).

Finally, a recent study from Ma's group demonstrated that the Cu(I)-catalyzed cycloisomerization of cyclopropenyl ketones **168** proceeded regioselectively, providing the corresponding 2,3,4-trisubstituted furans **169** in good to excellent yields (Scheme 8.65) [195]. A variety of functional groups could be tolerated under the

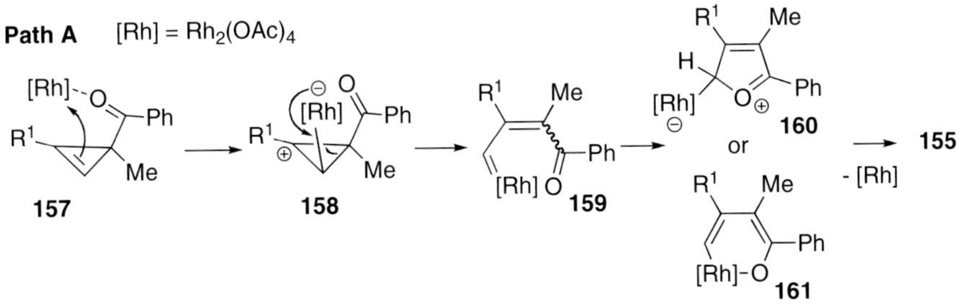

Path A [Rh] = $Rh_2(OAc)_4$

[Rh]--O, Ph, R^1, Me, **157** → [Rh], Ph, R^1, Me, **158** → R^1, Me, [Rh] O, Ph, **159** → R^1, Me, H, [Rh], Ph, **160** or R^1, Me, Ph, [Rh]-O, **161** → **155** (- [Rh])

Scheme 8.63

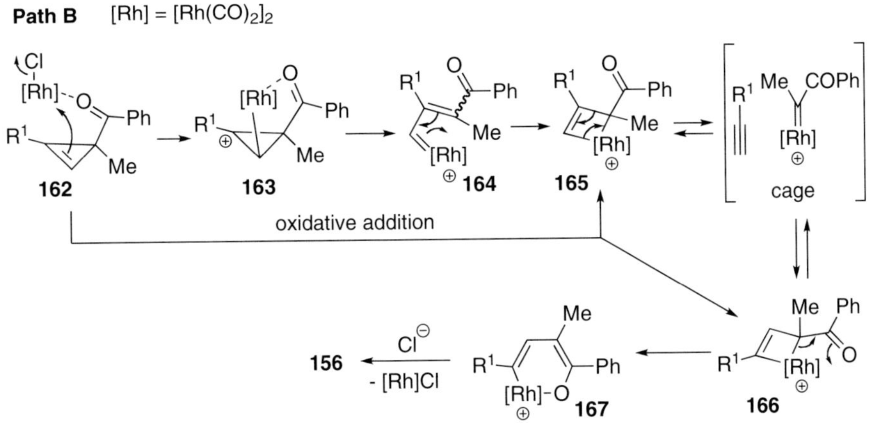

Path B [Rh] = $[Rh(CO)_2]_2$

Cl, [Rh]--O, Ph, R^1, Me, **162** → [Rh], O, Ph, R^1, Me, **163** → R^1, O, Ph, Me, [Rh], **164** → R^1, O, Ph, Me, [Rh], **165** ⇌ [R^1, Me, COPh, [Rh], cage]

oxidative addition

⇅ Me, Ph, R^1, [Rh], O, **166** → Me, R^1, Ph, [Rh]-O, **167** → **156** ($Cl^⊖$, - [Rh]Cl)

Scheme 8.64

R^2, R^3, O, R^1, **168** → (5 mol % CuI; MeCN, reflux) → R^1, R^2, O, R^3, **169**

TBSO, CO_2Et, Me: 83%

t-Bu, CO_2Et, Me: 80%

Ph, CO_2Et, Me: 89%

n-Bu, COMe, Me: 80%

n-C_5H_{11}, CO_2Et, Ph: 80%

TBSO, CO_2Et, Me: 85%

n-Bu, SO_2Ph, Me: 96%

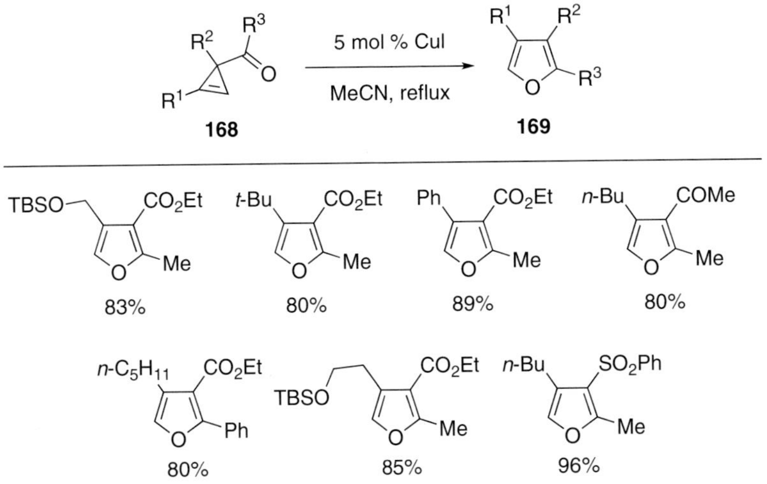

Scheme 8.65

reported reaction conditions. It should be noted that employment of the Pd(II)-catalyst completely changed the regioselectivity of this transformation, thus affording the isomeric 2,3,5-trisubstituted furans. According to the proposed mechanism, a regioselective iodocupration of the carbon–carbon double bond in cyclopropene **168** generates intermediate **170**. A subsequent β-decarbocupration gives the copper enolate **171**. Intramolecular *endo*-mode insertion of the carbon–carbon double bond into the copper–oxygen bond of **171** to generate **172** followed by the β-halide elimination affords product **169** (Scheme 8.66).

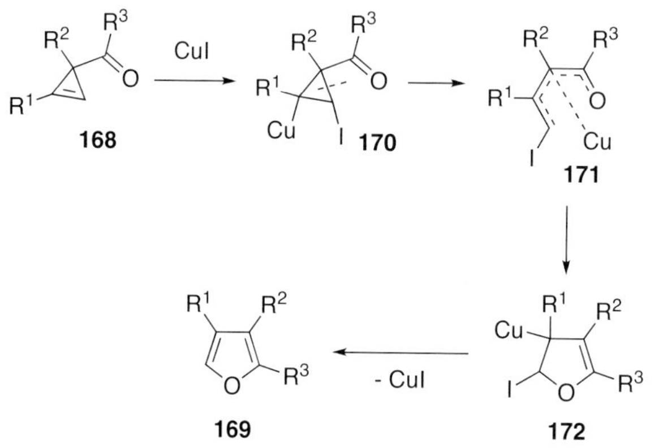

Scheme 8.66

8.2.1.2 Synthesis of Furans via "3 + 2" Cycloaddition Reactions

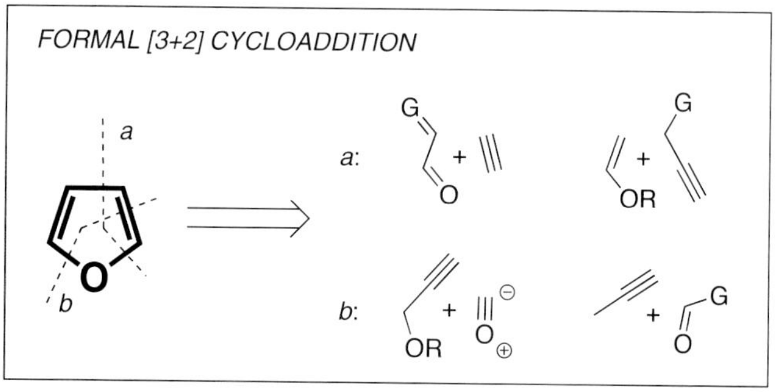

Among many various formal [3 + 2] approaches towards assembly of the furan core described in the literature, special attention was given to the Rh- or Cu-catalyzed reactions between alkynes and α-diazocarbonyl compounds [196–200]. This method quickly became very popular as a highly convenient and general tool for the construction of diversely substituted furans, and is often referred to as the

Rh-catalyzed [3 + 2] dipolar cycloaddition. Generally, this transformation is performed at elevated temperatures and affords furans directly in a single step without isolation or even observation of possible cyclopropene intermediate (Section 8.2.1.1). It was first demonstrated by D'yakonov that the Cu(II)-catalyzed reaction between α-diazoesters and internal alkynes provided the corresponding 2-alkoxyfurans in moderate yields [201–205]. Consequently, several research groups further elaborated on this transformation in the presence of different transition metal catalytic systems for a variety of differently substituted α-diazocarbonyl compounds and alkynes.

Davies reported that trisubstituted furans could be obtained via the Rh(II)-catalyzed formal [3 + 2] cycloaddition reaction between diazo-1,3-dicarbonyl compounds **173** and terminal alkynes **174** (Scheme 8.67) [189]. This protocol provided access to 2,3,5-trisubstituted furans **175** possessing various functional groups under relatively mild reaction conditions and low catalyst loading, albeit with generally moderate yields.

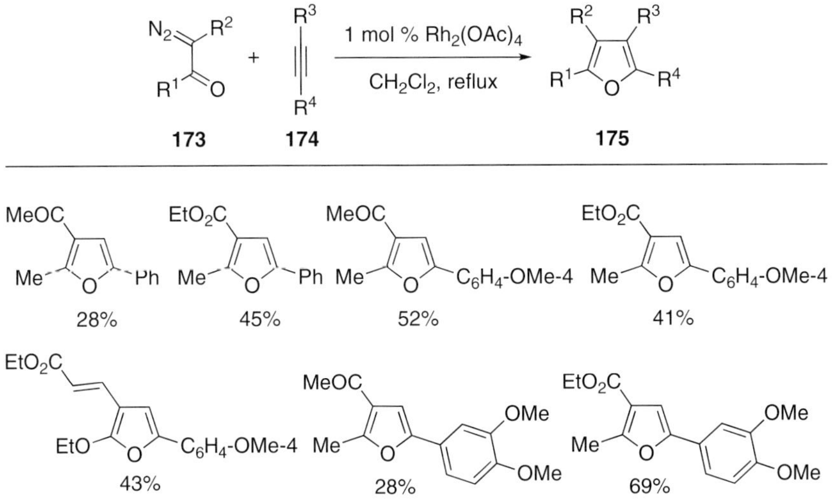

Scheme 8.67

Synthesis of tetrasubstituted furans via an intramolecular version of this methodology was reported by Padwa [206, 207]. Alkyne-tethered diazoketones **176** and **178** underwent the Rh(II)-catalyzed formal [3 + 2] cycloaddition affording fused furans **177** and **179** in good yields (Scheme 8.68).

Later, Pirrung investigated the employment of cyclic diazo-1,3-dicarbonyl compounds for the Rh(II)-catalyzed synthesis of fused furans (Scheme 8.69) [208]. Diazocyclohexane-1,3-dione **180** underwent a formal [3 + 2] cycloaddition with a variety of terminal alkynes **181** bearing sensitive functional groups in the presence of

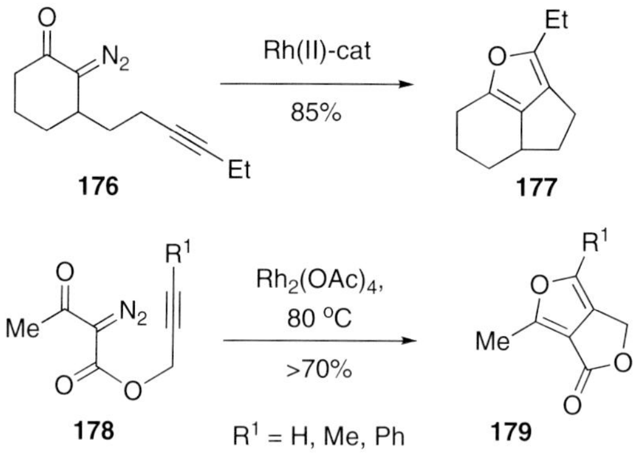

Scheme 8.68

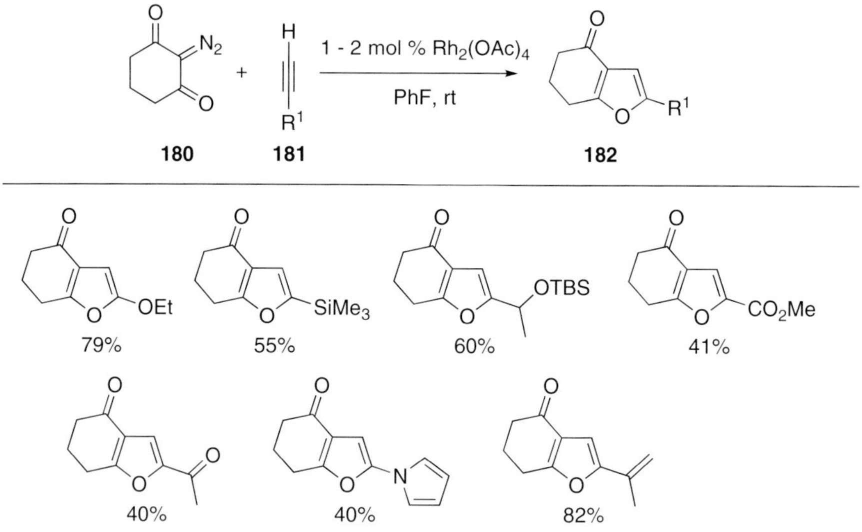

Scheme 8.69

Rh(II)-catalyst at room temperature to provide fused trisubstituted furans **182**. Apparently, employment of unsymmetrical diazodiones leads to the formation of mixtures of regioisomeric furan products.

Many research groups further investigated the scope of the Rh(II)-catalyzed synthesis of furans from alkynes and α-diazocarbonyl compounds [194, 209–221]. Some representative results are summarized in Scheme 8.70. Tri- and tetra-substituted furans possessing a vast array of functional groups and diverse substitution patterns could be efficiently accessed via this methodology.

Several mechanistic possibilities were proposed to account for the formation of furan products in the Rh(II)-catalyzed [3 + 2] cycloaddition reaction between

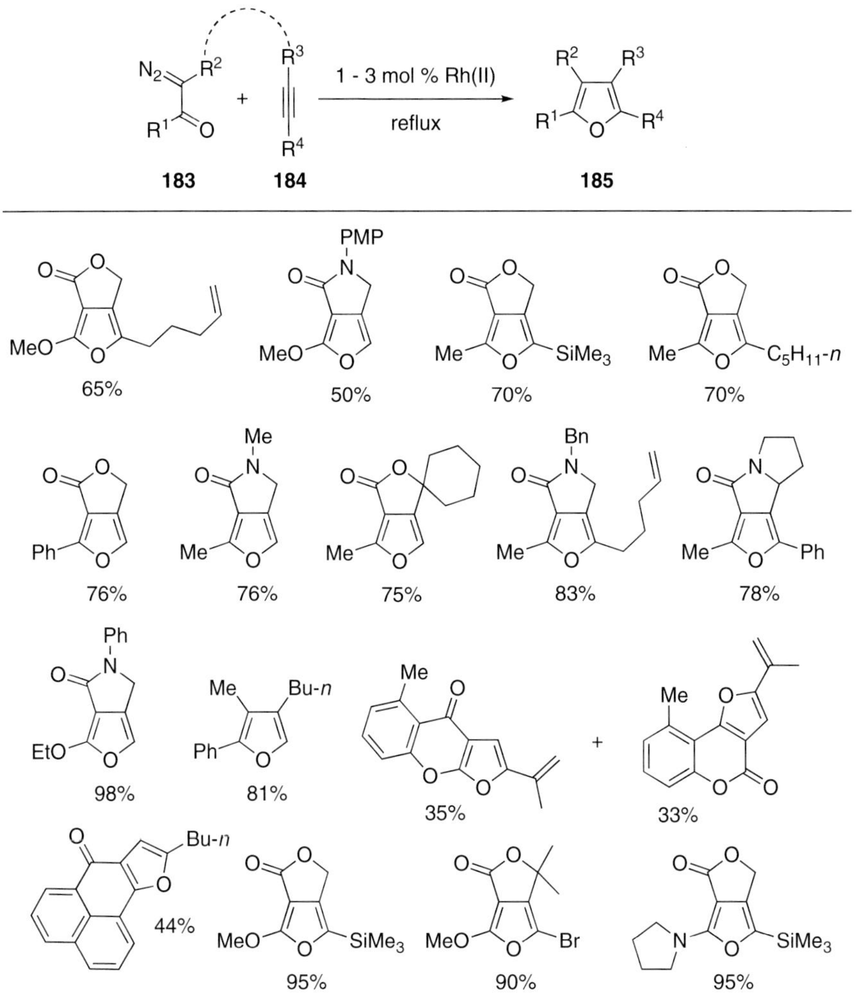

Scheme 8.70

α-diazocarbonyl compounds and alkynes (Scheme 8.71). First, diazo compound **186**, upon reaction with rhodium(II) carboxylate, generates the Rh-carbenoid species **187**. According to path A, a direct nucleophilic attack [222] of alkyne at **187** produces **188**, which then cyclizes to form furan **190** via a cyclic zwitterion **189**. Alternatively (Path B), a [2 + 2] cycloaddition of **187** and alkyne leads to the metallacyclobutene **191**, which can also be formed via cyclization of **188** [209]. Rhodacycle **191** then undergoes metathesis reaction to produce Rh carbenoid **192** which, upon 6π-electrocyclization and subsequent reductive elimination, furnishes product **190**. [2 + 1] Cycloaddition of **187** (Path C) leading to the cyclopropene **194** in the presence of rhodium(II) acetate could also account for the formation of furans via a subsequent Rh(II)-catalyzed cycloisomerization reaction [189, 207].

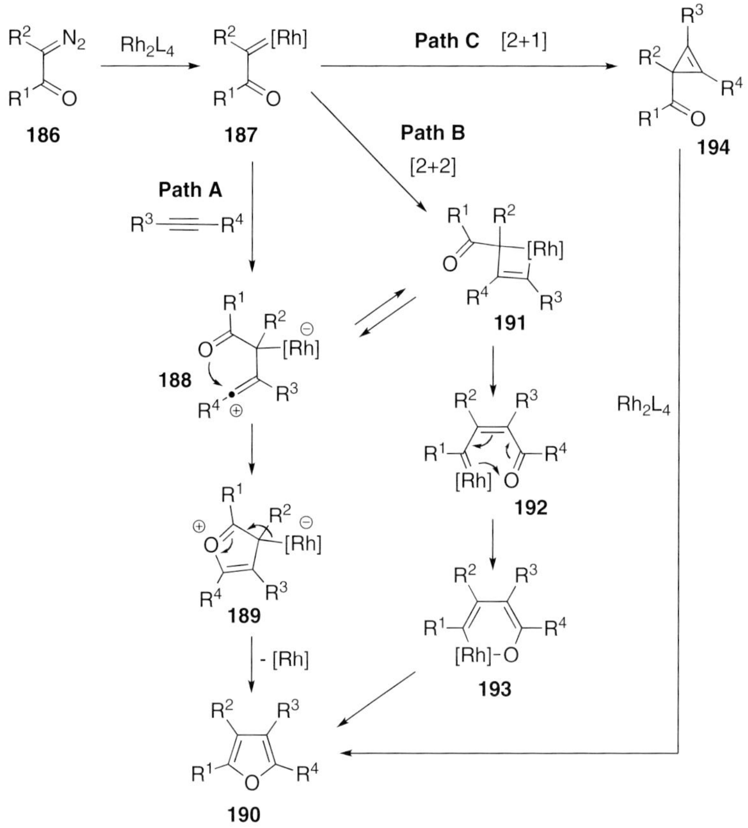

Scheme 8.71

Several highly efficient "3 + 2" approaches for the assembly of the furan heterocyclic core involve a transition metal-catalyzed propargylic substitution in propargylic alcohols or their derivatives with a variety of C-nucleophiles to access the key alkynyl ketones capable of further cycloisomerization. Hidai and Uemura reported that propargylic alcohols **195** produce the corresponding furans **197** upon reaction with cyclic and acyclic carbonyl compounds **196** in the presence of a bimetallic Ru and Pt(II) catalytic system (Scheme 8.72) [143]. The authors proposed that an initial Ru-catalyzed propargylic substitution [223] affords γ-ketoalkynes, like **87** which, upon further cycloisomerization catalyzed by Pt(II), produce furans **197**. Although good yields of tri- and tetra-substituted furans and functional group compatibility can be achieved in this "3 + 2" cycloaddition reaction, the synthetic usefulness of this protocol is questionable as a large excess of carbonyl compounds and relatively high catalyst loadings are required in this transformation.

An analogous Ru(II)-catalyzed transformation was investigated by Nebra and coworkers (Scheme 8.73) [144, 224]. Accordingly, "3 + 2" cycloaddition of pro-

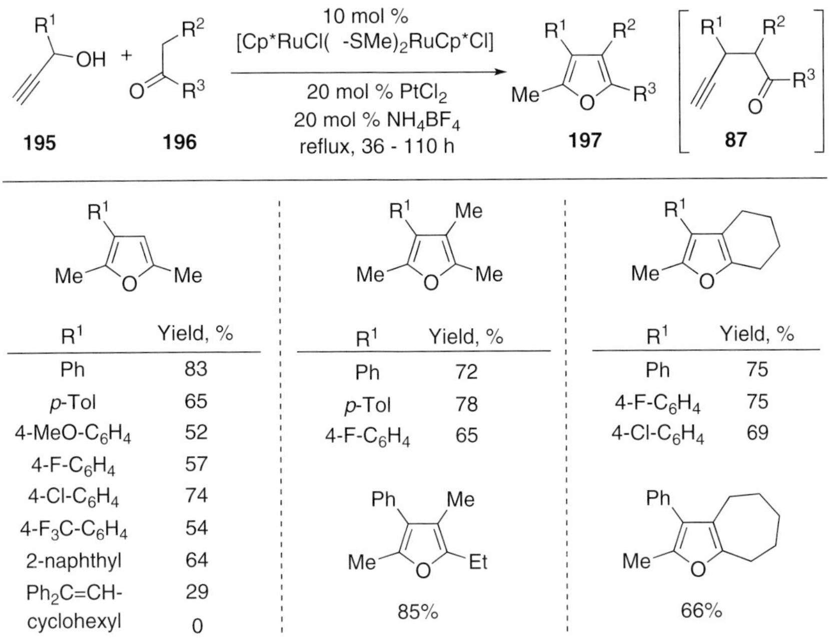

10 mol % [Cp*RuCl(-SMe)$_2$RuCp*Cl]; 20 mol % $PtCl_2$; 20 mol % NH_4BF_4; reflux, 36 - 110 h

195 + **196** → **197** [**87**]

R^1	Yield, %
Ph	83
p-Tol	65
4-MeO-C_6H_4	52
4-F-C_6H_4	57
4-Cl-C_6H_4	74
4-F_3C-C_6H_4	54
2-naphthyl	64
Ph_2C=CH-	29
cyclohexyl	0

R^1	Yield, %
Ph	72
p-Tol	78
4-F-C_6H_4	65

85%

R^1	Yield, %
Ph	75
4-F-C_6H_4	75
4-Cl-C_6H_4	69

66%

Scheme 8.72

pargylic alcohols **198** with 1,3-dicarbonyl compounds **199** afforded 3-acyl or -carbalkoxyfurans **200** in the presence of Ru(II) catalyst and substoichiometric amounts of trifluoroacetic acid. These new reaction conditions allowed achievement of a significantly extended scope and perfect functional group tolerance in this transformation. It was demonstrated that a variety of terminal and internal propargylic alcohols, cyclic and acyclic, aromatic and aliphatic 1,3-diketones and β-ketoesters could be efficiently employed in this transformation, providing up-to-tetrasubstituted furans **200** in good to excellent yields. The Cu(II)-triflate-catalyzed version of this methodology was elaborated by Zhan [225].

Zhan also utilized a similar approach for the "3 + 2" synthesis of furan compounds from propargylic esters and carbon-centered nucleophiles, overcoming the limitation of Hidai and Uemura's protocol for terminal alkynes (Scheme 8.74) [226]. Employment of a one-pot sequential Cu(II)-catalyzed nucleophilic substitution of propargylic acetates **201** with silyl enol ethers **202** to give the γ-alkynyl ketones allowed subsequent Cu(II)-catalyzed cycloisomerization of the latter upon addition of the *p*-toluenesulfonic acid co-promoter. Thus, tri-, and tetra-substituted, and even fused alkyl- and arylfurans **203** bearing various functionalities were synthesized in high yields. In addition, bisfuranylarenes could be accessed via this chemistry upon a double propargylic substitution/cycloisomerization sequence of bis(1-silyloxyvinyl) arenes. The same group also reported that this transformation could be efficiently catalyzed by Fe(III) [227–229] salts [230]. It was later demonstrated that the

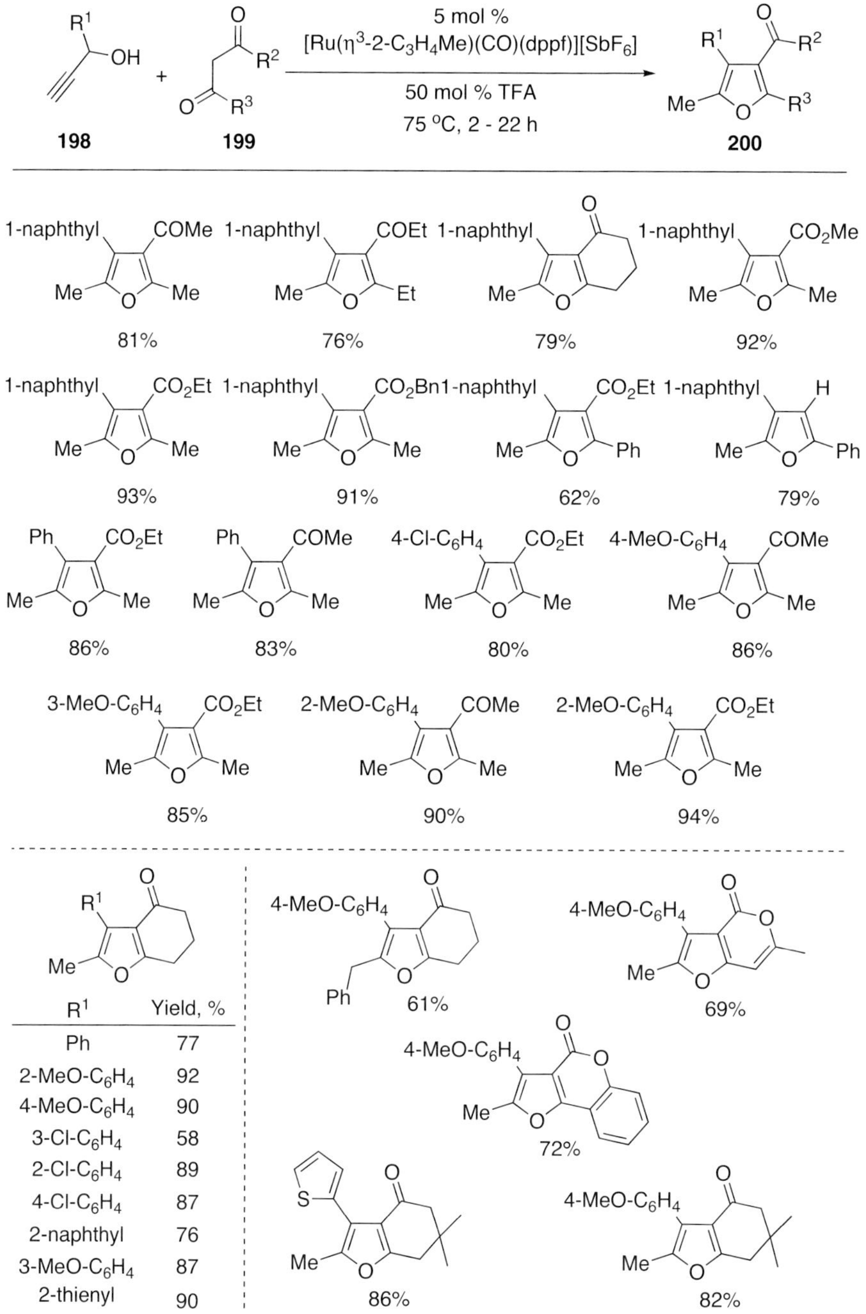

198 + **199** → **200**

5 mol % $[Ru(\eta^3\text{-}2\text{-}C_3H_4Me)(CO)(dppf)][SbF_6]$, 50 mol % TFA, 75 °C, 2 - 22 h

R^1	Yield, %
Ph	77
2-MeO-C_6H_4	92
4-MeO-C_6H_4	90
3-Cl-C_6H_4	58
2-Cl-C_6H_4	89
4-Cl-C_6H_4	87
2-naphthyl	76
3-MeO-C_6H_4	87
2-thienyl	90

Scheme 8.73

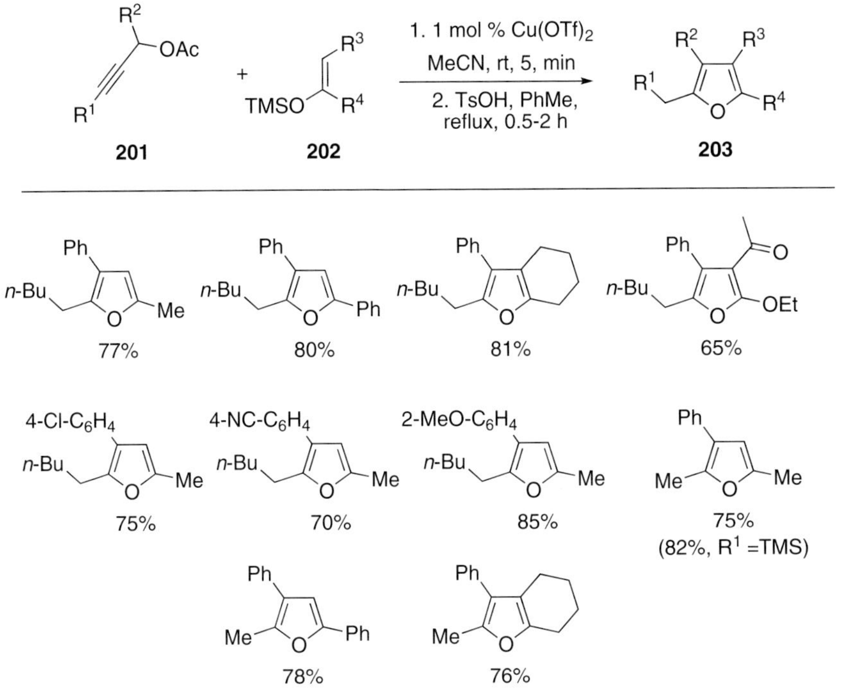

1. 1 mol % $Cu(OTf)_2$, MeCN, rt, 5, min
2. TsOH, PhMe, reflux, 0.5-2 h

201 + **202** → **203**

77% 80% 81% 65%

75% 70% 85% 75% (82%, R^1 =TMS)

78% 76%

Scheme 8.74

Cu(II)-catalyzed protocol does not require addition of the *p*-toluenesulfonic acid co-promoter [225].

Recently, Alper reported that primary propargylic alcohols **204** could serve as the C3-components in the synthesis of furans (Scheme 8.75) [231]. The Rh(II)-catalyzed hydroformylation [232] of **204** resulted in the formation of γ-hydroxyenals, which, upon subsequent cyclocondensation, provided 3-monosubstituted aryl- and hetarylfurans **205** possessing various functional groups in moderate yields.

Kim *et al.* described a two-component coupling–isomerization protocol for the synthesis of furans **208** from terminal propargylic ethers and amines **206** and acyl chlorides **207** in the presence of stoichiometric amounts of $ZnBr_2$ [233] promoter and tertiary amine base (Scheme 8.76) [234]. It is believed that the initial Zn(II)-mediated coupling reaction between alkynes and acyl chlorides provided the corresponding alkynones, which were further isomerized into the allenyl ketones under the basic conditions. The latter, upon Zn-activation, underwent a subsequent cycloisomerization to furnish the furan product. Furthermore, it was demonstrated that this approach could be applied to the synthesis of bis-furanyl-containing compounds and that the corresponding propargylic ethers or amines could be substituted with the simple alkynes.

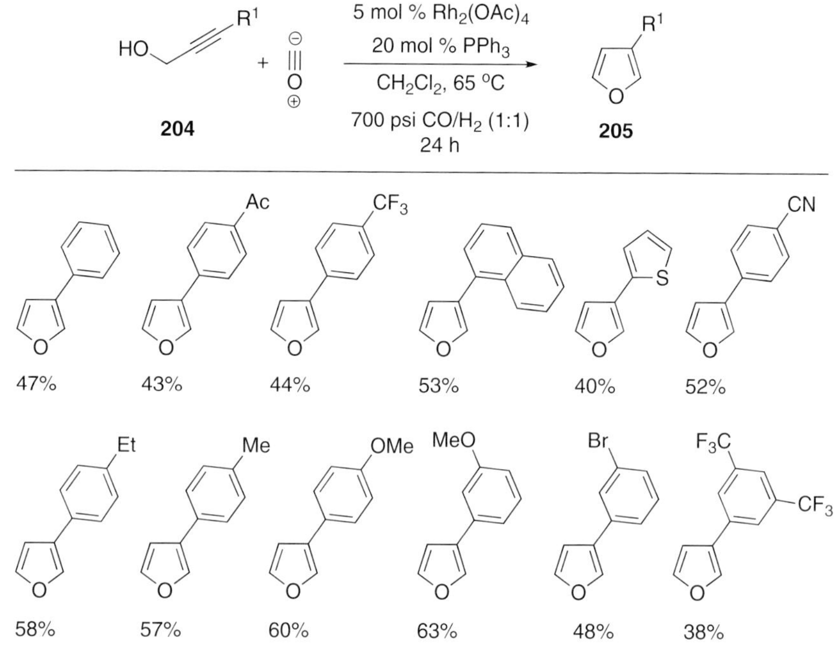

5 mol % $Rh_2(OAc)_4$, 20 mol % PPh_3, CH_2Cl_2, 65 °C, 700 psi CO/H_2 (1:1), 24 h

204 → 205

47% 43% 44% 53% 40% 52%

58% 57% 60% 63% 48% 38%

Scheme 8.75

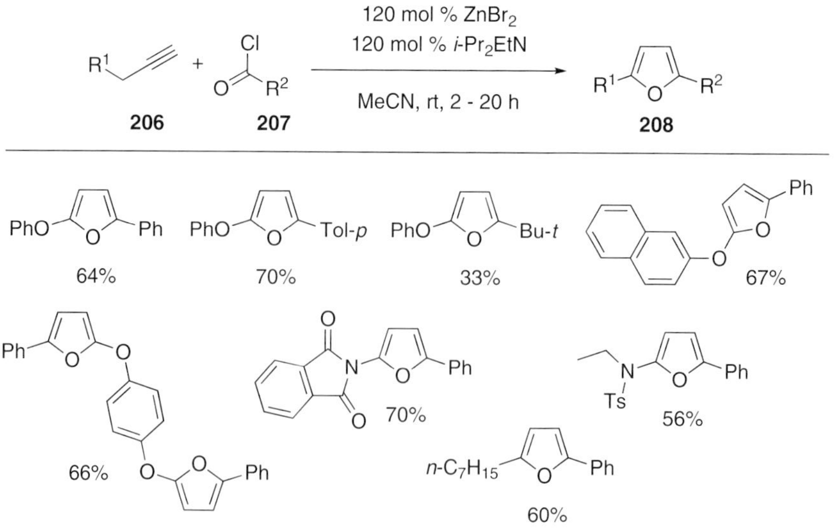

120 mol % $ZnBr_2$, 120 mol % i-Pr_2EtN, MeCN, rt, 2 - 20 h

206 + 207 → 208

64% 70% 33% 67%

66% 70% 56% 60%

Scheme 8.76

8.2.2 Pyrroles

Like furans, pyrroles represent a very important class of five-membered heterocyclic compounds, broadly found in naturally occurring and biologically active compounds [235–246], as well as in artificial materials [242, 247, 248]. In addition, the pyrrole framework represents some interest as a synthon, allowing access to various essential structures [66, 68, 249–254]. Quite expectedly, a great interest caused by the importance and valuable properties of pyrroles is reflected in a number of excellent reviews on the synthesis of the pyrrole framework appearing recently in the literature [83, 237, 242, 246, 248, 255–265]. Apparently, a rational design of general and chemo- and regioselective transition metal-catalyzed methodologies for the pyrrole syntheses [79] still remains a challenging task for many research groups. To some extent, advances in this area for construction of pyrroles were stimulated by the development of approaches for the assembly of furan cores. Many protocols involve transformations aimed at the generation of reactive intermediates similar to those employed in the furan syntheses. However, there are several methods that can be considered as characteristic to pyrroles only. Arguably, among all these transformations, cycloisomerizations and formal [4 + 1], [3 + 2], and [2 + 2 + 1] cycloaddition approaches are the most prominent methods for an efficient and convergent synthesis of pyrroles with a diverse substitution patterns and great functional group compatibility. Accordingly, the most important and recent contributions to these processes are discussed below.

8.2.2.1 Synthesis of Pyrroles via Cycloisomerization Reactions

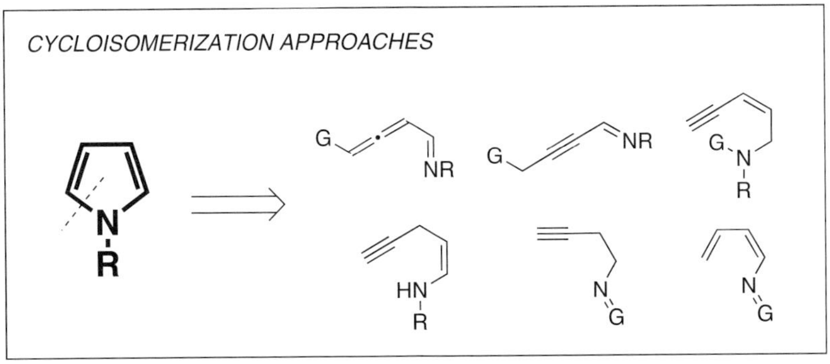

CYCLOISOMERIZATION APPROACHES

A single example of a transition metal-catalyzed cycloisomerization of allenyl imines, analogous to that reported for allenyl ketones, was described by Reißig. Iminoallene **209**, possessing a *tert*-butyl substituent, under Marshall's conditions with 20 mol% of $AgNO_3$ underwent cycloisomerization to give 2,3-disubstituted pyrrole **210** in moderate yield (Scheme 8.77) [266]. It should be mentioned that other iminoallenes were either completely inefficient in this transformation or provided inferior yields of the corresponding pyrroles, probably due to the high iminoallene instability. Despite the fact that this observation is severely limiting further

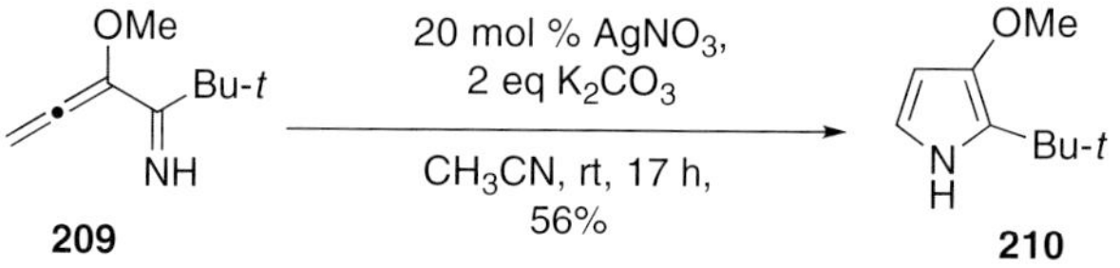

Scheme 8.77

application of this chemistry, iminoallenes represent very valuable and highly reactive intermediates, which could be easily generated *in situ* and used in subsequent transformations leading to multisubstituted pyrroles (*vide infra*).

An elegant solution to this iminoallene instability limitation was reported by Gevorgyan's group where easily accessible and far more stable conjugated alkynyl imines **211** served as surrogates to the corresponding allenes. It was shown that propargyl imines **211** underwent Cu(I)-catalyzed cycloisomerization in the presence of triethylamine as a tertiary amine base to provide 1,2-di- and 1,2,5-tri-substituted pyrroles **212** (Scheme 8.78) [267, 268]. A variety of sensitive groups could be tolerated under these reaction conditions. In addition, the authors demonstrated that 3-ethylbutyryl N1-substituent could be easily and highly efficiently deprotected under basic conditions. An anticipated mechanism is similar to that proposed for the Cu(I)-catalyzed cycloisomerization of allenyl ketones **46** into furans **47** (Scheme 8.21).

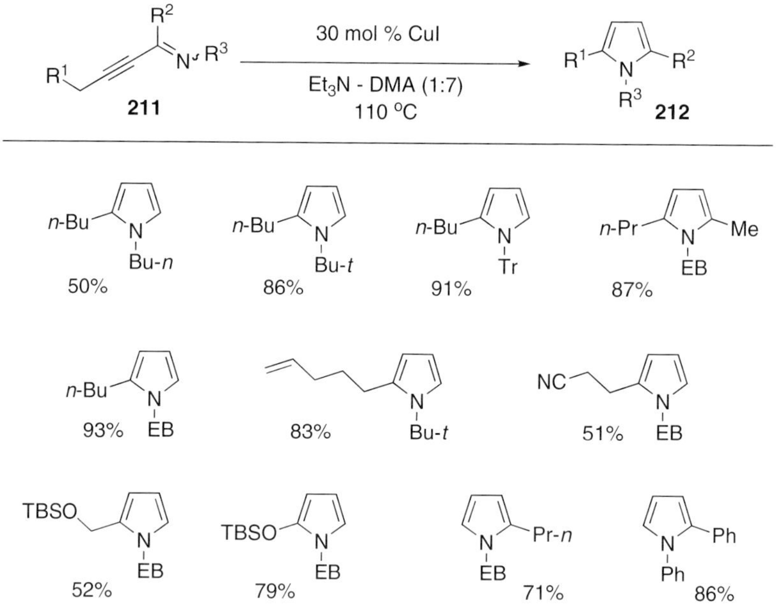

Scheme 8.78

The same group reported a novel approach toward the not so easily available 1,2,3-trisubstituted pyrroles (Scheme 8.79) [118, 124]. Accordingly, alkynyl imines **213**, possessing a sulfanyl group at the propargylic C4-position, underwent the Cu(I)-

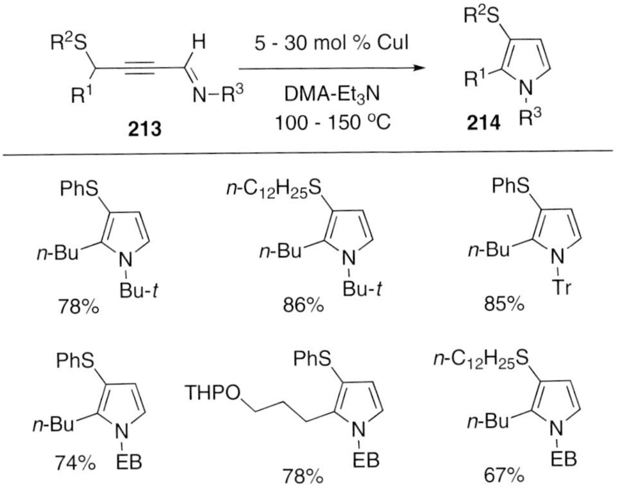

Scheme 8.79

catalyzed cycloisomerization proceeding with 1,2-migration of alkyl- and arylthio groups in the presence of amine base to afford 3-thiopyrroles **214** in high yields. It should be noted, that the alkylsulfanyl group migrated with efficiency comparable to that of its phenylsulfanyl analog, affording the corresponding pyrroles **214** in good yields.

In a recent report from the Gevorgyan group, this approach was adopted for an efficient synthesis of 1,2,3-trisubstituted 3-selenylpyrroles **216** proceeding via the Cu(I)-catalyzed cycloisomerization/1,2-Se migration cascade of propargyl selenides **215** (Scheme 8.80) [124]. In both cases, the proposed mechanism for the Cu(I)-catalyzed cycloisomerization of chalcogen-containing propargyl imines **213** and **215** is similar to that reported for an analogous migratory transformation of carbonyl compounds into furans.

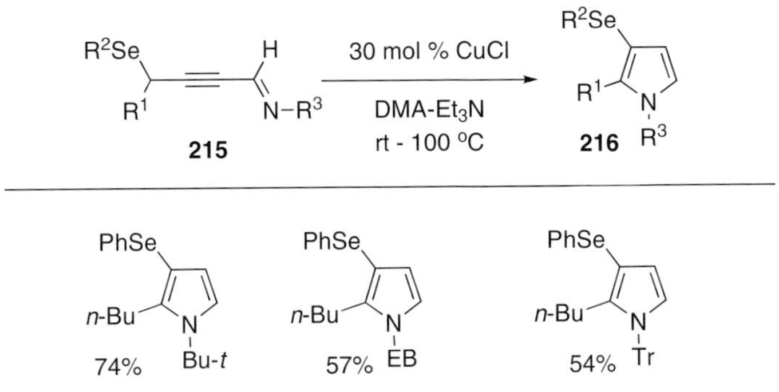

Scheme 8.80

Analogously to the precedent transition metal-catalyzed cycloisomerization of (*Z*)-pent-2-en-4-yn-1-ols like **113** into furan compounds (Scheme 8.45), Gabriele disclosed a general and a very convenient route leading to pyrroles **218** from nitrogen analogs of these reactive precursors (Scheme 8.81) [269]. Thus, di-, tri-, and tetra-substituted pyrroles **218** with different substitution patterns could be readily synthesized via the Cu(I)- or Cu(II)-catalyzed [270] cycloisomerization of (*Z*)-(2-en-4-ynyl)amines **217**. The mechanism of this transformation is similar to that proposed for the transition metal-catalyzed synthesis of furans (Scheme 8.46).

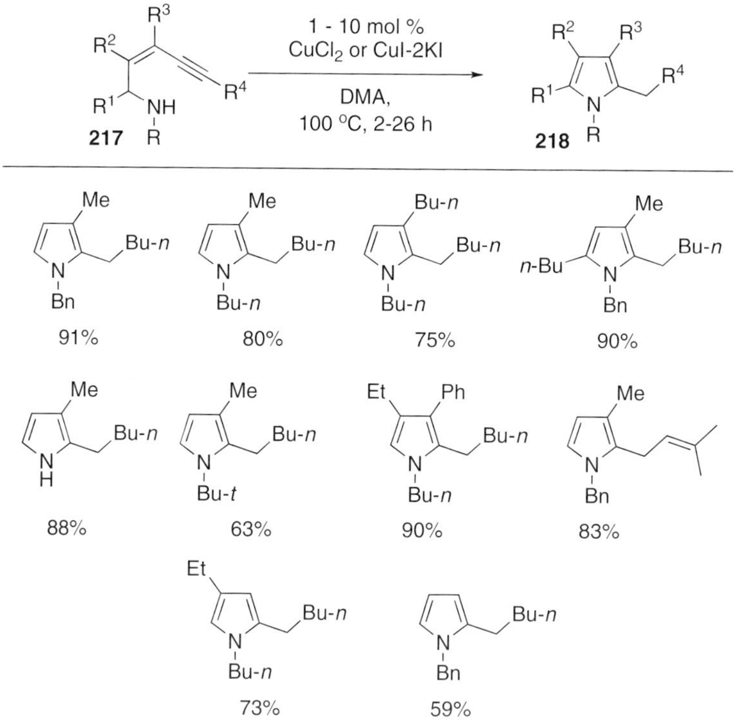

Scheme 8.81

The Au(I)-catalyzed version of Gabriele's protocol for the synthesis of pyrroles was recently reported by Gagosz *et al.* (Scheme 8.82) [271]. It was demonstrated that *N*-tosyl-protected (*Z*)-(2-en-4-ynyl)amines **219** could undergo a facile cycloisomerization into pyrroles **220** in the presence of cationic bis(trifluoromethanesulfonyl) imidate Au(I) complexes, very robust catalysts. Notably, tri- and tetra-substituted pyrroles **220** were synthesized in excellent yields and under very mild conditions within 5 min reaction time.

Furthermore, *N*,*N*-disubstituted (*Z*)-(2-en-4-ynyl)amines **221**, possessing an allyl group, underwent the Au(I)-catalyzed cycloisomerization with a 1,3-allyl shift affording tri- and tetra-substituted pyrroles **222** (Scheme 8.83) [271]. This transformation allowed efficient assembly of C2-homoallyl-substituted pyrroles bearing various

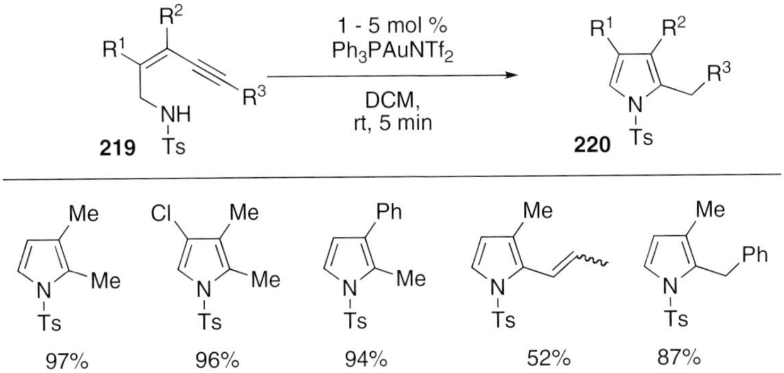

Scheme 8.82

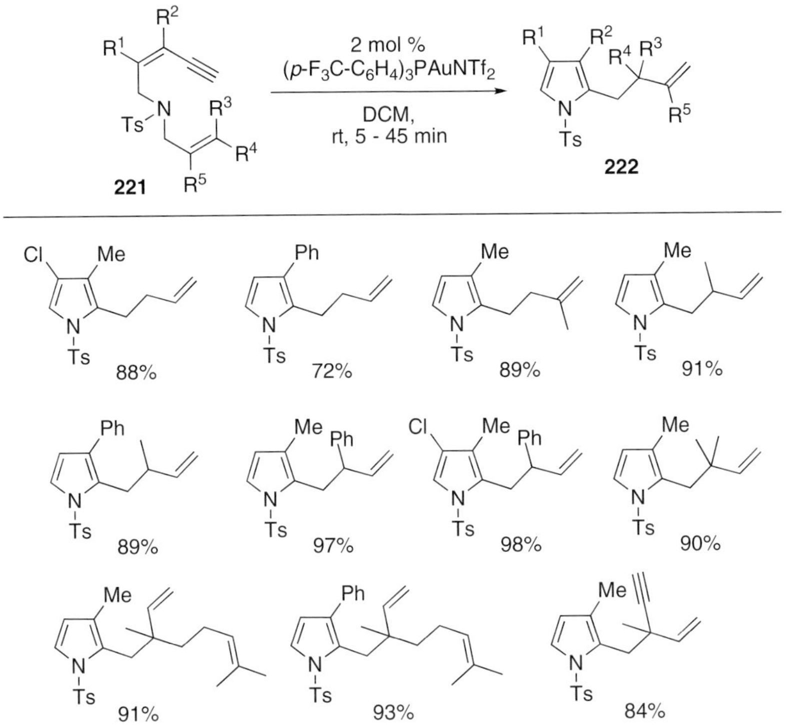

Scheme 8.83

functional groups in good to excellent yields. The proposed mechanism is depicted in Scheme 8.84 and involves the initial activation of the alkyne moiety of **221** toward an intramolecular nucleophilic attack by nitrogen atom (**223**) to give the cyclic vinylgold intermediate **224**. A subsequent Au(I)-catalyzed aza-Claisen type rearrangement produces an iminium intermediate **225**, which, upon proton elimination and protiodeauration of the generated **228**, furnishes the corresponding pyrrole **222**.

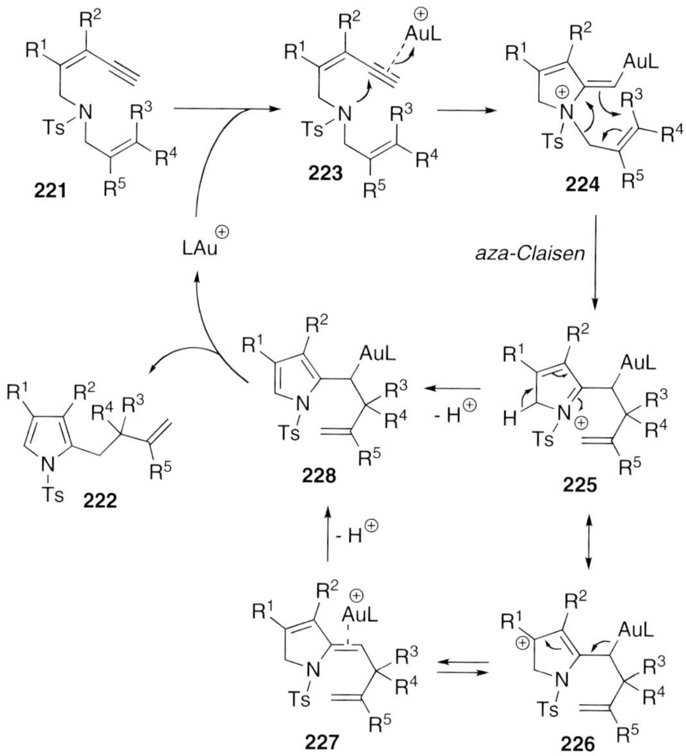

Scheme 8.84

An alternative route for the formation of pyrrolylmethylgold intermediate **228** engages a competitive elimination of the Au(I)-catalyst from **225** or **226** to form a Au(I)-complexed tosylenamide **227** followed by a proton loss.

An interesting approach toward the synthesis of N1-C2-fused pyrroles was recently elaborated by Zhang (Scheme 8.85) [272]. A variety of tri- and tetra-substituted pyrroles **230** were synthesized via the Au(I)-catalyzed cycloisomerization of (*Z*)-(2-en-4-ynyl)lactams **229** proceeding via a fragmentation of the former lactam moiety and followed by a subsequent carbocyclization [273]. The proposed mechanism is depicted in Scheme 8.86. Similarly to Gagosz's mechanistic proposal, *N,N*-disubstituted (*Z*)-(2-en-4-ynyl)amine derivatives **229** underwent a sequence of steps upon catalysis with the Au(I) complex, producing a key intermediate **232**. A lactam ring opening in the latter generated a vinylgold species **233** which, upon carbocyclization and aromatization, furnished the pyrrole **230**.

Dovey utilized propargyl enamines **236**, regioisomeric to (*Z*)-(2-en-4-ynyl)amines employed by Gabriele, in the Ag(I)-catalyzed synthesis of tetrasubstituted pyrroles **237** (Scheme 8.87) [274]. Thus, 3-acylpyrroles **237**, differently substituted at the N1-site, were obtained in generally high yields and within short reaction times using microwave irradiation [275–279]. This methodology was successfully applied to the synthesis of *N*-bridgehead pyrroles **239** (Scheme 8.88) [280]. It is believed that

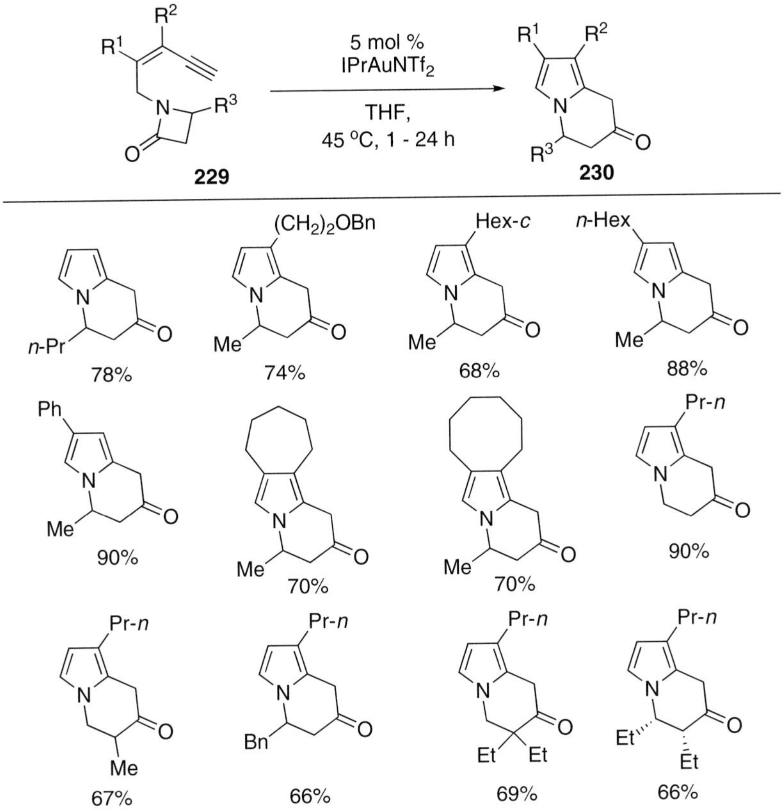

Scheme 8.85

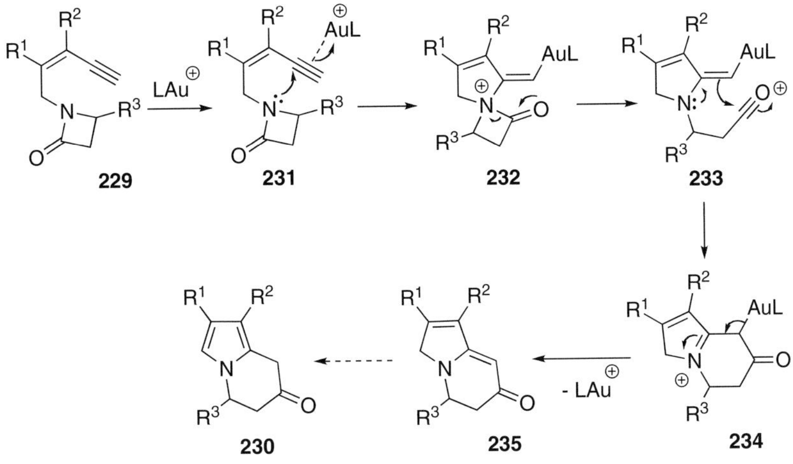

Scheme 8.86

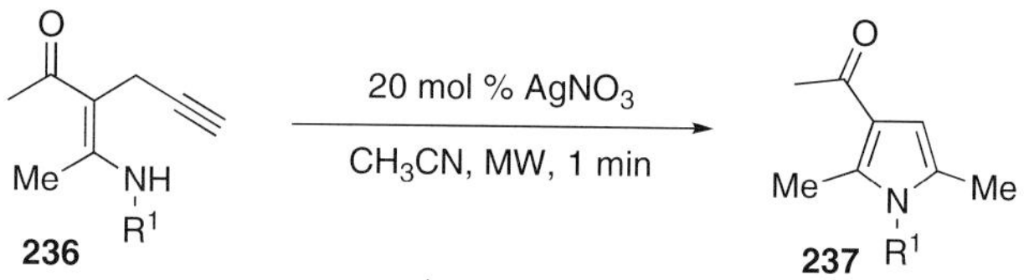

236 237

R^1	Yield, %
Ph	78
Me	93
n-Bu	91
c-Hex	96

Scheme 8.87

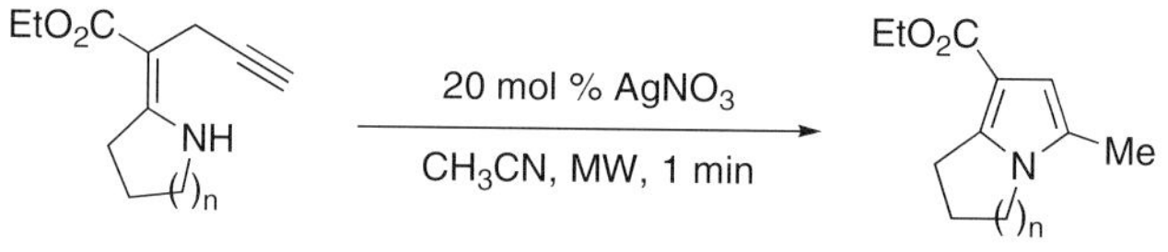

238 239

n	Yield, %
1	75
2	75
3	71

Scheme 8.88

mechanistically this transformation is similar to the general mechanism proposed for intramolecular nucleophilic addition of heteroatoms to transition metal-activated carbon–carbon multiple bonds (*vide supra*).

Among many recently introduced methods for assembly of heterocyclic N-containing frameworks, including aromatic pyrrole cores, employment of organic azides in transition metal-catalyzed cycloisomerization reactions received increasing attention over the last years. Toste first demonstrated that homopropargylic azides **240**, in the presence of a cationic Au(I) catalyst, could undergo an acetylenic Schmidt reaction providing pyrroles **241** in moderate to high yields (Scheme 8.89) [281]. This protocol allowed rapid assembly of up-to-tetrasubstituted and even fused *N*-unprotected pyrroles, possessing a variety of labile functional groups. A mechanistic hypothesis involving the Au(I)-induced activation of the alkyne moiety in **242** toward nucleophilic attack by the proximal nitrogen atom of the azide is outlined in Scheme 8.90. Subsequent fragmentation of **243** with the loss of dinitrogen produces a cationic intermediate **244**, which can be considered as a pyrrolydenegold carbenoid. The latter undergoes a formal 1,2-migration of the R^3-group to the gold carbenoid center to furnish the 2*H*-pyrrole **245** and regenerate an Au(I) catalyst. Finally, tautomerization of **245** gives 1*H*-pyrrole product **241** (Scheme 8.90).

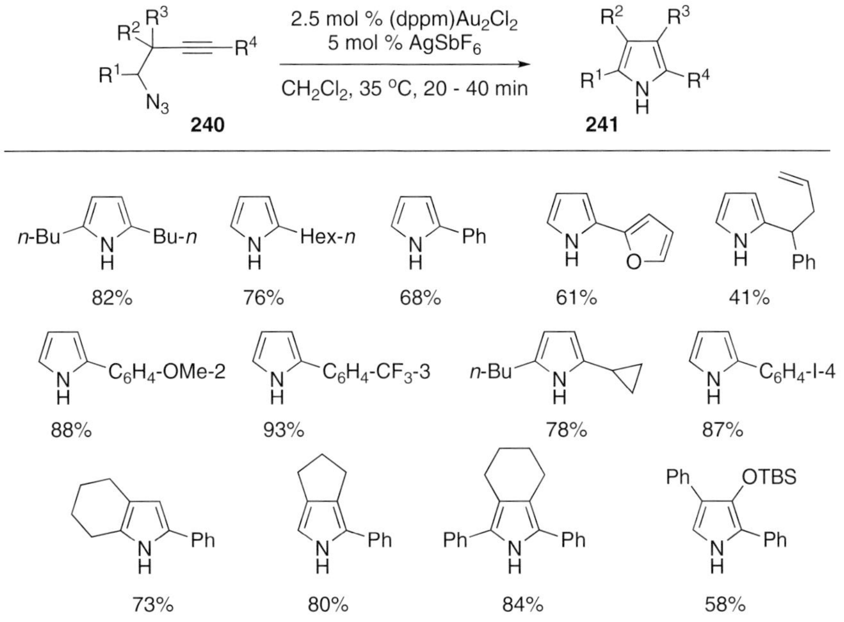

2.5 mol % (dppm)Au2Cl2
5 mol % AgSbF6
CH2Cl2, 35 °C, 20 - 40 min
240
241
82%
76%
68%
61%
41%
88%
93%
78%
87%
73%
80%
84%
58%

Scheme 8.89

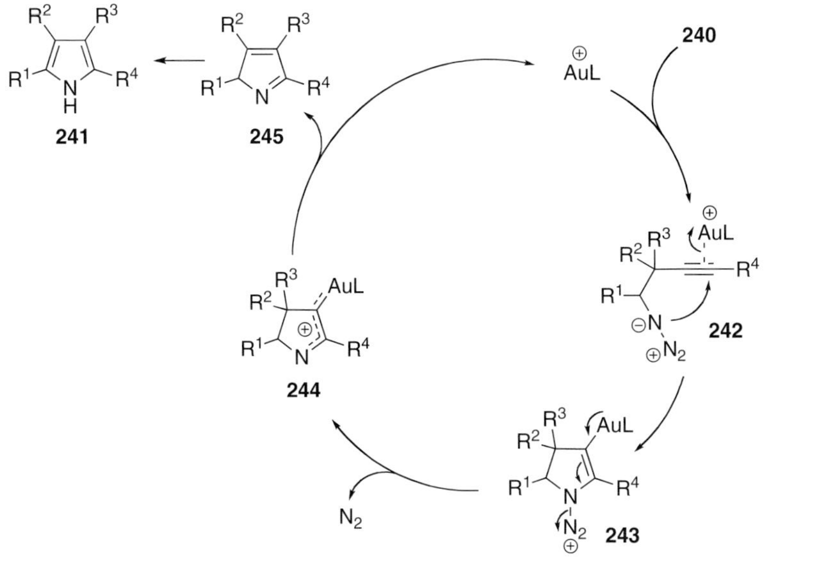

241
245
240
AuL
242
244
243
N2

Scheme 8.90

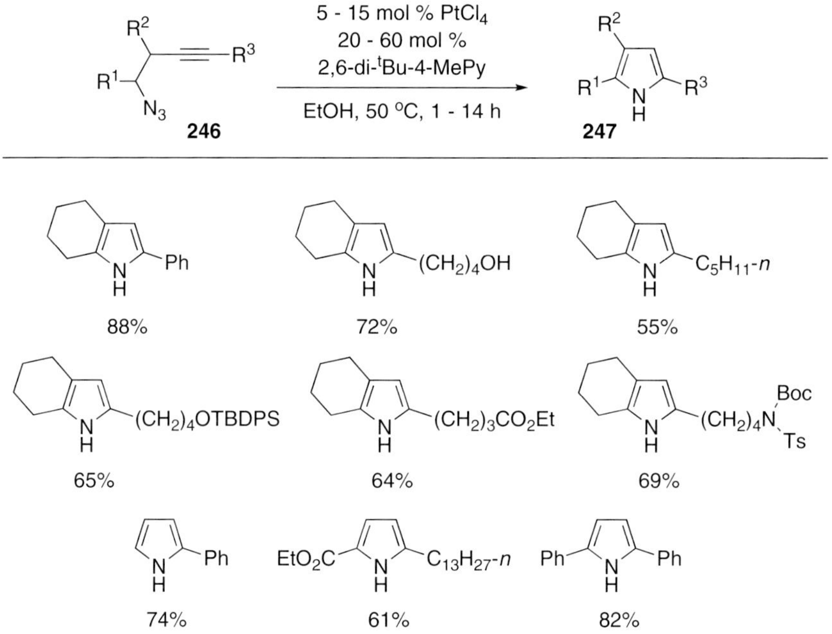

Scheme 8.91

Later, Hiroya reported that an analogous transformation of homopropargylic azides **246** leading to pyrroles **247** could be achieved with the Pt(IV)-pyridine ligand catalytic system (Scheme 8.91) [282]. Various mono-, di-, and tri-substituted pyrroles were obtained in moderate to high yields under these reaction conditions.

Recently, employment of 1,3-dienone azides **248** in a highly efficient Zn(II)- or Rh(II)-catalyzed synthesis of NH-pyrroles **249**, bearing ester or carbonyl groups at the C2 position, was reported by Driver and coworkers (Scheme 8.92) [283]. The reported mild reaction conditions for this transformation allowed the preparation of di- and tri-substituted alkyl- and arylpyrroles, possessing an array of different functional groups. It was demonstrated that this reaction could also be efficiently catalyzed by Cu(I)- and Cu(II) triflates. The authors proposed several mechanisms for the transformation of 1,3-dienyl azides **248** into pyrroles **249** (Scheme 8.93). Accordingly, the Rh(II)-catalyzed cycloisomerization proceeds via a nitrogen atom transfer through nitrenoid [13] **252** followed by the insertion into a vinylic C–H bond (**253**) (Path A). Alternatively, with the Lewis acidic Zn(II)-catalyst reaction begins with an initial activation of the azide in **251**, which then cyclizes into **253** with the loss of dinitrogen (Path B). In both routes, **253** liberates transition metal catalyst to produce 2*H*-pyrrole **250**, which further undergoes tautomerization to afford NH-pyrrole **249**. The observed higher reactivity trend for substrates bearing electron-donating C4-aryl groups, compared to that with electron-deficient substituents in the case of the Zn(II)-catalyst, supports involvement of the Schmidt-like mechanism B in this reaction (Scheme 8.93).

248 → **249**: 5-10 mol % ZnI_2 or 3 mol % $Rh_2(pfb)_4$; CH_2Cl_2 or PhMe, rt - 40 °C, 15 h

R^1	Yield, %
Ph	93
4-MeO-C_6H_4	83
2-MeO-C_6H_4	96
2,3-$(MeO)_2C_6H_3$	92
2-Me-C_6H_4	90
3-Me-C_6H_4	77
4-*t*-Bu-C_6H_4	90
4-Br-C_6H_4	97
2-Cl-C_6H_4	97
2,5-$Cl_2C_6H_3$	98
3-F-C_6H_4	98
4-F_3C-C_6H_4	85
2-naphthyl	71
2-furyl	21
2-thienyl	41

71% 86% 89% 72% 72% 99% 75%

Scheme 8.92

8.2.2.2 Synthesis of Pyrroles via "4 + 1" Cycloaddition Reactions

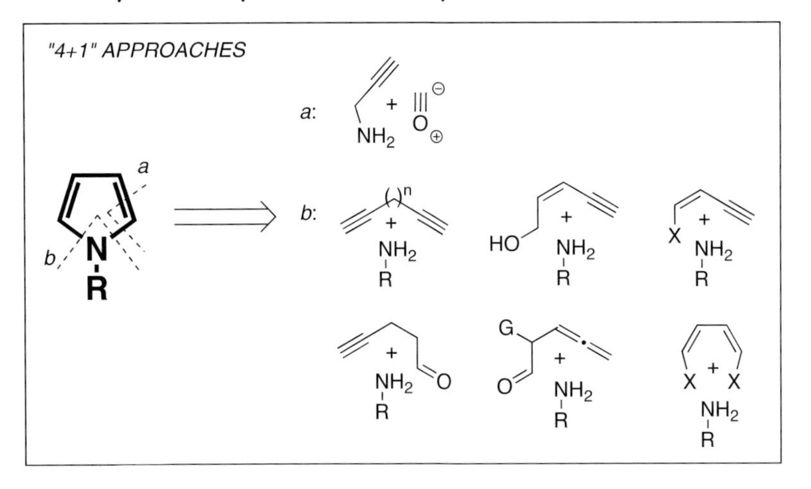

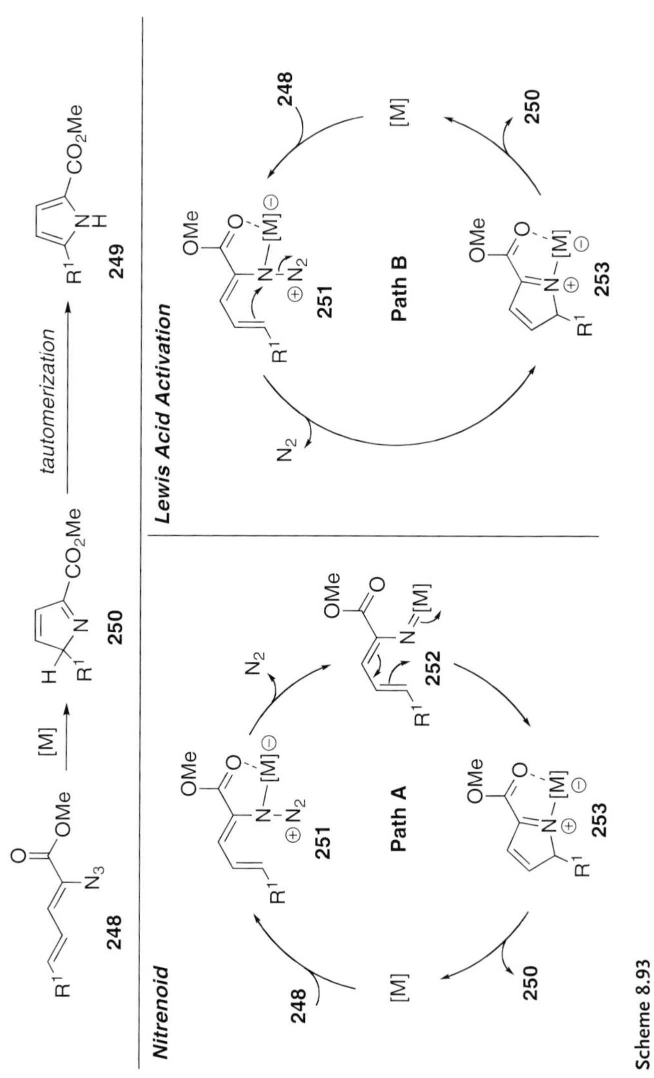

R1
O
OMe
N3
248
[M]
H
R1
N
CO2Me
250
tautomerization
R1
N
H
CO2Me
249
Nitrenoid
Lewis Acid Activation
248
[M]
250
251
Path A
252
253
N2
Path B

Scheme 8.93

A "4 + 1" approach for the synthesis of pyrroles **255** through Rh(II)-catalyzed hydroformylation of propargylic amines **254** was reported by Campi and coworkers. This transformation proceeded with an *in situ* generation of γ-aminoenals **256** which, upon cyclocondensation, provided the corresponding 3-mono- and 2,4-disubstituted aryl- and alkylpyrroles **255** (Scheme 8.94) [284].

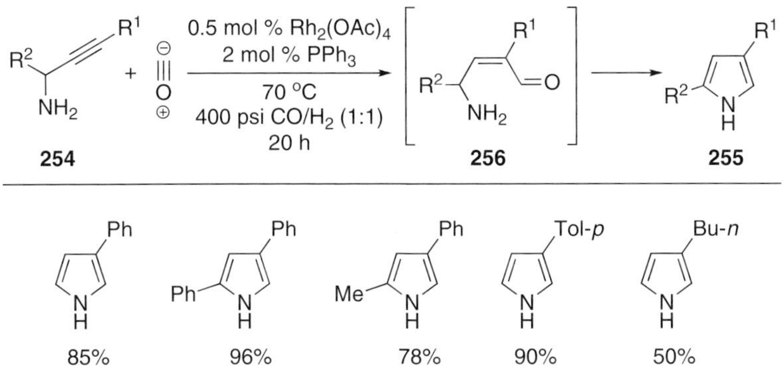

Scheme 8.94

Schulte [285, 286] and later Chalk [287] described the Cu(I)-catalyzed synthesis of symmetrical 2,5-diarylpyrroles **261** from conjugated diynes **257** and primary amines **258**. The reaction is believed to proceed via the transition metal-catalyzed hydroamination [33, 288–291] leading to tautomeric aminoenyne **259** or homopropargylic imine **260** intermediates, which further undergo 5-*endo-dig* cyclization to furnish pyrrole product **261** (Scheme 8.95).

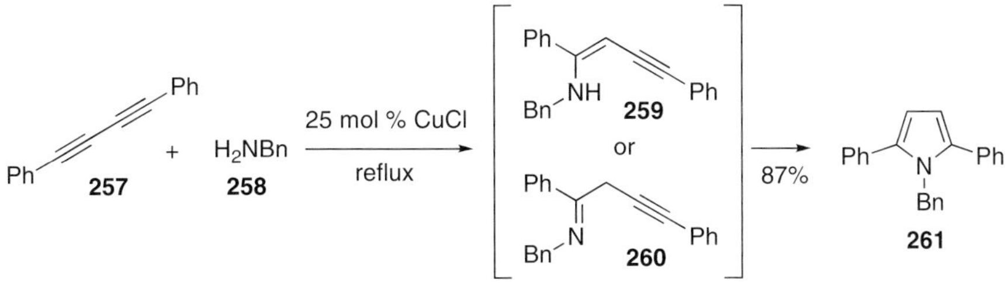

Scheme 8.95

Later, Odom extended this methodology to the Ti(IV)-catalyzed syntheses of pyrroles from skipped 1,*n*-diynes [292]. In the case of 1,4-diynes **262**, monohydroamination [288] occurred with Markovnikov selectivity, yielding intermediary propargylic imines **264** which were transformed into pyrroles **265** through a subsequent 5-*endo-dig* cyclization (Scheme 8.96). Similarly, reaction of 1,5-diynes **266** with primary amines **263** produced γ-iminoalkynes **267**. The Ti(IV)-catalyzed 5-*exo-dig* cyclization of the latter gave the corresponding pyrroles **268** (Scheme 8.97). Both routes tolerated employment of various primary amines and provided moderate to high yields of 1,2,5-trisubstituted pyrroles.

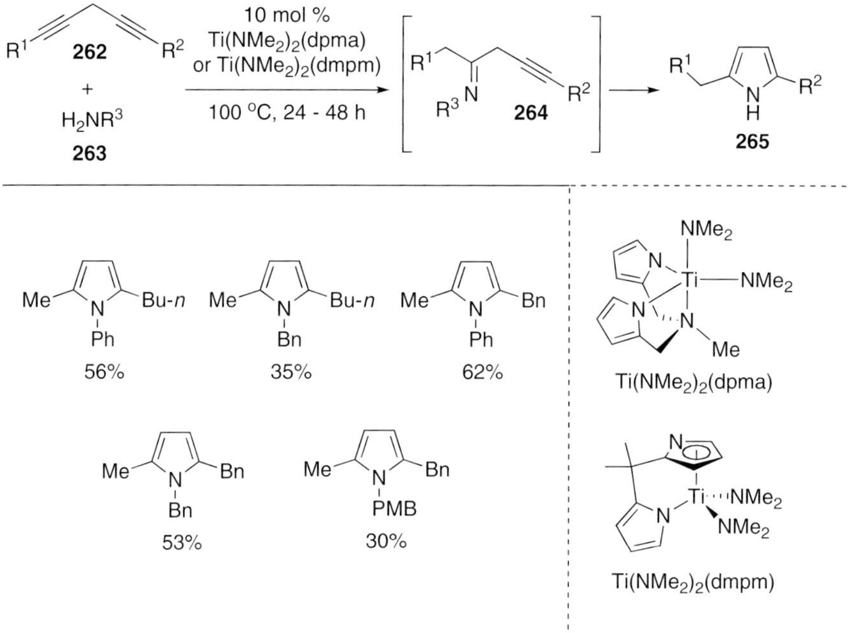

Scheme 8.96

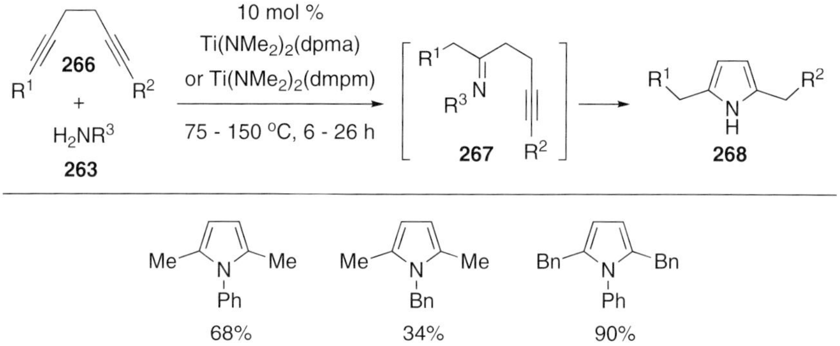

Scheme 8.97

Bertrand recently described the Au(I)-catalyzed version of these transformations involving 1,*n*-enynes leading to symmetrically 2,5-disubstituted pyrroles **269** and **270** (Scheme 8.98) [293].

In their recent work, Liang's group disclosed a convenient protocol for accessing complex pyrrole structures based on a transition metal-catalyzed S_N1 or S_N2'-like substitution reaction of allylpropargylic alcohols with primary amine derivatives (Scheme 8.99) [294]. Thus, easily accessible 1-en-4-yn-3-ols **272**, upon Au(III)-

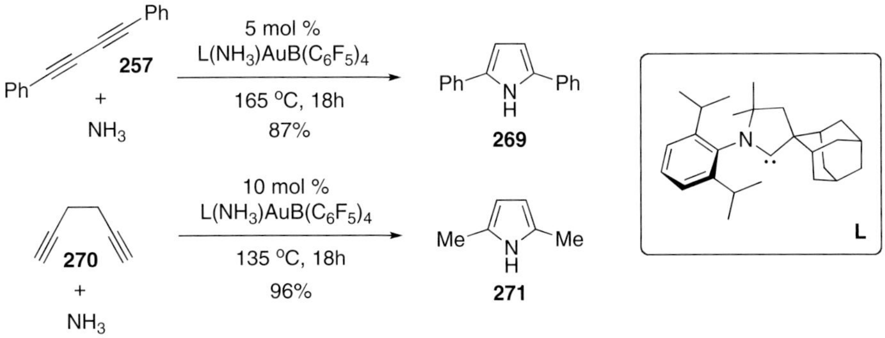

Scheme 8.98

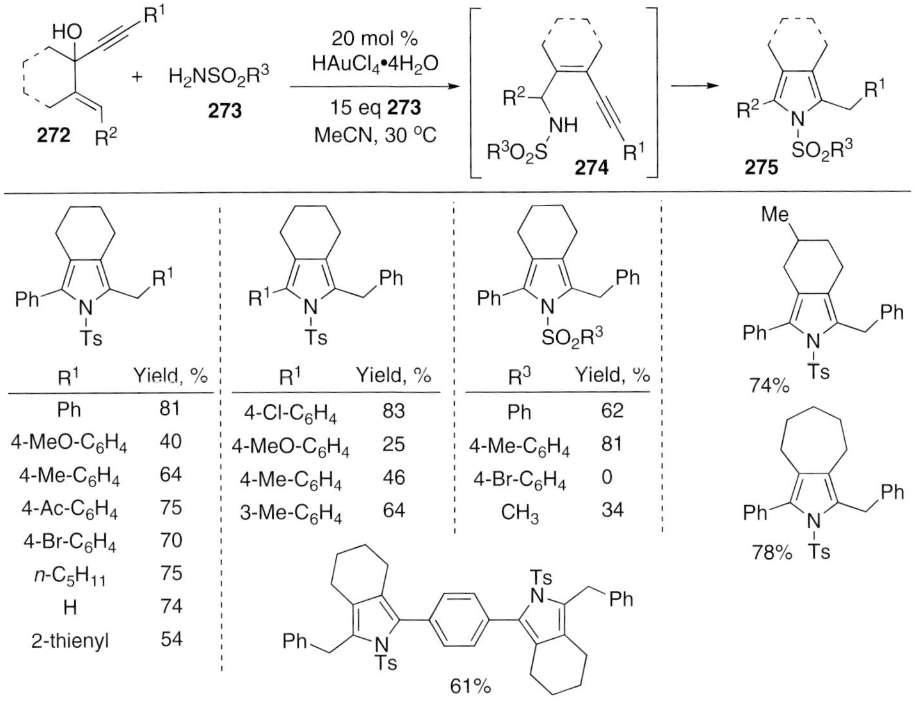

Scheme 8.99

R^1	Yield, %
Ph	81
4-MeO-C_6H_4	40
4-Me-C_6H_4	64
4-Ac-C_6H_4	75
4-Br-C_6H_4	70
n-C_5H_{11}	75
H	74
2-thienyl	54

R^1	Yield, %
4-Cl-C_6H_4	83
4-MeO-C_6H_4	25
4-Me-C_6H_4	46
3-Me-C_6H_4	64

R^3	Yield, %
Ph	62
4-Me-C_6H_4	81
4-Br-C_6H_4	0
CH_3	34

catalyzed amination with sulfonamides **273**, were transformed into the key (Z)-(2-en-4-ynyl)amines **274** (*vide supra*). The latter readily underwent Au(III)-catalyzed cycloisomerization into pyrroles **275**, possessing various functional groups. Although, the choice of N-nucleophiles was limited to sulfonamides. These soft nucleophiles possessed enough nucleophilicity, but did not deactivate the Au-catalyst, whereas other harder nucleophiles, such as amides and amines, were inefficient in this transformation.

Recently, Liu disclosed a more general "4 + 1" approach toward the synthesis of pyrroles, which combined an initial Au-catalyzed amination of (*Z*)-enynols **276** followed by cycloisomerization of *in situ* generated (*Z*)-(2-en-4-ynyl)amines **278** into pyrroles **279** in a cascade fashion (Scheme 8.100) [295]. This methodology provides a highly efficient and regioselective entry into fused, as well as acyclic tri-, tetra- and penta-substituted, pyrroles **279**. A variety of pyrroles bearing different functional groups could be accessed in moderate to high yields under very mild reaction conditions and from accessible starting materials. Unlike Liang's protocol, this cascade transformation proceeded in the presence of somewhat lower (fourfold) excess of different N-nucleophiles such as *p*-nitroaniline and tosylamine.

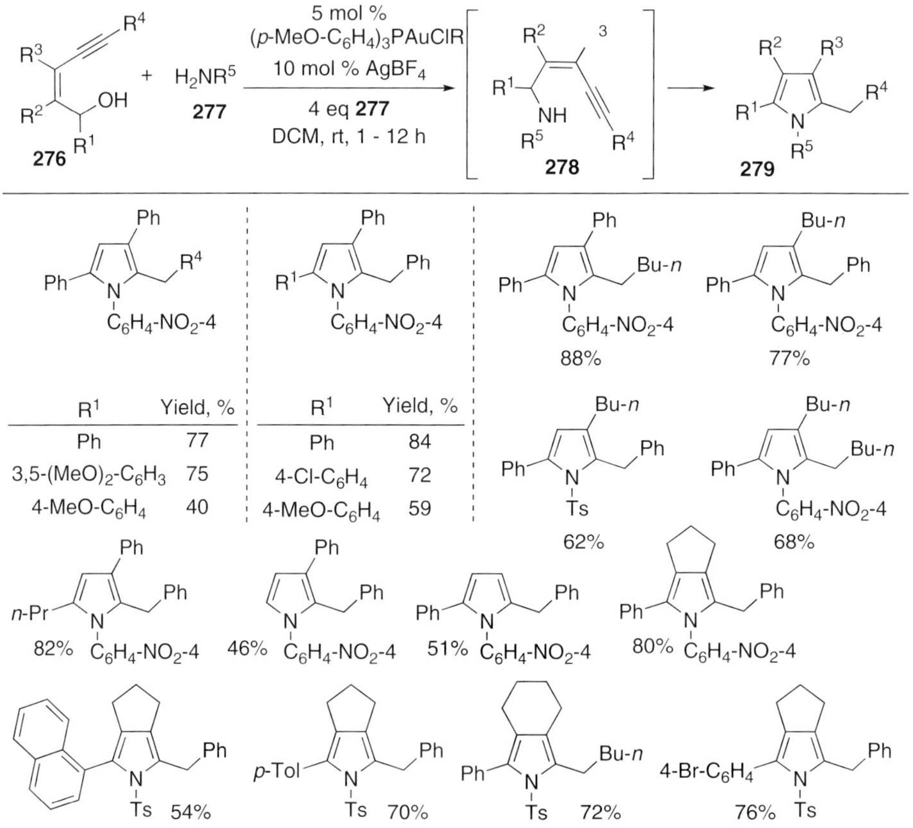
5 mol %
(p-MeO-C6H4)3PAuClR
10 mol % AgBF4
4 eq 277
DCM, rt, 1 - 12 h
276
277
278
279
H2NR5
R1 Yield, %
Ph 77
3,5-(MeO)2-C6H3 75
4-MeO-C6H4 40
R1 Yield, %
Ph 84
4-Cl-C6H4 72
4-MeO-C6H4 59
88%
77%
62%
68%
82%
46%
51%
80%
54%
70%
72%
76%

Scheme 8.100

Buchwald elaborated on a novel "4 + 1" Cu(I)-catalyzed protocol for the synthesis of variously substituted pyrroles **283** (Scheme 8.101) [296] utilizing 5-*endo-dig* cyclization of aminoenynes **282**, analogously to the protocol developed by Dovey (Schemes 8.87 and 8.88). The key reactive intermediates of this transformation,

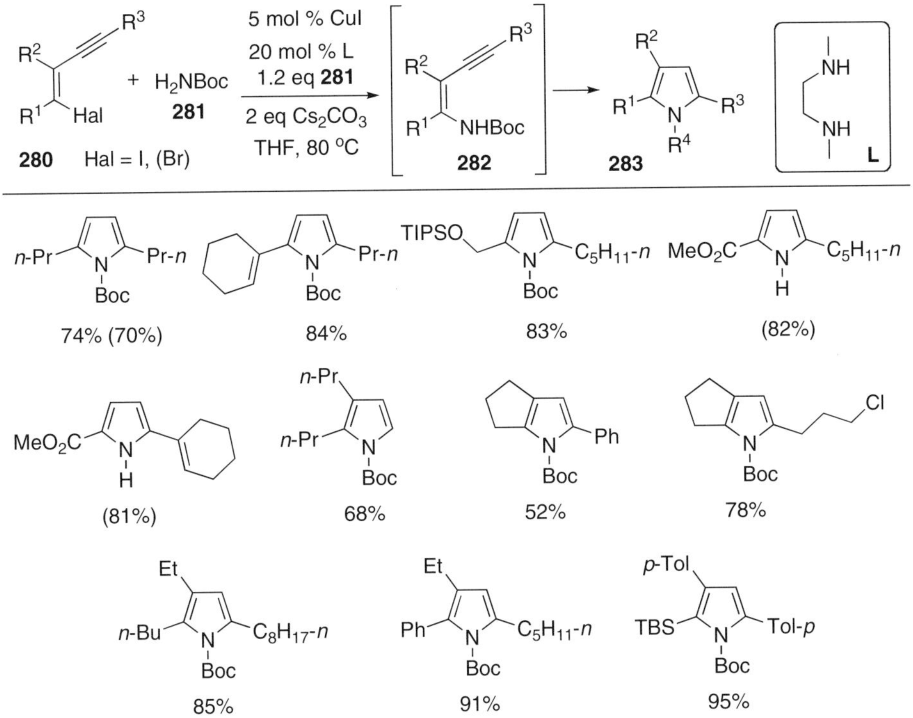

Scheme 8.101

enynes **282**, were accessed via the Cu(I)-catalyzed [7, 297–299] amidation of haloenynes **280** with *tert*-butyl carbamate **281**. The authors demonstrated that *N*-deprotected pyrroles could be accessed by performing the reaction in the presence of two equivalents of K_2CO_3 in toluene at more elevated temperatures (110 °C). In addition, both bromo- and iodo-enynes **280** could be employed as the cross-coupling partners in this transformation.

Arcadi *et al.* reported that 1,2,3,5-tetrasubstituted acyl- and carbethoxypyrroles **287**, possessing various functional groups, could be synthesized from homopropargylic 1,3-dicarbonyl compounds **284** possessing a terminal alkyne moiety and an array of primary amine derivatives **285** as N-nucleophiles under catalysis by gold salts (Scheme 8.102) [300, 301]. It was proposed that the Au(III)-catalyzed amination of 1,3-diketones or ketoesters **284** initially produced the corresponding imines or enamines **286**. Au(III)-catalyzed cycloisomerization of these reactive intermediates (*vide supra*) leads to pyrrole products **287** in moderate to quantitative yields. This domino reaction could also be catalyzed by Cu(I)- and Ag(I) salts, albeit requiring prolonged reaction times. Employment of enantiomerically pure chiral amines, β-amino alcohols, and α-amino esters **288** in this protocol gave the corresponding pyrroles **289** with perfect chirality transfer (Scheme 8.103) [301].

Dake extended this approach to the synthesis of pyrroles **293** from monocarbonyl-containing skipped propargylic ketones **290** catalyzed by Ag- or Au(I)-triflates

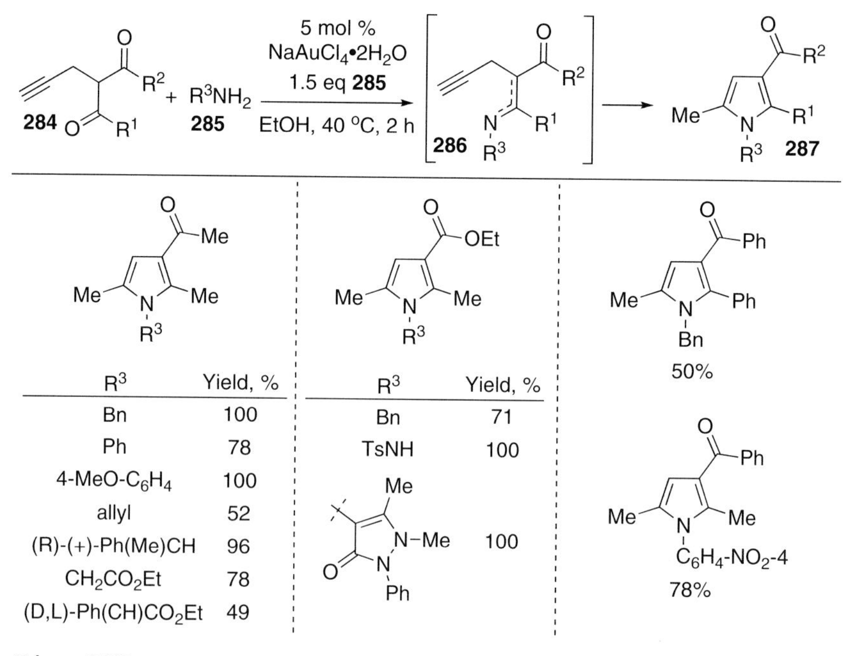

R^3	Yield, %
Bn	100
Ph	78
4-MeO-C_6H_4	100
allyl	52
(R)-(+)-Ph(Me)CH	96
CH_2CO_2Et	78
(D,L)-Ph(CH)CO_2Et	49

R^3	Yield, %
Bn	71
TsNH	100
(pyrazolone substituent)	100

Scheme 8.102

Scheme 8.103

(Scheme 8.104) [302]. Moderate to high yields of tri- and tetra-substituted, and even fused pyrroles **293** were obtained using this chemistry. Various amines were used as N-nucleophiles in this transformation, providing the corresponding *N*-functionalized pyrroles **293**. However, the reaction with sterically demanding amines **291** provided diminished yields of the target products, and in the case of *tert*-butylamine no product formation was observed. Notably, only stoichiometric amounts of ketone and amine components were employed in this reaction. Similarly, Hidai and Uemura reported that an analogous transformation could be performed in the presence of Pt(II) catalyst [143].

A recent report from Kirsch's group illustrated that vinyl propargyl ethers **36**, as surrogates of skipped allenyl ketones **37**, could be employed in a very efficient synthesis of densely substituted pyrroles **295** via the Ag/Au(I)-catalyzed

R^4	Yield, %
Bn	87
allyl	80
n-C_5H_{11}	77
Ph	59
c-C_6H_{11}	27
t-Bu	0
Ts	37
4-MeO-$C_6H_4CH_2$	74
4-MeO-C_6H_4	60

Scheme 8.104

condensation/cycloisomerization reaction (Scheme 8.105) [303] Hence, upon Ag-catalyzed rearrangement of **36** into **37**, the latter undergoes amination with a primary amine **294** to give the skipped allenyl imine or enamine intermediate which, upon Au(I)-catalyzed 5-*exo-dig*-cyclization, provides pyrrole **295**. Using this protocol, a variety of pyrroles possessing different labile groups could be rapidly obtained in good to high yields under mild reaction conditions. In addition, the reaction was demonstrated to be quite general with respect to the choice of an aromatic amine component **294**, though this transformation was not efficient with aliphatic amines (R^4 = Me, *i*-Pr, Bn). Moreover, this cascade transformation seems to be somewhat limited to propargyl vinyl ether substrates derived from primary propargylic alcohols possessing alkyl substituents at the triple bond.

Recently, Wang reported a novel approach for the assembly of up-to-pentasubstituted pyrroles **299** via an acid-catalyzed cascade transformation of skipped allenyl aldehydes **296** and anilines **297**. It was further demonstrated that this transformation could also be efficiently catalyzed by Au(I)- and Ag salts. Surprisingly, a sulfanyl group in **298** underwent an intramolecular 1,2-migration via the proposed thiiranium intermediate yielding the corresponding 2-thiopyrroles **299** (Scheme 8.106) [304].

A number of "4 + 1" protocols for the synthesis of pyrrole cores featuring the Cu(I)-catalyzed vinylation of primary amine derivatives as the key carbon–heteroatom bond forming reaction have been reported recently. Thus, Buchwald described an efficient Cu(I)-catalyzed synthesis of tri-, tetra-, and penta-substituted pyrroles **302** from 1,4-dihalo-1,3-dienes **300** and carbamates **281** (Scheme 8.107) [305]. This methodology displayed excellent functional group compatibility, providing good to

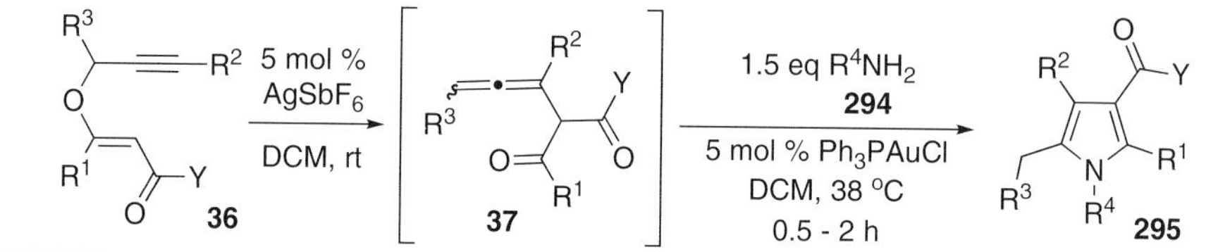

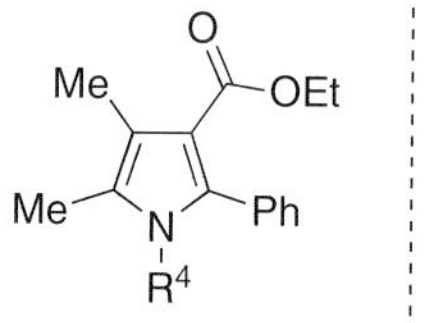

R^4	Yield, %
Ph	71
4-MeO-C_6H_4	75
4-*i*-Pr-C_6H_4	73
4-HO-C_6H_4	41
4-Br-C_6H_4	74
3-Cl-C_6H_4	83
2-*i*-Pr-C_6H_4	67
3-O_2N-C_6H_4	55
2-MeO_2C-3-thienyl	52
1-naphthyl	72
-C_6H_4-C_6H_4-	31

R^1	R^2	R^3	R^4	Yield, %
Me	Ph	H	Ph	75
Me	Ph	H	4-MeO-C_6H_4	75
Me	2-MeO-C_6H_4	H	Ph	83
n-C_5H_{11}	Me	H	Ph	68
Ph	Ph	H	Ph	72
Ph	3-thienyl	H	Ph	74
Ph	CH_2-C_6H_{11}-*c*	H	Ph	90
Ph	CH_2-C_6H_{11}-*c*	H	1-naphthyl	79
Ph	$(CH_2)_3$OTHP	H	Ph	70
Ph	H	H	Ph	52
Ph	TBS	H	Ph	77

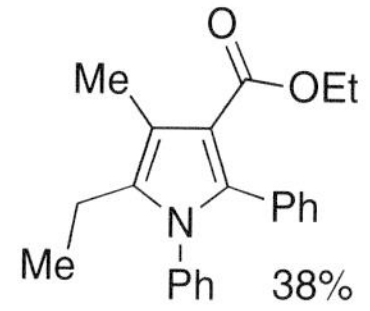

Scheme 8.105

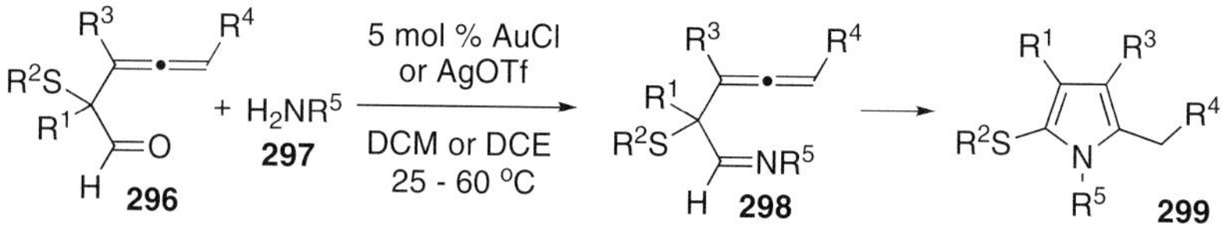

Scheme 8.106

excellent yields of the target pyrroles **302**, including fused ones. The reaction is believed to occur via two sequential Cu(I)-catalyzed inter- and intra-molecular coupling processes, as depicted in Scheme 8.107.

Further, Li reported an analogous Cu(I)-catalyzed double alkenylation of amides **304** with 1,4-diiodo-1,3-dienes **303** leading to the formation of tri- and tetra-substituted *N*-acylpyrroles **305** (Scheme 8.108) [306]. Interestingly, in contrast to Buchwald's approach, employment of alkylcarbamates as the N-nucleophile components in a catalytic reaction provided low yields of the pyrrole products. However, this limitation could be alleviated by performing the reaction in the presence of stoichiometric amounts of CuI and 1,2-diamine ligand **L**.

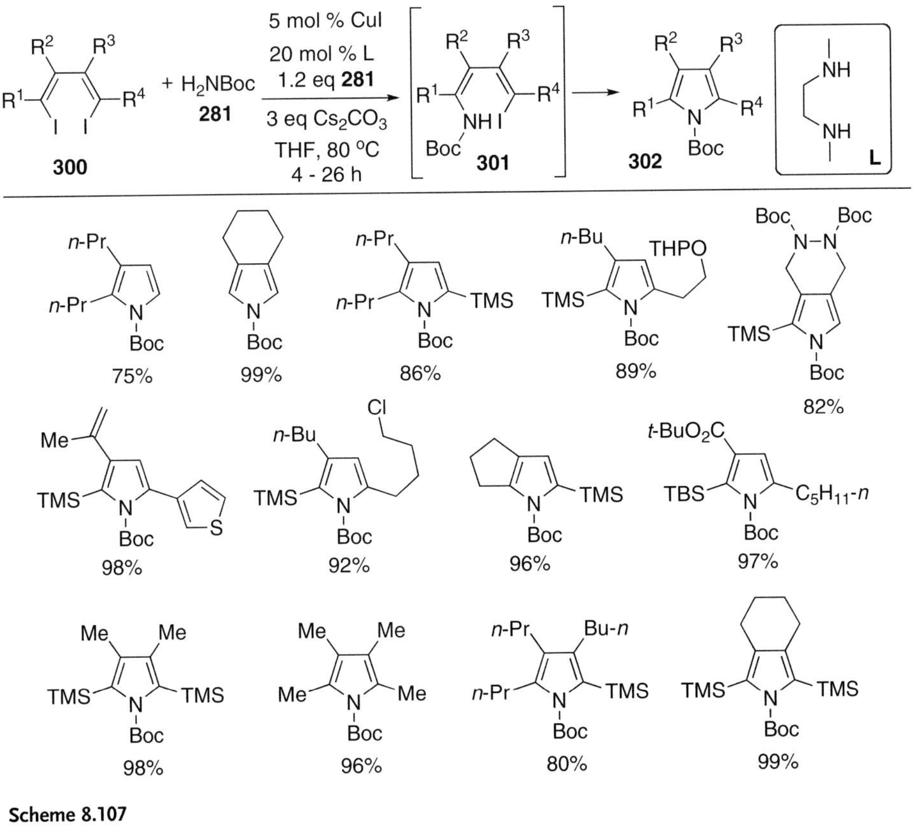

R^2 R^3 R^1 R^4 I I **300** + H_2NBoc **281**

5 mol % CuI
20 mol % L
1.2 eq **281**
3 eq Cs_2CO_3
THF, 80 °C
4 - 26 h

[R^2 R^3 R^1 R^4 NH I Boc **301**] → R^2 R^3 R^1 N R^4 Boc **302**

NH NH **L**

n-Pr, n-Pr, N, Boc — 75%

N, Boc — 99%

n-Pr, n-Pr, N, TMS, Boc — 86%

n-Bu, THPO, TMS, N, Boc — 89%

Boc, Boc, N–N, TMS, N, Boc — 82%

Me, TMS, N, Boc, S — 98%

Cl, n-Bu, TMS, N, Boc — 92%

N, TMS, Boc — 96%

t-BuO_2C, TBS, N, C_5H_{11}-n, Boc — 97%

Me, Me, TMS, N, TMS, Boc — 98%

Me, Me, Me, N, Me, Boc — 96%

n-Pr, Bu-n, n-Pr, N, TMS, Boc — 80%

TMS, N, TMS, Boc — 99%

Scheme 8.107

8.2.2.3 Synthesis of Pyrroles via "3 + 2" Cycloaddition Reactions

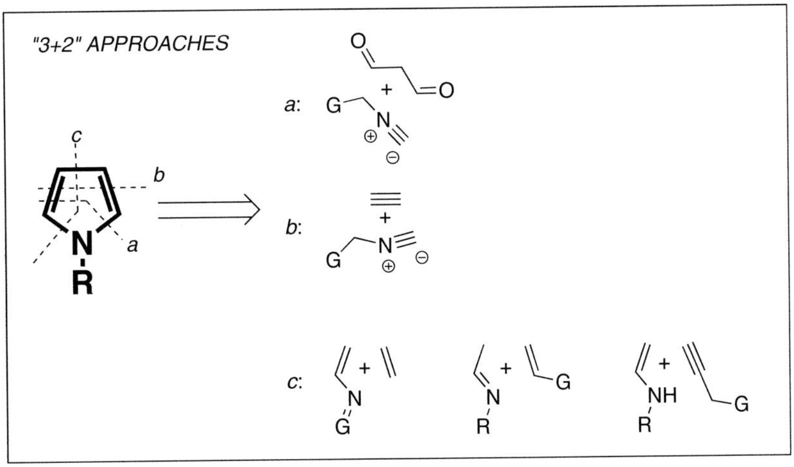

"3+2" APPROACHES

c, b, N, R, a

a: O, +, =O, G, N, ⊕, ⊖

b: +, G, N≡, ⊕, ⊖

c: +, N, G; +, N, R, G; +, NH, R, G

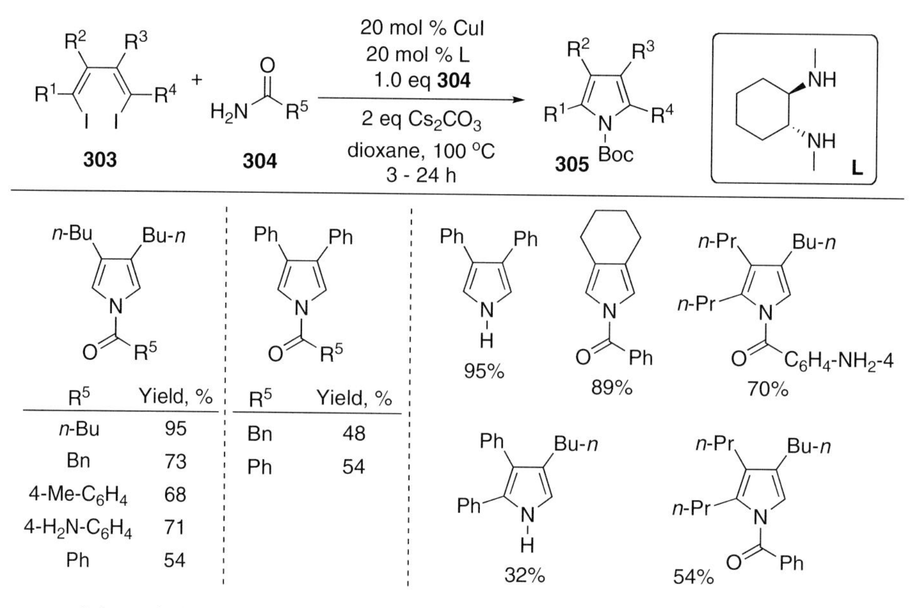

Scheme 8.108

Employment of isonitriles [307] in transition metal-catalyzed cascade transformations occupies a distinctive place among many recently introduced methodologies aimed at facile assembly of five-membered nitrogen-containing heterocycles, including multisubstituted pyrroles. Thus, an interesting Rh(0)-catalyzed approach toward pyrrole **308** from 1,3-dicarbonyl compounds **306** and isonitriles **307** was developed by Murahashi (Scheme 8.109) [308]. The proposed mechanism involves a Rh(0)-catalyzed α-C–H bond activation [85] of isonitriles (Scheme 8.110). Coordination of isonitriles **307** to the Rh(0)-catalyst gives complex **309**, C–H activation, of **309** yields isocyanoalkylrhodium hydride intermediate **310**. Addition of the latter to a dicarbonyl **306** is followed by either reductive elimination or protonation to afford **309** and **312**. The latter undergoes a dehydration–hydration sequence to give formamidine **313**, which, upon Rh-catalyzed decarbonylation (**314**) and cyclocondensation, produces the corresponding pyrrole **308** (Scheme 8.110). The regioselective addition of Rh-species **310** across unsymmetrical 1,3-dione **306** is dictated by stereo- and/or electronic effects of substituents at the carbonyl groups in 1,3-diketone **306**.

Yamamoto reported the Cu(I)-catalyzed synthesis of 2,4-di- and 2,3,4-tri-substituted pyrroles **317** (Scheme 8.111) [309]. Target products were obtained in modest to good yields via a "3 + 2" cycloaddition reaction between isonitriles **315** and activated alkynes **316**. Steric congestion around the triple bond of **316** decreases the reaction efficiency. It was also demonstrated that alkynes activated with electron-withdrawing groups other than carbalkoxy were somewhat inefficient in this cascade transformation. However, much better compatibility of electron-withdrawing substituents on isocyanide **315** was observed. The reaction mechanism is outlined in Scheme 8.112.

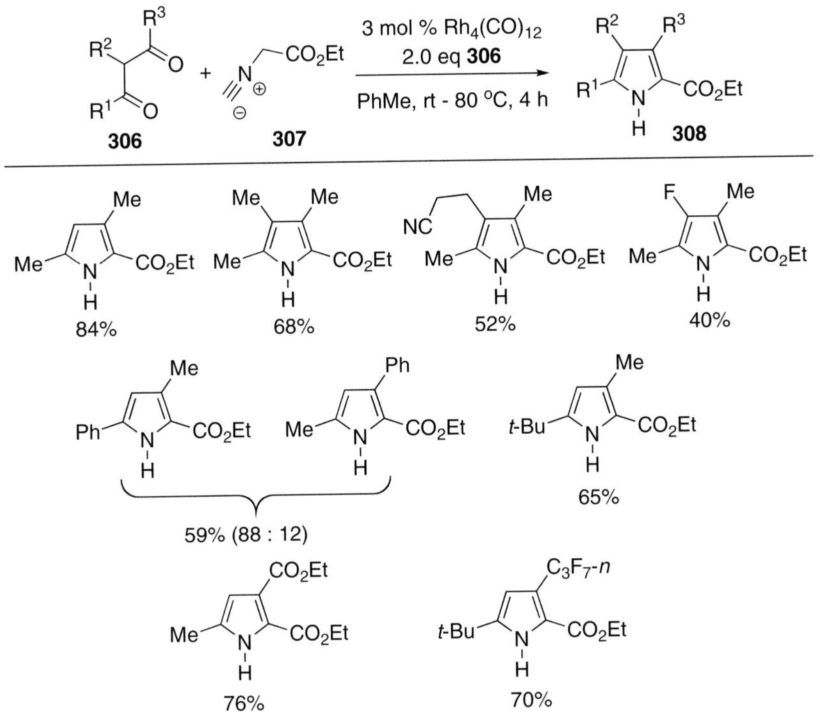

Scheme 8.109

It is believed that this transformation begins with a C–H activation of isocyanide by the Cu(I)-catalyst. Organocopper intermediate **318** or its tautomer **319** is generated from **315** and Cu_2O by extruding a water molecule. Michael addition of **318** and/or **319** to alkyne **316** produces a Cu-enolate, which undergoes an intramolecular cyclization to give the cyclic organocopper intermediate **320**. The latter species, upon protonation by isocyanide **315**, gives 2*H*-pyrrole **321** and returns active intermediates **318** to the catalytic cycle. A subsequent [1,5]-H shift in **321** furnishes pyrrole **317** (Scheme 8.112).

Analogous chemistry was independently described by de Meijere (Scheme 8.113) [310, 311]. The best results were achieved with Cu(I) benzenethiolate or preactivated nanosized copper powder (Cu^0-NP) catalysts. These conditions allowed better functional group compatibility and generally led to the formation of pyrroles **324** with a higher efficiency.

A major limitation of Yamamoto's and the early de Meijere's protocols is the employment of EWG-activated alkynes to achieve synthetically useful yields of pyrrole products. A recent work from de Meijere's group revealed that this issue can be resolved by employing stoichiometric amounts of a slightly modified Cu(I) system. Accordingly, terminal alkynes **326** were shown to undergo the Cu(I)-mediated formal [3 + 2] cycloaddition reaction with isocyanides **325**, affording the

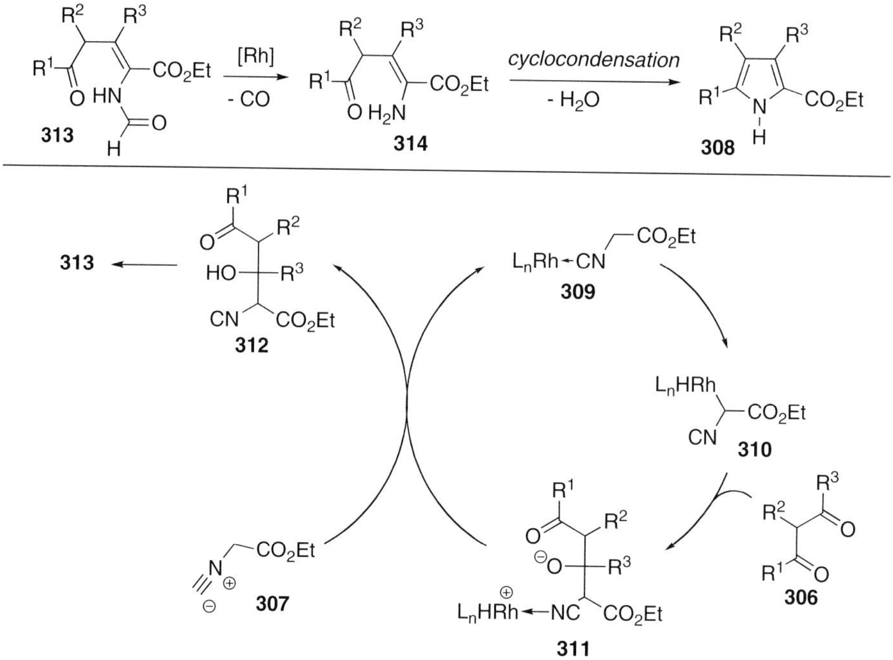

Scheme 8.110

corresponding 2,3-disubstituted pyrroles **327** in moderate to high yields (Scheme 8.114) [311].

Recently, two novel protocols for efficient syntheses of multisubstituted pyrroles employing vinyl azides were reported by Narasaka. Hence, it was shown that α–azidoacrylates **328**, upon Cu(II)-catalyzed reaction with 3 equiv of ethyl acetoacetate **329**, could be easily transformed into tri- and tetra-substituted pyrroles **330** (Scheme 8.115) [312]. A variety of functional groups were tolerated under the reaction conditions, though the presence of a carbalkoxy group, geminal to the azide function in **328**, was necessary to achieve an efficient pyrrole formation. According to the proposed mechanism, the reaction is initiated by the 1,4-addition of copper enolate **329** to the Cu(II)-activated vinyl azide **328**. A concomitant elimination of dinitrogen affords the alkylidene aminocopper intermediate **331**, which undergoes an intramolecular cyclization into the cyclic hemiaminal copper intermediate **332**. The latter, upon a dehydration–isomerization sequence, furnishes the pyrrole **330** (Scheme 8.116).

Later, Narasaka and coworkers reported that the scope of this transformation could be substantially improved with the use of Mn(III) catalysis [313]. These new reaction conditions allowed highly efficient employment of differently substituted simple non-activated alkyl-, aryl-, hetaryl- and even cyclic vinyl azides **328** (Scheme 8.117). In addition, previously unreactive 1,3-diketones **335** could serve as feasible 1,3-dicarbonyl components in this formal [3 + 2] cycloaddition reaction, affording 2,3,5-tri- and 2,3,4,5-tetra-substituted pyrroles **336** in moderate to excellent yields

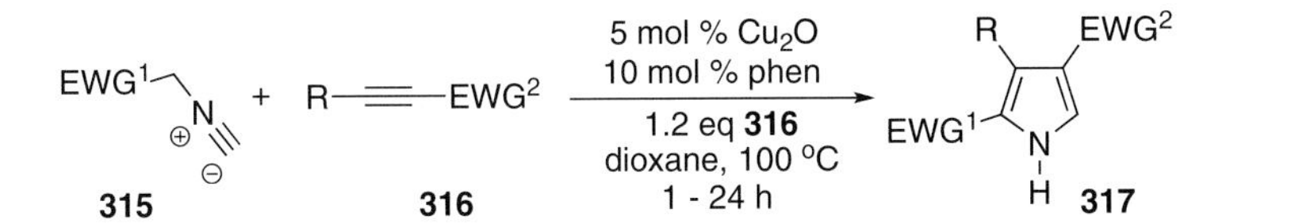

EWG^1–CH$_2$–$N^{\oplus}\equiv C^{\ominus}$ (**315**) + R–≡–EWG^2 (**316**) → [5 mol % Cu_2O, 10 mol % phen, 1.2 eq **316**, dioxane, 100 °C, 1 - 24 h] → **317**

R	EWG^1	EWG^2	Yield, %
Me	CO_2Et	CO_2Et	64
$HO(CH_2)_4$	CO_2Et	CO_2Et	65
c-C_6H_{11}	CO_2Et	CO_2Et	73
t-Bu	CO_2Et	CO_2Et	26
Ph	CO_2Et	CO_2Et	79
H	CO_2Et	CO_2Et	73
CO_2Et	CO_2Et	CO_2Et	43
Ph	CO_2Et	COMe	22
Ph	CO_2Et	$CONEt_2$	11
Ph	CO_2Et	CN	22
Ph	CO_2Et	SO_2Ph	13
Me	CO_2Me	CO_2Et	54
Me	CO_2Bu-t	CO_2Et	71
Me	$CONEt_2$	CO_2Et	75
Me	$P(O)(OEt)_2$	CO_2Et	59
Me	Ph	CO_2Et	46

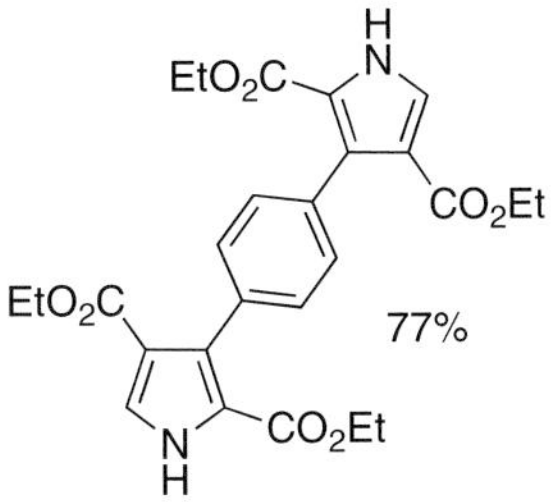

EtO_2C, CO_2Et, EtO_2C, CO_2Et (bis-pyrrole product) 77%

Scheme 8.111

(Scheme 8.118). The Mn(III)-catalyzed reaction is believed to occur via a radical mechanism.

Buchwald disclosed a domino approach to pyrroles based on the Piloty–Robinson synthesis involving [3,3]-rearrangement of divinylhydrazides **341**. The latter key intermediates were accessed via two sequential Cu(I)-catalyzed vinylations of bis-Boc-hydrazine **338** (Scheme 8.119) [314]. The scope of this transformation is displayed in Scheme 8.120. In general, when electron-withdrawing groups were attached to the pyrrole core, the carbamate group was deprotected *in situ*, affording the corresponding NH-pyrrole products. In some cases only partial cleavage of the Boc-group occurred. This synthetic route accommodates a variety of substituents and provides modular access to the pyrrole unit from simple and readily available starting materials.

Ishii reported the Sm(III)-catalyzed route to various tri- and tetra-substituted pyrroles **347** employing "3 + 2" cyclocondensation between imines **345** and nitroalkenes **346**. (Scheme 8.121) [315]. The use of aldimines and ketemines with steric congestion at the imine functionality led to formation of the corresponding pyrrole products in diminished yields. According to the mechanistic proposal, the samarium

Scheme 8.112

complex acts as a base and activates imine **345** to give enaminosamarium intermediate **348**. The latter undergoes 1,4-addition to nitroalkene **346** to form adduct **349**, followed by cyclization to yield **350**. Finally, the pyrrole precursor **350** is transformed into the corresponding product upon the elimination of HNO and H_2O molecules (Scheme 8.122).

Ila and Junjappa utilized a facile cycloisomerization of transient skipped allenyl imines **354** in the Cu(I)-catalyzed two-component "3 + 2" synthesis of 2-thiopyrroles **355** (Scheme 8.123) [316]. The key intermediates **354** were accessed via the CuBr-promoted S_N2' substitution reaction of propargyl bromide **353** with N,S-acetals **352**. A subsequent 5-*endo-dig* cyclization afforded a variety of 1,2,3,5-tetrasubstituted and even 1,2-fused pyrroles in moderate to good yields.

8.2.2.4 **Synthesis of Pyrroles via "2 + 2 + 1" Cycloaddition Reactions**

$R^1CH_2N^{\oplus}\equiv C^{\ominus}$ (**322**) + $R^2-\equiv-EWG$ (**323**) → (5 mol % CuSPh or Cu^0-NP; 1.1 eq **322**; DMF, 85 °C, 16 h) → **324**

R^1	R^2	EWG	Yield, %
CO_2Me	c-Pr	CO_2Me	93
CO_2Me	c-Pr	CO_2Bu-t	94
CO_2Bu-t	Me	CO_2Me	83
Ts	c-Pr	CO_2Me	64
Ph	c-Pr	CO_2Bu-t	87
Ts	c-Pr	CO_2Bu-t	91
Ts	Me	CO_2Me	83
CO_2Me	CH_2OMe	$P(O)(OEt)_2$	47
CO_2Bu-t	Ph	CO_2Et	78
CO_2Me	4-EtO-C_6H_4	CO_2Me	75
CO_2Me	4-F-C_6H_4	CO_2Me	78
CO_2Me	4-F_3C-C_6H_4	CO_2Me	70
CO_2Me	2-pyridyl	CO_2Me	68
CO_2Me	2-thienyl	CO_2Me	94
CO_2Me	CH(OMe)Me	CO_2Me	54
CO_2Et	H	CO_2Me	37
Ts	H	CO_2Me	30
4-O_2N-C_6H_4	H	CO_2Me	44
Ph	H	CO_2Me	25
CO_2Me	(cyclopropyl)C(Me)-≡-CO_2Me	CO_2Me	91

Scheme 8.113

In their recent study, Scheidt and coworkers devised the Rh(II)-catalyzed methodology employing imines, activated alkynes and diazoacetonitrile (DAN) in a convergent three-component "2 + 2 + 1" assembly reaction leading to pyrrolo-3,4-dicarboxylates **359** in moderate to high yields (Scheme 8.124) [317]. The authors proposed that DAN reacts with the Rh(II) catalyst to generate the corresponding metallocarbenoid **360**. The latter, upon reaction with imine **356**, produces a reactive transient azomethine ylide **361**, which is intercepted via a Huisgen [3 + 2]-cycloaddition with an activated alkynyl dipolarophile **358**. Next, 2,5-dihydropyrrole adduct **362** undergoes elimination to furnish the pyrrole **359** (Scheme 8.125).

Hidai and Uemura reported that the bimetallic Ru and Pt(II) two-catalyst system could be applied to the "2 + 2 + 1" synthesis of pyrroles **364** from propargylic alcohols **195**, enolizable ketones **196**, and anilines **363** (Scheme 8.126) [143]. Moderate yields of up-to-fully substituted pyrroles **364** could be achieved via this approach, though a large excess of carbonyl compound and aniline were required to achieve full conversion. This transformation is believed to proceed via the Ru-catalyzed propargylic substitution of **195** with **196** (*vide infra*) to give doubly skipped alkynyl ketone

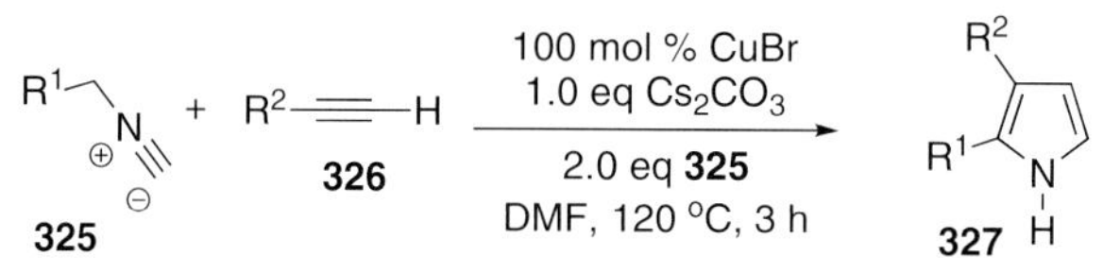

325 + 326 → 327 (100 mol % CuBr, 1.0 eq Cs_2CO_3, 2.0 eq 325, DMF, 120 °C, 3 h)

R^1	R^2	Yield, %
CO_2Et	Bu-*n*	70
CO_2Et	CH_2OMe	48
CO_2Et	CH(Me)OMe	74
CO_2Et	Ph	40
CO_2Et	*c*-Pr	88
CO_2Et	*t*-Bu	5
CO_2Et	2-Py	16
CO_2Et	*sec*-Bu	58
CO_2Et	CH_2CH_2OH	44
CO_2Bu-*t*	Bu-*n*	47
$4\text{-}O_2N\text{-}C_6H_4$	Bu-*n*	20
$CONEt_2$	Bu-*n*	0
$P(O)(OMe)_2$	Bu-*n*	0
Ts	Bu-*n*	0

Scheme 8.114

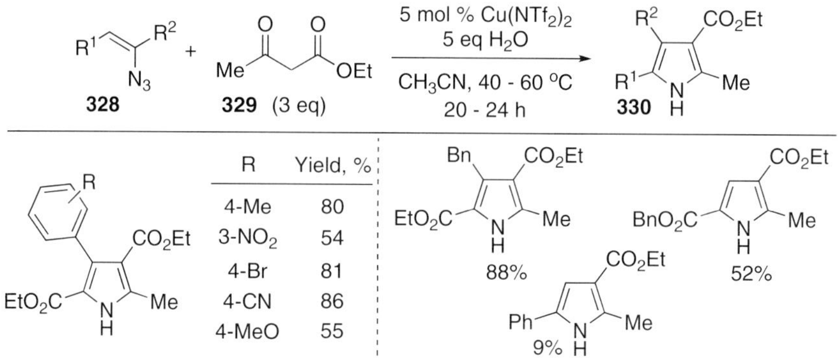

328 + 329 (3 eq) → 330 (5 mol % $Cu(NTf_2)_2$, 5 eq H_2O, CH_3CN, 40 - 60 °C, 20 - 24 h)

R	Yield, %
4-Me	80
$3\text{-}NO_2$	54
4-Br	81
4-CN	86
4-MeO	55

88% 52% 9%

Scheme 8.115

365, which, upon amination, furnishes the reactive intermediate **366**. The latter γ-iminoalkyne **366** undergoes Pt(II)-catalyzed cycloisomerization into the pyrrole product **364** (Scheme 8.127).

More recently, Nebra and Gimeno applied this concept to the synthesis of pentasubstituted 3-acylpyrroles **369** from 1,3-dicarbonyl compounds **368** [318]. The efficiency and the scope of this transformation were significantly improved by the use of the Ru(II) catalyst in the presence of trifluoroacetic acid co-promoter. The reaction tolerates 1,3-diketones, β-ketoesters, various secondary propargylic alcohols **195**

Scheme 8.116

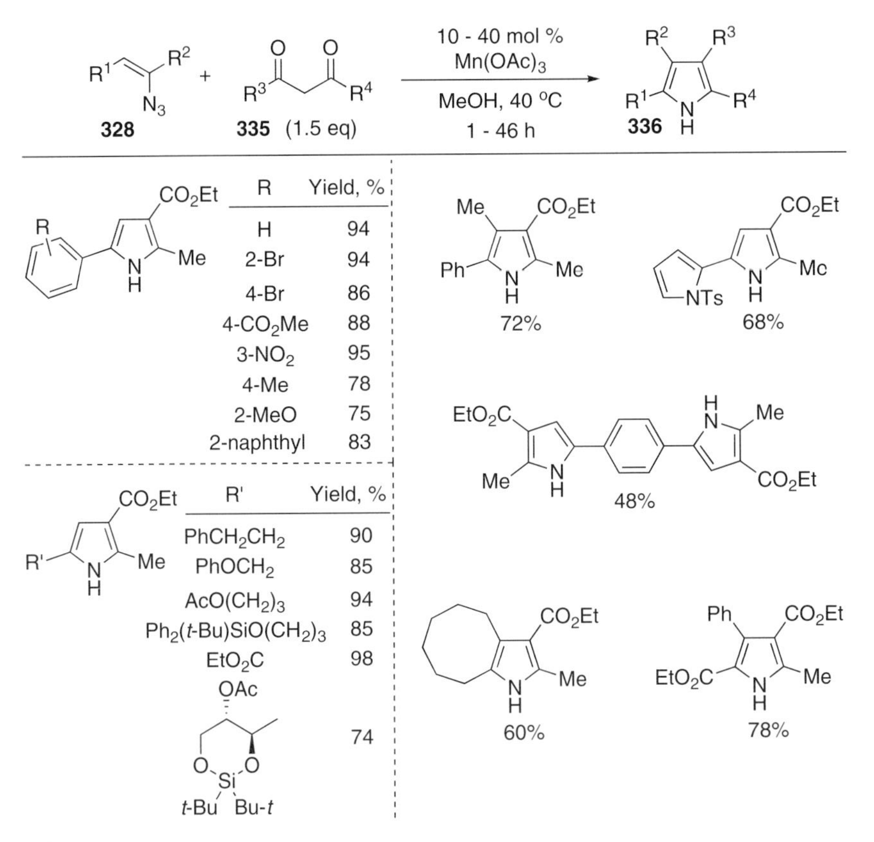

Scheme 8.117

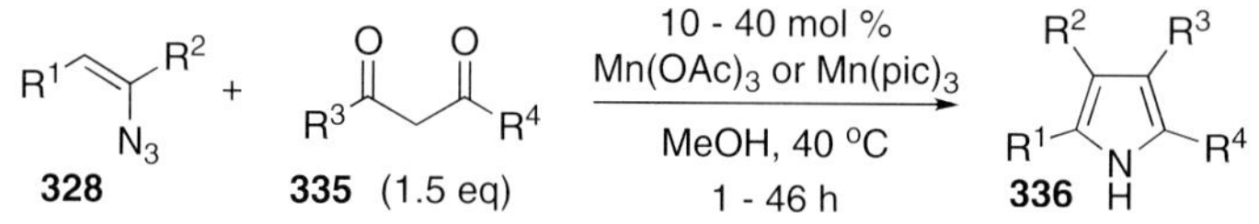

R^1	R^2	R^3	R^4	Yield, %
Ph	H	CO_2Et	Ph	63
CO_2Et	H	CO_2Et	Ph	72
Ph	H	CO_2Et	CH_2OEt	55
CO_2Et	H	CO_2Et	CH_2OEt	77
Ph	H	CO_2Et	*c*-Pr	56
CO_2Et	H	CO_2Et	*c*-Pr	72
Ph	H	COMe	Me	76
$3\text{-}O_2N\text{-}C_6H_4$	H	COMe	Me	80
$4\text{-Me-}C_6H_4$	H	COMe	Me	52
Ph	Me	COMe	Me	41
$AcO(CH_2)_3$	H	COMe	Me	71
Ph	H	COPh	Me	61

Scheme 8.118

(Scheme 8.128), and aromatic and aliphatic amines **367** (Scheme 8.129). In addition, the authors demonstrated that this protocol could be further extended to the syntheses of 1,2,3,5-tetrasubstituted- as well as NH-pyrroles by the *in situ* removal of various N- and C-protecting groups. According to the proposed mechanism, the Ru(II)-catalyzed three-component synthesis of pyrroles **369** proceeds via two competitive pathways occurring simultaneously and involving the amination reaction after or prior to the propargylic substitution step.

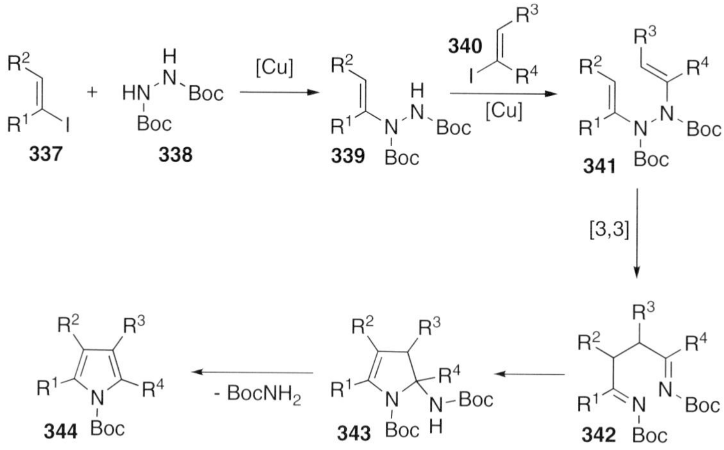

Scheme 8.119

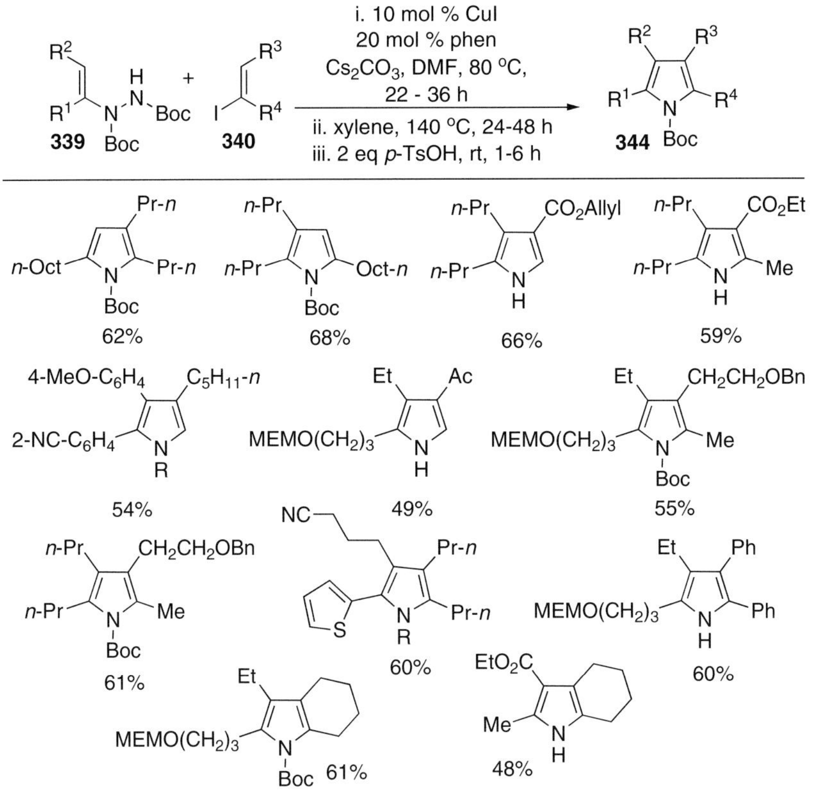

i. 10 mol % CuI, 20 mol % phen, Cs_2CO_3, DMF, 80 °C, 22 - 36 h
ii. xylene, 140 °C, 24-48 h
iii. 2 eq *p*-TsOH, rt, 1-6 h

339 + **340** → **344**

Scheme 8.120

8.3 Conclusion

The importance of aromatic heterocycles has long been recognized in organic chemistry from many perspectives. This chapter clearly indicates a growing interest in the development of novel general, highly efficient, convergent, and atom-economical approaches for the synthesis of furans and pyrroles. Evidently, employment of transition metals has revolutionized this area of organic chemistry allowing achievement of the desired new transformations in a highly chemo- and region-selective fashion and dramatically expanding the scope and applicability of the existing reactions. Throughout decades, employment of Ag, Cu, Hg, Pd, Rh, and Ru complexes allowed facile construction of furan and pyrrole heterocyclic frameworks with a diverse substitution patterns. In recent years, in addition to the continuing interest in these transition metals, the development of remarkably efficient Pt- and Au-catalyzed transformations has emerged as a new direction of research in this area of heterocycle synthesis.

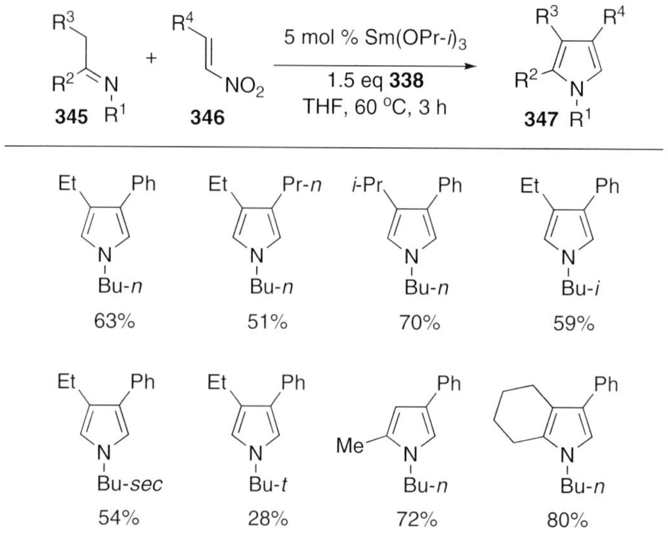

Scheme 8.121

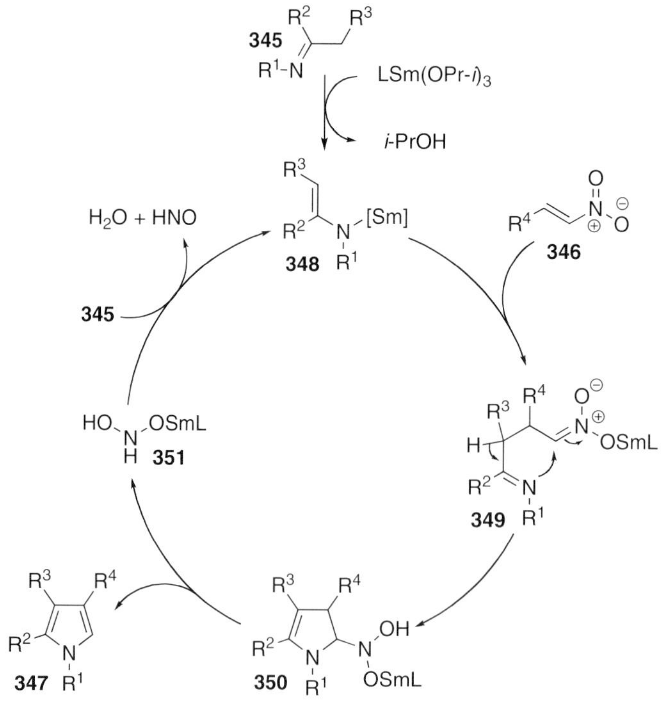

Scheme 8.122

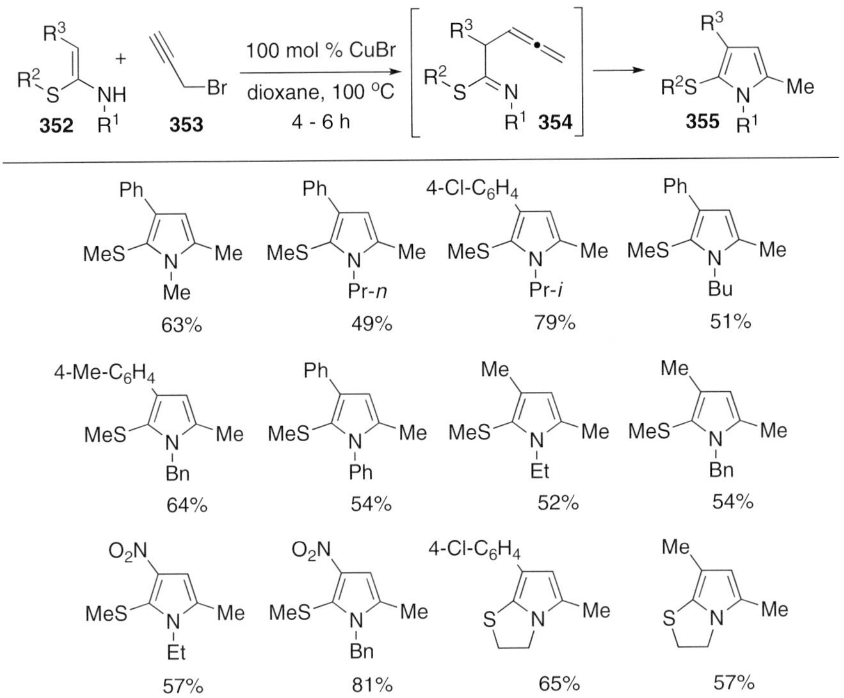

Scheme 8.123

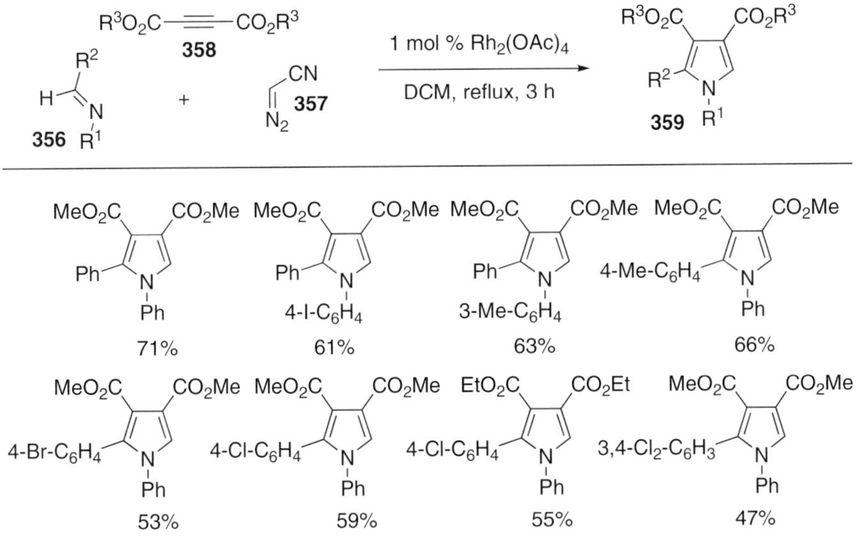

Scheme 8.124

Scheme 8.125

Scheme 8.126

Scheme 8.127

Despite the impressive progress of the transition metal-catalyzed chemistry of furans and pyrroles, undoubtedly, there is still a high demand for more general and efficient methodologies toward these heterocycles with more diverse substitution patterns.

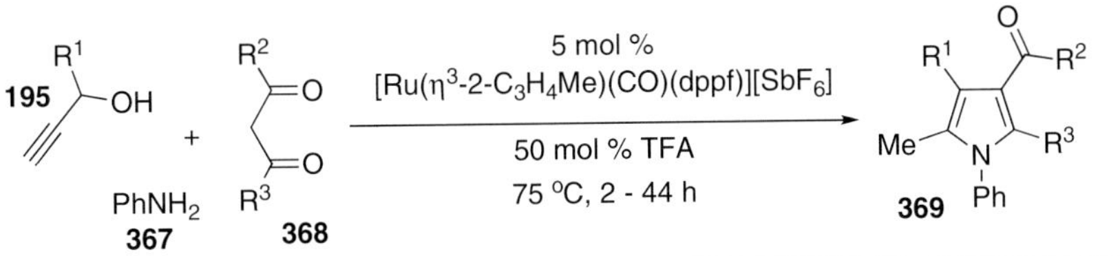

195 + $PhNH_2$ 367 + 368 → (5 mol % [Ru(η^3-2-C_3H_4Me)(CO)(dppf)][SbF_6], 50 mol % TFA, 75 °C, 2 - 44 h) → 369

R^1	R^2	R^3	Yield, %
Ph	OEt	Me	70
2-MeO-C_6H_4	OEt	Me	87
3-MeO-C_6H_4	OEt	Me	81
4-MeO-C_6H_4	OEt	Me	94
2-Cl-C_6H_4	OEt	Me	78
3-Cl-C_6H_4	OEt	Me	82
4-Cl-C_6H_4	OEt	Me	73
1-naphthyl	OEt	Me	72
2-naphthyl	OEt	Me	71
(*E*)-CH=CHPh	OEt	Me	68
2-thienyl	OEt	Me	94
4-MeO-C_6H_4	OMe	Me	93
4-MeO-C_6H_4	OBn	Me	87
4-MeO-C_6H_4	OBu-*t*	Me	84
4-MeO-C_6H_4	Me	Me	88
4-MeO-C_6H_4	Et	Et	83
4-MeO-C_6H_4	Ph	Ph	84
4-MeO-C_6H_4	Ph	Ph	91

Scheme 8.128

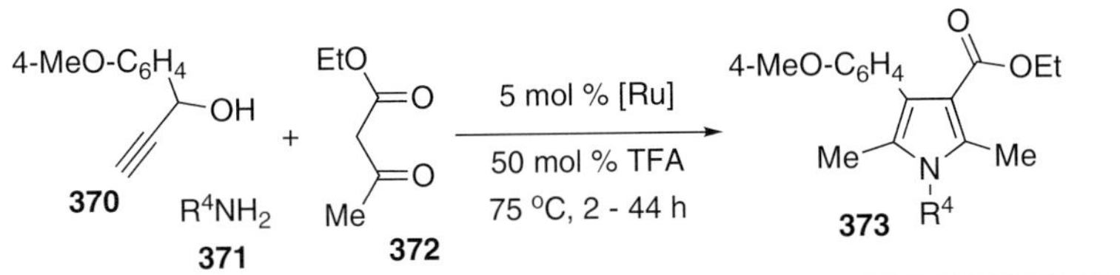

370 + R^4NH_2 371 + 372 → (5 mol % [Ru], 50 mol % TFA, 75 °C, 2 - 44 h) → 373

R^4	Yield, %	R^4	Yield, %	R^4	Yield, %
4-MeO-C_6H_4	70	4-EtO_2C-C_6H_4	70	Bn	70
3-MeO-C_6H_4	87	4-HO-C_6H_4	87	furfuryl	87
2-MeO-C_6H_4	81	3-Ac-C_6H_4	81	allyl	81
4-Cl-C_6H_4	94	4-H_2N(O)C-C_6H_4	94	propargyl	94
4-I-C_6H_4	78	2-H_2C=C(Me)-C_6H_4	78	(*S*)-CHMePh	78
4-O_2N-C_6H_4	82	Et	82	H	82

Scheme 8.129

8.4 Abbreviations

Ac	acetyl
aq	aqueous
BINOL	1,1′-bi-2-naphthol
Boc	*tert*-butylcarbonyl
cat.	catalytic
Cbz	carboxybenzyl
dap	α-(dimethylaminomethyl)pyrrolyl
DCE	1,2-dichloroethane
DCM	dichloromethane
DMA	*N,N*-dimethylacetamide
DMB	2,4-dimethoxybenzyl
DMEDA	*N,N*′-dimethylethylenediamine
DMF	*N,N*-dimethylformamide
dppf	1,1′-bis(diphenylphosphino)ferrocene
dppm	1,1-bis(diphenylphosphino)methane
dppp	1,3-bis(diphenylphosphino)propane
DTBMP	2,6-di-*tert*-butyl-4-methylpyridine
EB	3-ethylbutyryl
EDG	electron-donating group
EE	ethoxyethyl
eq	equivalent
EWG	electron-withdrawing
HFIP	hexafluoro-2-propanol
IPA	isopropanol
IPr	1,3-bis(2,6-diisopropylphenyl)imidazol-2-ylidene
Mes	mesityl
MOM	methoxymethyl
MPM	(*p*-methoxyphenyl)methyl
MS	molecular sieves
MW	microwave
NIS	*N*-iodosuccinimide
Ns	nosyl
pfb	perfluorobutyryl
Piv	pivalyl
phen	1,10-phenanthroline
PMB	*p*-methoxybenzyl
Py	pyridine
rt	room temperature
TBS	*tert*-butyldimethylsilyl
TBDMS	*tert*-butyldimethylsilyl
TBDPS	*tert*-butyldiphenylsilyl
TFA	trifluoroacetic acid

THF	tetrahydrofuran
THP	tetrahydropyranyl
TIPS	triisopropylsilyl
TMEDA	*N,N,N′,N′*-tetramethylethylenediamine
TMS	trimethylsilyl
TMSE	2-(trimethylsilyl)ethyl
Tol	tolyl
Tp	trispyrazolylborate
Ts	tosyl
Tf	triflyl

References

1 Katritzky, A.R. and Rees, C.W. (eds) (1984) *Comprehensive Heterocyclic Chemistry*, Pergamon, Oxford.
2 Katritzky, A.R., Scriven, E.F.V., and Rees, C.W. (eds) (1996) *Comprehensive Heterocyclic Chemistry II*, Elsevier, Oxford.
3 Katritzky, A.R., Ramsden, C.A., Scriven, E.F.V., and Taylor, R.J.K. (eds) (2008) *Comprehensive Heterocyclic Chemistry III*, Elsevier, Oxford.
4 Irie, M. (2000) *Chem. Rev.*, **100**, 1685.
5 Gribble, G.W., Saulnier, M.G., Pelkey, E.T., Kishbaugh, T.L.S., Yanbing, L., Jun, J., Trujillo, H.A., Keavy, D.J., Davis, D.A., Conway, S.C., Switzer, F.L., Roy, S., Silva, R.A., Obaza-Nutaitis, J.A., Sibi, M.P., Moskalev, N.V., Barden, T.C., Chang, L., Habeski, W.M., and Pelcman, B. (2005) *Curr. Org. Chem.*, **9**, 1493.
6 Schröter, S., Stock, C., and Bach, T. (2005) *Tetrahedron*, **61**, 2245.
7 Evano, G., Blanchard, N., and Toumi, M. (2008) *Chem. Rev.*, **108**, 3054.
8 Isambert, N. and Lavilla, R. (2008) *Chem. Eur. J.*, **14**, 8444.
9 Chinchilla, R., Najera, C., and Yus, M. (2004) *Chem. Rev.*, **104**, 2667.
10 Gilchrist, T.L. (1998) *J. Chem. Soc., Perkin Trans. 1*, 615.
11 Gilchrist, T.L. (1999) *J. Chem. Soc., Perkin Trans. 1*, 2849.
12 Ward, M.F. (2000) *Annu. Rep. Prog. Chem., Sect. B: Org. Chem.*, **96**, 157.
13 Söderberg, B.C.G. (2000) *Curr. Org. Chem.*, **4**, 727.
14 Ward, M.F. (2001) *Annu. Rep. Prog. Chem., Sect. B: Org. Chem.*, **97**, 143.
15 Gilchrist, T.L. (2001) *J. Chem. Soc., Perkin Trans. 1*, 2491.
16 Padwa, A. and Bur, S.K. (2007) *Tetrahedron*, **63**, 5341.
17 Stockman, R.A. (2002) *Annu. Rep. Prog. Chem., Sect. B: Org. Chem.*, **98**, 409.
18 Stockman, R.A. (2003) *Annu. Rep. Prog. Chem., Sect. B: Org. Chem.*, **99**, 161.
19 Stanovnik, B. and Svete, J. (2004) *Chem. Rev.*, **104**, 2433.
20 Stockman, R.A. (2005) *Annu. Rep. Prog. Chem., Sect. B: Org. Chem.*, **101**, 103.
21 Stockman, R.A. (2006) *Annu. Rep. Prog. Chem., Sect. B: Org. Chem.*, **102**, 81.
22 Stockman, R.A. (2007) *Annu. Rep. Prog. Chem., Sect. B: Org. Chem.*, **103**, 107.
23 Stockman, R.A. (2008) *Annu. Rep. Prog. Chem., Sect. B: Org. Chem.*, **104**, 106.
24 Cacchi, S. (1999) *J. Organomet. Chem.*, **576**, 42.
25 Rubin, M., Sromek, A.W., and Gevorgyan, V. (2003) *Synlett*, 2265.
26 Brandi, A., Cicchi, S., Cordero, F.M., and Goti, A. (2003) *Chem. Rev.*, **103**, 1213.
27 Tsoungas, P.G. and Diplas, A.I. (2004) *Curr. Org. Chem.*, **8**, 1607.
28 Tsoungas, P.G. and Diplas, A.I. (2004) *Curr. Org. Chem.*, **8**, 1579.
29 Kirsch, G., Hesse, S., and Comel, A. (2004) *Curr. Org. Synth.*, **1**, 47.
30 Barluenga, J., Santamaria, J., and Tomas, M. (2004) *Chem. Rev.*, **104**, 2259.
31 Deiters, A. and Martin, S.F. (2004) *Chem. Rev.*, **104**, 2199.

32 Diver, S.T. and Giessert, A.J. (2004) *Chem. Rev.*, **104**, 1317.
33 Alonso, F., Beletskaya, I.P., and Yus, M. (2004) *Chem. Rev.*, **104**, 3079.
34 Nakamura, I. and Yamamoto, Y. (2004) *Chem. Rev.*, **104**, 2127.
35 Zeni, G. and Larock, R.C. (2004) *Chem. Rev.*, **104**, 2285.
36 Wolfe, J.P. and Thomas, J.S. (2005) *Curr. Org. Chem.*, **9**, 625.
37 Maes, B.U.W. (2006) *Topics in Heterocyclic Chemistry*, vol. **1**, Springer-Verlag, Berlin, Heidelberg, pp. 155–211.
38 Zeni, G. and Larock, R.C. (2006) *Chem. Rev.*, **106**, 4644.
39 Conreaux, D., Bouyssi, D., Monteiro, N., and Balme, G. (2006) *Curr. Org. Chem.*, **10**, 1325.
40 Chopade, P.R. and Louie, J. (2006) *Adv. Synth. Catal.*, **348**, 2307.
41 Donohoe, T.J., Orr, A.J., and Bingham, M. (2006) *Angew. Chem. Int. Ed.*, **45**, 2664.
42 D'Souza, D.M. and Muller, T.J.J. (2007) *Chem. Soc. Rev.*, **36**, 1095.
43 Ferreira, V.F. (2007) *Curr. Org. Chem.*, **11**, 177.
44 Mihovilovic, M.D. and Stanetty, P. (2007) *Angew. Chem. Int. Ed.*, **46**, 3612.
45 Chemler, S.R. and Fuller, P.H. (2007) *Chem. Soc. Rev.*, **36**, 1153.
46 Arcadi, A. (2008) *Chem. Rev.*, **108**, 3266.
47 Álvarez-Corral, M., Muñoz-Dorado, M., and Rodríguez, G.I. (2008) *Chem. Rev.*, **108**, 3174.
48 Arndtsen, B.A. (2008) *Chem. Eur. J.*, **15**, 302.
49 Patil, N.T. and Yamamoto, Y. (2008) *Chem. Rev.*, **108**, 3395.
50 Weibel, J.-M., Blanc, A., and Pale, P. (2008) *Chem. Rev.*, **108**, 3149.
51 Donohoe, T.J., Fishlock, L.P., and Procopiou, P.A. (2008) *Chem. Eur. J.*, **14**, 5716.
52 Miyaura, N. and Suzuki, A. (1995) *Chem. Rev.*, **95**, 2457.
53 Duncton, M.A.J. and Pattenden, G. (1999) *J. Chem. Soc., Perkin Trans. 1*, 1235.
54 Yang, B.H. and Buchwald, S.L. (1999) *J. Organomet. Chem.*, **576**, 125.
55 Chemler, S.R., Trauner, D., and Danishefsky, S.J. (2001) *Angew. Chem. Int. Ed.*, **40**, 4544.
56 Pattenden, G. and Sinclair, D.J. (2002) *J. Organomet. Chem.*, **653**, 261.
57 Alberico, D., Scott, M.E., and Lautens, M. (2007) *Chem. Rev.*, **107**, 174.
58 Seregin, I.V. and Gevorgyan, V. (2007) *Chem. Soc. Rev.*, **36**, 1173.
59 Schlosser, M. (2007) *Synlett*, 3096.
60 Bellina, F., Cauteruccio, S., and Rossi, R. (2008) *Curr. Org. Chem.*, **12**, 774.
61 Donnelly, D.M.X. and Meegan, M.J. (1984) *Comprehensive Heterocyclic Chemistry*, vol. **4** (eds A.R. Katritzky and C.W. Rees), Pergamon, Oxford, p. 657.
62 Keay, B.A. and Dibble, P.W. (1996) *Comprehensive Heterocyclic Chemistry II*, vol. **2** (eds A.R. Katritzky, E.F.V. Scriven and C.W. Rees), Elsevier, Oxford, p. 395.
63 Keay, B.A., Hopkins, J.M., and Dibble, P.W. (2008) *Comprehensive Heterocyclic Chemistry III*, vol. **3** (eds A.R. Katritzky, C.A. Ramsden, E.F.V. Scriven, and R.J.K. Taylor), Elsevier, Oxford, pp. 571–623.
64 Hou, X.-L., Yang, Z., Yeung, K.-S., and Wong, H.N.C. (2007) *Progress in Heterocyclic Chemistry*, vol. **18** (eds G.W. Gribble and J.A. Joule), Elsevier, Oxford, pp. 187–217.
65 Hou, X.-L., Yang, Z., Yeung, K.-S., and Wong, H.N.C. (2008) *Progress in Heterocyclic Chemistry*, vol. **19** (eds G.W. Gribble and J.A. Joule), Elsevier, Oxford, pp. 176–207.
66 Lipshutz, B.H. (1986) *Chem. Rev.*, **86**, 795.
67 Wong, H.N.C., Yu, P., and Yick, C.Y. (1999) *Pure Appl. Chem.*, **71**, 1041.
68 Rassu, G., Zanardi, F., Battistini, L., and Casiraghi, G. (2000) *Chem. Soc. Rev.*, **29**, 109.
69 D'Auria, M., Emanuele, L., Racioppi, R., and Romaniello, G. (2003) *Curr. Org. Chem.*, **7**, 1443.
70 Lattanzi, A. and Scettri, A. (2004) *Curr. Org. Chem.*, **8**, 607.
71 Wright, D.L. (2005) *Progress in Heterocyclic Chemistry*, vol. **17** (eds G.W. Gribble and J.A. Joule), Elsevier, Oxford, pp. 1–32.
72 Lee, H.K., Chan, K.F., Hui, C.W., Yim, H.K., Wu, X.W., and Wong, H.N.C. (2005) *Pure Appl. Chem.*, **77**, 139.
73 Merino, P., Tejero, T., Delso, J.I., and Matute, R. (2007) *Curr. Org. Chem.*, **11**, 1076.

74 Brown, R.C.D. (2005) *Angew. Chem. Int. Ed.*, **44**, 850.
75 Graening, T. and Thrun, F. (2008) *Comprehensive Heterocyclic Chemistry III*, vol. **3** (eds A.R. Katritzky, C.A. Ramsden, E.F.V. Scriven, and R.J.K. Taylor), Elsevier, Oxford, pp. 497–569.
76 Hou, X.L., Cheung, H.Y., Hon, T.Y., Kwan, P.L., Lo, T.H., Tong, S.Y., and Wong, H.N.C. (1998) *Tetrahedron*, **54**, 1955.
77 Kirsch, S.F. (2006) *Org. Biomol. Chem.*, **4**, 2076.
78 König, B. (2000) *Science of Synthesis: Houben-Weyl Methods of Molecular Transformations*, vol. **9** (ed. G. Maas), Georg Thieme, Stuttgart, pp. 183–285.
79 Patil, N.T. and Yamamoto, Y. (2007) *ARKIVOC*, 121.
80 Wong, H.N.C., Yeung, K.S., and Yang, Z. (2008) *Comprehensive Heterocyclic Chemistry III*, vol. **3** (eds. A.R. Katritzky, C.A. Ramsden, E.F.V. Scriven, and R.J.K. Taylor), Elsevier, Oxford, pp. 407–496.
81 Keay, B.A. (1999) *Chem. Soc. Rev.*, **28**, 209.
82 Jeevanandam, A., Ghule, A., and Ling, Y.C. (2002) *Curr. Org. Chem.*, **6**, 841.
83 Balme, G., Bouyssi, D., and Monteiro, N. (2007) *Heterocycles*, **73**, 87.
84 Lautens, M., Klute, W., and Tam, W. (1996) *Chem. Rev.*, **96**, 49.
85 Ritleng, V., Sirlin, C., and Pfeffer, M. (2002) *Chem. Rev.*, **102**, 1731.
86 Nečas, D. and Kotora, M. (2007) *Curr. Org. Chem.*, **11**, 1566.
87 Marshall, J.A. and Robinson, E.D. (1990) *J. Org. Chem.*, **55**, 3450.
88 Ma, S. (2005) *Chem. Rev.*, **105**, 2829.
89 Bates, R.W. and Satcharoen, V. (2002) *Chem. Soc. Rev.*, **31**, 12.
90 Zimmer, R., Dinesh, C.U., Nandanan, E., and Khan, F.A. (2000) *Chem. Rev.*, **100**, 3067.
91 Marshall, J.A. and Wang, X.J. (1991) *J. Org. Chem.*, **56**, 960.
92 Marshall, J.A. and Wang, X.J. (1991) *J. Org. Chem.*, **56**, 6264.
93 Marshall, J.A. and Wang, X.J. (1992) *J. Org. Chem.*, **57**, 3387.
94 Marshall, J.A. and Bartley, G.S. (1994) *J. Org. Chem.*, **59**, 7169.
95 Marshall, J.A., Wallace, E.M., and Coan, P.S. (1995) *J. Org. Chem.*, **60**, 796.
96 Marshall, J.A., Bartley, G.S., and Wallace, E.M. (1996) *J. Org. Chem.*, **61**, 5729.
97 Marshall, J.A. and Sehon, C.A. (1997) *J. Org. Chem.*, **62**, 4313.
98 Marshall, J.A. and Liao, J. (1998) *J. Org. Chem.*, **63**, 5962.
99 Cong, X., Liu, K.-G., Liao, Q.-J., and Yao, Z.-J. (2005) *Tetrahedron Lett.*, **46**, 8567.
100 Hashmi, A.S.K. (1995) *Angew. Chem. Int. Ed. Engl.*, **34**, 1581.
101 Hashmi, A.S.K., Ruppert, T.L., Knofel, T., and Bats, J.W. (1997) *J. Org. Chem.*, **62**, 7295.
102 Hashmi, A.S.K., Schwarz, L., and Bats, J.W. (2000) *J. Prakt. Chem.*, **342**, 40.
103 Fürstner, A. and Davies, P.W. (2007) *Angew. Chem. Int. Ed.*, **46**, 3410.
104 Hashmi, A.S.K. and Hutchings, G.J. (2006) *Angew. Chem. Int. Ed.*, **45**, 7896.
105 Jiménez-Núñez, E. and Echavarren, A.M. (2007) *Chem. Commun.*, 333.
106 Shen, H.C. (2008) *Tetrahedron*, **64**, 3885.
107 Shen, H.C. (2008) *Tetrahedron*, **64**, 7847.
108 Li, Z., Brouwer, C., and He, C. (2008) *Chem. Rev.*, **108**, 3239.
109 Hashmi, A.S.K. (2007) *Chem. Rev.*, **107**, 3180.
110 Hashmi, A.S.K., Frost, T.M., and Bats, J.W. (2000) *J. Am. Chem. Soc.*, **122**, 11553.
111 Hashmi, A.S.K., Schwarz, L., Choi, J.-H., and Frost, T.M. (2000) *Angew. Chem. Int. Ed.*, **39**, 2285.
112 Hashmi, A.S.K., Schwarz, L., Rubenbauer, P., and Blanco, M.C. (2006) *Adv. Synth. Catal.*, **348**, 705.
113 Leclerc, E. and Tius, M.A. (2003) *Org. Lett.*, **5**, 1171.
114 Ménard, D., Vidal, A., Barthomeuf, C., Lebreton, J., and Gosselin, P. (2006) *Synlett*, 57.
115 Zhou, C.Y., Chan, P.W.H., and Che, C.M. (2006) *Org. Lett.*, **8**, 325.
116 Kel'in, A.V. and Gevorgyan, V. (2002) *J. Org. Chem.*, **67**, 95.
117 Sromek, A.W. and Gevorgyan, V. (2007) *Top. Curr. Chem.*, **274**, 77.
118 Kim, J.T., Kel'in, A.V., and Gevorgyan, V. (2003) *Angew. Chem. Int. Ed.*, **42**, 98.
119 Sromek, A.W., Rubina, M., and Gevorgyan, V. (2005) *J. Am. Chem. Soc.*, **127**, 10500.
120 Gorin, D.J., Sherry, B.D., and Toste, F.D. (2008) *Chem. Rev.*, **108**, 3351.

121 Xia, Y., Dudnik, A.S., Gevorgyan, V., and Li, Y. (2008) *J. Am. Chem. Soc.*, **130**, 6940.
122 Crone, B. and Kirsch, S.F. (2008) *Chem. Eur. J.*, **14**, 3514.
123 Dudnik, A.S. and Gevorgyan, V. (2007) *Angew. Chem. Int. Ed.*, **46**, 5195.
124 Dudnik, A.S., Sromek, A.W., Rubina, M., Kim, J.T., Kel'in, A.V., and Gevorgyan, V. (2008) *J. Am. Chem. Soc.*, **130**, 1440.
125 Suhre, M.H., Reif, M., and Kirsch, S.F. (2005) *Org. Lett.*, **7**, 3925.
126 Naota, T., Takaya, H., and Murahashi, S.-I. (1998) *Chem. Rev.*, **98**, 2599.
127 Trost, B.M., Frederiksen, M.U., and Rudd, M.T. (2005) *Angew. Chem. Int. Ed.*, **44**, 6630.
128 Trost, B.M., Toste, F.D., and Pinkerton, A.B. (2001) *Chem. Rev.*, **101**, 2067.
129 Arisawa, M., Terada, Y., Theeraladanon, C., Takahashi, K., Nakagawa, M., and Nishida, A. (2005) *J. Organomet. Chem.*, **690**, 5398.
130 Conrad, J.C. and Fogg, D.E. (2006) *Curr. Org. Chem.*, **10**, 185.
131 Faller, J. and Parr, J. (2006) *Curr. Org. Chem.*, **10**, 151.
132 Nishibayashi, Y. and Uemura, S. (2006) *Curr. Org. Chem.*, **10**, 135.
133 Peng, L., Zhang, X., Ma, M., and Wang, J. (2007) *Angew. Chem. Int. Ed.*, **46**, 1905.
134 Kirsch, S.F. (2008) *Synthesis*, 3183.
135 Omae, I. (2008) *Appl. Organomet. Chem.*, **22**, 149.
136 Marion, N. and Nolan, S.P. (2007) *Angew. Chem. Int. Ed.*, **46**, 2750.
137 Marco-Contelles, J. and Soriano, E. (2007) *Chem. Eur. J.*, **13**, 1350.
138 Correa, A., Marion, N., Fensterbank, L., Malacria, M., Nolan, S.P., and Cavallo, L. (2008) *Angew. Chem. Int. Ed.*, **47**, 718.
139 Sromek, A.W., Kel'in, A.V., and Gevorgyan, V. (2004) *Angew. Chem. Int. Ed.*, **43**, 2280.
140 Schwier, T., Sromek, A.W., Yap, D.M.L., Chernyak, D., and Gevorgyan, V. (2007) *J. Am. Chem. Soc.*, **129**, 9868.
141 Shapiro, N.D. and Toste, F.D. (2007) *J. Am. Chem. Soc.*, **129**, 4160.
142 Imagawa, H., Kurisaki, T., and Nishizawa, M. (2004) *Org. Lett.*, **6**, 3679.
143 Nishibayashi, Y., Yoshikawa, M., Inada, Y., Milton, M.D., Hidai, M., and Uemura, S. (2003) *Angew. Chem. Int. Ed.*, **42**, 2681.
144 Cadierno, V., Gimeno, J., and Nebra, N. (2007) *Adv. Synth. Catal.*, **349**, 382.
145 Sniady, A., Durham, A., Morreale, M.S., Wheeler, K.A., and Dembinski, R. (2007) *Org. Lett.*, **9**, 1175.
146 Sniady, A., Durham, A., Morreale, M.S., Marcinek, A., Szafert, S., Lis, T., Brzezinska, K.R., Iwasaki, T., Ohshima, T., Mashima, K., and Dembinski, R. (2008) *J. Org. Chem.*, **73**, 5881.
147 Esho, N., Desaulniers, J.-P., Davies, B., Chui, H.M.P., Rao, M.S., Chow, C.S., Szafert, S., and Dembinski, R. (2005) *Bioorg. Med. Chem.*, **13**, 1231.
148 Aucagne, V., Amblard, F., and Agrofoglio, L.A. (2004) *Synlett*, 2406.
149 Amblard, F., Aucagne, V., Guenot, P., Schinazi, R.F., and Agrofoglio, L.A. (2005) *Bioorg. Med. Chem.*, **13**, 1239.
150 Hudson, R.H.E. and Moszynski, J.M. (2006) *Synlett*, 2997.
151 Aubert, C., Buisine, O., and Malacria, M. (2002) *Chem. Rev.*, **102**, 813.
152 Lloyd-Jones, G.C. (2003) *Org. Biomol. Chem.*, **1**, 215.
153 Michelet, V., Toullec, P.Y., and Genêt, J.-P. (2008) *Angew. Chem. Int. Ed.*, **47**, 4268.
154 Yao, T., Zhang, X., and Larock, R.C. (2004) *J. Am. Chem. Soc.*, **126**, 11164.
155 Yao, T., Zhang, X., and Larock, R.C. (2005) *J. Org. Chem.*, **70**, 7679.
156 Patil, N.T., Wu, H., and Yamamoto, Y. (2005) *J. Org. Chem.*, **70**, 4531.
157 Martins, M.A.P., Frizzo, C.P., Moreira, D.N., Zanatta, N., and Bonacorso, H.G. (2008) *Chem. Rev.*, **108**, 2015.
158 Liu, X.Y., Pan, Z.L., Shu, X.Z., Duan, X.H., and Liang, Y.M. (2006) *Synlett*, 1962.
159 Oh, C.H., Reddy, V.R., Kim, A., and Rhim, C.Y. (2006) *Tetrahedron Lett.*, **47**, 5307.
160 Zhang, J. and Schmalz, H.-G. (2006) *Angew. Chem. Int. Ed.*, **45**, 6704.
161 Zhang, G., Huang, X., Li, G., and Zhang, L. (2008) *J. Am. Chem. Soc.*, **130**, 1814.
162 Seiller, B., Bruneau, C., and Dixneuf, P.H. (1994) *J. Chem. Soc., Chem. Commun.*, 493.
163 Seiller, B., Bruneau, C., and Dixneuf, P.H. (1995) *Tetrahedron*, **51**, 13089.
164 Kücükbay, H., Cetinkaya, B., Guesmi, S., and Dixneuf, P.H. (1996) *Organometallics*, **15**, 2434.

165 Marshall, J.A. and Sehon, C.A. (1995) *J. Org. Chem.*, **60**, 5966.
166 Díaz-Álvarez, A.E., Crochet, P., Zablocka, M., Duhayon, C., Cadierno, V., Gimeno, J., and Majoral, J.P. (2006) *Adv. Synth. Catal.*, **348**, 1671.
167 Liu, Y., Song, F., Song, Z., Liu, M., and Yan, B. (2005) *Org. Lett.*, **7**, 5409.
168 Kim, S. and Lee, P.H. (2008) *Adv. Synth. Catal.*, **350**, 547.
169 Miller, D. (1969) *J. Chem. Soc. C*, 12.
170 McDonald, F.E. and Schultz, C.C. (1994) *J. Am. Chem. Soc.*, **116**, 9363.
171 McDonald, F.E. (1999) *Chem. Eur. J.*, **5**, 3103.
172 Bruneau, C. and Dixneuf, P.H. (1999) *Acc. Chem. Res.*, **32**, 311.
173 Wakatsuki, Y. (2004) *J. Organomet. Chem.*, **689**, 4092.
174 Varela, J.A. and Saá, C. (2006) *Chem. Eur. J.*, **12**, 6450.
175 Lo, C.-Y., Guo, H., Lian, J.-J., Shen, F.-M., and Liu, R.-S. (2002) *J. Org. Chem.*, **67**, 3930.
176 Hashmi, A.S.K. and Sinha, P. (2004) *Adv. Synth. Catal.*, **346**, 432.
177 Yoshida, M., Al-Amin, M., Matsuda, K., and Shishido, K. (2008) *Tetrahedron Lett.*, **49**, 5021.
178 Marson, C.M., Harper, S., and Wrigglesworth, R. (1879) *J. Chem. Soc., Chem. Commun.*, **1994**.
179 Marson, C.M. and Harper, S. (1998) *J. Org. Chem.*, **63**, 9223.
180 Shu, X.-Z., Liu, X.-Y., Xiao, H.-Q., Ji, K.-G., Guo, L.-N., Qi, C.-Z., and Liang, Y.-M. (2007) *Adv. Synth. Catal.*, **349**, 2493.
181 Ji, K.-G., Shen, Y.-W., Shu, X.-Z., Xiao, H.-Q., Bian, Y.-J., and Liang, Y.-M. (2008) *Adv. Synth. Catal.*, **350**, 1275.
182 Dai, L.-Z. and Shi, M. (2008) *Tetrahedron Lett.*, **49**, 6437.
183 Wen, S.G., Liu, W.M., and Liang, Y.M. (2007) *Synthesis*, 3295.
184 Rubin, M., Rubina, M., and Gevorgyan, V. (2007) *Chem. Rev.*, **107**, 3117.
185 Rubin, M., Rubina, M., and Gevorgyan, V. (2006) *Synthesis*, 1221.
186 Tomilov, Y.V., Shapiro, E.A., Protopopova, M.N., Ioffe, A.I., Dolgii, I.E., and Nefedov, O.M. (1985) *Izv. Akad. Nauk SSSR, Ser. Khim.*, 631.
187 Davies, H.M.L. and Hedley, S.J. (2007) *Chem. Soc. Rev.*, **36**, 1109.
188 Cho, S.H. and Liebeskind, L.S. (1987) *J. Org. Chem.*, **52**, 2631.
189 Davies, H.M.L. and Romines, K.R. (1988) *Tetrahedron*, **44**, 3343.
190 Müller, P., Pautex, N., Doyle, M.P., and Bagheri, V. (1990) *Helv. Chim. Acta*, **73**, 1233.
191 Müller, P. and Gränicher, C. (1993) *Helv. Chim. Acta*, **76**, 521.
192 Müller, P. and Gränicher, C. (1995) *Helv. Chim. Acta*, **78**, 129.
193 Padwa, A., Kassir, J.M., and Xu, S.L. (1991) *J. Org. Chem.*, **56**, 6971.
194 Padwa, A., Kassir, J.M., and Xu, S.L. (1997) *J. Org. Chem.*, **62**, 1642.
195 Ma, S. and Zhang, J. (2003) *J. Am. Chem. Soc.*, **125**, 12386.
196 Ye, T. and McKervey, M.A. (1994) *Chem. Rev.*, **94**, 1091.
197 Padwa, A. and Austin, D.J. (1994) *Angew. Chem. Int. Ed. Engl.*, **33**, 1797.
198 Padwa, A. and Weingarten, M.D. (1996) *Chem. Rev.*, **96**, 223.
199 Padwa, A. (2001) *J. Organomet. Chem.*, **617–618**, 3.
200 Muthusamy, S. and Krishnamurthi, J. (2008) *Topics in Heterocyclic Chemistry*, vol. **12**, Springer-Verlag, Berlin, Heidelberg, pp. 147–192.
201 D'yakonov, I.A. and Komendantov, M.I. (1959) *Zh. Obshch. Khim.*, **29**, 1749.
202 D'yakonov, I.A. and Komendantov, M.I. (1961) *Zh. Obshch. Khim.*, **31**, 3483.
203 D'yakonov, I.A. and Komendantov, M.I. (1961) *Zh. Obshch. Khim.*, **31**, 3881.
204 D'yakonov, I.A., Komendantov, M.I., and Korshunov, S.P. (1962) *Zh. Obshch. Khim.*, **32**, 923.
205 D'yakonov, I.A. and Komendantov, M.I. (1963) *Zh. Obshch. Khim.*, **33**, 2448.
206 Padwa, A., Chiacchio, U., Garreau, Y., Kassir, J.M., Krumpe, K.E., and Schoffstall, A.M. (1990) *J. Org. Chem.*, **55**, 414.
207 Kinder, F.R. and Padwa, A. (1990) *Tetrahedron Lett.*, **31**, 6835.
208 Pirrung, M.C., Zhang, J., and Morehead, A.T. (1994) *Tetrahedron Lett.*, **35**, 6229.
209 Hoye, T.R., Dinsmore, C.J., Johnson, D.S., and Korkowski, P.F. (1990) *J. Org. Chem.*, **55**, 4518.

210 Wee, A.G.H., Liu, B., and Zhang, L. (1992) *J. Org. Chem.*, **57**, 4404.
211 Padwa, A. and Kinder, F.R. (1993) *J. Org. Chem.*, **58**, 21.
212 Padwa, A., Dean, D.C., Fairfax, D.J., and Xu, S.L. (1993) *J. Org. Chem.*, **58**, 4646.
213 Brown, D.S., Elliott, M.C., Moody, C.J., Mowlem, T.J., Marino, J.P.J., and Padwa, A. (1994) *J. Org. Chem.*, **59**, 2447.
214 Lee, Y.R., Suk, J.Y., and Kim, B.S. (1999) *Tetrahedron Lett.*, **40**, 6603.
215 Lee, Y.R. and Suk, J.Y. (2000) *Tetrahedron Lett.*, **41**, 4795.
216 Padwa, A. and Straub, C.S. (2000) *Org. Lett.*, **2**, 2093.
217 Padwa, A. (2000) *J. Organomet. Chem.*, **610**, 88.
218 Tollari, S., Palmisano, G., Cenini, S., Cravotto, G., Giovenzana, G.B., and Penoni, A. (2001) *Synthesis*, 735.
219 Padwa, A. and Straub, C.S. (2003) *J. Org. Chem.*, **68**, 227.
220 Ma, S., Lu, L., and Lu, P. (2005) *J. Org. Chem.*, **70**, 1063.
221 Müller, P., Allenbach, Y.F., and Bernardinelli, G. (2003) *Helv. Chim. Acta*, **86**, 3164.
222 Padwa, A., Austin, D.J., Price, A.T., Semones, M.A., Doyle, M.P., Protopopova, M.N., Winchester, W.R., and Tran, A. (1993) *J. Am. Chem. Soc.*, **115**, 8669.
223 Ljungdahl, N. and Kann, N. (2009) *Angew. Chem. Int. Ed.*, **48**, 642.
224 Cadierno, V., Díez, J., Gimeno, J., and Nebra, N. (2008) *J. Org. Chem.*, **73**, 5852.
225 Pan, Y.-m., Zhao, S.-y., Ji, W.-h., and Zhan, Z.-p. (2009) *J. Comb. Chem.*, **11**, 103.
226 Zhan, Z.-p., Wang, S.-p., Cai, X.-b., Liu, H.-j., Yu, J.-l., and Cui, Y.-y. (2007) *Adv. Synth. Catal.*, **349**, 2097.
227 Bolm, C., Legros, J., Le Paih, J., and Zani, L. (2004) *Chem. Rev.*, **104**, 6217.
228 Díaz, D.D., Miranda, P.O., Padrón, J.I., and Martín, V.S. (2006) *Curr. Org. Chem.*, **10**, 457.
229 Bauer, E.B. (2008) *Curr. Org. Chem.*, **12**, 1341.
230 Zhan, Z.-p., Cai, X.-b., Wang, S.-p., Yu, J.-l., Liu, H.-j., and Cui, Y.-y. (2007) *J. Org. Chem.*, **72**, 9838.
231 Nanayakkara, P. and Alper, H. (2006) *Adv. Synth. Catal.*, **348**, 545.
232 Eilbracht, P., Bärfacker, L., Buss, C., Hollmann, C., Kitsos-Rzychon, B.E., Kranemann, C.L., Rische, T., Roggenbuck, R., and Schmidt, A. (1999) *Chem. Rev.*, **99**, 3329.
233 Gossage, R.A. (2006) *Curr. Org. Chem.*, **10**, 923.
234 Lee, K.Y., Lee, M.J., and Kim, J.N. (2005) *Tetrahedron*, **61**, 8705.
235 O'Hagan, D. (1997) *Nat. Prod. Rep.*, **14**, 637.
236 O'Hagan, D. (2000) *Nat. Prod. Rep.*, **17**, 435.
237 Bellina, F. and Rossi, R. (2006) *Tetrahedron*, **62**, 7213.
238 Boger, D.L., Boyce, C.W., Labroli, M.A., Sehon, C.A., and Jin, Q. (1999) *J. Am. Chem. Soc.*, **121**, 54.
239 Fan, H., Peng, J., Hamann, M.T., and Hu, J.-F. (2008) *Chem. Rev.*, **108**, 264.
240 Fürstner, A. (2003) *Angew. Chem. Int. Ed.*, **42**, 3582.
241 Gribble, G.W. (2003) *Progress in Heterocyclic Chemistry*, vol. **15** (eds G.W. Gribble and T.L. Gilchrist), Elsevier, Oxford, pp. 58–74.
242 Janosik, T., Bergman, J., and Pelkey, E.T. (2005) *Progress in Heterocyclic Chemistry*, vol. **16** (eds G.W. Gribble and J.A. Joule), Elsevier, Oxford, pp. 128–155.
243 Gupton, J.T. (2006) *Topics in Heterocyclic Chemistry*, vol. **2**, Springer-Verlag, Berlin, Heidelberg, pp. 53–92.
244 Walsh, C.T., Garneau-Tsodikova, S., and Howard-Jones, A.R. (2006) *Nat. Prod. Rep.*, **23**, 517.
245 Wood, T.E. and Thompson, A. (2007) *Chem. Rev.*, **107**, 1831.
246 Biava, M., Porretta, G.C., Poce, G., Supino, S., and Sleiter, G. (2007) *Curr. Org. Chem.*, **11**, 1092.
247 d'Ischia, M., Napolitano, A., and Pezzella, A. (2008) *Comprehensive Heterocyclic Chemistry III*, vol. **3** (eds A.R. Katritzky, C.A. Ramsden, E.F.V. Scriven, and R.J.K. Taylor), Elsevier, Oxford, pp. 353–388.
248 Guernion, N.J.L. and Hayes, W. (2004) *Curr. Org. Chem.*, **8**, 637.
249 Trofimov, B.A. and Nedolya, N.A. (2008) *Comprehensive Heterocyclic Chemistry III*, vol. **3** (eds A.R. Katritzky, C.A. Ramsden,

E.F.V. Scriven, and R.J.K. Taylor), Elsevier, Oxford, pp. 45–268.
250 Banwell, M.G., Goodwin, T.E., Ng, S., Smith, J.A., and Wong, D.J. (2006) *Eur. J. Org. Chem.*, 3043.
251 Banwell, M.G., Beck, D.A.S., Stanislawski, P.C., Sydnes, M.O., and Taylor, R.M. (2005) *Curr. Org. Chem.*, **9**, 1589.
252 Jefford, C.W. (2000) *Curr. Org. Chem.*, **4**, 205.
253 Trofimov, B.A., Sobenina, L.N., Demenev, A.P., and Mikhaleva, A.I. (2004) *Chem. Rev.*, **104**, 2481.
254 Saracoglu, N. (2007) *Topics in Heterocyclic Chemistry*, vol. **11**, Springer-Verlag, Berlin, Heidelberg, pp. 1–61.
255 Sundberg, R.J. (1996) *Comprehensive Heterocyclic Chemistry II*, vol. **2** (eds A.R. Katritzky, E.F.V. Scriven, and C.W. Rees), Elsevier, Oxford, pp. 119–206.
256 Pelkey, E.T. (2005) *Progress in Heterocyclic Chemistry*, vol. **17** (eds G.W. Gribble and J.A. Joule), Elsevier, Oxford, pp. 109–141.
257 Pelkey, E.T. (2007) *Progress in Heterocyclic Chemistry*, vol. **18** (eds G.W. Gribble and J.A. Joule), Elsevier, Oxford, pp. 150–186.
258 Pelkey, E.T. and Russel, J.S. (2008) *Progress in Heterocyclic Chemistry*, vol. **19** (eds G.W. Gribble and J.A. Joule), Elsevier, Oxford, pp. 135–175.
259 Black, D.S. (2000) *Science of Synthesis: Houben-Weyl Methods of Molecular Transformations*, vol. **9** (ed. G. Maas), Georg Thieme, Stuttgart, pp. 441–552.
260 Bergman, J. and Janosik, T. (2008) *Comprehensive Heterocyclic Chemistry III*, vol. **3** (eds A.R. Katritzky, C.A. Ramsden, E.F.V. Scriven, and R.J.K. Taylor), Elsevier, Oxford, pp. 269–351.
261 Agarwal, S., Cämmerer, S., Filali, S., Fröhner, W., Knöll, J., Krahl, M.P., Reddy, K.R., and Knölker, H.-J. (2005) *Curr. Org. Chem.*, **9**, 1601.
262 Balme, G. (2004) *Angew. Chem. Int. Ed.*, **43**, 6238.
263 Ferreira, V.F., De Souza, M.C.B.V., Cunha, A.C., Pereira, L.O.R., and Ferreira, M.L.G. (2001) *Org. Prep. Proced. Int.*, **33**, 411.
264 Joshi, U., Pipelier, M., Naud, S., and Dubreuil, D. (2005) *Curr. Org. Chem.*, **9**, 261.
265 Schmuck, C. and Rupprecht, D. (2007) *Synthesis*, 3095.
266 Flögel, O., Dash, J., Brüdgam, I., Hartl, H., and Reißig, H.-U. (2004) *Chem. Eur. J.*, **10**, 4283.
267 Kel'in, A.V., Sromek, A.W., and Gevorgyan, V. (2001) *J. Am. Chem. Soc.*, **123**, 2074.
268 Sromek, A.W., Rheingold, A.L., Wink, D.J., and Gevorgyan, V. (2006) *Synlett*, 2325.
269 Gabriele, B., Salerno, G., and Fazio, A. (2003) *J. Org. Chem.*, **68**, 7853.
270 Csende, F. and Stájer, G. (2005) *Curr. Org. Chem.*, **9**, 1737.
271 Istrate, F.M. and Gagosz, F. (2007) *Org. Lett.*, **9**, 3181.
272 Peng, Y., Yu, M., and Zhang, L. (2008) *Org. Lett.*, **10**, 5187.
273 Ojima, I., Tzamarioudaki, M., Li, Z., and Donovan, R.J. (1996) *Chem. Rev.*, **96**, 635.
274 Robinson, R.S., Dovey, M.C., and Gravestock, D. (2004) *Tetrahedron Lett.*, **45**, 6787.
275 Bagley, M.C. and Lubinu, M.C. (2006) *Topics in Heterocyclic Chemistry*, vol. **1**, Springer-Verlag, Berlin, Heidelberg, pp. 31–58.
276 Besson, T. and Thiéry, V. (2006) *Topics in Heterocyclic Chemistry*, vol. **1**, Springer-Verlag, Berlin, Heidelberg, pp. 59–78.
277 Crosignani, S. and Linclau, B. (2006) *Topics in Heterocyclic Chemistry*, vol. **1**, Springer-Verlag, Berlin, Heidelberg, pp. 129–154.
278 Erdélyi, M. (2006) *Topics in Heterocyclic* Chemistry, vol. **1**, Springer-Verlag, Berlin, Heidelberg, pp. 79–128.
279 Rodriquez, M. and Taddei, M. (2006) *Topics in Heterocyclic* Chemistry, vol. **1**, Springer-Verlag, Berlin, Heidelberg, pp. 213–266.
280 Robinson, R.S., Dovey, M.C., and Gravestock, D. (2005) *Eur. J. Org. Chem.*, 505.
281 Gorin, D.J., Davis, N.R., and Toste, F.D. (2005) *J. Am. Chem. Soc.*, **127**, 11260.
282 Hiroya, K., Matsumoto, S., Ashikawa, M., Ogiwara, K., and Sakamoto, T. (2006) *Org. Lett.*, **8**, 5349.
283 Dong, H., Shen, M., Redford, J.E., Stokes, B.J., Pumphrey, A.L., and Driver, T.G. (2007) *Org. Lett.*, **9**, 5191.

284 Campi, E.M., Jackson, R.W., and Nilsson, Y. (1991) *Tetrahedron Lett.*, **32**, 1093.

285 Schulte, K.E. and Reisch, J. (1961) *Angew. Chem.*, **73**, 241.

286 Schulte, K.E., Reisch, J., and Walker, H. (1965) *Chem. Ber.*, **98**, 98.

287 Chalk, A.J. (1972) *Tetrahedron Lett.*, **13**, 3487.

288 Müller, T.E. and Beller, M. (1998) *Chem. Rev.*, **98**, 675.

289 Pohlki, F. and Doye, S. (2003) *Chem. Soc. Rev.*, **32**, 104.

290 Widenhoefer, R.A. and Han, X. (2006) *Eur. J. Org. Chem.*, 4555.

291 Müller, T.E., Hultzsch, K.C., Yus, M., Foubelo, F., and Tada, M. (2008) *Chem. Rev.*, **108**, 3795.

292 Ramanathan, B., Keith, A.J., Armstrong, D., and Odom, A.L. (2004) *Org. Lett.*, **6**, 2957.

293 Lavallo, V., Frey, G.D., Donnadieu, B., Soleilhavoup, M., and Bertrand, G. (2008) *Angew. Chem. Int. Ed.*, **47**, 5224.

294 Shu, X.-Z., Liu, X.-Y., Xiao, H.-Q., Ji, K.-G., Guo, L.-N., and Liang, Y.-M. (2008) *Adv. Synth. Catal.*, **350**, 243.

295 Lu, Y., Fu, X., Chen, H., Du, X., Jia, X., and Liu, Y. (2009) *Adv. Synth. Catal.*, **351**, 129.

296 Martín, R., Rivero, M.R., and Buchwald, S.L. (2006) *Angew. Chem. Int. Ed.*, **45**, 7079.

297 Ley, S.V. and Thomas, A.W. (2003) *Angew. Chem. Int. Ed.*, **42**, 5400.

298 Beletskaya, I.P. and Cheprakov, A.V. (2004) *Coord. Chem. Rev.*, **248**, 2337.

299 Monnier, F. and Taillefer, M. (2008) *Angew. Chem. Int. Ed.*, **47**, 3096.

300 Arcadi, A., Di Giuseppe, S., Marinelli, F., and Rossi, E. (2001) *Adv. Synth. Catal.*, **343**, 443.

301 Arcadi, A., Di Giuseppe, S., Marinelli, F., and Rossi, E. (2001) *Tetrahedron: Asymmetry*, **12**, 2715.

302 Harrison, T.J., Kozak, J.A., Corbella-Pane, M., and Dake, G.R. (2006) *J. Org. Chem.*, **71**, 4525.

303 Binder, J.T. and Kirsch, S.F. (2006) *Org. Lett.*, **8**, 2151.

304 Peng, L., Zhang, X., Ma, J., Zhong, Z., and Wang, J. (2007) *Org. Lett.*, **9**, 1445.

305 Martín, R., Larsen, C.H., Cuenca, A., and Buchwald, S.L. (2007) *Org. Lett.*, **9**, 3379.

306 Yuan, X., Xu, X., Zhou, X., Yuan, J., Mai, L., and Li, Y. (2007) *J. Org. Chem.*, **72**, 1510.

307 Zhu, J. (2003) *Eur. J. Org. Chem.*, 1133.

308 Takaya, H., Kojima, S., and Murahashi, S.-I. (2001) *Org. Lett.*, **3**, 421.

309 Kamijo, S., Kanazawa, C., and Yamamoto, Y. (2005) *J. Am. Chem. Soc.*, **127**, 9260.

310 Larionov, O.V. and de Meijere, A. (2005) *Angew. Chem. Int. Ed.*, **44**, 5664.

311 Lygin, A.V., Larionov, O.V., Korotkov, V.S., and de Meijere, A. (2009) *Chem. Eur. J.*, **15**, 227.

312 Chiba, S., Wang, Y.-F., Lapointe, G., and Narasaka, K. (2008) *Org. Lett.*, **10**, 313.

313 Wang, Y.-F., Toh, K.K., Chiba, S., and Narasaka, K. (2008) *Org. Lett.*, **10**, 5019.

314 Rivero, M.R. and Buchwald, S.L. (2007) *Org. Lett.*, **9**, 973.

315 Shiraishi, H., Nishitani, T., Nishihara, T., Sakaguchi, S., and Ishii, Y. (1999) *Tetrahedron*, **55**, 13957.

316 Gupta, A.K., Reddy, K.R., Ila, H., and Junjappa, H. (1995) *J. Chem. Soc., Perkin Trans. 1*, 1725.

317 Galliford, C.V. and Scheidt, K.A. (2007) *J. Org. Chem.*, **72**, 1811.

318 Cadierno, V., Gimeno, J. and Nebra, N. (2007) *Chem. Eur. J.*, **13**, 9973.

9
Transition Metal-Catalyzed Synthesis of Fused Five-Membered Aromatic Heterocycles

Alexander S. Dudnik and Vladimir Gevorgyan

9.1
Introduction

Benzo[*b*]furans, benzo[*b*]thiophenes, 1*H*-indoles, isoindoles, and indolizines are key elements of an array of biologically active compounds [1–3]. In addition, these heterocycles have found application as highly potent pharmaceuticals with a broad spectrum of activities and as pesticides, insecticides, and herbicides in agriculture [1–3]. Among other examples of applications of these fused heterocycles are dyes, polymers, electronic materials, solvents [4] and advanced intermediates in organic synthesis [5–9]. The significance of benzo[*b*]furans, benzo[*b*]thiophenes, 1*H*-indoles, isoindoles, and indolizines in applied and fundamental areas of chemistry placed them at the forefront of contemporary synthetic organic chemistry. A variety of synthetic routes toward these heterocycles have been reported [1–3, 10–23]. Nonetheless, many research groups still actively pursue the development of novel and more general methods for the synthesis of these heterocyclic cores. Employment of transition metal-catalyzed transformations is arguably the most attractive approach toward this objective [24–51]. This review covers transition metal-catalyzed syntheses of selected fused five-membered aromatic heterocycles with one heteroatom, including benzo[*b*]furans, benzo[*b*]thiophenes, 1*H*-indoles, isoindoles, and indolizines. The main organization of this review is based on a heterocycle, wherein syntheses of a particular heterocyclic core are categorized by the type of transformation and substrates employed. We tried to discuss equally both synthetic applicability and mechanistic aspects of the described transformations. It should be noted that the most important, recent, and conceptionally interesting catalytic reactions are discussed in this chapter. Syntheses of benzo[*b*]furans, benzo[*b*]thiophenes, 1*H*-indoles, isoindoles, and indolizines via functionalization of the preexisting heterocyclic cores [37, 52–60] are not covered and, thus, only transformations in which an assembly of a heterocyclic ring occurred are described in this chapter. In addition, synthesis of heterocycles under Pd catalysis is not covered herein, as this is covered in Chapter 6.

Catalyzed Carbon-Heteroatom Bond Formation. Edited by Andrei K. Yudin

ISBN: 978-3-527-32428-6

9.2
Fused Five-Membered Heterocycles

9.2.1
Benzofurans

The benzo[*b*]furan moiety bearing different functional groups represents one of the most important heterocyclic pharmacophores and is often incorporated in many biologically active compounds and pharmaceutical agents as a core structural motif [61–63]. Furthermore, the benzo[*b*]furan skeleton is found in a variety of artificial materials with highly valuable properties and can be employed in the synthesis of diverse important synthons and final products [6, 9, 61, 64–66]. As a result, this heterocyclic unit continues to attract extensive synthetic efforts aimed at the development of selective and general methodologies for the facile assembly of the benzo[*b*] furan core. In particular, among a variety of synthetic protocols, recent research has been focused on the establishment of catalytic approaches from easily accessible precursors under mild reaction conditions [61, 62, 65, 67–74]. Special attention was given to transformations leading to the benzofuran heterocycle catalyzed by transition metals. Within this area, cycloisomerization, intramolecular arylation, and formal [4 + 1] and [3 + 2] cycloaddition approaches arguably represent the most important routes toward an efficient construction of multisubstituted benzo[*b*]furan units. Hence, recent contributions to these methods are discussed in detail below.

9.2.1.1 Synthesis of Benzofurans via Cycloisomerization Reactions

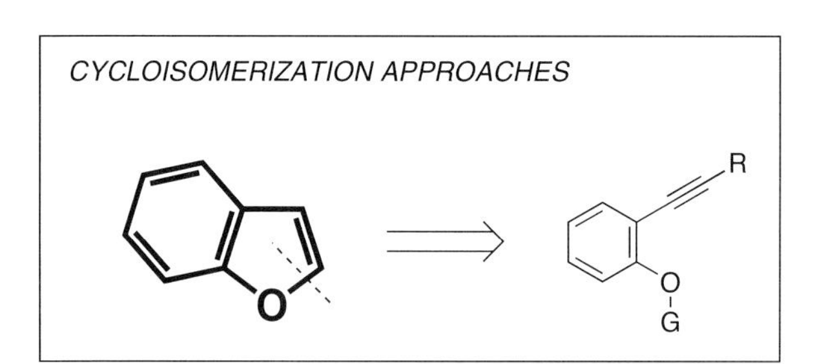

The synthesis of benzofurans and related compounds from *ortho*-iodophenol and copper(I) acetylides via a coupling–cyclization reaction (*vide infra*) was introduced by Castro in the 1960s [75–77] and stimulated the development of many contemporary transition metal-catalyzed synthetic approaches toward the assembly of the benzofuran unit. In a subsequent report from Castro's group, it was proposed that the formation of the benzofuran core occurs in a stepwise fashion with the intermediacy of *ortho*-alkynylphenol, which, upon activation by the Cu(I) salt promoter, undergoes subsequent cycloisomerization [77].

Following the original proposal by Castro, many other research groups have utilized *ortho*-alkynylphenols as convenient precursors in transition metal-catalyzed cycloisomerizations toward a highly efficient synthesis of functionalized benzofurans.

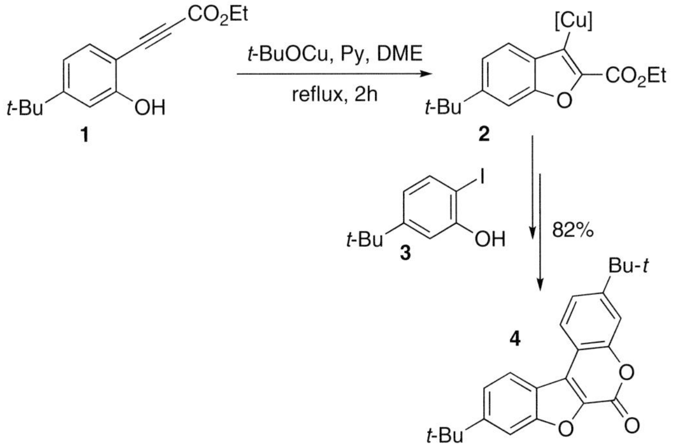

Scheme 9.1

Thus, Nilsson reported an interesting approach for the assembly of tetracyclic benzofuran derivatives via the Cu(I)-mediated cyclization–annulation cascade reaction of *o*-alkynylphenol **1** and *o*-iodophenol **3** (Scheme 9.1). The formation of benzofuran **4** was rationalized by the proposed involvement of a nucleophilic benzofuranylcopper intermediate **2**. The latter, upon coupling and subsequent lactonization with **3**, furnished the corresponding polycyclic product **4** [78].

Recently, an analogous stoichiometric transformation, involving the utilization of a putative organocopper(I) intermediate similar to **2**, was described by Nakamura (Scheme 9.2) [79, 80]. Accordingly, *o*-alkynylphenols **5**, upon activation with a stoichiometric organometallic base, undergo the Zn(II)-mediated cyclization into the corresponding 3-zinciobenzofuran **6**, which upon transmetalation with the CuCN-LiCl complex produces the key cuprate **7**. The latter, upon interception with a suitable electrophile, such as proton, activated alkyl halides, aldehydes, acyl chlorides, or α,β-enones, furnishes 2,3-disubstituted benzofuran **8**. Although this approach allows facile introduction of a C3-substitutent, providing an elegant solution to the limitation of Castro's chemistry to the synthesis of C3-unfunctionalized benzofurans only, the synthetic utility of this method is somewhat limited by the necessity of using stoichiometric amounts of base, Zn(II)- and Cu(I) reagents. Moreover, C2-unsubstituted benzofurans cannot be accessed via this protocol, as terminal *o*-alkynylphenols are not tolerated under these reaction conditions.

An extension of Hashmi's Au(III)-catalyzed phenol synthesis [81] to furan substrates **9** bearing an additional alkyne moiety allowed the preparation of C6-C7-heterofused benzofuran **11** (Scheme 9.3) [82]. According to the proposed mechanism, the Au(III)-catalyzed arene formation reaction generates *o*-alkynylphenol **10**. A subsequent Au(III)-catalyzed cycloisomerization of the latter, following the general mechanism for an intramolecular nucleophilic addition of heteroatom to transition metal-activated carbon–carbon multiple bonds, gives **11** (Scheme 9.3).

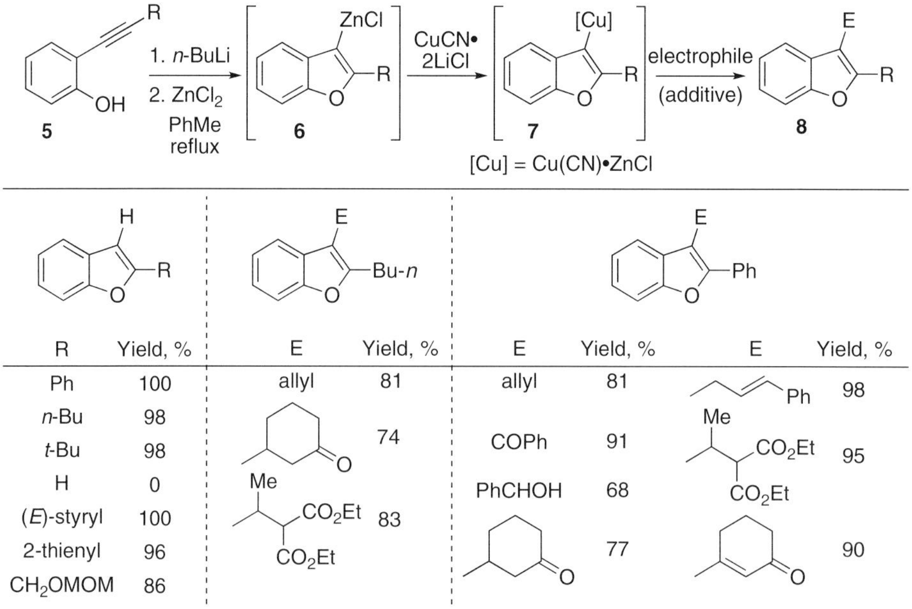

R	Yield, %
Ph	100
n-Bu	98
t-Bu	98
H	0
(E)-styryl	100
2-thienyl	96
CH_2OMOM	86

Scheme 9.2

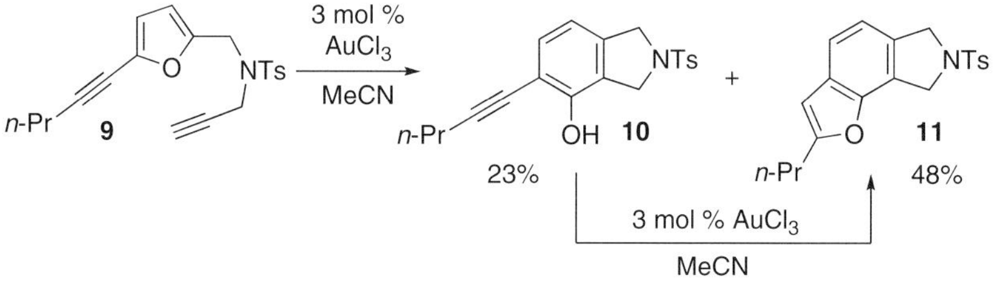

Scheme 9.3

Kabalka and coworkers disclosed a successful incorporation of the Mannich condensation into the cascade Cu(I)-promoted microwave-assisted cycloisomerization of *o*-ethynylphenols **12** into 2-(aminomethyl)benzofuran derivatives **15** in the presence of amines and *para*-formaldehyde (Scheme 9.4) [83, 84]. A variety of secondary alkyl and aryl amines **14** was employed in this transformation, affording the corresponding products within short (5 min) reaction times. Although, a threefold excess of CuI/alumina was required to achieve high levels of conversion, this solid-supported [85–88] promoter can be recycled without a noticeable loss of its catalytic activity. In addition, employment of the bidentate secondary amines, for instance piperazine, allows a double-Mannich/cyclization reaction, providing bis (benzofuranyl)-containing derivatives. The proposed mechanism (Scheme 9.5) involves the initial Mannich condensation of the alkyne **12** with an *in situ* generated

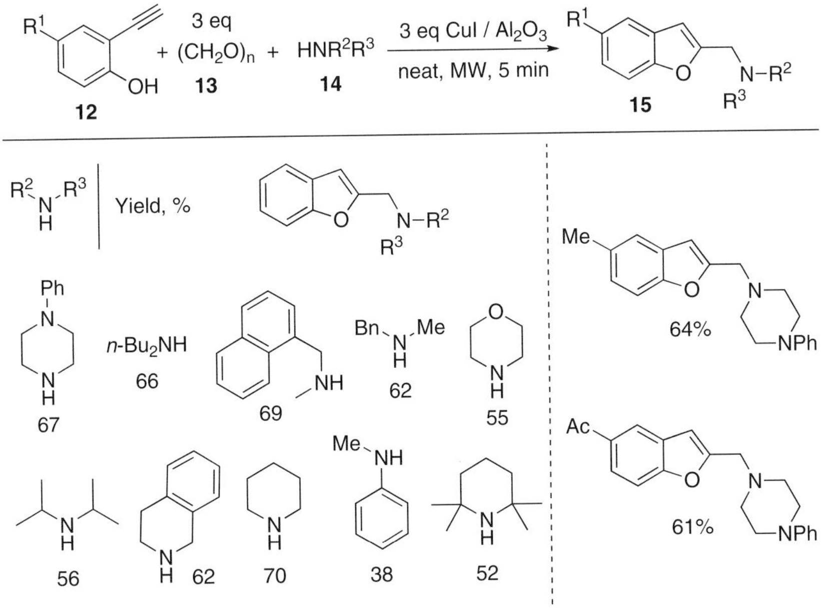

Scheme 9.4

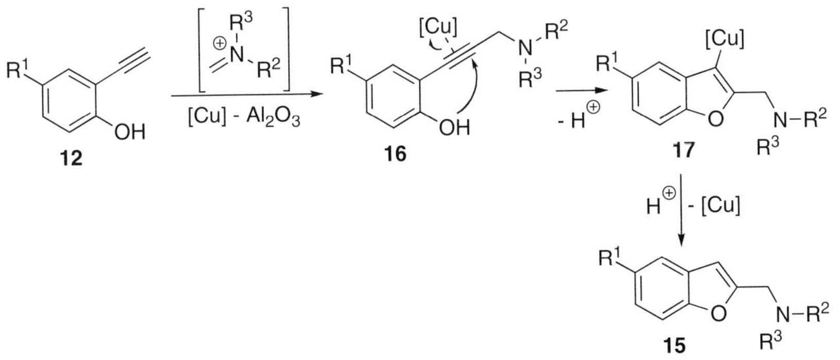

Scheme 9.5

iminium salt to give a reactive *o*-alkynylphenol intermediate **16**. The latter, upon activation with the Cu(I)-promoter, cyclizes into the corresponding benzofuran **15**.

Recently, Fürstner revealed that $PtCl_2$ could serve as a highly efficient catalyst for the synthesis of C2-monosubstituted benzofurans **19**. Thus, cyclization of *o*-alkynylphenols **18** occurred smoothly to provide a variety of benzofurans in excellent yields (Scheme 9.6) [89].

Later, many other research groups reported that other transition metals could serve as feasible catalysts for a facile cycloisomerization of **18** into the corresponding C2-substituted benzofuran cores **19** (Scheme 9.7). Along this line, Hiroya illustrated that

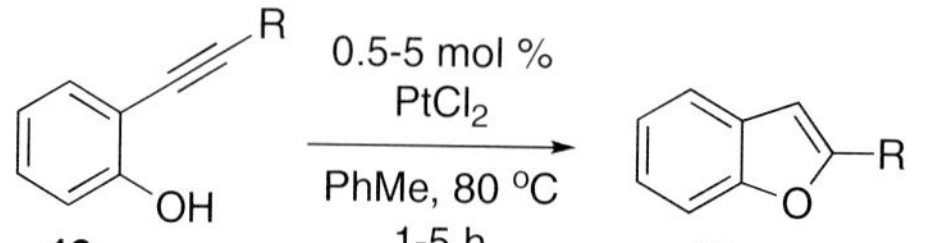

18 → **19**: 0.5-5 mol % $PtCl_2$, PhMe, 80 °C, 1-5 h

R^1	Yield, %
n-Pr	88
n-C_5H_{11}	98
c-Pr	98
CH_2CH_2Ph	98
$CH_2CH(CO_2Me)_2$	92
Ph	97
p-MeO-C_6H_4	95
m-F_3C-C_6H_4	94

Scheme 9.6

this transformation could successfully occur under catalysis with Cu(II) salts in aqueous alcohol media [90]. Crabtree's group utilized a cationic Ir(III) hydride complex for the cyclization of aryl- and alkyl-substituted alkyne precursors [91]. Catalysis with Au(III) salts for this transformation was employed by Krause (Scheme 9.7) [92]. Trost disclosed the Rh(II)-catalyzed cycloisomerization of terminal *o*-alkynylphenols **18** (R = H) into C2-C3-unsubstituted benzofurans involving the alkyne–vinylidene isomerization as an alkyne activation mode, similar to that utilized by McDonald in the synthesis of indoles (*vide infra*, Scheme 9.40) [93].

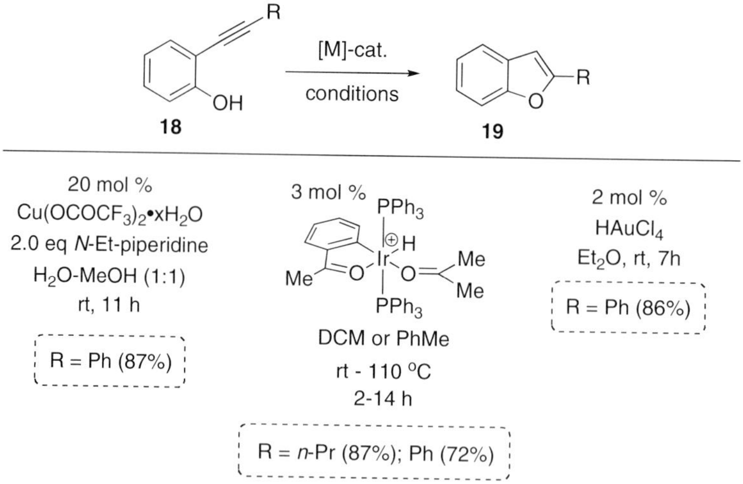

18 → **19**: [M]-cat., conditions

20 mol % $Cu(OCOCF_3)_2 \cdot xH_2O$, 2.0 eq *N*-Et-piperidine, H_2O-MeOH (1:1), rt, 11 h — R = Ph (87%)

3 mol % [Ir complex: PPh_3, H, Me, Me, PPh_3], DCM or PhMe, rt - 110 °C, 2-14 h — R = *n*-Pr (87%); Ph (72%)

2 mol % $HAuCl_4$, Et_2O, rt, 7h — R = Ph (86%)

Scheme 9.7

Ying and coworkers demonstrated that employment of cationic Au(I) complexes allowed a highly efficient cyclization of *o*-alkynylphenols **20** into an array of mono- and

Scheme 9.8

di-substituted benzofurans **21** (Scheme 9.8) [94, 95]. A facile cycloisomerization reaction proceeded under very mild reaction conditions in the presence of 2 mol% Au(I)-catalyst.

Further, Fiandanese reported the Cu(II)-catalyzed cycloisomerization/oxidative coupling of *ortho*-hydroxyphenyl-substituted diynes, allowing the preparation of conjugated bis-furyl-containing diynes [96]. Hsung illustrated that *O*-protected alkynylphenols, namely *ortho*-anisyl ynamides, could serve as surrogates for the corresponding reactive *ortho*-alkynylphenols in the synthesis of 2-amido benzofurans catalyzed by Rh(I) complex via the demethylation–cyclization sequence [97].

Despite high levels of efficiency and functional group compatibility achieved in the synthesis of the benzofuran framework via the cycloisomerization approach, utilization of alkynylphenols in this transformation clearly sets limits to the preparation of C3-unsubstituted cores only. This fact prompted development of alternative transition metal-catalyzed routes toward a straightforward assembly of far more valuable C3-functionalized benzofurans with a diverse substitution pattern.

Thus, Yamamoto developed the Pt(II)-catalyzed migratory cycloisomerization, namely carboalkoxylation, of *o*-alkynylphenyl acetals **22** leading to the synthesis of 2,3-disubstituted benzofurans **23**, where the former phenolic substituent resides at the C3-site (Scheme 9.9) [98, 99]. It was demonstrated that this Pt(II)-catalyzed cycloisomerization could be applied to substrates with a variety of alkynyl substituents, providing the corresponding benzofurans in moderate to excellent yields. In addition, different migrating oxyalkyl groups, such as 1-ethoxyethyl, MOM, BOM, and TBS-oxymethyl, can be effectively incorporated into this cascade transformation,

22 → **23**: 2 mol % $PtCl_2$, 8 mol % COD, PhMe, 30 °C, 1-96 h

R^1	R^2	R^3	Yield, %
n-Hex	Et	Me	91
$(CH_2)_4Cl$	Et	Me	92
c-Hex	Et	Me	94
t-Bu	Et	Me	trace
p-MeO-C_6H_4	Et	Me	90
Ph	Et	Me	88
p-F_3C-C_6H_4	Et	Me	61
n-Pr	Me	H	92
Ph	Me	H	73
n-Pr	Bn	H	94
Ph	Bn	H	83
Ph	TBS	H	61

Scheme 9.9

whereas the THP-group was shown to be completely inefficient. A plausible mechanism proposed by the authors for the Pt(II)-catalyzed reaction of **22** is illustrated in Scheme 9.10. Accordingly, $PtCl_2$ coordinates to the triple bond of the alkyne **22** and activates it toward the nucleophilic attack (**24**) by the oxygen atom of the acetal moiety to form the cyclized zwitterionic intermediate **25**. An intramolecular

Scheme 9.10

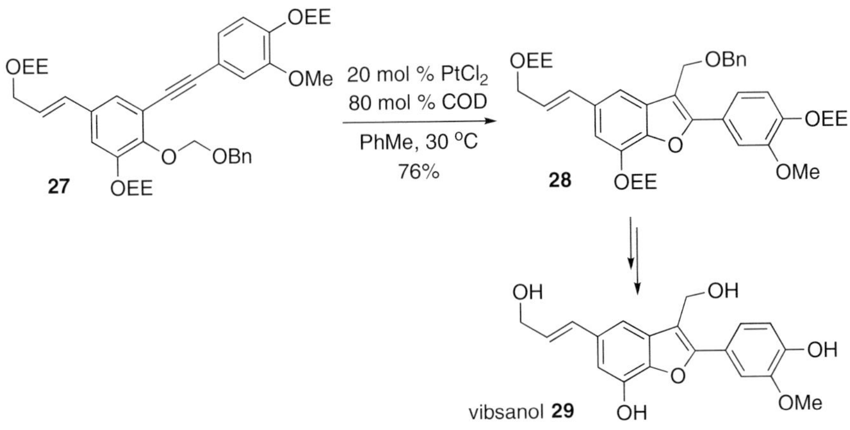

Scheme 9.11

1,3-migration of the alkoxyalkyl group in the latter to the C-Pt site produces the oxonium intermediate **26**. A subsequent elimination of the Pt(II) catalyst provides product **23**. The synthetic potential of this methodology was further demonstrated by the successful application of this carboalkoxylation reaction to a concise total synthesis of vibsanol **29**, an inhibitor of lipid peroxidation (Scheme 9.11).

Furthermore, Fürstner independently reported that *O*-functionalized *ortho*-alkynylphenols **30**, bearing cation-stabilizing groups other than alkoxyalkyl, underwent the Pt(II)-catalyzed cycloisomerization involving a formal 1,3-migration of these groups to give 2,3-disubstituted benzofurans **31** (Scheme 9.12) [89]. In addition, the migratory

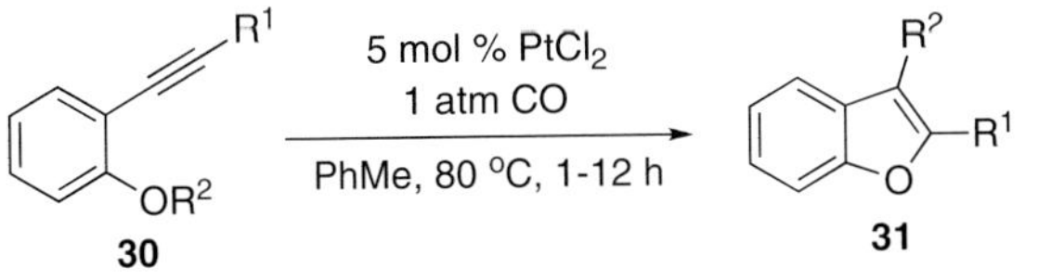

R^1	R^2	Yield, %
n-Pr	allyl	88
CH_2CH_2Ph	allyl	94
m-MeO-C_6H_4	allyl	98
m-F_3C-C_6H_4	allyl	94
n-C_5H_{11}	-CH_2(Me)C=CH_2	73
CH_2CH_2Ph	-CH_2(Me)C=CH_2	71
n-C_5H_{11}	-CH_2(Br)C=CH_2	54
n-C_5H_{11}	-CH_2CH=CH-Ph-(*E*)	68
n-C_5H_{11}	Bn	66
n-C_5H_{11}	PMB	76
m-F_3C-C_6H_4	PMB	78
c-Pr	PMB	77

Scheme 9.12

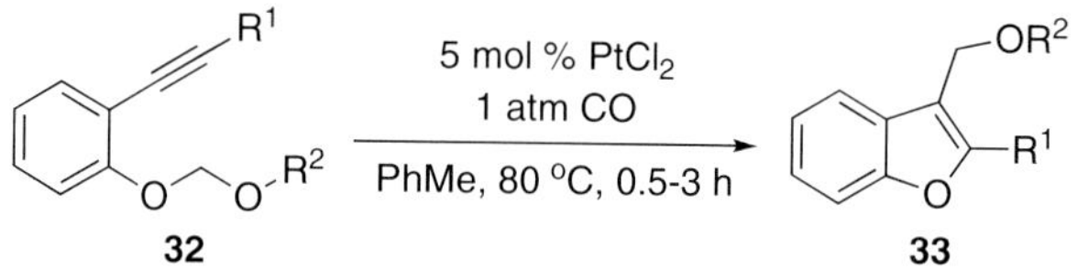

32 → 33: 5 mol % $PtCl_2$, 1 atm CO, PhMe, 80 °C, 0.5-3 h

R^1	R^2	Yield, %
n-C_5H_{11}	Me	91
CH_2CH_2Ph	Me	84
CH_2CH_2Ph	Bn	78
CH_2CH_2Ph	$CH_2CH_2SiMe_3$	81
c-Pr	Me	84
Ph	Me	95
m-MeO-C_6H_4	Me	62
m-F_3C-C_6H_4	Me	74
m-F_3C-C_6H_4	Bn	84
o-(i-PrO)-C_6H_4	Bn	94
o-Br-C_6H_4	$CH_2CH_2SiMe_3$	75

Scheme 9.13

cycloisomerization of alkoxymethyl analogs **32** proceeded equally efficiently, yielding benzofuran derivatives **33**, analogously to that disclosed by the Yamamoto's group (Scheme 9.13). A variety of functional groups could be tolerated in these cascade transformations, affording the corresponding benzofuran derivatives **31** and **33** in moderate to excellent yields. The mechanistic rationale proposed by the authors is similar to that outlined in Scheme 9.10. In the case of allyl migrating groups, a formal 1,3-migration was proposed to occur via the π–allylplatinum intermediate. A practical significance of the developed chemistry was later confirmed by its application to an efficient synthesis of the pterocarpane skeleton of the antibiotic Erypoegin H **36** and its derivatives (Scheme 9.14) [100].

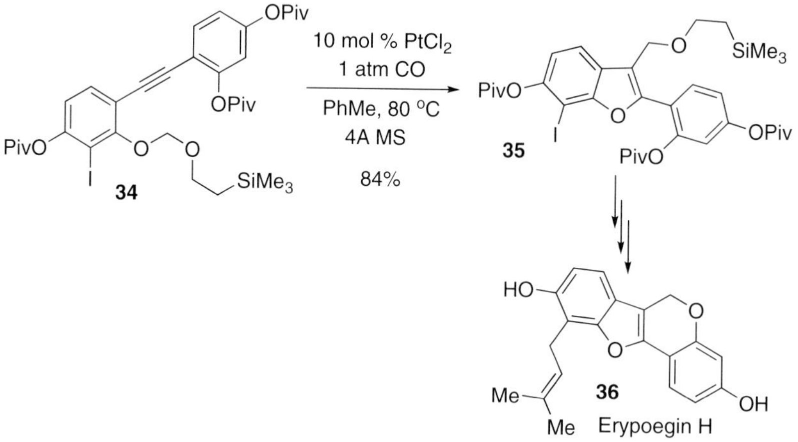

34 → 35: 10 mol % $PtCl_2$, 1 atm CO, PhMe, 80 °C, 4A MS, 84%; 35 → 36 Erypoegin H

Scheme 9.14

9.2.1.2 Synthesis of Benzofurans via Intramolecular Arylation Reactions

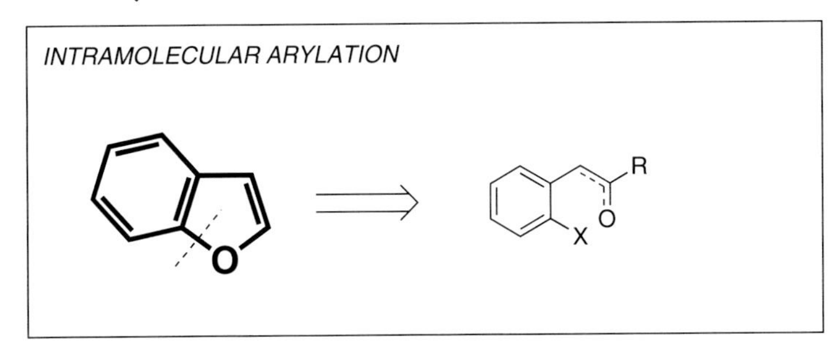

Employment of intramolecular arylation of *o*-halophenyl ethanones for the construction of a substituted benzofuran skeleton in the presence of Cu(0) powder was first reported by Grimshaw [101, 102].

Recently, Chen and coworkers revealed that a wide range of mono- and di-substituted benzofurans **38** could be efficiently synthesized via the Cu(I)-catalyzed ring closure of 2-haloaromatic ketones **37**, which proceeded through the Cu(I)-catalyzed intramolecular O-arylation [103] of enolates generated from the corresponding carbonyl compounds (Scheme 9.15) [104]. In addition, it was shown that both iodo- and bromo-ketones **37** could be employed in the Cu(I)-catalyzed reaction, whereas closure of 2-chloro derivatives could not be achieved under these conditions. Aromatic ketones **37** underwent cyclization into the corresponding benzofurans more readily than their alkyl analogs, presumably due to a more facile requisite enolization in the presence of an adjacent aromatic group.

Analogous Cu(I)-catalyzed transformation of 2-bromophenyl ketones into benzofuran derivatives was also described by Ackermann [105]. Li observed a similar cyclization in the presence of the CuI–DMEDA–Cs_2CO_3 catalytic system [106].

Following these reports, SanMartin and Domínguez developed a recyclable CuI–TMEDA catalytic system for the ring-closing transformation of a variety of 2-bromoaryl ethanones **39**, which allowed facile assembly of mono- and tri-substituted, and even fused benzofurans **40** in aqueous media, normally, in good to excellent yields (Scheme 9.16) [107, 108].

Ackermann further elaborated on this chemistry and found that the previously unreactive chlorides **41** could be cyclized into the corresponding benzofurans **42** in the presence of 10 mol% of CuI and Me_2NCH_2COOH as a stabilizing ligand (Scheme 9.17) [105].

Willis reported that alkenyl triflates **43** could be employed in the Cu(I)-catalyzed synthesis of benzofurans **44** as surrogates of the respective 2-bromoaryl ketones analogs **37** used by Chen (Scheme 9.18) [109]. This reaction was proposed to proceed via the initial hydrolysis of triflates **43** to give reactive enolates which, upon Cu-catalyzed intramolecular arylation, were transformed into various polycyclic fused benzofuran derivatives **44**.

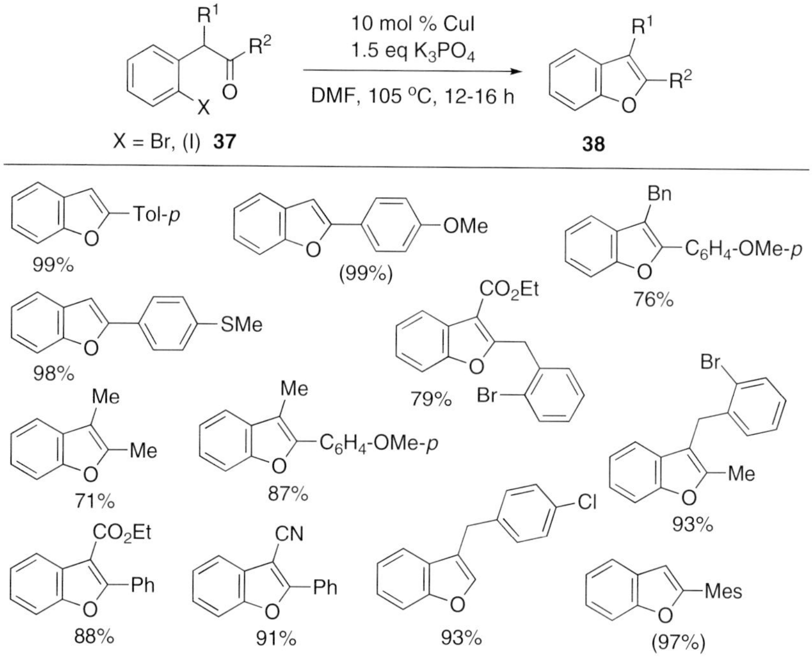

R^1, R^2, X; X = Br, (I) **37** → 10 mol % CuI, 1.5 eq K_3PO_4, DMF, 105 °C, 12-16 h → R^1, R^2, O **38**

Tol-*p* 99%; OMe (99%); Bn, C_6H_4-OMe-*p* 76%; SMe 98%; CO_2Et, Br 79%; Me, Me 71%; Me, C_6H_4-OMe-*p* 87%; Br, Me 93%; CO_2Et, Ph 88%; CN, Ph 91%; Cl 93%; Mes (97%)

Scheme 9.15

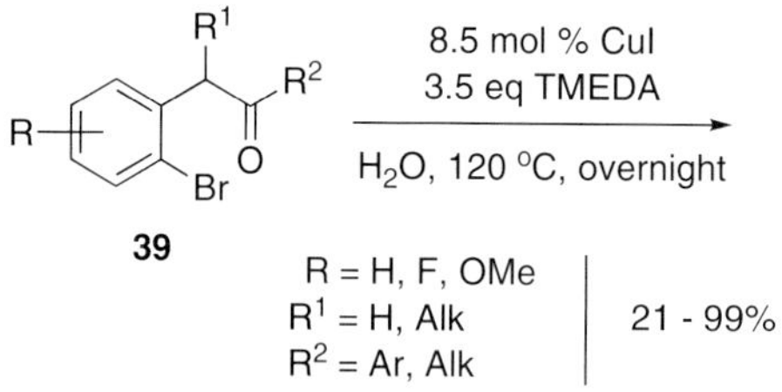

R^1, R^2, R, Br, O **39** → 8.5 mol % CuI, 3.5 eq TMEDA, H_2O, 120 °C, overnight → R^1, R^2, R, O **40**

R = H, F, OMe; R^1 = H, Alk; R^2 = Ar, Alk | 21 - 99%

Scheme 9.16

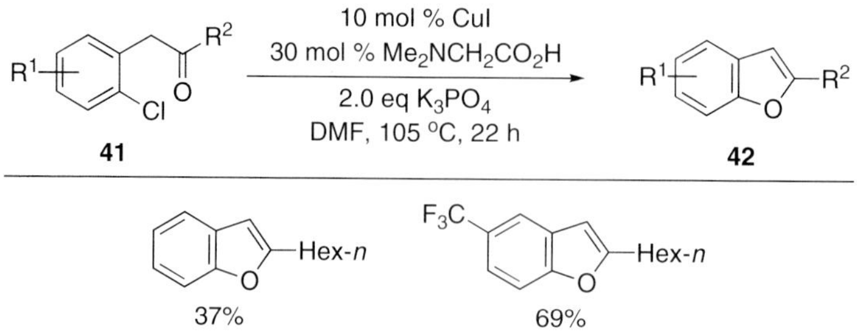

R^1, R^2, Cl, O **41** → 10 mol % CuI, 30 mol % $Me_2NCH_2CO_2H$, 2.0 eq K_3PO_4, DMF, 105 °C, 22 h → R^1, R^2, O **42**

Hex-*n* 37%; F_3C, Hex-*n* 69%

Scheme 9.17

Scheme 9.18

9.2.1.3 Synthesis of Benzofurans via "4 + 1" Cycloaddition Reactions

FORMAL [4+1] CYCLOADDITION

A highly efficient three-component coupling–cycloisomerization reaction, leading to the assembly of 3-aminobenzofuran derivatives, was recently reported by Sakai (Scheme 9.19) [110]. Substituted salicylaldehydes **45** underwent a facile cascade transformation in the presence of secondary amines **46**, alkynyl silanes **47**, and the bimetallic Cu(I)–Cu(II) catalytic system, producing an array of different di- and tri-substituted benzofurans **48**. According to the proposed mechanism, the reaction begins with the exchange process between the alkynyl silane **47** and the Cu(I)-catalyst to generate the copper(I) acetylide **49**. In parallel, a more Lewis-acidic Cu(II)-catalyst triggers the formation of an iminium intermediate **51** from a secondary amine **46** and aldehyde **45**. A subsequent nucleophilic addition of the organocopper species **49** to **51** produces the key skipped *o*-propargylphenol **52**. This reactive intermediate undergoes Cu(II)-catalyzed 5-*exo-dig* cyclization and transforms into the cyclic vinylcopper species **53**. Finally, a protiodemetalation–aromatization sequence of the latter furnishes the benzofuran product **48** (Scheme 9.20).

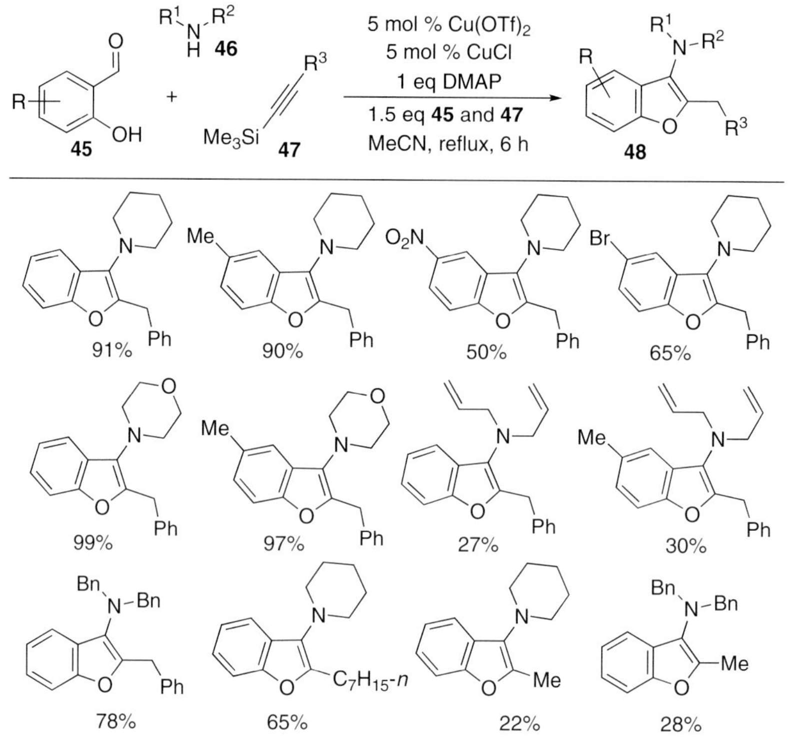

R–(C₆H₃)(CHO)(OH) **45** + R^1–NH–R^2 **46** + Me_3Si–C≡C–R^3 **47**

5 mol % $Cu(OTf)_2$
5 mol % CuCl
1 eq DMAP
1.5 eq **45** and **47**
MeCN, reflux, 6 h

→ **48** (3-NR^1R^2-2-CH_2R^3-benzofuran)

91%, 90%, 50%, 65%

99%, 97%, 27%, 30%

78%, 65%, 22%, 28%

Me, O_2N, Br, N, O, Ph, Bn, C_7H_{15}-*n*, Me

Scheme 9.19

X = DMAP or $\overset{\ominus}{OH}$

Ar = R–(C₆H₃)–OH

R^1, N, R^2, $Cu(OTf)_2$, R^3, R, O, H, X- -H, **52**

- XH

R^1–N–R^2, $\overset{\ominus}{Cu}(OTf)_2$, R^3, **53**

48

$Cu(OTf)_2$

45 + **46**

NR^1R^2, Ar, O, H, $Cu(OTf)_2$, **50**

$\overset{\oplus}{N}R^1R^2$, Ar, **51**

CuCl

Me_3Si–C≡C–R^3 **47**

Cu–C≡C–R^3 **49**

Me_3SiCl

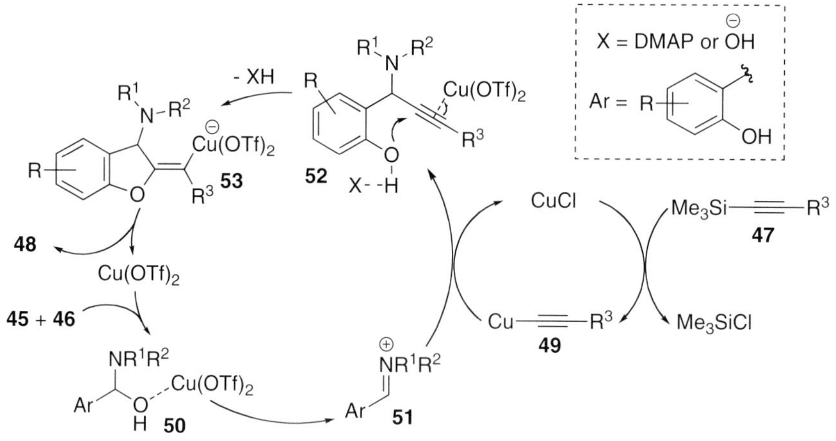

Scheme 9.20

9.2.1.4 Synthesis of Benzofurans via "3 + 2" Cycloaddition Reactions

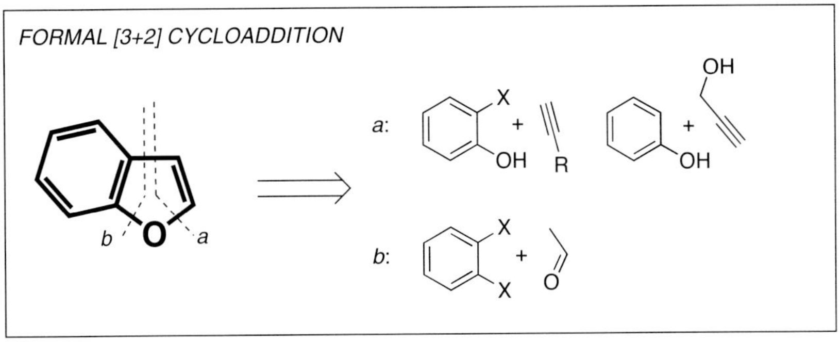

FORMAL [3+2] CYCLOADDITION

As mentioned earlier (Chapter 9.2.1.1), a formal [3 + 2] approach toward the synthesis of benzofurans **57** and related compounds involving a coupling–cyclization reaction between *ortho*-halophenol **54** and copper(I) acetylides **55** was first established by Castro and coworkers (Scheme 9.21) [75–77, 111]. It was proposed that the formation of the benzofuran **57** is a stepwise reaction involving formation of the *ortho*-alkynylphenol intermediate **56** [77]. The authors demonstrated that a variety of substituted benzofurans could be synthesized in generally high yields via this stoichiometric approach.

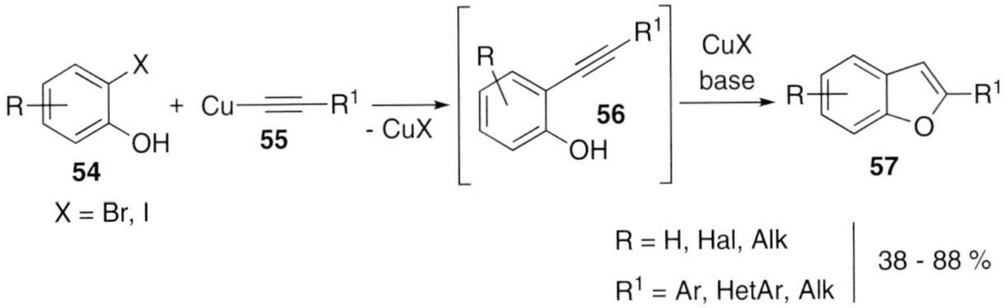

X = Br, I

R = H, Hal, Alk | 38 - 88 %
R^1 = Ar, HetAr, Alk

Scheme 9.21

Despite its obvious limitation to the requisite preparation of copper(I) acetylides, this seminal work greatly influenced progress in the development of transition metal-catalyzed approaches for the preparation of benzofuran units, as well as other heterocyclic cores. Furthermore, since its original discovery, this protocol was featured in a number of syntheses of an array of multisubstituted benzofuran derivatives, including the preparation of various biologically active molecules and valuable intermediates [112–116].

This methodology was later extended by Owen [117] and Nilsson [78] to the employment of *o*-iodo- and -bromophenols and terminal acetylenes in the presence of stoichiometric amounts of Cu(I) oxide in pyridine or Cu(I) *tert*-butoxide to generate acetylide **55**, allowing the achievement of typically higher or comparable yields (60–82%) of the corresponding benzofurans **57**. Ohemeng [118] and Scammells [119] effectively employed the first catalytic system in the synthesis of biologically active benzofuran derivatives, whereas Nilsson succeeded in the assembly of

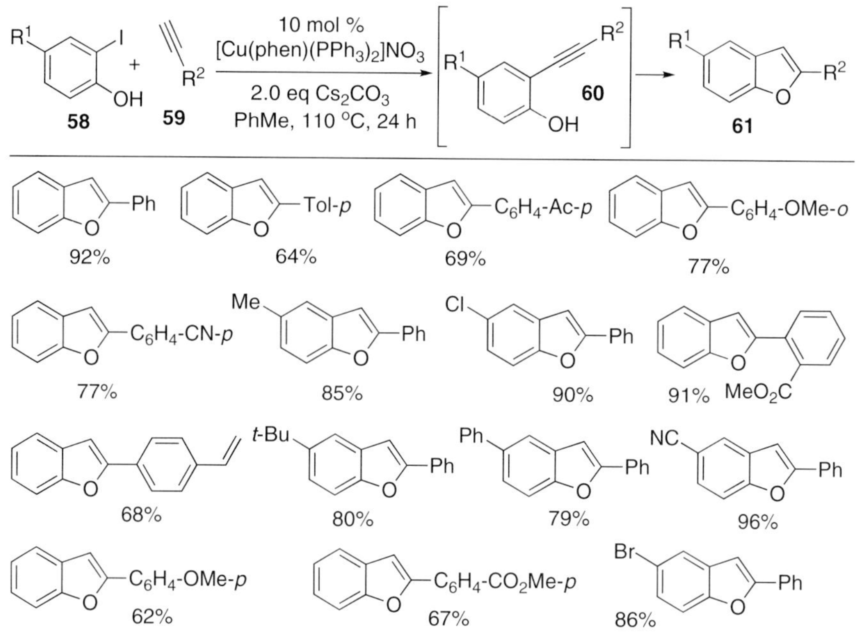

Scheme 9.22

isocoumestan **4** from *o*-iodophenol **3** and ethyl propiolate using the second catalyst (*vide supra*, Scheme 9.1) [78].

The synthesis of mono- and di-substituted benzofurans **61** utilizing a catalytic-in-copper version of this transformation was recently described by Venkataraman (Scheme 9.22) [120, 121]. The Cu(I)-catalyzed protocol tolerates a wide range of sensitive functional groups and is general in terms of its applicability toward both electron-rich and electron-deficient alkynes **59** and *o*-iodophenol **58** coupling components.

More recently, Bolm reported that the "3 + 2" synthesis of benzofurans from terminal alkynes and *ortho*-iodophenols could be achieved with a moderate efficiency in the presence of the $FeCl_3$–DMEDA [122] catalytic system [123].

Liu described an interesting "3 + 2" approach toward benzofuran compounds **65**, involving the cascade Zn(II)-catalyzed Markovnikov hydroxylation of terminal propargylic alcohols **63** with phenols **62**, followed by the electrophilic cyclocondensation of an *in situ* generated α-aryloxy carbonyl compound **64**. (Scheme 9.23) [124].

Ma and coworkers recently utilized a double arylation strategy for an efficient synthesis of benzofuran-3-carboxylates **69** (Scheme 9.24) [125]. In the first step of this domino reaction, the Cu(I)-catalyzed chemoselective C-arylation of β-ketoesters **67** with *ortho*-bromoiodobenzenes **66** provided the key *o*-bromoaryl-substituted enolates **68**. These reactive intermediates (*vide supra*) readily underwent the second Cu(I)-catalyzed intramolecular *O*-arylation, furnishing the desired benzofurans **69**. The reaction tolerates a variety of substituents on both coupling partners **66** and **67**, allowing for the preparation of tri- and tetra-substituted frameworks in good yields.

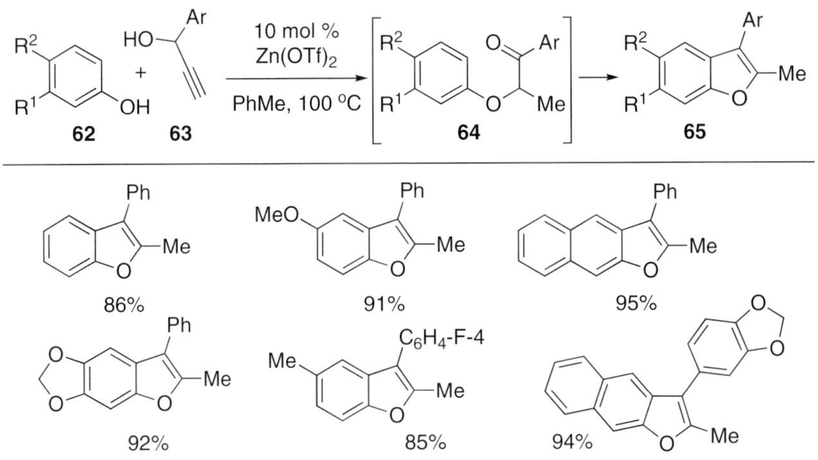

Scheme 9.23

9.2.2
Benzothiophenes

Synthesis of functionalized benzo[*b*]thiophenes is of substantial interest as this heterocyclic framework is often found in biologically active compounds and

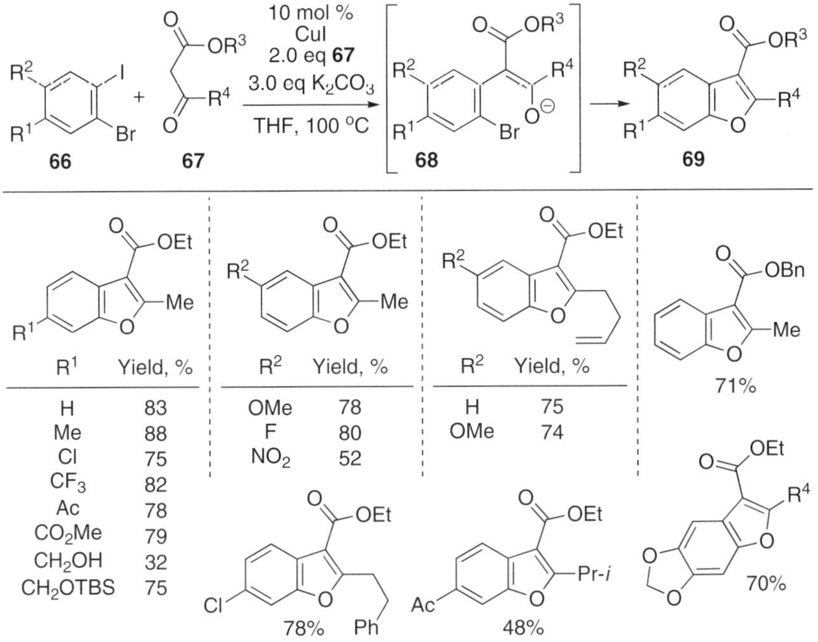

R^1	Yield, %
H	83
Me	88
Cl	75
CF_3	82
Ac	78
CO_2Me	79
CH_2OH	32
CH_2OTBS	75

R^2	Yield, %
OMe	78
F	80
NO_2	52

R^2	Yield, %
H	75
OMe	74

Scheme 9.24

incorporated into a variety of important synthetic materials [126–128]. In addition, the benzothiophene core represents an interesting synthon, which has found some applications in the syntheses of advanced intermediates [6, 9, 64, 129–132]. A variety of methodologies and different protocols for the synthesis of benzothiophenes and their derivatives have been reported [126, 133–138]. However, relatively few of them involve the employment of transition metal catalysts or promoters for the construction of the benzothiophene ring from acyclic precursors. Among the limited number of the latter protocols, increasing attention has been focused recently on the development of cycloisomerization approaches, whereas a few methodologies featuring formal [4 + 1] and [3 + 2] cycloaddition processes have also been reported. Accordingly, a detailed discussion of the recent contributions to the area of benzo[*b*]thiophene synthesis via these three routes is provided below.

9.2.2.1 **Synthesis of Benzothiophenes via Cycloisomerization Reactions**

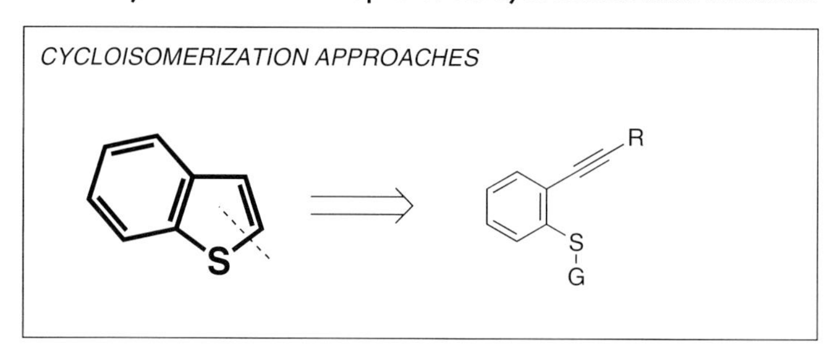

During their studies on the benzo[*b*]thiophene core synthesis via the "3 + 2" cycloaddition, namely the coupling–cyclization route of copper(I) acetylides or their surrogates and *ortho*-bromothiophenol (*vide infra*, Chapter 9.2.2.3), Castro and coworkers suggested that the formation of the title heterocyclic core occurred through the generation of *ortho*-alkynylthiophenols [139]. Unfortunately, these attractive substrates are inaccessible via the Sonogashira coupling [140] of *ortho*-halobenzenethiols, as the mercapto group poisons Pd-catalysts. Consequently, an array of existing cycloisomerization methodologies developed for the assembly of benzo[*b*]furan and 1*H*-indole systems found very limited application toward the synthesis of the benzo[*b*]thiophene core.

An elegant solution to this limitation was recently developed by Yamamoto, where highly efficient syntheses of diversely C2–C3 functionalized benzothiophenes became available via a set of transition metal-catalyzed migratory cycloisomerization approaches utilizing easily accessible *S*-protected substrates. Considering recent developments of such processes for the construction of benzofuran (*vide supra*) and indole (*vide infra*) cores, Yamamoto's group first disclosed the Au(I)-catalyzed cycloisomerization of alkoxymethyl-substituted *o*-alkynylthiophenols **70** into the corresponding 2,3-disubstituted benzothiophenes **71**, proceeding via the key intramolecular 1,3-migration of the former *S*-substituent (Scheme 9.25) [141]. The authors demonstrated that this carbothiolation reaction could be equally efficiently catalyzed by Au(III) and Pt(II) salts. Other groups capable of stabilizing positive

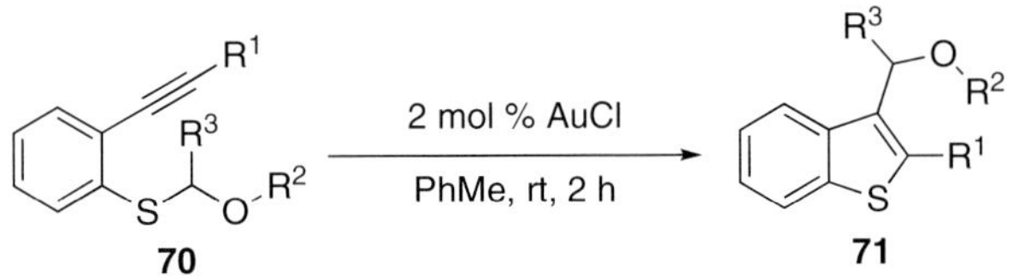

70 → 71 (2 mol % AuCl, PhMe, rt, 2 h)

R^1	R^2	R^3	Yield, %
n-Pr	Me	H	93
c-Hex	Me	H	92
t-Bu	Me	H	96
Ph	Me	H	99
p-F_3C-C_6H_4	Me	H	100
p-MeO-C_6H_4	Me	H	96
CO_2Et	Me	H	85
Ph	TBS	H	99
Ph	MPM	H	95
Ph	TMSE	H	92
n-Pr	Et	Me	92
Ph	Et	Me	98
n-Pr	$-(CH_2)_4-$		98
Ph	$-(CH_2)_4-$		93

Scheme 9.25

charge migration, such as allyl and *p*-methoxybenzyl, were also efficient in this cascade transformation, providing the target 3-allyl- or 3-benzyl benzothiophene derivatives **73**, respectively (Scheme 9.26). It was further demonstrated that substrates bearing a variety of alkyne substituents could readily undergo the Au(I)-catalyzed cycloisomerization, affording the corresponding benzothiophenes in excellent yields. Remarkably, different 1,3-migrating oxyalkyl groups, including Si-containing and cyclic tetrahydropyranyl substituent, can be easily incorporated into this cascade transformation. The proposed mechanism is outlined in Scheme 9.27 and is similar to that suggested for the Pt(II)-catalyzed benzofuran synthesis in Scheme 9.10.

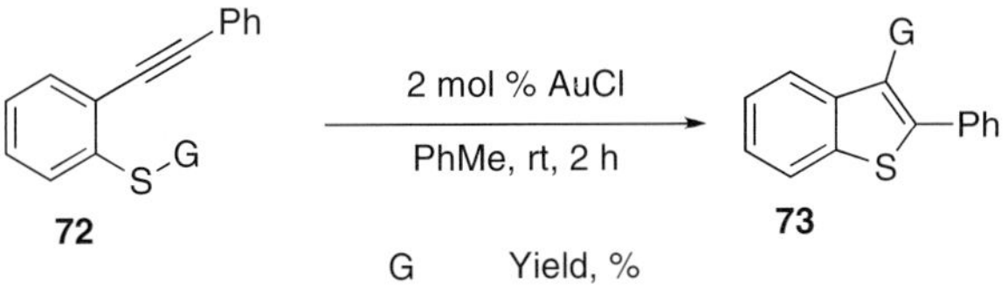

72 → 73 (2 mol % AuCl, PhMe, rt, 2 h)

G	Yield, %
allyl	98
PMB	93

Scheme 9.26

Scheme 9.27

Later, the same group extended this methodology to the synthesis of 3-silyl-containing benzothiophenes **78** via the Au(I)-catalyzed cyclization–1,3-Si-migration cascade transformation of silyl-protected substrates **77** (Scheme 9.28) [142]. A variety of di-, tri-, and tetra-substituted, as well as fused 3-benzothiophene silanes **77** can be efficiently synthesized via this protocol in excellent yields. Lower yields were achieved for substrates with a substantial congestion at the alkyne moiety, strong electron-withdrawing substituents, or less nucleophilic Si-migrating groups. Interestingly, the authors observed that a 1,3-migration of a silyl substituent proceeded intermolec-

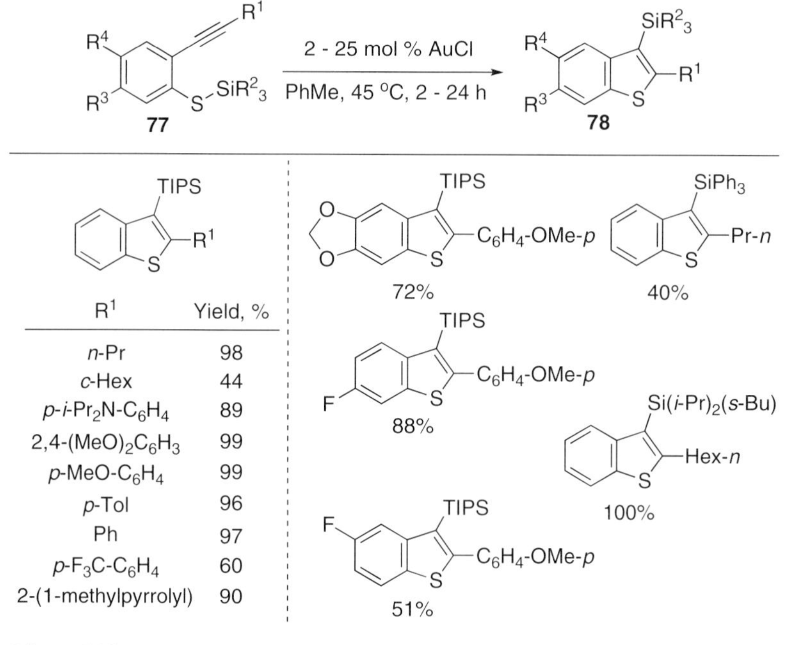

R^1	Yield, %
n-Pr	98
c-Hex	44
p-*i*-Pr_2N-C_6H_4	89
2,4-$(MeO)_2C_6H_3$	99
p-MeO-C_6H_4	99
p-Tol	96
Ph	97
p-F_3C-C_6H_4	60
2-(1-methylpyrrolyl)	90

Scheme 9.28

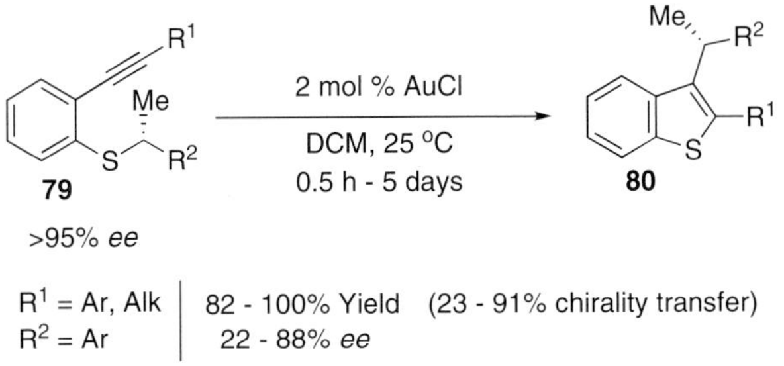

R1
Me
S
R2
79
>95% ee
2 mol % AuCl
DCM, 25 °C
0.5 h - 5 days
Me
R2
R1
S
80
R1 = Ar, Alk
R2 = Ar
82 - 100% Yield (23 - 91% chirality transfer)
22 - 88% ee

Scheme 9.29

ularly, presumably due to the much lower migratory ability of the Si-group, relative to that of the C-migrating group, and, consequently, a longer lifetime for the silylsulfonium intermediate analogous to **75**.

Finally, Yamamoto reported that the Au-catalyzed carbothiolation reaction of optically active *o*-alkynylphenyl sulfides **79**, bearing chiral *S*-substituents, proceeded with predominant retention of the configuration at the 1-arylethyl migrating group in **80** (Scheme 9.29) [143]. This observation indicated that the 1,3-migration of the 1-arylethyl group proceeded through the formation of the contact ion pair **82**, followed by C–C bond formation to give the corresponding benzothiophene **80** (Scheme 9.30).

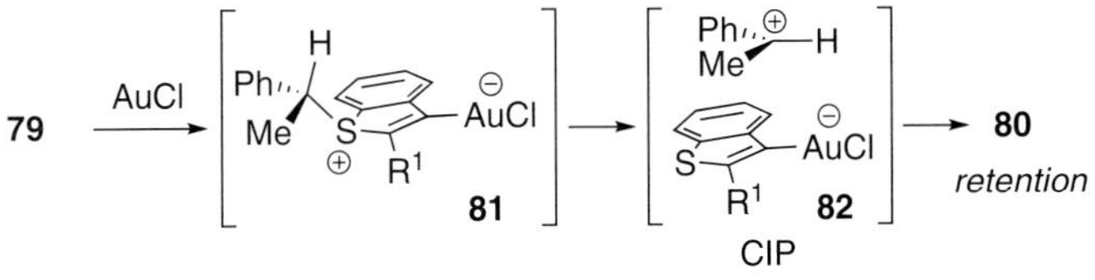

79
AuCl
H
Ph
Me
S
R1
AuCl
81
Ph
H
Me
S
AuCl
R1
82
CIP
80
retention

Scheme 9.30

9.2.2.2 **Synthesis of Benzothiophenes via "4 + 1" Cycloaddition Reactions**

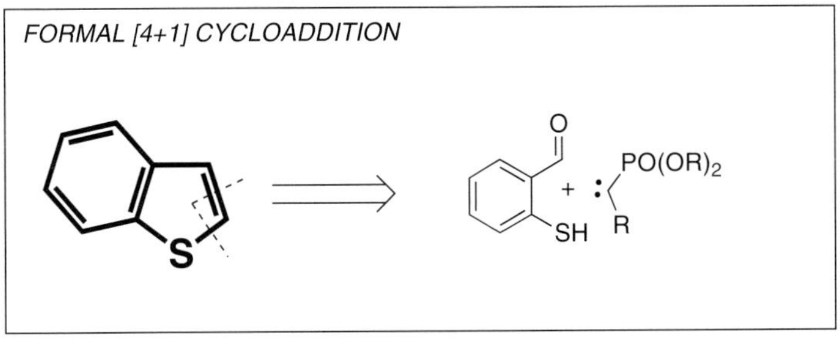

FORMAL [4+1] CYCLOADDITION
S
O
SH
PO(OR)2
R

Nakamura reported a single example of the fairly efficient Rh(II)-catalyzed "4 + 1" benzothiophene synthesis, utilizing a one-pot reaction initiated by the S–H insertion of thiophenol **83** into the metallocarbenoid center derived from α-diazophosphonate **84**. A subsequent base-mediated intramolecular Horner–Wadsworth–Emmons

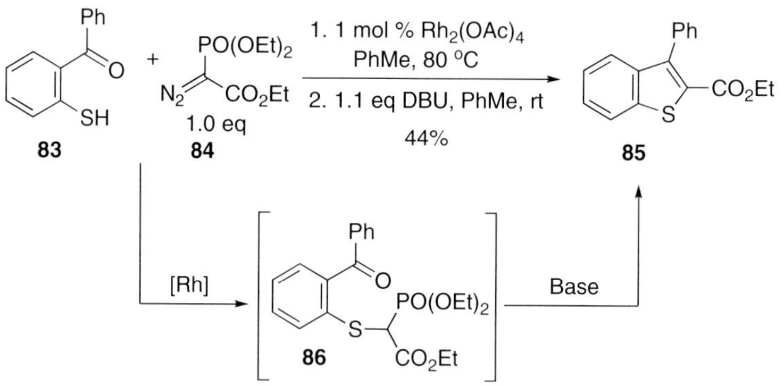

Scheme 9.31

reaction [144] of the *in situ* generated phosphonate **86** furnished the 3-phenylbenzothiophene-2-carboxylate **85** (Scheme 9.31) [145].

9.2.2.3 Synthesis of Benzothiophenes via "3 + 2" Cycloaddition Reactions

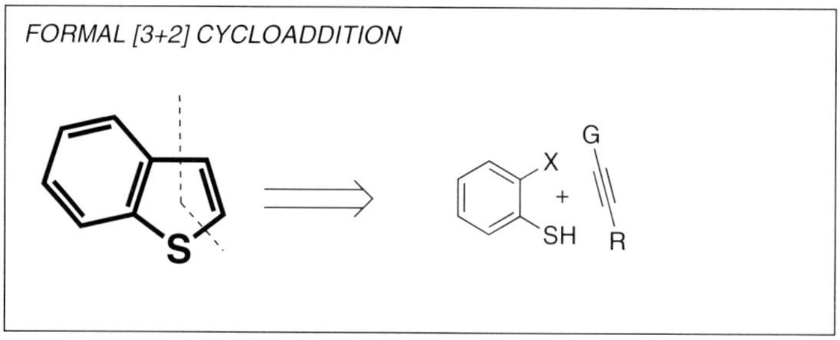

FORMAL [3+2] CYCLOADDITION

Castro introduced a stoichiometric-in-copper synthesis of the benzothiophene **90** framework via the formal [3 + 2] cycloaddition route, involving the coupling–cyclization cascade of stable copper(I) acetylides **88** and *ortho*-bromothiophenol **87** (Scheme 9.32) [139]. Later, the same researchers suggested that this reaction proceeded through the formation of a reactive *ortho*-alkynylthiophenol intermediate **89**, which, upon the concomitant cyclization, accomplished the assembly of the

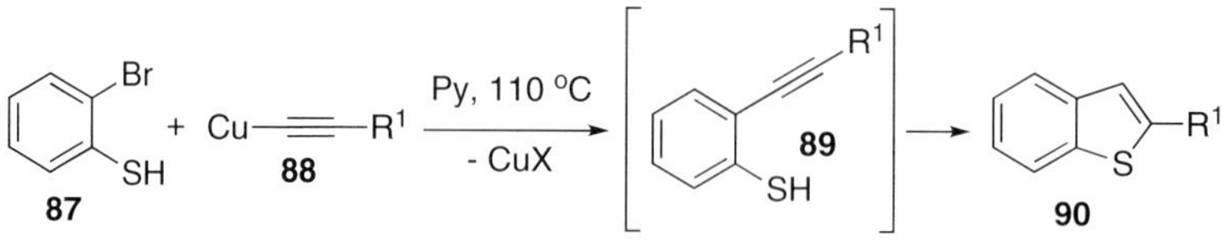

R^1	Yield, %
Ph	90
n-Bu	80
n-Pr	80

Scheme 9.32

benzo[*b*]thiophene skeleton. In the case of unstable Cu(I) acetylides, the latter can be generated *in situ* from terminal alkynes **91** in the presence of stoichiometric amounts of CuI and tertiary amine base. However, these reaction conditions provide the corresponding benzothiophenes in low yields (Scheme 9.33).

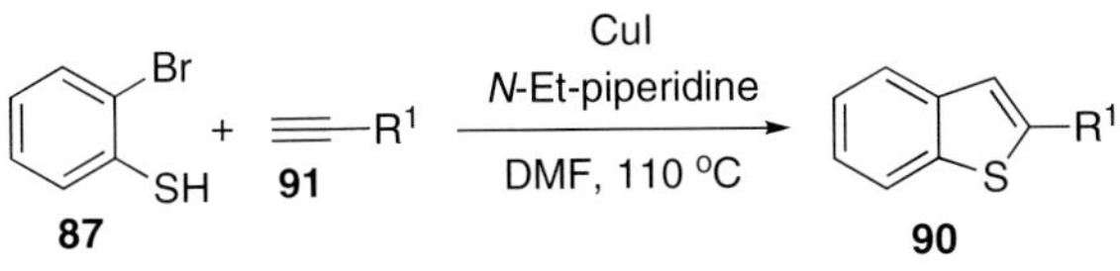

R^1	Yield, %
CO_2Et	35
CH_2OH	10

Scheme 9.33

Effenberger reported the Ag-mediated synthesis of trisubstituted benzothiophenes **95** via a formal [3 + 2] cycloaddition reaction of arylsulfenyl chlorides **92** and acetylenes **93** (Scheme 9.34) [146]. This transformation involved the initial *trans*-sulfenyltriflation of alkynes **93** to give highly reactive vinyl triflates **94**, which, upon a subsequent Bischler-type cyclization under the reaction conditions, produced benzothiophenes **95** in variable yields.

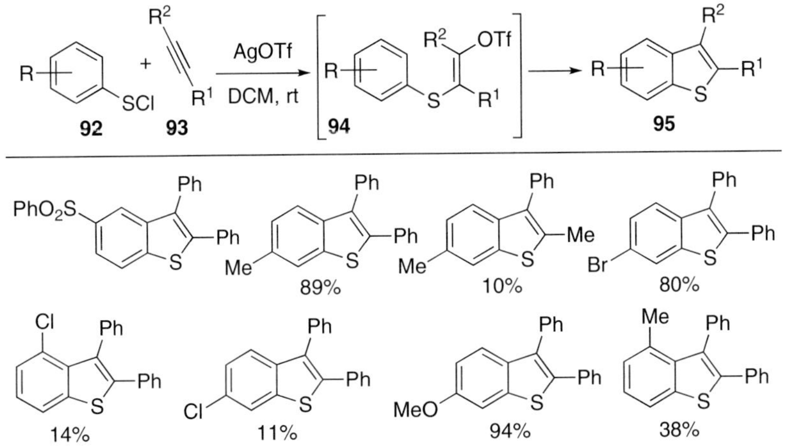

Scheme 9.34

9.2.3
Indoles

Indoles represent a very important and unique class of fused five-membered heterocyclic compounds and are probably the most prevalent among other heterocyclic frameworks broadly found in nature. The indole ring system is incorporated into a vast number of structurally diverse biologically active natural and synthetic compounds [147–157]. Consequently, the indole ring system has become an essential, or so called "privileged," structural motif in many pharmaceutical agents [156, 158]. Furthermore, indole-containing structures have found widespread application as materials with an array of valuable properties [5, 156, 158, 159], as well as reactive intermediates in the synthesis of fine chemicals [6, 9, 64, 160–164]. Not surprisingly, investigation of the chemistry of this fascinating heterocycle has been sustained to be one of the most important objectives of heterocyclic chemistry for over 100 years. Great attention directed towards the synthesis of functionalized indole derivatives stimulated the establishment of a myriad of methods for their preparation [165–177]. Nonetheless, the development of novel general transition metal-catalyzed processes for the synthesis and functionalization of the indole ring system in a highly chemo- and region-selective fashion continues to be of significant importance [148, 163, 167, 176, 178–181]. Numerous innovative and practical methodologies using transition metal complexes as either catalysts or reagents have been developed over the last three decades. Arguably, among all the advances in this area, cycloisomerization, intramolecular arylation, and formal [4 + 1] and [3 + 2] cycloaddition approaches demonstrate the highest potential in the facile construction of diversely functionalized indoles. Accordingly, the most important and recent contributions to these processes are included in the discussion provided below.

9.2.3.1 **Synthesis of Indoles via Cycloisomerization Reactions**

CYCLOISOMERIZATION APPROACHES

Castro first documented cycloisomerization of *ortho*-alkynylanilines **96** to give the 2-phenyl indole **97** in the presence of substoichiometric amounts of Cu(I) catalyst in excellent yield (Scheme 9.35) [77, 111]. Mechanistically, it is believed that 5-*exo-dig* cyclization follows the generally accepted mechanism proposed for an intramolecular addition of various nucleophilic entities to transition metal-activated carbon–carbon multiple bonds.

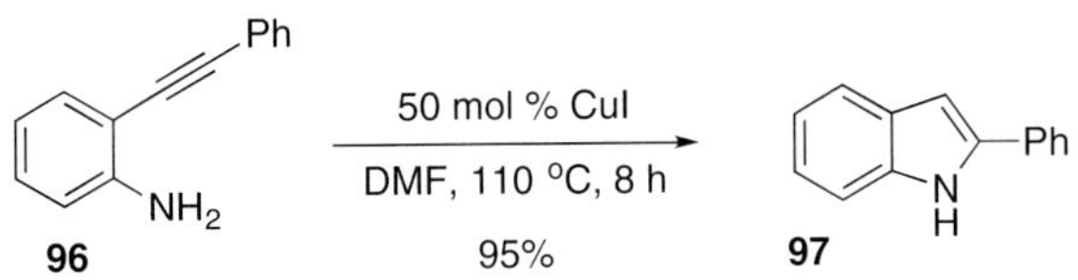

Ph
50 mol % CuI
DMF, 110 °C, 8 h
95%
NH_2
96
97
N H
Ph

Scheme 9.35

Later, Yamamoto employed this chemistry for very efficient synthesis of functionalized indole derivatives **99** (Scheme 9.36). Substantially higher yields were achieved upon addition of $CaCO_3$, whereupon Si-substituted alkynes **98** underwent a concomitant deprotection to give C-2 unsubstituted indoles in good to excellent yields [182].

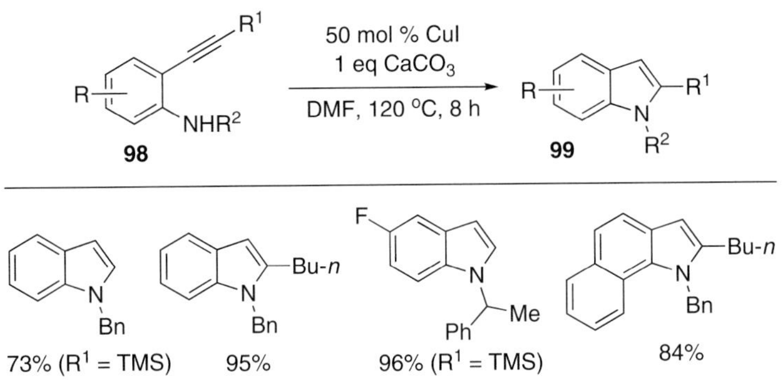

R^1
50 mol % CuI
1 eq $CaCO_3$
DMF, 120 °C, 8 h
NHR^2
98
99
R^2

73% (R^1 = TMS) 95% 96% (R^1 = TMS) 84%

Scheme 9.36

An analogous cycloisomerization, involving a subsequent Si-deprotection, was utilized by Lamas and Barluenga for the synthesis of 5,7-disubstituted NH-indoles [183]. Although good to excellent yields of the desired indole products and an excellent functional group compatibility were achieved in this transformation, two equivalents of CuI promoter were required for this cyclization. Later, Duncton described that 6-chloro-5-fluoroindole derivatives could be prepared using this approach [184]. Cook employed the same Cu(I)-mediated protocol in the synthesis of biologically important tryptophan analogs (Scheme 9.37) [185]. Similarly, Castro's conditions were efficiently applied to the kilogram scale synthesis of Gonanotropin

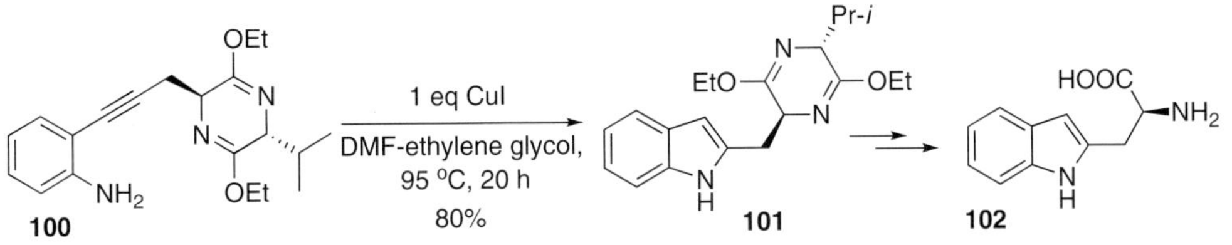

1 eq CuI
DMF-ethylene glycol, 95 °C, 20 h
80%
100
101
102

Scheme 9.37

releasing hormone (GnRH) antagonists, possessing functionalized indole cores [186]. The Cu(I)-promoted cycloisomerization–oxidative coupling cascade of silyl-protected *ortho*-diyne-substituted anilines into bis(indolyl)diynes was recently reported by Fiandanese and coworkers [96].

Catalytic-in-Cu synthesis of indoles was reported by Yamanaka [187], Villemin [188], Yamamoto (Scheme 9.38) [189], and then Cacchi (Scheme 9.39) [190]. In contrast to the previous examples, no loss of the TMS-group was observed under these reaction conditions.

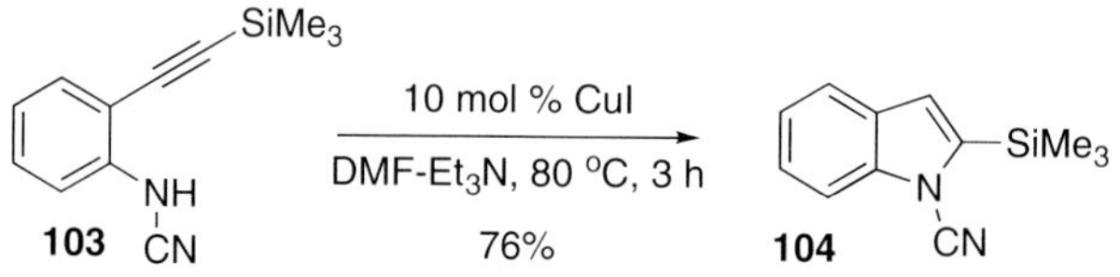
SiMe3
10 mol % CuI
DMF-Et3N, 80 °C, 3 h
76%
NH
CN
103
104
SiMe3
N
CN

Scheme 9.38

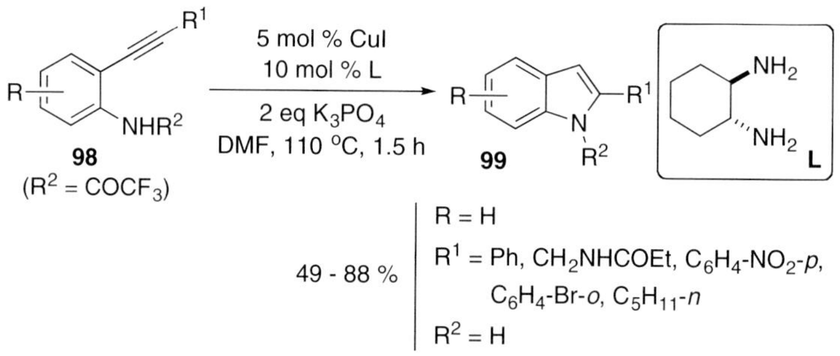
R1
R
NHR2
98
(R2 = COCF3)
5 mol % CuI
10 mol % L
2 eq K3PO4
DMF, 110 °C, 1.5 h
R
R1
N
R2
99
NH2
NH2
L
49 - 88 %
R = H
R1 = Ph, CH2NHCOEt, C6H4-NO2-p,
C6H4-Br-o, C5H11-n
R2 = H

Scheme 9.39

Another mode of activation of *o*-alkynylanilines **105**, involving an alkyne–vinylidene isomerization, was reported by Mc-Donald (Scheme 9.40) [191]. High yields of 1-monosubstituted indoles **106** were obtained upon cycloisomerization of terminal *o*-alkynylanilines **105** using an *in situ* generated $Et_3N{:}Mo(CO)_5$ catalyst. The authors proposed a mechanism involving the initial alkyne–vinylidene isomerization [192–194] of a terminal acetylene **105** into the reactive carbenoid **107**. The

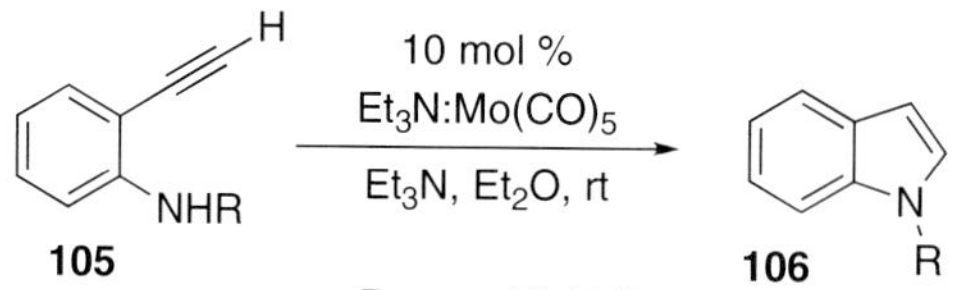
H
NHR
105
10 mol %
Et3N:Mo(CO)5
Et3N, Et2O, rt
N
R
106

R	Yield,%
H	79
Boc	50

Scheme 9.40

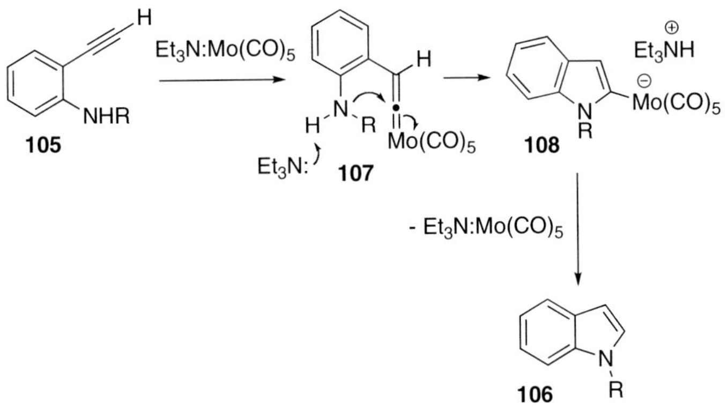

Scheme 9.41

latter underwent intramolecular cyclization to give the indolylmolybdenum zwitterionic intermediate **108**. A subsequent protiodemetalation afforded indole **106** and regenerated the catalyst (Scheme 9.41).

Recently, an analogous Rh(I)-catalyzed transformation of terminal *o*-alkynylanilines **109** into indoles **111** was reported by Trost (Scheme 9.42) [93]. Mechanistically, this cycloisomerization proceeded via an alkyne–vinylidene isomerization of a terminal alkyne **109**, leading to the Rh-vinylidene **110**, similar to that proposed by McDonald (Scheme 9.41). Consequently, internal *o*-alkynylanilines were shown to be completely unreactive in this transformation.

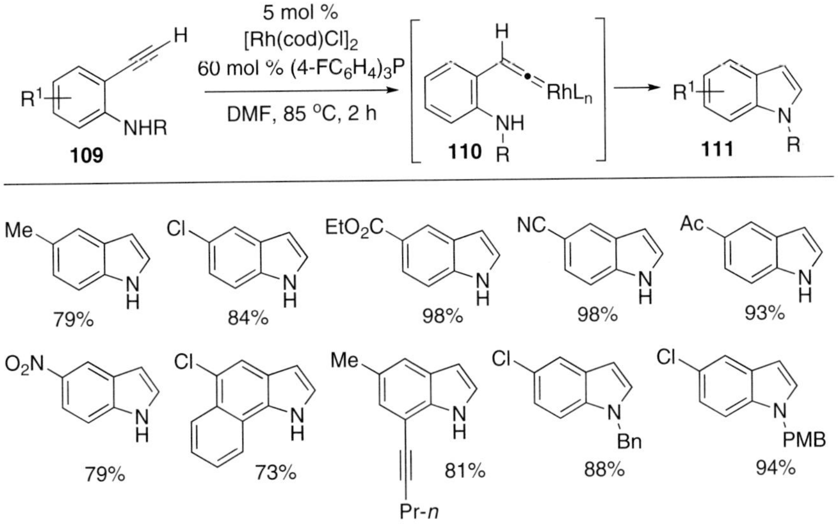

Scheme 9.42

Following the original report by Castro (*vide supra*), many other research groups utilized *ortho*-alkynylanilines as versatile substrates in transition metal-catalyzed cycloisomerization reactions, leading to a highly efficient assembly of multisubstituted indole cores. Thus, Field and Messerle achieved the synthesis of indoles **113** via the cycloisomerization reaction of **112** catalyzed by the cationic Rh(I) dicarbonyl complex $[Rh((mim)_2CH_2)(CO)_2{}^+ BPh_4{}^-]$ [195]. Müller observed indole formation in the presence of a cationic Cu(I) catalyst [196]. Mitsudo [197] and Rutjes [198] employed Ru(0)- and Ag-catalysts, respectively, for this transformation. Finally, Crabtree [91] and Majumdar [199] demonstrated that Ir(III) and Au(III) complexes, accordingly, could be effective catalysts for the intramolecular hydroamination leading to indoles (Scheme 9.43).

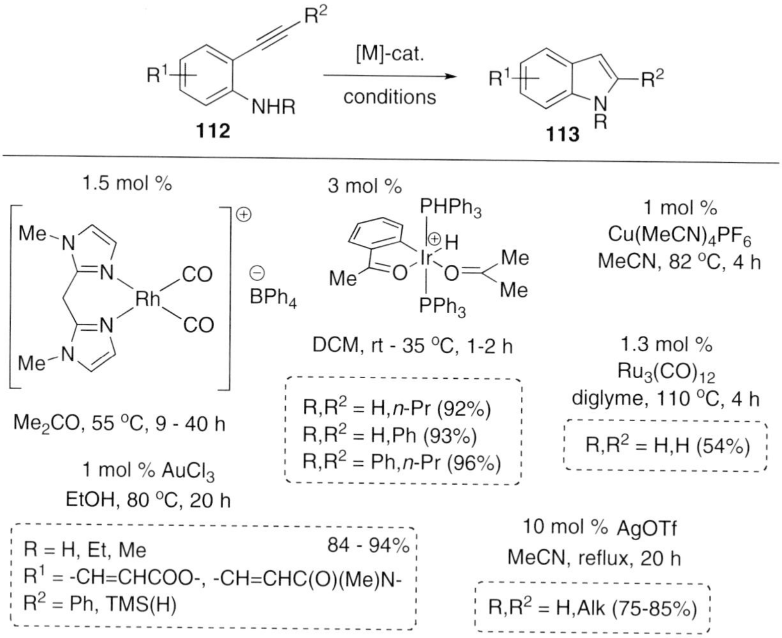

Scheme 9.43

Cycloisomerization of *o*-alkynylanilines **114** into multisubstituted *N*-protected as well as NH-indoles **115** in the presence of various Cu(II) catalysts was extensively investigated by Hiroya *et al.* (Scheme 9.44) [200, 201]. The authors, demonstrated that, in addition to Cu-catalysts, Ti(IV) and Zn(II) salts could also trigger this cyclization to some extent. Further, it was shown that $Cu(OTf)_2$ catalyst displayed good activities in the case of primary aniline derivatives, whereas $Cu(OAc)_2$ was better suited for secondary aniline derivatives. In both cases, moderate to excellent yields of up-to-trisubstituted indoles **115**, bearing a variety of labile functional groups, can be synthesized using this procedure. A cascade cyclization–intramolecular C3-alkylation leading to C2–C3 fused indoles was also demonstrated.

Furthermore, the same group illustrated that this protocol could be successfully applied for the construction of the key indole core of lycorine alkaloid Hippadine **118** (Scheme 9.45) [201]. Later, it was also shown that this transformation could be

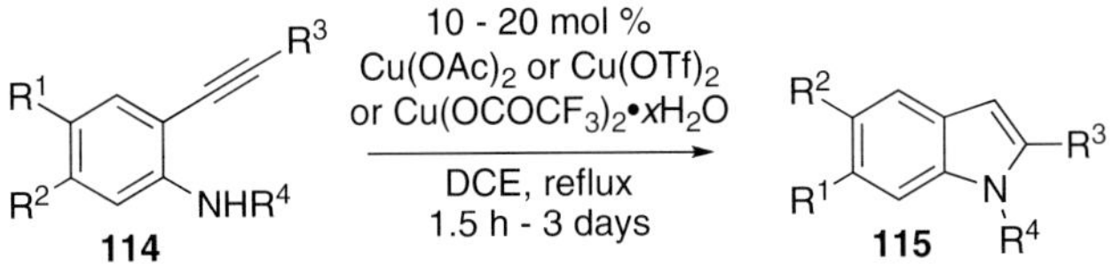

10 - 20 mol % $Cu(OAc)_2$ or $Cu(OTf)_2$ or $Cu(OCOCF_3)_2 \cdot xH_2O$, DCE, reflux, 1.5 h - 3 days; **114** → **115**

R^1	R^2	R^3	R^4	Yield, %
H	H	Ph	Ms	94
H	H	Ph	CO_2Et	88
H	H	*n*-Bu	Ms	92
H	H	*n*-Bu	CO_2Et	81
H	H	H	Ms	87
H	H	H	CO_2Et	87
H	H	Ph	H	68
H	H	*t*-Bu	Ms	22
H	H	TMS	Ms	9
Br	H	Ph	Ms	76
H	OMe	Ph	Ms	95
CN	H	Ph	Ms	74
Me	H	Ph	Ms	95
H	H	CO_2Me	Ms	79
Br	H	Ph	H	83
H	OMe	Ph	H	58
Me	H	Ph	H	71
H	CN	Ph	H	84
H	H	Ph	PMB	92
H	H	H	PMB	89
H	H	$(CH_2)_4OTBS$	*i*-Pr	100

Scheme 9.44

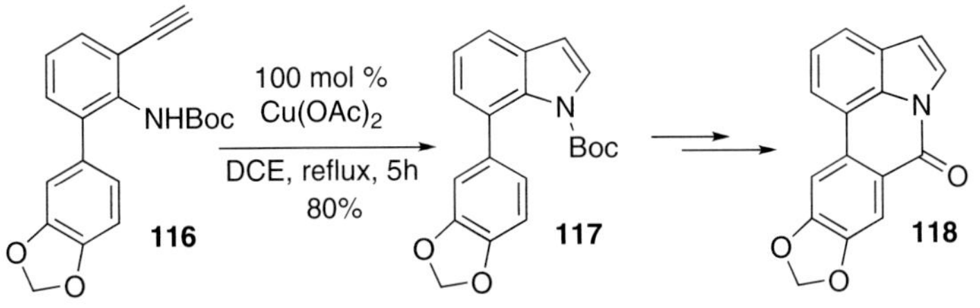

116 → (100 mol % $Cu(OAc)_2$, DCE, reflux, 5h, 80%) → **117** (N-Boc) → → **118**

Scheme 9.45

efficiently performed in an aqueous alcohol medium, which allowed multiple recycling of the Cu(II) trifluoroacetate catalyst without noticeable change to its catalytic activity [90].

Following these reports, Dai and coworkers disclosed a microwave-assisted solid-phase-supported synthesis of C5-substituted indole [168] derivatives using Hiroya's Cu(II)-conditions [202].

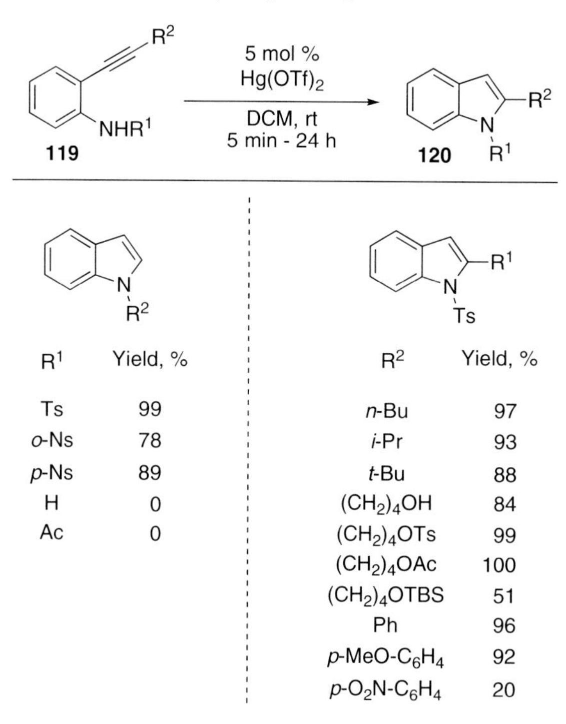

Scheme 9.46

Hg(II)-salts were reported by Nishizawa to effectively catalyze the cyclization of *o*-alkynylanilines **119** into indoles **120** (Scheme 9.46) [203]. However, strongly electron-deficient anilines **119** provided substantially decreased yields of the target products **120**.

Similarly, excellent yields of indole products were achieved when employing catalytic amounts (5 mol%) of phenylmercuric triflate. This finding led to the development of a solid-supported silaphenylmercuric triflate catalyst, which allowed indole synthesis with high efficiency in a flow reactor [204].

Catalysis of an analogous transformation by gold complexes was first reported by Utimoto [205] and further investigated by Marinelli [206] (Scheme 9.47). Generally, moderate to excellent yields of the expected NH-indoles **122**, possessing different functional groups, were obtained under these mild reaction conditions. A one-pot Au(III)-catalyzed cyclization–halogenation cascade process yielding the corresponding 3-bromo- and -iodoindoles was demonstrated.

Kobayashi reported that the microencapsulated Au(III) salt supported on silica gel could serve as a very efficient catalyst for this transformation, providing good to excellent yields of differently C2-monosubstituted indoles at room temperature with little or no detectable gold leaching [207].

Arcadi and coworkers further elaborated on the Au(III)-catalyzed cyclization of *o*-alkynylanilines **123** in the presence of enones **124** (Scheme 9.48) [208]. A sequential

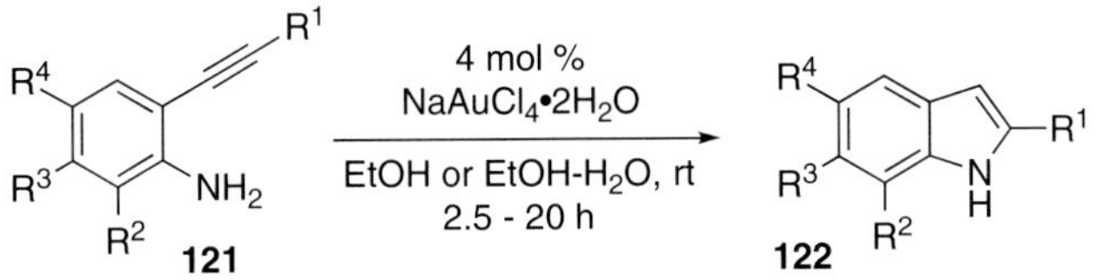

4 mol % $NaAuCl_4 \cdot 2H_2O$, EtOH or EtOH-H_2O, rt, 2.5 - 20 h; **121** → **122**

R^1	R^2	R^3	R^4	Yield, %
p-Tol	H	H	H	94
n-Bu	H	H	H	88
(*E*)-Ph-CH=CH-	H	H	H	92
4-*t*-Bu-1-cyclohexenyl	H	H	H	81
2-thienyl	H	H	H	87
p-Cl-C_6H_4	H	H	H	87
3,4-dihydro-1-naphthyl	Cl	H	Cl	68
p-MeO-C_6H_4	Cl	H	Cl	22
(*E*)-Ph-CH=CH-	Cl	H	NO_2	9
H	Cl	H	Cl	76
H	Cl	H	NO_2	95
H	H	CF_3	H	74
H	F	H	F	95

Scheme 9.47

cyclization–alkylation of **123** and **124** provided a straightforward route for the synthesis of β-indolyl ketones **126**. In addition, a domino indole synthesis, involving Au(III)-catalyzed amination, cyclization, and alkylation processes, was realized by the employment of 2 equiv of enone **124** at elevated temperatures. It is believed that this transformation occurred through the initial Au(III)-catalyzed 5-*endo-dig* cyclization leading to the formation of the corresponding C3-unsubstituted indole products **125**, followed by the electrophilic aromatic substitution of indole involving the conjugate addition-type alkylation with the activated enone **124**.

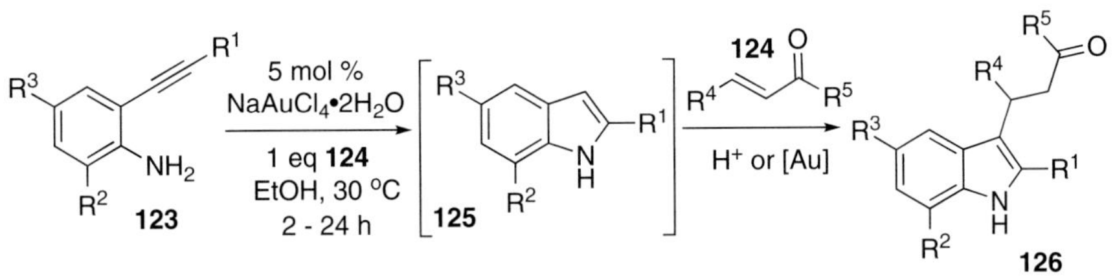

5 mol % $NaAuCl_4 \cdot 2H_2O$, 1 eq **124**, EtOH, 30 °C, 2 - 24 h; **123** → [**125**] → (**124**, H^+ or [Au]) → **126**

R^1 = Ph, 2-naphthyl, 4-Cl-C_6H_4-, 2-thienyl, *n*-Bu, H, 4-Ph-1-cyclohexenyl, 3,4-dihydro-1-naphthyl

R^2 = H, Cl; R^4 = H, Ph, Alk

R^3 = H, Cl, Me; R^5 = Ph, Alk

45 - 95 %

Scheme 9.48

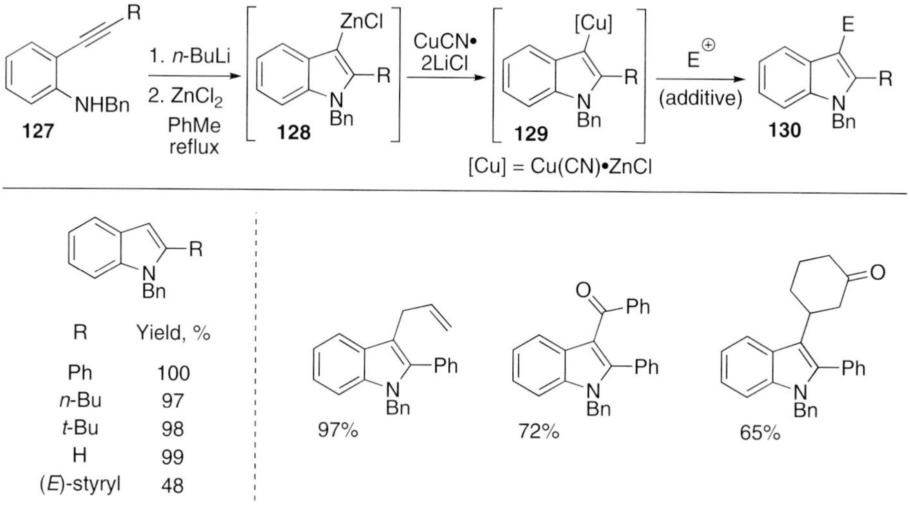

127 → 1. n-BuLi, 2. $ZnCl_2$, PhMe, reflux → [128] → CuCN•2LiCl → [129] → $E^{\oplus}$ (additive) → 130

[Cu] = Cu(CN)•ZnCl

R	Yield, %
Ph	100
n-Bu	97
t-Bu	98
H	99
(E)-styryl	48

97% 72% 65%

Scheme 9.49

Similarly to the synthesis of C3-functionalized benzo[*b*]furans (Scheme 9.2), Nakamura employed the Zn/Cu-mediated cyclization–substitution strategy for a facile assembly of 2,3-disubstituted indoles **130** (Scheme 9.49) [79, 80]. This reaction involves the initial formation of 3-zincindole **128**, which undergoes transmetalation with CuCN–LiCl complex to give the corresponding cuprate **129**. The latter, upon reaction with a suitable electrophile, furnishes indole **130**.

Analogous Et_2Zn-mediated approach for an efficient synthesis of 3-bromo- and -acylindoles was reported by Zhao and coworkers [209, 210]. In addition, the same authors demonstrated that catalytic amounts of diethylzinc could promote a common 5-*endo-dig* cyclization of *N*-sulfonylated *ortho*-alkynylanilines, affording C3-unfunctionalized indoles in good to excellent yields.

Recently, Fujii and Ohno, incorporated a Mannich-type aminomethylation of alkynes **131** in the presence of *para*-formaldehyde **132** and secondary amines **133** into a cascade cyclization leading to 2-aminomethyl indole derivatives **134** (Scheme 9.50) [211]. The proposed mechanism (Scheme 9.51) is similar to that suggested by Kabalka for the analogous benzo[*b*]furan synthesis (Scheme 9.5).

More recently, Corma reported that this aminomethylation–cyclization transformation could be successfully achieved with a highly active and recoverable heterogeneous catalyst, consisting of nanoparticles of Au supported on a nanocrystalline ZrO_2 [212]. Another cascade reaction, involving a domino coupling–cycloisomerization of terminal *o*-ethynyl-acetanilides and aryl iodides in the presence of 10 mol% $[Cu(phen)(PPh_3)_2]NO_3$, was developed by Cacchi and Venkataraman for the synthesis of C2-aryl-functionalized indoles [121, 190].

Yamamoto was first to report that *ortho*-alkynylanilines, which are fully substituted at the N-site, could undergo a transition metal-catalyzed cycloisomerization reaction to give indoles via a formal 1,3-migration of a suitable migrating group from the

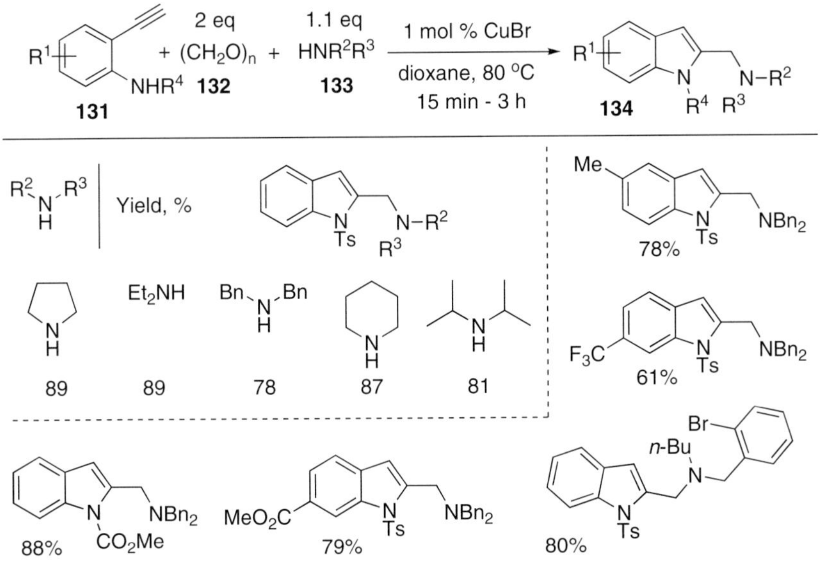

Scheme 9.50

N-atom to the C3-position of the formed indole skeleton. This strategy represents a significant advantage over classical cycloisomerization protocols as most of them provide C3-unsubstituted indole derivatives only. Accordingly, it was demonstrated that *N*-acyl-substituted alkynylanilines **138** underwent the Pt(II)-catalyzed aminoacylation reaction, affording the corresponding indoles **139** (Scheme 9.52) [213]. Although excellent yields of indoles **139** and a notable functional group compatibility were observed for this reaction, in many cases the formation of the target products **139** was accompanied by significant amounts of the corresponding deacylated indole derivatives. The proposed mechanism (Scheme 9.53) involves the initial coordination of **138** with the Pt(II)-catalyst to give the π-complex **140**. Intramolecular attack of the *ortho*-nitrogen atom at the activated alkyne in **140** produces the zwitterionic cyclic intermediate **141** which, upon a subsequent intramolecular

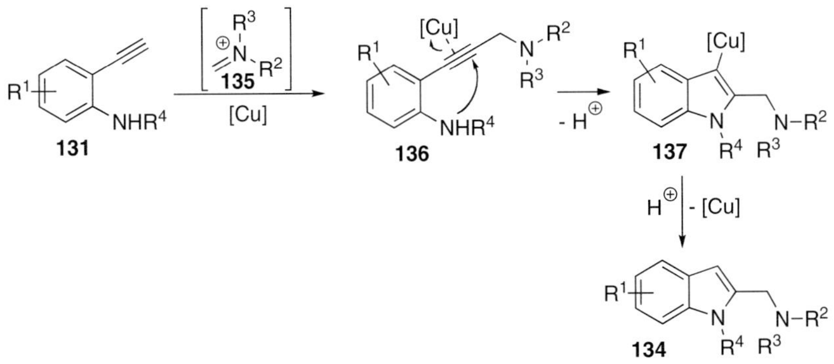

Scheme 9.51

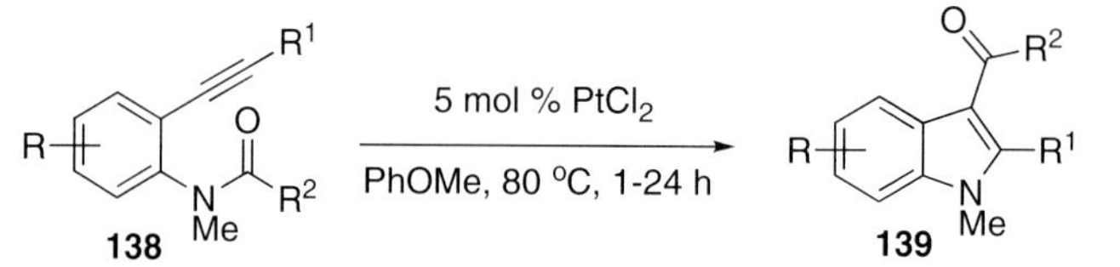

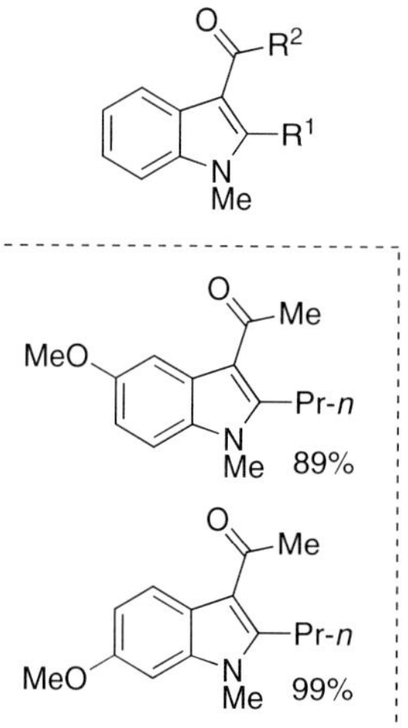

138 → 139: 5 mol % $PtCl_2$, PhOMe, 80 °C, 1-24 h

R^1	R^2	Yield, %
n-Pr	Me	96
n-Pr	Me	93
c-Hex	Me	98
t-Bu	Me	91
p-MeO-C_6H_4	Me	81
p-Tol	Me	97
Ph	Me	94
p-F-C_6H_4	Me	88
p-F_3C-C_6H_4	Me	93
n-Pr	H	74
n-Pr	Ph	75
n-Pr	Bn	99
Ph	Bn	87
n-Pr	CF_3	>99

5-MeO, 2-Pr-*n*, 3-acetyl, N-Me indole: 89%

6-MeO, 2-Pr-*n*, 3-acetyl, N-Me indole: 99%

Scheme 9.52

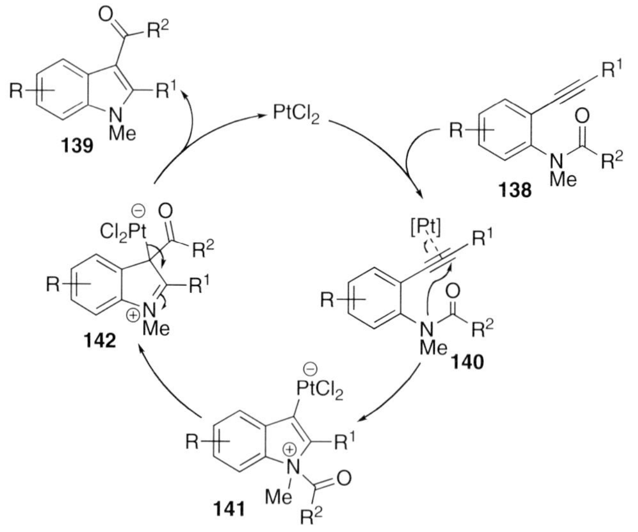

Scheme 9.53

1,3-migration of the acyl group, yields the iminium intermediate **142**. Elimination of the Pt-catalyst in the latter provides the corresponding indole product.

An analogous formal 1,3-allyl group migration was reported by Fürstner in the synthesis of indole **145** (Scheme 9.54) [89]. Supposedly, a 1,3-migration occurred through the π-allylplatinum intermediate **144**. Zhang observed a 1,3-migration of the benzyl group during the Au(I)-catalyzed cycloisomerization of *N,N*-disubstituted *ortho*-alkynylanilines [214]. Fensterbank and Malacria reported a 1,3-migration of the allyl group, proceeding via the aza-Claisen type rearrangement, in the Pt(II)-catalyzed cycloisomerization of the skipped *ortho*-alkynylanilines, leading to indoles [215].

143 → 144 → 145: 5 mol % $PtCl_2$, 1 atm CO, PhMe, 80 °C, 1-12 h, 59%

Scheme 9.54

Recently, Yamamoto extended the cycloisomerization–1,3-migration approach toward the construction of indole cores possessing a C3-sulfonyl group. Thus, cycloisomerization of *ortho*-alkynylsulfonamides **146** occurred in the presence of Au(III)-catalyst to produce 3-sulfonylindoles **147** via a 1,3-migration of a sulfanyl group (Scheme 9.55) [216]. The proposed mechanism is similar to that reported for

146 → 147: 10 mol % $AuBr_3$, PhMe, 80 °C, 1 h

R^1	R^2	R^3	Yield, %
n-Pr	Me	Me	95
c-Hex	Me	Me	62
t-Bu	Me	Me	38
Ph	Me	Me	92
p-Tol	Me	Me	87
p-MeO-C_6H_4	Me	Me	81
p-F_3C-C_6H_4	Me	Me	51
H	Me	Me	71
CO_2Et	Me	Me	0
n-Pr	Me	Bn	44
n-Pr	Me	*i*-Pr	60
n-Pr	Ph	Me	52
n-Pr	*p*-MeO-C_6H_4	Me	85
n-Pr	*p*-Ac-C_6H_4	Me	80

Scheme 9.55

the synthesis of indoles via the 1,3-acyl group migration (*vide supra*). Surprisingly, employment of Lewis acid catalyst, $InBr_3$, triggered cyclization of the substrates **146** with a subsequent 1,7-migration of the sulfanyl group to produce the regioisomer to **147**, 6-sulfonylindoles, as the major product in good to high yields.

Zhang recently reported an interesting double migration approach for the synthesis of N1–C2-fused indole structures (Scheme 9.56) [214]. A variety of tri- and tetra-substituted indoles **149** were synthesized via the Pt(IV)-catalyzed cycloisomerization of alkynyl-containing lactams **148**, proceeding via a fragmentation of the former lactam moiety followed by a subsequent carbocyclization [217]. It was shown that lactams with different ring sizes can be easily accommodated in this transformation. However, in most cases, small quantities of 1,7-fused indole products were observed as a result of the competing C7-acylation reaction. The proposed mechanism is depicted in Scheme 9.57. Accordingly, *ortho*-alkynylaniline derivatives **148** underwent a sequence of steps upon catalysis with Au(I) complex, producing a spiro key intermediate **151**. A lactam ring opening in the latter generated a vinylplatinum species **152**, which, upon a subsequent carbocyclization and 1,2-migration steps, furnished indole **149**.

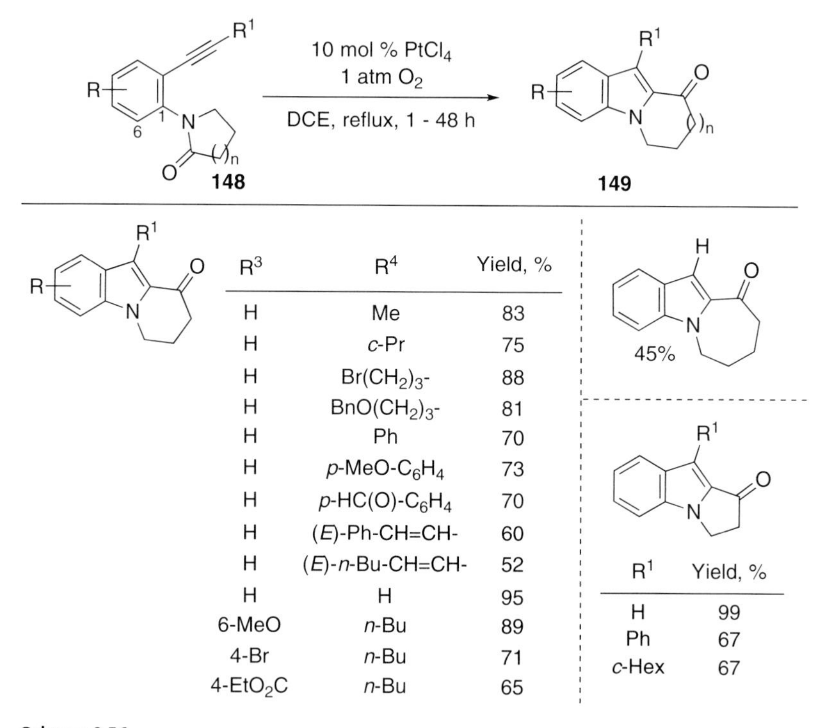

R^3	R^4	Yield, %
H	Me	83
H	*c*-Pr	75
H	$Br(CH_2)_3$-	88
H	$BnO(CH_2)_3$-	81
H	Ph	70
H	*p*-MeO-C_6H_4	73
H	*p*-HC(O)-C_6H_4	70
H	(*E*)-Ph-CH=CH-	60
H	(*E*)-*n*-Bu-CH=CH-	52
H	H	95
6-MeO	*n*-Bu	89
4-Br	*n*-Bu	71
4-EtO_2C	*n*-Bu	65

R^1	Yield, %
H	99
Ph	67
c-Hex	67

Scheme 9.56

A somewhat similar double 1,2-migration cascade transformation of *N,N*-disubstituted *o*-alkynylanilines **154** was recently disclosed by Iwasawa (Scheme 9.58) [218].

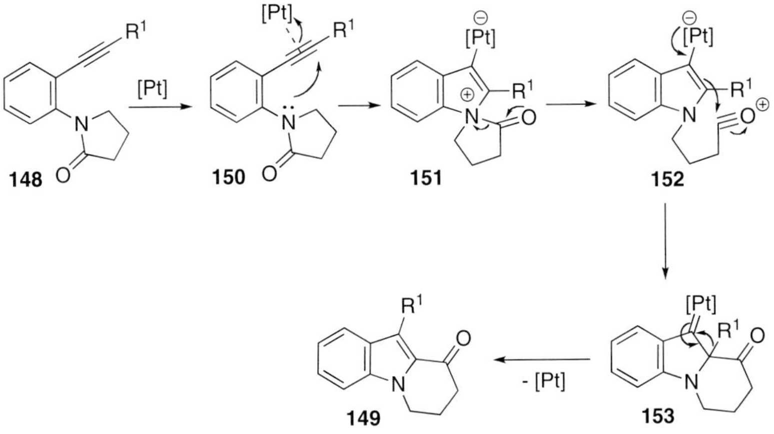

Scheme 9.57

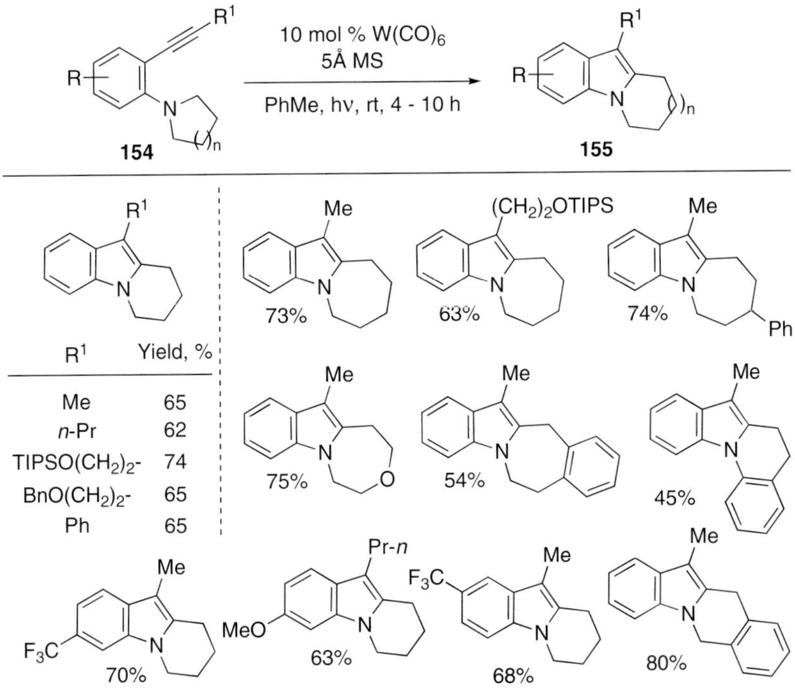

Scheme 9.58

R^1	Yield, %
Me	65
n-Pr	62
$TIPSO(CH_2)_2$-	74
$BnO(CH_2)_2$-	65
Ph	65

A variety of differently fused indoles **155** can be efficiently assembled via the W(0)-catalyzed cycloisomerization of *o*-alkynylaniline derivatives **154** under photoirradiation conditions. The proposed mechanistic rationale featured a 1,2 Stevens-type rearrangement in the metal-containing ammonium ylide **157**, followed by a 1,2-alkyl migration to the metallocarbenoid center to generate intermediate **158** (Scheme 9.59).

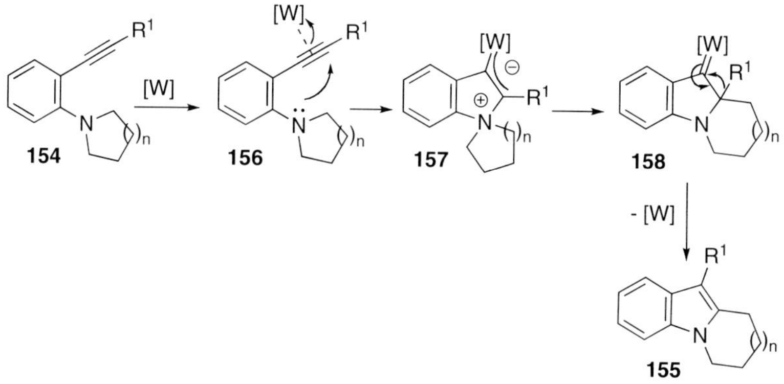

Scheme 9.59

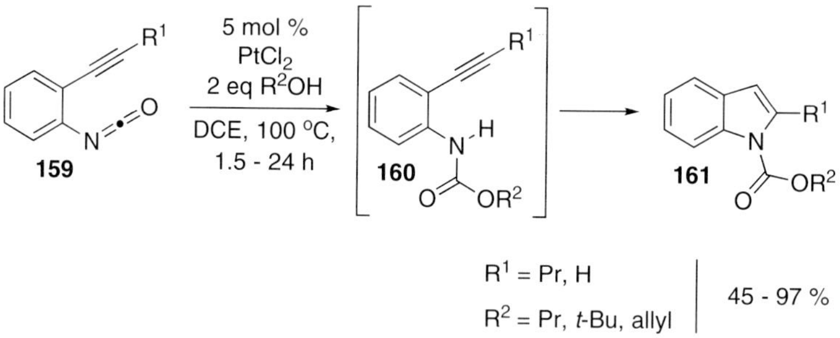

Scheme 9.60

Furthermore, Yamamoto and coworkers illustrated that *o*-(alkynyl)phenylisocyanates **159** could also be efficiently employed in a similar Pt(II)-catalyzed cycloisomerization reaction, serving as surrogates of the corresponding carbamate derivatives **160**, to provide *N*-(alkoxycarbonyl)indoles **161** in moderate to excellent yields (Scheme 9.60) [219]. It is believed that a dual-role catalysis with the Pt(II) salt first triggered the initial intermolecular nucleophilic addition of alcohol to isocyanate **159**, affording the key transient carbamate **160**, which, upon a subsequent Pt(II)-catalyzed 5-*endo-dig* cyclization, generated the desired product **161**.

A similar concept was applied to the two-component synthesis of indoles **164** via cycloisomerization of alkynyl imines **162** (Scheme 9.61) [220]. *N*-(Alkoxymethyl) indoles **164** were readily prepared through the Cu(I)-catalyzed cyclization–hydroxylation reaction of the imine precursors **162** in the presence of a variety of external O-nucleophiles **163**, such as alcohols and phenols, with good functional group compatibility. The authors believe that this domino reaction begins with the Cu(I)-activation of alkyne **162** toward 5-*endo-dig* cyclization to form a cyclic metal-containing azomethine

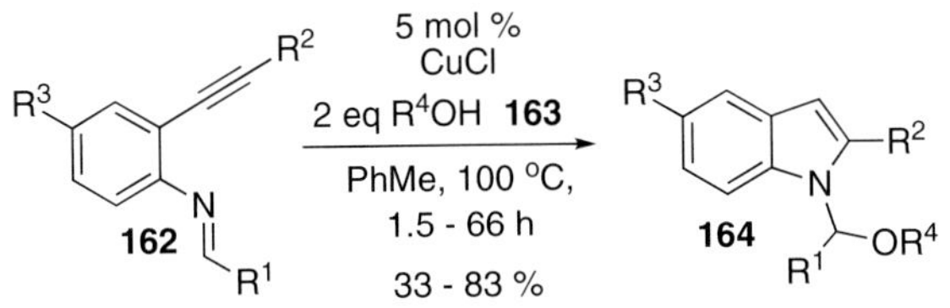

R^1 = *p*-Tol, *m*-Tol, *o*-Tol, *p*-MeO-C_6H_4-, *p*-MeS-C_6H_4-, Ph, *t*-Bu
p-Br-C_6H_4-, *p*-Cl-C_6H_4-, *p*-O_2N-C_6H_4-, (*E*)-CH=CH-Ph

R^2 = *n*-Pr, *c*-Hex, *t*-Bu, H, Ph, *p*-MeO-C_6H_4-, *p*-MeOCO-C_6H_4, TMS-C≡C-

R^3 = H, CN, MeO

R^4 = Me, *n*-Pr, *i*-Pr, *t*-Bu, Ph, *p*-MeO-C_6H_4-
CF_3CH_2, CH_2=CH-CH_2-, Bn, TMS-CH_2CH_2-

Scheme 9.61

ylide **166**, which, upon interception with alcohol **163** and a subsequent protiodemetalation, provides the target indole product (Scheme 9.62).

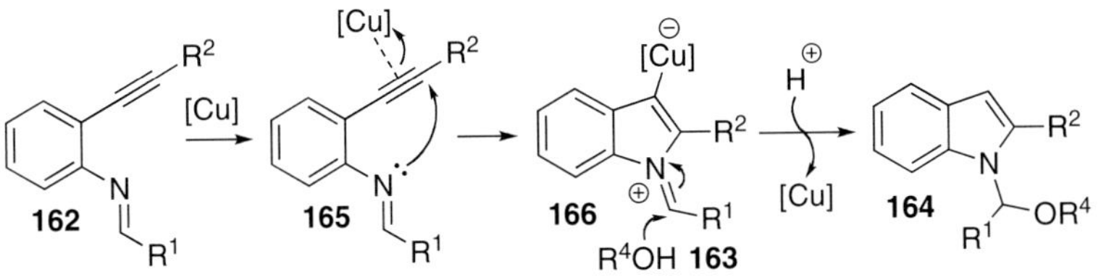

Scheme 9.62

The Au(III)-catalyzed double hydroamination cascade reaction of *ortho*-alkynylanilines **167** with terminal alkynes **168** affording *N*-vinylindoles **169** was reported by Li (Scheme 9.63) [221]. In the case of alkyl-substituted acetylenes, this protocol provided mixtures of isomeric *N*-vinylindoles with both terminal and internal double bonds. This transformation is believed to occur via the Au(III)-catalyzed cycloisomerization of transient key alkynyl imines, similar to **162** utilized by Yamamoto, which were generated via the initial Au(III)-catalyzed hydroamination of the corresponding anilines **167** with alkynes **168**.

Recently, Iwasawa established a set of transition metal-catalyzed protocols for an efficient construction of N1–C2-fused polycyclic indole skeletons via a cycloisomerization–cycloaddition domino reaction of alkynyl imines **172** [222–224]. It was shown that the latter substrates, upon activation with transition metal catalysts, such as W(0), Pt(II), and Au(III), generate reactive azomethine ylide intermediates **174** similar to **166** (Scheme 9.64). Interception of such ylides with a variety of suitably substituted alkenes **175** via a [3 + 2]-cycloaddition affords fused indole products **177** through a transient formation of the corresponding metallocarbenoids **176**. Transformation of terminal alkynyl imines proceeds with a 1,2-H shift in the **176**, whereas

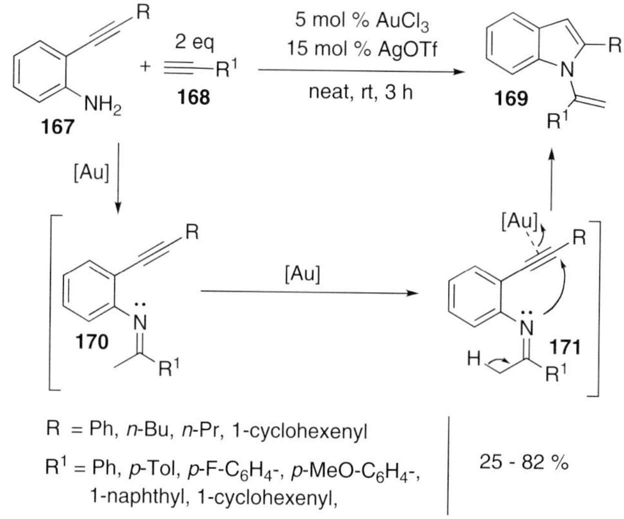
5 mol % $AuCl_3$
15 mol % AgOTf
neat, rt, 3 h
2 eq
167
168
169
170
171
[Au]
R = Ph, n-Bu, n-Pr, 1-cyclohexenyl
R^1 = Ph, p-Tol, p-F-C_6H_4-, p-MeO-C_6H_4-, 1-naphthyl, 1-cyclohexenyl,
25 - 82 %

Scheme 9.63

internal alkynes undergo a 1,2-alkyl migration of the R-group in the latter intermediate (Scheme 9.65) [222, 223]. Moreover, cycloisomerization of vinyl-substituted substrates **178**, namely enynyl imines, in the presence of bulky ketene acetals **179** proceeded via a formal [5 + 2]-cycloaddition in **180** to give the indoloazepinone derivatives **181** (Scheme 9.66) [224]. In contrast, employment of smaller ketene acetal dipolarophiles triggers the [3 + 2]-cycloaddition pathway involving a 1,2-vinyl group migration. The authors also demonstrated that the use of imines as dipolarophiles in this transformation allowed assembly of diazepine **184**, while a related

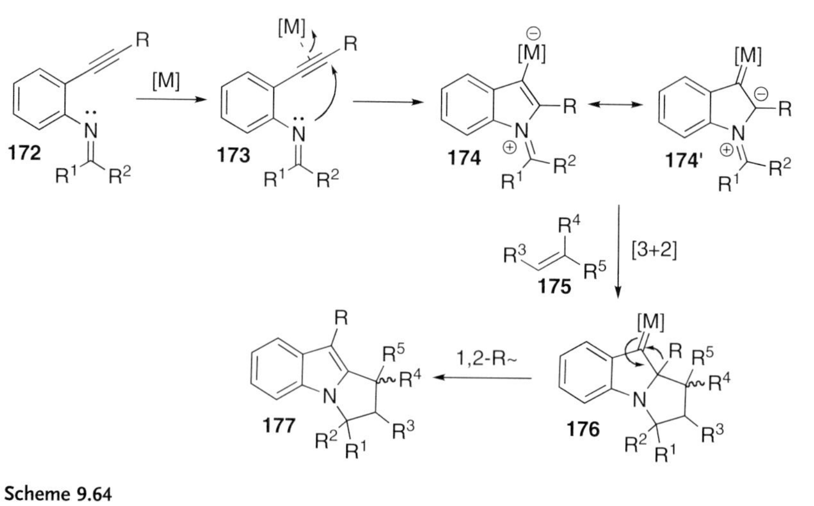
[M]
172
173
174
174'
175
176
177
[3+2]
1,2-R~

Scheme 9.64

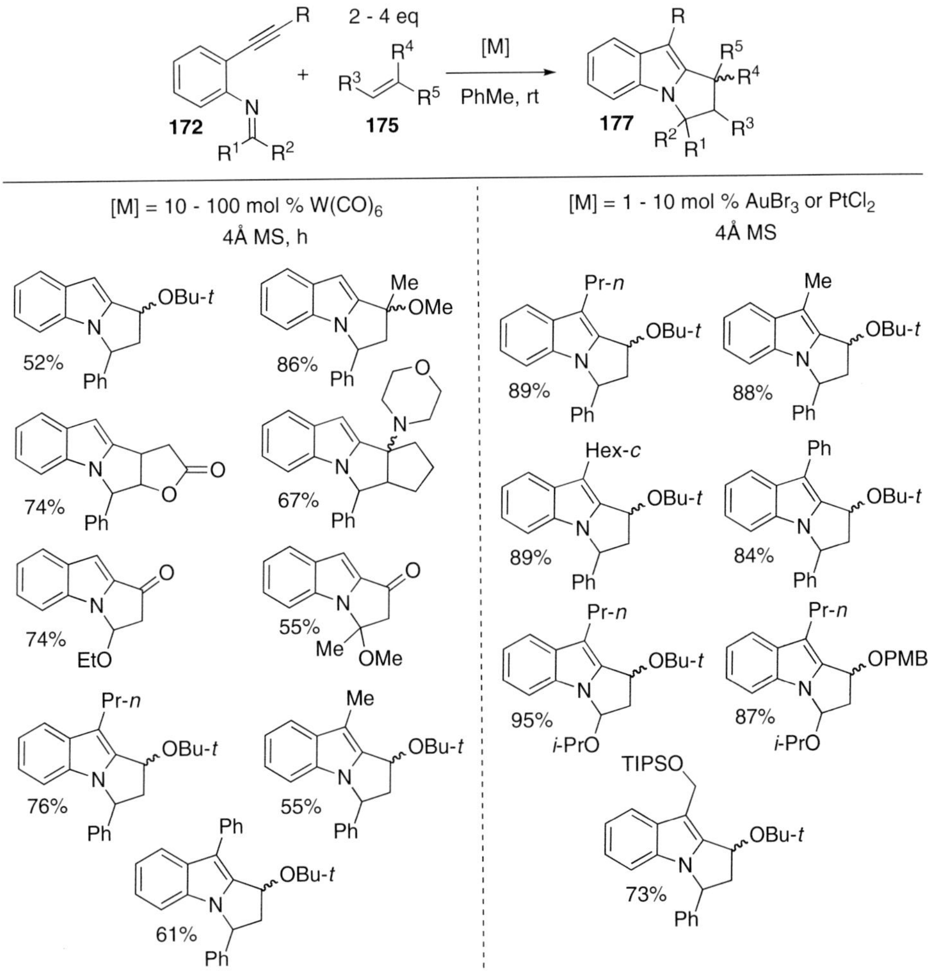

Scheme 9.65

azepino[1,2-*a*]indole core **187** could be accessed via the W(0)-catalyzed ring closure of *C*,*N*-divinyl-substituted imines **185** (Scheme 9.67) [224].

More recently, Saito and Hanzawa demonstrated that a variety of fused, tri-, tetra- and penta-substituted indoles **190** could be efficiently accessed via the Rh(I)-catalyzed cycloisomerization of *N*-propargyl anilines **188** (Scheme 9.68) [225, 226]. This transformation presumably involves the initial Rh(I)-catalyzed amino-Claisen rearrangement of **188** to produce a putative reactive allenyl aniline intermediate **189**, which is smoothly converted into the indole core, as exemplified by the cycloisomerization of an independently prepared allene **191** (Scheme 9.69). On the other hand, the latter transformation of allenes can be achieved under thermal reaction conditions, albeit with a diminished efficiency.

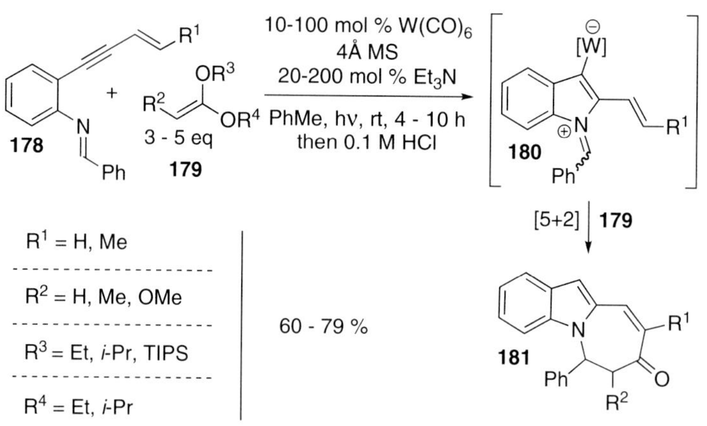

Scheme 9.66

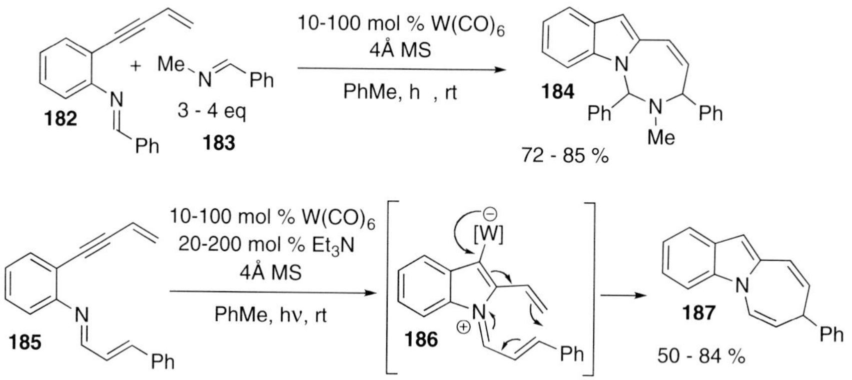

Scheme 9.67

Cenini was the first to observe that a transition metal-catalyzed reduction of *o*-nitrostyrenes **193** afforded indoles **196** or **197** (Scheme 9.70) [227]. It was suggested that nitrocompounds **193**, upon reaction with Fe-, Ru-, and Rh-carbonyls, underwent reduction, forming a highly reactive nitrene [13] **194** or the corresponding metallonitrenoid **195** intermediates. A subsequent C–H insertion of the latter furnishes indole **196**, whereas β-disubstituted styrenes **193** were isomerized into 1,2-disubstituted indoles **197** through a 1,2-shift of a suitable migrating group (Scheme 9.70).

Later, an analogous Mo(VI)-catalyzed reductive cyclization of *o*-nitrostyrenes **198** was utilized by Arnáiz in the synthesis of multisubstituted indoles **199**, possessing various functional groups (Scheme 9.71) [228]. Similarly to Cenini's report, the authors proposed a mechanism involving cyclization of a reactive nitrenoid like **195**.

More recently, several reports on the utilization of a facile cyclization of analogous to **195** Rh-nitrenoids for the assembly of multisubstituted indole cores appeared in the literature. Along this line, Narasaka revealed that 2*H*-azirines **200** could serve as versatile surrogates of these highly reactive nitrene–metal complexes **201**

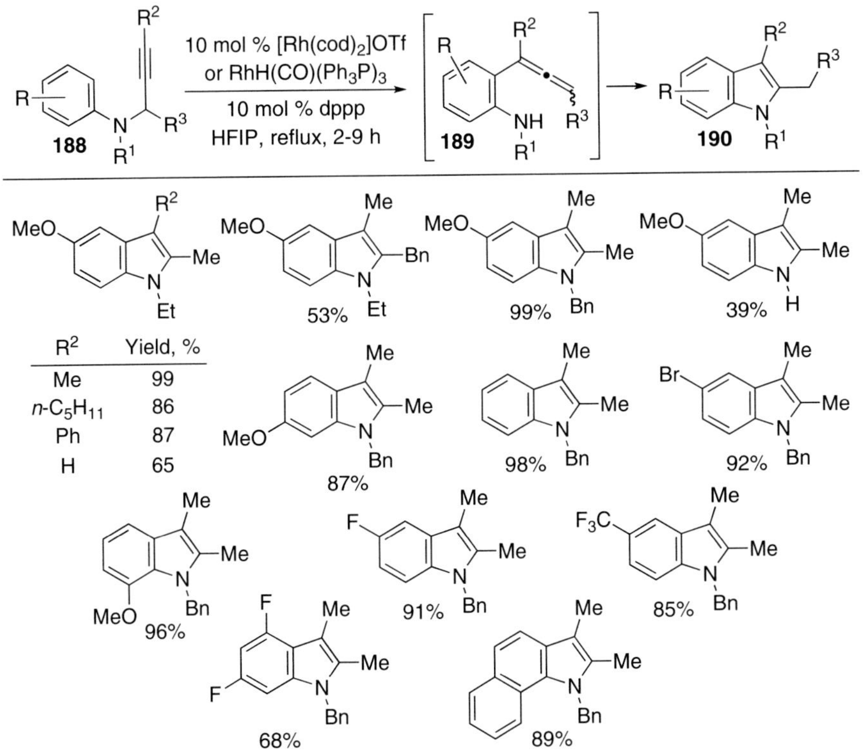

R^2	Yield, %
Me	99
n-C_5H_{11}	86
Ph	87
H	65

Scheme 9.68

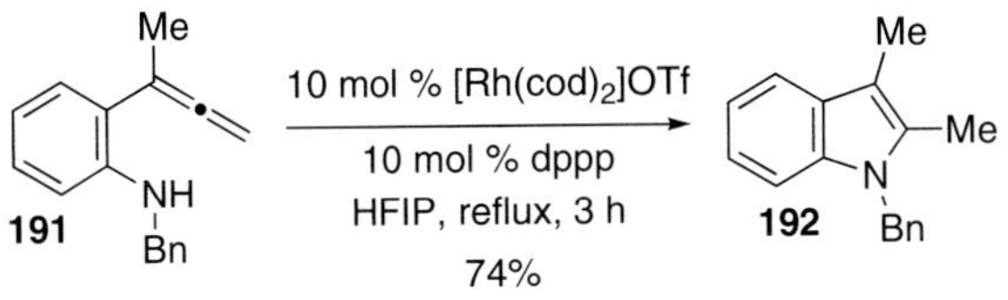

Scheme 9.69

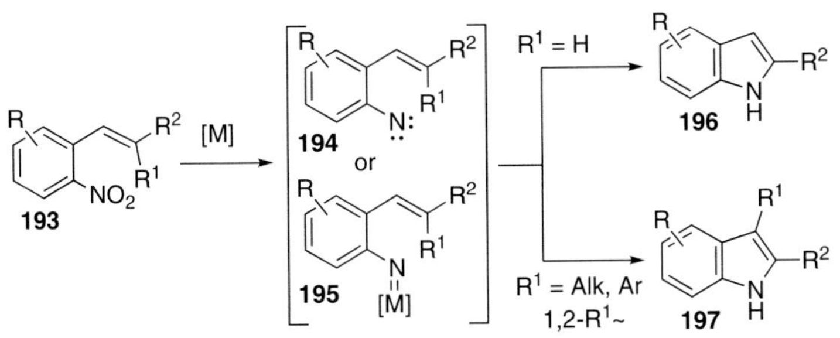

[M] = $Fe(CO)_5$, $Ru_3(CO)_{12}$, $Rh_6(CO)_{16}$

Scheme 9.70

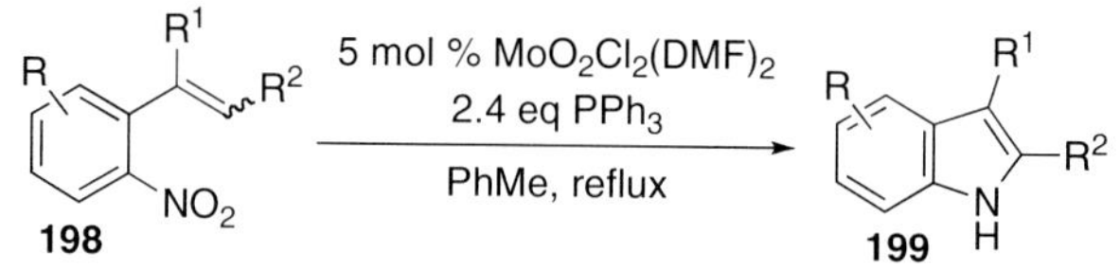

R = H, 4-Cl, 5-Cl, 6-Cl 5,6-(OCH_2O)-, 7-OMe	15 - 93 %
R^1 = H, Me	
R^2 = Ph, n-C_5H_{11}, Me, CO_2Et, Ac	

Scheme 9.71

(Scheme 9.72). Thus, it was shown that a variety of disubstituted indoles could be prepared via the Rh(II)-catalyzed cycloisomerization of trisubstituted aryl azirines **200** (Scheme 9.72) [229] Despite the fact that good functional group compatibility and high yields were attained in this transformation, C2-monosubstituted 2*H*-azirines were unreactive under these reaction conditions, whereas alkyl-aryl C2-disubstituted substrates **200** provided indoles **202** in generally lower yields than for the diaryl-substituted azirines. In the case of unsymmetrical C2-aryl substituents, C–H insertion of the Rh-nitrenoid **201** occurs into both aryl rings, providing a mixture of indole products. In the latter case, the electronic effect of the aryl group substituents did not have any significant influence on the regioselectivity of the C–H insertion process.

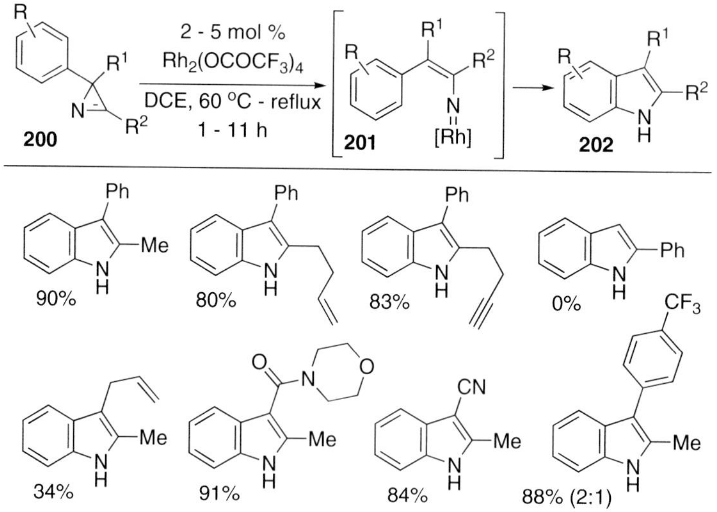

Scheme 9.72

Later, Driver demonstrated that the key nitrenoid species could be effectively generated from organic vinyl azides **203** (Scheme 9.73) [230]. Accordingly, aryl-substituted azidoacrylates **203** were smoothly converted into the corresponding functionalized indole-2-carboxylates **204** in the presence of the Rh(II)-catalyst. The authors demonstrated that this cascade reaction easily tolerates substrates with both electron-donating and electron-withdrawing aryl substituents. Meta-substituted azides **203** provided mixtures of indoles, with a moderate to high regioselectivity toward the formation of 5-substituted regioisomer. It is believed that this reaction proceeds via the rhodium(II)-mediated nitrogen atom transfer. Accordingly, the initial coordination of the dirhodium(II) carboxylate with the α-nitrogen of azide **203** produces complex **205**, which, upon loss of dinitrogen, leads to the formation of the rhodium nitrenoid **206**. A subsequent stepwise electrophilic aromatic substitution via arenium ion **207** accomplishes the assembly of indole core **204** and regenerates the catalyst (Scheme 9.74).

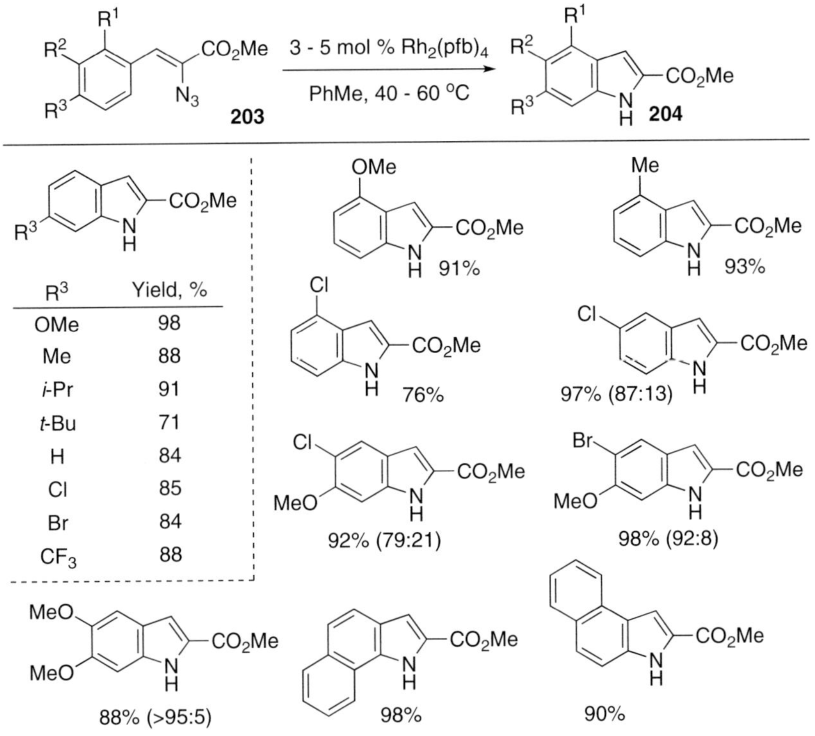

3 - 5 mol % $Rh_2(pfb)_4$, PhMe, 40 - 60 °C; **203** → **204**

R^3	Yield, %
OMe	98
Me	88
i-Pr	91
t-Bu	71
H	84
Cl	85
Br	84
CF_3	88

91%; 93%; 76%; 97% (87:13); 92% (79:21); 98% (92:8); 88% (>95:5); 98%; 90%

Scheme 9.73

The limitation of the previous methodology to indole-2-carboxylates somewhat diminishes the synthetic value of this approach. The same group recently provided a possible solution to this problem by developing a very efficient complementary approach toward the indole ring synthesis, where aryl azides were employed as electrophilic nitrogen sources. In this way, *ortho*-vinyl-substituted aryl azides **208**,

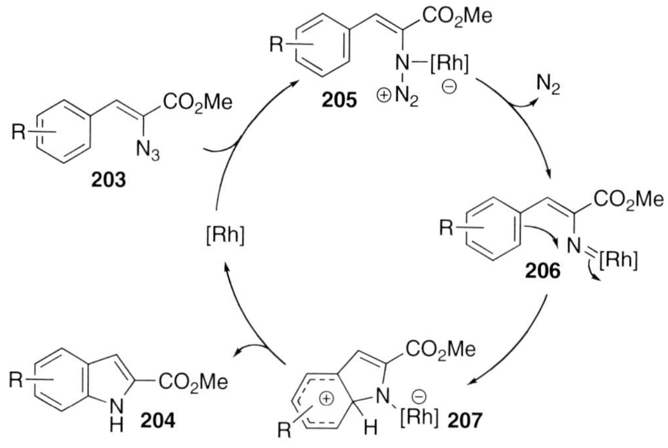

Scheme 9.74

upon the Rh(II)-catalyzed vinyl C–H bond amination, were readily cycloisomerized into differently functionalized indoles **209** (Scheme 9.75) [231]. However, transformation of unfunctionalized substrates, as well as substrates containing alkyl substituents at the vinylic β-position **208** afforded lower conversions and yields of the target indole products. In addition, the authors demonstrated that *gem*-diphenyl-substituted aryl azide **210** underwent the Rh(II)-catalyzed nitrogen transfer reaction with a concomitant 1,2-phenyl group migration, presumably via intermediate **211**, providing 2,3-disubstituted indole **212** in good yield (Scheme 9.76). The proposed mechanism outlined in Scheme 9.77 features cyclization of the generated nitrenoid **214** on the pendant *ortho*-vinylic substituent via a stepwise electrophilic aromatic substitution process (**215**).

9.2.3.2 Synthesis of Indoles via Intramolecular Arylation Reactions

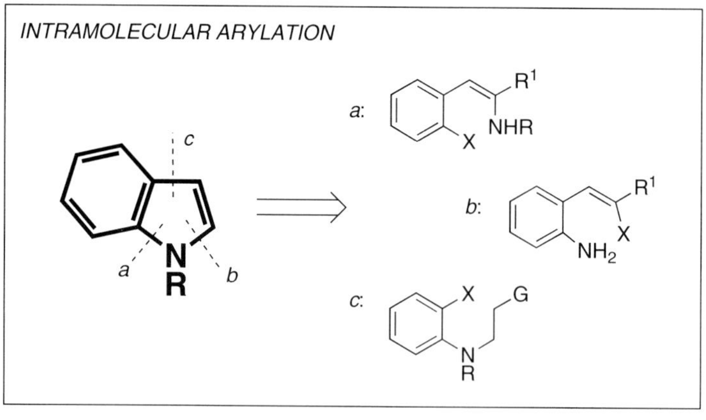

INTRAMOLECULAR ARYLATION

The Cu(I)-catalyzed intramolecular arylation approach as an efficient route toward the construction of multisubstituted indole cores was first demonstrated by Cusack.

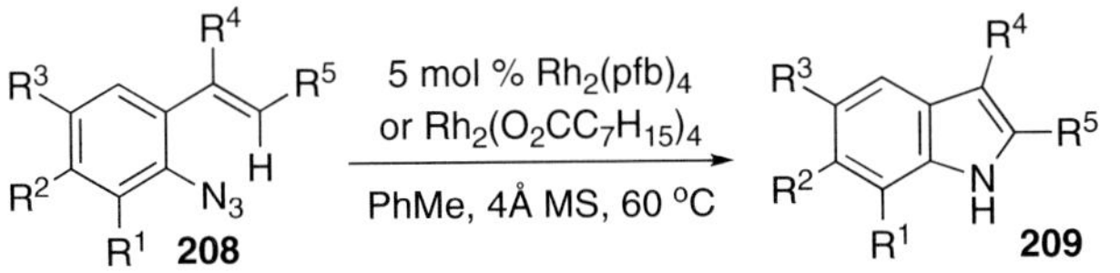

208 → 209: 5 mol % $Rh_2(pfb)_4$ or $Rh_2(O_2CC_7H_{15})_4$, PhMe, 4Å MS, 60 °C

R^1	R^2	R^3	R^4	R^5	Yield, %
H	H	OMe	H	Ph	98
H	H	Me	H	Ph	89
H	H	H	H	Ph	94
H	H	F	H	Ph	99
H	H	OCF_3	H	Ph	95
H	H	CF_3	H	Ph	82
H	OMe	H	H	Ph	88
H	Me	H	H	Ph	89
H	CF_3	H	H	Ph	89
Me	H	F	H	Ph	92
H	H	H	Ph	H	39
H	H	H	H	*p*-Tol	96
H	H	H	H	*p*-F-C_6H_4	95
H	H	H	H	*m*-F-C_6H_4	91
H	H	H	H	*p*-Cl-C_6H_4	84
H	H	H	H	*p*-F_3C-C_6H_4	95
H	H	OMe	H	*p*-F_3C-C_6H_4	90
H	H	F	H	*p*-F_3C-C_6H_4	89
H	H	OCF_3	H	*p*-F_3C-C_6H_4	91
H	H	CF_3	H	Me	75
H	H	H	H	*n*-Hex	68

Scheme 9.75

It was shown that the intramolecular N-arylation of ene-carbamates **216** in the presence of the Cu(I)–L-proline catalytic system and an external inorganic base could provide *N*-protected indole-3-carboxylate derivatives **217** (Scheme 9.78) [232]. In the case of substrates bearing electron-deficient substituents on the aryl ring, partial deprotection of the Cbz-group accompanied the formation of the corresponding products. General utility of this protocol was further demonstrated by an efficient synthesis of the 2,4,5-trisubstituted indole core **219** of the antihypertensive agent U86192A **220** (Scheme 9.79).

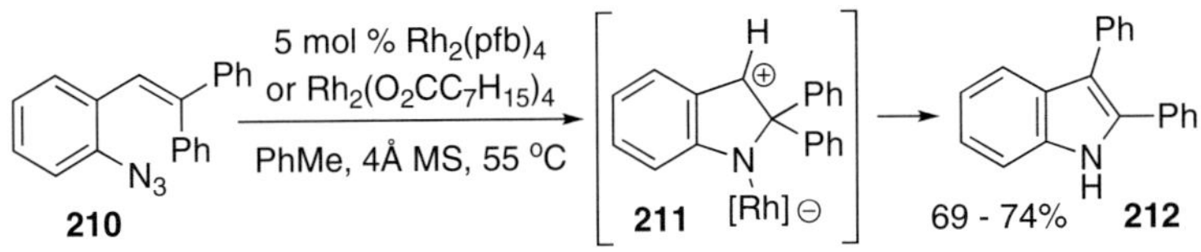

210 → [211] → 212: 5 mol % $Rh_2(pfb)_4$ or $Rh_2(O_2CC_7H_{15})_4$, PhMe, 4Å MS, 55 °C; 69 - 74%

Scheme 9.76

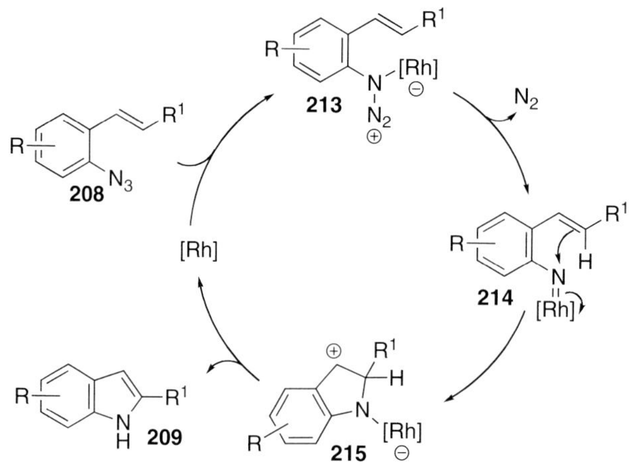

Scheme 9.77

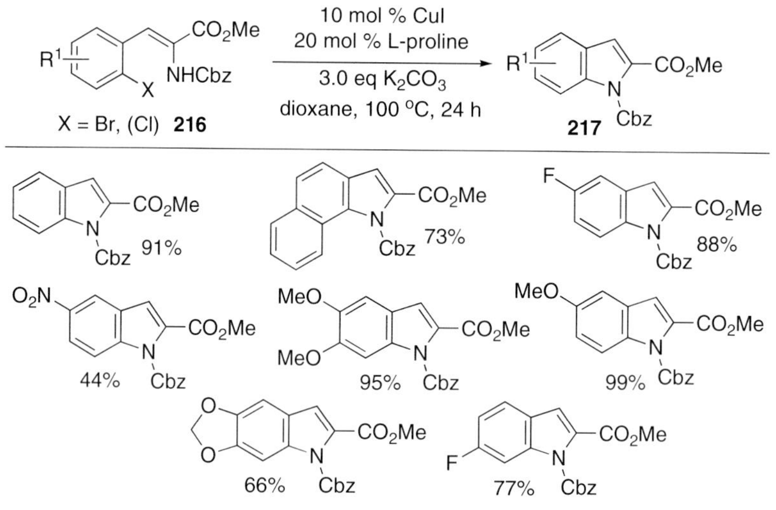

Scheme 9.78

Moreover, Fukuyama successfully utilized the same chemistry for the assembly of indole-2-carboxylic acid fragments **222**, **226**, **228**, and **230** during a highly convergent total synthesis of potent antitumor antibiotics, (+)-duocarmycins A **223** and SA **224** (Scheme 9.80), and (+)-yatakemycin **231**, respectively (Scheme 9.81) [233, 234].

A similar Cu(I)-catalyzed arylation protocol leading to the formation of differently *N*-substituted indole-3-carboxylates **233** was recently reported by Karchava and coworkers (Scheme 9.82) [235]. The corresponding indole precursors **232** can be easily accessed in nearly quantitative yields by the amination of the respective common 1,3-dicarbonyl compound. A variety of primary alkyl- and aryl-amines, including sterically hindered ones, can be employed in this reaction, affording indole products **233** in good to high yields.

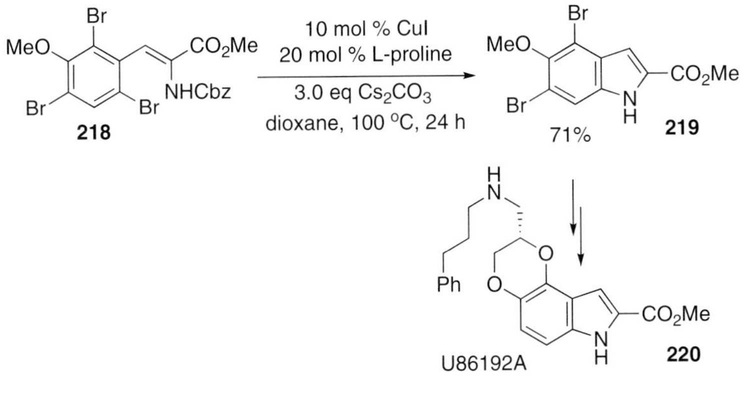

Scheme 9.79

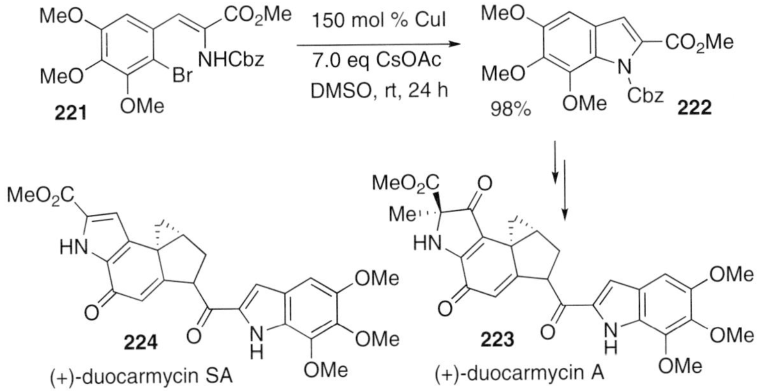

Scheme 9.80

More recently, the same group extended this approach to a facile Cu(I)-catalyzed assembly of pharmacologically important 1-amino- and 1-hydroxyindoles [152] **235** from easily available enehydrazines **234** (Scheme 9.83) [236]. In the case of benzyl-substituted substrates, arylation reaction was accompanied by an oxidation process leading to the formation of *N*-benzyledene-1-aminoindole-3-carboxylates.

Lautens developed an efficient route for the preparation of optically active 1,2-C5-fused indoles **237**, namely imidazoindolone cores, by taking advantage of the Cu(I)-catalyzed double amidation tandem reaction of *gem*-dihalovinylanilides **236** (Scheme 9.84) [237]. A variety of alkyl substituents at the imidazolone ring can be tolerated, providing products **237**, generally in good yields. The preservation of chirality originating from the amino acid substituent of **236** varied from moderate to very high.

Finally, Ma reported a general strategy for the synthesis of pharmacologically important 3-acyloxindole derivatives utilizing a C3–C3a disconnection approach.

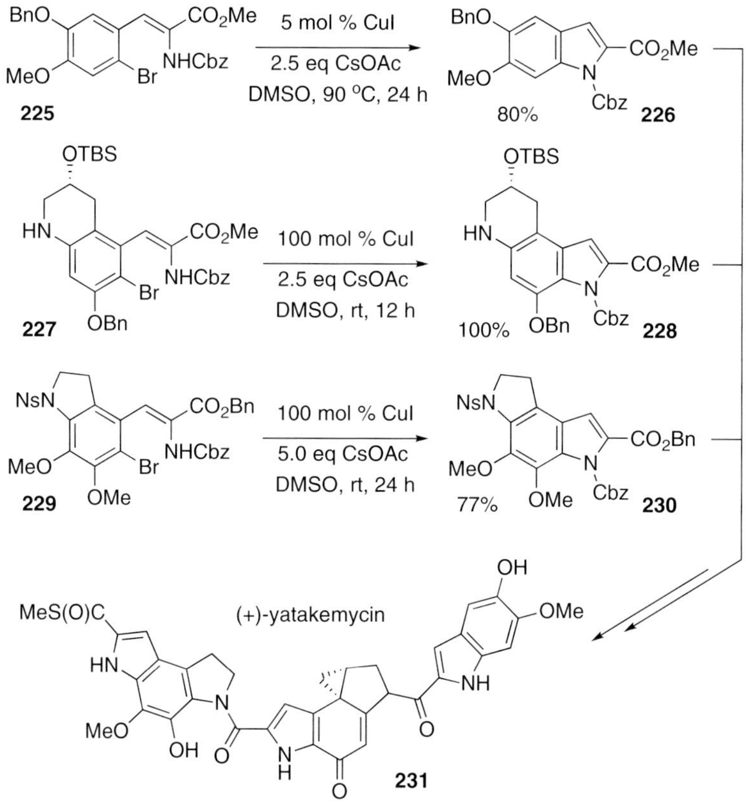

5 mol % CuI, 2.5 eq CsOAc, DMSO, 90 °C, 24 h: 225 → 226 (80%)

100 mol % CuI, 2.5 eq CsOAc, DMSO, rt, 12 h: 227 → 228 (100%)

100 mol % CuI, 5.0 eq CsOAc, DMSO, rt, 24 h: 229 → 230 (77%)

(+)-yatakemycin 231

Scheme 9.81

5 mol % CuI, 0 - 2 eq $(CH_2OH)_2$, 2.0 eq K_3PO_4, DMF or IPA, 75 °C, 4-20 h: 232 → 233

92%, 80%, 68%, 32%, 78%, 57%, 65%, 74%, 77%, 94%

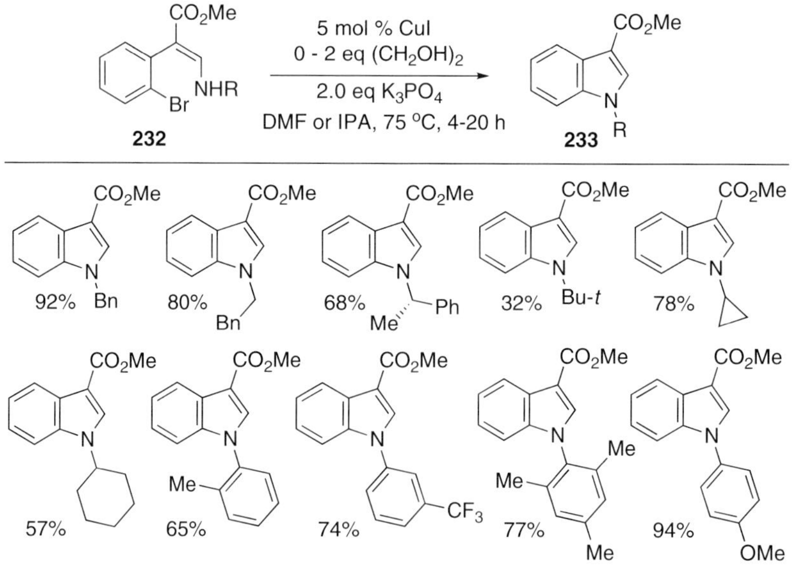

Scheme 9.82

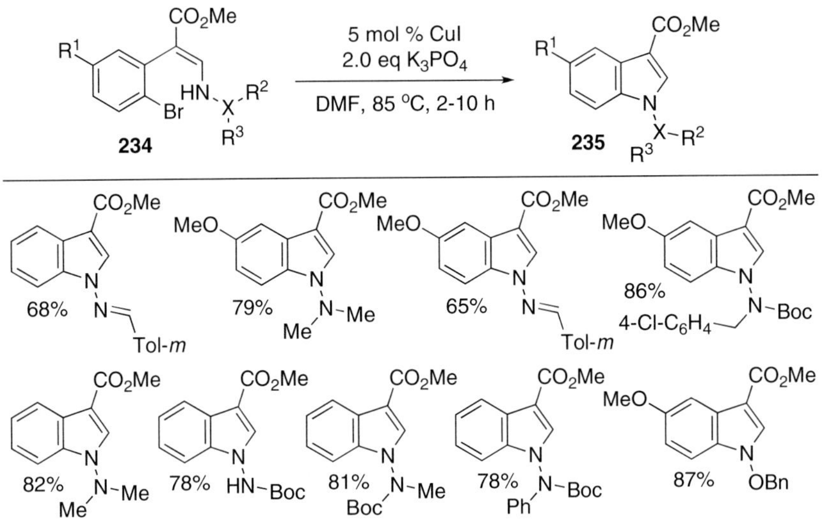

Scheme 9.83

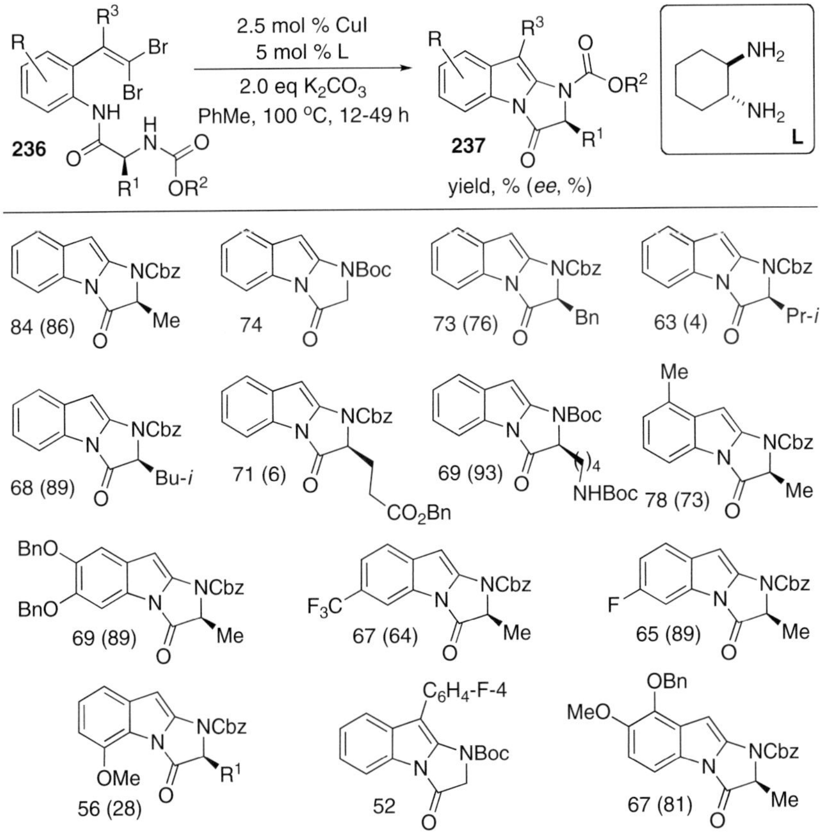

Scheme 9.84

Accordingly, employment of β-keto anilides **238** in an intramolecular arylation reaction, catalyzed by the Cu(I)–L-proline system, led to an efficient assembly of the target indole derivatives **239** (Scheme 9.85) [238, 239]. The authors demonstrated that electronic effects of substituents on the aromatic ring have very little influence on this arylation reaction, as complete conversion was observed for substrates possessing both electron-donating and electron-withdrawing groups.

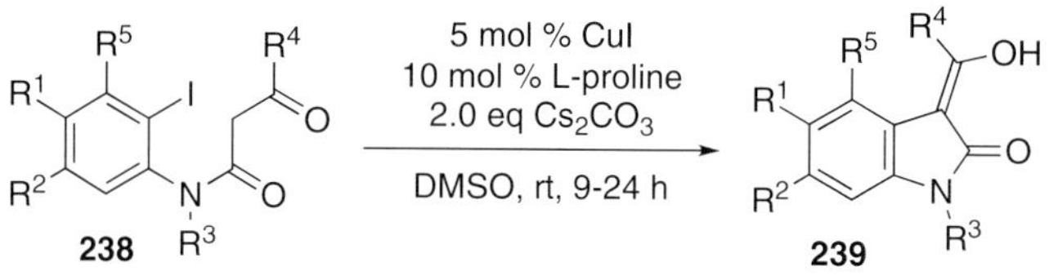

5 mol % CuI, 10 mol % L-proline, 2.0 eq Cs_2CO_3, DMSO, rt, 9-24 h: **238** → **239**

Substituents	Yield
R^1 = H, F, Me	66 - 83 %
R^2 = H, Cl, MeO	
R^3 = Me, Bn, DMB	
R^4 = Me, Ph, 4-Cl-C_6H_4-, 2-furyl, 5-(2-Cl-Py)-, 2-thienyl, 2-benzo[*b*]thienyl	
R^5 = H, Me	

Scheme 9.85

9.2.3.3 **Synthesis of Indoles via "4 + 1" Cycloaddition Reactions**

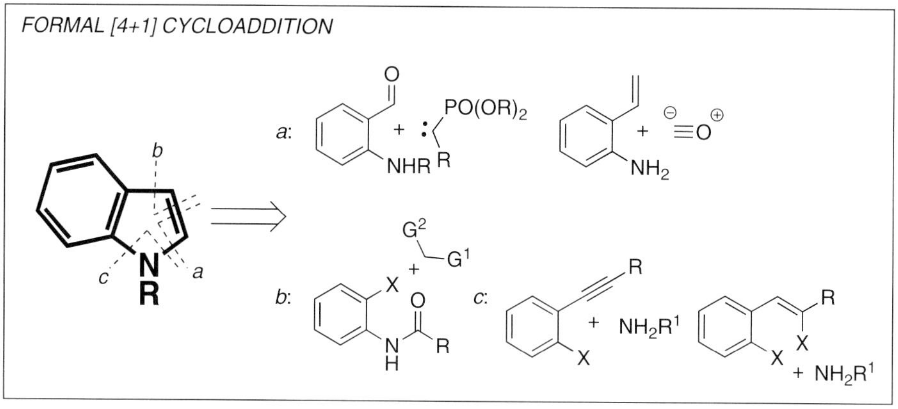

FORMAL [4+1] CYCLOADDITION

An interesting "4 + 1" approach based on the utilization of α-diazophosphonate as a one-carbon synthon toward the construction of a functionalized indole framework was illustrated by Nakamura. This was achieved by combining the Rh(II)-catalyzed N–H insertion reaction with the base-mediated intramolecular Horner–

Wadsworth–Emmons process [144], namely, in this case, a modified Madelung-type indole synthesis, into a one-pot transformation. First, reaction between *ortho*-carbonyl-functionalized aniline **240** and diazocompound **241** in the presence of the Rh(II)-catalyst generated the phosphonate **243**, which, upon treatment with base, was transformed into multisubstituted indoles **242** in moderate to high yields (Scheme 9.86) [145].

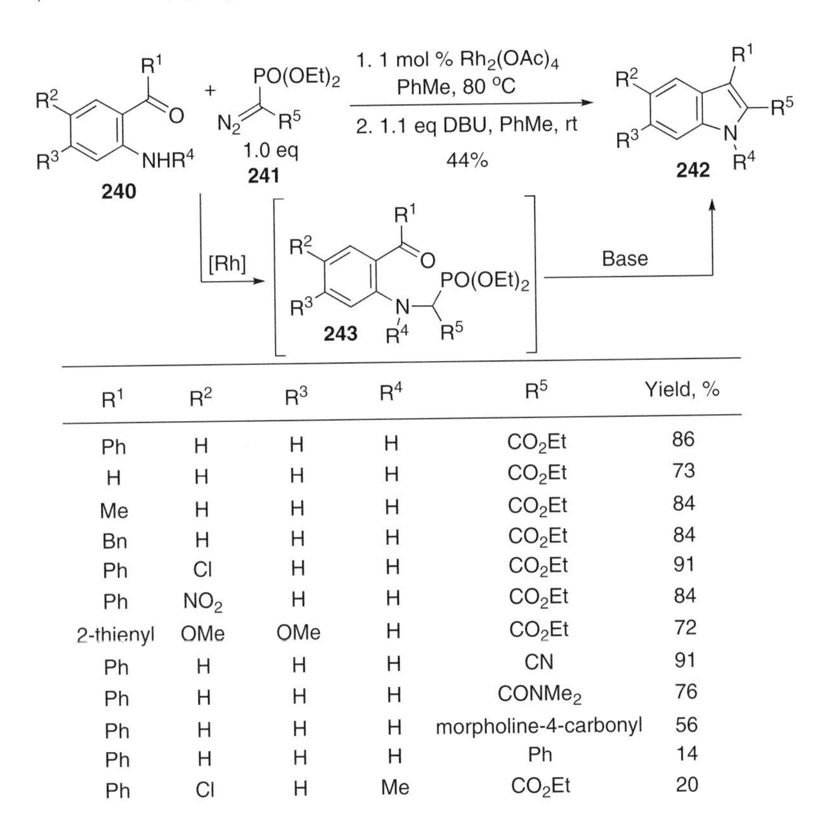

R^1	R^2	R^3	R^4	R^5	Yield, %
Ph	H	H	H	CO_2Et	86
H	H	H	H	CO_2Et	73
Me	H	H	H	CO_2Et	84
Bn	H	H	H	CO_2Et	84
Ph	Cl	H	H	CO_2Et	91
Ph	NO_2	H	H	CO_2Et	84
2-thienyl	OMe	OMe	H	CO_2Et	72
Ph	H	H	H	CN	91
Ph	H	H	H	$CONMe_2$	76
Ph	H	H	H	morpholine-4-carbonyl	56
Ph	H	H	H	Ph	14
Ph	Cl	H	Me	CO_2Et	20

Scheme 9.86

Dong developed the Rh(I)-catalyzed "4 + 1" synthesis of tryptamines and tryptophols **245** based on a regioselective hydroformylation of *ortho*-aminostyrenes **244**, wherein carbon monoxide served as a one-carbon synthon (Scheme 9.87) [240]. This hydroformylation–cyclocondensation sequence, occurring via the formation of the key *o*-aminophenyl aldehyde intermediate **246**, allows the preparation of a variety of substituted indoles with unfunctionalized N- and C2-sites in modest to good yields (Scheme 9.87). However, indoles possessing substituents at the C4 position could not be prepared via this protocol.

Ma reported the Cu(I)-catalyzed cascade coupling–condensation–deacylation process for the synthesis of 2-(trifluoromethyl)indoles **249** (Scheme 9.88) [239, 241]. Thus, a formal [3 + 2] annulation of trifluoroacetanilides **247** with β-ketoesters **248**

244 → 245: 5 mol % $HRh(CO)(PPh_3)_3$, 300 psi H_2/CO (1:1), PPh_3, PhMe, 70 °C, 70 h

244 → [Rh] → [246] → $- H_2O$ → 245

R^1 = H, F, Cl, Me

R^2 = H, CF_3, MeO

R^3 = H, F, Br

R^4 = NHTs, $N(Boc)_2$, OH

32 - 73 %

Scheme 9.87

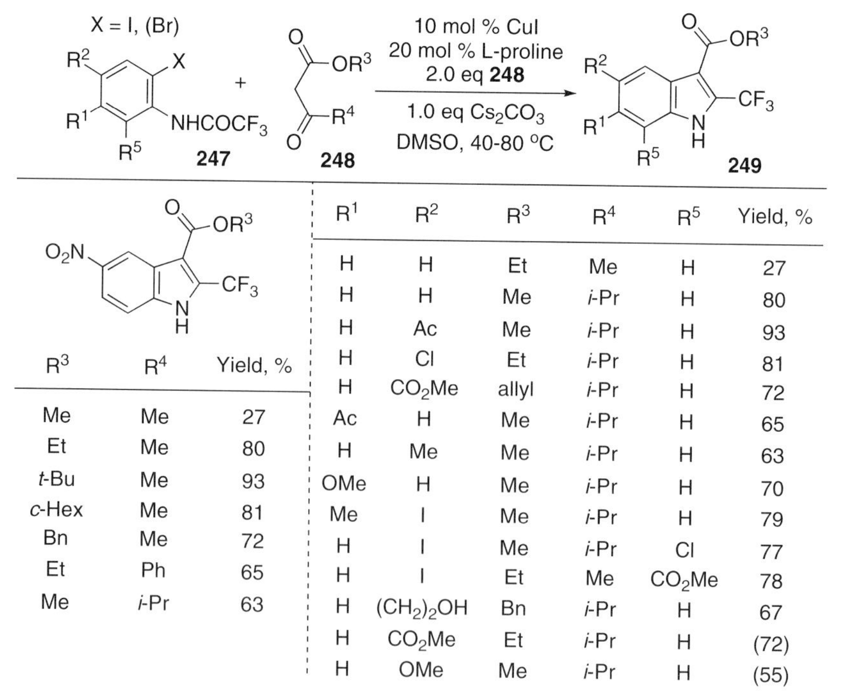

R^3	R^4	Yield, %
Me	Me	27
Et	Me	80
t-Bu	Me	93
c-Hex	Me	81
Bn	Me	72
Et	Ph	65
Me	*i*-Pr	63

R^1	R^2	R^3	R^4	R^5	Yield, %
H	H	Et	Me	H	27
H	H	Me	*i*-Pr	H	80
H	Ac	Me	*i*-Pr	H	93
H	Cl	Et	*i*-Pr	H	81
H	CO_2Me	allyl	*i*-Pr	H	72
Ac	H	Me	*i*-Pr	H	65
H	Me	Me	*i*-Pr	H	63
OMe	H	Me	*i*-Pr	H	70
Me	I	Me	*i*-Pr	H	79
H	I	Me	*i*-Pr	Cl	77
H	I	Et	Me	CO_2Me	78
H	$(CH_2)_2OH$	Bn	*i*-Pr	H	67
H	CO_2Me	Et	*i*-Pr	H	(72)
H	OMe	Me	*i*-Pr	H	(55)

Scheme 9.88

in the presence of the L-proline stabilized Cu(I)-catalyst under requisite anhydrous conditions provided the corresponding indole-3-carboxylates **249** with perfect functional group compatibility. Generally, electron-deficient substrates **247** were more reactive and gave better yields than those possessing electron-donating substituents. According to the proposed mechanism, the Cu(I)-catalyzed arylation of β-ketoester **248** leads to the enolate **250**, in which the carbanion attack at the carbonyl group of the trifluoroacetamide moiety produces cyclic dihydroindolyl alkoxide **251**. A subsequent intramolecular attack of the alkoxide in the latter leads to the formation of the 4-membered ring in **252**, which, upon fragmentation (**253**) followed by deacylation, furnishes the corresponding indole **249** (Scheme 9.89).

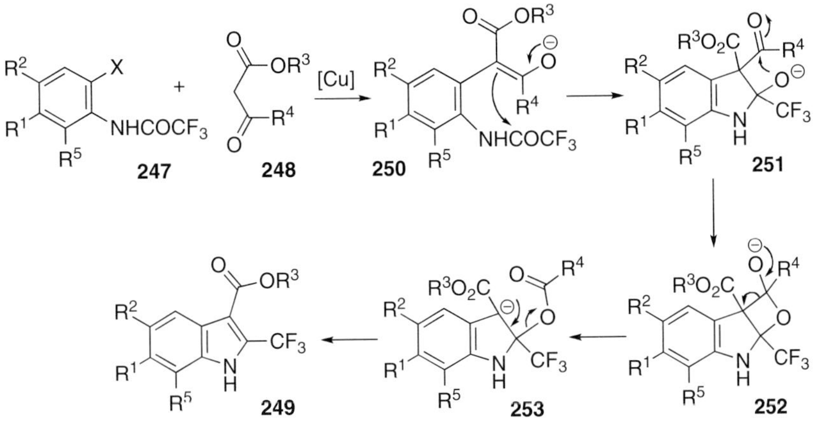

Scheme 9.89

Synthesis of 1,2-disubstituted indole frameworks **257** via a formal "4 + 1" cycloaddition between a 4-carbon unit and a primary amine was recently developed by Ackermann (Scheme 9.90) [242]. Reactive intermediates, 2-(*o*-haloaryl)enamines **256** were generated via the Cu(I)-catalyzed hydroamination of the *ortho*-halo-substituted phenylacetylenes **254** with primary amines **255**. A subsequent Cu(I)-catalyzed intramolecular enamine arylation reaction gave the corresponding indoles **257** in good yields. The authors demonstrated that alkynyl chlorides **254** could also participate in this cascade double amination process, albeit with a substantially diminished efficiency.

Another double amination route to indoles **260** was reported by Willis (Scheme 9.91) [243]. Accordingly, tandem inter-/intramolecular-bisamination of 2-(2-haloalkenyl)-aryl halides **258** with an array of N-nucleophiles **259**, such as anilines, amides, and carbamates, proceeded in the presence of different Cu(I) catalysts, providing *N*-functionalized indoles **260** in moderate to high yields. Conversely, employment of simple alkyl amines was far less efficient.

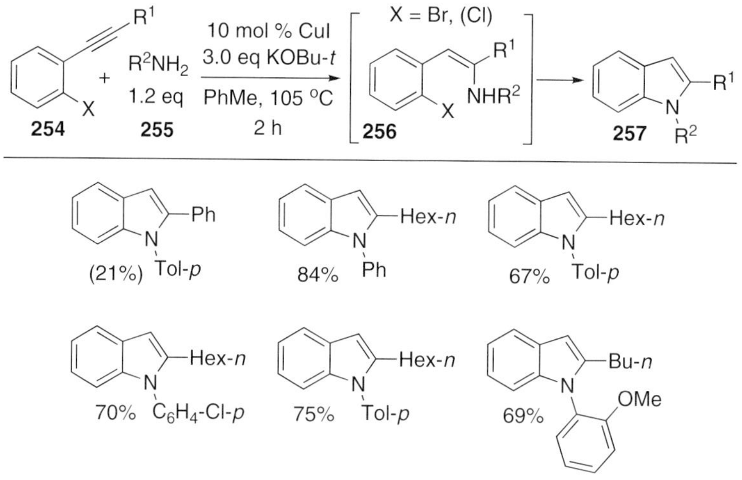

Scheme 9.90

In addition, this reaction seems to be limited to the corresponding dibromides as coupling components, since attempts to employ the chloro analogs provided either diminished yields of the products **260** or a complete lack of reactivity.

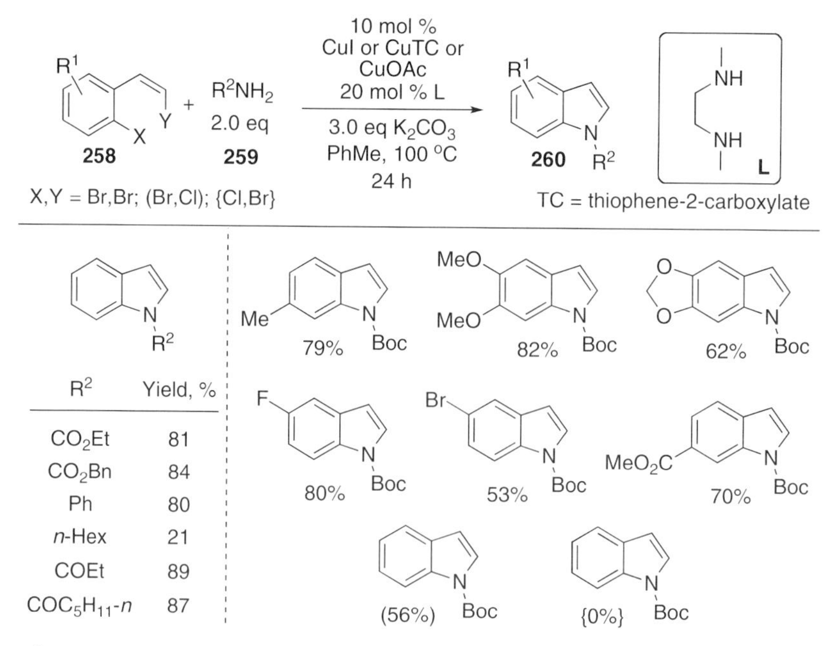

Scheme 9.91

9.2.3.4 Synthesis of Indoles via "3 + 2" Cycloaddition Reactions

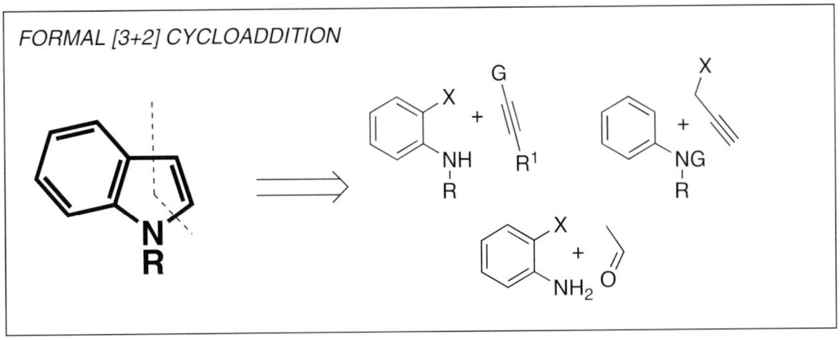

FORMAL [3+2] CYCLOADDITION

A formal [3 + 2] cycloaddition route leading to the preparation of di- and trisubstituted indoles **264** from *ortho*-haloanilines **261** and copper(I) acetylides **262** was introduced by Castro (Scheme 9.92) [75, 76, 111], simultaneously with the mentioned earlier coupling–cyclization reaction *en route* to benzofurans **57** (Scheme 9.21). It is believed that this stoichiometric in copper cascade reaction proceeded with the generation of an intermediary *ortho*-alkynylaniline **263** [77]. Attempts to turn this reaction into a more practical catalytic process, using basic conditions to generate Cu(I)-acetylide from a terminal alkyne in the presence of CuI, provided poor yields of the respective indoles **264**.

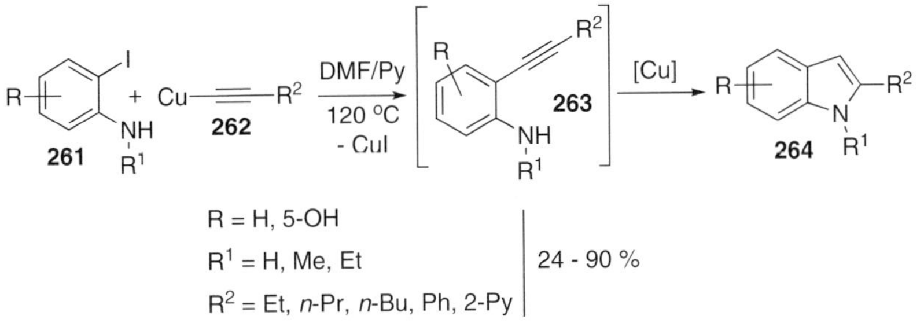

DMF/Py, 120 °C, - CuI; [Cu]

R = H, 5-OH
R^1 = H, Me, Et
R^2 = Et, *n*-Pr, *n*-Bu, Ph, 2-Py
24 - 90 %

Scheme 9.92

Nonetheless, this stoichiometric methodology was successfully applied by others for the syntheses of a variety of complex indole derivatives. For instance, Katritzky utilized this chemistry to obtain benzotriazolyl-substituted indoles, valuable intermediates in the synthesis of fused and polycyclic indole-derived structures [244, 245]. Shvartsberg used Castro's protocol for the assembly of benz[g]indolediones [246], whereas the synthesis of pharmacologically important 1-aminoindole derivatives was reported by Prikhod'ko [247, 248].

Recently, a general Cu(I)-catalyzed version leading to 2-aryl and 2-heteroaryl indoles **267** was reported by Cacchi (Scheme 9.93) [190]. Thus, *o*-iodotrifluoroacetanilides **265** and terminal alkynes **266** underwent the Cu(I)-catalyzed domino coupling–cycloisomerization reaction with a concomitant deprotection of the trifluoroacetyl group, providing NH-indoles **267**. Moderate to excellent yields can be

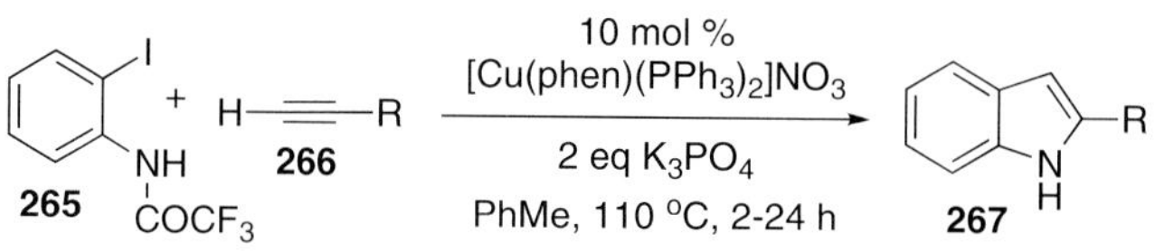

265 + $H-\equiv-R$ 266 → (10 mol % $[Cu(phen)(PPh_3)_2]NO_3$, 2 eq K_3PO_4, PhMe, 110 °C, 2-24 h) → 267

R	Yield, %
Ph	78
p-Ac-C_6H_4	96
p-Cl-C_6H_4	80
p-MeCONH-C_6H_4	71
p-MeO-C_6H_4	81
m-MeO-C_6H_4	92
o-MeO-C_6H_4	62
3,5-Me_2-C_6H_3	68
p-MeO_2C-C_6H_4	93
p-OHC-C_6H_4	57
p-O_2N-C_6H_4	62
n-Bu	11
2-Py	65
3-quinolyl	89

Scheme 9.93

achieved in the case of aryl- and hetaryl-substituted substrates, whereas alkynes, possessing alkyl groups, provided indoles only in low yields.

Later, Venkataraman and coworkers reported an analogous "3 + 2" coupling–cyclization protocol for the synthesis of 5-mono and 5,7-dihalosubstituted indoles [121]. Slough demonstrated that the same transformation can be performed with a comparable efficiency employing a solid-supported Cu(I) catalyst [249].

Ma applied this chemistry to the Cu(I)-catalyzed indolization of 2-bromotrifluoroacetanilides **268** with a variety of terminal alkynes **269** (Scheme 9.94) [250]. A variety of substituted NH-indoles **270** were synthesized via this procedure in generally good to excellent yields. Most importantly, employment of protected propargylic alcohols allowed the efficient introduction of a C2-alkyl substituent into the indole ring, which was shown to be problematic under Cacchi's conditions. However, a coupling–cyclization cascade of alkynes **269**, bearing simpler alkyl groups, for instance *n*-pentyl, led to a diminished yield of the target indole product.

Recently, Fagnou reported a very interesting, atom-economical route to the 1,2,3-trisubstituted indole derivatives **273** via the Rh(II)-catalyzed oxidative coupling–indolization reaction (Scheme 9.95) [251]. Accordingly, simple acetanilides **271**, upon a directed C–H activation with the Rh(II)-catalyst [252] followed by a subsequent carborhodation–indolization sequence of alkyne **272**, gave *N*-acylated indoles **273**. Both electron-rich and electron-deficient acetanilides **271**, possessing different functionalities were perfectly tolerated under these reaction conditions. In the case of unsymmetrical alkyl-alkyl-substituted acetylenes, a mixture of indole products

1.1 eq **268** + H—≡—R^5 **269** → (20 mol % CuI, 60 mol % L-proline, 2 eq K_2CO_3, DMF, 80 °C, 24 h) → **270**

Substituents	Yield
R^1 = H, OMe	25 - 94 %
R^2 = H, Et, Cl, MeO, NO_2, CO_2Me	
R^3 = Me, $CH_2N(Me)Bn$, CH_2NMe_2, CO_2Me	
R^4 = H, Cl	
R^5 = Ph, p-Cl-C_6H_4, p-MeO_2C-C_6H_4, p-MeO-C_6H_4, m-AcO-C_6H_4, n-C_5H_{11}, CH_2OBn, CH_2OTHP	

Scheme 9.94

with the C2–C3 regioselectivity favoring the formation of indole where the larger substituent resides at C2 was obtained.

Tokunaga and Wakatsuki developed a "3 + 2" transition metal-catalyzed route to multisubstituted indoles **277** which relies on the Bischler indolization reaction (Scheme 9.96) [253]. The key 2-arylaminoketone intermediates **276** and **276′** were accessed via tandem Ru(0)-catalyzed hydroamination–isomerization of propargylic alcohols **275** with substituted anilines **274** under solvent-free conditions.

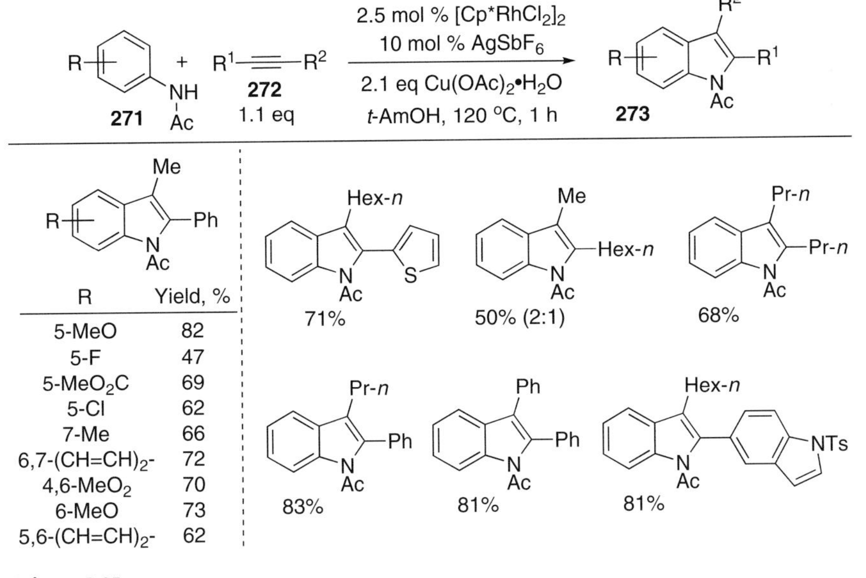

R	Yield, %
5-MeO	82
5-F	47
5-MeO_2C	69
5-Cl	62
7-Me	66
6,7-$(CH{=}CH)_2$-	72
4,6-MeO_2	70
6-MeO	73
5,6-$(CH{=}CH)_2$-	62

Scheme 9.95

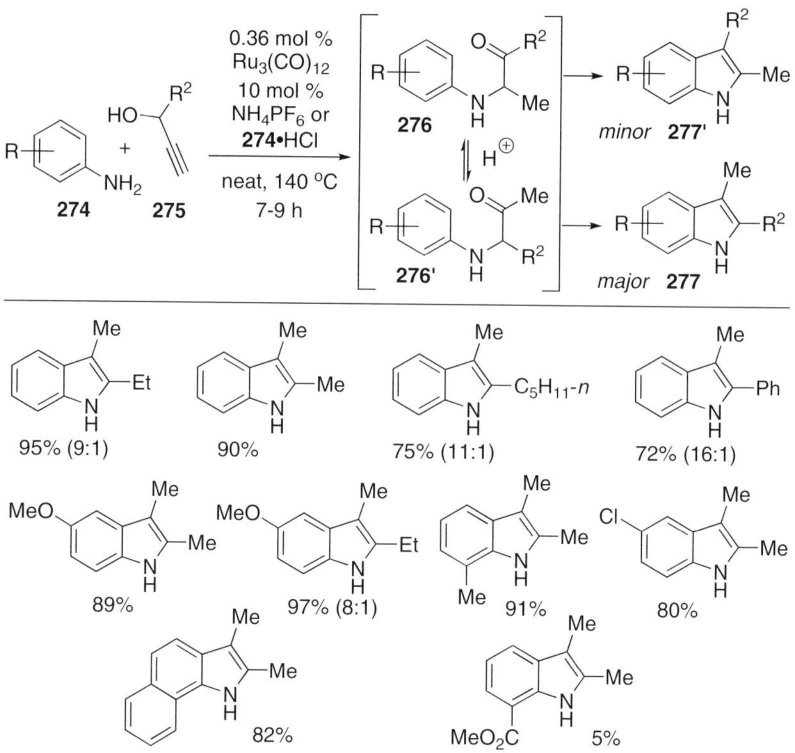

Scheme 9.96

An apparent fast interconversion of the regioisomeric 2-aminoketones **276** and **276′** in the presence of Brønsted acid led to the formation of mixtures, in some cases, of C2–C3 regioisomeric indoles with a generally good selectivity toward 3-methyl derivatives.

Later, an analogous highly regioselective transformation in the presence of Zn(II) or Cu(II) triflate catalysts in toluene was reported by Liu [124]. Notably, the amination–indolization sequence of **274** and **275** under these modified reaction conditions provided the corresponding multisubstituted indole derivatives **277** as sole C3-methyl-substituted regioisomers.

Saito and Hanzawa extended their cycloisomerization route, involving the amino-Claisen rearrangement–cyclization cascade reaction (Scheme 9.68), to a one-pot procedure toward the functionalized 2,3-dimethylindole derivatives **281** (Scheme 9.97) [226]. The key propargyl anilines **280** were generated *in situ* via the propargylation of substituted anilines **278** with the propargyl bromide **279**.

Recently, transition metal-catalyzed hydroamination reaction of alkynes with hydrazines was used to generate aryl hydrazones, highly reactive and versatile intermediates in a subsequent metal-catalyzed Fischer indole synthesis. Thus, Odom first reported a "3 + 2" synthesis of 1,7-fused, di-, and tri-substituted indoles **285** featuring the Ti(IV)-catalyzed hydroamination of various alkynes **283** with

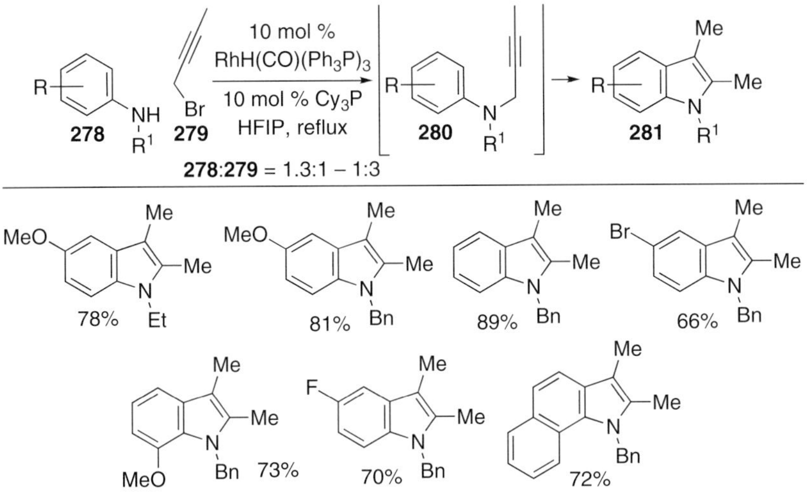

Scheme 9.97

arylhydrazines **282** (Scheme 9.98) [254]. The Fischer cyclization of *in situ* generated aryl hydrazones **284** was promoted by stoichiometric amounts of $ZnCl_2$. The regioselectivity of the reaction appears to be influenced by alkyne electronics, favoring the Markovnikov addition to terminal alkyl acetylenes and *anti*-Markovnikov addition to aryl-substituted alkynes.

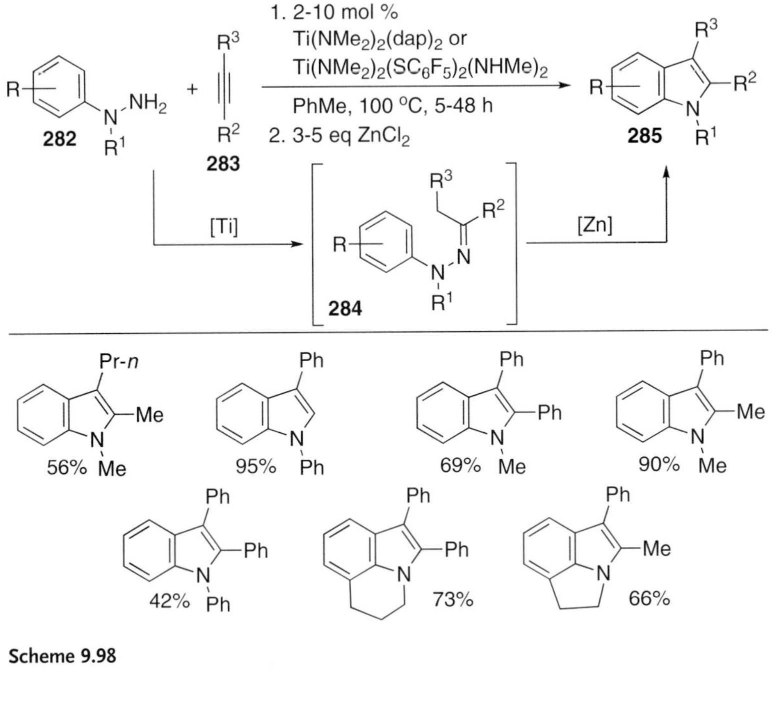

Scheme 9.98

Later, an entirely catalytic version of a similar domino process, involving Grandberg tryptamine synthesis, was elaborated on by Beller and coworkers (Scheme 9.99) [255]. Moderate to high yields of indoles **288** were achieved via the Ti(IV)-catalyzed amination–indolization reaction of differently substituted arylhydrazones **286** and chloro-substituted terminal alkynes **287**.

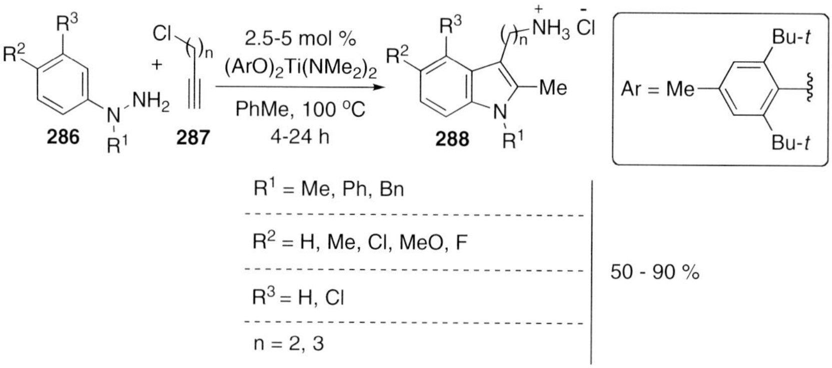

R^1 = Me, Ph, Bn

R^2 = H, Me, Cl, MeO, F

R^3 = H, Cl

n = 2, 3

50 - 90 %

Scheme 9.99

Along this line, several research groups reported similar transition metal-catalyzed transformations of differently substituted alkynes and hydrazines, leading to the construction of multisubstituted indole cores. Ackermann disclosed that the $TiCl_4$–*t*-$BuNH_2$ system as a single catalyst could efficiently catalyze the assembly of 1,2,3-trisubstituted indoles [256]. Later, Beller elaborated on the Ti(IV)-catalyzed hydrohydrazination–Zn(II)-promoted Fischer cyclization *en route* to tryptophols, their homologs [257], and 3-silyloxyindoles [258]. The same group performed a general study of the Ti(II)-catalyzed version of this reaction, and later investigated the Zn(II)-mediated synthesis of indole-2,3-carboxylates [259]. A substantially improved scope of this reaction in the case of terminal alkynes was achieved when stoichiometric amounts of Zn(II) triflate were used to promote both amination and indole ring formation processes [260]. Further, the Ti(IV) catalyst with pyrrolyl-containing ligands was recently utilized by Odom in the synthesis of 2,3-disubstituted NH-indoles [261]. Finally, Beller reported the Ti(IV)/Zn(II)-mediated approach toward pharmacologically important 3-hydroxyindole-containing 5-HT_6 receptor ligands [262].

In addition, a number of reports on the indole synthesis utilizing a similar concept of the *in situ* generation of the key reactive arylhydrazones, similar to **284**, with the aid of transition metal catalysts recently appeared in the literature. Among them, the domino Rh-catalyzed hydroformylation [263] of alkenes leading to aldehydes followed by a subsequent hydrazone formation and Fischer indolization process, which was introduced by Eilbracht [264–267] and later investigated by Beller [268], represents an attractive and efficient "3 + 1 + 1" approach for a one-pot construction of complex indoles.

Finally, a number of efficient indole syntheses utilizing a transition metal catalyzed C-arylation of 1,3-dicarbonyl compounds with *ortho*-iodoaniline derivatives followed by the N1–C2 indole ring closure via a cyclocondensation reaction were elaborated on by several research groups.

An early report on this chemistry from Suzuki's group described the employment of stable Na-enolates **290** as coupling components in the synthesis of disubstituted, and even fused, indoles **292** (Scheme 9.100) [269]. The Cu(I)-mediated α-arylation of **290** with *o*-iodoaniline **289** initially provided α–*o*-aminophenyl-ketones **291**, whereupon a subsequent cyclocondensation reaction furnished the corresponding indole nucleus.

289 + **290** → (150 mol % CuI; DMF, 120-130 °C; 6-7 h) → [**291**] → **292**

R^1 = Me, Et, CH_2CO_2Et, $-CH_2C(Me)_2CH_2CO-$; R^2 = COMe, COPh, CO_2Et | 60 - 80 %

Scheme 9.100

Later, Miura found that this reaction could be efficiently performed in a catalytic fashion with a carbonyl compound possessing an active methylene group in the presence of 5 mol% of CuI catalyst and an external base [269, 270].

Recently, Tanimori demonstrated that a variety of 2-functionalized indole-3-carboxylates **292** could be readily accessed through the Cu(I)–BINOL-catalyzed arylation–cyclocondensation sequence of *o*-iodoaniline **289** and β-ketoesters **293** under mild reaction conditions (Scheme 9.101) [271].

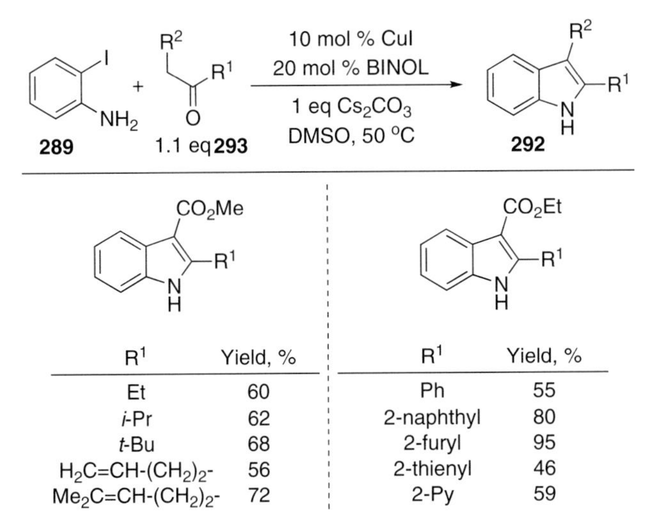

R^1 (CO_2Me)	Yield, %
Et	60
i-Pr	62
t-Bu	68
$H_2C=CH-(CH_2)_2-$	56
$Me_2C=CH-(CH_2)_2-$	72

R^1 (CO_2Et)	Yield, %
Ph	55
2-naphthyl	80
2-furyl	95
2-thienyl	46
2-Py	59

Scheme 9.101

Finally, Ma et al. further elaborated on this chemistry toward the development of an efficient cascade process for the assembly of 2,3-disubstituted indoles **296** (Scheme 9.102) [239, 272]. The authors showed that an array of aliphatic and aromatic β-ketoesters and ketoamides **295** could be utilized in the "3 + 2" annulation reaction with anilides **294** providing the corresponding indoles **296** in moderate to excellent yields. In addition, this reaction was applied to 2-bromotrifluoroacetanilides, albeit requiring slightly elevated temperatures to achieve complete conversion. It is believed that cyclocondensation of aminoketone **298** accomplished the indole ring

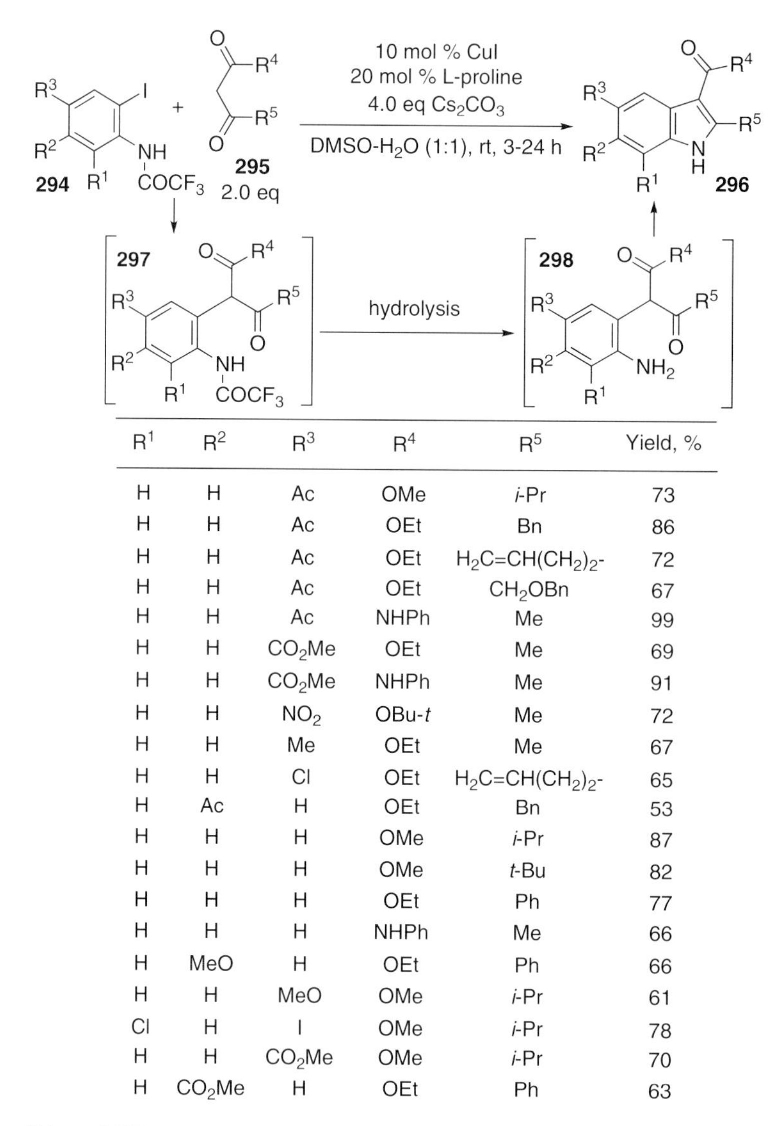

R^1	R^2	R^3	R^4	R^5	Yield, %
H	H	Ac	OMe	*i*-Pr	73
H	H	Ac	OEt	Bn	86
H	H	Ac	OEt	$H_2C{=}CH(CH_2)_2$-	72
H	H	Ac	OEt	CH_2OBn	67
H	H	Ac	NHPh	Me	99
H	H	CO_2Me	OEt	Me	69
H	H	CO_2Me	NHPh	Me	91
H	H	NO_2	OBu-*t*	Me	72
H	H	Me	OEt	Me	67
H	H	Cl	OEt	$H_2C{=}CH(CH_2)_2$-	65
H	Ac	H	OEt	Bn	53
H	H	H	OMe	*i*-Pr	87
H	H	H	OMe	*t*-Bu	82
H	H	H	OEt	Ph	77
H	H	H	NHPh	Me	66
H	MeO	H	OEt	Ph	66
H	H	MeO	OMe	*i*-Pr	61
Cl	H	I	OMe	*i*-Pr	78
H	H	CO_2Me	OMe	*i*-Pr	70
H	CO_2Me	H	OEt	Ph	63

Scheme 9.102

formation process, where a requisite hydrolysis of the arylation product, *N*-protected aminoketone **297**, occurred *in situ* under the basic reaction conditions. Alternatively, in some cases, addition of an acid promoter was necessary to trigger this process.

9.2.4 Isoindoles

2*H*-Isoindole is a fairly well-used heterocyclic unit in organic chemistry [164, 273, 274]. The isoindole core is found in natural and synthetic products with a wide range of applications in medicinal chemistry, due to their notable antibiotic and antitumor activities and other biological properties [156, 158], as well as in materials [274]. Despite the fact that a variety of syntheses of isoindoles and their derivatives [165, 170, 171, 173–175, 274, 275] from acyclic precursors have been developed, only a few reports describe the use of transition metal catalysts or promoters. Within this group, transformations involving disconnections of the isoindole core via a cycloisomerization and a formal [4 + 1] cycloaddition reaction are described in detail below.

9.2.4.1 Synthesis of Isoindoles via Cycloisomerization Reactions

CYCLOISOMERIZATION APPROACHES

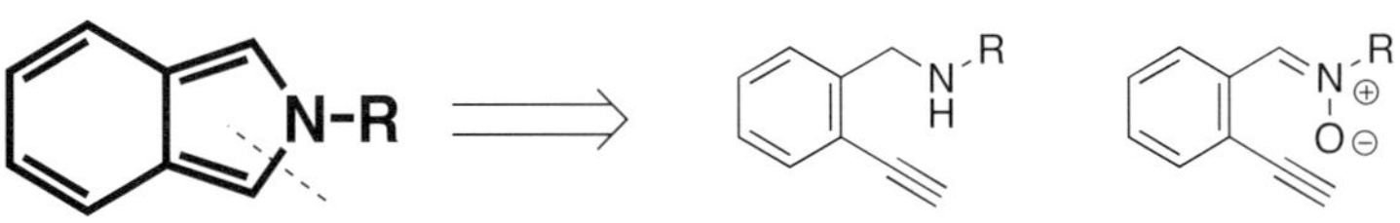

N-R ⟹ N(H)-R ; N(⊕)(O⊖)-R

Dyker provided two examples of the isoindole synthesis based on a transition metal-catalyzed 5-*exo-dig* cyclization. Thus *ortho*-alkynyl benzylamines **299** were cycloisomerized into the isoindole-1-carboxamides **300** in the presence of Au(III)-catalyst, providing the corresponding products in modest yields (Scheme 9.103) [276]. A

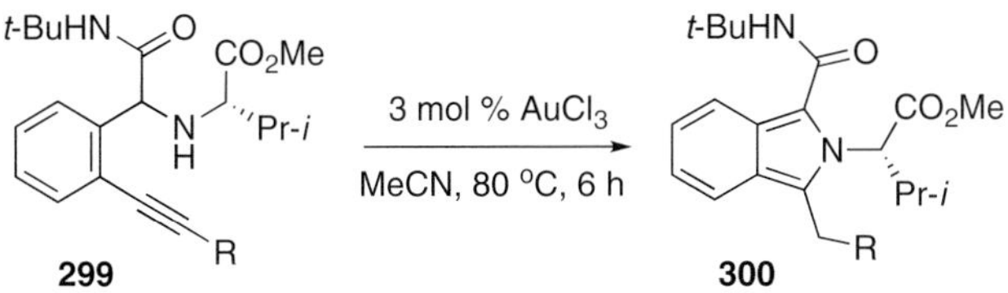

t-BuHN, O, CO_2Me, N, H, Pr-*i*, R, **299**

3 mol % $AuCl_3$

MeCN, 80 °C, 6 h

t-BuHN, O, N, CO_2Me, Pr-*i*, R, **300**

R	Yield, %
n-Bu	49%
Ph	38%

Scheme 9.103

competitive 6-*endo-dig* cyclization of **299** into dihydroisoquinoline derivatives, accompanying this reaction, diminished the efficiency of the isoindole synthesis.

Wang reported that an analogous transformation of *ortho*-alkynyl benzylamines **301** into isoindoles **302** could be efficiently catalyzed by the Cu(I) complex (Scheme 9.104) [277]. The *para*-toluidine additive was essential to ensure a high selectivity toward the desired 5-*exo-dig* versus the competitive 6-*exo-dig* cyclization path leading to the dihydroisoquinoline framework.

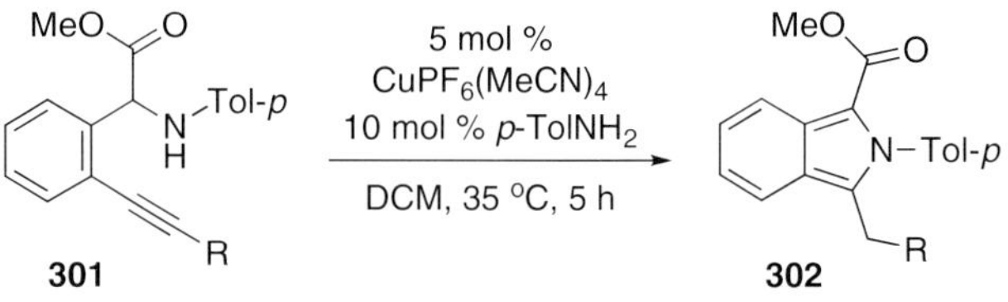

5 mol % $CuPF_6(MeCN)_4$
10 mol % *p*-$TolNH_2$
DCM, 35 °C, 5 h

301 → **302**

R	Yield, %
Ph	90%
o-Cl-C_6H_4	85%
o-Br-C_6H_4	>99%

Scheme 9.104

Recently, Shin reported an interesting Au(I)-catalyzed cycloisomerization of alkyne tethered nitrones **303** leading to 2,3-disubstituted isoindoles **304** in moderate yields (Scheme 9.105) [278]. The proposed mechanistic rationale is outlined in Scheme 9.106. First, nitrone **303** undergoes the Au(I)-catalyzed 7-*endo-dig* cyclization to give zwitterionic intermediate **305** through an intramolecular attack of the O-atom of the nitrone at the alkyne moiety. A subsequent N–O bond cleavage in **305** leads to the α-oxo Au-carbenoid **306**, which, upon addition of the imine to the Au-carbenoid center, provides the cyclic intermediate **307**. Elimination of the Au-catalyst in the latter affords the corresponding isoindole **304**. Interestingly, the authors documented that the use of a bulky and relatively electron-rich Au(I) catalyst produces isoindole chemoselectively, whereas employment of an Au(III) complex completely diverts the reaction path to a competitive 6-*exo-dig* cyclization process.

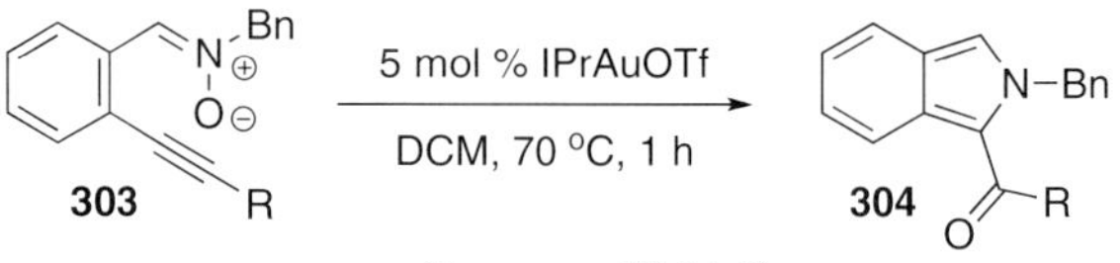

5 mol % IPrAuOTf
DCM, 70 °C, 1 h

303 → **304**

R	Yield, %
H	55%
Me	53%
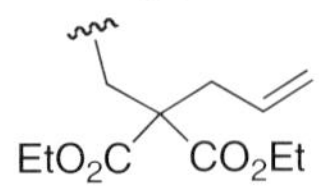 $CH_2C(CO_2Et)_2CH_2CH=CH_2$	66%

Scheme 9.105

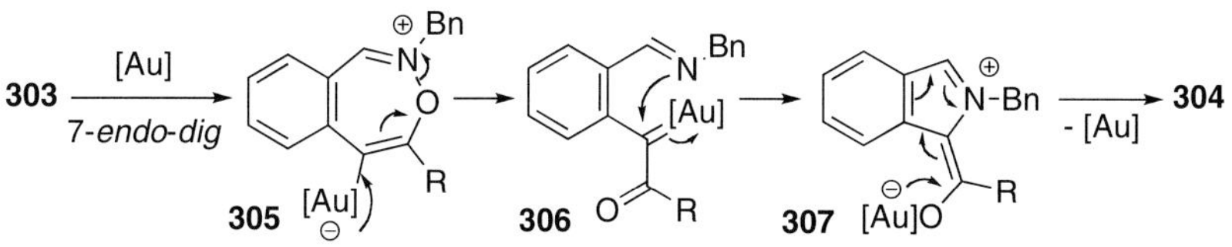

Scheme 9.106

9.2.4.2 **Synthesis of Isoindoles via "4 + 1" Cycloaddition Reactions**

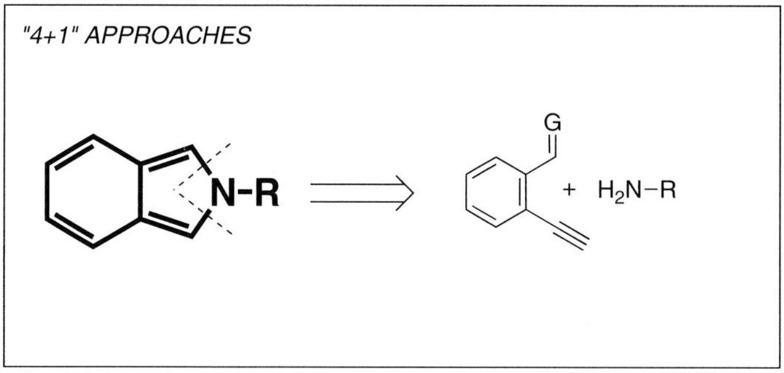

Recently, Wang designed a "4 + 1" route to isoindoles based on the *in situ* generation of reactive *ortho*-alkynyl benzylamine intermediates like **301**. Accordingly, *o*-alkynyl-substituted phenyldiazoacetates **308** were smoothly converted into isoindole-1-carboxylates **310** in the presence of a variety of alkyl- or aryl-amines **309** and the Cu(I)-catalyst under mild reaction conditions (Scheme 9.107) [277]. The proposed mechanism is shown in Scheme 9.108. The Cu(I) carbenoid **311** is generated upon treatment of diazocompound **308** with the $Cu(MeCN)_4PF_6$ catalyst. A concomitant intermolecular N–H insertion reaction with a primary amine **309** produces the key reactive alkynylamine **312** suited for the Cu(I)-catalyzed 5-*exo-dig* cyclization. Consequently, the later process in **312** proceeds via the amine nitrogen atom attack at the activated triple bond, affording a zwitterionic vinylcopper(I) intermediate **313**. A subsequent protiodemetalation of this organocopper species gives 3-exomethyleneisoindoline **314**, which further tautomerizes into isoindole **310**.

9.2.5
Indolizines

The diverse pharmacological potential of indolizines and related derivatives stimulates a continuous interest in utilizing molecules containing this core in synthetic biologically active compounds [279–287]. Furthermore, indolizines, due to their intriguing molecular properties, found applications in the field of material science and in a range of artificial materials [279, 280]. Not surprisingly, development of effective synthetic methods [173, 174, 279, 280, 288, 289] toward a facile assembly of the indolizine core and related structures with diverse substitution patterns has received increasing attention over the last two decades. In particular, recent efforts in

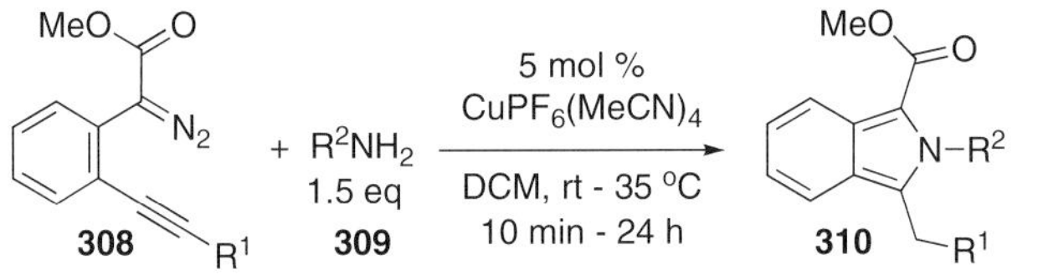

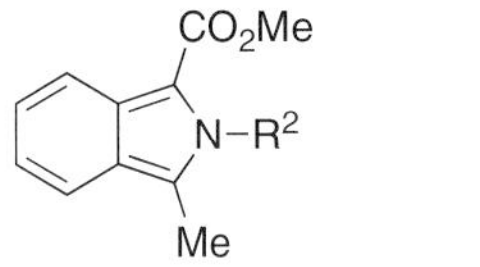

308 + R^2NH_2 (1.5 eq) **309** → (5 mol % $CuPF_6(MeCN)_4$, DCM, rt - 35 °C, 10 min - 24 h) → **310**

N-R^2 products, R^1 = Me (1-methyl, 2-R^2 series):

R^2	Yield, %
Ph	90
p-Tol	95
p-MeO-C_6H_4	93
p-O_2N-C_6H_4	90
2,4-$Cl_2C_6H_3$	97
p-Br-C_6H_4	92
Bn	28
t-Bu	trace
Ts	0

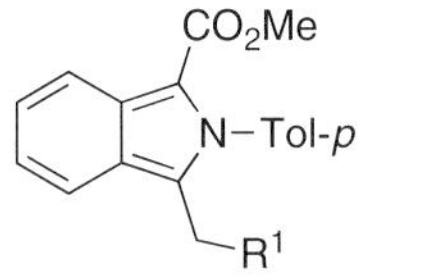

N-Tol-*p* products (1-CH_2R^1 series):

R^2	Yield, %
Ph	85
p-Br-C_6H_4	94
p-Tol	78
o-Cl-C_6H_4	87
o-Br-C_6H_4	87
m-Tol	93
p-MeO-C_6H_4	57
Me	66

Scheme 9.107

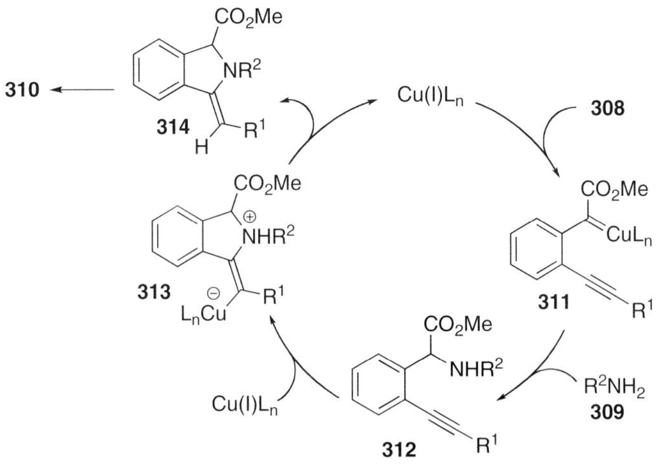

Scheme 9.108

this field have been focused on the design of catalytic transformations, with an emphasis placed on the processes catalyzed by transition metals. A detailed discussion of the most important approaches for the construction of the indolizine framework, including cycloisomerization and formal [3 + 2] cycloaddition reaction, is provided in the following section.

9.2.5.1 Synthesis of Indolizines via Cycloisomerization Reactions

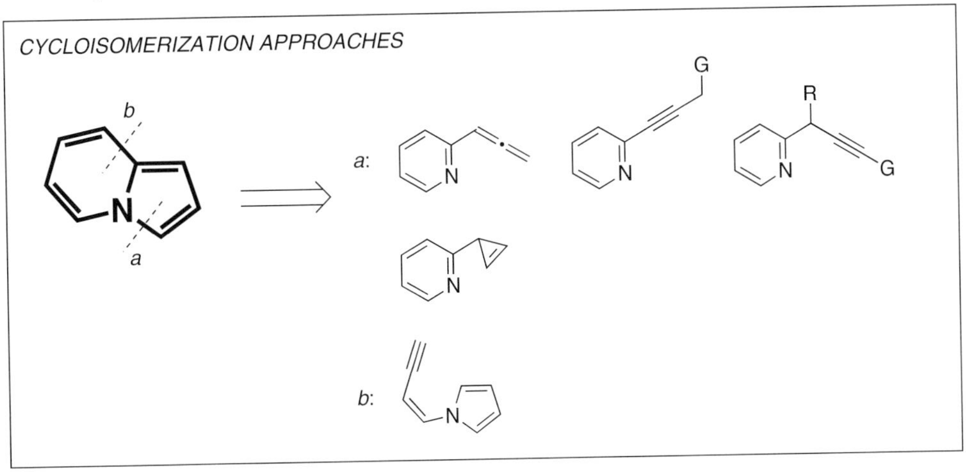

CYCLOISOMERIZATION APPROACHES

a:

b:

Until very recently, no examples of a transition metal-catalyzed cycloisomerization of allenylpyridines into indolizines had been reported. A well-known instability of these seemingly versatile precursors prevented any significant use in the construction of the indolizine nucleus. The only precedent of this transformation was documented by Gevorgyan and coworkers. Thus, 1,3-disubstituted alkyl- and aryl-indolizines **316** were synthesized via the Cu(I)-catalyzed 5-*endo-trig* cyclization of the corresponding disubstituted allenylpyridines **315**, whereas the cycloisomerization of *gem*-disubstituted allene failed to produce the expected monosubstituted indolizine (Scheme 9.109) [290].

Addressing the allenylpyridine instability problem, Gevorgyan developed the Cu(I)-catalyzed cycloisomerization of conjugated alkynyl imines **317**, perfect allene **318** surrogates, as an alternative route for the construction of indolizine and other pyrrole-fused cores **319** (Scheme 9.110) [291–293]. This protocol proved to be very general and efficient for the synthesis of a variety of diversely functionalized C3-substituted heterocycles **319**. On the other hand, this approach was not applicable

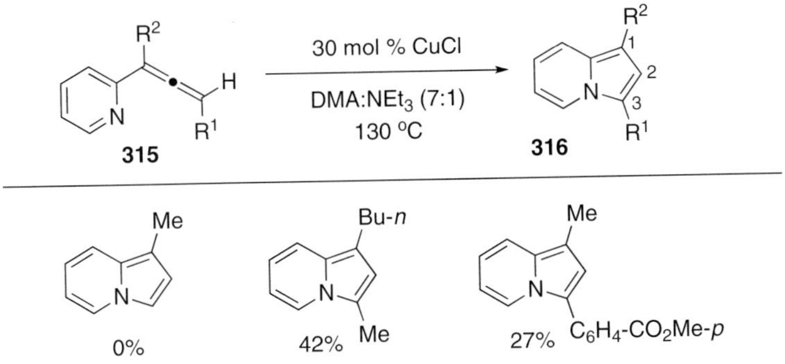

315 → **316**: 30 mol % CuCl; DMA:NEt_3 (7:1); 130 °C

0% (R² = Me); 42% (R² = Bu-*n*, R¹ = Me); 27% (R² = Me, R¹ = C_6H_4-CO_2Me-*p*)

Scheme 9.109

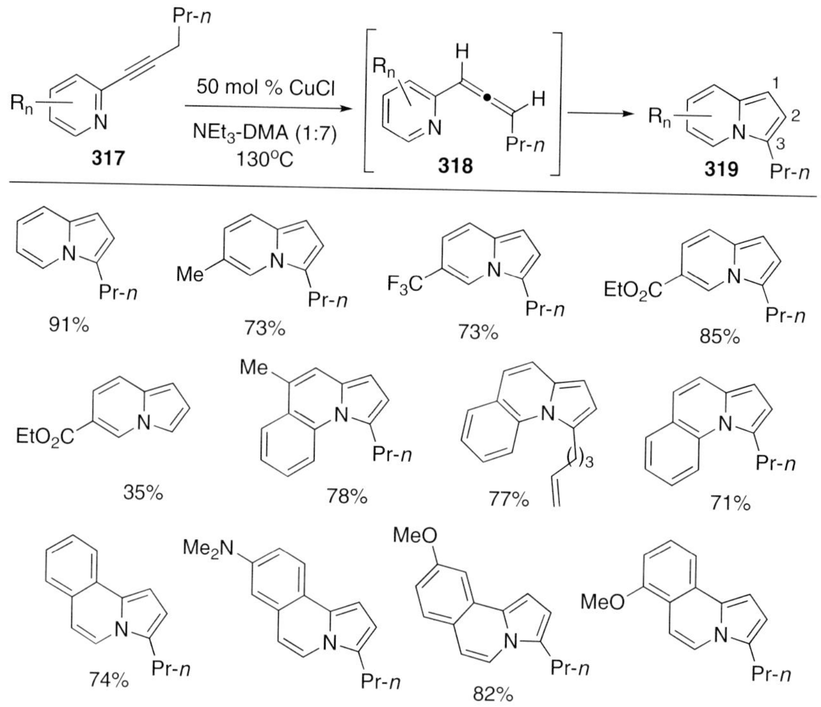

Scheme 9.110

for the preparation of C2- and/or C1-substituted indolizines. The authors proposed that this cascade transformation involved a base-induced propargyl–allenyl isomerization of **317** into the reactive allene species **318**, which was subsequently cyclized into the indolizine product, following similar elementary processes suggested for the Cu(I)-catalyzed cycloisomerization of allenyl ketones [294] and imines [291] (See Scheme 8.21 in the previous chapter).

As a further demonstration of the synthetic utility of this Cu(I)-catalyzed cycloisomerization methodology, the same group described its application for an expeditious and highly diastereoselective synthesis of (±)-monomorine **322** (Scheme 9.111) [291].

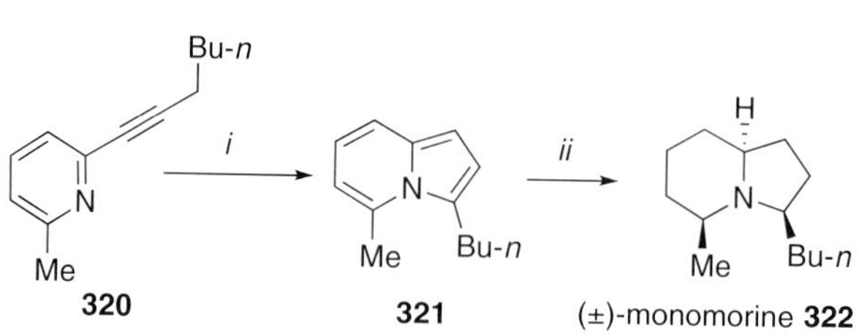

Scheme 9.111

Recently, Sun utilized this chemistry for a straightforward and highly efficient assembly of the indolizine core **324** in the synthesis of a microtubule inhibitor STA-5312 **325**, possessing potent antitumor activity against multidrug-resistant-type cancers (Scheme 9.112) [295].

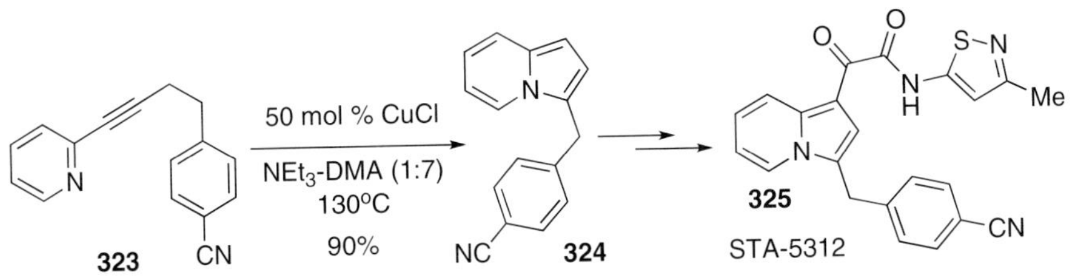

50 mol % CuCl
NEt$_3$-DMA (1:7)
130°C
90%
323 CN NC 324 325 STA-5312 CN Me S–N N H O O

Scheme 9.112

A very mild and highly efficient two-component Cu(I)-mediated substitution–cycloisomerization approach for the modular synthesis of not so easily available 1,3-disubstituted indolizine structures was recently devised by Gevorgyan. This route features generation of the reactive allenylpyridines via a propargylic substitution in conjugated pyridyl-substituted propargylic mesylates **326** with various organocopper reagents **327**. A subsequent facile cycloisomerization of these allenes in the presence of stoichiometric amounts of the *in situ* generated Cu(I) salt by-product gives the expected indolizines **328** (Scheme 9.113) [290]. The authors suggested two mechanisms for this cascade transformation which are illustrated in Scheme 9.114. Mechanistic investigation suggested that the cyclization of the transient allene intermediate **329** proceeded through the deprotonation–protonation sequence of the cyclic vinylcopper species **330**.

The same group designed the Cu(I)-assisted cycloisomerization of conjugated 3-thiopropargyl pyridine **332** for the efficient synthesis of scarcely available 2-substituted indolizine **334**. The corresponding product was formed, presumably via the 5-*endo-trig* cyclization of the *in situ* generated allene intermediate **333** under basic reaction conditions, whereupon a subsequent 1,2-thio group migration allowed the introduction of a C2-substitutent (Scheme 9.115) [296].

The Gevorgyan group recently disclosed a highly efficient synthesis of 3-phosphatyloxy indolizines **337** via the Cu(I)-catalyzed formal 1,3-phosphatyloxy group migration–cycloisomerization of a series of conjugated propargyl pyridines **335** (Scheme 9.116) [297]. Thorough mechanistic studies of the course of this transformation with the aid of ^{17}O-labeled substrates **337** revealed that this cycloisomerization proceeded via the initial [3]-sigmatropic rearrangement of the latter into the reactive allene species **336**. It should be noted that phosphatyloxy-containing indolizines **337** are versatile synthons, as various alkyl and aryl groups can be introduced at the C-2 position of indolizine via the Kumada cross-coupling reaction [297].

Shortly after, a similar Pt(II)-catalyzed transformation of alkynylpyridines **338** providing 2-acyloxy-substituted indolizines **339/340** via a 1,2/1,3-acyloxy group migration was disclosed by Sarpong (Scheme 9.117) [298]. Although a variety of

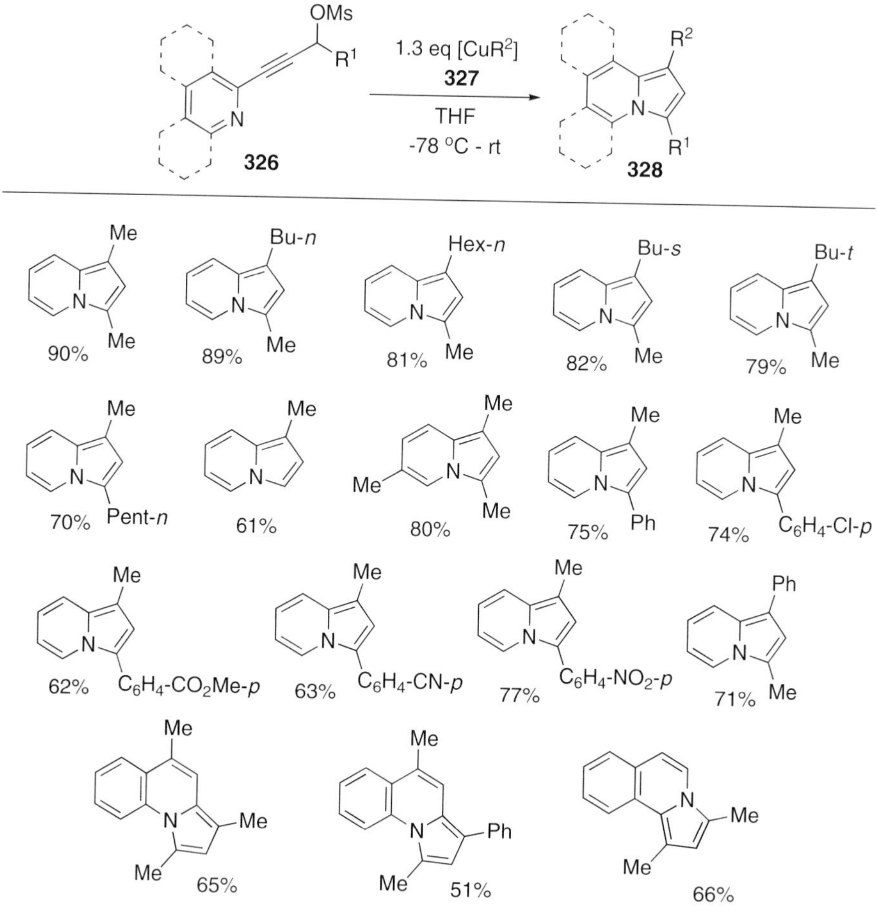

1.3 eq $[CuR^2]$ **327**, THF, -78 °C - rt; **326** → **328**

90%, 89%, 81%, 82%, 79%, 70%, 61%, 80%, 75%, 74%, 62%, 63%, 77%, 71%, 65%, 51%, 66%

Scheme 9.113

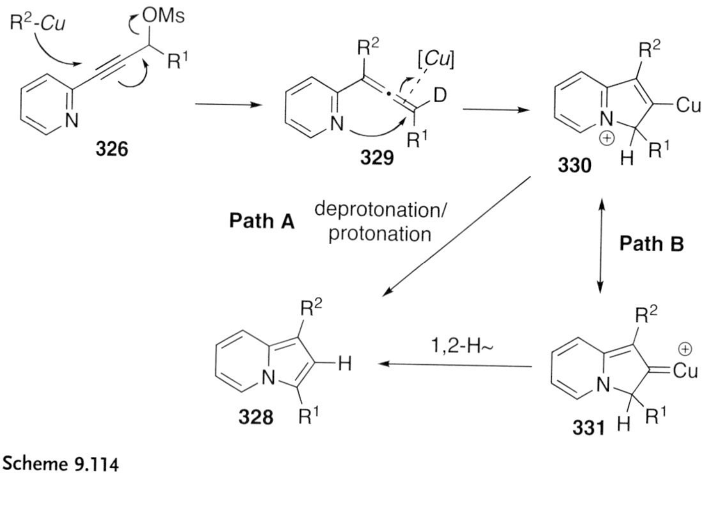

Path A: deprotonation/protonation; Path B; 1,2-H~ (**326** → **329** → **330** → **328**; **330** ⇌ **331** → **328**)

Scheme 9.114

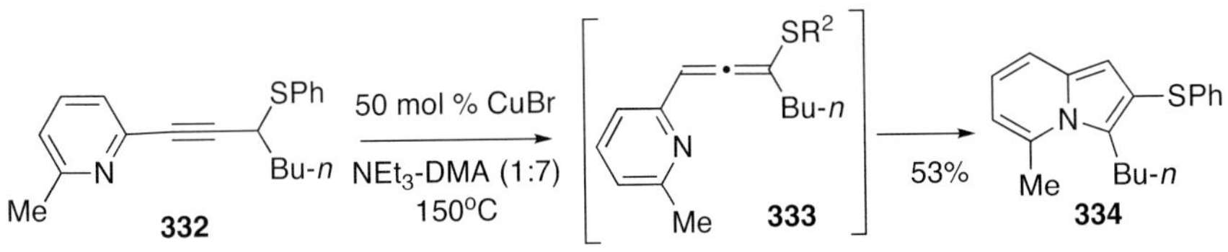

Scheme 9.115

functional groups could be tolerated under the reaction conditions, this cascade transformation is not highly regioselective, as it yields mixtures of 2- and 3-acyloxyindolizines **339** and **340** in variable ratios, always favoring the preferential formation of the former regioisomer. According to the proposed mechanism (Scheme 9.118), **338** undergoes cycloisomerization proceeding via initial 5-*exo-dig* cyclization to form intermediate **341**, which further isomerizes into a carbenoid species **342** (path **A**). Cyclization of the latter (**343**), followed by a proton transfer, furnishes a C-2 substituted indolizine **339**. Alternatively, in the case of electron-withdrawing R^1 groups, the reaction could also occur via a competitive path **B**, involving a formal [3, 3]-acyloxy group migration to give the allenic intermediate **344**, which is set for a transition metal-catalyzed cycloisomerization into a C-3 substituted indolizine **340** (*vide supra*).

Aiming for the development of transition metal-catalyzed protocols, which would allow a regioselective synthesis of C2-substituted indolizine cores, Gevorgyan reported a highly efficient Au(III)-catalyzed cascade cycloisomerization of skipped propargylpyridines **346** into indolizines **347** (Scheme 9.119) [299, 300]. This cascade transformation involved a facile 1,2-migration of silyl-, stannyl-, and even germyl

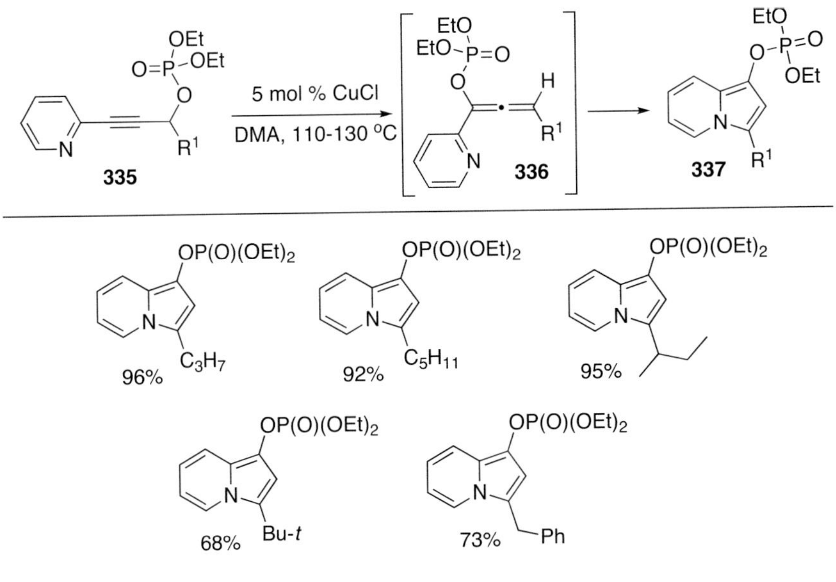

Scheme 9.116

5 mol % PtI_2, 20 mol % $P(C_6F_5)_3$, PhH, 100 °C, 24 h: **338** → **339** + **340**

R^1	R^2	Yield, %	Ratio **339** : **340**
t-Bu	*t*-Bu	11	4 : 1
c-Pr	*t*-Bu	49	3 : 1
1-cyclohexenyl	*t*-Bu	74	14 : 1
(Me, Me-substituted alkenyl)	*t*-Bu	54	14 : 1
(Ph-substituted alkenyl)	*t*-Bu	56	>20 : 1
1-furyl	*t*-Bu	34	>20 : 1
2-naphthyl	*t*-Bu	73	11 : 1
(benzodioxolyl)	*t*-Bu	67	13 : 1
3,5-$(MeO)_2$-C_6H_3	*t*-Bu	76	7 : 1
2-Br-C_6H_4	*t*-Bu	72	10 : 1
2,6-Me_2-C_6H_3	*t*-Bu	91	>20 : 1
Ph	*t*-Bu	75	13 : 1
Ph	4-MeO-C_6H_4	67	>20 : 1
4-MeO-C_6H_4	*t*-Bu	74	13 : 1
4-F_3C-C_6H_4	*t*-Bu	58	2 : 1
4-F_3C-C_6H_4	4-MeO-C_6H_4	56	13 : 1
4-MeO_2C-C_6H_4	*t*-Bu	78	6 : 1
4-NC-C_6H_4	*t*-Bu	31	3 : 1

Scheme 9.117

groups via an alkyne–vinylidene isomerization of substrates **346** into the reactive organogold species **348**. A subsequent cyclization of the latter intermediate followed by a series of 1,2-hydride shifts in **349** furnished the corresponding indolizine **347** (Scheme 9.120).

The same group demonstrated that a variety of mono- and disubstituted 1-oxyindolizine derivatives **352** could be readily synthesized via a facile Ag-catalyzed cycloisomerization of skipped propargylpyridines **351** (Scheme 9.121) [300, 301]. It was suggested that this Au-catalyzed reaction involved a 5-*endo-dig* cyclization of the alkyne **351** activated by a π-philic metal. Formation of the indolizine product **352** was accomplished via a subsequent proton transfer in cyclic vinylmetal zwitterion **354** (Scheme 9.122). It should be noted that a variety of transition metals, such as Au(I), Au(III), Cu(I), Pt(II), and Pd(II), were shown to catalyze this transformation with variable degrees of efficiency.

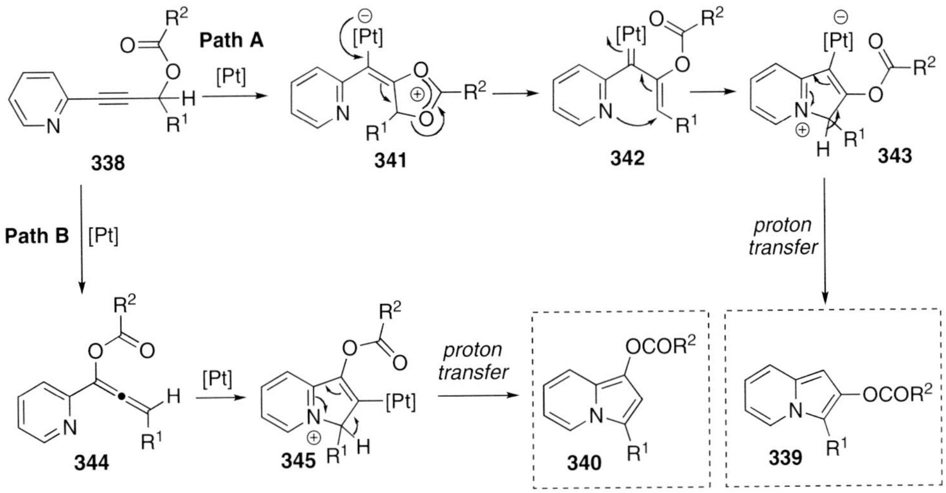

Scheme 9.118

Very recently, the Pt(II)-catalyzed version of an analogous cycloisomerization of skipped propargylpyridines **351** into *N*-fused heterocycles **352** was reported by Sarpong (Scheme 9.123) [302, 303].

Almost simultaneously, Liu disclosed that the Cu(I)–NEt_3 catalytic system could also efficiently trigger this useful transformation (Scheme 9.124) [304].

A regiodivergent transition metal-catalyzed synthesis of differently substituted indolizines was recently elaborated on by Gevorgyan and coworkers. Thus, pyridyl-substituted cyclopropenes **356** were smoothly cycloisomerized in the presence of

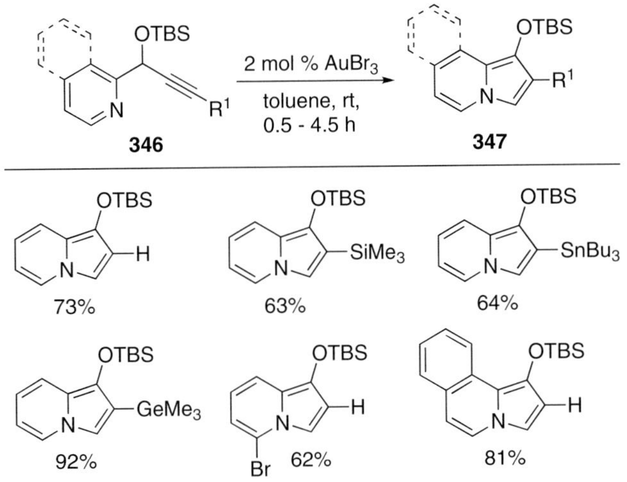

Scheme 9.119

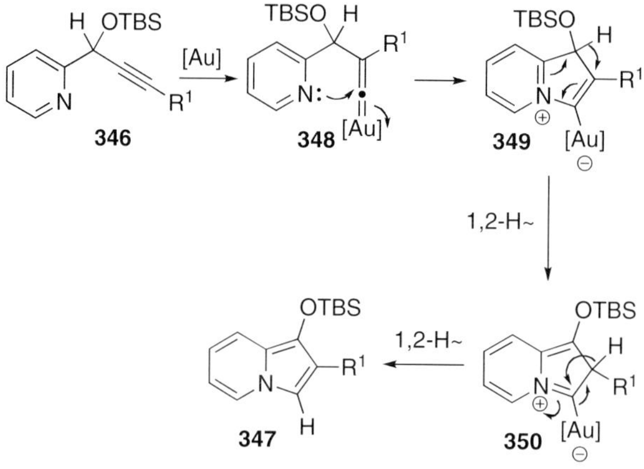

Scheme 9.120

the Rh(I) catalyst to provide the corresponding 1,3-disubstituted indolizines **357** regioselectively (Scheme 9.125) [305]. Conversely, employment of the Cu(I) catalyst led to the exclusive formation of the regioisomeric 1,2-disubstituted indolizines **358** in high yields (Scheme 9.126). The proposed mechanistic rationale for the regiodivergent rearrangement of iminocyclopropenes **356** into fused pyrroloheterocycles **357**/**358** (Scheme 9.127) involves ring opening of the former in the presence of Rh(I) complex to give the most substituted carbenoid **359**. A subsequent cyclization through ylide **360** or rhodacycle **361** followed by elimination of the metal furnishes regioisomer **357**. In contrast, when Cu(I) catalyst is used, a similar cyclization of a less substituted carbenoid **362** leads to the product **358** selectively.

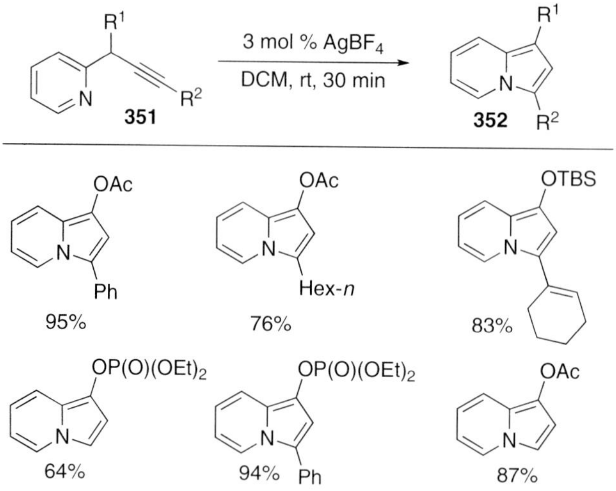

Scheme 9.121

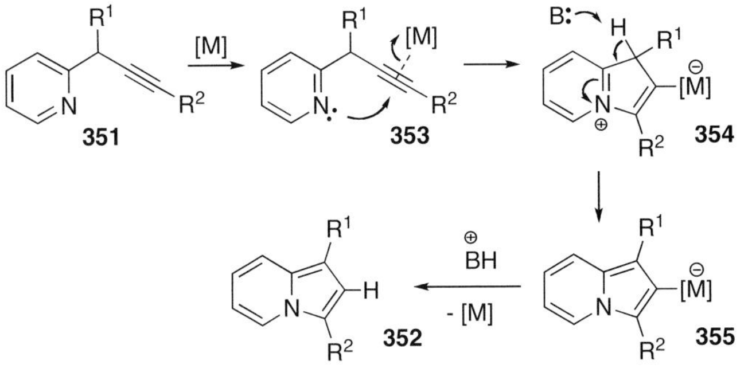

Scheme 9.122

Alternatively, regioisomeric azametalacycles **361** and **364** may arise directly from a regioselective oxidative addition process in cyclopropene **356**.

An interesting Cu(I)-catalyzed annulation of tethered pyrroloalkynes **365** and **367** into amino-substituted *N*-fused pyrroloheterocycles **366** and **368**, respectively, was recently demonstrated by Chang (Scheme 9.128) [306]. This reaction was proposed to proceed through a ketenimine intermediate **370**, which in turn was generated *in situ* from the 1,2,3-triazole cycloadduct **369** upon extrusion of dinitrogen (Scheme 9.129).

Finally, Fürstner provided a single example of the Pt(II)-catalyzed [307] annulation of the same substrate **365** into the fused indolizine derivative **372**, proceeding through a facile cyclization of Pt-vinylidene **371** (Scheme 9.130) [308, 309].

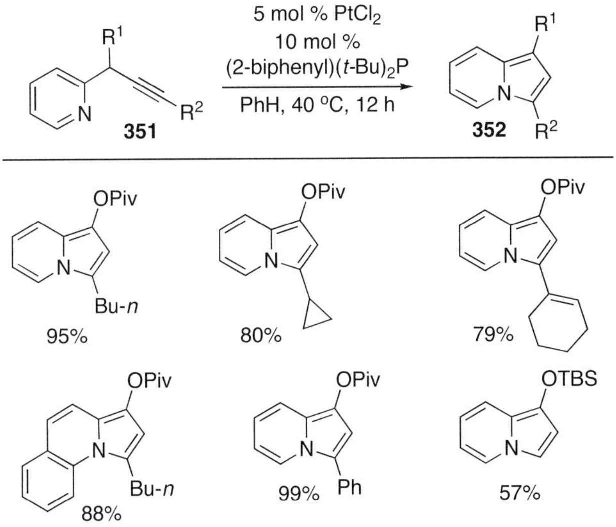

Scheme 9.123

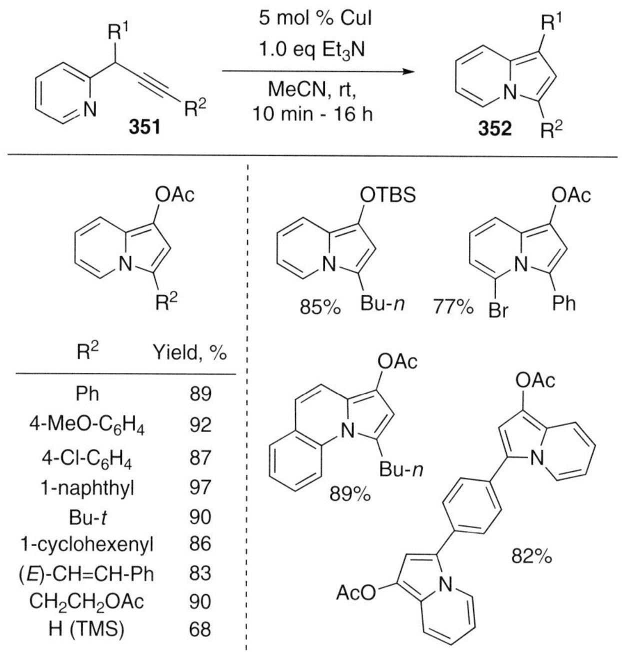

351 → **352**: 5 mol % CuI, 1.0 eq Et_3N, MeCN, rt, 10 min - 16 h

R^2	Yield, %
Ph	89
4-MeO-C_6H_4	92
4-Cl-C_6H_4	87
1-naphthyl	97
Bu-*t*	90
1-cyclohexenyl	86
(*E*)-CH=CH-Ph	83
CH_2CH_2OAc	90
H (TMS)	68

OTBS, Bu-*n*: 85%; OAc, Br, Ph: 77%; OAc, Bu-*n*: 89%; bis-indolizine (AcO): 82%

Scheme 9.124

356 → **357**: 1-2 mol % $RhCl(PPh_3)_3$, DMF, rt

85%; 91%; 79%; 87%; 81%; 93%; 88%; 84%

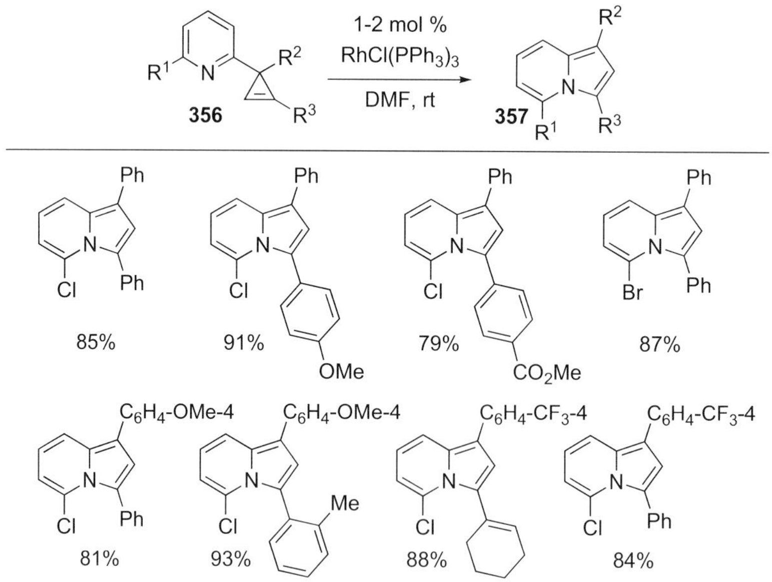

Scheme 9.125

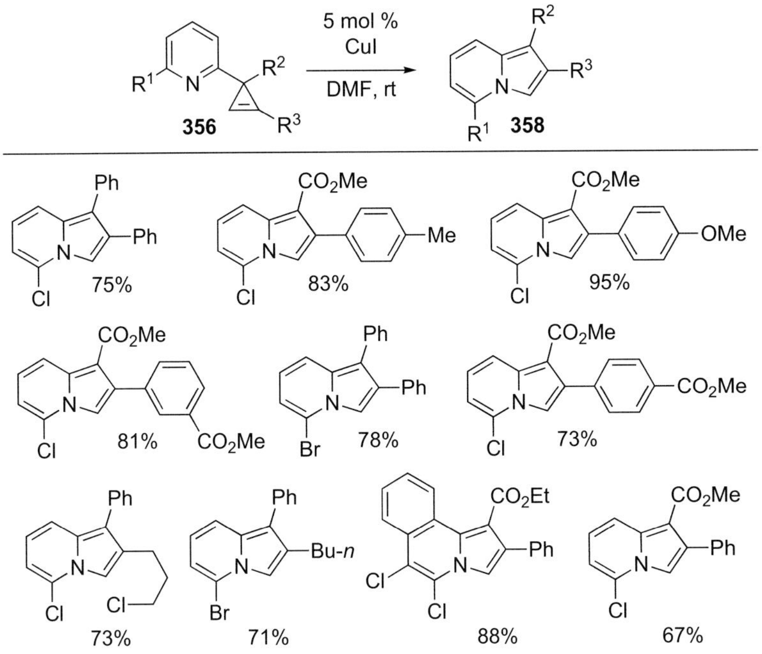

5 mol % CuI, DMF, rt

356 → 358

75% 83% 95% 81% 78% 73% 73% 71% 88% 67%

Scheme 9.126

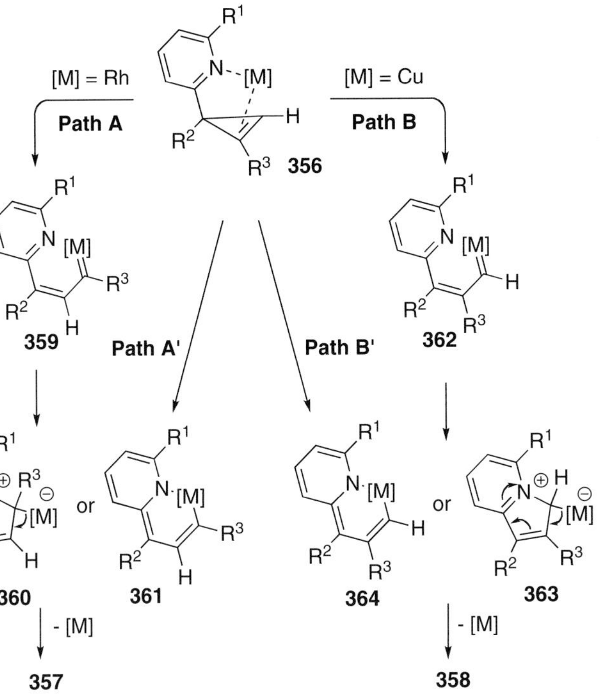

[M] = Rh Path A; [M] = Cu Path B; Path A'; Path B'

356, 359, 360, 361, 362, 363, 364, - [M], 357, 358

Scheme 9.127

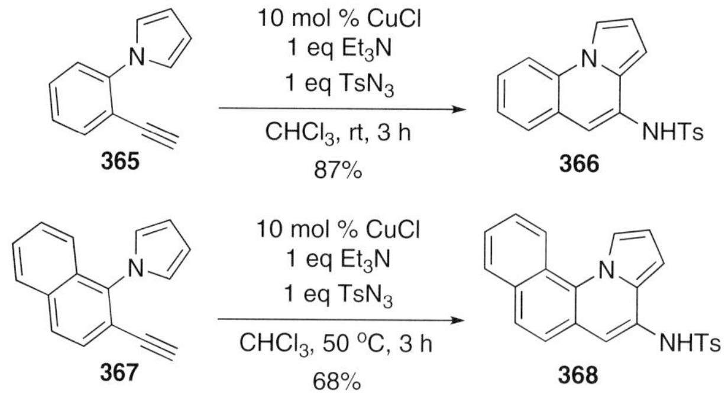

N
365
10 mol % CuCl
1 eq Et_3N
1 eq TsN_3
$CHCl_3$, rt, 3 h
87%
N
NHTs
366

N
367
10 mol % CuCl
1 eq Et_3N
1 eq TsN_3
$CHCl_3$, 50 °C, 3 h
68%
N
NHTs
368

Scheme 9.128

N
365
[Cu]
TsN_3
N
CuL_n
N-Ts
N=N
369
- N_2
N
N
Ts
370
X
X = [Cu] or H
N
NHTs
366

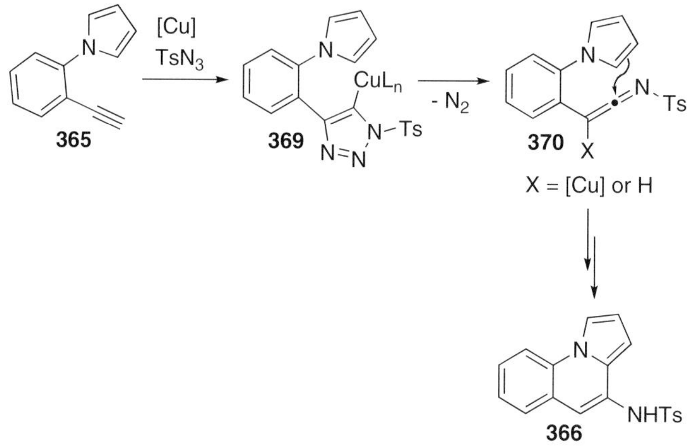

Scheme 9.129

9.2.5.2 Synthesis of Indolizines via "3 + 2" Cycloaddition Reactions

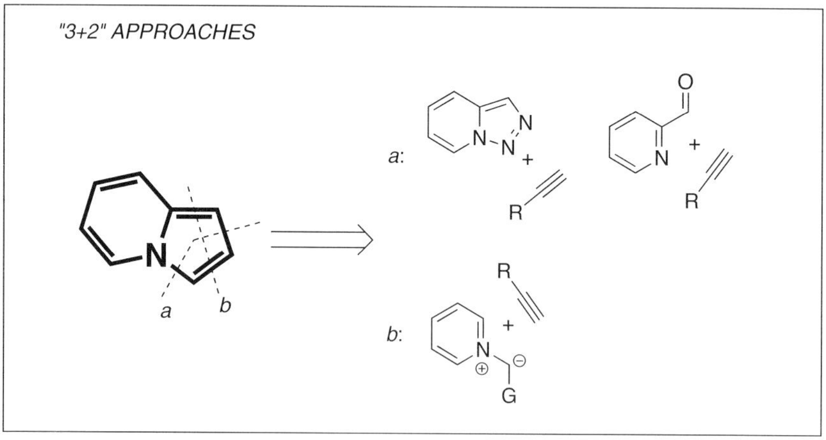

"3+2" APPROACHES

N
a
b

a:
N-N
N
+
R
O
N
+
R

b:
R
+
N
⊕
⊖
G

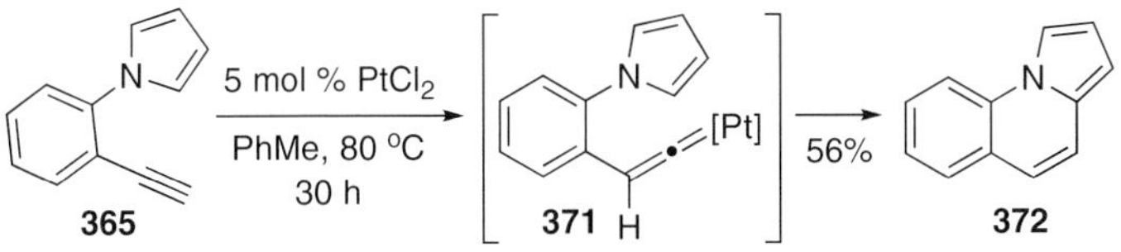

Scheme 9.130

A remarkably efficient Rh(II)-catalyzed formal [3 + 2] cycloaddition reaction, namely transannulation, between 7-halo-substituted *N*-fused triazole **373** and a variety of terminal alkynes **374**, leading to the regioselective assembly of 1,3-disubsituted indolizines **376**, was recently developed by Gevorgyan (Scheme 9.131) [310]. Formation of indolizines **376** was accompanied by small quantities of the corresponding cyclopropenes **375**, which, however, could not be converted into indolizine products **376** under these reaction conditions, thereby suggesting an independent route for the formation of the latter. It is believed that pyridotriazole **373** undergoes a closed–open form equilibrium, serving as a stable and convenient surrogate of the respective pyridyl diazocompound **377** (Scheme 9.132). Being produced in small amounts, the latter, upon reaction with the Rh(II) carboxylate, generates the Rh-carbenoid species **378**. Further feasible transformations (paths **A** and **B**,

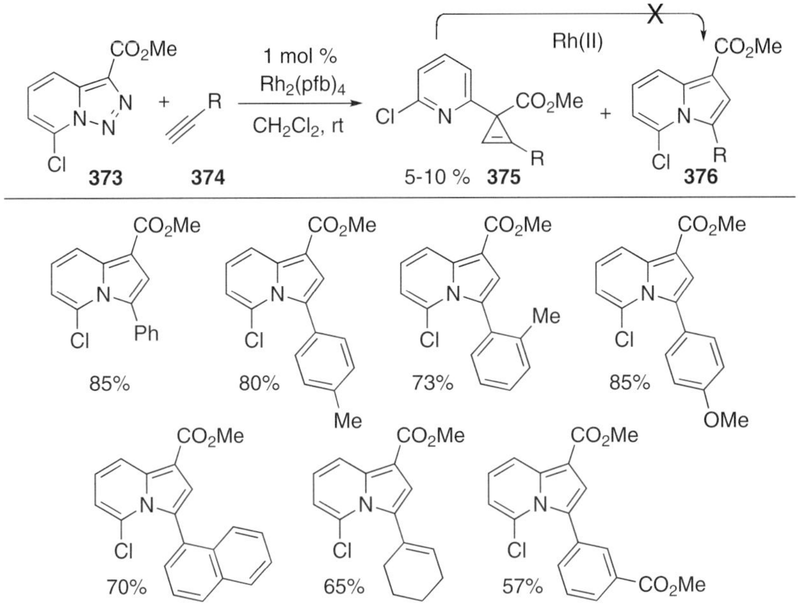

Scheme 9.131

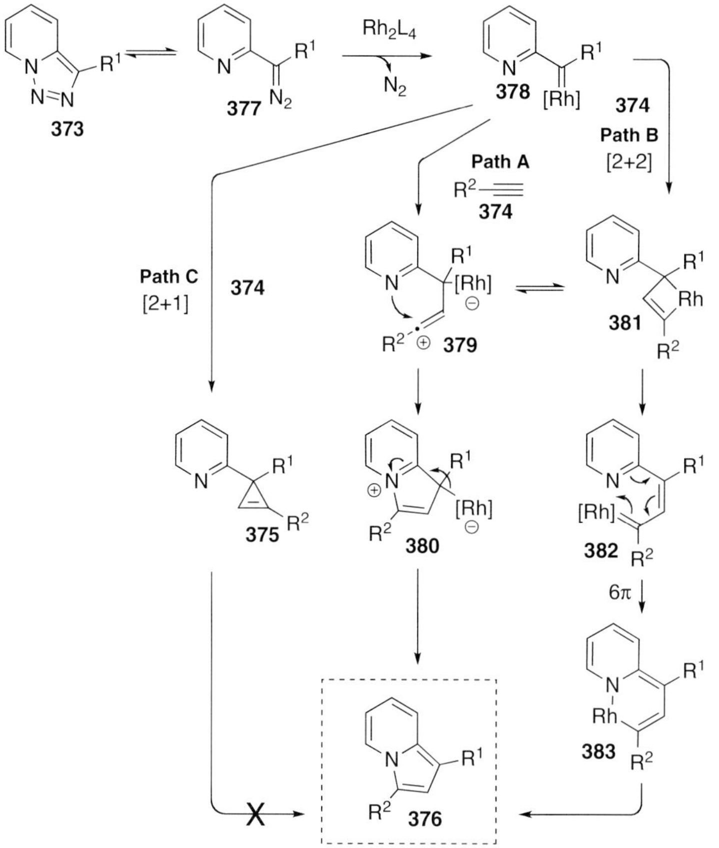

Scheme 9.132

Scheme 9.132) of these aza-analogs of oxacarbenoids resemble those suggested for the formation of furan products in the Rh(II)-catalyzed [3 + 2] cycloaddition reaction between α-diazocarbonyl compounds and alkynes (See Scheme 8.71 in the previous chapter).

Further, Liu recently disclosed a highly efficient "3 + 2" coupling–cycloisomerization reaction leading to the assembly of 1-aminoindolizine derivatives **388** (Scheme 9.133) [311]. It was found, that the reactive skipped propargylpyridine **387** could be generated *in situ* via the Au(III)-catalyzed Mannich-type three-component reaction of picolyl aldehyde **384**, secondary amine **385**, and terminal acetylene **386**. The subsequent Au(III)-catalyzed cycloisomerization of **387** furnishes indolizines **388** in moderate to excellent yields. In addition, the authors demonstrated that the coupling of optically active amino acids produced the corresponding indolizine derivatives with complete preservation of the stereogenic centers.

Finally, Porco described a straightforward Ag-catalyzed domino approach toward the efficient construction of diversely substituted Lamellarin-type indolizine

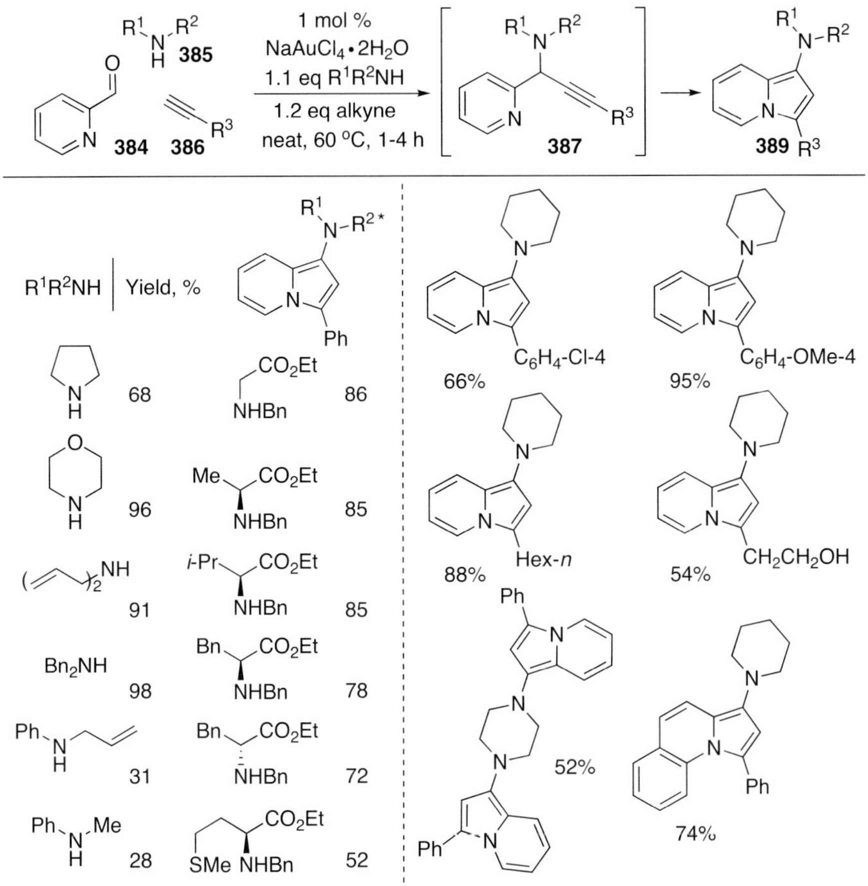

Scheme 9.133

frameworks **391** from alkynyl imines **389** and activated alkynes **390** (Scheme 9.134) [312]. It was suggested that the initial Ag-catalyzed cycloisomerization of alkynyl imines **389** followed by a proton transfer in **392** generated the key azomethine ylide **393**. A subsequent [3 + 2] cycloaddition of the latter with **390**, upon a concomitant oxidation–aromatization sequence, gave the target fused indolizine **391** (Scheme 9.135).

9.3 Conclusion

Currently, there is growing interest in the development of novel, general, highly efficient, convergent, and practical approaches for the synthesis of fused five-membered heterocycles. Throughout decades, employment of Cu and Pd complexes

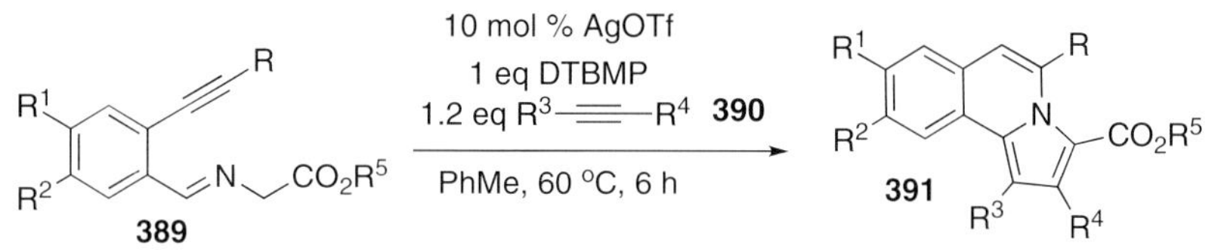

10 mol % AgOTf, 1 eq DTBMP, 1.2 eq R^3—≡—R^4 **390**, PhMe, 60 °C, 6 h; **389** → **391**

R	R^1	R^2	R^3	R^4	R^5	Yield, %
n-C_7H_{15}	H	H	CO_2Me	CO_2Me	Me	72
n-C_7H_{15}	H	F	CO_2Me	CO_2Me	Me	59
Me	H	H	CO_2Et	CO_2Et	Me	83
Ph	H	H	CO_2Me	CO_2Me	Me	78
n-Hex	OMe	OMe	CO_2Me	CO_2Me	Me	78
c-Pr	H	H	CO_2Me	CO_2Me	Me	63
CH_2OMe	-OCH_2O-		CO_2Me	CO_2Me	Bu-t	68
$SiMe_3$	OMe	OMe	CO_2Me	CO_2Me	Me	70
H	OMe	OMe	CO_2Me	CO_2Me	Me	42
(E)-Me-CH=CH-	H	H	CO_2Me	CO_2Me	Bu-t	29
n-C_7H_{15}	H	H	CHO	Bu-n	Me	46
n-C_7H_{15}	H	H	COMe	Ph	Me	56
n-C_7H_{15}	H	H	CO_2Me	Ph	Me	57
Me	H	H	CO_2Me	H	Me	52
n-C_7H_{15}	H	H	CO_2Et	$SiMe_3$	Me	28

Scheme 9.134

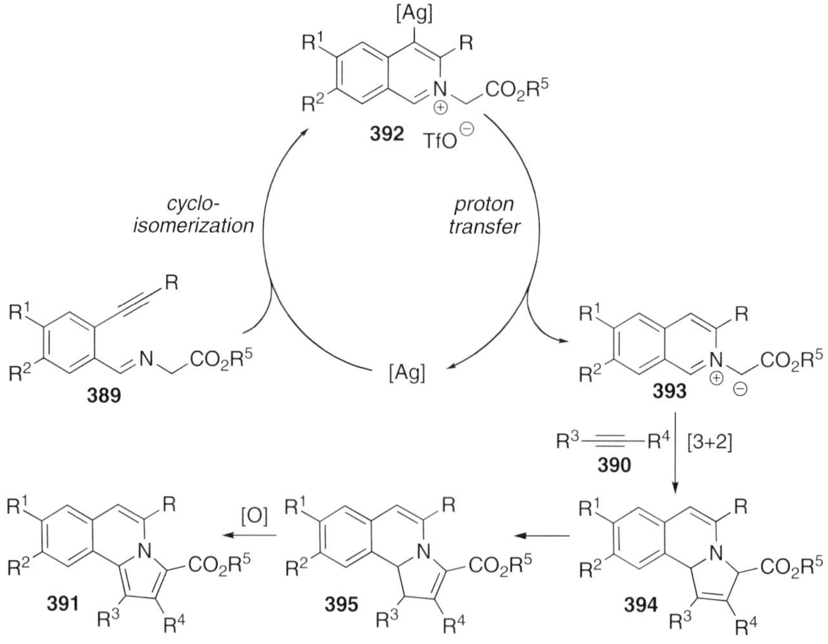

[Ag], **392** TfO⁻, cyclo-isomerization, proton transfer, **389**, [Ag], **393**, R^3—≡—R^4 **390** [3+2], **394**, **395**, [O], **391**

Scheme 9.135

clearly dominated in the syntheses of indoles, benzo[*b*]furans, isoindoles, and indolizines. In recent years, in addition to the continuing attention to these “classical” transition metals, focus has been shifted to employment of Pt-, Ag-, Au-, Ru-, and Rh-catalysts, allowing facile construction of fused heterocycles with diverse substitution patterns. Novel approaches aimed for the synthesis of differently functionalized heterocycles, including novel complex cascade cycloisomerization reactions, more efficient cross-coupling protocols of previously unreactive substrates, as well as C–H activation processes, are being extensively developed. Although remarkable progress has been achieved in the field of transition metal-catalyzed chemistry of aromatic heterocycles, undoubtedly, development of more general and efficient transformations toward indoles, benzo[*b*]furans, isoindoles, benzo[*b*]thiophenes, and indolizines is still highly warranted.

9.4 Abbreviations

Ac	acetyl
aq	aqueous
BINOL	1,1′-bi-2-naphthol
Boc	*tert*-butylcarbonyl
cat.	catalytic
Cbz	carboxybenzyl
dap	α-(dimethylaminomethyl)pyrrolyl
DCE	1,2-dichloroethane
DCM	dichloromethane
DMA	*N,N*-dimethylacetamide
DMB	2,4-dimethoxybenzyl
DMEDA	*N,N′*-dimethylethylenediamine
DMF	*N,N*-dimethylformamide
dppf	1,1′-bis(diphenylphosphino)ferrocene
dppm	1,1-bis(diphenylphosphino)methane
dppp	1,3-bis(diphenylphosphino)propane
DTBMP	2,6-di-*tert*-butyl-4-methylpyridine
EB	3-ethylbutyryl
EDG	electron-donating group
EE	ethoxyethyl
eq	equivalent
EWG	electron-withdrawing
HFIP	hexafluoro-2-propanol
IPA	isopropanol
IPr	1,3-bis(2,6-diisopropylphenyl)imidazol-2-ylidene
Mes	mesityl
MOM	methoxymethyl
MPM	(*p*-methoxyphenyl)methyl

MS	molecular sieves
MW	microwave
NIS	*N*-iodosuccinimide
Ns	nosyl
pfb	perfluorobutyryl
Piv	pivalyl
phen	1,10-phenanthroline
PMB	*p*-methoxybenzyl
Py	pyridine
rt	room temperature
TBS	*tert*-butyldimethylsilyl
TBDMS	*tert*-butyldimethylsilyl
TBDPS	*tert*-butyldiphenylsilyl
TFA	trifluoroacetic acid
THF	tetrahydrofuran
THP	tetrahydropyranyl
TIPS	triisopropylsilyl
TMEDA	*N,N,N′,N′*-tetramethylethylenediamine
TMS	trimethylsilyl
TMSE	2-(trimethylsilyl)ethyl
Tol	tolyl
Tp	trispyrazolylborate
Ts	tosyl
Tf	triflyl

References

1 Katritzky, A.R. and Rees, C.W. (eds) (1984) *Comprehensive Heterocyclic Chemistry*, Pergamon, Oxford.

2 Katritzky, A.R., Scriven, E.F.V., and Rees, C.W. (eds) (1996) *Comprehensive Heterocyclic Chemistry II*, Elsevier, Oxford.

3 Katritzky, A.R., Ramsden, C.A., Scriven, E.F.V., and Taylor, R.J.K. (eds) (2008) *Comprehensive Heterocyclic Chemistry III*, Elsevier, Oxford.

4 Irie, M. (2000) *Chem. Rev.*, **100**, 1685.

5 Gribble, G.W., Saulnier, M.G., Pelkey, E.T., Kishbaugh, T.L.S., Yanbing, L., Jun, J., Trujillo, H.A., Keavy, D.J., Davis, D.A., Conway, S.C., Switzer, F.L., Roy, S., Silva, R.A., Obaza-Nutaitis, J.A., Sibi, M.P., Moskalev, N.V., Barden, T.C., Chang, L., Habeski, W.M., and Pelcman, B. (2005) *Curr. Org. Chem.*, **9**, 1493.

6 Schröter, S., Stock, C., and Bach, T. (2005) *Tetrahedron*, **61**, 2245.

7 Evano, G., Blanchard, N., and Toumi, M. (2008) *Chem. Rev.*, **108**, 3054.

8 Isambert, N. and Lavilla, R. (2008) *Chem. Eur. J.*, **14**, 8444.

9 Chinchilla, R., Najera, C., and Yus, M. (2004) *Chem. Rev.*, **104**, 2667.

10 Gilchrist, T.L. (1998) *J. Chem. Soc., Perkin Trans. 1*, 615.

11 Gilchrist, T.L. (1999) *J. Chem. Soc., Perkin Trans. 1*, 2849.

12 Ward, M.F. (2000) *Annu. Rep. Prog. Chem., Sect. B: Org. Chem.*, **96**, 157.

13 Söderberg, B.C.G. (2000) *Curr. Org. Chem.*, **4**, 727.

14 Ward, M.F. (2001) *Annu. Rep. Prog. Chem., Sect. B: Org. Chem.*, **97**, 143.

15 Gilchrist, T.L. (2001) *J. Chem. Soc., Perkin Trans. 1*, 2491.

16 Padwa, A. and Bur, S.K. (2007) *Tetrahedron*, **63**, 5341.
17 Stockman, R.A. (2002) *Annu. Rep. Prog. Chem., Sect. B: Org. Chem.*, **98**, 409.
18 Stockman, R.A. (2003) *Annu. Rep. Prog. Chem., Sect. B: Org. Chem.*, **99**, 161.
19 Stanovnik, B. and Svete, J. (2004) *Chem. Rev.*, **104**, 2433.
20 Stockman, R.A. (2005) *Annu. Rep. Prog. Chem., Sect. B: Org. Chem.*, **101**, 103.
21 Stockman, R.A. (2006) *Annu. Rep. Prog. Chem., Sect. B: Org. Chem.*, **102**, 81.
22 Stockman, R.A. (2007) *Annu. Rep. Prog. Chem., Sect. B: Org. Chem.*, **103**, 107.
23 Stockman, R.A. (2008) *Annu. Rep. Prog. Chem., Sect. B: Org. Chem.*, **104**, 106.
24 Cacchi, S. (1999) *J. Organomet. Chem.*, **576**, 42.
25 Rubin, M., Sromek, A.W., and Gevorgyan, V. (2003) *Synlett*, 2265.
26 Brandi, A., Cicchi, S., Cordero, F.M., and Goti, A. (2003) *Chem. Rev.*, **103**, 1213.
27 Tsoungas, P.G. and Diplas, A.I. (2004) *Curr. Org. Chem.*, **8**, 1607.
28 Tsoungas, P.G. and Diplas, A.I. (2004) *Curr. Org. Chem.*, **8**, 1579.
29 Kirsch, G., Hesse, S., and Comel, A. (2004) *Curr. Org. Synth.*, **1**, 47.
30 Barluenga, J., Santamaria, J., and Tomas, M. (2004) *Chem. Rev.*, **104**, 2259.
31 Deiters, A. and Martin, S.F. (2004) *Chem. Rev.*, **104**, 2199.
32 Diver, S.T. and Giessert, A.J. (2004) *Chem. Rev.*, **104**, 1317.
33 Alonso, F., Beletskaya, I.P., and Yus, M. (2004) *Chem. Rev.*, **104**, 3079.
34 Nakamura, I. and Yamamoto, Y. (2004) *Chem. Rev.*, **104**, 2127.
35 Zeni, G. and Larock, R.C. (2004) *Chem. Rev.*, **104**, 2285.
36 Wolfe, J.P. and Thomas, J.S. (2005) *Curr. Org. Chem.*, **9**, 625.
37 Maes, B.U.W. (2006) *Topics in Heterocyclic Chemistry*, vol. 1, Springer-Verlag, Berlin, Heidelberg, pp. 155–211.
38 Zeni, G. and Larock, R.C. (2006) *Chem. Rev.*, **106**, 4644.
39 Conreaux, D., Bouyssi, D., Monteiro, N., and Balme, G. (2006) *Curr. Org. Chem.*, **10**, 1325.
40 Chopade, P.R. and Louie, J. (2006) *Adv. Synth. Catal.*, **348**, 2307.
41 Donohoe, T.J., Orr, A.J., and Bingham, M. (2006) *Angew. Chem. Int. Ed.*, **45**, 2664.
42 D'Souza, D.M. and Muller, T.J.J. (2007) *Chem. Soc. Rev.*, **36**, 1095.
43 Ferreira, V.F. (2007) *Curr. Org. Chem.*, **11**, 177.
44 Mihovilovic, M.D. and Stanetty, P. (2007) *Angew. Chem. Int. Ed.*, **46**, 3612.
45 Chemler, S.R. and Fuller, P.H. (2007) *Chem. Soc. Rev.*, **36**, 1153.
46 Arcadi, A. (2008) *Chem. Rev.*, **108**, 3266.
47 Álvarez-Corral, M., Muñoz-Dorado, M., and Rodríguez, G.I. (2008) *Chem. Rev.*, **108**, 3174.
48 Arndtsen, B.A. (2008) *Chem. Eur. J.*, **15**, 302.
49 Patil, N.T. and Yamamoto, Y. (2008) *Chem. Rev.*, **108**, 3395.
50 Weibel, J.-M., Blanc, A., and Pale, P. (2008) *Chem. Rev.*, **108**, 3149.
51 Donohoe, T.J., Fishlock, L.P., and Procopiou, P.A. (2008) *Chem. Eur. J.*, **14**, 5716.
52 Miyaura, N. and Suzuki, A. (1995) *Chem. Rev.*, **95**, 2457.
53 Duncton, M.A.J. and Pattenden, G. (1999) *J. Chem. Soc., Perkin Trans. 1*, 1235.
54 Yang, B.H. and Buchwald, S.L. (1999) *J. Organomet. Chem.*, **576**, 125.
55 Chemler, S.R., Trauner, D., and Danishefsky, S.J. (2001) *Angew. Chem. Int. Ed.*, **40**, 4544.
56 Pattenden, G. and Sinclair, D.J. (2002) *J. Organomet. Chem.*, **653**, 261.
57 Alberico, D., Scott, M.E., and Lautens, M. (2007) *Chem. Rev.*, **107**, 174.
58 Seregin, I.V. and Gevorgyan, V. (2007) *Chem. Soc. Rev.*, **36**, 1173.
59 Schlosser, M. (2007) *Synlett*, 3096.
60 Bellina, F., Cauteruccio, S., and Rossi, R. (2008) *Curr. Org. Chem.*, **12**, 774.
61 Donnelly, D.M.X. and Meegan, M.J. (1984) *Comprehensive Heterocyclic Chemistry*, vol. 4 (eds A.R. Katritzky and C.W. Rees), Pergamon, Oxford, p. 657.
62 Ziegert, R.E., Torang, J., Knepper, K., and Bräse, S. (2005) *J. Comb. Chem.*, **7**, 147.
63 Keay, B.A., Hopkins, J.M., and Dibble, P.W. (2008) *Comprehensive Heterocyclic Chemistry III*, vol. 3 (eds A.R. Katritzky, C.A. Ramsden, E.F.V. Scriven, and R.J.K. Taylor), Elsevier, Oxford, pp. 571–623.

64 Lipshutz, B.H. (1986) *Chem. Rev.*, **86**, 795.
65 Cacchi, S., Fabrizi, G., and Goggiamani, A. (2006) *Curr. Org. Chem.*, **10**, 1423.
66 Wong, H.N.C., Yeung, K.S., and Yang, Z. (2008) *Comprehensive Heterocyclic Chemistry III*, vol. 3 (eds A.R. Katritzky, C.A. Ramsden, E.F.V. Scriven, and R.J.K. Taylor), Elsevier, Oxford, pp. 407–496.
67 Keay, B.A. and Dibble, P.W. (1996) *Comprehensive Heterocyclic Chemistry II*, vol. 2 (eds A.R. Katritzky, E.F.V. Scriven, and C.W. Rees), Elsevier, Oxford, p. 395.
68 Kadieva, M. and Oganesyan, É. (1997) *Chem. Heterocycl. Compd.*, **33**, 1245.
69 Dell, C.P. (2000) *Science of Synthesis: Houben-Weyl Methods of Molecular Transformations*, vol. 10 (ed. G. Maas), Georg Thieme, Stuttgart, pp. 11–86.
70 Cacchi, S., Fabrizi, G., and Goggiomani, A. (2002) *Heterocycles*, **56**, 613.
71 Jeevanandam, A., Ghule, A., and Ling, Y.C. (2002) *Curr. Org. Chem.*, **6**, 841.
72 Hou, X.-L., Yang, Z., Yeung, K.-S., and Wong, H.N.C. (2007) *Progress in Heterocyclic Chemistry*, vol. 18 (eds G.W. Gribble and J.A. Joule), Elsevier, Oxford, pp. 187–217.
73 Graening, T. and Thrun, F. (2008) *Comprehensive Heterocyclic Chemistry III*, vol. 3 (eds A.R. Katritzky, C.A. Ramsden, E.F.V. Scriven, and R.J.K. Taylor), Elsevier, Oxford, pp. 497–569.
74 Hou, X.-L., Yang, Z., Yeung, K.-S., and Wong, H.N.C. (2008) *Progress in Heterocyclic Chemistry*, vol. 19 (eds G.W. Gribble and J.A. Joule), Elsevier, Oxford, pp. 176–207.
75 Stephens, R.D. and Castro, C.E. (1963) *J. Org. Chem.*, **28**, 3313.
76 Castro, C.E. and Stephens, R.D. (1963) *J. Org. Chem.*, **28**, 2163.
77 Castro, C.E., Havlin, R., Honwad, V.K., Malte, A.M., and Moje, S.W. (1969) *J. Am. Chem. Soc.*, **91**, 6464.
78 Haglund, O. and Nilsson, M. (1991) *Synlett*, 723.
79 Nakamura, M., Ilies, L., Otsubo, S., and Nakamura, E. (2006) *Angew. Chem. Int. Ed.*, **45**, 944.
80 Nakamura, M., Ilies, L., Otsubo, S., and Nakamura, E. (2006) *Org. Lett.*, **8**, 2803.
81 Hashmi, A.S.K., Frost, T.M., and Bats, J.W. (2000) *J. Am. Chem. Soc.*, **122**, 11553.
82 Hashmi, A.S.K., Frost, T.M., and Bats, J.W. (2001) *Org. Lett.*, **3**, 3769.
83 Kabalka, G.W., Wang, L., and Pagni, R.M. (2001) *Tetrahedron Lett.*, **42**, 6049.
84 Kabalka, G.W., Zhou, L.-L., Wang, L., and Pagni, R.M. (2006) *Tetrahedron*, **62**, 857.
85 Bräse, S., Gil, C., and Knepper, K. (2002) *Bioorg. Med. Chem.*, **10**, 2415.
86 Krchňák, V. and Holladay, M.W. (2002) *Chem. Rev.*, **102**, 61.
87 Crosignani, S. and Linclau, B. (2006) *Topics in Heterocyclic Chemistry*, vol. 1, Springer-Verlag, Berlin, Heidelberg, pp. 129–154.
88 Erdélyi, M. (2006) *Topics in HeterocyclicChemistry*, vol. 1, Springer-Verlag, Berlin, Heidelberg, pp. 79–128.
89 Fürstner, A. and Davies, P.W. (2005) *J. Am. Chem. Soc.*, **127**, 15024.
90 Hiroya, K., Itoh, S., and Sakamoto, T. (2005) *Tetrahedron*, **61**, 10958.
91 Li, X., Chianese, A.R., Vogel, T., and Crabtree, R.H. (2005) *Org. Lett.*, **7**, 5437.
92 Belting, V. and Krause, N. (2006) *Org. Lett.*, **8**, 4489.
93 Trost, B.M. and McClory, A. (2007) *Angew. Chem. Int. Ed.*, **46**, 2074.
94 Zhang, Y., Zhi-Jun, X.I.N., Ji-Jun, X.U.E., and Ying, L.I. (2008) *Chin. J. Chem.*, **26**, 1461.
95 Zhang, Y., Xue, J., Xin, Z., Xie, Z., and Li, Y. (2008) *Synlett*, 940.
96 Fiandanese, V., Bottalico, D., Marchese, G., and Punzi, A. (2008) *Tetrahedron*, **64**, 53.
97 Oppenheimer, J., Johnson, W.L., Tracey, M.R., Hsung, R.P., Yao, P.-Y., Liu, R., and Zhao, K. (2007) *Org. Lett.*, **9**, 2361.
98 Nakamura, I., Mizushima, Y., and Yamamoto, Y. (2005) *J. Am. Chem. Soc.*, **127**, 15022.
99 Nakamura, I., Mizushima, Y., Yamagishi, U., and Yamamoto, Y. (2007) *Tetrahedron*, **63**, 8670.
100 Fürstner, A., Heilmann, E.K., and Davies, P.W. (2007) *Angew. Chem. Int. Ed.*, **46**, 4760.
101 Grimshaw, J. and Thompson, N. (1987) *J. Chem. Soc., Chem. Commun.*, 240.
102 Baumgartner, M.T., Jimenez, L.B., Pierini, A.B., and Rossi, R.A. (2002) *J. Chem. Soc., Perkin Trans. 2*, 1092.

103 Ley, S.V. and Thomas, A.W. (2003) *Angew. Chem. Int. Ed.*, **42**, 5400.
104 Chen, C.-y. and Dormer, P.G. (2005) *J. Org. Chem.*, **70**, 6964.
105 Ackermann, L. and Kaspar, L.T. (2007) *J. Org. Chem.*, **72**, 6149.
106 Fang, Y. and Li, C. (2006) *J. Org. Chem.*, **71**, 6427.
107 Carril, M., SanMartin, R., Tellitu, I., and Domínguez, E. (2006) *Org. Lett.*, **8**, 1467.
108 Carril, M., SanMartin, R., Domínguez, E., and Tellitu, I. (2007) *Green Chem.*, **9**, 219.
109 Tadd, A.C., Fielding, M.R., and Willis, M.C. (2007) *Tetrahedron Lett.*, **48**, 7578.
110 Sakai, N., Uchida, N., and Konakahara, T. (2008) *Tetrahedron Lett.*, **49**, 3437.
111 Castro, C.E., Gaughan, E.J., and Owsley, D.C. (1966) *J. Org. Chem.*, **31**, 4071.
112 Schneiders, G.E. and Stevenson, R. (1980) *Synth. Commun.*, **10**, 699.
113 Yang, Z., Hon, P.M., Chui, K.Y., Xu, Z.L., Chang, H.M., Lee, C.M., Cui, Y.X., Wong, H.N.C., Poon, C.D., and Fung, B.M. (1991) *Tetrahedron Lett.*, **32**, 2061.
114 Yang, Z., Liu, H.B., Lee, C.M., Chang, H.M., and Wong, H.N.C. (1992) *J. Org. Chem.*, **57**, 7248.
115 Mzhelskaya, M.A., Ivanchikova, I.D., Polyakov, N.E., Moroz, A.A., and Shvartsberg, M.S. (2004) *Russ. Chem. Bull.*, **53**, 2798.
116 Yang, X.F. and Kong, L.Y. (2007) *Chin. Chem. Lett.*, **18**, 380.
117 Doad, G.J.S., Barltrop, J.A., Petty, C.M., and Owen, T.C. (1989) *Tetrahedron Lett.*, **30**, 1597.
118 Ohemeng, K.A., Appollina, M.A., Nguyen, V.N., Schwender, C.F., Singer, M., Steber, M., Ansell, J., Argentieri, D., and Hageman, W. (1994) *J. Med. Chem.*, **37**, 3663.
119 Lütjens, H. and Scammells, P.J. (1998) *Tetrahedron Lett.*, **39**, 6581.
120 Bates, C.G., Saejueng, P., Murphy, J.M., and Venkataraman, D. (2002) *Org. Lett.*, **4**, 4727.
121 Saejueng, P., Bates, C.G., and Venkataraman, D. (2005) *Synthesis*, 1706.
122 Díaz, D.D., Miranda, P.O., Padrón, J.I., and Martín, V.S. (2006) *Curr. Org. Chem.*, **10**, 457.
123 Carril, M., Correa, A., and Bolm, C. (2008) *Angew. Chem. Int. Ed.*, **47**, 4862.
124 Kumar, M.P. and Liu, R.-S. (2006) *J. Org. Chem.*, **71**, 4951.
125 Lu, B., Wang, B., Zhang, Y., and Ma, D. (2007) *J. Org. Chem.*, **72**, 5337.
126 Zhang, T.Y., O'Toole, J., and Proctor, C.S. (1999) *Sulfur Rep.*, **22**, 1.
127 Russell, R.K. and Press, J.B. (1996) *Comprehensive Heterocyclic Chemistry II*, vol. 2 (eds A.R. Katritzky, E.F.V. Scriven, and C.W. Rees), Elsevier, Oxford, pp. 679–729.
128 Schatz, J., Brendgen, T., and Schühle, D. (2008) *Comprehensive Heterocyclic Chemistry III*, vol. 3 (eds A.R. Katritzky, C.A. Ramsden, E.F.V. Scriven, and R.J.K. Taylor), Elsevier, Oxford, pp. 931–974.
129 Bianchini, C. and Meli, A. (1997) *Synlett*, 643.
130 Pradhan, T.K. and De, A. (2005) *Heterocycles*, **65**, 1491.
131 Rajappa, S. and Natekar, M.V. (1996) *Comprehensive Heterocyclic Chemistry II*, vol. 2 (eds A.R. Katritzky, E.F.V. Scriven, and C.W. Rees), Elsevier, Oxford, pp. 491–605.
132 Rajappa, S. and Deshmukh, A.R. (2008) *Comprehensive Heterocyclic Chemistry III*, vol. 3 (eds A.R. Katritzky, C.A. Ramsden, E.F.V. Scriven, and R.J.K. Taylor), Elsevier, Oxford, pp. 741–841.
133 Nakayama, J. (1996) *Comprehensive Heterocyclic Chemistry II*, vol. 2 (eds A.R. Katritzky, E.F.V. Scriven, and C.W. Rees), Elsevier, Oxford, pp. 607–677.
134 Rayner, C.M. and Graham, M.A. (2000) *Science of Synthesis: Houben-Weyl Methods of Molecular Transformations*, vol. 10 (ed. G. Maas), Georg Thieme, Stuttgart, pp. 155–184.
135 Sato, O. and Nakayama, J. (2008) *Comprehensive Heterocyclic Chemistry III*, vol. 3 (eds A.R. Katritzky, C.A. Ramsden, E.F.V. Scriven, and R.J.K. Taylor), Elsevier, Oxford, pp. 843–930.
136 Janosik, T. and Bergman, J. (2005) *Progress in Heterocyclic Chemistry*, vol. 17 (eds G.W. Gribble and J.A. Joule), Elsevier, Oxford, pp. 84–108.
137 Janosik, T. and Bergman, J. (2007) *Progress in Heterocyclic Chemistry*, vol. 18 (eds G.W. Gribble and J.A. Joule), Elsevier, Oxford, pp. 126–149.

138 Janosik, T. and Bergman, J. (2008) *Progress in Heterocyclic Chemistry*, vol. 19 (eds G.W. Gribble and J.A. Joule), Elsevier, Oxford, pp. 112–134.

139 Malte, A.M. and Castro, C.E. (1967) *J. Am. Chem. Soc.*, **89**, 6770.

140 Chinchilla, R. and Najera, C. (2007) *Chem. Rev.*, **107**, 874.

141 Nakamura, I., Sato, T., and Yamamoto, Y. (2006) *Angew. Chem. Int. Ed.*, **45**, 4473.

142 Nakamura, I., Sato, T., Terada, M., and Yamamoto, Y. (2007) *Org. Lett.*, **9**, 4081.

143 Nakamura, I., Sato, T., Terada, M., and Yamamoto, Y. (2008) *Org. Lett.*, **10**, 2649.

144 Hajós, G. and Nagy, I. (2008) *Curr. Org. Chem.*, **12**, 39.

145 Nakamura, Y. and Ukita, T. (2002) *Org. Lett.*, **4**, 2317.

146 Effenberger, F. and Russ, W. (1982) *Chem. Ber.*, **115**, 3719.

147 Gribble, G.W. (2003) *Progress in Heterocyclic Chemistry*, vol. 15 (eds G.W. Gribble and T.L. Gilchrist), Elsevier, Oxford, pp. 58–74.

148 Agarwal, S., Cämmerer, S., Filali, S., Fröhner, W., Knöll, J., Krahl, M.P., Reddy, K.R., and Knölker, H.-J. (2005) *Curr. Org. Chem.*, **9**, 1601.

149 Takayama, H., Kitajima, M., and Kogure, N. (2005) *Curr. Org. Chem.*, **9**, 1445.

150 Lewis, S.E. (2006) *Tetrahedron*, **62**, 8655.

151 Sánchez, C., Méndez, C., and Salas, J.A. (2006) *Nat. Prod. Rep.*, **23**, 1007.

152 Somei, M. (2006) *Topics in Heterocyclic Chemistry*, vol. 6, Springer-Verlag, Berlin, Heidelberg, pp. 77–111.

153 Higuchi, K. and Kawasaki, T. (2007) *Nat. Prod. Rep.*, **24**, 843.

154 Süzen, S. (2007) *Topics in Heterocyclic Chemistry*, vol. 11, Springer-Verlag, Berlin, Heidelberg, pp. 145–178.

155 Gupta, L., Talwar, A., and Chauhan, P.M.S. (2007) *Curr. Med. Chem.*, **14**, 1789.

156 d'Ischia, M., Napolitano, A., and Pezzella, A. (2008) *Comprehensive Heterocyclic Chemistry III*, vol. 3 (eds A.R. Katritzky, C.A. Ramsden, E.F.V. Scriven, and R.J.K. Taylor), Elsevier, Oxford, pp. 353–388.

157 Patil, S.A., Patil, R., and Miller, D.D. (2008) *Curr. Org. Chem.*, **12**, 691.

158 Gribble, G.W. (1996) *Comprehensive Heterocyclic Chemistry II*, vol. 2 (eds A.R. Katritzky, E.F.V. Scriven, and C.W. Rees), Elsevier, Oxford, pp. 207–257.

159 Gribble, G.W. (2003) *Pure Appl. Chem.*, **75**, 1417.

160 Joule, J.A. (1999) *Progress in Heterocyclic Chemistry*, vol. 11 (eds G.W. Gribble and T.L. Gilchrist), Elsevier, Oxford, pp. 45–65.

161 Sapi, J. and Laronze, J.-Y. (2004) *ARKIVOC*, 208.

162 Saracoglu, N. (2007) *Topics in Heterocyclic Chemistry*, vol. 11, Springer-Verlag, Berlin, Heidelberg, pp. 1–61.

163 Patil, S. and Patil, R. (2007) *Curr. Org. Synth.*, **4**, 201.

164 Trofimov, B.A., and Nedolya, N.A. (2008) *Comprehensive Heterocyclic Chemistry III*, vol. 3 (eds A.R. Katritzky, C.A. Ramsden, E.F.V. Scriven, and R.J.K. Taylor), Elsevier, Oxford, pp. 45–268.

165 Sundberg, R.J. (1996) *Comprehensive Heterocyclic Chemistry II*, vol. 2 (eds A.R. Katritzky, E.F.V. Scriven, and C.W. Rees), Elsevier, Oxford, pp. 119–206.

166 Joule, J.A. (2000) *Science of Synthesis: Houben-Weyl Methods of Molecular Transformations*, vol. 10 (eds G. Maas), Georg Thieme, Stuttgart, pp. 361–652.

167 Gribble, G.W. (2000) *J. Chem. Soc., Perkin Trans. 1*, 1045.

168 Tois, J., Franzén, R., and Koskinen, A. (2003) *Tetrahedron*, **59**, 5395.

169 Dalpozzo, R. and Bartoli, G. (2005) *Curr. Org. Chem.*, **9**, 163.

170 Janosik, T., Bergman, J., and Pelkey, E.T. (2005) *Progress in Heterocyclic Chemistry*, vol. 16 (eds G.W. Gribble and J.A. Joule), Elsevier, Oxford, pp. 128–155.

171 Pelkey, E.T. (2005) *Progress in Heterocyclic Chemistry*, vol. 17 (eds G.W. Gribble and J.A. Joule), Elsevier, Oxford, pp. 109–141.

172 Campo, J., García-Valverde, M., Marcaccini, S., Rojo, M.J., and Torroba, T. (2006) *Org. Biomol. Chem.*, **4**, 757.

173 Pelkey, E.T. (2007) *Progress in Heterocyclic Chemistry*, vol. 18 (eds G.W. Gribble and J.A. Joule), Elsevier, Oxford, pp. 150–186.

174 Pelkey, E.T. and Russel, J.S. (2008) *Progress in Heterocyclic Chemistry*, vol. 19 (eds G.W. Gribble and J.A. Joule), Elsevier, Oxford, pp. 135–175.

175 Bergman, J. and Janosik, T. (2008) *Comprehensive Heterocyclic Chemistry III*,

vol. 3 (eds A.R. Katritzky, C.A., Ramsden, E.F.V., Scriven, and R.J.K. Taylor), Elsevier, Oxford, pp. 269–351.
176 Krüger, K., Tillack, A., and Beller, M. (2008) *Adv. Synth. Catal.*, **350**, 2153.
177 Schmidt, M.A. and Movassaghi, M. (2008) *Synlett*, 313.
178 Battistuzzi, G., Cacchi, S., and Fabrizi, G. (2002) *Eur. J. Org. Chem.*, 2671.
179 Cacchi, S. and Fabrizi, G. (2005) *Chem. Rev.*, **105**, 2873.
180 Humphrey, G.R. and Kuethe, J.T. (2006) *Chem. Rev.*, **106**, 2875.
181 Patil, S. and Buolamwini, J.K. (2006) *Curr. Org. Synth.*, **3**, 477.
182 Fujiwara, J., Fukutani, Y., Sano, H., Maruoka, K., and Yamamoto, H. (1983) *J. Am. Chem. Soc.*, **105**, 7177.
183 Ezquerra, J., Pedregal, C., Lamas, C., Barluenga, J., Pérez, M., García-Martín, M.A., and González, J.M. (1996) *J. Org. Chem.*, **61**, 5804.
184 Adams, D.R., Duncton, M.A.J., Roffey, J.R.A., and Spencer, J. (2002) *Tetrahedron Lett.*, **43**, 7581.
185 Ma, C., Liu, X., Li, X., Flippen-Anderson, J., Yu, S., and Cook, J.M. (2001) *J. Org. Chem.*, **66**, 4525.
186 Farr, R.N., Alabaster, R.J., Chung, J.Y.L., Craig, B., Edwards, J.S., Gibson, A.W., Ho, G.-J., Humphrey, G.R., Johnson, S.A., and Grabowski, E.J.J. (2003) *Tetrahedron: Asymmetry*, **14**, 3503.
187 Sakamoto, T., Kondo, Y., Iwashita, S., Nagano, T., and Yamanaka, H. (1988) *Chem. Pharm. Bull.*, **36**, 1305.
188 Villemin, D. and Goussu, D. (1989) *Heterocycles*, **29**, 1255.
189 Kamijo, S., Jin, T., and Yamamoto, Y. (2002) *Angew. Chem. Int. Ed.*, **41**, 1780.
190 Cacchi, S., Fabrizi, G., and Parisi, L.M. (2003) *Org. Lett.*, **5**, 3843.
191 McDonald, F.E. and Chatterjee, A.K. (1997) *Tetrahedron Lett.*, **38**, 7687.
192 Bruneau, C. and Dixneuf, P.H. (1999) *Acc. Chem. Res.*, **32**, 311.
193 Wakatsuki, Y. (2004) *J. Organomet. Chem.*, **689**, 4092.
194 Varela, J.A. and Saá, C. (2006) *Chem. Eur. J.*, **12**, 6450.
195 Burling, S., Field, L.D., and Messerle, B.A. (2000) *Organometallics*, **19**, 87.
196 Müller, T.E., Grosche, M., Herdtweck, E., Pleier, A.K., Walter, E., and Yan, Y.K. (2000) *Organometallics*, **19**, 170.
197 Kondo, T., Okada, T., Suzuki, T., and Mitsudo, T.-a. (2001) *J. Organomet. Chem.*, **622**, 149.
198 van Esseveldt, B.C.J., van Delft, F.L., Smits, J.M.M., de Gelder, R., Schoemaker, H.E., and Rutjes, F.P.J.T. (2004) *Adv. Synth. Catal.*, **346**, 823.
199 Majumdar, K.C., Chattopadhyay, B., and Samanta, S. (2009) *Synthesis*, 311.
200 Hiroya, K., Itoh, S., Ozawa, M., Kanamori, Y., and Sakamoto, T. (2002) *Tetrahedron Lett.*, **43**, 1277.
201 Hiroya, K., Itoh, S., and Sakamoto, T. (2004) *J. Org. Chem.*, **69**, 1126.
202 Dai, W.-M., Guo, D.-S., Sun, L.-P., and Huang, X.-H. (2003) *Org. Lett.*, **5**, 2919.
203 Kurisaki, T., Naniwa, T., Yamamoto, H., Imagawa, H., and Nishizawa, M. (2007) *Tetrahedron Lett.*, **48**, 1871.
204 Yamamoto, H., Sasaki, I., Hirai, Y., Namba, K., Imagawa, H., and Nishizawa, M. (2009) *Angew. Chem. Int. Ed.*, **48**, 1244.
205 Iritani, K., Matsubara, S., and Utimoto, K. (1988) *Tetrahedron Lett.*, **29**, 1799.
206 Arcadi, A., Bianchi, G., and Marinelli, F. (2004) *Synthesis*, 610.
207 Miyazaki, Y. and Kobayashi, S. (2008) *J. Comb. Chem.*, **10**, 355.
208 Alfonsi, M., Arcadi, A., Aschi, M., Bianchi, G., and Marinelli, F. (2005) *J. Org. Chem.*, **70**, 2265.
209 Yin, Y., Ma, W., Chai, Z., and Zhao, G. (2007) *J. Org. Chem.*, **72**, 5731.
210 Yin, Y., Chai, Z., Ma, W.-Y., and Zhao, G. (2008) *Synthesis*, 4036.
211 Ohno, H., Ohta, Y., Oishi, S., and Fujii, N. (2007) *Angew. Chem. Int. Ed.*, **46**, 2295.
212 Zhang, X. and Corma, A. (2008) *Angew. Chem. Int. Ed.*, **47**, 4358.
213 Shimada, T., Nakamura, I., and Yamamoto, Y. (2004) *J. Am. Chem. Soc.*, **126**, 10546.
214 Li, G., Huang, X., and Zhang, L. (2008) *Angew. Chem. Int. Ed.*, **47**, 346.
215 Cariou, K., Ronan, B., Mignani, S., Fensterbank, L., and Malacria, M. (2007) *Angew. Chem. Int. Ed.*, **46**, 1881.
216 Nakamura, I., Yamagishi, U., Song, D., Konta, S., and Yamamoto, Y. (2007) *Angew. Chem. Int. Ed.*, **46**, 2284.

217 Ojima, I., Tzamarioudaki, M., Li, Z., and Donovan, R.J. (1996) *Chem. Rev.*, **96**, 635.

218 Takaya, J., Udagawa, S., Kusama, H., and Iwasawa, N. (2008) *Angew. Chem. Int. Ed.*, **47**, 4906.

219 Kamijo, S. and Yamamoto, Y. (2003) *J. Org. Chem.*, **68**, 4764.

220 Kamijo, S., Sasaki, Y., and Yamamoto, Y. (2004) *Tetrahedron Lett.*, **45**, 35.

221 Zhang, Y., Donahue, J.P., and Li, C.J. (2007) *Org. Lett.*, **9**, 627.

222 Kusama, H., Takaya, J., and Iwasawa, N. (2002) *J. Am. Chem. Soc.*, **124**, 11592.

223 Kusama, H., Miyashita, Y., Takaya, J., and Iwasawa, N. (2006) *Org. Lett.*, **8**, 289.

224 Kusama, H., Suzuki, Y., Takaya, J., and Iwasawa, N. (2006) *Org. Lett.*, **8**, 895.

225 Saito, A., Kanno, A., and Hanzawa, Y. (2007) *Angew. Chem. Int. Ed.*, **46**, 3931.

226 Saito, A., Oda, S., Fukaya, H., and Hanzawa, Y. (2009) *J. Org. Chem.*, **74**, 1517.

227 Crotti, C., Cenini, S., Rindone, B., Tollari, S., and Demartin, F. (1986) *J. Chem. Soc., Chem. Commun.*, 784.

228 Sanz, R., Escribano, J., Pedrosa, M.R., Aguado, R., and Arnáiz, F.J. (2007) *Adv. Synth. Catal.*, **349**, 713.

229 Chiba, S., Hattori, G., and Narasaka, K. (2007) *Chem. Lett.*, **36**, 52.

230 Stokes, B.J., Dong, H., Leslie, B.E., Pumphrey, A.L., and Driver, T.G. (2007) *J. Am. Chem. Soc.*, **129**, 7500.

231 Shen, M., Leslie, B.E., and Driver, T.G. (2008) *Angew. Chem. Int. Ed.*, **47**, 5056.

232 Barberis, C., Gordon, T.D., Thomas, C., Zhang, X., and Cusack, K.P. (2005) *Tetrahedron Lett.*, **46**, 8877.

233 Yamada, K., Kurokawa, T., Tokuyama, H., and Fukuyama, T. (2003) *J. Am. Chem. Soc.*, **125**, 6630.

234 Okano, K., Tokuyama, H., and Fukuyama, T. (2006) *J. Am. Chem. Soc.*, **128**, 7136.

235 Melkonyan, F.S., Karchava, A.V., and Yurovskaya, M.A. (2008) *J. Org. Chem.*, **73**, 4275.

236 Melkonyan, F., Topolyan, A., Yurovskaya, M., and Karchava, A. (2008) *Eur. J. Org. Chem.*, 5952.

237 Yuen, J., Fang, Y.Q., and Lautens, M. (2006) *Org. Lett.*, **8**, 653.

238 Lu, B. and Ma, D. (2006) *Org. Lett.*, **8**, 6115.

239 Ma, D. and Cai, Q. (2008) *Acc. Chem. Res.*, **41**, 1450.

240 Dong, Y. and Busacca, C.A. (1997) *J. Org. Chem.*, **62**, 6464.

241 Chen, Y., Wang, Y., Sun, Z., and Ma, D. (2008) *Org. Lett.*, **10**, 625.

242 Ackermann, L. (2005) *Org. Lett.*, **7**, 439.

243 Hodgkinson, R.C., Schulz, J., and Willis, M.C. (2009) *Org. Biomol. Chem.*, **7**, 432.

244 Katritzky, A.R., Li, J., and Stevens, C.V. (1995) *J. Org. Chem.*, **60**, 3401.

245 Katritzky, A.R., Fali, C.N., and Li, J. (1997) *J. Org. Chem.*, **62**, 4148.

246 Yakovleva, E., Ivanchikova, I., and Shvartsberg, M. (2005) *Russ. Chem. Bull.*, **54**, 421.

247 Prikhod'ko, T.A. and Vasilevsky, S.F. (1998) *Mendeleev Commun.*, **8**, 149.

248 Prikhodko, T.A. and Vasilevsky, S.F. (2001) *Russ. Chem. Bull.*, **50**, 1268.

249 Slough, G.A., Krchnak, V., Helquist, P., and Canham, S.M. (2004) *Org. Lett.*, **6**, 2909.

250 Liu, F. and Ma, D. (2007) *J. Org. Chem.*, **72**, 4844.

251 Stuart, D.R., Bertrand-Laperle, M., Burgess, K.M.N., and Fagnou, K. (2008) *J. Am. Chem. Soc.*, **130**, 16474.

252 Ritleng, V., Sirlin, C., and Pfeffer, M. (2002) *Chem. Rev.*, **102**, 1731.

253 Tokunaga, M., Ota, M., Haga, M.-a., and Wakatsuki, Y. (2001) *Tetrahedron Lett.*, **42**, 3865.

254 Cao, C., Shi, Y., and Odom, A.L. (2002) *Org. Lett.*, **4**, 2853.

255 Khedkar, V., Tillack, A., Michalik, M., and Beller, M. (2004) *Tetrahedron Lett.*, **45**, 3123.

256 Ackermann, L. and Born, R. (2004) *Tetrahedron Lett.*, **45**, 9541.

257 Khedkar, V., Tillack, A., Michalik, M., and Beller, M. (2005) *Tetrahedron*, **61**, 7622.

258 Schwarz, N., Alex, K., Sayyed, I.A., Khedkar, V., Tillack, A., and Beller, M. (2007) *Synlett*, 1091.

259 Sayyed, I.A., Alex, K., Tillack, A., Schwarz, N., Michalik, D., and Beller, M. (2007) *Eur. J. Org. Chem.*, 4525.

260 Alex, K., Tillack, A., Schwarz, N., and Beller, M. (2008) *Angew. Chem. Int. Ed.*, **47**, 2304.

261 Banerjee, S., Barnea, E., and Odom, A.L. (2008) *Organometallics*, **27**, 1005.

262 Alex, K., Schwarz, N., Khedkar, V., Sayyed, I.A., Tillack, A., Michalik, D., Holenz, J., Diaz, J.L., and Beller, M. (2008) *Org. Biomol. Chem.*, **6**, 1802.

263 Eilbracht, P., Bärfacker, L., Buss, C., Hollmann, C., Kitsos-Rzychon, B.E., Kranemann, C.L., Rische, T., Roggenbuck, R., and Schmidt, A. (1999) *Chem. Rev.*, **99**, 3329.

264 Köhling, P., Schmidt, A.M., and Eilbracht, P. (2003) *Org. Lett.*, **5**, 3213.

265 Schmidt, A.M. and Eilbracht, P. (2005) *Org. Biomol. Chem.*, **3**, 2333.

266 Schmidt, A.M. and Eilbracht, P. (2005) *J. Org. Chem.*, **70**, 5528.

267 Köhling, P.L., Schmidt, A.M., and Eilbracht, P. (2006) *Org. Biomol. Chem.*, **4**, 302.

268 Ahmed, M., Jackstell, R., Seayad, A.M., Klein, H., and Beller, M. (2004) *Tetrahedron Lett.*, **45**, 869.

269 Suzuki, H., Thiruvikraman, S.V., and Osuka, A. (1984) *Synthesis*, 616.

270 For an analogous reaction leading to indoles and proceeding via SRN1 mechanism, see: Baumgartner, M.T., Nazareno, M.A., Murguia, M.C., Pierini, A.B., and Rossi, R.A. (1999) *Synthesis*, 2053.

271 Tanimori, S., Ura, H., and Kirihata, M. (2007) *Eur. J. Org. Chem.*, 3977.

272 Chen, Y., Xie, X., and Ma, D. (2007) *J. Org. Chem.*, **72**, 9329.

273 Babichev, F.S., Kovtunenko, V.A., and Tyltin, A.K. (1981) *Usp. Khim.*, **50**, 2073.

274 Kovtunenko, V.A. and Voitenko, Z.V. (1994) *Russ. Chem. Rev.*, **63**, 997.

275 Donohoe, T.J. (2000) *Science of Synthesis: Houben-Weyl Methods of Molecular Transformations*, vol. 10 (ed. G. Maas), Georg Thieme, Stuttgart, pp. 653–692.

276 Kadzimirsz, D., Hildebrandt, D., Merz, K., and Dyker, G. (2006) *Chem. Commun.*, 661.

277 Peng, C., Cheng, J., and Wang, J. (2008) *Adv. Synth. Catal.*, **350**, 2359.

278 Yeom, H.-S., Lee, J.-E., and Shin, S. (2008) *Angew. Chem. Int. Ed.*, **47**, 7040.

279 Flitsch, W. (1996) *Comprehensive Heterocyclic Chemistry II*, vol. 8 (eds A.R. Katritzky, E.F.V., Scriven, and C.W. Rees), Elsevier, Oxford, p. 237.

280 Brandi, A. and Cicchi, S. (2008) *Comprehensive Heterocyclic Chemistry III*, vol. 11 (eds A.R. Katritzky, C.A. Ramsden, E.F.V. Scriven, and R.J.K. Taylor), Elsevier, Oxford, pp. 367–408.

281 Michael, J.P. (2000) *Nat. Prod. Rep.*, **17**, 579.

282 Michael, J.P. (2002) *Nat. Prod. Rep.*, **19**, 719.

283 Michael, J.P. (2004) *Nat. Prod. Rep.*, **21**, 625.

284 Michael, J.P. (2005) *Nat. Prod. Rep.*, **22**, 603.

285 Michael, J.P. (2007) *Nat. Prod. Rep.*, **24**, 191.

286 Michael, J.P. (2008) *Nat. Prod. Rep.*, **25**, 139.

287 Fan, H., Peng, J., Hamann, M.T., and Hu, J.-F. (2008) *Chem. Rev.*, **108**, 264.

288 Uchida, T. and Matsumoto, K. (1976) *Synthesis*, 209.

289 Shipman, M. (2000) *Science of Synthesis: Houben-Weyl Methods of Molecular Transformations*, vol. 10 (ed. G. Maas), Georg Thieme, Stuttgart, pp. 745–788.

290 Chernyak, D., Gadamsetty, S.B., and Gevorgyan, V. (2008) *Org. Lett.*, **10**, 2307.

291 Kel'in, A.V., Sromek, A.W., and Gevorgyan, V. (2001) *J. Am. Chem. Soc.*, **123**, 2074.

292 Kim, J.T. and Gevorgyan, V. (2005) *J. Org. Chem.*, **70**, 2054.

293 Walker, D.P., Wishka, D.G., Piotrowski, D.W., Jia, S., Reitz, S.C., Yates, K.M., Myers, J.K., Vetman, T.N., Margolis, B.J., Jacobsen, E.J., Acker, B.A., Groppi, V.E., Wolfe, M.L., Thornburgh, B.A., Tinholt, P.M., Cortes-Burgos, L.A., Walters, R.R., Hester, M.R., Seest, E.P., Dolak, L.A., Han, F., Olson, B.A., Fitzgerald, L., Staton, B.A., Raub, T.J., Hajos, M., Hoffmann, W.E., Li, K.S., Higdon, N.R., Wall, T.M., Hurst, R.S., Wong, E.H.F., and Rogers, B.N. (2006) *Bioorg. Med. Chem*, **14** 8219.

294 Kel'in, A.V. and Gevorgyan, V. (2002) *J. Org. Chem.*, **67**, 95.

295 Li, H., Xia, Z., Chen, S., Koya, K., Ono, M., and Sun, L. (2007) *Org. Process Res. Dev.*, **11**, 246.

296 Dudnik, A.S., Sromek, A.W., Rubina, M., Kim, J.T., Kel'in, A.V., and Gevorgyan, V. (2008) *J. Am. Chem. Soc.*, **130**, 1440.

297 Schwier, T., Sromek, A.W., Yap, D.M.L., Chernyak, D., and Gevorgyan, V. (2007) *J. Am. Chem. Soc.*, **129**, 9868.

298 Hardin, A.R. and Sarpong, R. (2007) *Org. Lett.*, **9**, 4547.

299 Seregin, I.V. and Gevorgyan, V. (2006) *J. Am. Chem. Soc.*, **128**, 12050.

300 Seregin, I.V., Schammel, A.W., and Gevorgyan, V. (2008) *Tetrahedron*, **64**, 6876.

301 Seregin, I.V., Schammel, A.W., and Gevorgyan, V. (2007) *Org. Lett.*, **9**, 3433.

302 Smith, C.R., Bunnelle, E.M., Rhodes, A.J., and Sarpong, R. (2007) *Org. Lett.*, **9**, 1169.

303 Bunnelle, E.M., Smith, C.R., Lee, S.K., Singaram, S.W., Rhodes, A.J., and Sarpong, R. (2008) *Tetrahedron*, **64**, 7008.

304 Yan, B., Zhou, Y., Zhang, H., Chen, J., and Liu, Y. (2007) *J. Org. Chem.*, **72**, 7783.

305 Chuprakov, S. and Gevorgyan, V. (2007) *Org. Lett.*, **9**, 4463.

306 Cho, S.H. and Chang, S. (2008) *Angew. Chem. Int. Ed.*, **47**, 2836.

307 Vedernikov, A.N. (2007) *Curr. Org. Chem.*, **11**, 1401.

308 Fürstner, A. and Mamane, V. (2002) *J. Org. Chem.*, **67**, 6264.

309 Mamane, V., Hannen, P., and Fürstner, A. (2004) *Chem. Eur. J.*, **10**, 4556.

310 Chuprakov, S., Hwang, F.W., and Gevorgyan, V. (2007) *Angew. Chem. Int. Ed.*, **46**, 4757.

311 Yan, B. and Liu, Y. (2007) *Org. Lett.*, **9**, 4323.

312 Su, S. and Porco, J.A. Jr. (2007) *J. Am. Chem. Soc.*, **129**, 7744.

10
Carbon–Heteroatom Bond Formation by Rh^{I}-Catalyzed Ring-Opening Reactions

Matthew J. Fleming and Mark Lautens

10.1
Introduction

Transition metal-catalyzed allylic nucleophilic substitution (ANS) represents an important reaction for the construction of useful chiral building blocks. One class of ANS reactions is the transition metal-catalyzed ring-opening of strained bridged bicyclic allylic ethers and amines of general structure **1** by nucleophiles to give the corresponding cyclohexene adducts **2** (Scheme 10.1). As the leaving group remains within the product, this reaction allows the possibility of generating defined contiguous stereogenic centers within the product when starting from a *meso*-substrate and carrying out the reaction under asymmetric conditions.

The majority of research focus has involved developing methodologies for the ring-opening of strained allylic ethers and amines of general structure **1** by carbon or hydride nucleophiles forming carbon–carbon or carbon–hydrogen bonds during the process. A number of research groups [1], including our own [2], have demonstrated that a variety of transition metal complexes can catalyze this reaction, and, depending on the combination of metal catalyst and nucleophile, different diastereoisomers of the product can be preferentially formed. However, less attention has been applied to the ring-opening of compounds of general structure **1** by heteroatom nucleophiles, which would form valuable carbon–heteroatom bonds. To date, Rh^{I}-complexes have been the most extensively explored catalysts for this type of reaction. The first reported example was by Hogeveen and Middelkoop, who in 1973 demonstrated that $[Rh(CO)_2Cl]_2$ could catalyze the ring-opening of **4** by MeOH to give one diastereoisomer of **5** (Scheme 10.2) [3]. It was later determined by Ashworth and Berchtold that the methoxy and hydroxy substituents have a cis-relationship by analysis of the coupling constant between H_a and H_b in the ^1H NMR spectrum of cycloadduct **6** [4]. While mechanistic details were not discussed, Brønsted acid catalysis seemed unlikely as reaction of **4** with protic catalysts resulted in the aromatized product **3**.

Since the seminal report by Hogeveen and Middelkoop, our group and others have examined the synthetic utility and mechanistic pathway of the Rh^{I}-catalyzed ring-opening of strained bridged bicyclic allylic ethers and amines by heteroatom

Catalyzed Carbon-Heteroatom Bond Formation. Edited by Andrei K. Yudin

ISBN: 978-3-527-32428-6

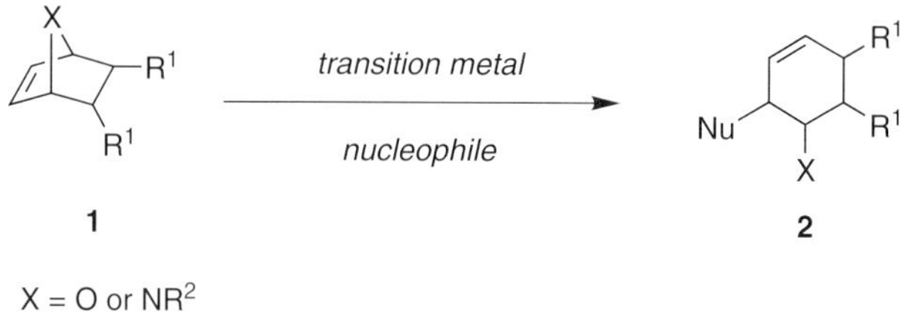

transition metal

nucleophile

1

2

X = O or NR^2

Scheme 10.1

nucleophiles. This chapter describes the advances made in this reaction using oxygen-, nitrogen- and sulfur-based nucleophiles to form the corresponding carbon–heteroatom bonds, with particular emphasis on the asymmetric variant of this reaction.

10.2
Ring-Opening *meso*-Oxabicyclic Alkenes with Oxygen-Based Nucleophiles

Considerable attention has been focused on the desymmetrization of oxobenzonorbornadiene **7** by heteroatom nucleophiles as the asymmetric ring-opening (ARO) methodology enables access to the hydronaphthalene core **8**, a privileged structural motif in medicinal chemistry [5]. In 2000 we reported the Rh^I-catalyzed carbon–oxygen bond formation by the ring-opening of oxobenzonorbornadienes with aliphatic alcohol nucleophiles (Scheme 10.3) [6]. While the conditions reported by Hogeveen and Middelkoop failed to induce the reaction, it was discovered that ring-opening of **7** with MeOH could be achieved with $[Rh(COD)Cl]_2$ in combination with bidentate bisphosphine ligands. Optimal results were obtained with bisphosphines derived

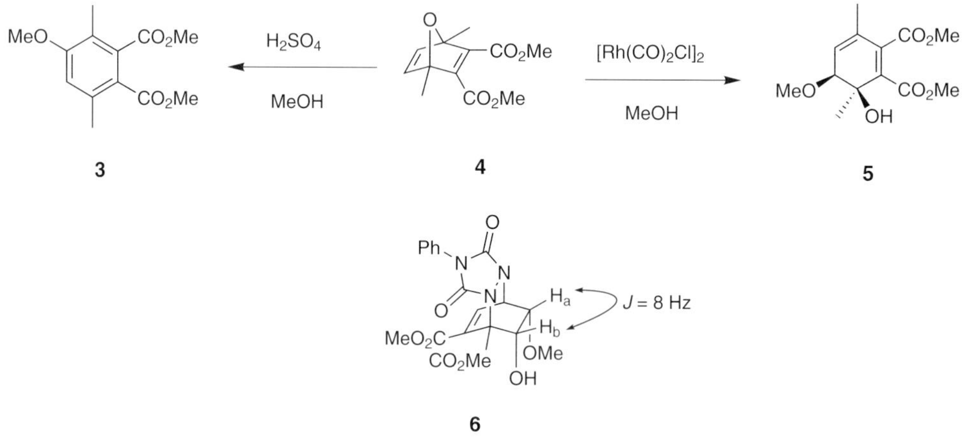

H_2SO_4 / MeOH; $[Rh(CO)_2Cl]_2$ / MeOH

3 4 5

6

J = 8 Hz

Scheme 10.2

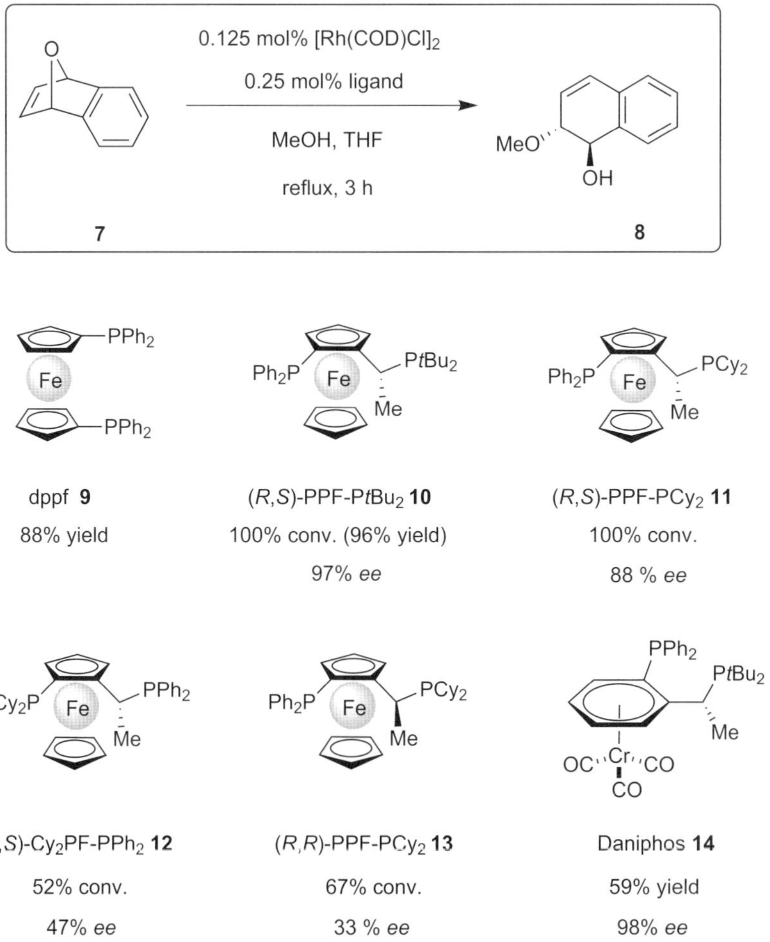

0.125 mol% $[Rh(COD)Cl]_2$
0.25 mol% ligand
MeOH, THF
reflux, 3 h

7 → 8

dppf **9**
88% yield

(*R*,*S*)-PPF-P*t*Bu_2 **10**
100% conv. (96% yield)
97% *ee*

(*R*,*S*)-PPF-PCy_2 **11**
100% conv.
88 % *ee*

(*R*,*S*)-Cy_2PF-PPh_2 **12**
52% conv.
47% *ee*

(*R*,*R*)-PPF-PCy_2 **13**
67% conv.
33 % *ee*

Daniphos **14**
59% yield
98% *ee*

Scheme 10.3

from ferrocene, such as dppf **9**. Of the chiral analogues surveyed, ligands based on the electron-rich Josiphos scaffold proved to be the most promising. A systematic study of the structure of this ligand class was conducted to optimize reaction efficiency and enantioselectivity [7]. To maximize enantioselectivity it was necessary to increase the steric bulk of the benzylic phosphine (comparing ligands **10** with **11**), and for the larger phosphine to be situated at the benzylic position and not on the ferrocene ring (comparing **10** with **12**). The Josiphos skeleton contains planar and carbon-centered chirality. It was discovered that the optimal diastereoisomer contains (*R*)-central and (*S*)-planar chirality (or the enantiomeric (*S*,*R*)-pair), while the (*R*,*R*) variant **13** showed lower yield and enantioselectivity. The optimal ligand, PPF-P*t*Bu_2**10** gave the corresponding dihydronaphthalene product **8** in 96% isolated yield and in 97% *ee*. Salzer has demonstrated that the Daniphos diphosphine ligand

14, based on the arene chromium tricarbonyl scaffold, can also be applied to the Rh^I-catalyzed ARO reaction, generating the products in high enantioselectivity, albeit in lower yield [8].

Key features of this methodology include low catalyst loadings (typically 0.25 mol% of catalytically active rhodium species) and neutral reaction conditions. No activation of the nucleophile is required and in fact no reaction occurs if a deprotonated nucleophilic species is used. In contrast to Hogeveen and Middelkoop's result, the newly incorporated methoxy functionality was determined to be trans to the hydroxy group; suggesting the reaction is following a different mechanistic pathway than when using the oxabicyclic alkene **4**. The reaction was demonstrated to work effectively for electron-rich and electron-poor oxabicyclic alkenes with a wide range of aliphatic alcohols (Scheme 10.4). In each case, the reaction was highly efficient and enantioselective (>95% *ee*) with only one regio- and diastereo-isomer being observed. Interestingly even the poorly nucleophilic (and more acidic) hexafluoroisopropanol (HFIP) proved to be a viable nucleophile. In fact, when a competitive experiment was conducted whereby equimolar amounts of HFIP

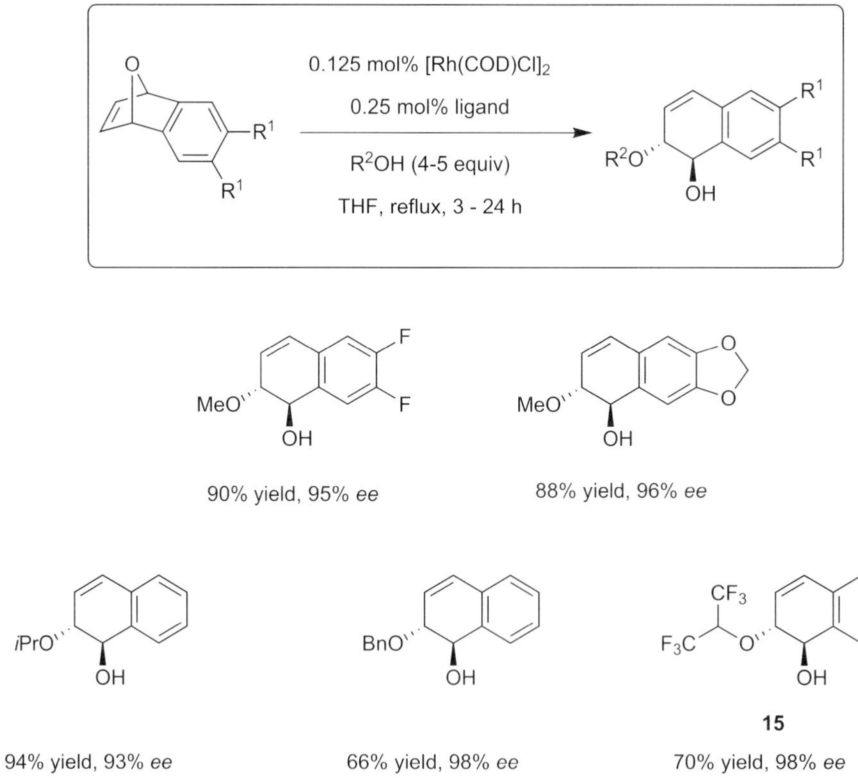

0.125 mol% $[Rh(COD)Cl]_2$
0.25 mol% ligand
R^2OH (4-5 equiv)
THF, reflux, 3 - 24 h

90% yield, 95% *ee*

88% yield, 96% *ee*

94% yield, 93% *ee*

66% yield, 98% *ee*

15
70% yield, 98% *ee*

Scheme 10.4

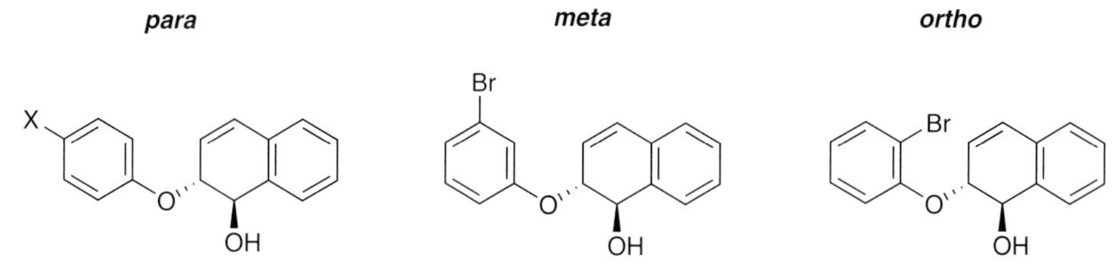

para *meta* *ortho*

X = H 91% yield, 99% *ee*
X = I 92% yield, 98% *ee*
X = C(O)Me 91% yield, 99% *ee*
X = OMe 85% yield, 95% *ee*

94% yield, 98% *ee*

17% yield, 97% *ee* after 24 h (using $[Rh(COD)Cl]_2$)
92% yield, 97% *ee* after 24 h (using $[Rh(CO)_2Cl]_2$)

conditions: **7**, 0.5 mol% $[Rh(COD)Cl]_2$, 1 mol% PPF-P*t*Bu_2, phenol (5 equiv), THF, 80 °C, 6 - 24 h

Scheme 10.5

and *i*PrOH were reacted with **7** only the product **15** formed by HFIP addition was observed [7].

Other oxygen-based nucleophiles have demonstrated efficient reactivity under our reported conditions. *Para*- and *meta*-substituted phenols were shown to add in high yield and excellent enantioselectivity (Scheme 10.5) [9]. The reaction proceeded well, even when aryl iodides were used, indicating that the rhodium insertion into the aryl iodide bond is slow compared to ring-opening.

Phenols containing either π-donating or -withdrawing groups were both compatible, however, there was significant difference in the relative rates of the reaction. Competition experiments revealed that 4-hydroxyacetophenone underwent addition to **7** 16 times faster than 4-hydroxyanisole, confirming that the presence of an electron-withdrawing group on the aromatic ring accelerates the rate of addition. *Ortho*-substituted phenols, while adding in high enantioselectivity, reacted less efficiently under these conditions. For example, 2-bromophenol added in only 17% yield after prolonged reaction times. The ability of 2-halophenols to form a chelate with rhodium through the halide and hydroxy groups was proposed to explain the diminished reactivity [10]. An increase in the reaction efficiency was achieved by changing the catalyst from $[Rh(COD)Cl]_2$ to $[Rh(CO)_2Cl]_2$ which gave the product in 92% yield and 97% *ee* after 24 h. It was reasoned that by replacing a COD with a more tightly bound CO, occupation of an open coordination site with a carbonyl ligand disfavors the formation of a chelate with 2-halophenols.

As more acidic oxygen nucleophiles added preferentially to **7**, it was reasoned that carboxylic acids would be good candidates for the ring-opening oxabicyclic alkenes to form the corresponding allylic esters. While certain transition metals, most notably palladium, are known to react with allylic esters to form π-allyl intermediates, rhodium-phosphine complexes are less reactive [11]. The product allylic ester formed

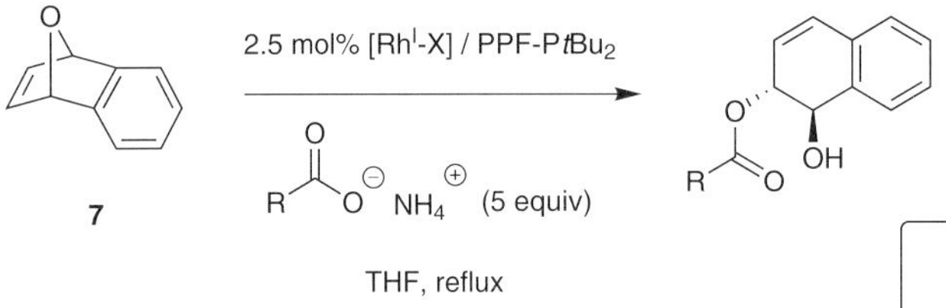

2.5 mol% [Rh^{I}-X] / PPF-PtBu$_2$

$RCO_2^{\ominus}$ $NH_4^{\oplus}$ (5 equiv)

THF, reflux

7

halide-exchange protocol

$[Rh(COD)Cl]_2$ (0.5 mol%)
PPF-tBu$_2$ (1.5 mol%)

1) THF, 5 min
2) AgOTf (1.5 mol%), 5 min
3) Bu_4NI (2 mol%), 5 min

[Rh^{I}-I] / PPF-PtBu$_2$

R	Ligand	X	Yield (%)	*ee* (%)
Me	dppf	Cl	84	-
Me	PPF-PtBu$_2$	Cl	81	61
Me	PPF-PtBu$_2$	I	93	92
Bn	PPF-PtBu$_2$	Cl	72	31
Bn	PPF-PtBu$_2$	I	91	92

Scheme 10.6

from ring-opening an oxabicyclic alkene with a carboxylic acid should therefore be stable to the standard ring-opening reaction conditions. While acetic acid and sodium acetate were unreactive, ammonium acetate proved to be a viable nucleophile to ring-open oxobenzonorbornadiene **7** when using $[Rh(COD)Cl]_2$ as the catalyst precursor and dppf as the phosphine ligand to give racemic dihydronaphthalene in 84% yield (Scheme 10.6) [12]. The asymmetric variant employing PPF-PtBu$_2$ as the chiral ligand gave only a moderate 61% *ee*. While ligand screening failed to improve the enantioselectivity of the reaction, it was discovered that changing the halide ligand had a dramatic effect on the enantioselectivity of the reaction [13]. Optimal results are obtained as one moves down the halide group with best enantioselectivities obtained with an iodide ligand. The rhodium iodide species can be readily prepared *in situ* by first adding AgOTf to a solution of commercially available $[Rh(COD)Cl]_2$ and PPF-PtBu$_2$ and then, after 5 min of stirring, adding an iodide source, such as Bu_4NI to give the rhodium iodide catalyst, as evidenced by ^{31}P NMR. Carrying out the ARO reaction using this [Rh^{I}-I]/PPF-tBu$_2$ catalyst with ammonium acetate as the nucleophile gave the allylic acetate product in an improved 93% yield and 92% *ee*. It was demonstrated that benzoate is also compatible, despite its lower nucleophilicity and the higher propensity of the allylic benzoate moiety to undergo ionization with transition metals. Employing the halide-exchange protocol again proved to be essential to obtain a high yield and enantioselectivity of the product.

Oxabicyclic alkenes other than oxobenzonorbornadiene have been examined as meso-substrates for the asymmetric ring-opening reaction with oxygen nucleophiles. Oxabicyclo[2.2.1.]heptenes proved to be less reactive and failed to open under any of the conditions described, probably due to reduced strain within the molecule. To overcome this barrier, a higher reaction temperature and concentration

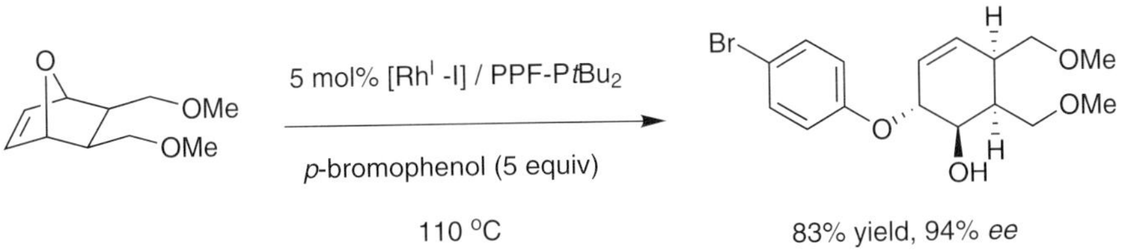

5 mol% [Rh^I-I] / PPF-P*t*Bu_2

p-bromophenol (5 equiv)

110 °C

83% yield, 94% *ee*

16 **17**

Scheme 10.7

were employed. Heating oxabicyclo[2.2.1.]heptene **16** at 110 °C with 5 equiv of the nucleophile in the absence of solvent, and using the [Rh^I-I]/PPF-*t*Bu_2 catalyst provided the desired ring-opened product **17** as a single diastereoisomer in 94% *ee* (Scheme 10.7) [14]. As with oxobenzonorbornadienes, only *trans*-1,2-cyclohexenol was formed. The same regio- and stereo-chemical outcome for both substrates is of mechanistic significance as it can be concluded that the outcome for ring-opening oxobenzonorbornadienes is not governed by the conjugation between the newly formed alkene and the aromatic moiety present in the 1,2-dihydronaphthalene products. This reaction demonstrates the potential of this methodology as starting from a meso-compound four contiguous defined stereogenic centers can be obtained within the product.

10.3 Ring-Opening *meso*-Oxabicyclic Alkenes with Nitrogen-Based Nucleophiles

It proved possible to form carbon–nitrogen bonds by the asymmetric ring-opening of oxobenzonorbornadiene with nitrogen nucleophiles under Rh^I catalysis. Using $[Rh(COD)Cl]_2$ as the catalyst precursor and PPF-*t*Bu_2 as the chiral ligand, sulfonamides could be added in high yield and enantioselectivity (Scheme 10.8) [6a]. While other electron deficient nitrogen sources, such as phthalimide and anilines, added efficiently under the same conditions, the enantioselectivity was lower. As in the case of ammonium carboxylate nucleophiles, altering the rhodium halide ligand proved highly effective in increasing the enantioselectivity of the reaction, with the iodide ligand providing the products with the highest *ee*. Carrying out the halogen-exchange protocol prior to the addition of reagents when using phthalimide or anilines as the nucleophile gave ring-opened products in >90% *ee* [15].

The majority of aliphatic secondary amines investigated proved to be unreactive under the standard conditions. Competition experiments revealed that the lack of reactivity was due to deactivation of the catalyst by the amine. As amines are known to be good ligands for rhodium, it is likely that the catalyst poisoning may be due to a binding interaction of the basic amine to the metal center [16]. A solution to this problem was found by a combination of altering the rhodium halide ligand and adding a proton source. Using the [Rh^I-I]/PPF-*t*Bu_2 catalyst in combination with

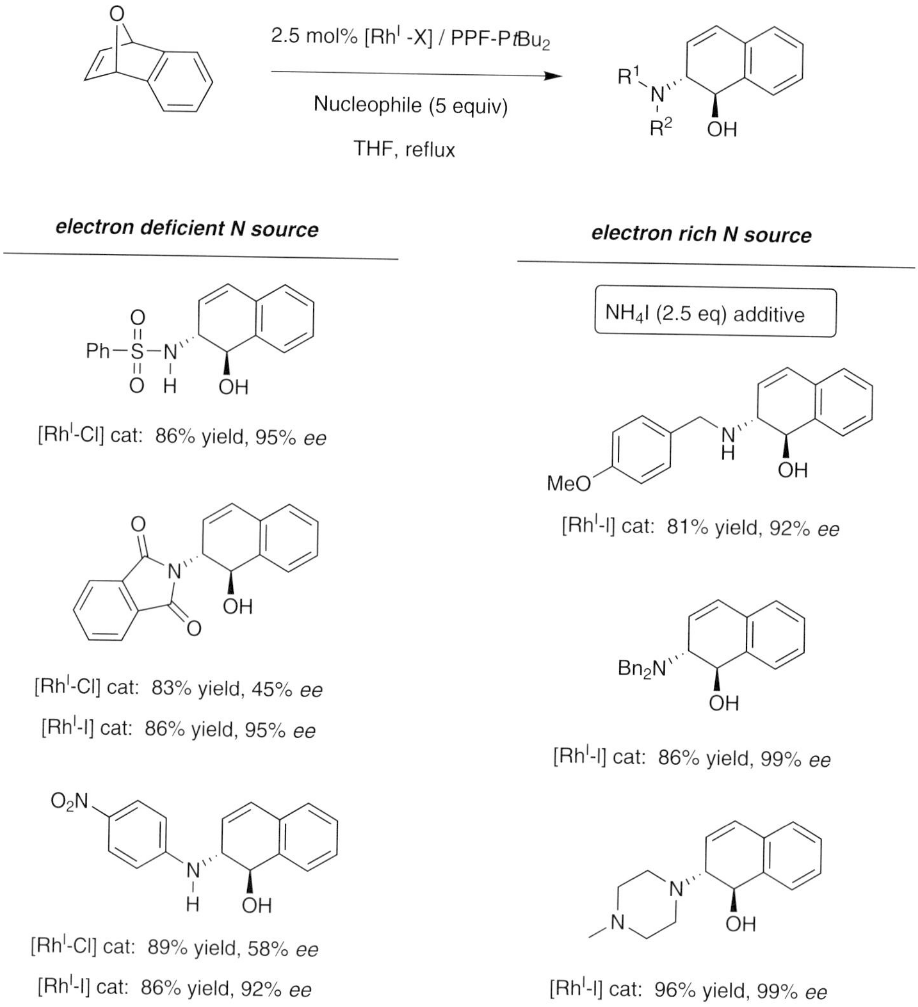

2.5 mol% [Rh^I-X] / PPF-PtBu$_2$
Nucleophile (5 equiv)
THF, reflux

electron deficient N source

[Rh^I-Cl] cat: 86% yield, 95% *ee*

[Rh^I-Cl] cat: 83% yield, 45% *ee*
[Rh^I-I] cat: 86% yield, 95% *ee*

[Rh^I-Cl] cat: 89% yield, 58% *ee*
[Rh^I-I] cat: 86% yield, 92% *ee*

electron rich N source

NH_4I (2.5 eq) additive

[Rh^I-I] cat: 81% yield, 92% *ee*

[Rh^I-I] cat: 86% yield, 99% *ee*

[Rh^I-I] cat: 96% yield, 99% *ee*

Scheme 10.8

NH_4I alleviated catalyst poisoning allowing the possibility of ring-opening with aliphatic amine nucleophiles. While a variety of acids were examined, the use of NH_4I consistently gave the highest yields and enantioselectivities, and it was also noted that the work-up procedure was simplified as NH_4I is water soluble.

The less reactive oxabicyclo[2.2.1.]heptene **18** can also be ring-opened by nitrogen nucleophiles, such as *N*-methylaniline, to give the amino alcohol **19** as a single diastereoisomer in 93% yield and 95% *ee* (Scheme 10.9) [15]. As with alcohol nucleophiles, the [Rh^I-I]/PPF-tBu$_2$ catalyst in solvent-less conditions was required for efficient ring-opening.

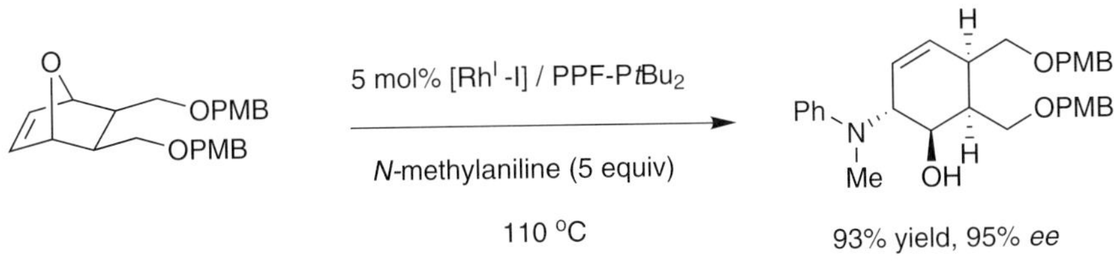

5 mol% [Rh^I-I] / PPF-P*t*Bu_2

N-methylaniline (5 equiv)

110 °C

93% yield, 95% *ee*

18 → 19

Scheme 10.9

10.4
Ring-Opening *meso*-Azabicyclic Alkenes with Nitrogen-Based Nucleophiles

The ARO of azabenzonorbornadienes by nitrogen nucleophiles is an important reaction as it allows access to a regio- and stereo-defined diaminotetralin core. These diamines not only exhibit biological activity, but can also be used as ligands in asymmetric transformations. There are two major differences, and, therefore, obstacles to overcome when comparing the ring-opening of oxa- and azabicyclic alkenes. (i) Nitrogen is a poorer leaving group than oxygen, therefore azabicyclic alkenes are less prone to ring-opening. (ii) During the course of the reaction a nucleophilic nitrogen anion is generated, which could potentially lead to the formation of oligomerization by-products or poisoning of the catalyst. We found that azabenzonorbornadienes could be ring-opened by nitrogen nucleophiles using Rh^I-catalysts with the nitrogen activating group (R in Scheme 10.10) playing an important role in determining the success of the reaction [17]. While azabenzonorbornadienes with *N*-alkyl activating groups, such as methyl or benzyl, proved to be unreactive, it was demonstrated that with electron withdrawing groups on nitrogen,

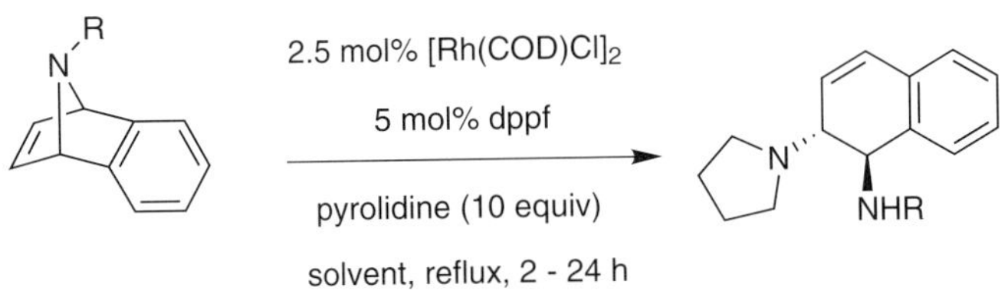

2.5 mol% $[Rh(COD)Cl]_2$

5 mol% dppf

pyrolidine (10 equiv)

solvent, reflux, 2 - 24 h

R	solvent	additive	yield, %
Tos	THF	$Et_3N.HCl$	91
Nos	THF	NH_4I	94
Boc	THP	$Et_3N.HCl$	82

Scheme 10.10

such as carbamates and sulfonamides the products could be formed with good conversion when ring-opening with pyrrolidine using the $[Rh(COD)Cl]_2$/dppf catalyst in combination with a protic additive (Scheme 10.10). In the case of the *N*-Boc azabenzonorbornadiene it was necessary to switch from THF to the higher boiling solvent tetrahydropyran (THP) and increase the reaction temperature to obtain high conversion. It was reasoned that the addition of the ammonium salt prevents catalyst poisoning, thereby lowering catalyst loading and the number of equivalents of amine nucleophile required for efficient ring-opening.

The development of an asymmetric version concentrated on the *N*-Boc-containing substrate as the removal of this group to liberate the free amine is usually straightforward and high yielding. Screening experiments showed that Ferriphos ligand **20** in combination with a triethylammonium halide gave the best enantioselectivity (Scheme 10.11). The nature of the halide of the ammonium salt again influenced the enantioselectivity of the reaction. Interestingly, the *ee* increased in the order Cl>Br>I, which is the inverse relationship relative to the ring-opening of oxabicyclic alkenes with amines. It was also noted that the Rh^{I} to ligand ratio also proved to be important in terms of enantioselectivity. The optimal ratio of Rh:ligand was found to be 1:2.2. It was reasoned that the excess chiral ligand in solution minimizes catalyst–nucleophile interactions which presumably cause lower enantioselectivity during the enantiodiscriminating step. This methodology was applied in the synthesis of the κ-opioid agonist **22** [17a], chiral 1,2-diamine ligands, such as Trost-type ligand **23** [17c] and to prepare 1-aminotetralin-type scaffolds that were used to make libraries of amines (such as **24**) to screen against opioid receptors [17d].

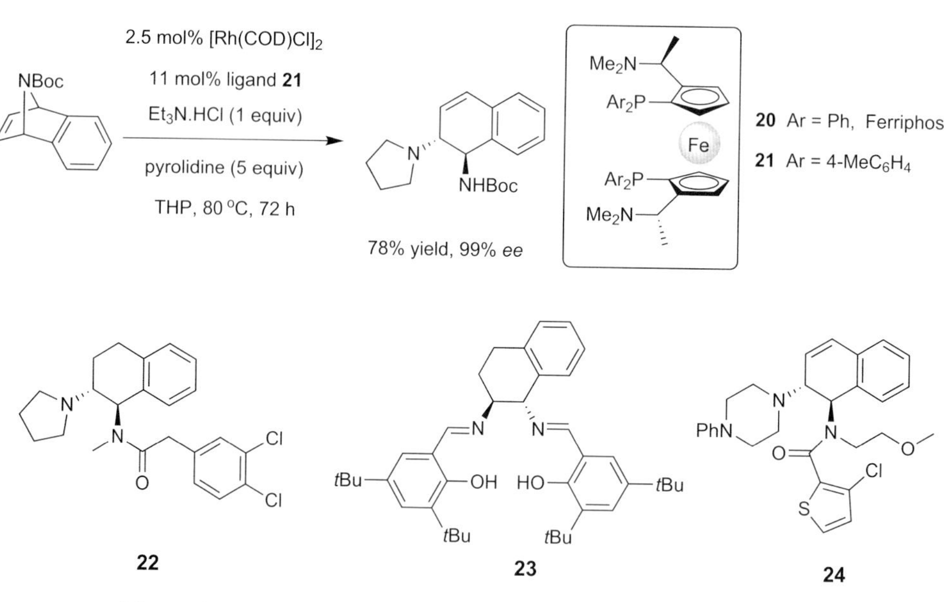

Scheme 10.11

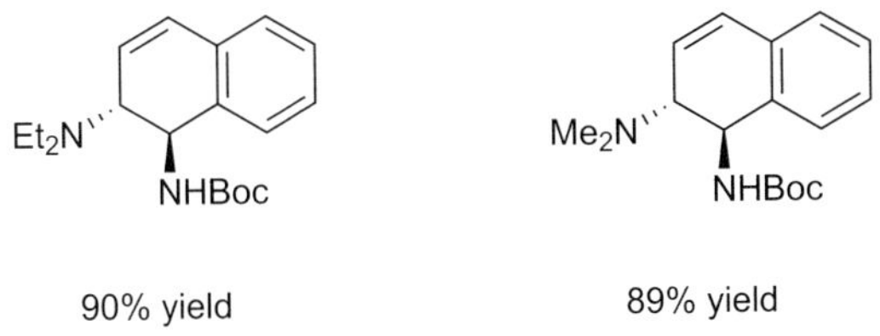

Et_2N NHBoc — 90% yield

Me_2N NHBoc — 89% yield

conditions: 1 mol% $[Rh(COD)Cl]_2$, 3 mol% dppf, R_2NH_2I (2 equiv), $i Pr_2NEt$, dioxane, 110 °C, 14 h

Scheme 10.12

Volatile aliphatic secondary amines such as diethylamine (bp 55 °C) and dimethylamine (bp 7 °C) have been used to ring-open azabenzonorbornadienes (Scheme 10.12). To simplify the practical procedure these amines were added as their corresponding hydrogen iodide salts [17b].

We later found that primary and secondary amines can ring-open azabenzonorbornadienes under solventless conditions (Scheme 10.13) [17b]. In these cases it was not necessary to include any additives to obtain high conversion and enantioselectivity. Both 6,7- and 5,8-disubstituted azabenzonorbornadienes could be used as substrates for this reaction, though longer reaction times were required for the 5,8- disubstituted analogues, probably due to greater steric hindrance.

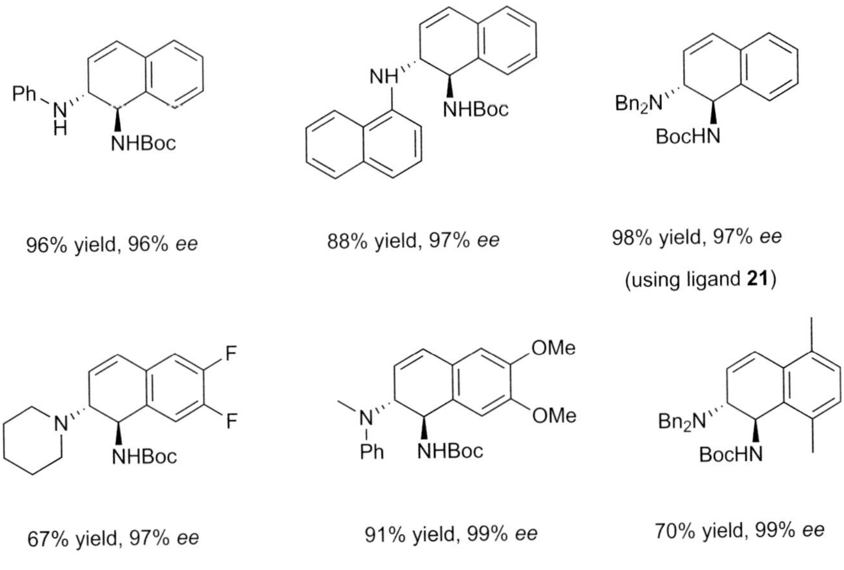

Ph NH, NHBoc — 96% yield, 96% *ee*

NH, NHBoc — 88% yield, 97% *ee*

Bn_2N, BocHN — 98% yield, 97% *ee* (using ligand **21**)

N, F, F, NHBoc — 67% yield, 97% *ee*

N, Ph, NHBoc, OMe, OMe — 91% yield, 99% *ee*

Bn_2N, BocHN — 70% yield, 99% *ee*

conditions: 2.5 mol% $[Rh(COD)Cl]_2$, 11 mol% **20**, nucleophile (5 equiv), 80 °C, 24 - 72 h

Scheme 10.13

Chiral amines derived from (*S*)-proline were used to ring-open azabenzonorbornadienes using an achiral $[Rh(COD)Cl]_2$/dppf catalyst (Scheme 10.14) [18]. While only marginal diastereoselectivity was observed the resulting enantiopure diastereoisomers could be readily separated and then converted into unique cyclic

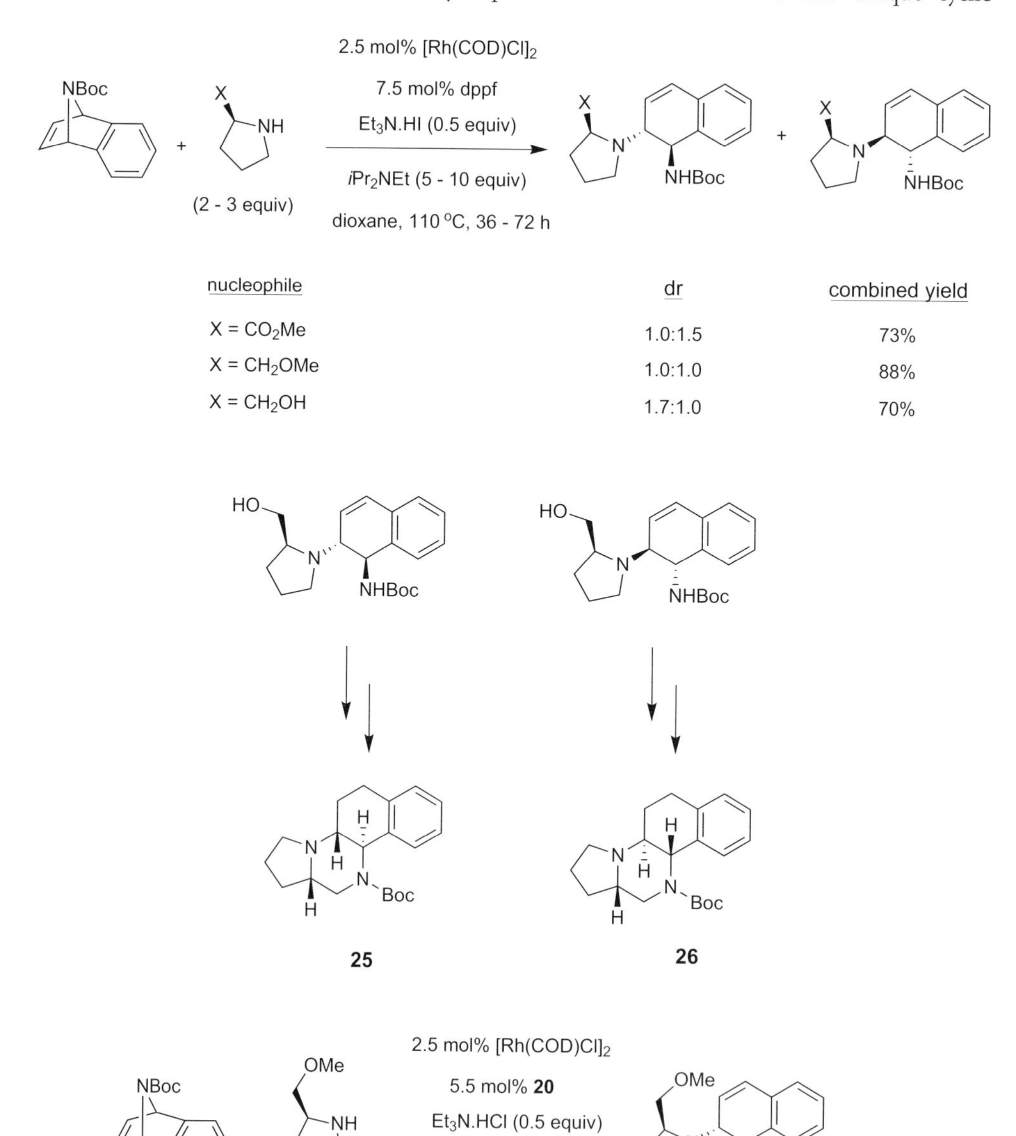

Scheme 10.14

compounds, such as **25** and **26**. Interestingly, in the case of (*S*)-prolinol, complete chemoselectivity was observed between the amino and hydroxy nucleophilic centers. When carrying out the reaction with the enantiomerically pure Ferriphos ligand **20** only isomer **27** was observed in 93% yield.

10.5 Ring-Opening *meso*-Oxabicyclic Alkenes with Sulfur-Based Nucleophiles

In contrast to oxygen- and nitrogen-based nucleophiles, the high polarizability and redox potential of sulfur-containing nucleophiles make them more prone to catalyst poisoning and background reactions. This was demonstrated in the Rh^{I}-catalyzed ring-opening of oxobenzonorbornadiene with thiophenol whereby the resulting dihydronaphthalene was formed in only 50% yield along with by-products 1-naphthol, sulfide **28** and disulfide **29** (Scheme 10.15) [19]. As with carboxylate- and nitrogen-based nucleophiles, halide counterions and protic additives had a dramatic influence on the *ee* of the product. The highest enantioselectivity was observed using the $[Rh^{I}\text{-}I]$/PPF-*t*Bu_2 catalyst in combination with the protic additive NH_4I, giving the ring-opened product in 94% *ee*.

Suppression of the background reactions was achieved by a number of modifications to the reaction procedure. These included adding the thiol to the reaction flask prior to the catalyst and then slowly adding the substrate to the reaction mixture. The addition of a radical inhibitor, such as galvinoxyl, further improved the yield of the ring-opened product. By employing these changes it was demonstrated that electron rich, electron deficient and hindered aryl sulfides could be added in high yield (up to 92%) and high enantioselectivity (>92% *ee*, Scheme 10.16). While it was demonstrated that some functionalized alkyl thiols successfully ring-opened oxobenzonorbornadiene, the majority of alkyl thiols examined failed to induce ring-opening.

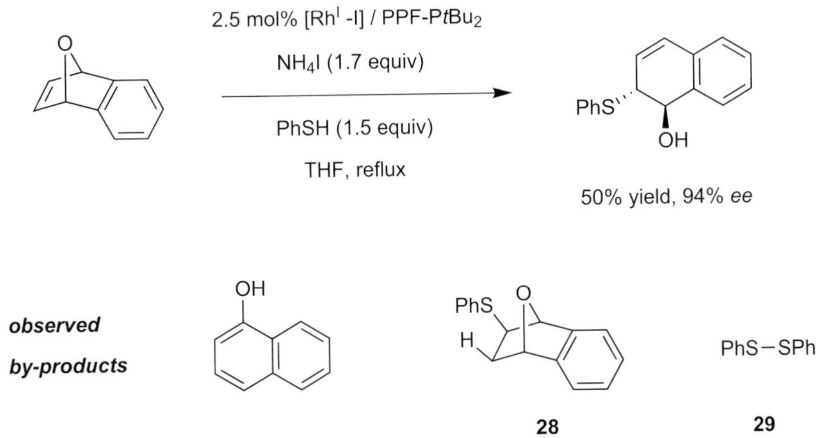

2.5 mol% [Rh^{I} -I] / PPF-P*t*Bu$_2$

NH$_4$I (1.7 equiv)

PhSH (1.5 equiv)

THF, reflux

PhS OH

50% yield, 94% *ee*

observed by-products

OH

PhS O H

28

PhS–SPh

29

Scheme 10.15

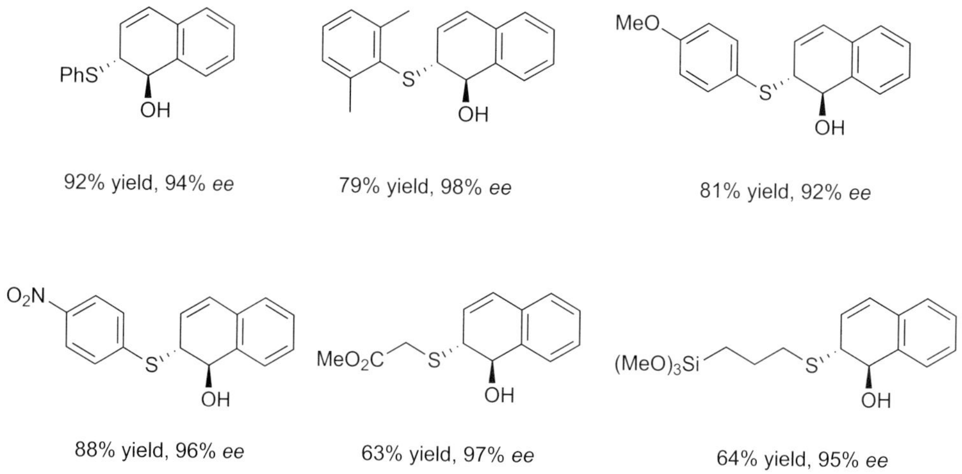

PhS OH — 92% yield, 94% ee

79% yield, 98% ee

MeO — 81% yield, 92% ee

O_2N — 88% yield, 96% ee

MeO_2C — 63% yield, 97% ee

$(MeO)_3Si$ — 64% yield, 95% ee

conditions: **7**, 2.5 mol% $[Rh(COD)Cl]_2$, 6 mol% PPF-P*t*Bu_2, NH_4I (1.7 equiv), thiol (1.5 equiv), THF, reflux, 6-24 h, 5 mol% Galvinoxyl or slow addition of **7**

Scheme 10.16

Competition experiments ruled out catalyst poisoning in these cases as competent nucleophiles, such as thiophenol, gave the desired product in the presence of the unreactive alkyl thiols.

10.6 Mechanistic Model

The proposed catalytic cycle for ring-opening oxa- or aza-bicyclic alkenes with heteroatom nucleophiles using the $[Rh(COD)Cl]_2$/bidentate phosphine (usually PPF-P*t*Bu_2**10**) catalyst is shown in Scheme 10.17. Initially the dimeric rhodium precursor **30** is cleaved by solvent, to give the monomeric rhodium species **31**. Reversible coordination of **31** to the least hindered exo face of the double bond of oxa- or aza-bicyclic alkene **32** gives rise to the Rh^I species **33**. This mode of binding may be further strengthened by rhodium metal coordination to the heteroatom of the oxa- or aza-bicyclic alkene. Oxidative insertion into one of the two enantiotopic C–X bonds with retention of stereochemistry results in the σ-Rh^{III} complex **34**. This step is likely to be irreversible due to the release of ring-strain. σ-Rh^{III} species **34** may be in equilibrium with the π-allyl **35** or σ-rhodium species **36**. The latter seems less likely due to the ring-strain associated with the [4.2.0] structure. Protonation of the oxa- or aza-bicyclic alkene's heteroatom by the nucleophile increases the electrophilic character of the organorhodium species and simultaneously increases the nucleophility of the heteroatom nucleophile. Ring-opening of **37** in an S_N2' fashion relative

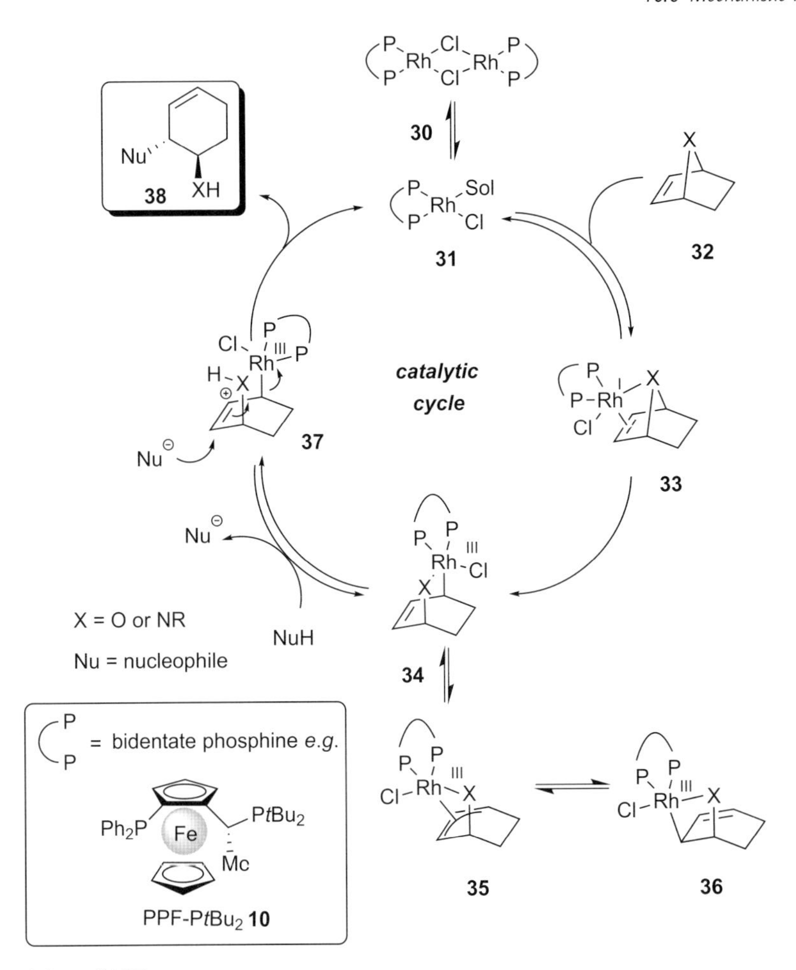

Scheme 10.17

to the rhodium metal with inversion of stereochemistry liberates the observed product **38** and regenerates the catalyst. The protonation of the oxa- or aza-bicyclic alkene's heteroatom by the nucleophile during the catalytic cycle explains the results in competitive reactions whereby the more acidic alcohol (e.g., HFIP over *i*PrOH) reacts preferentially.

For ring-opening when protic additives are required, the rhodium dimer is proposed to enter two different catalytic cycles (Scheme 10.18). During the productive cycle (catalytic cycle A) the dimer **39**, after solvation and complexation, oxidatively inserts into one of the two carbon–heteroatom bonds of the oxa- or aza-bicyclic alkene. A proton transfer from the nucleophile to the organorhodium species **40** gives the electrophilic species **41**, which can then be ring-opened in an S_N2' fashion by the nucleophile to give the observed product **42**. Alternatively, rhodium dimer **39** is cleaved by the nucleophile to give the stable rhodium–nucleophile complex **43**

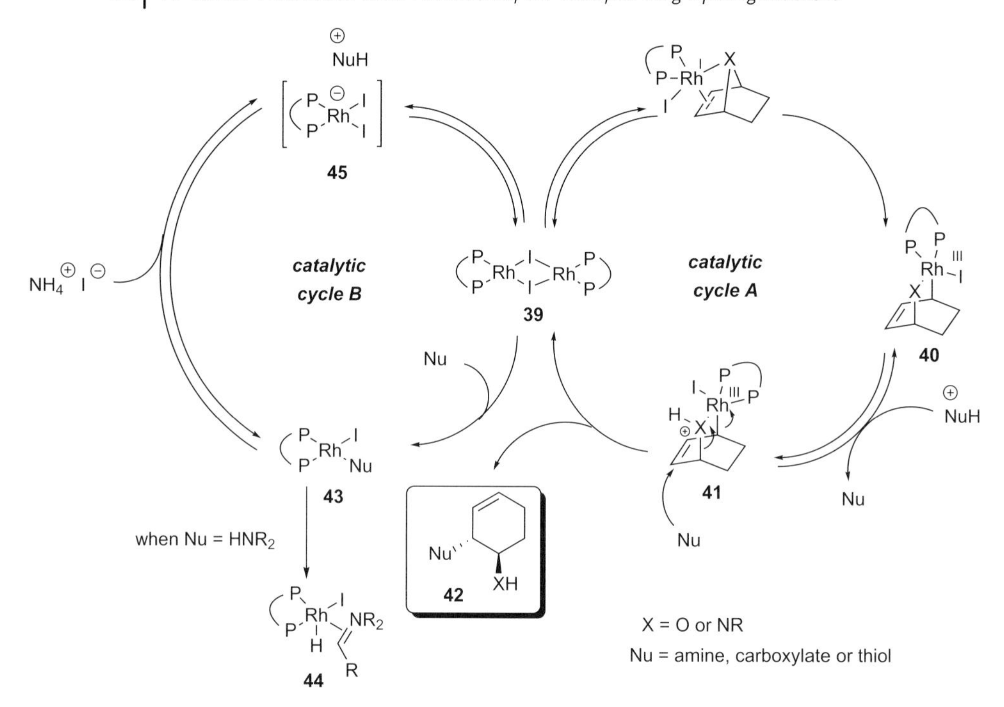

Scheme 10.18

(catalytic cycle B). In the absence of a protic additive, **43** would remain intact, inhibiting catalyst turnover. Amine–rhodium complexes are also known to undergo β-elimination to form imine complexes such as **44**. If this was to occur, catalyst poisoning would result. In the presence of a protic additive, rhodium–nucleophile complex **43** can covert to dihalorhodate **45**, which can then reform the dimeric complex **39** by reaction with another monomer or go on to react with another substrate molecule with loss of one of the halide ligands. Regeneration of the active catalyst by a proton source is supported by Vallarino's observation that HCl reacts with rhodium–amine complexes to generate anionic dihalorhodium species [20].

The proposed enantiodiscriminating step is the oxidative insertion of Rh^I into one of the two bridgehead carbon–heteroatom bonds of the substrate. As this step occurs prior to nucleophilic attack, the nature of the nucleophile should not influence the enantioselectivity. However, when using the $[Rh(COD)Cl]_2$/PPF-tBu$_2$ catalyst, the *ee* of the product can vary greatly, depending on the nucleophile used. To explain this phenomenon it is reasonable to propose a nucleophile–catalyst interaction (Scheme 10.19). If the nucleophile is bound to the catalyst during the enatiodiscriminating step, the resulting perturbation of the coordination sphere can impact the chiral space affecting the stereoselectivity of the reaction. Changing the catalyst halide ligand and the addition of protic additives may maintain the catalyst's integrity and therefore uphold high enantioselectivity.

Y = halide or excess ligand

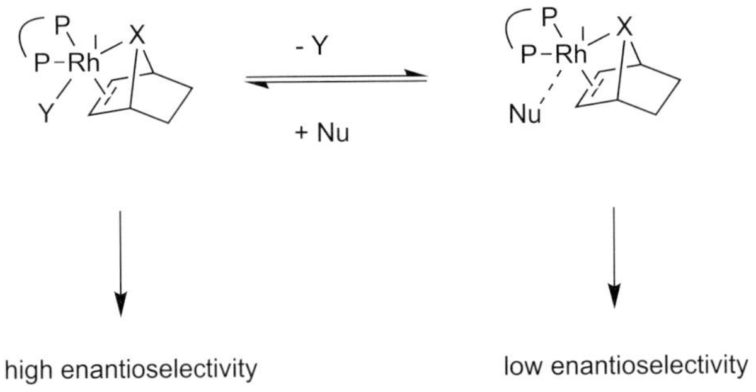

- Y

+ Nu

high enantioselectivity

low enantioselectivity

Scheme 10.19

10.7 Ring-Opening Unsymmetrical Oxa- and Aza-bicyclic Alkenes with Heteroatom Nucleophiles

Ring-opening a symmetrical *meso*-oxabicyclic alkene catalyzed by an achiral rhodium–phosphine complex ultimately leads to a racemic product, as the achiral rhodium catalyst does not distinguish between either of the two enantiotopic carbon–oxygen bonds during the oxidative insertion step. When an achiral rhodium–phosphine complex catalyzes the ring-opening of an unsymmetrically substituted chiral oxabicycle of generic structure **46**, depending on which carbon–oxygen bond rhodium inserts into, two regioisomers **47** and **48** can be potentially formed (Scheme 10.20).

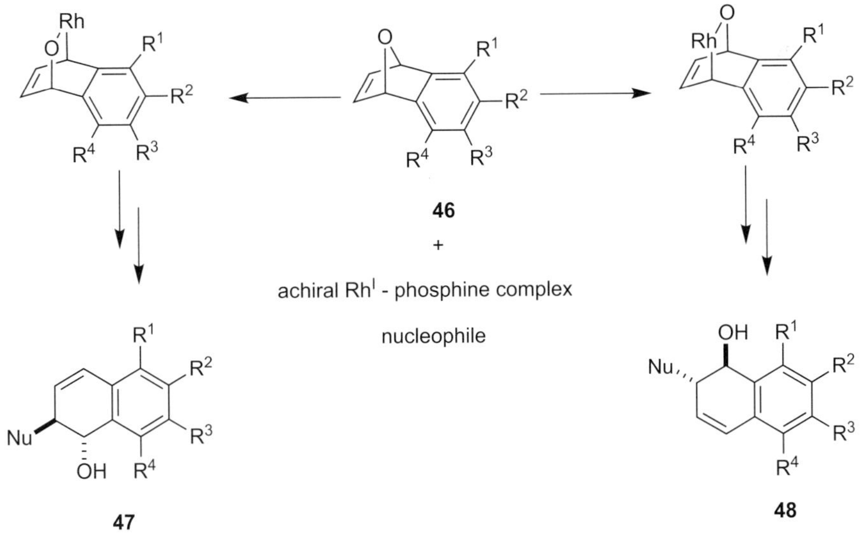

46

\+

achiral Rh^{I} - phosphine complex

nucleophile

47

48

Scheme 10.20

Table 10.1 Ring opening unsymmetrical oxabicyclical alkenes with heteroatom nucleophiles using an achiral Rh^I-catalyst.

2.5 mol% $[Rh(COD)Cl]_2$, 5 mol% dppf, NuH (5 equiv), THF, reflux, 1 - 3 h; **46** → **47** + **48**

Entry	R^1	R^2	R^3	R^4	NuH	Yield 47 + 48, %	Ratio, 47:48
1	OMe	H	OMe	H	MeOH	80	>25 : 1
2[a]	OMe	H	OMe	H	Bn_2NH	77	11 : 1
3[a]	H	H	$N(C_4H_8)$	H	Ph(Me)NH	86	>25 : 1
4	H	H	N(Me)Ph	H	MeOH	58	>25 : 1
5	H	H	OMe	H	MeOH	89	12 : 1
6	OMe	H	H	H	MeOH	81	3.5 : 1
7	H	Ac	H	H	MeOH	83	1.05 : 1
8	H	H	H	Me	MeOH	86	1.05 : 1
9	H	H	CF_3	H	MeOH	84	1.05 : 1

a) NH_4I (2.5 equiv) as additive.

In 2002 we reported that the nature of the substituents on the aromatic ring can greatly influence which of the two potential product isomers is preferentially formed [21]. Ring-opening substrates with strong π-donating substituents, containing either oxygen or nitrogen atoms, using alcohol or amine nucleophiles led to high regioselectivity (Table 10.1, entries 1–7). In certain cases, we found that only one regioisomer product was observed (entries 1, 3 and 4). Due to greater steric crowding, the regioselective bias towards the electronically favored product is reduced when the π-donating substituent is in close proximity to the carbon–oxygen bond that Rh^I will insert into (comparing entries 5 and 6). The regioselectivity of the ring-opening suggests that ionization of the carbon–oxygen bond is a key step in the catalytic cycle. Substrates with π-withdrawing groups (acetyl, entry 7), σ-withdrawing groups (CF_3 group, entry 8) and σ-donating groups (methyl group, entry 9) resulted in near equal amounts of each regioisomer.

The results obtained when using unsymmetrical substrates support the proposed mechanism. The remote electronic effects mediated by an aromatic π-system favor the cleavage of carbon–oxygen bond *b* in substrate **49** (Scheme 10.21). Bond *b* should be more prone to cleavage as a result of the stabilization of the positive charge that results following ionization. The resulting organorhodium species **50** leads to the observed regioisomer **51**.

When ring-opening racemic oxabicyclic alkene **52** with phthalimide catalyzed by a chiral rhodium complex, the inherent substrate bias was overridden by the chiral rhodium catalyst, resulting in near equal mixture of regioisomers **53** and **54**, both formed in >95% *ee* (Scheme 10.22) [22]. Regioisomer **53**, a potential precursor to

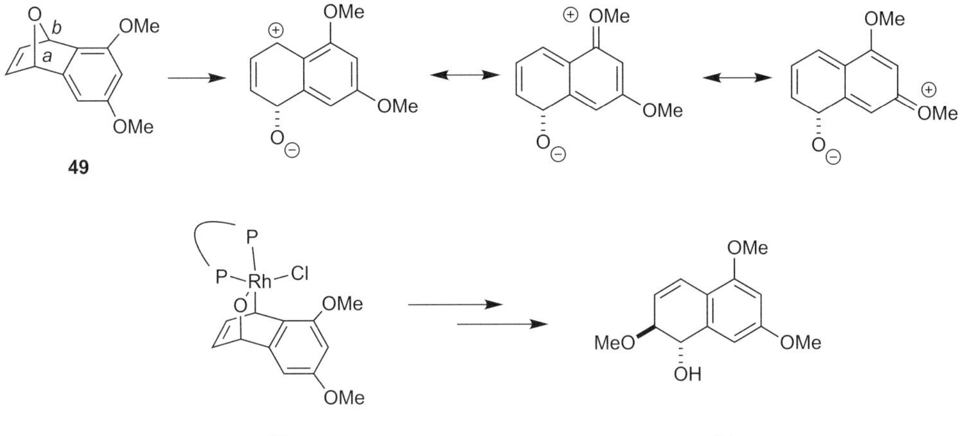

Scheme 10.21

a variety of active pharmaceutical ingredients, could be conveniently isolated from the reaction mixture by selective precipitation.

The effect of bridgehead substitution was also probed by reacting methyl-substituted oxa- and aza-bicyclic alkenes **55** and **56** with heteroatom nucleophiles (Scheme 10.23) [7, 17b]. In both cases the products resulting from carbon–oxygen or carbon–nitrogen bond cleavage at the more highly substituted bridgehead carbon were observed. The ionization of this carbon–heteroatom bond should be preferred from an electronic point of view since the tertiary carbocation will have greater stability. The oxidative insertion of the catalyst into a bridgehead carbon–heteroatom bond is considered the regiodetermining step in the catalytic cycle. As shown in the regioselective ring-opening of **56**, the formation of **57** (via **59**) supports oxidative insertion prior to nucleophilic attack.

Despite the dominant role exerted by bridgehead substituted oxabicyclic alkene substrates in determining ring-opening regioselectivity, it was demonstrated that a chiral Rh^{I} catalyst can override this bias (Scheme 10.24) [23]. Simply by switching

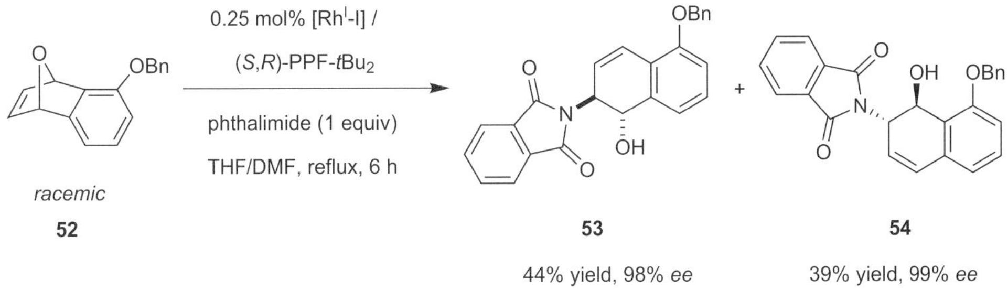

Scheme 10.22

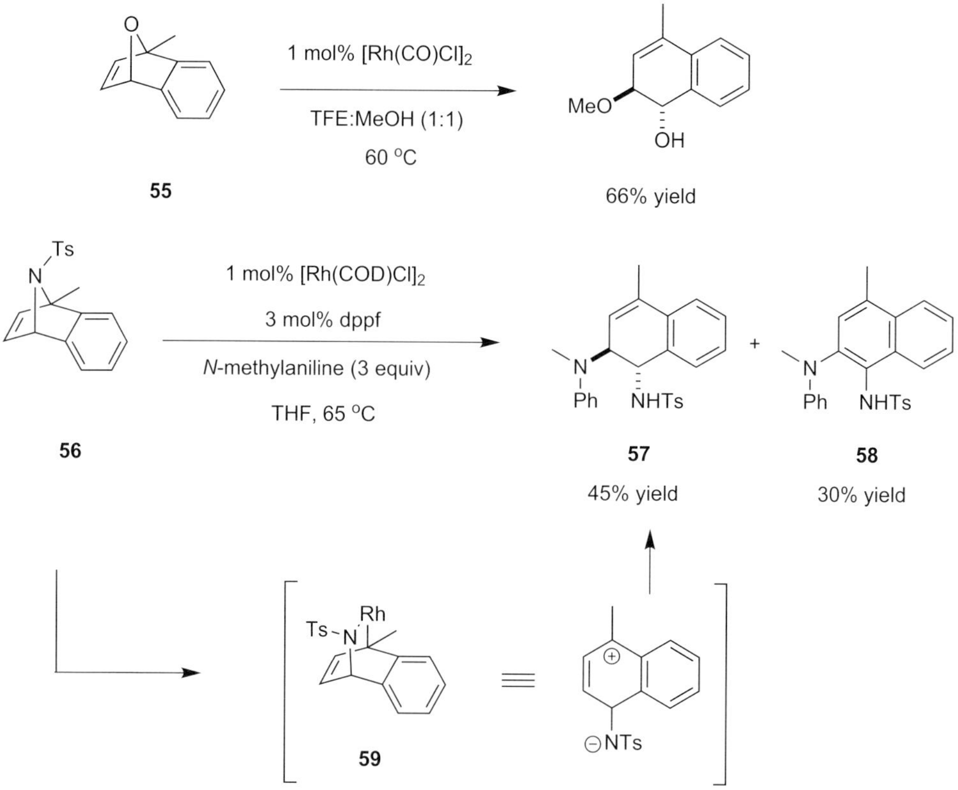

1 mol% $[Rh(CO)Cl]_2$
TFE:MeOH (1:1)
60 °C
55
MeO
OH
66% yield

1 mol% $[Rh(COD)Cl]_2$
3 mol% dppf
N-methylaniline (3 equiv)
THF, 65 °C
Ts
N
56
N
Ph
NHTs
57
45% yield
+
N
Ph
NHTs
58
30% yield
Ts
N
Rh
59
NTs

Scheme 10.23

5 mol% $Rh(COD)_2OTf$
5.5 mol% (*R*,*S*)-PPF-P*t*Bu_2
MeOH, THF, 80 °C, 12 h
OTBS
O
OH
61
75% yield, 95% *ee*
O
OTBS
60
95% *ee*
5 mol% $Rh(COD)_2OTf$
5.5 mol% (*S*,*R*)-PPF-P*t*Bu_2
MeOH, THF, 80 °C, 12 h
HO
OTBS
O
62
90% yield, 94% *ee*

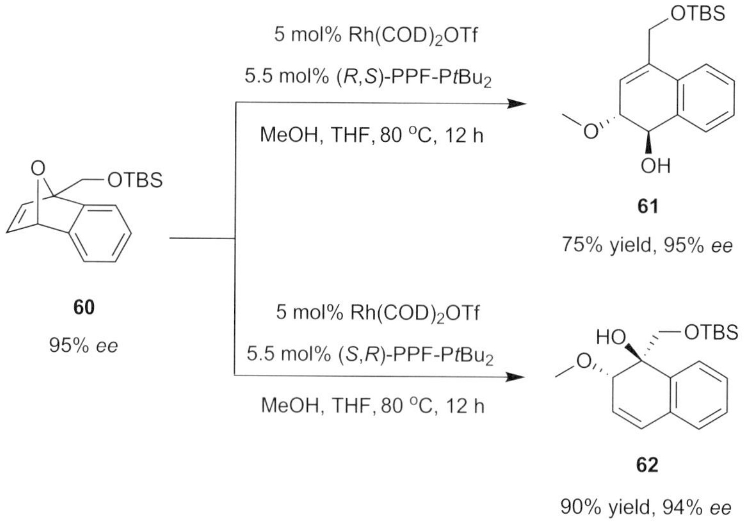

Scheme 10.24

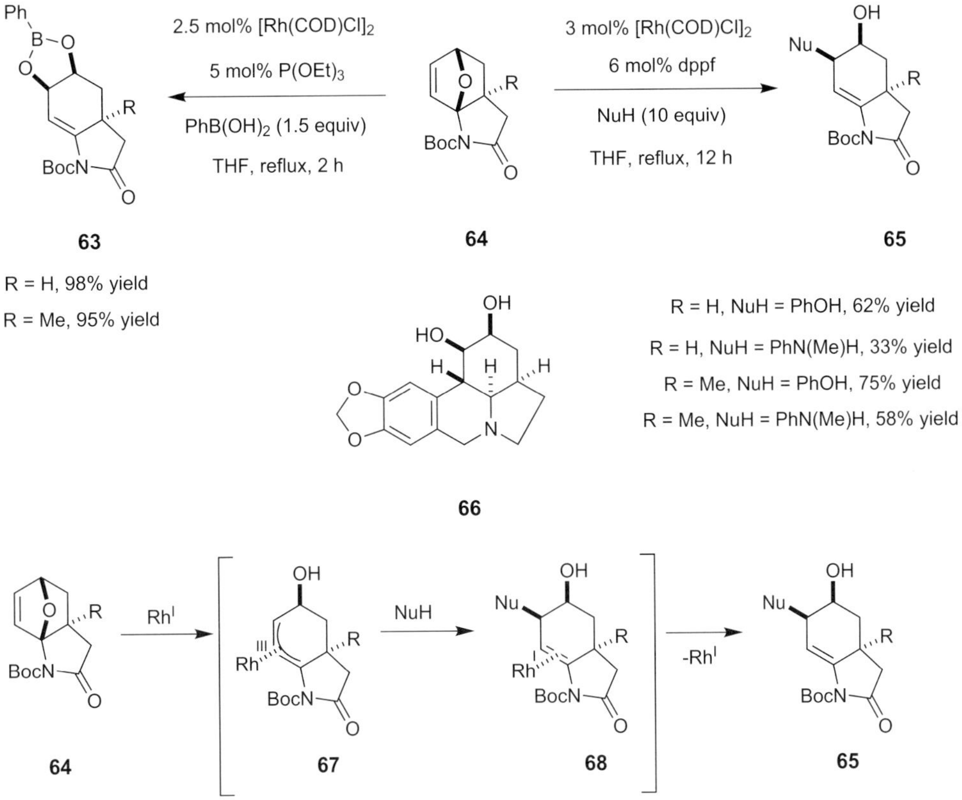

2.5 mol% $[Rh(COD)Cl]_2$, 5 mol% $P(OEt)_3$, $PhB(OH)_2$ (1.5 equiv), THF, reflux, 2 h: **64** → **63**

3 mol% $[Rh(COD)Cl]_2$, 6 mol% dppf, NuH (10 equiv), THF, reflux, 12 h: **64** → **65**

63: R = H, 98% yield; R = Me, 95% yield

65: R = H, NuH = PhOH, 62% yield; R = H, NuH = PhN(Me)H, 33% yield; R = Me, NuH = PhOH, 75% yield; R = Me, NuH = PhN(Me)H, 58% yield

66

64 → (Rh^I) → [**67** → (NuH) → **68**] → ($-Rh^I$) → **65**

Scheme 10.25

the absolute configuration of the chiral ligand when ring-opening enantioenriched oxabicyclic alkene **60** resulted in two different regiomeric products **61** and **62**. Evidently there is strong catalyst control during the mismatched reaction. The use of a cationic Rh^I triflate catalyst proved to be essential for this mode of reactivity.

During studies toward the synthesis of *epi*-zephyranthine **66**, Padwa also demonstrated that unsymmetrical oxabicyclic alkenes can be regioselectively ring-opened with heteroatom nucleophiles under rhodium catalysis (Scheme 10.25) [24]. Oxabicyclic alkene **64** was ring-opened with either phenol or aniline forming the corresponding carbon–oxygen or carbon–nitrogen bond. Interestingly, employing phenylboronic acid as the nucleophile resulted in boronate **63**. Rhodium has previously been reported to catalyze the ring-opening of oxabicyclic alkenes by boronic acids to form the corresponding carbon–carbon bonds via a transmetallation step [25]; however, in this case, due to the absence of base the boronic acid acts as an oxygen-based nucleophile. An X-ray structure established the cis-relationship between the nucleophile and hydroxy groups. The stereochemical outcome of these reactions is therefore opposite to that reported by us for the Rh^I-catalyzed aminolysis and alcoholysis of oxabenzonorbornadiene. It was proposed that rhodium coordinates to the alkenyl π-bond followed by nitrogen-assisted cleavage of the carbon–oxygen bond to furnish the π-allyl Rh^{III} species **67**. Subsequent nucleophilic addition occurs at the least hindered terminus of **67** and on the opposite side of the rhodium complex giving, after proton exchange, **68** and then ultimately the observed cis-isomer **63** or **65**.

10.8
Ring-Opening of Vinyl Epoxides with Heteroatom Nucleophiles

The transition metal-catalyzed ring-opening of vinyl epoxides by nucleophiles is a well established reaction. Most notably under palladium catalysis, the preference is for the nucleophile to add in a 1,4-manner *syn* to the leaving group, resulting in net S_N2' reaction with retention of stereochemistry [26]. In 2000 we demonstrated that rhodium can also catalyze the ring-opening of vinyl epoxides by heteroatom nucleophiles. However, to further highlight rhodium's unique reactivity with strained allylic systems, only products resulting from 1,2-addition were observed (Scheme 10.26) [27].

Although rhodium could be reacting as a Lewis acid, this possibility was discounted by control experiments which demonstrated that substrates such as cyclohexene oxide and styrene oxide were unreactive; indeed, this established that the alkene was essential for reactivity. Ultimately, the vinyl functionality activates the epoxide and acts as a directing group to influence the regiochemistry. While the previously reported ARO conditions failed to show reactivity, employing catalytic $[Rh(CO)_2Cl]_2$ without the need for phosphine ligands worked best. Both alcohols and aromatic amines can be added under neutral conditions and at room temperature to give only one regio- and diastereo-isomer of the corresponding *trans*-1,2-amino alcohol or alkoxy alcohol (Scheme 10.27).

The proposed catalytic cycle starts by coordination of rhodium to both the oxygen and alkene component of the vinyl epoxide, forming complex **69**, followed by oxidative insertion of rhodium into the allylic carbon–oxygen bond with retention of stereochemistry (Scheme 10.28). The initially formed Rh^{III} species **70** undergoes isomerization via the π-allyl Rh^{III} complex **71** to the less strained species **72**. Subsequent intermolecular nucleophilic S_N2' displacement with inversion affords the observed product **73** with concomitant release of Rh^I.

Additional support for this working model has been obtained by examining the intramolecular variant of this reaction (Scheme 10.29). In an attempt to prepare

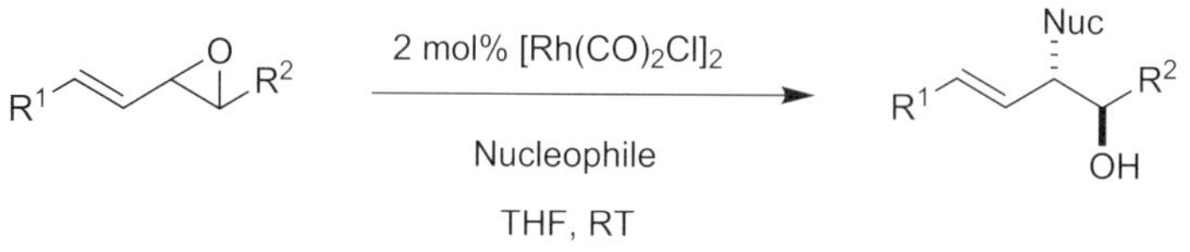

R1 … O … R2 —[2 mol% $[Rh(CO)_2Cl]_2$ / Nucleophile / THF, RT]→ R1 … Nuc … R2 … OH

Scheme 10.26

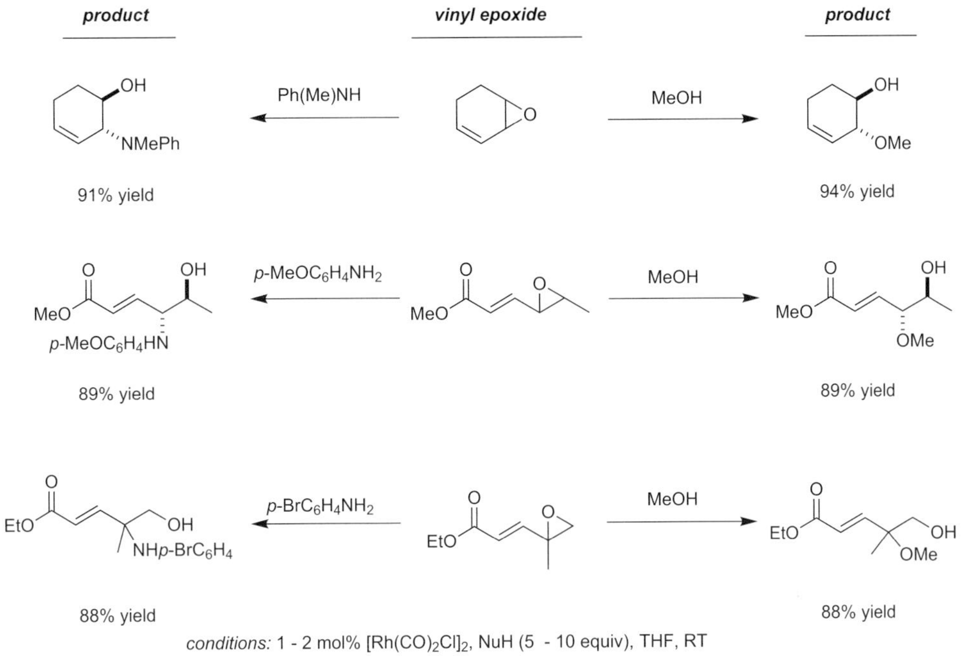

product		vinyl epoxide		product
OH / NMePh (cyclohexenol), 91% yield	Ph(Me)NH ←	cyclohexadiene monoepoxide	→ MeOH	OH / OMe (cyclohexenol), 94% yield
MeO, O, OH, p-$MeOC_6H_4HN$, 89% yield	p-$MeOC_6H_4NH_2$ ←	MeO, O, O	→ MeOH	MeO, O, OH, OMe, 89% yield
EtO, O, OH, NHp-BrC_6H_4, 88% yield	p-$BrC_6H_4NH_2$ ←	EtO, O, O	→ MeOH	EtO, O, OH, OMe, 88% yield

conditions: 1 - 2 mol% $[Rh(CO)_2Cl]_2$, NuH (5 - 10 equiv), THF, RT

Scheme 10.27

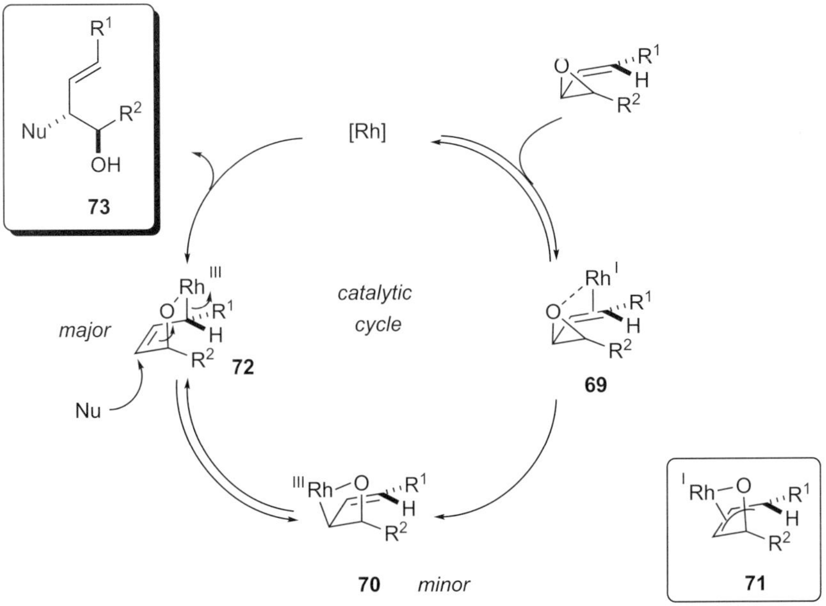

Scheme 10.28

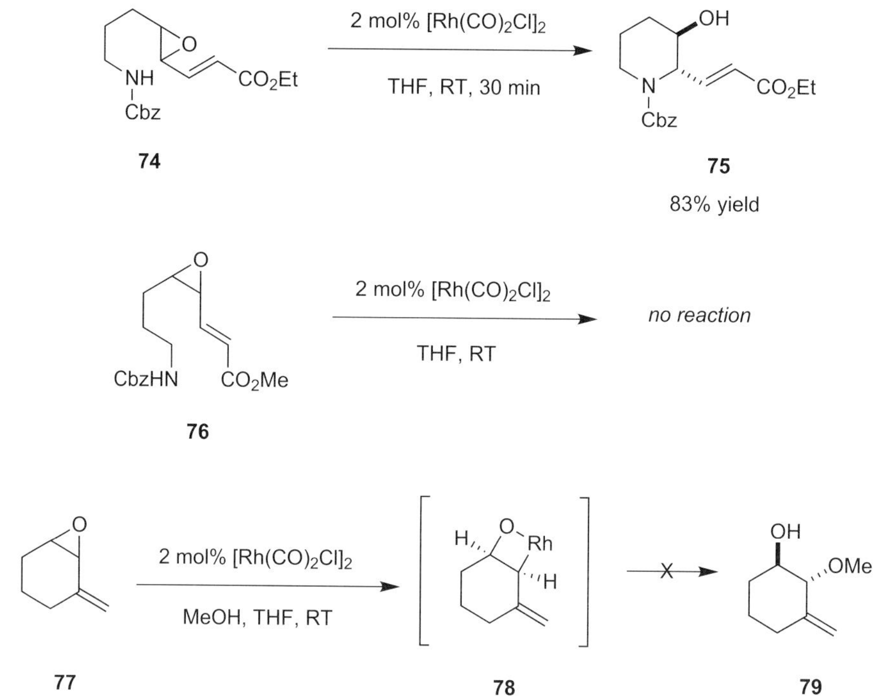

Scheme 10.29

5- and 6-membered heterocycles by this approach, Ha noted that *trans, trans*-vinyl epoxide **74** undergoes intramolecular addition in the presence of $[Rh(CO)_2Cl]_2$ to afford piperidine **75**, whereas the corresponding *cis, trans*-vinyl epoxide **76** failed to cyclize under the same reaction conditions [28]. The latter was attributed to the inability of rhodium to simultaneously coordinate the alkene and oxygen atom with the requisite geometry for the oxidative addition step. The position of the alkene was shown to be important as the intermolecular reaction of *exo*-methylene cyclohexene oxide **77** with MeOH failed to generate **79**, and resulted in intractable mixtures. Presumably, the more rigid framework of **78** does not allow isomerization to the oxametallacycle.

10.9 Conclusion

Rh^I-complexes demonstrate the ability to catalyze the ring-opening of strained allylic ethers and amines by heteroatom nucleophiles. The reaction produces either a new carbon–oxygen, carbon–nitrogen or carbon–sulfur bond via an intermolecular allylic displacement of an oxygen or nitrogen, which establishes the regioselectivity as well as the relative and absolute stereochemistry in one step. When ring-opening oxa- and aza-bicyclic alkenes it has been demonstrated that, with manipulation of the catalyst and reaction conditions, a large range of nucleophiles can react, usually leading to the *trans*-1,2-product. However, alteration of the substrate structure can lead to the *cis*-1,2-product. Meso-substrates can be ring-opened in high enantioselectivity using a Rh^I precursor in combination with a chiral diphosphine such as Josiphos or Ferriphos. Interesting reactivity patterns have also been observed ring-opening unsymmetrical oxa- and aza-bicyclic alkenes with heteroatom nucleophiles. The use of an achiral Rh^I–diphosphine catalyst can result in high regioselectively, whereby the electronics of the substrate control the product distribution, whilst this substrate bias can be overruled using a chiral Rh^I–diphosphine catalyst, resulting in a mixture of regioisomers formed in high enantiomeric excess or a single regioisomer, depending on the level of optical purity of the substrate. This methodology has been applied successfully in the enantioselective synthesis of several natural products, bioactive molecules and ligands that contain a tetralin core that are not readily available using classical transformations. Furthermore, the regio- and stereo-selective Rh^I-catalyzed ring-opening of vinyl epoxides by alcohols and amines readily leads to the *trans*-1,2-amino alcohol or alkoxy alcohol motif under neutral reaction conditions.

Acknowledgment

We thank the coworkers named in the references for their intellectual and experimental contributions. Tomislav Rovis and Keith Fagnou deserve special mention for their seminal discoveries. We thank Andrew Martins and Hans-Ulrich Blaser (Solvias AG) for helpful discussions when preparing this manuscript. We thank AstraZeneca, the ORDCF, the University of Toronto and NSERC for financial support.

References

1 (a) Chen, C.-L. and Martin, S.F. (2006) *J. Org. Chem.*, **71**, 4810; (b) Arrayas, R.G., Cabrera, S., and Carretero, J.C. (2006) *Synthesis*, **7**, 1205; (c) Rayabarapu, D.K. and Cheng, C.-H. (2007) *Acc. Chem. Res.*, **40**, 971.

2 For a review see: Lautens, M., Fagnou, K., and Hiebert, S. (2003) *Acc. Chem. Res.*, **36**, 48.

3 Hogeveen, H. and Middelkoop, T.B. (1973) *Tetrahedron Lett.*, 3671.

4 Ashworth, R.W. and Berchtold, G.A. (1977) *Tetrahedron Lett.*, **18**, 339.

5 (a) Johnson, B.M. and Chang, P.–T.L. (1996) *Analytical Profiles of Drug Substances and Excipients*, Elsevier, vol. 24, 443; (b) Snyder, S.E. (1995) *J. Med. Chem.*, **38**, 2395; (c) Kamal, A. and Gayatri, L. (1996) *Tetrahedron Lett.*, **37**, 3359; (d) Kim, K., Guo, Y., and Sulikowski, G.A. (1995) *J. Org. Chem.*, **60**, 6866; (e) Perrone, R. (1995) *J. Med. Chem.*, **38**, 942.

6 (a) Lautens, M., Fagnou, K., and Rovis, T. (2000) *J. Am. Chem. Soc.*, **122**, 5650; (b) Lautens, M., Fagnou, K., Taylor, M., and Rovis, T. (2001) *J. Organomet. Chem.*, **624**, 259.

7 (a) Lautens, M. and Fagnou, K. (2004) *Proc. Natl. Acad. Sci.*, **101**, 5455. For additional information on the mechanism see (b) Preetz, A., Kohrt, C., Prexter, H.-J., Torrens, A., Buschmann. H., Garcia

Lopez, M., Heller, D., *Adv. Syn. Catal.* DOI:10.1002/adsc.2000.
8 Braun, W., Müller, W., Calmuschi, B., and Salzer, A. (2005) *J. Organomet. Chem.*, **690**, 1166.
9 Lautens, M., Fagnou, K., and Taylor, M. (2000) *Org. Lett.*, **2**, 1677.
10 For similar binding patterns with transition metals, see: (a) Kitamura, M., Ohkuma, T., Inoue, S., Sayo, N., Kumobayashi, H., Akutagawa, S., Ohta, T., Takaya, H., and Noyori, R. (1988) *J. Am. Chem. Soc.*, **110**, 629; (b) Brenchly, G., Fedouloff, M., Merrifield, E., and Wills, M. (1996) *Tetrahedron: Asymmetry*, **7**, 2809.
11 Takeuchi, R. and Kitamura, N. (1998) *New J. Chem.*, **22**, 659.
12 Lautens, M. and Fagnou, K. (2001) *Tetrahedron*, **57**, 5067.
13 For a review of halide effects on enantioselectivity, see: Lautens, M. and Fagnou, K. (2002) *Angew. Chem. Int. Ed.*, **41**, 26.
14 Lautens, M. and Fagnou, K. (2001) *J. Am. Chem. Soc.*, **123**, 7170.
15 Lautens, M., Fagnou, K., and Yang, D. (2003) *J. Am. Chem. Soc.*, **125**, 14884.
16 Wilkinson, G. (1982) *Comprehensive Organometallic Chemistry*, Pergamon Press, vol. 5.
17 (a) Lautens, M., Fagnou, K., and Zunic, V. (2002) *Org. Lett.*, **4**, 3465; (b) Cho, Y.-h., Zunic, V., Senboku, H., Olsen, M., and Lautens, M. (2006) *J. Am. Chem. Soc.*, **128**, 6837; (c) Cho, Y.-h., Fayol, A., and Lautens, M. (2006) *Tetrahedron: Asymmetry*, **17**, 416; (d) Dockendorff, C., Jin, S., Olsen, M., Lautens, M., Coupal, M., Hodzic, L., Spear, N., Payza, K., Walpole, C., and Tomaszewski, M.J. (2009) *Bioorg. Med. Chem. Lett.*, **19**, 1228. For recent examples of carbon heteroatom bond formation catalysed by iridium see (e) Yang, D., Long, Y., Wang, H., and Zang, Z (2008) *Org. Lett.*, **10**, 4723; (f) Yang, D., Ping, H., Long, Y., Wu, Y., Zeng, H., Wang, H., and Zuo, X. (2009) Beilstein *J. Org. Chem.*, **5**, No. 53.
18 Cho, Y.-h., Tseng, N.-W., Senboku, H., and Lautens, M. (2008) *Synthesis*, **15**, 2467.
19 Lautens, M. and Leong, P. (2004) *J. Org. Chem.*, **69**, 2194.
20 Vallarino, L.M. and Sheargold, S.W. (1979) *Inorg. Chim. Acta.*, **36**, 243.
21 Lautens, M., Schmid, G.A., and Chau, A. (2002) *J. Org. Chem.*, **67**, 8043.
22 Fleming, M.J., Lautens, M., Thommen, M., and Spielvogel, D. (2009) *PCT Int. Appl.* WO 2009109648.
23 Webster, R., Böing, C., and Lautens, M. (2009) *J. Am. Chem. Soc.*, **131**, 444.
24 (a) Wang, Q. and Padwa, A. (2004) *Org. Lett.*, **6**, 2189; (b) Wang, Q. and Padwa, A. (2006) *J. Org. Chem.*, **71**, 3210.
25 Lautens, M., Dockendorff, C., Fagnou, K., and Malicki, A. (2002) *Org. Lett.*, **4**, 1311.
26 For recent reviews of catalytic asymmetric allylic substitution, see: (a) Trost, B.M. and Crawely, M.L. (2003) *Chem. Rev.*, **103**, 2921; (b) Pfaltz, A. and Lautens, M. (1999) Chapter 24, in *Comprehensive Asymmetric Catalysis*, vol. 2 (eds E.N. Jacobsen, A. Pfaltz, and H. Yamamoto), Springer, Berlin.
27 Fagnou, K. and Lautens, M. (2000) *Org. Lett.*, **2**, 2319.
28 Ha, J.D., Shin, E.Y., Kang, S.K., Ahn, J.H., and Choi, J.-K. (2004) *Tetrahedron. Lett.*, **45**, 4193.

11
Gold-Catalyzed Addition of Nitrogen and Sulfur Nucleophiles to C–C Multiple Bonds

Ross A. Widenhoefer and Feijie Song

11.1
Introduction

The catalytic addition of nitrogen nucleophiles to electronically unactivated C–C multiple bonds represents an attractive approach to the synthesis of functionalized amines, carboxamide derivatives and nitrogen heterocycles, with potential applications to target-oriented synthesis, pharmaceutical development, and fine chemical synthesis [1]. Although a number of transition metal complexes have been employed as catalysts for the amination of C–C multiple bonds, soluble gold complexes played little role in the development of this chemistry prior to 2001. Since this time, however, gold(III) salts and, to a greater extent, cationic gold(I) complexes have emerged as effective and highly versatile catalysts for the amination of C–C multiple bonds. The utility of gold complexes as catalysts for the amination of C–C multiple bonds stems largely from the pronounced carbophilic behavior of these complexes, which activates C–C bonds toward nucleophilic attack by amines and amine derivatives [2]. Even more recently, this gold-based methodology has been extended to the addition of sulfur nucleophiles to C–C multiple bonds. Herein we review the application of gold complexes as catalysts for the addition of nitrogen and sulfur nucleophiles to C–C multiple bonds. This review covers the literature through 2008 and is restricted to transformations that form a C–N or C–S bond via addition of the nucleophile to an electronically unactivated C–C multiple bond catalyzed by a well-defined, soluble gold complex.

11.2
Addition of Nitrogen Nucleophiles to Alkynes

11.2.1
Hydroamination

11.2.1.1 Intramolecular Processes

In one of the first examples of gold-catalyzed hydroamination, Utimoto reported the gold(III)-catalyzed intramolecular hydroamination of 5-alkynyl amines to form

Catalyzed Carbon-Heteroatom Bond Formation. Edited by Andrei K. Yudin

ISBN: 978-3-527-32428-6

tetrahydropyridine derivatives [3]. As an example, reaction of 5-dodecynylamine with a catalytic amount of sodium tetrachloroaurate dihydrate in refluxing acetonitrile for 1 h led to 6-*exo* cyclization to form the cyclic imine **1** in 80% isolated yield (Eq. (11.1)). The procedure was also effective for the intramolecular hydroamination of 5-alkynyl amines that possessed alkyl substitution at the C1 position of the 5-alkynyl chain and for the hydroamination of terminal alkynes. In addition, 4-alkynylamines and 3-alkynylamines underwent 5-*exo* and 5-*endo* cyclization, respectively, in the presence of a catalytic amount of $NaAuCl_4$ to form dihydropyrroles in good yield (Eq. (11.2)).

$NaAuCl_4 \cdot 2H_2O$ (5 mol %), CH_3CN, reflux, 1 h, 80%; R = *n*-hexyl; **1** (11.1)

$NaAuCl_4 \cdot 2H_2O$ (5 mol %), CH_3CN, reflux, ≥98%; R = *n*-C_7H_{15} (11.2)

Utimoto has also shown that simple gold(III) salts catalyze the intramolecular hydroamination of alkynes with arylamines [4]. For example, treatment of 2-(3,3-dimethyl-1-butynyl)aniline with a catalytic amount of sodium tetrachloroaurate in refluxing THF for 30 min led to isolation of 2-*t*-butylindole in 90% yield (Eq. (11.3)). Marinelli has modified and expanded the scope of Utimoto's procedure through employment of ethanol, ethanol/water [5], or ionic liquids [6] solvents. As an example of this modified protocol, treatment of 2-alkynylaniline **2** with a catalytic amount of $NaAuCl_4$ dihydrate in ethanol at room temperature for 6 h led to isolation of 2-(4-chlorophenyl)indole **3** in 92% yield (Eq. (11.4)). A similar hydroamination protocol employing $AuCl_3$ as a catalyst has been recently reported by Majumdar [7].

$NaAuCl_4$ (5 mol %), THF, reflux, 30 min, 90% (11.3)

$NaAuCl_4 \cdot 2H_2O$ (4 mol %), EtOH, 25 °C, 6 h, 92%; **2** → **3** (11.4)

Arcadi has reported the gold(III)-catalyzed hydroamination of alkynes with enaminones [8]. For example, treatment of **4** with a catalytic amount of $NaAuCl_4$ dihydrate in ethanol at 60 °C for 2 h led to cyclization and subsequent isolation of

1,2,3,5-tetrasubstituted pyrrole **5** in 75% yield (Eq. (11.5)). Alternatively, treatment of a 1 : 1.5 mixture of 2-(3-propynyl)-2,4-pentanedione and benzylamine with a catalytic amount of $NaAuCl_4$ dihydrate led to isolation of **4** in quantitative yield via sequential condensation/hydroamination (Eq. (11.5)). Dyker has reported the gold(III)-catalyzed intramolecular reaction of *o*-secondary benzyl amines with alkynes to form isoindoles and dihydroisoquinolines [9].

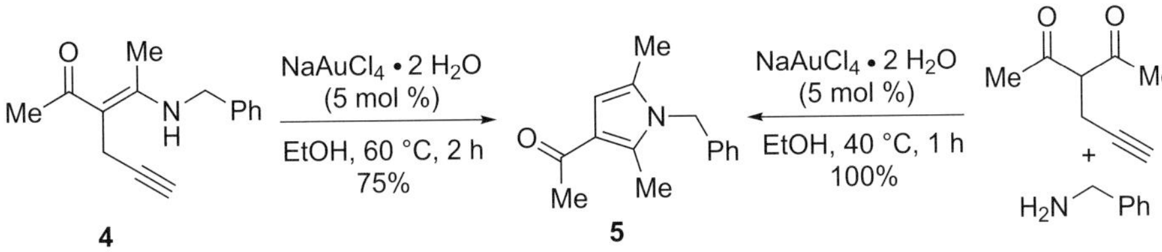

(11.5)

Shin has reported an effective gold(I)-catalyzed method for the intramolecular *exo*-hydroamination of propargylic and homopropargylic trichloroacetimidates to form five- and six-membered heterocycles, respectively [10]. As an example of the former, treatment of propargylic trichloroacetimidate **6** with a catalytic amount of $[P(C_6F_5)_3]AuCl$ and $AgSbF_6$ in 1,2-dichloroethane at 0 °C for 3 h led to isolation of 4-methylene-4,5-dihydrooxazole **7** in 87% yield (Eq. (11.6)). As an example of the latter transformation, treatment of homopropargylic trichloroacetimidate **8** with a catalytic amount of $(PPh_3)AuCl$ and $AgBF_4$ in DCE at 0 °C for 30 min led to isolation of the heterocycle **9** in 91% yield (Eq. (11.7)). The 6-*exo*, but not the 5-*exo*, transformation tolerated substitution at the terminal alkyne carbon atom and both transformations tolerated a range of substitution at the oxygen-bound sp^3 carbon atom. Similar transformations were reported near the same time by Hashmi [11].

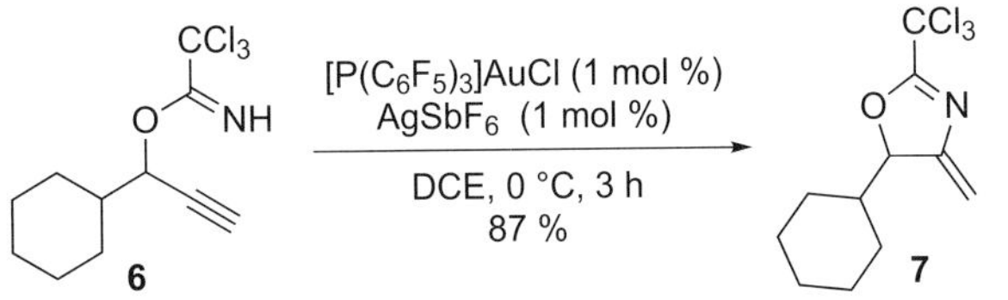

(11.6)

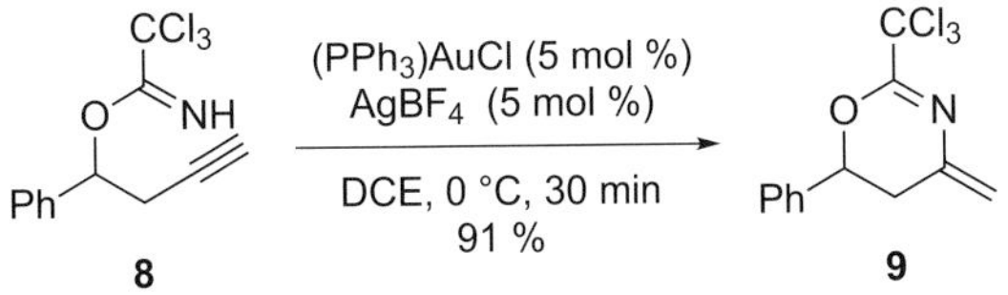

(11.7)

Carbamates are also effective nucleophiles for the gold(I)-catalyzed intramolecular hydroamination of alkynes [12]. For example, treatment of *O*-propargylic carbamate **10** with a catalytic 1 : 1 mixture of $(PPh_3)AuCl$ and AgOTf in dichloromethane at room temperature led to isolation of 2,5-dihydroisoxazole **11** in 88% yield (Eq. (11.8)). The transformation was effective for alkyl-, alkenyl- and aryl-substituted internal alkynes and terminal alkynes and for *N*–Boc, Cbz, and Ts derivatives. A similar transformation was developed by Schmalz [13]. As an example of this procedure, treatment of *O*-propargylic carbamate **12** with a catalytic 1 : 1 mixture of AuCl and NEt_3 in dichloromethane at room temperature for 1 h led to isolation of 4-methylene oxazolidinone **13** in 99% yield (Eq. (11.9)). Takemoto has reported a gold(I)-catalyzed protocol for the intramolecular hydroamination of alkynes with carbamates as an efficient route to the synthesis of hydroisoquinolines [14].

Boc, NH, O, *n*-Hexyl, **10** — $(PPh_3)AuCl$ (5 mol %), AgOTf (5 mol %), CH_2Cl_2, rt, 88 % → O, N–Boc, *n*-Hexyl, **11** (11.8)

O, O, N–Ts, H, **12** — AuCl (5 mol %), NEt_3 (5 mol %), CH_2Cl_2, rt, 99 % → O, O, N–Ts, **13** (11.9)

11.2.1.2 Intermolecular Processes

Tanaka has reported a gold(I)-catalyzed procedure for the intermolecular hydroamination of alkynes with aryl amines to form imines [15], which is modeled after the catalyst system optimized for the gold-catalyzed intermolecular hydration of alkynes [16]. As an example of Tanaka's procedure, treatment of a neat 1 : 1.1 mixture of phenylacetylene and *p*-bromoaniline with a catalytic 1 : 5 mixture of $(PPh_3)AuMe$ and $H_3PW_{12}O_{40}$ at 70 °C for 2 h formed ketimine **14** in 94% NMR yield (Eq. (11.10)). Both gold and acid were required for effective hydroamination and the reaction was quite sensitive to the nature of the acid promoter. The protocol tolerated a range of alkynes and anilines, but did not tolerate alkyl amines. The catalyst was highly stable and turnover numbers approaching 9000 were realized for the reaction of phenylacetylene with either 4-cyanoaniline or 4-bromoaniline.

+ H_2N, Br — $(PPh_3)AuMe$ (0.1 mol %), $H_3PW_{12}O_{40}$ (0.5 mol %), 70 °C, 2 h, 94% → Br, N, CH_3, **14** (11.10)

Li has reported that cationic gold(III) complexes also catalyze the intermolecular hydroamination of terminal alkynes with aniline derivatives [17]. For example, reaction of a neat 1 : 1.5 mixture of phenylacetylene and aniline catalyzed by $AuCl_3$ at room temperature followed by reduction with sodium borohydride led to isolation of phenyl-(1-phenylethyl)amine in 88% yield (Eq. (11.11)). The gold(III)-catalyzed intermolecular hydroamination/reduction of terminal alkynes was effective for both alkyl and aryl alkynes and for both electron-rich and electron-deficient anilines. Recently, Bertrand has demonstrated that cationic gold(I) complexes that contain a cyclic alkyl(amino)carbene (CAAC) ligand such as the Werner complex **15** catalyze the hydroamination of alkynes and allenes with ammonia at elevated temperatures (Eq. (11.12)) [18].

1) $AuCl_3$ (5 mol %) neat, 25 °C; 2) $NaBH_4$, 88% (11.11)

15 (5 mol %), 110 °C, 18 h, 60% (11.12)

15 (Ar = 2,6-diisopropylphenyl)

11.2.2
Acetylenic Schmidt Reaction

Toste has reported a gold(I)-catalyzed protocol for the intramolecular addition of an alkyl azide to an alkyne with loss of dinitrogen (acetylenic Schmidt reaction) to form pyrrole derivatives [19]. For example, treatment of homopropargyl azide **16** with a catalytic 1 : 2 mixture of (dppm)Au_2Cl_2 (dppm = 1,2-bis(diphenylphosphino)methane) and $AgSbF_6$ in methylene chloride at 35 °C led to isolation of 2,5-dibutylpyrrole in 82% yield (Eq. (11.13)). The method was also effective for the cyclization of α-unsubstituted homopropargyl azides and tolerated both alkyl and aryl substitution of the alkyne.

(dppm)Au_2Cl_2 (2.5 mol %), $AgSbF_6$ (5 mol %), CH_2Cl_2, 35 °C, 82% (11.13)

11.2.3
Tandem C–N/C–C Bond Forming Processes

Liu has reported the gold(III)-catalyzed three-component coupling of a heteroaromatic aldehyde, an amine, and an alkyne to form an aminoindolizine [20]. For example, heating a neat 1 : 1.1 : 1.2 mixture of pyridine-2-carboxyaldehyde, piperidine, and 4-methoxyphenylacetylene with a catalytic amount of $NaAuCl_4{\bullet}2H_2O$ at 60 °C for 4 h led to isolation of aminoindolizine **17** in 95% yield (Scheme 11.1). The transformation presumably proceeds via initial formation of propargylic amine derivative **18**, generated via a Mannich–Grignard reaction involving a gold acetylide species, followed by gold(I)-catalyzed amination of the C≡C bond with the pendant pyridine nitrogen atom (Scheme 11.1). A similar transformation was reported by Gevorgyan near the same time [21]. Liu's protocol required an aryl-substituted internal alkyne, but was effective for both dialkyl and alkyl/aryl substituted secondary amines. Three-component coupling employing enantiomerically enriched secondary amines occurred with retention of configuration.

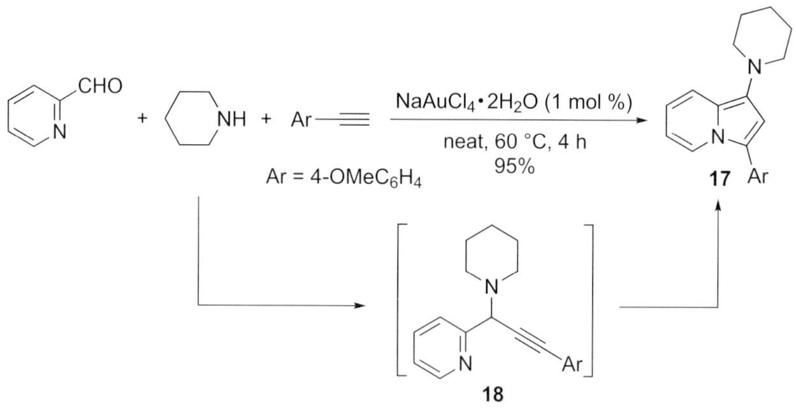

CHO + NH + Ar—≡ → $NaAuCl_4{\cdot}2H_2O$ (1 mol %), neat, 60 °C, 4 h, 95% → **17**

Ar = 4-OMeC_6H_4

18

Scheme 11.1

Li has reported a gold(I)-catalyzed procedure for the annulation of 2-tosylaminobenzaldehyde **19** with terminal aromatic alkynes to form aza-isoflavanones although extremely forcing conditions were required [22]. As an example, heating a toluene solution of **19** and phenylacetylene with a catalytic mixture of AuCN and PBu_3 in a sealed tube at 150 °C for 2.5 days led to isolation of azaisoflavanone **20** in 65% yield (Eq. (11.14)).

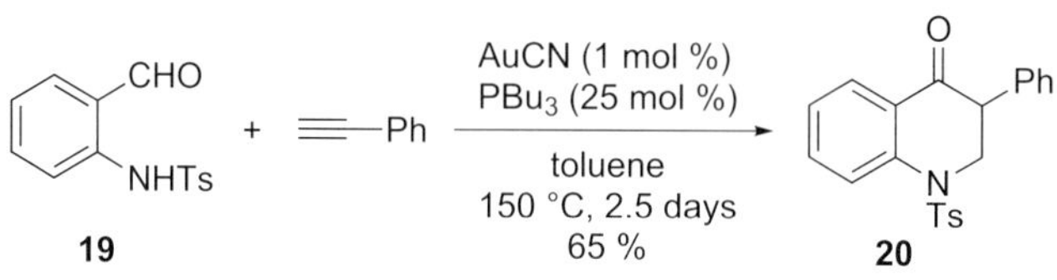

19 (CHO, NHTs) + ≡—Ph → AuCN (1 mol %), PBu_3 (25 mol %), toluene, 150 °C, 2.5 days, 65 % → **20** (Ph, N Ts) (11.14)

Che has reported the tandem hydroamination/hydroarylation of aromatic amines wirh terminal alkynes to form dihydroquinolines in which 1 equiv of aniline combines with 2 equiv of alkyne [23]. For example, reaction of 3-methoxyaniline with phenylacetylene (5 equiv) and a catalytic 1 : 1 mixture of the gold(I) *N*-heterocyclic carbene complex (IPr)AuCl (IPr = 1,3-bis(2,6-diisopropylphenyl)imidazol-2-ylidine) and AgOTf at 150 °C under microwave irradiation led to isolation of dihydroquinoline **21** in 82 % yield (Eq. (11.15)). Alternatively, reaction of *o*-acetylaniline with phenylacetylene catalyzed by a mixture of (IPr)AuCl and AgOTf at 150 °C led to isolation of the quinoline derivative **22** in 93% yield via incorporation of a single equivalent of alkyne (Eq. (11.16)). Arcadi has reported the gold(III)-catalyzed hydroamination/hydroarylation of 2-alkynylanilines with α,β-enones to form C3-alkyl indoles [24]. As an example of this transformation, treatment of 2-(phenylethynyl)aniline with 4-phenyl-3-buten-2-one and a catalytic amount of sodium tetrachloroaurate dihydrate in ethanol at 30 °C formed 1,2,3-trisubstituted indole **23** in 88% yield (Eq. (11.17)).

MeO, NH_2 + Ph—≡ → (IPr)AuCl (5 mol %), AgOTf (5 mol %), NH_4PF_6 (15 mol %); CH_3CN, 150 °C, MW, 25 min, 82% → MeO, H, N, Me, Ph, Ph **21**

IPr = i-Pr, i-Pr, N, N, i-Pr, i-Pr

(11.15)

NH_2, Me, O + Ph—≡ → (IPr)AuCl (5 mol %), AgOTf (5 mol %), NH_4PF_6 (15 mol %); CH_3CN, 150 °C, MW, 30 min, 93% → MeO, N, Ph, **22** Me (11.16)

Ph, NH_2 + O, Me, Ph → $NaAuCl_4 \cdot 2\,H_2O$ (5 mol %); EtOH, 30 °C, 88% → O, Me, Ph, Ph, N, Me **23**

(11.17)

Kirsch has reported the conversion of propargyl vinyl ethers to form pyrroles via a multi-step transformation involving the silver-catalyzed propargyl Claisen rearrangement to form the α-allenyl β-keto ester, condensation with a primary aryl amine,

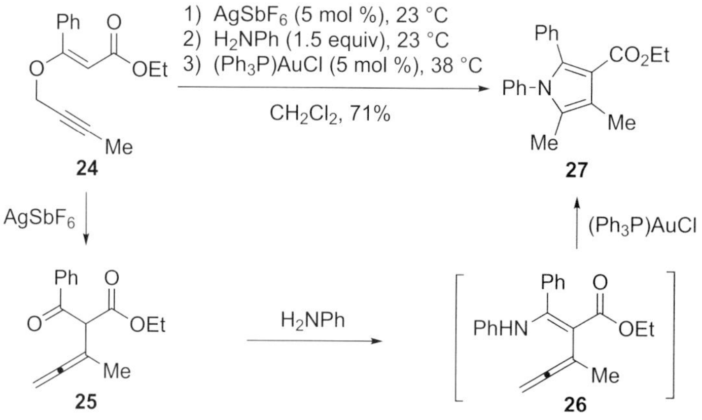

1) $AgSbF_6$ (5 mol %), 23 °C
2) H_2NPh (1.5 equiv), 23 °C
3) $(Ph_3P)AuCl$ (5 mol %), 38 °C
CH_2Cl_2, 71%

24 → 27; 24 —$AgSbF_6$→ 25 —H_2NPh→ [26] —$(Ph_3P)AuCl$→ 27

Scheme 11.2

and gold(I)-catalyzed *exo*-hydroamination [25]. For example, treatment of propargyl ether **24** with a catalytic amount of $AgSbF_6$ at room temperature led to propargyl Claisen rearrangement with formation of the stable α-allenyl β-keto ester **25** that was treated sequentially with aniline and a catalytic amount of $(PPh_3)AuCl$ to form pyrrole **27** in 71% isolated yield (Scheme 11.2). In comparison, sequential treatment of propargyl vinyl ethers that possessed a propargylic ethyl or phenyl group underwent condensation/cyclization to form 1,2-dihydropyridines in modest yield (Eq. (11.18)).

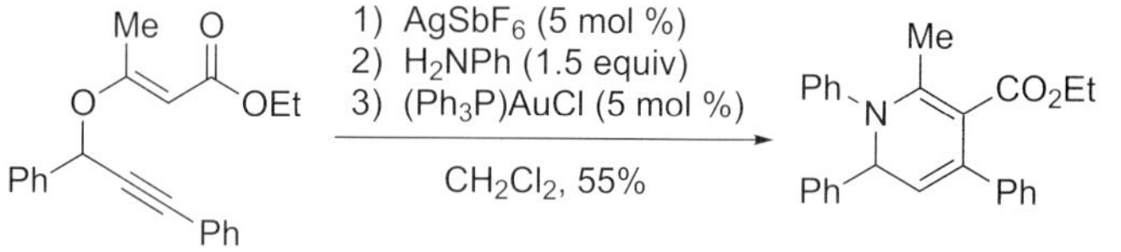

1) $AgSbF_6$ (5 mol %)
2) H_2NPh (1.5 equiv)
3) $(Ph_3P)AuCl$ (5 mol %)
CH_2Cl_2, 55%
(11.18)

Iwasawa has reported the gold(III)-catalyzed reaction of *N*-(*o*-ethynylphenyl)imines with electron-rich alkenes to form polycyclic indole derivatives [26]. As an example, reaction of *N*-[1-(1-pentynyl)phenyl]imine **28** and *tert*-butyl vinyl ether with a catalytic amount of $AuBr_3$ in toluene at room temperature led to isolation of the polycyclic indole **29** in 80% yield as a mixture of diastereomers (Scheme 11.3). Conversion of **28** to **29** presumably occurs via initial intramolecular hydroamination to form the gold carbene containing azomethine ylide **30** that undergoes intermolecular [3 + 2] cycloaddition with *tert*-butyl vinyl ether to form the carbene complex **31**. 1,2-Migration of the *n*-propyl group to the metal-bound carbon atom coupled with deauration then forms **29**. This transformation is also catalyzed efficiently by $PtCl_2$ [26].

Genêt and Michelet have reported the gold(I)-catalyzed cyclization/amination of 1,6-enynes with carboxamides and electron-deficient aromatic amines [27]. Reaction of 1,6-enyne **32** with *o*-cyanoaniline and a catalytic 1 : 1 mixture of $(PPh_3)AuCl$ and $AgSbF_6$ in dioxane at room temperature for 20 h led to isolation of methylene cyclopentane **33** in 93% yield as a single diastereomer (Eq. (11.19)). Gold(I)-catalyzed

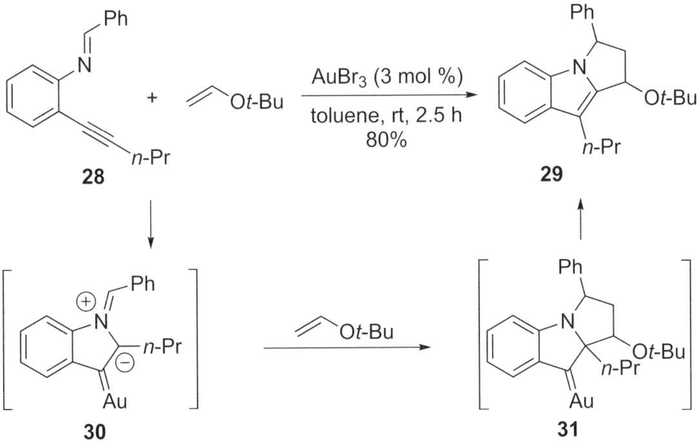

Scheme 11.3

cyclization/amination required aromatic substitution at the terminal alkenyl carbon atom but was also effective for the cyclization/amination of allyl propargyl ethers to form substituted oxygen-containing heterocycles.

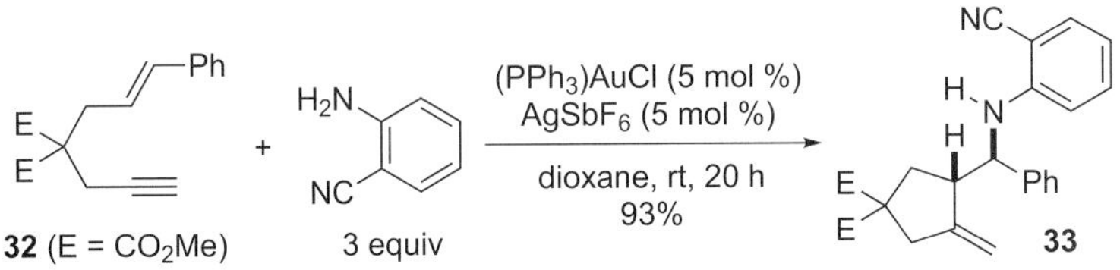

(11.19)

Zhang has shown that the gold(I) *N*-heterocyclic carbene complex (IPr)AuNTf$_2$ catalyzes the intramolecular amination/ring-expansion of *N*-(pent-2-en-4-ynyl)-β-lactams to form 5,6-dihydro-8*H*-indolizin-7-ones (Eq. (11.20)) [28]. Barluenga and Aguilar have reported an unusual gold(I)-catalyzed hetero-cyclization of dienynes triggered by addition of nitrile to the C–C triple bond [29]. As an example, treatment of a mixture of dienyne **34** and acrylonitrile with a catalytic 1 : 1 mixture of (Et$_3$P)AuCl and AgSbF$_6$ in DCE at 85 °C led to isolation of 2-vinyl pyridine derivative **35** in 75% yield (Scheme 11.4). Complexation of gold to the C–C triple bond followed by outer-sphere, regioselective addition of acrylonitrile would form the vinyl gold intermediate **36**. Cyclization and elimination of gold would then form **35** (Scheme 11.4).

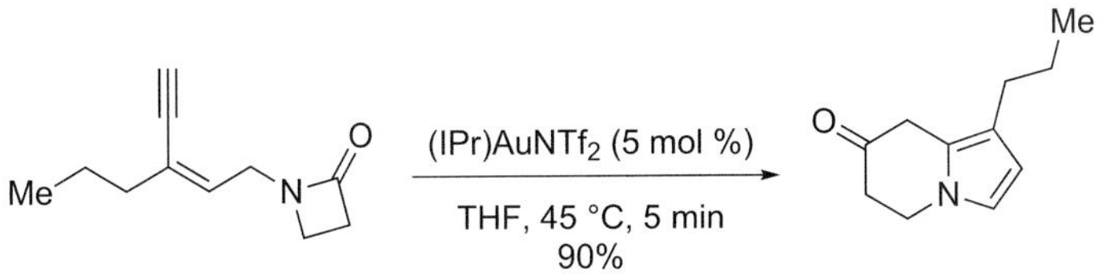

(11.20)

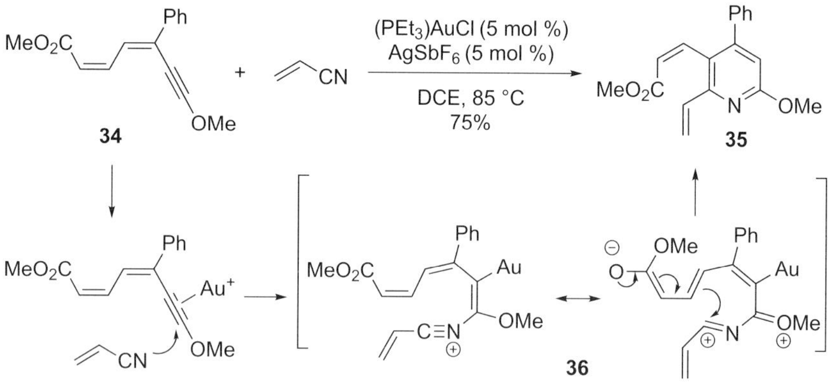

Scheme 11.4

11.2.4
Tandem C–N/C–X Bond Forming Processes

Li has reported the gold(III)-catalyzed tandem intermolecular/intramolecular hydroamination of *o*-alkynylanilines with terminal alkynes to form *N*-vinyl indoles [30]. For example, treatment of a mixture of 2-(phenylethynyl)aniline and phenylacetylene with a catalytic 1 : 3 mixture of $AuCl_3$ and AgOTf without solvent at room temperature for 2 h led to isolation of 1-(1-phenylethenyl)-2-phenylindole **37** in 88% yield (Scheme 11.5). The method tolerated aromatic, alkenyl, and alkyl substitution of the *o*-alkynylaniline, but was largely restricted to aromatic terminal alkynes for the non-aniline component, although both electron-rich and electron-poor 1-aryl alkynes functioned effectively. Interestingly, treatment of 2-phenylindole with phenyl acetylene and a catalytic 1 : 3 mixture of $AuCl_3$ and AgOTf did not lead to formation of **37** (Scheme 11.5). This observation points to a mechanism of gold(III)-catalyzed double hydroamination involving initial intermolecular hydroamination followed by intramolecular hydroamination of the resulting imine intermediate.

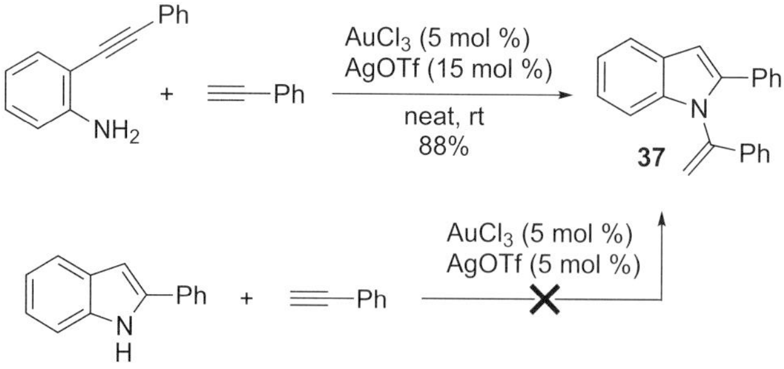

Scheme 11.5

Dake has reported a gold(I)-catalyzed protocol for the condensation/hydroamination of γ-alkynyl ketones with primary amines to form pyrrole derivatives [31]. For example, treatment of the carbamate-protected 1-amino-7-octyn-4-one **38** with a catalytic 1 : 1 : 1 mixture of AuCl, AgOTf, and PPh_3 in DCE at 50 °C for 19 h led to isolation of the pyrrole **39** in 75% yield (Eq. (11.21)). This transformation also tolerated internal alkynes, and employment of AgOTf alone typically produced results inferior to those obtained with cationic gold(I) complexes. Similarly, Liang has reported the gold(III)-catalyzed amination/intramolecular hydroamination of 1-en-4-yne-3-ols with sulfonamides to form substituted pyrroles (Eq. (11.22)) [32].

BocHN **38** + H_2NBn → AuCl (5 mol %), AgOTf (5 mol %), PPh_3 (5 mol %), DCE, 50 °C, 75% → **39**

(11.21)

+ H_2NTs (15 equiv) → $NaAuCl_4 \cdot 4\ H_2O$ (20 mol %), CH_3CN, 30 °C, 8 h, 74% (11.22)

Gevorgyan has reported the gold(III)-catalyzed tandem amination/1,2-migration of *o*-propargylic pyridine derivatives that contain a silicon, tin, or germanium group at the terminal alkynyl carbon atom to form fused pyrrole derivatives [33]. As an example, reaction of the propargyl pyridine derivative **40** with a catalytic amount of $AuBr_3$ in toluene at 50 °C led to isolation of the heterobicyclic compound **41** in 63% yield (Eq. (11.23)). Nakamura has reported the gold(III)-catalyzed synthesis of 3-sulfonylindoles from *o*-alkynyl-*N*-sulfonylanilines via net addition of the N–S bond across the pendant C≡C bond [34]. For example, heating a solution of sulfonamide **42** with a catalytic amount of $AuBr_3$ at 80 °C led to isolation of the indole **43** in 95% yield (Eq. (11.24)). Although cross-over experiments established the intramolecular transfer of a sulfonyl group to the C3 carbon atom of the indole, gold(III)-catalyzed cyclization of a number of *o*-alkynyl-*N*-sulfonylanilines often led to formation of mixtures of 3-, 4-, and 6-sulfonylindoles.

40 → $AuBr_3$ (2 mol %), toluene, 50 °C, 63% → **41** (11.23)

42 → $AuBr_3$ (10 mol %), toluene, 80 °C, 95% → **43** (11.24)

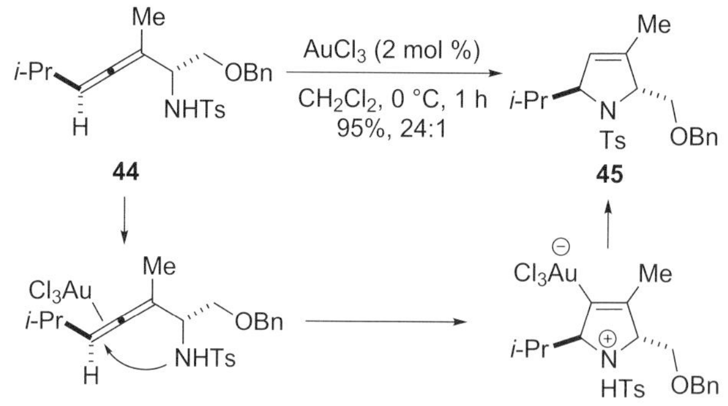

Scheme 11.6

11.3
Hydroamination of Allenes

11.3.1
Intramolecular Processes

Krause has shown that gold(III) salts catalyze the intramolecular *endo*-hydroamination of *N*-protected α-aminoallenes [35]. For example, treatment of the diasteromerically pure α-allenyl sulfonamide **44** with a catalytic amount of $AuCl_3$ in dichloromethane at 0 °C for 1 h formed the pyrroline derivative **45** in 95% yield with 96% diastereomeric purity (Scheme 11.6). The protocol tolerated aryl and alkyl substitution of the distal allenyl carbon atom and was also effective for the hydroamination of *N*-unprotected α-allenylamines although these latter transformations required considerably longer reaction time. In a similar manner, Lee has reported the gold(III)-catalyzed *endo*-hydroamination of 4-allenyl-2-azetidinone **46** to form bicyclic β-lactams **47** (Eq. (11.25)) [36].

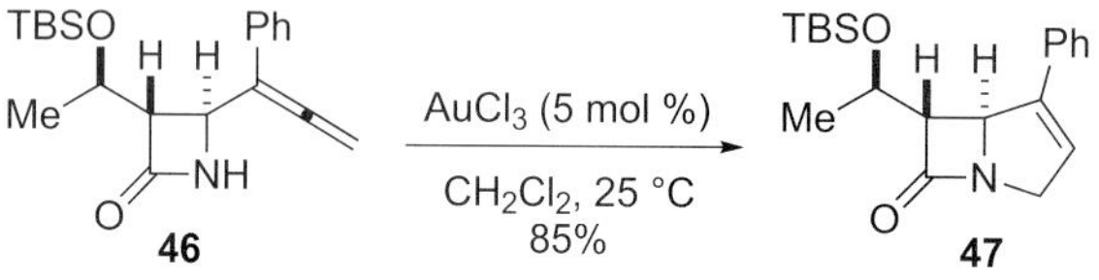

(11.25)

Widenhoefer and coworkers have reported that the gold(I) phosphine complex [P(*t*-Bu)$_2$(*o*-biphenyl)]AuCl is a highly active and selective precatalyst for the intramolecular *exo*-hydroamination of *N*-γ- and δ-allenyl carbamates [37]. As an example, treatment of the *N*-δ-allenyl carbamate **48** with a catalytic 1 : 1 mixture of [P(*t*-Bu)$_2$*o*-biphenyl]AuCl and AgOTf (5 mol%) in dioxane at room temperature for 22 h led to isolation of piperidine **49** in 92% yield as a 7.0 : 1 mixture of *cis*:*trans* diasteromers (Eq. (11.26)). Gold(I)-catalyzed hydroamination of *N*-γ- and δ-allenyl carbamates tolerated substitution at both the internal and terminal allenyl carbon atoms and the transformation displayed modest selectivity for the transfer of chirality from an axially chiral 1,3-disubstituted allenyl moiety to a tetrahedral stereogenic carbon atom. For example, gold(I)-catalyzed cyclization of chiral, non-racemic *N*-γ-allenyl carbamate (*S*)-**50** (≤84% *ee*) led to isolation of the 2-alkenylpyrroldine derivative (*R*)-**51** in 96% yield with 74% *ee* (Eq. (11.27)).

Ph, Ph, NHCbz, Me, **48** → [P(*t*-Bu)$_2$(*o*-biphenyl)]AuCl (5 mol %), AgOTf (5 mol %); dioxane, 25 °C, 22 h; 92% (7.0:1) → Ph, Ph, NCbz, Me, **49** (11.26)

NHCbz, Me, (*S*)–**50** (≤84% ee) → [P(*t*-Bu)$_2$(*o*-biphenyl)]AuCl (5 mol %), AgOTf (5 mol %); dioxane, 25 °C; 96% → Cbz, N, H, Me, (*R*)–**51** (74% ee) (11.27)

Yamamoto has reported the intramolecular *exo*-hydroamination of *N*-allenyl sulfonamides and carbamates catalyzed by simple, unligated gold(I) and gold(III) salts. Noteworthy was that cyclization of *N*-allenyl sulfonamide derivatives that possessed an axially chiral allenyl moiety occurred with highly selective transfer of chirality to the newly formed tetrahedral stereogenic carbon atom [38]. For example, treatment of enantiomerically enriched γ-allenyl tosylamide **52** (96% *ee*) with a catalytic amount of AuCl in THF at room temperature led to isolation of (*E*)-2-(1-heptenyl)pyrroldine (*E*)-**53** in 99% yield with 94% *ee* (Eq. (11.28)).

NHTs, C_5H_{11}, **52** (96% ee) → AuCl (1 mol %); THF, RT; 99% → Ts, N, C_5H_{11}, (*E*)-**53** (94% ee) (11.28)

11.3.2 Intermolecular Processes

Yamamoto has reported the gold(III)-catalyzed intermolecular hydroamination of allenes with aromatic amines to form allylic amines [39]. As an example, reaction of a 1 : 2 mixture of 1-phenyl-1,2-propadiene with aniline catalyzed by $AuBr_3$ in THF at 30 °C formed the allylic amine **54** in 78% isolated yield as a single regio- and stereo-isomer (Eq. (11.29)). Monosubstituted alkyl allenes and 1,3-disubstituted allenes also underwent gold-catalyzed hydroamination in good yields. In the case of chiral, non-racemic 1,3-disubstituted allenes, hydroamination occurred with modest selectivity for the transfer of chirality to the newly formed tetrahedral stereogenic carbon atom. For example, the gold(III)-catalyzed reaction of (*R*)-1-

phenyl-1,2-butadiene (94% *ee*) with aniline formed allylic amine (*S,E*)-**55** in 68% yield with 88% *ee* (Eq. (11.30)).

Ph-allene + H_2NPh → [$AuBr_3$ (10 mol %), THF, 30 °C, 78%] → Ph–CH=CH–CH$_2$–NHPh **54** (11.29)

(94% ee) 1-phenyl-1,2-butadiene + H_2NPh → [$AuBr_3$ (10 mol %), THF, 30 °C, 68%] → (*S,E*)-**55** (88% ee) (11.30)

Yamamoto has also demonstrated the intermolecular hydroamination of monosubstituted arylallenes with morpholine [40]. For example, treatment of 1-*p*-tolyl-1,2-propadiene with morpholine and a catalytic 1 : 1 mixture of [PPh_2(*o*-tolyl)]AuCl and AgOTf in toluene at 80 °C for 24 h led to isolation of the allylic amine **56** in 83% yield as a single regio- and stereo-isomer (Eq. (11.31)). 1-Alkylallenes, 1,1-, and 1,3-disubstituted allenes also underwent gold(I)-catalyzed hydroamination with morpholine, albeit with diminished efficiency and/or regioselectivity.

1-*p*-tolyl-1,2-propadiene + morpholine → [[PPh_2(*o*-tolyl)]AuCl (10 mol %), AgOTf (10 mol %), toluene, 80 °C, 24 h, 83%] → **56** (11.31)

Widenhoefer has reported the intermolecular hydroamination of allenes with *N*-unsubstituted carbamates catalyzed by a gold(I) *N*-heterocyclic carbene complex [41]. For example, reaction of 3-methyl-1,2-butadiene **57** with benzyl carbamate catalyzed by a 1 : 1 mixture of (IPr)AuCl and AgOTf in dioxane at 23 °C for 24 h formed benzyl 1,1-dimethyl-2-propenylcarbamate in 93% isolated yield as a single regioisomer resulting from attack of benzyl carbamate at the more hindered terminus of **57** (Eq. (11.32)). 1,3-Disubstituted, trisubstituted, and tetrasubstituted allenes also underwent efficient and selective hydroamination in the presence of a catalytic mixture of (IPr)AuCl and AgOTf. Differentially 1,3-disubstituted allenes, such as **58**, underwent gold(I)-catalyzed hydroamination with exclusive delivery of the carbamate to the more electron-rich allene terminus (Eq. (11.33)). However, gold(I)-catalyzed intermolecular hydroamination of enantiomerically enriched 1,3-disubstituted allenes with carbamates occurred with no detectable transfer of chirality to the newly formed tetrahedral stereogenic carbon atom.

57 + H_2NCbz → [(IPr)AuCl (5 mol %), AgOTf (5 mol %), 23 °C, 24 h, 93%] → CH_2=CH–C(Me)(Me)–NHCbz (11.32)

BzO 58 Me + H_2NCbz → (IPr)AuCl (5 mol %), AgOTf (5 mol %), dioxane, 23 °C, 24 h, 84% → BzO NHCbz Me

(11.33)

11.3.3
Enantioselective Processes

Toste has described the intramolecular enantioselective hydroamination of γ- and δ-allenyl sulfonamides catalyzed by enantiomerically pure bis(gold) phosphine complexes [42]. For example, treatment of the terminally-disubstituted γ-allenyl sulfonamide **59** with a catalytic amount of [(*R*)-3,5-xylyl-binap]$(AuOPNB)_2$ (OPNB = *p*-nitrobenzoate) formed protected pyrrolidine **60** in 88% yield with 98% *ee* (Eq. (11.34)). Likewise, treatment of δ-allenyl sulfonamide **61** with a catalytic amount of [(*R*)-Cl-MeObiphep]$(AuOPNB)_2$ in nitromethane at 50 °C for 24 h formed 2-alkenyl piperidine **62** in 70% isolated yield with 98% *ee* (Eq. (11.35)). Realization of high enantioselectivity in this protocol required employment of both a terminally disubstituted allene and a sulfonamide nucleophile.

NHTs 59 → [(*R*)-3,5-xylyl-binap]$(AuOPNB)_2$ (3 mol %), DCE, 23 °C, 17 h, 88%, 98% ee (OPNB = *p*-nitrobenzoate) → Ts N 60

PAr_2 PAr_2 Ar = Me Me (*R*)-3,5-xylyl-binap

(11.34)

Me Me NHTs Me Me 61 → [(*R*)-Cl-MeObiphep]$(AuOPNB)_2$ (5 mol %), $MeNO_2$, 50 °C, 24 h, 70%, 98% *ee* → Me Me NTs Me Me 62

Cl MeO PPh_2 MeO PPh_2 Cl (*R*)-Cl-MeObiphep

(11.35)

Toste has developed an alternative approach to the enantioselective hydroamination of *N*-allenyl sulfonamides that employs a chiral or achiral gold complex in combination with a chiral phosphonate anion [43]. As an example of this approach, treatment of γ-allenyl sulfonamide **63** with a catalytic 1 : 1 mixture of the achiral mono (gold) phosphine complex (PMe_2Ph)AuCl and chiral, enantiomerically pure silver phosphonate Ag-(*R*)-**64** (5 mol%) in benzene at room temperature for 48 h formed 2-alkenyl pyrrolidine **65** in 84% isolated yield with 99% *ee* (Eq. (11.36)).

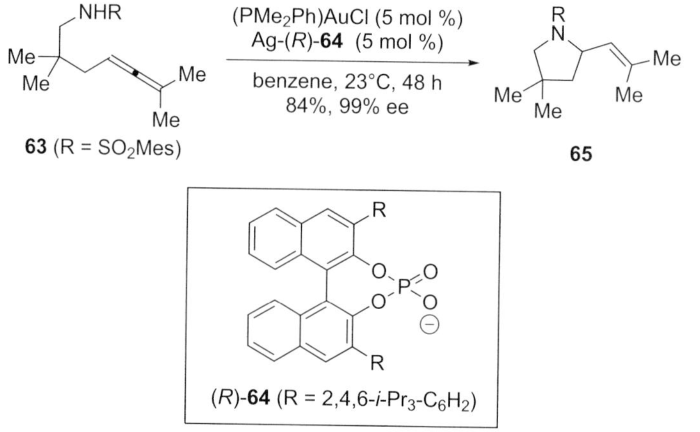

(11.36)

Widenhoefer and Bender have reported the enantioselective intramolecular hydroamination of *N*-γ-allenyl carbamates [44]. As an example, treatment of terminally disubstituted *N*-γ-allenyl carbamate **66** with a catalytic 1 : 2 mixture of [(*S*)-**67**]Au_2Cl_2 [(*S*)-**67** = (*S*)-DTBM-MeObiphep] and $AgClO_4$ in *m*-xylene at 0–23 °C for 48 h formed pyrrolidine **68** in 83% isolated yield with 91% *ee* (Eq. (11.37)). Terminally unsubstituted *N*-γ-allenyl carbamates underwent hydroamination under these conditions with higher yield but lower enantioselectivity. Widenhoefer and coworkers have exploited the ability of cationic gold(I) complexes to racemize allenes [45] to achieve the dynamic kinetic enantioselective hydroamination (DKEH) of *N*-γ-allenyl carbamates that possessed a trisubstituted allenyl group [46]. For example, reaction of *N*-γ-allenyl carbamate **69** with a catalytic 1 : 2 mixture of [(*S*)-**67**]Au_2Cl_2 (2.5 mol%) and $AgClO_4$ (5 mol%) in *m*-xylene at room temperature for 24 h led to isolation of a 10.1 : 1 mixture of (*Z*)-**70** and (*E*)-**70** in 99% combined yield with 91 and 9% *ee*, respectively (Eq. (11.38)).

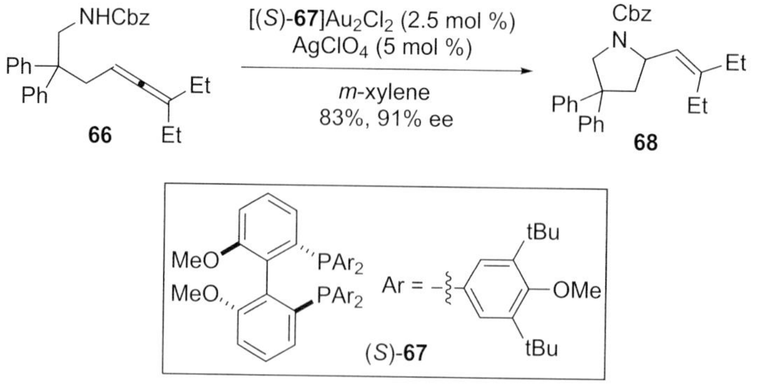

(11.37)

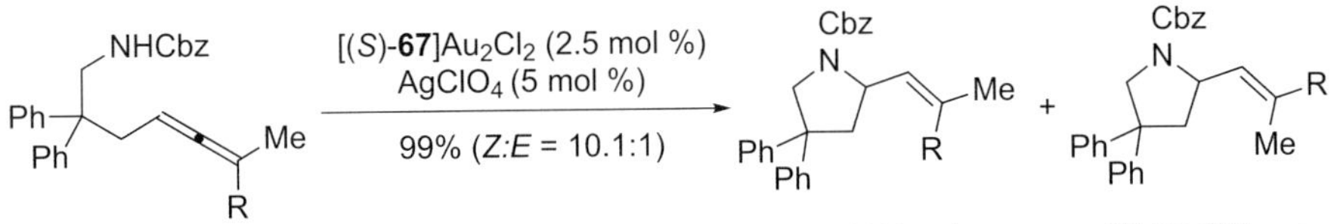
NHCbz
Ph
Ph
Me
R
[(S)-67]Au2Cl2 (2.5 mol %)
AgClO4 (5 mol %)
99% (Z:E = 10.1:1)
Cbz
N
Me
Ph
Ph
R
+
Cbz
N
R
Ph
Ph
Me

69 (R = n-hexyl) (Z)-70 (91% ee) (E)-70 (9% ee)

(11.38)

11.4 Hydroamination of Alkenes and Dienes

11.4.1 Unactivated Alkenes

11.4.1.1 Sulfonamides as Nucleophiles

He has reported that $(PPh_3)AuOTf$, generated *in situ* from $(PPh_3)AuCl$ and AgOTf, catalyzes the intermolecular hydroamination of unactivated alkenes with sulfonamides [47]. As an example, reaction of a 1 : 4 mixture of H_2NTs and 4-(2-propenyl) anisole catalyzed by a 1 : 1 mixture of $(PPh_3)AuCl$ and AgOTf (5 mol%) in toluene at 85 °C for 14 h led to isolation of sulfonamide **71** in 95% yield (Eq. (11.39)). Likewise, *N*-γ-alkenyl sulfonamides underwent gold-catalyzed intramolecular hydroamination to form pyrrolidine derivatives. For example, treatment of the cyclohexenyl sulfonamide derivative **72** with a catalytic 1 : 1 mixture of $(PPh_3)AuCl$ and AgOTf (5 mol%) led to isolation of **73** in 95% yield (Eq. (11.40)). Che has reported the gold(I)-catalyzed hydroamination of unactivated alkenes with sulfonamides under both thermal and microwave conditions [48]. As an example, treatment of sulfonamide **74** with a 1 : 1 mixture of $(PPh_3)AuCl$ and AgOTf in toluene at 100 °C for 24 h led to isolation of heterobicycle **75** in 95% yield (Eq. (11.41)). More recently, Nájera has shown that $[P(OPh)_3]AuCl$ is a particularly effective precatalyst for the intermolecular hydroamination of sulfonamides with cyclic alkenes and 1,3-dienes under thermal or microwave-assisted conditions (Eq. (11.42)) [49].

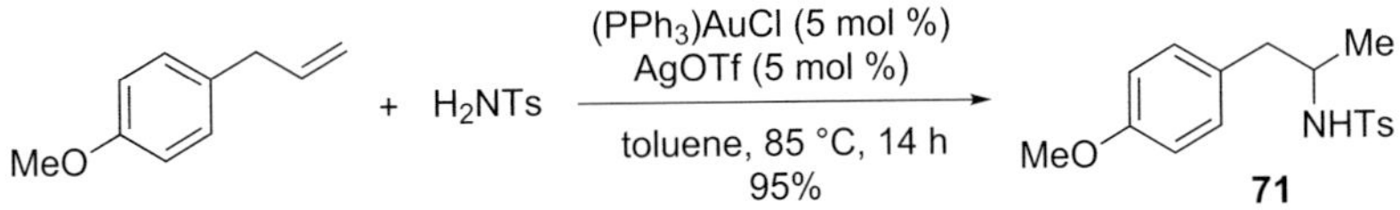
MeO
+ H2NTs
(PPh3)AuCl (5 mol %)
AgOTf (5 mol %)
toluene, 85 °C, 14 h
95%
Me
NHTs
MeO

71

(11.39)

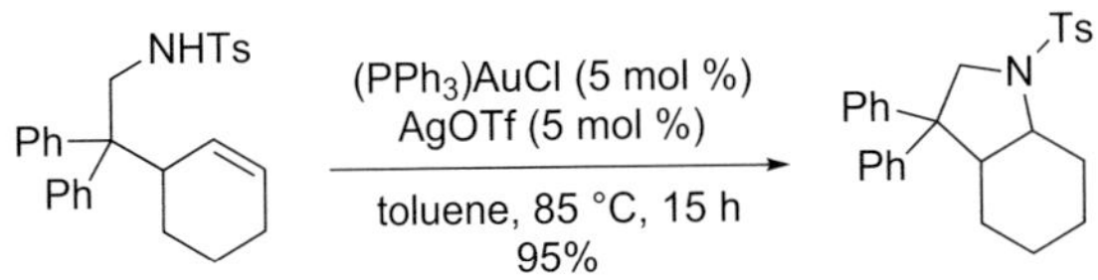
NHTs
Ph
Ph
(PPh3)AuCl (5 mol %)
AgOTf (5 mol %)
toluene, 85 °C, 15 h
95%
Ts
N
Ph
Ph

72 73

(11.40)

74 → 75: $(PPh_3)AuCl$ (5 mol %), AgOTf (5 mol %), 100 °C, 24 h, 95% (11.41)

cyclooctene + H_2NTs → cyclooctyl-NHTs: $[(P(OPh)_3]AuCl$ (1 mol %), AgOTf (1 mol %), 85 °C, 24 h, 75% (11.42)

There has been growing concern that the apparent gold(I)/AgOTf-catalyzed hydroamination of alkenes with sulfonamides is actually due to triflic acid generated under reaction conditions. In 2002, Hartwig demonstrated that triflic acid (20 mol%) catalyzes the intramolecular hydroamination of unactivated alkenes with sulfonamides in toluene at 100 °C [50]. In 2006, Hartwig reported that triflic acid catalyzes the intermolecular hydroamination of alkenes and dienes with sulfonamides and provided direct comparisons of these triflic acid-catalyzed transformations to the corresponding reactions initiated by $(PPh_3)AuCl$ and AgOTf [51]. In all cases examined, triflic acid-catalyzed hydroamination occurred with rates, yields, and selectivities comparable to those obtained in the presence of catalytic amounts of $(PPh_3)AuCl$ and AgOTf. Furthermore, control experiments that would rule out the presence of an acid-catalyzed reaction pathway in the hydroamination of alkenes with sulfonamides in the presence of catalytic amounts of $(PPh_3)AuCl$ and AgOTf were not reported [47–49].

11.4.1.2 **Carboxamide Derivatives as Nucleophiles**

Widenhoefer and Han have reported an effective protocol for the intramolecular hydroamination of unactivated C=C bonds with carbamates [52]. As an example of this protocol, treatment of the *N*-γ-alkenyl carbamate **76** with a catalytic 1 : 1 mixture of $[P(t\text{-}Bu)_2(o\text{-biphenyl})]AuCl$ and AgOTf (5 mol%) in dioxane at 60 °C for 22 h formed pyrrolidine **77** in 91% isolated yield as a 3.6 : 1 mixture of diastereomers (Eq. (11.43)). The protocol tolerated substitution at the internal olefinic carbon atom and along the alkyl backbone and the method was applied to the synthesis of both heterobicyclic compounds and piperidine derivatives. This protocol was subsequently expanded to include the intramolecular hydroamination of *N*-alkenyl carboxamides including 2-allyl aniline derivatives (Eq. (11.44)) [53].

76 → 77: $[P(t\text{-}Bu)_2(o\text{-biphenyl})]AuCl$ (5 mol %), AgOTf (5 mol %), dioxane, 60 °C, 22 h, 91% (3.6:1) (11.43)

NHAc

[P(*t*-Bu)$_2$(*o*-biphenyl)]AuCl (5 mol %)
AgOTf (5 mol %)

dioxane, 80 °C, 22 h
99%

Ac
N
Me

(11.44)

Bender and Widenhoefer have reported the room temperature intramolecular hydroamination of unactivated alkenes with *N,N′*-disubstituted ureas catalyzed by a gold(I) *N*-heterocyclic carbene complex [54]. For example, treatment of 2-isopropyl-4-pentenyl urea **78** with a catalytic 1 : 1 mixture of (IPr)AuCl and AgOTf at room temperature for 22 h led to isolation of pyrroldine **79** in 98% yield as a 5.5 : 1 mixture of diastereomers (Eq. (11.45)). Gold(I)-catalyzed hydroamination of *N*-4-pentenyl ureas tolerated substitution at the C1 or C4 carbon atoms and was effective for the cyclization of unsubstituted 4-pentenyl ureas and 5-hexenyl ureas. Conversely, the method did not tolerate substitution at the terminal alkenyl carbon atom.

i-Pr
O
N
H
NHPh
78

(IPr)AuCl (5 mol %)
AgOTf (5 mol %)

MeOH, rt, 22 h
98% (5.5:1)

i-Pr
N
O
NHPh
Me
79

(11.45)

In contrast to the hydroamination of alkenes with sulfonamides, the potential of an acid-catalyzed reaction pathway in the hydroamination of alkenes with carboxamide derivatives appears less likely. Hartwig found that the intramolecular hydroamination of alkenes with *N*-arylcarboxamides was only realized in the presence of stoichiometric amounts of triflic acid [50]. In contrast, He reported that triflic acid catalyzes the intramolecular hydroamination of an *N*-4-methyl-4-pentenyl carbamate in toluene at 85 °C [55]. However, in the corresponding gold(I)-catalyzed transformation, the intramolecular hydroamination of an *N*-4-methyl-4-pentenyl carbamate was markedly slower than was the intramolecular hydroamination of an *N*-4-pentenyl carbamate [52], which is inconsistent with the anticipated behavior of an acid-catalyzed pathway. Furthermore, control experiments firmly ruled out the presence of an acid-catalyzed reaction pathway in the gold(I)-catalyzed intramolecular hydroamination of alkenes with carboxamide derivates and ureas [53, 54].

11.4.1.3 **Ammonium Salts as Nucleophiles**

Widenhoefer and Bender have recently reported the gold(I)-catalyzed intramolecular hydroamination of unactivated C=C bonds with alkyl ammonium salts [56]. As an example, treatment of the HBF_4 salt of 2,2-diphenyl-4-pentenyl amine (**80**•HBF_4) with a catalytic 1 : 1 mixture of the gold(I) *o*-biphenyl phosphine complex (**81**)AuCl (**81** = 2-dicyclohexylphosphino-2′,6′-dimethoxy-1,1′-biphenyl) and AgOTf in toluene at 80 °C for 24 h followed by basification with NaOH led to isolation of pyrrolidine **82** in 94% yield (Eq. (11.46)). Although the transformation was of limited scope, gold(I)-catalyzed intramolecular hydroamination was also effective for primary 5-hexenyl ammonium salts and secondary 4-pentenyl ammonium salts.

$80 \cdot HBF_4$ → 1) (81)AuCl (5 mol %), AgOTf (5 mol %), 80 °C, 24 h; 2) NaOH → 82

81: Cy_2P-biaryl with OMe, OMe

(11.46)

11.4.2
Methylenecyclopropanes, Vinylcyclopropanes, and Dienes

Shi has reported a novel ring-opening, ring-closing hydroamination of methylenecyclopropanes (MCPs) with sulfonamides catalyzed by mixtures of $(PPh_3)AuCl$ and AgOTf [57]. For example, treatment of MCP **83** with tosylamide and a catalytic 1 : 1 mixture of $(PPh_3)AuCl$ and AgOTf in toluene at 85 °C for 8 h led to isolation of pyrrolidine **84** in 68% yield (Eq. (11.47)). Alkyl and aryl mono- and di-substituted MCPs underwent effective gold(I)-catalyzed ring-opening/ring-closing hydroamination. Likewise, Togni has demonstrated the gold(I)-catalyzed ring-opening hydroamination of vinylcyclopropanes (VCPs) with sulfonamides [58]. For example, reaction of VCP **85** with methylsulfonamide and a catalytic 1 : 1 mixture of $(PPh_3)AuCl$ and AgOTf in toluene at 50 °C for 24 h led to isolation of 4-phenyl-3-pentenyl sulfonamide **86** in 97% yield (Eq. (11.48)). It should also be noted that compound **86** was isolated in 92% yield from reaction of **85** and methylsulfonamide catalyzed by HOTf at 50 °C for 24 h [58].

83 (Ar = 2-MeC_6H_4) + H_2NTs → $(PPh_3)AuCl$ (5 mol %), AgOTf (5 mol %), 85 °C, 8 h, 68% → **84** (11.47)

85 + HMeNTs → $(PPh_3)AuCl$ (10 mol %), AgOTf (10 mol %), toluene, 50 °C, 24 h, 97% → **86** (11.48)

He has reported the gold(I)-catalyzed intermolecular hydroamination of 1,3-dienes with carbamates and sulfonamides to form allylic amines [59]. As an example, reaction of benzyl carbamate with a slight excess of 1,3-cyclohexadiene catalyzed by a 1 : 1 mixture of $(PPh_3)AuCl$ and AgOTf in 1,2-dichloroethane at 50 °C overnight led to isolation of benzyl (2-cyclohexenyl)carbamate **87** in 87% yield (Eq. (11.49)). In addition, 2-oxazolidinone, methyl carbamate, and *p*-toluenesulfonamide were effective nucleophiles and 1,2-pentadiene and cyclohexadiene were effective dienes for gold-catalyzed intermolecular hydroamination. The mechanism of the gold(I)-catalyzed intermolecular hydroamination of conjugated dienes with carbamates has been investigated employing computational analysis [60].

+ H_2NCbz —(PPh3)AuCl (5 mol %), AgOTf (5 mol %), DCE, 50 °C, 87%→ NHCbz **87** (11.49)

11.5 Addition of Sulfur Nucleophiles to C–C Multiple Bonds

11.5.1 Alkynes

Wang has reported that simple gold(I) and gold(III) salts catalyze the migration/hydroarylation of propargylic sulfides and dithioacetals that possess an aryl group bound to the propargylic carbon atom [61]. As an example of the gold(I)-catalyzed sulfide migration of a propargyl sulfide, treatment of **88** with a catalytic amount of AuCl (5 mol%) in toluene at 80 °C led to isolation of a 7 : 1 mixture of regioisomeric 2-thioindenes **89** and **90** in 70% combined yield (Eq. (11.50)). Employment of $AuCl_3$ as a catalyst under similar conditions formed a 5 : 1 mixture of **89** and **90** in 70% yield. As an example of gold-catalyzed sulfide migration of a propargylic dithioacetal, treatment of a toluene solution of **91** with a catalytic amount of AuCl at 80 °C led to isolation of indene **92** in 95% yield (Eq. (11.51)).

Me SR **88** —AuCl (5 mol %), toluene, 80 °C, 70 % (**89**:**90** = 7:1)→ Me, SR **89** + Me, SR **90** (11.50)

S S, Ph **91** —AuCl (5 mol %), toluene, 80 °C, 95 %→ S, S, Ph **92** (11.51)

Nakamura has reported the gold(I)-catalyzed cyclization/migration of (α-alkoxyalkyl) (*o*-alkynylphenyl)sulfides to form 2,3-disubstituted benzothiophenes [62] and the gold(I)-catalyzed cyclization/migration of (ortho-alkynylphenylthio)silanes to form 3-silylbenzothiophenes [63]. As an example of the former procedure, treatment of 2-(1-pentynyl)phenyl sulfide **93** with a catalytic amount of AuCl in toluene at room temperature led to isolation of benzothiophene **94** in 93% yield (Scheme 11.7). The transformation presumably occurs via *endo*-attack of the sulfur atom on the gold(I)-complexed alkyne to form zwitterion **95** followed by migration of the α-alkoxyalkyl group to the gold-bound carbon atom and subsequent elimination of gold chloride. Cyclization/migration of enantiomerically enriched *o*-alkynylphenyl

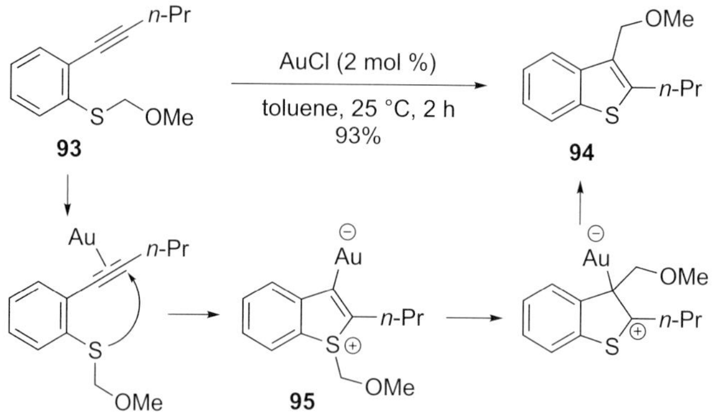

Scheme 11.7

1-arylethylsulfides occurred with predominant retention of configuration (Eq. (11.52)) [64].

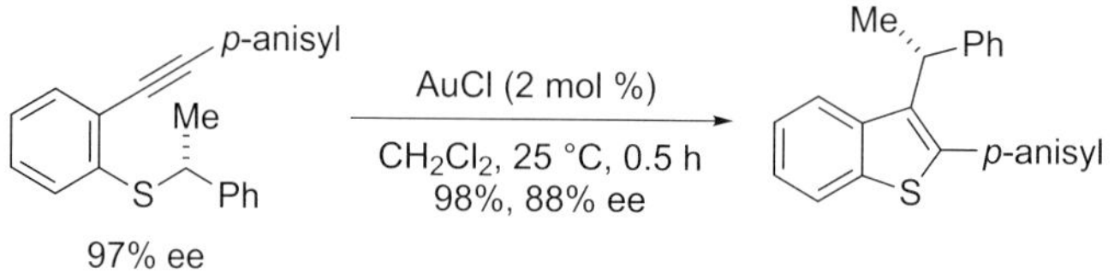

(11.52)

Several additional transformations involving the addition of sulfur nucleophiles to alkynes have recently been disclosed. Corma has reported the intermolecular dihydrothiolation of alkynes with dithiols to form thioacetals [65]. Ohe has reported the gold(I)-catalyzed rearrangement/ring-opening alkylation of *N,N'*-dimethylthiocarbamate with 2-methoxyfuran to form the conjugated trienyl ester **96** as a mixture of diastereomers (Eq. (11.53)) [66]. The transformation is presumably initiated via the [3,3]-sigmatropic rearrangement of the propargylic thiocarbamate moiety, reminiscent of the gold(I)-catalyzed [3,3]-sigmatropic rearrangement of propargyl carboxylates to form allenyl carboxylates.

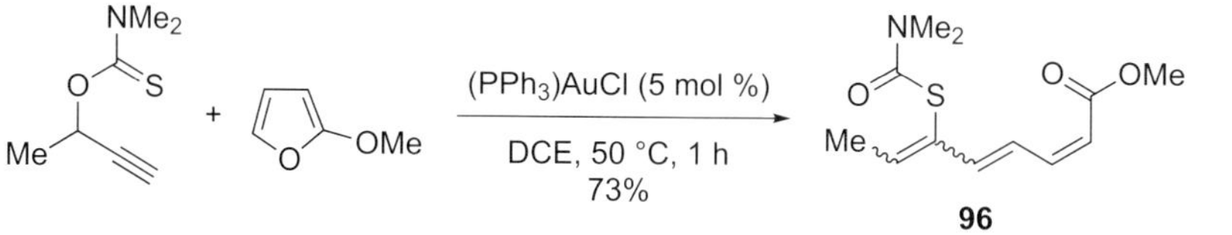

(11.53)

11.5.2 Allenes and Dienes

Krause has reported a gold(I)-catalyzed protocol for the intramolecular *endo*-hydrothiolation of α-thioallenes to form 2,5-dihydrothiophenes [67]. As an example, treatment of diastereomerically pure α-thioallene **97** with a catalytic amount of AuCl in dichloromethane at 20 °C for 1.5 h led to isolation of 2,5-dihydrothiophene **98** in 86% yield as a 95 : 5 mixture of diastereomers (Eq. (11.54)). He has reported the gold (I)-catalyzed hydrothiolation of conjugated dienes [68] employing a catalyst system similar to that used for the intermolecular hydroamination of carbamates with dienes [59]. As an example, reaction of 3-methyl-1,3-pentadiene with thiophenol and a catalytic amount of $(PPh_3)AuOTf$ in DCE at room temperature overnight led to isolation of sulfide **99** in 90% yield via addition of the thiol to the internal carbon atom of the less-substituted C=C bond of the diene (Eq. (11.55)). This latter protocol was effective for both electron-rich and electron-poor thiols, although employment of simple alkyl thiols led to diminished yield for both conjugated dienes and vinyl arenes.

97 → **98**: AuCl (5 mol %); CH_2Cl_2, 20 °C, 1.5 h; 86%, 95:5 (11.54)

3-methyl-1,3-pentadiene + PhSH → **99**: $(PPh_3)AuOTf$ (5 mol %); DCE, RT, overnight; 90% (11.55)

References

1 Müller, T.E., Hultzsch, K.C., Yus, M., Foubelo, F., and Tada, M. (2008) *Chem. Rev.*, **108**, 3795–3892.

2 Puddephatt, R.J. (1982) *Comprehensive Organometallic Chemistry*, vol. 2 (eds G. Wilkinson, F.G.A. Stone, and E.W. Abel), Pergamon Press, Oxford, pp. 762–821.

3 (a) Fukuda, Y., Utimoto, K., and Nozaki, H. (1987) *Heterocycles*, **25**, 297–300; (b) Fukuda, Y. and Utimoto, K. (1991) *Synthesis*, 975–978.

4 Iritani, K., Matsubara, S., and Utimoto, K. (1988) *Tetrahedron Lett.*, **29**, 1799–1802.

5 Arcadi, A., Bianchi, G., and Marinelli, F. (2004) *Synthesis*, 610–618.

6 Ambrogio, I., Arcadi, A., Cacchi, S., Fabrizi, G., and Marinelli, F. (2007) *Synlett*, 1775–1779.

7 Majumdar, K.C., Samanta, S., and Chattopadhyay, B. (2008) *Tetrahedron Lett.*, **49**, 7213–7216.

8 (a) Arcadi, A., Giuseppe, S.D., Marinelli, F., and Rossi, E. (2001) *Adv. Synth. Catal.*, **343**, 443–446; (b) Arcadi, A., Giuseppe, S.D., Marinelli, F., and Rossi, E. (2001) *Tetrahedron: Asymmetry*, **12**, 2715–2720.

9 Kadzimirsz, D., Hildebrandt, D., Merz, K., and Dyker, G. (2006) *Chem. Commun.*, 661–662.

10 Kang, J.-E., Kim, H.-B., Lee, J.-W., and Shin, S. (2006) *Org. Lett.*, **8**, 3537–3540.

11 Hashmi, A.S.K., Rudolph, M., Schymura, S., Visus, J., and Frey, W. (2006) *Eur. J. Org. Chem.*, 4905–4909.

12 Yeom, H.-S., Lee, E.-S., and Shin, S. (2007) *Synlett*, 2292–2294.

13 Ritter, S., Horino, Y., Lex, J., and Schmalz, H.-G. (2006) *Synlett*, 3309–3313.
14 Enomoto, T., Obika, S., Yasui, Y., and Takemoto, Y. (2008) *Synlett*, 1647–1650.
15 Mizushima, E., Hayashi, T., and Tanaka, M. (2003) *Org. Lett.*, **5**, 3349–3352.
16 Teles, J.H., Brode, S., and Chabanas, M. (1998) *Angew Chem. Int. Ed.*, **37**, 1415–1418.
17 Luo, Y., Li, Z., and Li, C.-J. (2005) *Org. Lett.*, **7**, 2675–2678.
18 Lavallo, V., Frey, G.D., Donnadieu, B., Soleihavoup, M., and Bertrand, G. (2008) *Angew Chem. Int. Ed.*, **47**, 5224–5228.
19 Gorin, D.J., Davis, N.R., and Toste, F.D. (2005) *J. Am. Chem. Soc.*, **127**, 11260–11261.
20 Yan, B. and Liu, Y. (2007) *Org. Lett.*, **9**, 4323–4326.
21 Seregin, I.V., Schammel, A.W., and Gevorgyan, V. (2007) *Org. Lett.*, **9**, 3433–3436.
22 Skouta, R. and Li, C.-J. (2007) *Synlett*, 1759–1762.
23 Liu, X.-Y., Ding, P., Huang, J.-S., and Che, C.-M. (2007) *Org. Lett.*, **9**, 2645–2648.
24 Alfonsi, M., Arcadi, A., Aschi, M., Bianchi, G., and Marinelli, F. (2005) *J. Org. Chem.*, **70**, 2265–2273.
25 Binder, J.T. and Kirsch, S.F. (2006) *Org. Lett.*, **8**, 2151–2153.
26 Kusama, H., Miyashita, Y., Takaya, J., and Iwasawa, N. (2006) *Org. Lett.*, **8**, 289–292.
27 Leseurre, L., Toullec, P.Y., Genêt, J.-P., and Michelet, V. (2007) *Org. Lett.*, **9**, 4049–4052.
28 Peng, Y., Yu, M., and Zhang, L. (2008) *Org. Lett.*, **10**, 5187–5190.
29 Barluenga, J., Fernández-Rodríquez, M.A., García-Garcia, P., and Aguilar, E. (2008) *J. Am. Chem. Soc.*, **130**, 2764–2765.
30 Zhang, Y., Donahue, J.P., and Li, C.-J. (2007) *Org. Lett.*, **9**, 627–630.
31 Harrison, T.J., Kozak, J.A., Corbella-Pané, M., and Dake, G.R. (2006) *J. Org. Chem.*, **71**, 4525–4529.
32 Shu, X.-Z., Liu, X.-Y., Xial, H.-Q., Ji, K.-G., Guo, L.-N., and Liang, Y.-M. (2008) *Adv. Synth. Catal.*, **350**, 243–248.
33 Seregin, I.V. and Gevorgyan, V. (2006) *J. Am. Chem. Soc.*, **128**, 12050–12051.
34 Nakamura, I., Yamagishi, U., Song, D., Konta, S., and Yamamoto, Y. (2007) *Angew Chem. Int. Ed.*, **46**, 2284–2287.
35 (a) Morita, N. and Krause, N. (2004) *Org. Lett.*, **6**, 4121–4123; (b) Morita, N. and Krause, N. (2006) *Eur. J. Org. Chem.*, 4634–4641.
36 Lee, P.H., Kim, H., Lee, K., Kim, M., Noh, K., Kim, H., and Seomoon, D. (2005) *Angew Chem. Int. Edit.*, **44**, 1840–1843.
37 Zhang, Z., Liu, C., Kinder, R.E., Han, H., Qian, H., and Widenhoefer, R.A. (2006) *J. Am. Chem. Soc.*, **128**, 9066–9073.
38 Patil, N.T., Lutete, L.M., Nishina, N., and Yamamoto, Y. (2006) *Tetrahedron Lett.*, **47**, 4749–4751.
39 Nishina, N. and Yamamoto, Y. (2006) *Angew Chem. Int. Ed.*, **45**, 3314–3317.
40 Nishina, N. and Yamamoto, Y. (2007) *Synlett*, 1767–1770.
41 Kinder, R.E., Zhang, Z., and Widenhoefer, R.A. (2008) *Org. Lett.*, **10**, 3157–3159.
42 LaLonde, R.L., Sherry, B.D., Kang, E.J., and Toste, F.D. (2007) *J. Am. Chem. Soc.*, **129**, 2452–2453.
43 Hamilton, G.L., Kang, E.J., Mba, M., and Toste, F.D. (2007) *Science*, **317**, 496–499.
44 Zhang, Z., Bender, C.F., and Widenhoefer, R.A. (2007) *Org. Lett.*, **9**, 2887–2889.
45 Sherry, B.D. and Toste, F.D. (2004) *J. Am. Chem. Soc.*, **126**, 15978–15979.
46 Zhang, Z., Bender, C.F., and Widenhoefer, R.A. (2007) *J. Am. Chem. Soc.*, **129**, 14148–14149.
47 Zhang, J., Yang, C.-G., and He, C. (2006) *J. Am. Chem. Soc.*, **128**, 1798–1799.
48 Liu, X.-Y., Li, C.-H., and Che, C.-M. (2006) *Org. Lett.*, **8**, 2707–2710.
49 Giner, X. and Nájera, C. (2008) *Org. Lett.*, **10**, 2919–2922.
50 Schlummer, B. and Hartwig, J.F. (2002) *Org. Lett.*, **4**, 1471–1474.
51 Rosenfeld, D.C., Shekhar, S., Takemiya, A., Utsunomiya, M., and Hartwig, J.F. (2006) *Org. Lett.*, **8**, 4179–4182.
52 Han, X. and Widenhoefer, R.A. (2006) *Angew Chem. Int. Ed.*, **45**, 1747–1749.
53 Bender, C.F. and Widenhoefer, R.A. (2006) *Chem. Commun.*, 4143–4144.
54 Bender, C.F. and Widenhoefer, R.A. (2006) *Org. Lett.*, **8**, 5303–5305.
55 Li, Z., Zhang, J., Brouwer, C., Yang, C.-G., Reich, N.W., and He, C. (2006) *Org. Lett.*, **8**, 4175–4178.

56 Bender, C.F. and Widenhoefer, R.A. (2008) *Chem. Commun.*, 2741–2743.

57 Shi, M., Liu, L.-P., and Tang, J. (2006) *Org. Lett.*, **8**, 4043–4046.

58 Shi, W.-J., Liu, Y., Butti, P., and Togni, A. (2007) *Adv. Synth. Catal.*, **349**, 1619–1623.

59 Brouwer, C. and He, C. (2006) *Angew Chem. Int. Ed.*, **45**, 1744–1747.

60 Kovács, G., Ujaque, G., and Lledós, A. (2008) *J. Am. Chem. Soc.*, **130**, 853–864.

61 Peng, L., Zhang, X., Zhang, S., and Wang, J. (2007) *J. Org. Chem.*, **72**, 1192–1197.

62 Nakamura, I., Sato, T., and Yamamoto, Y. (2006) *Angew Chem. Int. Ed.*, **45**, 4473–4475.

63 Nakamura, I., Sato, T., Terada, M., and Yamamoto, Y. (2007) *Org. Lett.*, **9**, 4081–4083.

64 Nakamura, I., Sato, T., Terada, M., and Yamamoto, Y. (2008) *Org. Lett.*, **10**, 2649–2651.

65 Santos, L.L., Ruiz, V.R., Sabater, M.J., and Corma, A. (2008) *Tetrahedron*, **64**, 7902–7909.

66 Ikeda, Y., Murai, M., Abo, T., Miki, K., and Ohe, K. (2007) *Tetrahedron Lett.*, **48**, 6651–6654.

67 Morita, N. and Krause, N. (2006) *Angew Chem. Int. Ed.*, **45**, 1897–1899.

68 Brouwer, C., Rahaman, R., and He, C. (2007) *Synlett*, 1785–1789.

12
Gold-Catalyzed Addition of Oxygen Nucleophiles to C–C Multiple Bonds

Ross A. Widenhoefer and Feijie Song

12.1
Introduction

The direct, catalytic addition of an oxygen nucleophile to a C–C multiple bond represents an attractive approach to the formation of C–O bonds from simple, readily available precursors. For this reason, there has been considerable interest in the identification of new and effective catalysts for the addition of oxygen nucleophiles to C–C multiple bonds. One of the most significant developments in homogenous catalysis over the past 5–10 years has been the emergence of soluble gold complexes as catalysts for organic transformations [1]. Although a number of synthetically useful transformations have been developed, gold complexes have demonstrated particular utility as catalysts for the functionalization of C–C multiple bonds with heteroatom nucleophiles, including oxygen nucleophiles such as water, carbinols, carbonyl compounds, and ethers. Early efforts in this area focused primarily on the application of simple gold(III) salts as catalysts, but interest has increasingly shifted toward the application of cationic gold(I) complexes supported by a phosphine or *N*-heterocyclic carbene ligands as catalysts for C–O bond formation. These cationic gold(I) complexes are electrophilic, highly carbophilic compounds that are particularly suitable for π-activation catalysis. Herein we review the application of soluble gold(I) and gold(III) complexes as catalysts for the addition of oxygen nucleophiles to C–C multiple bonds through 2008. This review is restricted to transformations that involve addition of an oxygen nucleophile to an electronically unactivated C–C multiple bond catalyzed by a well-defined, soluble gold species.

Catalyzed Carbon-Heteroatom Bond Formation. Edited by Andrei K. Yudin

ISBN: 978-3-527-32428-6

12.2
Addition to Alkynes

12.2.1
Carbinols as Nucleophiles

12.2.1.1 Intermolecular Processes

Building upon the earlier work of Thomas [2], Utimoto demonstrated that simple gold(III) complexes catalyze the hydration of both terminal and internal alkynes to form ketones [3]. As an example, reaction of phenyl acetylene with a catalytic amount of sodium tetrachloroaurate (2 mol%) in refluxing aqueous methanol for 1 h led to isolation of acetophenone in 91% yield (Scheme 12.1). Both alkyl- and aryl-substituted terminal alkynes underwent gold(III)-catalyzed hydration to form methyl ketones, but unsymmetric internal alkynes formed mixtures of regioisomeric alkyl ketones. The gold(III)-catalyzed hydration of alkynes has also been conducted in ionic liquids [4]. Utimoto [3], and, more recently, Laguna [5] have also shown that gold(III) complexes catalyze the hydroalkoxylation of terminal and internal alkynes with anhydrous methanol to form dimethyl acetals in excellent yield (Scheme 12.1).

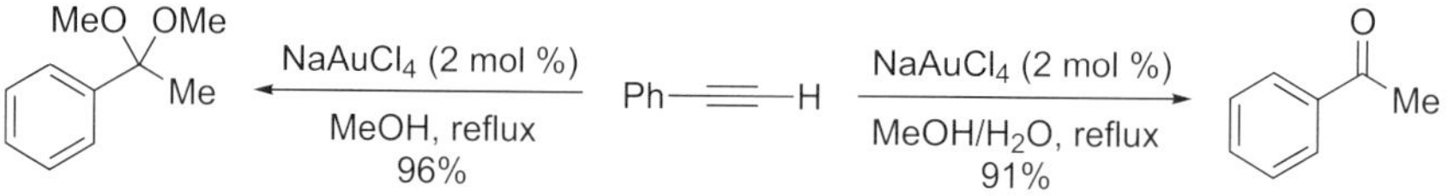

MeO OMe, Me ← $NaAuCl_4$ (2 mol %), MeOH, reflux, 96% — Ph—≡—H — $NaAuCl_4$ (2 mol %), $MeOH/H_2O$, reflux, 91% → O, Me

Scheme 12.1

In 1998, Teles reported that cationic gold(I) complexes, generated *in situ* via protonolysis of methyl(triphenylphosphine)gold, catalyzed the intermolecular hydroalkoxylation of alkynes with alcohols (Eq. (12.1)) [6]. These gold(I) catalysts were active under mild conditions ($\leq 50\,^{\circ}C$) with turnover numbers up to 1×10^5 and turnover frequencies up to $5400\,h^{-1}$. Tanaka extended this methodology to include the gold(I)-catalyzed hydration of alkynes [7]. For example, treatment of 1-hexyne with a mixture of $(PPh_3)AuMe$ (0.2 mol%) and sulfuric acid (50 mol%) in aqueous methanol at 60 °C for 2 h led to formation of 2-hexanone in 99% GC yield (Eq. (12.2)). In related work, gold(I)-catalyzed reaction of alkynes with diols has been employed in the synthesis of cyclic acetals [8], $BF_3{\cdot}OEt_2$ has been employed as the co-catalyst activator for alkyne hydration [9], and gold (I) carbene complexes have been employed as catalysts for alkyne hydration [10]. Gold(I)-catalyzed alkyne hydration has also been applied to good effect in the synthesis of Pterosines B and C [11].

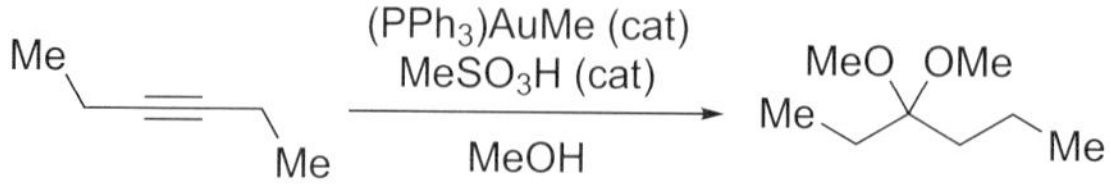

Me, Me — $(PPh_3)AuMe$ (cat), $MeSO_3H$ (cat), MeOH → MeO OMe, Me, Me (12.1)

Me–(CH₂)₃–C≡C–H → Me–(CH₂)₃–C(=O)–Me

(PPh$_3$)AuMe (0.2 mol %)
H_2SO_4 (50 mol %)
H_2O/MeOH, 60 °C
99% (GC)

(12.2)

12.2.1.2 **Intramolecular Processes**

Michelet and Genêt have reported the gold(I)- and gold(III)-catalyzed tandem intramolecular dihydroalkoxylation of bis(homopropargylic) diols to form strained bicyclic ketals [12]. As an example, treatment of bis(homopropargylic) diol **1** with a catalytic amount of AuCl in methanol at room temperature for 30 min led to isolation of bicyclic ketal **2** in 99% yield (Eq. (12.3)). De Brabander has reported the regioselective, gold(I)-catalyzed intramolecular 5-*exo*-hydroalkoxylation of 4-alkynols to form acetals [13]. As an example, treatment of 4-undecyne-1-ol with a catalytic 1 : 1 mixture of (PPh$_3$)AuMe and AgPF$_6$ in diisopropyl ether at room temperature, followed by treatment with a mixture of PPTS, methanol, and trimethyl orthoformate, led to isolation of a 6 : 1 mixture of tetrahydrofuran **3** and tetrahydropyran **4** in 90% combined yield (Eq. (12.4)). Similarly, Krause has shown that treatment of aryl-substituted propargyl alcohol **5** with a catalytic 1 : 1 : 5 mixture of (PPh$_3$)AuCl, AgBF$_4$, and *p*-toluenesulfonic acid in ethanol at room temperature for 1 h led to isolation of cyclic acetal **6** in 76% yield (Eq. (12.5)) [14].

1 → **2**
AuCl (2 mol %)
MeOH, rt
99%

(12.3)

HO–(CH₂)₃–C≡C–*n*-hexyl → **3** + **4** (6:1)
1. (PPh$_3$)AuMe (5 mol %)
AgPF$_6$ (5 mol %)
i-Pr$_2$O, 23 °C
2. PPTS, MeOH, HC(OMe)$_3$
90 %

(12.4)

5 → **6**
(PPh$_3$)AuCl (2 mol %)
AgBF$_4$ (2 mol %)
p-TsOH (10 mol %)
EtOH, rt
76 %

(12.5)

Building on the earlier work of Hashmi [15], Liu has reported the gold-catalyzed cycloisomerization and oxidative cyclization of (*Z*)-enynols [16, 17]. As an example of the former transformation, treatment of the tetraphenyl enynol **7** with a catalytic amount of AuCl$_3$ in dichloromethane at room temperature led to isolation of

(*Z*)-5-benzylidene-2,5-dihydrofuran **8** in 91% yield (Scheme 12.2) [16]. As an example of oxidative cyclization, treatment of **7** with a catalytic amount of $(PPh_3)AuCl$ and AgOTf in oxygen-saturated THF at 50 °C for 17 h led to isolation of butenolide **9** in 97% yield and benzoic acid in 64% yield (Scheme 12.2) [17].

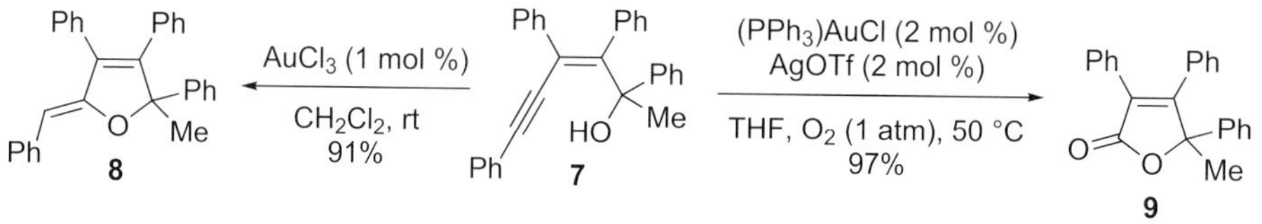

Ph Ph Ph Me O Ph **8** ← $AuCl_3$ (1 mol %), CH_2Cl_2, rt, 91% — Ph Ph Ph Me HO Ph **7** → $(PPh_3)AuCl$ (2 mol %), AgOTf (2 mol %), THF, O_2 (1 atm), 50 °C, 97% — Ph Ph O O Ph Me **9**

Scheme 12.2

A number of permutations and applications of gold-catalyzed alkyne hydroalkoxylation have been reported. Gold(I)-catalyzed double intramolecular hydroalkoxylation has been employed in the synthesis of the bisbenzannelated spiroketal core of the Rubromycins [18]. Gouverneur has recently described a gold(I)-catalyzed protocol for the alkoxyhalogenation of β-hydroxy-α,α-difluroynones initiated by intramolecular alkyne hydroalkoxylation [19]. Floreancig has described the synthesis of oxygen and nitrogen heterocycles via the gold(I)-catalyzed hydrative cyclization of alkynyl ethers [20]. Forsyth has applied gold(I)-catalyzed bis-spiroketalization in the synthesis of the A–D ring subunit of Azaspiracid [21]. Gold(I) complexes also catalyze the rearrangement of propargylic alcohols to form α,β-unsaturated ketones [22].

12.2.1.3 Tandem C–O/C–C Bond Forming Processes

A number of gold-catalyzed processes that combine alkyne hydroalkoxylation with carbon–carbon bond formation have also been reported. Li has described a gold(III)-catalyzed protocol for the annulation of salicylicaldehydes with aryl acetylenes to form isoflavanones [23]. Barluenga has reported the gold(III)-catalyzed tandem cycloisomerization/Prins-type cyclization of ω-alkynols that possess an appropriately positioned alkenyl group to form eight-membered carbocycles [24]. As an example, treatment of diallyl ω-alkynol **10** with a catalytic amount of $AuCl_3$ in propanol at 80 °C for 12 h led to isolation of bicyclic ether **11** in 94% yield (Eq. (12.6)). Hayashi has reported the gold(I)-catalyzed hydrative cyclization of diynes to form cyclohexenones [25] and Liu has reported the hydrative carbocyclization of oxo diynes and aromatic 1,6-diyne-4-en-3-ols [26]. As an example of the latter, treatment of 1,6-diyne **12** with a catalytic mixture of $(PPh_3)AuCl$ and AgOTf led to isolation of naphthyl derivative **13** in 90% yield (Eq. (12.7)).

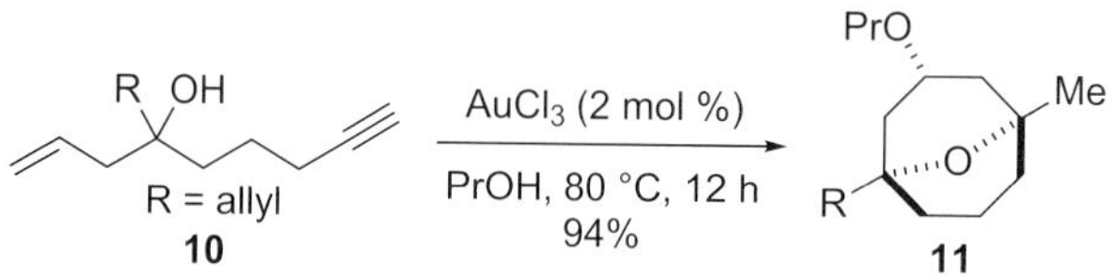

R OH, R = allyl, **10** → $AuCl_3$ (2 mol %), PrOH, 80 °C, 12 h, 94% → PrO, Me, O, R, **11** (12.6)

OH Ph *p*-tolyl **12** — $(PPh_3)AuCl$ (2 mol %), AgOTf (2 mol %), CH_2Cl_2, rt, 3 h, 90% → Ph, O, *p*-tolyl **13** (12.7)

12.2.2 Ketones as Nucleophiles

In 2000, Hashmi reported that reaction of 5-octyne-3-one with a catalytic amount of $AuCl_3$ in acetonitrile formed 2,5-diethylfuran in quantitative yield within minutes at room temperature (Eq. (12.8)) [15]. As an extension of this protocol, Liu has reported the gold(III)-catalyzed tandem cyclization/alkoxylation of 2-oxo-3-butynoic esters and 1,2-diones with alcohols to form 2,2-disubstituted 3(2*H*)-furanones [27]. For example, treatment of 1,6-diphenyl-1,5-hexadiyne-3,4-dione with 3 equiv of methanol and a catalytic amount of $AuCl_3$ in dichloromethane at room temperature led to isolation of furanone **14** in 88% yield (Eq. (12.9)). Attack of methanol at the distal carbonyl carbon atom may occur in concert with attack of the carbonyl oxygen atom on the gold-complexed alkyne or, more likely, may occur on the oxocarbenium ion generated via carbonyl attack on the gold-coordinated alkyne.

Et, O, Et — $AuCl_3$ (0.1 mol %), CH_3CN, rt, quant → Et, O, Et (12.8)

Ph, O, O, Ph — $AuCl_3$ (2 mol %), MeOH (3 equiv), CH_2Cl_2, rt, 88 % → Ph, O, O, Ph, OMe **14** (12.9)

Larock has developed an effective gold(III)-catalyzed method for the conversion of 2-(1-alkynyl)-2-alken-1-ones to substituted furans [28]. As an example, treatment of cyclohexenone **15** with methanol and a catalytic amount of $AuCl_3$ in dichloromethane at room temperature for 1 h led to isolation of furan **16** in 88% yield (Eq. (12.10)). Liang has demonstrated the suitability of ionic liquids as solvents for this transformation [29].

O, Ph **15** — $AuCl_3$ (1 mol %), MeOH (1.5 equiv), CH_2Cl_2, rt, 88 % → Ph, O, OMe **16** (12.10)

A number of interesting gold-catalyzed processes involving the heterocyclization/1,2-alkyl migration of α-alkynyl ketones have appeared recently. For example, Schmalz has shown that treatment of 1-(1-alkynyl)cyclopropyl ketone **17** with a catalytic amount of $(PPh_3)AuOTf$ in dichloromethane that contained methanol (2 equiv) led to isolation of fused bicyclic furan **18** in 91% yield [30]. Conversion of **17** to **18** presumably occurs via attack of the carbonyl oxygen atom on the gold(I)-alkyne complex **19** coupled with 1,2-alkyl migration followed by attack of methanol at the resulting carbocation and protodeauration (Scheme 12.3). Alternatively, Kirsch has shown that treatment of α-hydroxy-α-alkynylcyclohexanone **20** with a catalytic amount of $AuCl_3$ in CH_2Cl_2 at room temperature led to isolation of spirocyclic furanone **21** in 81% yield (Scheme 12.4) [31]. The conversion of **20** to **21** is presumably initiated by 5-*endo* attack of the ketone carbonyl group on a gold-complexed alkyne, followed by alkyl migration and protodeauration (Scheme 12.4).

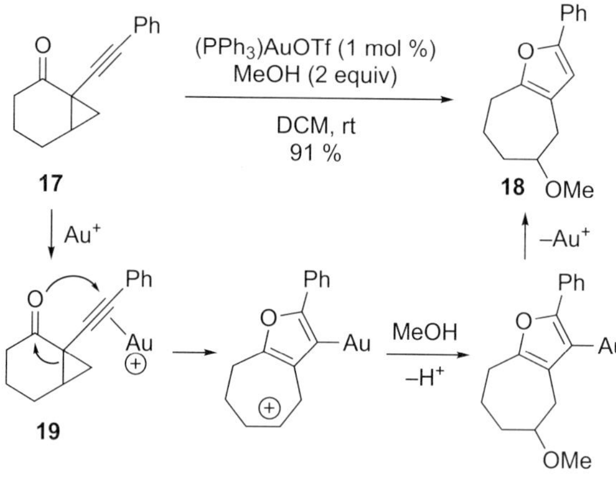

Scheme 12.3

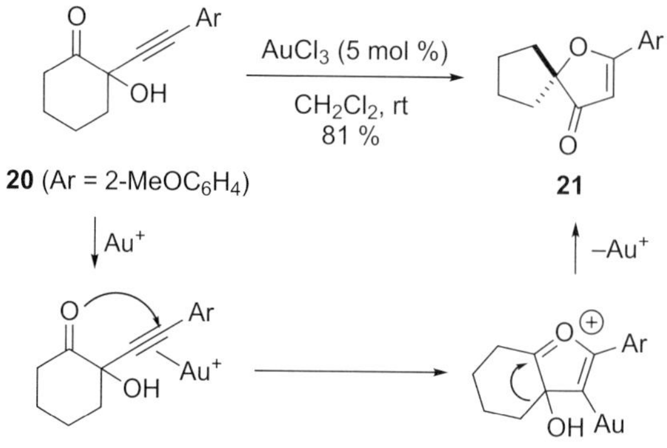

Scheme 12.4

A number of extensions and modification of the gold-catalyzed cyclization/1,2-alkyl migration of α-alkynyl ketones have been developed. Zhang has recently reported the gold(I)-catalyzed methods for the [4 + 2] annulation [32] and 1,3-dipolar cycloaddition [33] of 1-(2-phenylethynyl)cyclopropyl methyl ketones to form carbo/heterocyclic compounds and bicyclo[3.2.0]heptanes, respectively (Scheme 12.5). Kirsch has extended his heterocyclization/1,2-alkyl migration methodology to include the synthesis of 4-iodo-3-furanones through the NIS-induced, gold(III)-catalyzed cyclization/1,2-migration of silylated α-hydroxy-α-alkynyl ketones [34].

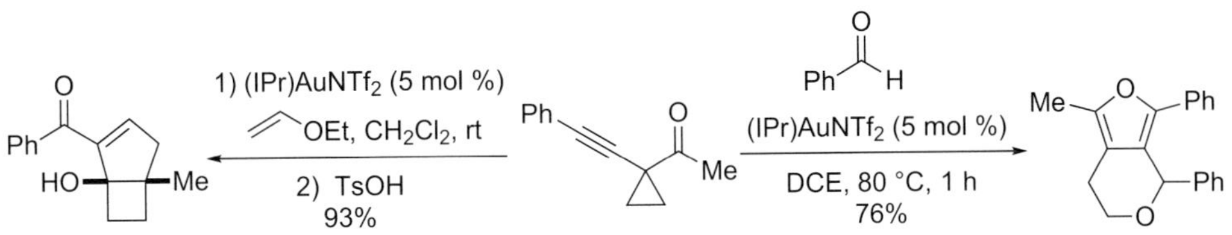

1) (IPr)AuNTf$_2$ (5 mol %), OEt, CH$_2$Cl$_2$, rt; 2) TsOH, 93%

PhCHO, (IPr)AuNTf$_2$ (5 mol %), DCE, 80 °C, 1 h, 76%

Scheme 12.5

12.2.3
Aldehydes as Nucleophiles

The gold(I)-catalyzed cyclization of *o*-alkynylbenzaldehydes has emerged as an efficient approach to the synthesis of aromatic and heteroaromatic compounds. In 2002, Yamamoto reported the gold(III)-catalyzed benzannulation of *o*-alkynylbenzaldehydes with alkynes to form naphthyl ketone derivatives [35]. As an example, treatment of 2-(2-phenylethynyl)benzaldehyde **22** with 1-phenyl-1-propyne and a catalytic amount of AuCl$_3$ in DCE at 80 °C led to isolation of 1,2,3-trisubstituted naphthyl derivative **23** in 89% yield as a single regioisomer (Scheme 12.6). The conversion of **22** to **23** presumably proceeds via initial formation of benzopyrylium zwitterion **24** via addition of the carbonyl oxygen atom to the gold-coordinated alkyne, followed by [4 + 2]-cycloaddition with 1-phenyl-1-propyne to form **25**. Retrocycliza-

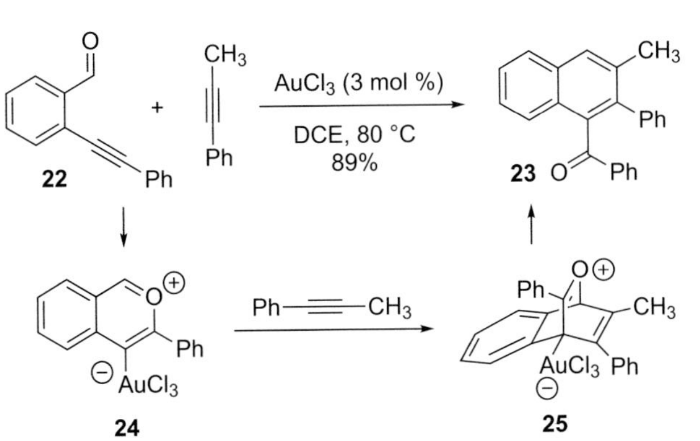

AuCl$_3$ (3 mol %), DCE, 80 °C, 89%

Scheme 12.6

tion/deauration would form **23**. [4 + 2]-Trapping of the benzopyrylium ion with alkynes has also been realized with tethered alkynes, leading to formation of aromatic polycyclic derivatives [36]. Asao has reported similar intermolecular [4 + 2]-benzannulation reactions through employment of an enol [37] or a benzyne [38] as the dieneophile component, and Dyker has described gold(III)-catalyzed domino reactions of the isobenzopyrylium cations similar to **24** with olefins and electron-rich heteroarenes [39].

Porco has reported the gold(III)-catalyzed cyclization of *o*-alkynylbenzaldehydes as a route to azaphilones and related compounds [40]. Toward this goal, treatment of *o*-alkynylbenzaldehyde **26** with a catalytic amount of $Au(OAc)_3$ in a 10 : 1 DCE/TFA mixture led to quantitative formation of the benzopyrylium salt **27** within 1 min at room temperature (Scheme 12.7). In a preparative-scale reaction, gold(III)-catalyzed cyclization of **26** followed by oxidation with IBX led to isolation of azaphilone **28** in 84% yield (Scheme 12.7). Similarly, the two-step cyclization/oxidation protocol was applied to the synthesis of a number of unnatural azaphilones.

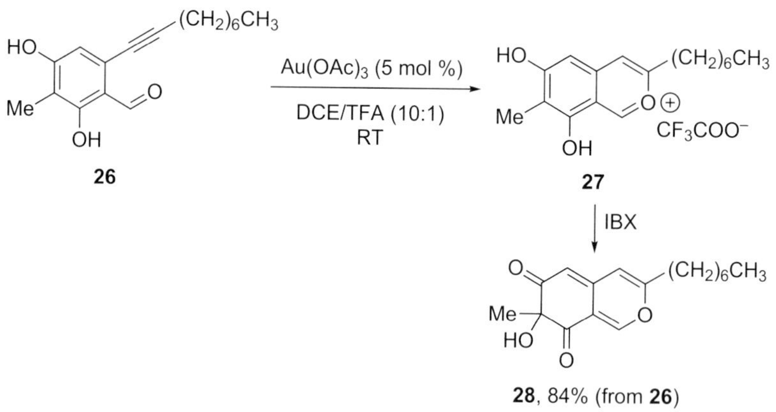

HO, Me, OH, $(CH_2)_6CH_3$, O — **26**; $Au(OAc)_3$ (5 mol %), DCE/TFA (10:1), RT → HO, Me, OH, $(CH_2)_6CH_3$, $O^{\oplus}$ CF_3COO^- — **27**; IBX → O, Me, HO, O, $(CH_2)_6CH_3$, O — **28**, 84% (from **26**)

Scheme 12.7

Li has reported the water-promoted, gold(I)-catalyzed cascade addition/cyclization of terminal alkynes with *o*-alkynylbenzaldehyde derivatives to form 1-alkynyl-1*H*-isochromenes [41]. For example, reaction of 1-(2-phenylethynyl)benzaldehyde with phenylacetylene catalyzed by a 1 : 4 mixture of $(PMe_3)AuCl$ and Hunig's base in a water/toluene mixture at 70 °C for 1 day led to isolation of isochromene **29** in 81% yield (Eq. (12.11)). The transformation is presumably initiated by gold-catalyzed addition of acetylide to the C=O bond of the aldehyde moiety followed by addition of the resulting alkoxide across the pendant C≡C triple bond.

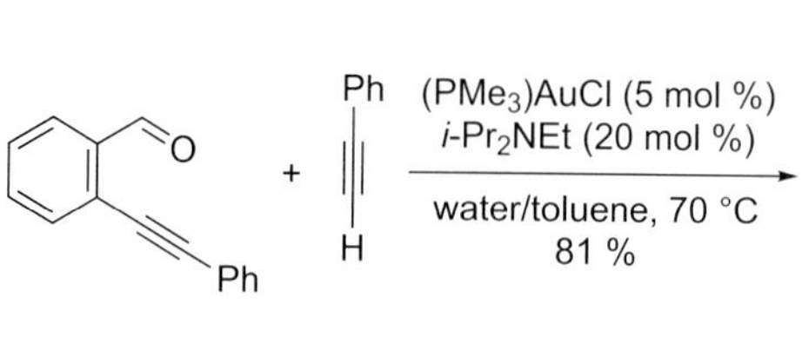

O, Ph + Ph, H; $(PMe_3)AuCl$ (5 mol %), *i*-Pr_2NEt (20 mol %), water/toluene, 70 °C, 81 % → Ph, O, Ph — **29** (12.11)

12.2.4
Carboxylic Acids as Nucleophiles

Gold complexes have also been shown to catalyze the addition of the O–H group of a carboxylic acid across a C≡C triple bond (hydroalkoxycarbonylation). Pale has developed an efficient and selective protocol for the intramolecular hydroalkoxycarbonylation of alkynes that is catalytic in both gold and base [42]. In an optimized protocol, treatment of 5-phenyl-4-pentynoic acid with a catalytic 1 : 1 mixture of AuCl and potassium carbonate in acetonitrile at 20 °C for 2 h led to isolation of the γ-benzylidine lactone **30** in 96% yield with exclusive formation of the *Z*-stereoisomer (Eq. (12.12)). Likewise, Genêt and Michelet have developed a gold(I)-catalyzed method for the intramolecular hydroalkoxycarbonylation of propargylmalonate derivaties [43]. As an example, treatment of ethyl propargylmalonate derivative **31** with a catalytic amount of AuCl in acetonitrile at room temperature led to isolation of γ-benzylidene lactone **32** in 84% yield with exclusive formation of the *Z*-stereoisomer (Eq. (12.13)). 2-(1-Alkynyl)benzoic acid derivatives also undergo intramolecular hydroalkoxycarbonylation at room temperature in the presence of a catalytic 1 : 1 mixture of AuCl and potassium carbonate to form aromatic lactones [44].

AuCl (10 mol %), K_2CO_3 (10 mol %), CH_3CN, 20 °C, 96% → **30** (12.12)

31 (Ar = 4-CNC_6H_4) → AuCl (5 mol %), CH_3CN, rt, 84% → **32** (12.13)

12.2.5
Rearrangements of Propargylic Carboxylates

12.2.5.1 Acyl Migration Followed by Nucleophilc Attack

A number of efficient gold(I)-catalyzed tandem transformations initiated by the 1,2- or 1,3-acyl migration of propargylic carboxylates to form a gold carbenoid or gold allene species, respectively, have appeared in recent years [45]. One group of these transformations involves acyl migration followed by addition of a heteroatom nucleophile to the resulting carbene or allene intermediate. Gagosz has reported the gold(I)-catalyzed tandem rearrangement/hydroalkoxylation of α-benzyloxy-α′--hydroxyalkynes to form functionalized 2,5-dihydrofurans [46]. As an example, treatment of the diastereomerically and enantiomerically enriched butynediol monobenzoate **33** with a catalytic amount of $(PPh_3)AuNTf_2$ in dichloromethane at room temperature formed 2,5-dihydrofuran **34** in 97% yield with complete preservation of relative and absolute configuration. Compound **34** is presumably formed via an initial 1,3-acyl migration followed by 5-*endo*-trig hydroalkoxylation of the resulting α-hydroxy allenyl carboxylate **35** (Scheme 12.8).

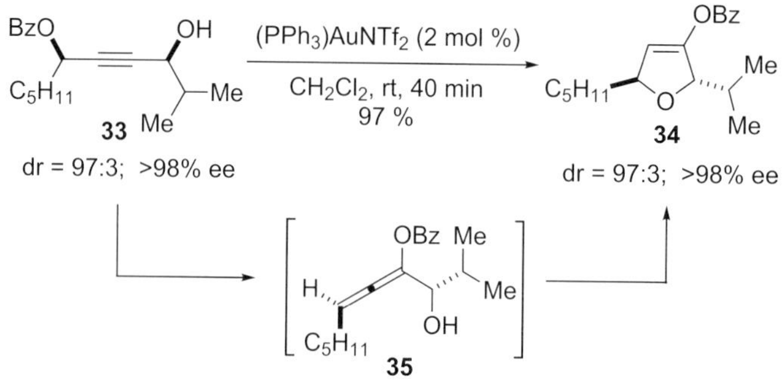

BzO OH C_5H_{11} Me **33** Me dr = 97:3; >98% ee
$(PPh_3)AuNTf_2$ (2 mol %), CH_2Cl_2, rt, 40 min, 97 %
OBz C_5H_{11} O Me Me **34** dr = 97:3; >98% ee
[H OBz Me Me OH C_5H_{11} **35**]

Scheme 12.8

Shin [47] and Gevorgyan [48] have reported the syntheses of furan derivatives via the gold-catalyzed migration/cyclization of propargylic carboxylates that possess a pendant hydroxy or carbonyl group, respectively. Similarly, De Brabander has described the gold(I)-catalyzed cycloetherification of ω-hydroxy propargylic esters to form oxacyclic enolesters [49]. Kato and Akita have reported the hydrative cyclization of 1,1-diethynyl acetate to form lactones [50]. Davies has reported the intermolecular thiolation of propargylic carboxylates with sulfides [51] and Toste [52] has reported the synthesis of azepines by a gold(III)-catalyzed annulation of a propargylic carboxylate with a vinyl imine. The latter transformation is likely initiated by a 1,2-acyl migration. Echavarren has shown that propargylic carboxylates react with carbon-based nucleophiles such as 1,3-dicarbonyl compounds or electron-rich arenes to form enol carboxylates. These transformations presumably occur via attack of the nucleophile on the reactive gold species (gold carbenoid or gold allene complex) generated from substrate-dependent 1,2- or 1,3-acyl migration (Scheme 12.9) [53].

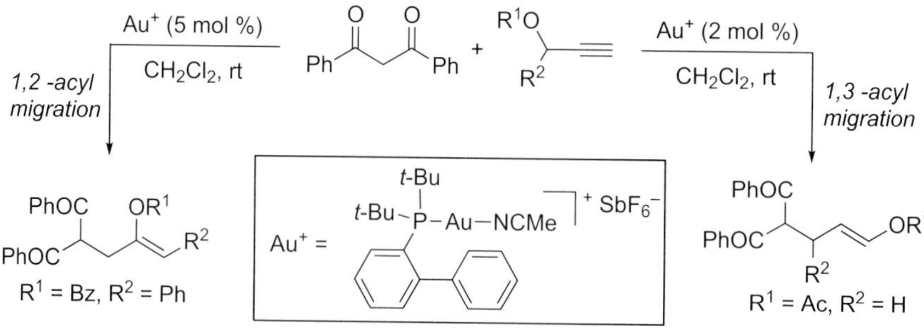

Ph(CO)CH_2(CO)Ph + R^1O, R^2 propargyl
Au^+ (5 mol %), CH_2Cl_2, rt — *1,2-acyl migration* → PhOC, PhOC, OR^1, R^2; R^1 = Bz, R^2 = Ph
Au^+ (2 mol %), CH_2Cl_2, rt — *1,3-acyl migration* → PhOC, PhOC, R^2, OR^1; R^1 = Ac, R^2 = H
Au^+ = [t-Bu(t-Bu)P(biphenyl)–Au–NCMe]$^+$ SbF_6^-

Scheme 12.9

12.2.5.2 Acyl Migration Followed by C=C/C≡C Addition

A number of gold-catalyzed transformations involving the 1,2- or 1,3-acyl migration of a propargylic carboxylate followed by reaction of the resulting gold carbenoid or allene intermediate with an alkene have been documented. Gagosz has described the gold(I)-catalyzed isomerization of 5-en-2-yn-1-ol acetates to form acetoxy bicyclo[3.1.0]hexenes [54]. As an example, treatment of enynyl acetate **36** with a catalytic amount of [P(*t*-Bu)$_2$*o*-biphenyl]AuNTf$_2$ (1 mol %) in dichloromethane at room temperature for 5 min led to isolation of bicycle **37** in 97% yield (Eq. (12.14)) [54]. In comparison, Zhang has shown that 4-en-2-yn-1-ol acetates undergo tandem [3]-rearrangement/Nazarov cyclization in the presence of cationic gold(I) complexes to form substituted cyclopentenes (Eq. (12.15)) [55]. Fürstner has shown that β-alkenyl propargyl acetates undergo rearrangement/cyclization to form bicyclo[3.1.0]hexane derivatives in the presence of a gold(I) phosphine complex and has applied closely related transformations to the synthesis of biologically active terpernes [56]. As an example of the former transformation, treatment of enynyl acetate **38** with a catalytic mixture of (PPh$_3$)AuCl and AgSbF$_6$ at room temperature followed by deacylation with potassium carbonate in methanol led to isolation of bicycle **39** in 74% yield from **38** (Eq. (12.16)) [56]. Nolan [57], Hanna [58], and Toste [59] have reported similar cyclopropane-forming transformations. Toste has also documented the gold(I)-catalyzed rearrangement/cyclization of 1-ethynyl-2-propenyl pivalates to form cyclopentenones [60] and Zhang has reported the tandem gold(I)-catalyzed rearrangement/[2 + 2]-cycloaddition of indole-derived propargylic esters **40** to form indoline-fused cyclobutanes **41** (Eq. (12.17)) [61].

OAc, Ph, Me, **36** → [P(*t*-Bu)$_2$*o*-biphenyl]AuNTf$_2$ (1 mol %), CH$_2$Cl$_2$, rt, 5 min, 97% → OAc, Ph, Me, **37**

t-Bu, P–*t*-Bu, P(*t*-Bu)$_2$*o*-biphenyl

(12.14)

OAc, Me → (PPh$_3$)AuCl (1 mol %), AgSbF$_6$ (1 mol %), wet CH$_2$Cl$_2$, rt, 30 min, 95% → O, Me

(12.15)

OAc, Ph, **38** → (PPh$_3$)AuCl (2 mol %), AgSbF$_6$ (2 mol %), CH$_2$Cl$_2$, rt → AcO, Ph, H → K$_2$CO$_3$, MeOH → O, Ph, H, **39** (74 %)

(12.16)

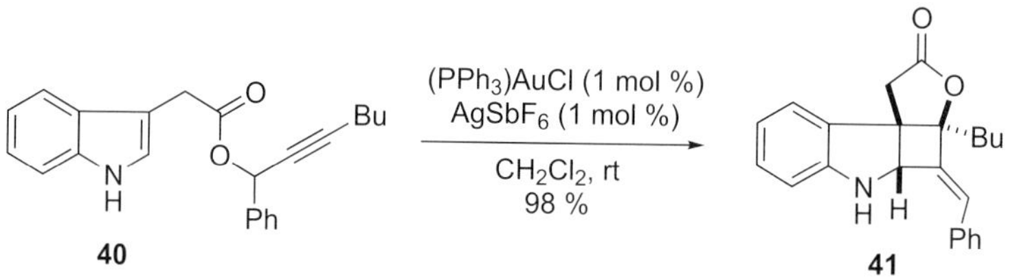

(12.17)

Alkynyl propargylic carboxylates have also been employed to good effect in cascade C–O/C–C bond forming processes initiated by 1,2- or 1,3-acyl migration. For example, Schreiber has reported the synthesis of α-pyrones via the gold-catalyzed rearrangement/cyclization of propargyl α-alkynoates (Eq. (12.18)) [62]. Oh has reported the gold(I)-catalyzed rearrangement/cyclization of alkynyl propargyl acetates to form naphthyl ketone derivatives [63] and 2,3-bis(alkylidiene)cycloalkanones [64]. As an example of the former transformation, treatment of propargylic acetate **42** with a catalytic mixture of sodium tetrachloroaurate and triphenylphosphine in DCE at room temperature for 30 min led to isolation of naphthyl ketone **43** in 90% yield (Eq. (12.19)) [63]. The gold(I)-catalyzed conversion of aromatic alkynyl propargyl carboxylates to naphthyl ketones was also reported by Toste, who showed that silver phosphine complexes also catalyzed this transformation [65].

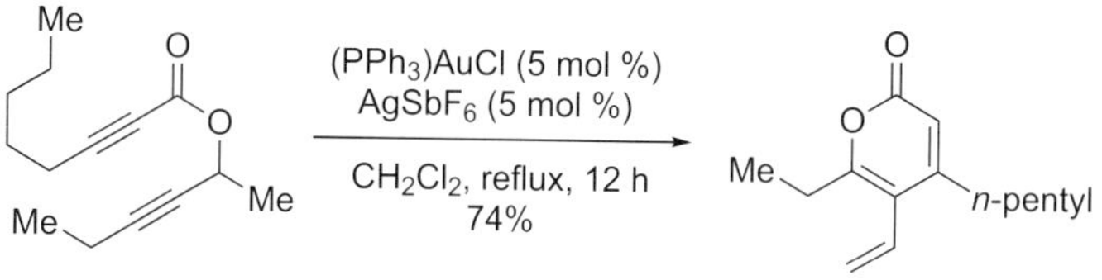

(12.18)

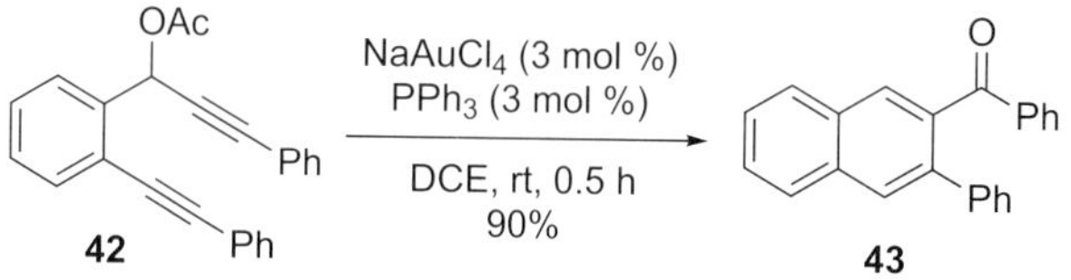

(12.19)

12.2.5.3 **Acyl Migration Leading to Diene/Ketone Formation**

Gold complexes catalyze the rearrangement of propargyl acetates to form dienes and/or unsaturated ketones. Zhang has shown that gold(I) complexes catalyze the isomerization of propargylic pivalates [66] or trimethylsilyl-substituted propargyl acetates [67] to form 1,3-butadien-2-ol esters (Eq. (12.20)). In comparison, gold(III) complexes catalyze the isomerization of propargyl acetates to form α-ylidene-β--diketones (Eq. (12.21)) [68]. Zhang has also demonstrated the gold(I)-catalyzed rearrangement/deoxygenation of propargyl acetates to form α,β-unsaturated ketones under neutral [69] or oxidative conditions [70]. Similar to the former transformation, cationic gold(I) complexes that contain an *N*-heterocyclic carbene ligand catalyze the conversion of allylic acetates to enones and enals [71].

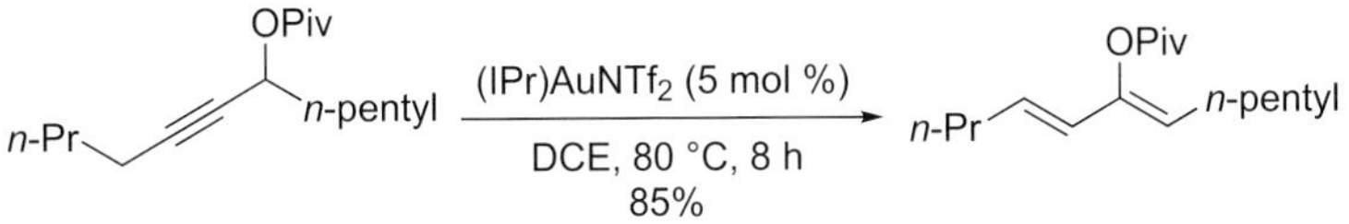

(12.20)

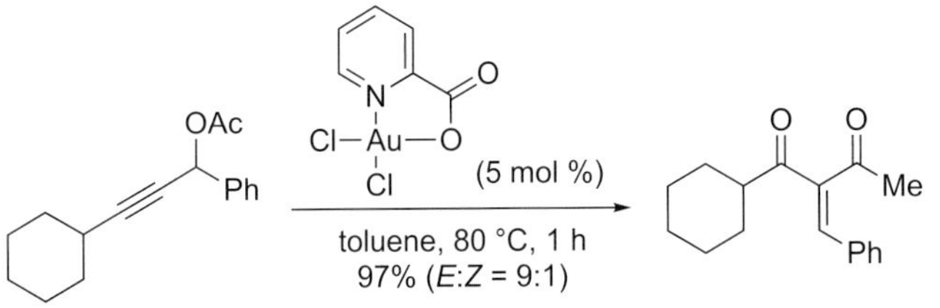

(12.21)

12.2.6
Carbonates and Carbamates as Nucleophiles

Shin has reported a gold(I)-catalyzed protocol for the cyclization of homopropargyl *tert*-butyl carbonates to form cyclic enol carbonates [72]. As an example, treatment of propargyl carbonate **44** with a catalytic 1 : 1 mixture of the electron-deficient gold(I) complex $[P(C_6F_5)_3]AuCl$ and $AgSbF_6$ in dichloroethane at 0 °C to room temperature led to isolation of cyclic carbonate **45** in 80% yield (Eq. (12.22)). Gagosz reported a similar transformation that employed the preformed gold(I) complex $(PPh_3)AuNTf_2$ as a catalyst (Eq. (12.23)) [73]. Asao has utilized *o*-alkynylbenzoates as alkylating agents for Friedel–Crafts alkylations in a transformation initiated by intramolecular attack of the carbonyl oxygen atom at the proximal C≡C bond [74].

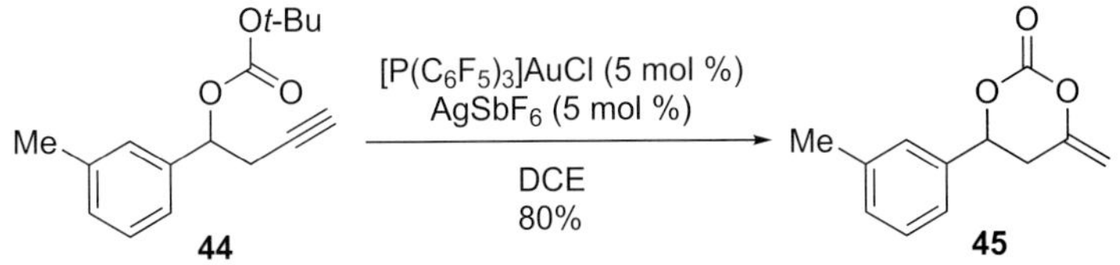

(12.22)

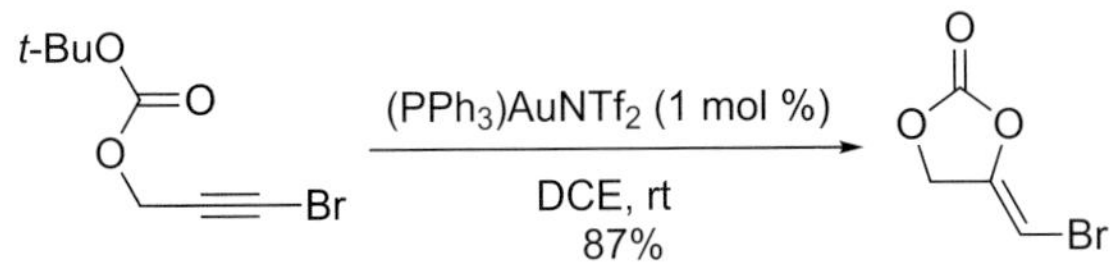

(12.23)

Gagosz has also described an effective gold(I)-catalyzed method for the cyclization of *N*-propargyl carbamates to form 5-methylene-1,3-oxazolidin-2-ones [75]. As an example, treatment of carbamate **46** with a catalytic amount of $(PPh_3)AuNTf_2$ in dichloromethane at room temperature for 5 min led to isolation of 1,3-oxazolidin-2-one **47** in 99% yield (Eq. (12.24)). Similarly, mixtures of $(PPh_3)AuCl$ and either $AgSbF_6$ [76] or AgOTf [77] catalyze the cyclization of *N*-propargyl carbamates to form

5-methylene-1,3-oxazolidin-2-ones. The former procedure was also effective for the cyclization of *N*-3-butynylcarbamates to form 2-oxazinones via 6-*exo* cyclization. Similar to these transformations, Hashmi has reported the gold(I)-catalyzed 5-*endo*-dig cyclization of *N*-alkynyl carbamates to form oxazolinones at or below room temperature [78]. For example, reaction of *tert*-butyl *N*-alkynylcarbamate **48** with a catalytic amount of $(PPh_3)AuNTf_2$ in deuterated chloroform at room temperature led to isolation of oxazolinone **49** in 93% yield (Eq. (12.25)). Hashimi [79] and Uemura [80] have independently reported the gold(III)-catalyzed cyclization of *N*-propargylcarboxamides to form oxazoles (Eq. (12.26)).

46 → **47**: $(PPh_3)AuNTf_2$ (1 mol %), CH_2Cl_2, rt, 99% (12.24)

48 → **49**: $(PPh_3)AuNTf_2$ (5 mol %), $CDCl_3$, rt, 93% (12.25)

$AuCl_3$ (5 mol %), CH_2Cl_2, rt, 98%; R = 1-adamantyl (12.26)

12.2.7
Ethers and Epoxides as Nucleophiles

Hashmi has shown that gold(III) complexes catalyze the conversion of alkynyl epoxides to furans [81]. As an example, treatment of the alkynyl epoxide **50** with a catalytic amount of $AuCl_3$ in acetonitrile at room temperature led to isolation of the 2,4-disubstituted furan **51** in 84% yield (Eq. (12.27)). A number of extensions of this basic transformation have been developed. Hashmi has reported the gold(I)-catalyzed synthesis of 3-acylindenes from 2-alkynylaryl epoxides [82]. Shi has described gold(I)-catalyzed procedures for the hydrative cyclization of epoxy alkynes to form cyclic ketals [83]. Similarly, Liang has reported the gold(III)-catalyzed tandem cyclization/1,2-alkyl migration of 1-alkynyl-2,3-epoxy alcohols to form pyranone derivatives [84] and the tandem cycloisomerization/dimerization of 1-alkynyl-2,3-epoxy acetates to form difurylmethane derivatives [85]. Similar to this latter transformation, Shi has reported a gold(I)-catalyzed procedure for the tandem cycloisomerization/dimerization of 1-alkynyl-2,3-epoxy alcohols to form difurylmethane derivatives [86].

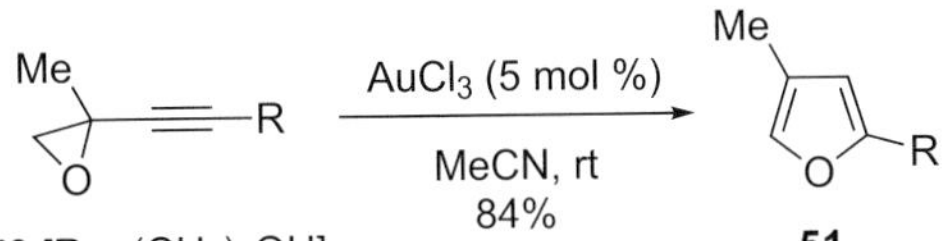

Me, $AuCl_3$ (5 mol %), MeCN, rt, 84%, Me, R, O, R, O, **50** [R = $(CH_2)_4OH$], **51** (12.27)

Toste has reported the synthesis of indenyl ethers via the gold(I)-catalyzed intramolecular carboalkoxylation of aromatic alkynes, presumably via attack of the ether oxygen atom on a gold(I)-alkyne complex followed by electrophilic deauration of the resulting vinyl gold species with the pendant benzylic carbocation (Scheme 12.10) [87]. Rhee has reported the gold(I)-catalyzed cycloisomerization of 3-methoxy-1,6-enynes to form 1-methoxy-1,4-cycloheptadienes (Eq. (12.28)) [88]. The course of the gold(I)-catalyzed cyclization of 3-silyloxy-1,6-enynes depends on the nature of the supporting ligand. For example, treatment of 3-silyloxy-1,6-enynes with a catalytic amount of $[P(t\text{-}Bu)_2o\text{-biphenyl}]AuCl/AgSbF_6$ leads predominantly to tandem carbocyclization/pinacol rearrangement to form methylene cyclopentane whereas reaction of 3-silyloxy-1,6-enynes with a catalytic amount of $[P(C_6F_5)_3]AuCl/AgSbF_6$ leads to tandem carbocylization/Claisen rearrangement to form 4-cycloheptenone [89].

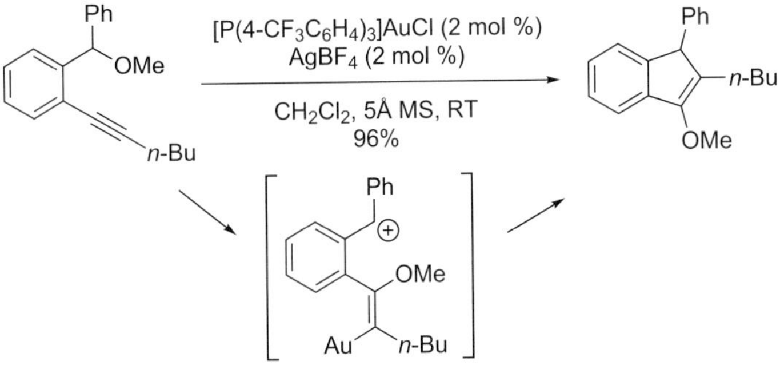

Ph, OMe, n-Bu, $[P(4\text{-}CF_3C_6H_4)_3]AuCl$ (2 mol %), $AgBF_4$ (2 mol %), CH_2Cl_2, 5Å MS, RT, 96%, Ph, n-Bu, OMe, Ph, ⊕, OMe, Au, n-Bu

Scheme 12.10

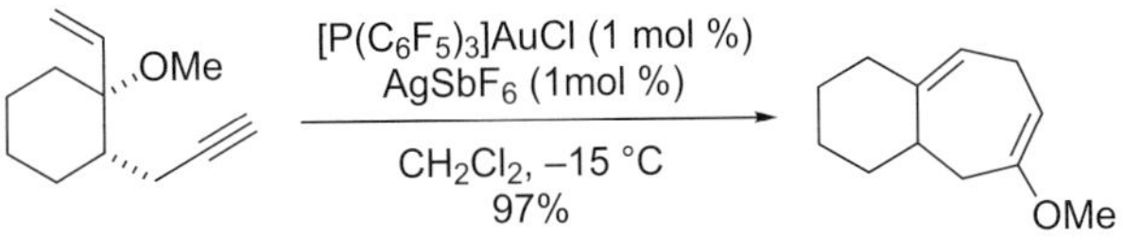

OMe, $[P(C_6F_5)_3]AuCl$ (1 mol %), $AgSbF_6$ (1mol %), CH_2Cl_2, −15 °C, 97%, OMe (12.28)

12.2.8
Additional Nucleophiles

Toste has reported the cyclization/rearrangement of aryl γ-alkynyl sulfoxides catalyzed by a gold(I) *N*-heterocyclic carbene complex [90]. For example, treatment of phenyl 3-butynyl sulfoxide **52** with a catalytic 1 : 1 mixture of (IMes)AuCl [IMes = 1,3-bis(2,4,6-trimethylphenyl)imidazol-2-ylidene] and $AgSbF_6$ led to isolation of

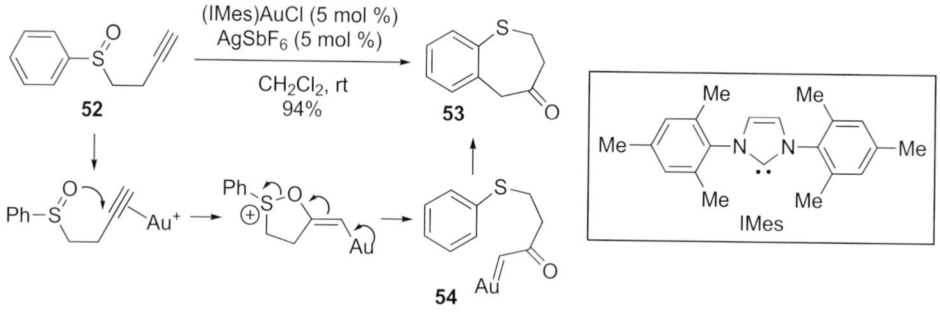

Scheme 12.11

1-benzothiepin-4-one **53** in 94% yield (Scheme 12.11). Conversion of **52** to **53** presumably occurs via addition of the sulfoxide carbonyl group to the pendant C≡C bond, followed by ring opening to form the gold-carbenoid complex **54**. C–H activation/arylation then forms **53** (Scheme 12.11). Zhang has shown that the conversion of **52** to **53** is also accomplished in 93% yield through employment of the gold(III) complex dichloro(pyridine-2-carboxylate)gold as a catalyst [91]. Asao and Yamamoto have reported the gold(III)-catalyzed cyclization of *o*-alkynylnitrobenzenes as a route to isatogens and anthranils [92]. As an example, treatment of 1-(2-phenylethynyl)nitrobenzene with a catalytic amount of $AuBr_3$ in DCE at room temperature formed isatogen **55** and anthranil **56** in 53% and 34% yield, respectively (Eq. (12.29)). In a related transformation, Shin has reported the generation of azomethine ylides from gold-catalyzed addition of the oxygen atom of a nitrone to an alkyne [93].

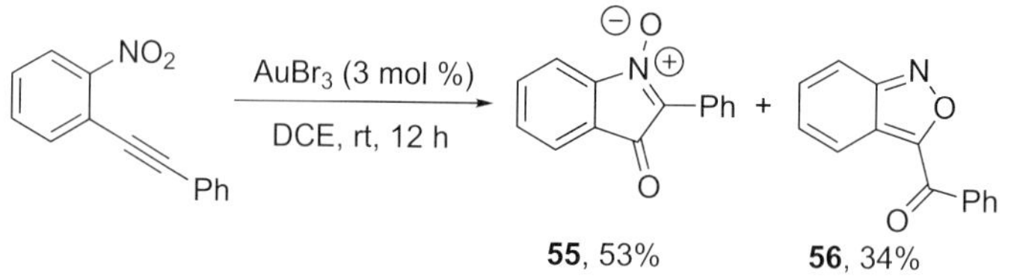

(12.29)

12.3 Addition to Allenes

12.3.1 Carbinols as Nucleophiles

12.3.1.1 Intramolecular Processes

In 2001, Krause reported that gold(III) salts catalyzed the stereoselective *endo*-hydroalkoxylation of α-hydroxyallenes to form 2,5-dihydrofurans [94]. For example, treatment of the diastereomerically enriched α-hydroxyallene (dr = 9 : 1) **57** with a catalytic amount of $AuCl_3$ in dichloromethane at room temperature led to isolation of 2,5-dihydrofuran **58** in 74% yield as a 9 : 1 mixture of diastereomers (Eq. (12.30)). This transformation was also realized through catalysis with HCl or Amerlyst 15.

57 (dr = 9:1) → [$AuCl_3$ (5 mol %), CH_2Cl_2, rt, 74%] → 58 (dr = 9:1) (12.30)

A number of extensions and applications of the gold-catalyzed *endo*-hydroalkoxylation of α-allenyl alcohols have been reported [95–102]. Krause has demonstrated the effectiveness of ionic liquids as solvents for the gold(III)-catalyzed *endo*-hydroalkoxylation of α-hydroxy allenes [96]. Krause has also reported a method for the stereoselective cyclization of phenyl-substituted α-hydroxyallenes, which epimerized under the originally developed conditions [97]. Hashmi has shown that the gold(III)-catalyzed cyclization of tertiary α-hydroxyallenes is accompanied by the formation of oxidative coupling products, which points to the *in situ* reduction of gold(III) to gold(I) under the reaction conditions [98]. Consistent with this hypothesis, Hegedus reported that treatment of the α-hydroxy allenamide **59** with a catalytic 1 : 1 mixture of $(PPh_3)AuCl$ and $AgBF_4$ at room temperature led to isolation of 2,5-dihydrofuran **60** in 80% yield as a single stereoisomer (Eq. (12.31)) [99]. Reissig has applied gold(I)-catalyzed cyclization of α-hydroxyallenes to the synthesis of 3-alkoxy-2,5-dihydrofurans [100]. Krause applied the gold(III)-catalyzed cycloisomerization of α-hydroxyallenes to the stereoselective synthesis of (−)-isocyclocapitelline and (−)-isochrysotricine [101], and to the synthesis of furanomycin derivatives [102].

59 → [$(PPh_3)AuCl$ (5 mol %), $AgBF_4$ (5 mol %), CH_2Cl_2, rt, 80%] → 60 (12.31)

Krause has reported the gold-catalyzed intramolecular *endo*-hydroalkoxylation of β-hydroxyallenes to form dihydropyrans [103]. For example, treatment of a 70 : 30 diastereomeric mixture of β-hydroxyallene **61** with a catalytic 1 : 1 mixture of AuCl and pyridine in dichloromethane led to isolation of dihydropyran **62** in 84% yield as a 70 : 30 mixture of diastereomers (Eq. (12.32)). This transformation was also catalyzed effectively by a 1 : 1 mixture of $(PPh_3)AuCl$ and $AgBF_4$ in toluene. β-Hydroxyallenes that possessed substitution at the internal allenyl carbon atom also underwent gold-catalyzed cycloisomerization with selective transfer of chirality from the allenyl moiety to the newly formed tetrahedral stereocenter, but dihydropyran formation was sometimes accompanied by formation of 2,5-dihydrofurans as side products.

AuCl (5 mol %), pyridine (5 mol %), CH_2Cl_2, rt, 84%

61 (dr = 70:30) → **62** (dr = 70:30) (12.32)

In 2006, Widenhoefer reported an effective gold(I)-catalyzed protocol for the *exo*-hydroalkoxylation of γ- and δ-hydroxy allenes to form 2-vinyl tetrahydrofurans and 2-vinyl tetrahydropyrans, respectively [104]. For example, treatment of 1-phenyl-5,6-heptadienol with a catalytic 1 : 1 mixture of [P(*t*-Bu)$_2$*o*-biphenyl]AuCl and AgOTs in toluene at room temperature led to isolation of 2-phenyl-6-vinyltetrahydropyran in 96% yield as a 7.2 : 1 mixture of diastereomers (Eq. (12.33)). This gold(I)-catalyzed hydroalkoxylation protocol tolerated substitution at the terminal allenyl carbon atoms and along the alkyl chain that tethered the hydroxy group to the allenyl moiety and was also effective for the 5-*exo* hydroalkoxylation of γ-hydroxy allenes. Alcaide and Almendros have shown that gold(III) also catalyzes the 5-*exo* hydroalkoxylation of γ-allenyl alcohols in modest yields (Eq. (12.34)) [105].

Au[P(*t*-Bu)$_2$(*o*-biphenyl)]Cl (5 mol %), AgOTs (5 mol %), toluene, 25 °C, 96% (7.2:1) (12.33)

$AuCl_3$ (5 mol %), CH_2Cl_2, rt, 57% (12.34)

12.3.1.2 Enantioselective Processes

In early 2007, Widenhoefer and Zhang reported the gold(I)-catalyzed intramolecular enantioselective hydroalkoxylation of γ- and δ-hydroxyallenes [106]. For example, reaction of 2,2-diphenyl-4,5-hexadienol with a catalytic 1 : 2 mixture of [(*S*)-**63**]Au_2Cl_2 [(*S*)-**63** = (*S*)-DTBM-MeObiphep] and AgOTs at −20 °C in toluene for 18 h led to isolation of 4,4-diphenyl-2-vinyltetrahydrofuran in 67% yield with 93% *ee* (Eq. (12.35)). This protocol was also effective for the enantioselective 6-*exo*-hydroalkoxylation of δ-hydroxyallenes to form tetrahydropyrans. Gold(I)-catalyzed cyclization of γ-hydroxyallenes that possessed an axially chiral 1,3-disubstituted allenyl moiety occurred with high enantioselectivity/low diastereoselectivity in a catalyst-

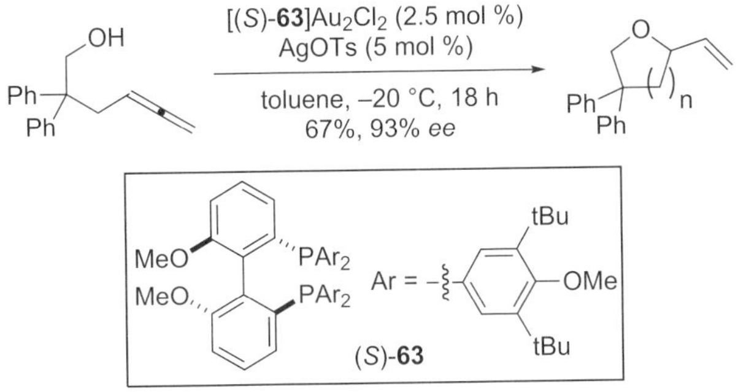

[(*S*)-**63**]Au_2Cl_2 (2.5 mol %)
AgOTs (5 mol %)
toluene, −20 °C, 18 h
67%, 93% *ee*

OH
Ph
Ph
O
Ph
Ph
n

MeO
MeO
PAr$_2$
PAr$_2$
Ar =
tBu
OMe
tBu
(*S*)-**63**

(12.35)

controlled process. As an example, reaction of racemic allenyl alcohol **64** with a catalytic 1 : 2 mixture of [(*S*)-**63**]Au_2Cl_2 and AgOTs led to isolation of a homochiral 1 : 1 mixture of (*E*)-**65** (>95% *ee*) and (*Z*)-**65** (>95% *ee*) in 94% combined yield (Scheme 12.12). Hydrogenation of this mixture confirmed that (*E*)-**65** and (*Z*)-**65** possessed the same absolute configuration

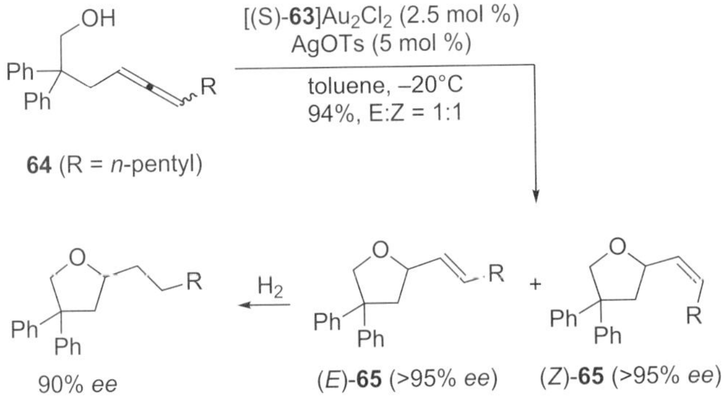

OH
Ph
Ph
R
64 (R = *n*-pentyl)

[(S)-**63**]Au_2Cl_2 (2.5 mol %)
AgOTs (5 mol %)
toluene, −20°C
94%, E:Z = 1:1

O
R
Ph
Ph
90% *ee*
H_2
(*E*)-**65** (>95% *ee*)
+
(*Z*)-**65** (>95% *ee*)

Scheme 12.12

Toste and coworkers have developed effective gold(I)-catalyzed protocols for the intramolecular enantioselective hydroalkoxylation of γ- and δ-hydroxy allenes employing chiral, enantiomerically pure silver salts [107]. For example, treatment of γ-hydroxy allene **66** with a catalytic 1 : 2 mixture of the achiral bis(gold) complex (dppm)Au_2Cl_2 [dppm = bis(diphenylphosphino)methane] and chiral silver phosphonate Ag-(*R*)-**67** in benzene at room temperature led to isolation of 2-alkenyl tetrahydrofuran **68** in 90% yield with 97% *ee* (Eq. (12.36)). A combination of chiral bis(gold) complex with a chiral silver salt proved effective for terminally unsubstituted allenyl alcohols. For example, reaction of 5,6-heptadienol catalyzed by a mixture of [(*S*,*S*)-DIPAMP]Au_2Cl_2 [DIPAMP = 1,2-ethanediylbis[(2-methoxyphenyl) phenylphosphine] and Ag-(*R*)-**67** gave 2-vinyltetrahydropyran **69** in 96% yield with 92% *ee* (Eq. (12.37)).

66 → 68: dppm(AuCl)$_2$ (2.5 mol %), Ag-(*R*)-**67** (5 mol %), benzene, 23°C, 1 h, 90%, 97% ee

(*R*)-**67** (R = 2,4,6-*i*-Pr$_3$-C$_6$H$_2$)

(12.36)

→ **69**: [(*S*,*S*)-DIPAMP]Au$_2$Cl$_2$ (2.5 mol %), Ag-(*R*)-**67** (5 mol %), benzene, 23 °C, 96%, 92% ee

(*S*,*S*)-DIPAMP

(12.37)

12.3.1.3 Intermolecular Processes

Zhang and Widenhoefer have reported a highly regio- and diastereo-selective method for the intermolecular hydroalkoxylation of allenes with alcohols [108]. As an example, reaction of 1-phenyl-1,2-butadiene with 2-phenyl-1-ethanol catalyzed by a 1 : 1 mixture of (IPr)AuCl [IPr = 1,3-bis(2,6-diisopropylphenyl)imidazol-2-ylidene] and AgOTf in toluene at room temperature led to isolation of (*E*)-(3-phenethoxy-1-butenyl)benzene in 96% yield as a single regio- and stereo-isomer (Eq. (12.38)). The protocol was effective for primary and secondary alcohols and for monosubstituted, 1,1- and 1,3-disubstituted, trisubstituted, and tetrasubstituted allenes. Transfer of chirality from the allenyl moiety to the newly formed tetrahedral stereocenter in the hydroalkoxylation of axially chiral 1,3-disubstituted allenes ranged from 0 to 81%, depending on the nature of the allene and the concentration of the alcohol.

(IPr)AuCl (5 mol %), AgOTf (5 mol %), toluene, rt, 1 h, 96%

IPr

(12.38)

Nishina and Yamamoto have also reported the gold(I)-catalyzed intermolecular hydroalkoxylation of allenes with alcohols [109]. As an example, treatment of a neat mixture of *p*-tolyl allene and isopropanol with a catalytic 1 : 1 mixture of $(PPh_3)AuCl$ and AgOTf at 30 °C led to isolation of isopropyl (*E*)-3-(4-tolyl)-2-propenyl ether in 98% yield (Eq. (12.39)). The protocol was most effective for monosubstituted and 1,3-disubstituted allenes and gave no transfer of chirality for the hydroalkoxylation of 1-phenyl-1,2-butadiene. Horino has reported the gold(I)-catalyzed intermolecular addition of alcohols to the allenyl moiety of 4-vinylidene-2-oxazolidinones [110].

$(PPh_3)AuCl$ (5 mol %)
AgOTf (5 mol %)
neat, 30 °C
98%

(12.39)

12.3.2 Ketones as Nucleophiles

In 2000, Hashmi reported the gold(III)-catalyzed cyclization of α-allenyl ketones to form substituted furans [15]. As an example, treatment of allenyl *p*-methoxybenzyl ketone **70** with a catalytic amount of $AuCl_3$ formed a 2 : 1 mixture of the cycloisomerization product **71** and the cycloisomerization/hydroarylation product **72** in 91% combined yield (Eq. (12.40)). Marshall had previously demonstrated the Ag(I)-catalyzed conversion of allenyl ketones to furans [111]. The Ag(I)-catalyzed transformation was four orders of magnitude slower than was the gold(III)-catalyzed transformation, but the silver-catalyzed process forms the 2-substituted furan as the exclusive product without the competing secondary hydroarylation. The selective conversion of α-allenyl ketones to substituted furans is also catalyzed by gold(III)-porphyrin complexes [112].

$AuCl_3$ (1 mol %)
CH_3CN

70 (R = CH_2-4-$MeOC_6H_4$) **71** (60%) **72** (31%)

(12.40)

Gevorgyan has reported the gold(III)-catalyzed cycloisomerization of haloallenyl ketones to form halofurans [113]. As an example, treatment of 3-bromo-1,2-propadienyl ketone **73** with a catalytic amount of $AuCl_3$ in toluene at 70 °C led to formation of 3-bromofuran **74** with 98% regioselectivity. Gevorgyan has also reported the gold(I)-catalyzed cycloisomerization of haloallenyl ketones to form halofurans that occurs with halide migration [113]. As an example of this protocol, treatment of **73** with a catalytic amount of $(PEt_3)AuCl$ in toluene at room temperature led to formation

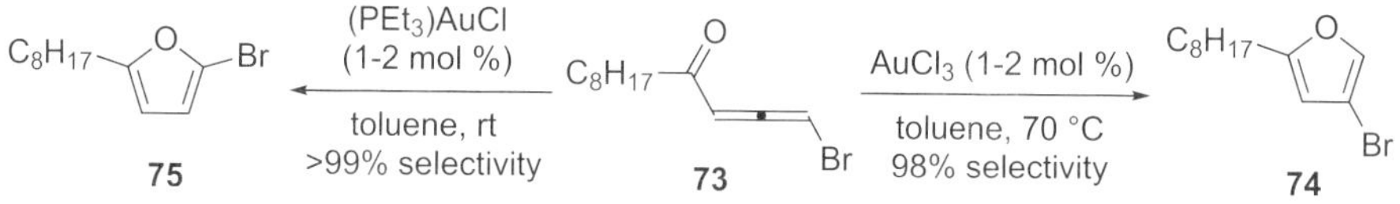

Scheme 12.13

of 2-bromofuran **75** with >99% regioselectivity (Scheme 12.13). Cationic gold(I) complexes catalyze the closely related cyclization/1,2-migration of 1,1-disubstituted allenyl ketones to form polysubstituted furan derivatives [114].

12.3.3 Carboxylic Acid Derivatives as Nucleophiles

Bäckvall has reported the gold(III)-catalyzed cycloisomerization of allene-substituted malonic esters to form β,γ-unsaturated δ-lactones [115]. For example, treatment of allenyl malonate derivative **76** with a catalytic 1 : 3 mixture of $AuCl_3$ and $AgSbF_6$ in acetic acid at 70 °C led to isolation of δ-lactone **77** in 99% yield (Eq. (12.41)). The transformation was restricted to substrates that contained a terminally disubstituted allenyl moiety and presumably occurs via nucleophilic attack of the carbonyl oxygen atom on a gold-complexed allene followed by acetate-mediated demethylation and protodeauration.

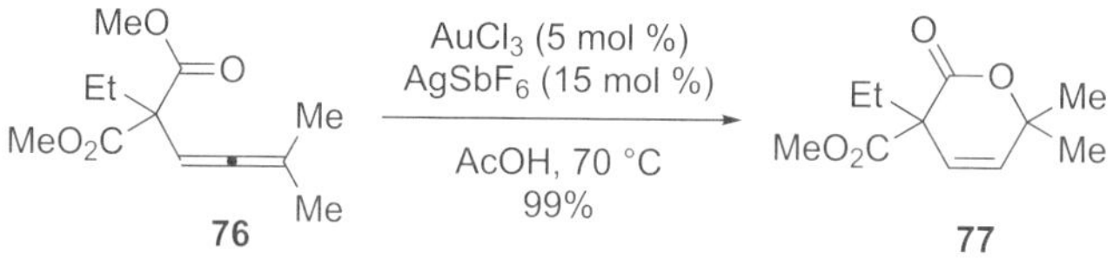

(12.41)

Shin has reported the gold(III)-catalyzed 5-*endo* cyclization of *tert*-butyl allenoates to form 2,4-disubstituted butenolides [116]. As an example, treatment of *tert*-butyl 2-benzyl-5-phenyl-2,3-pentadienoate with a catalytic amount of $AuCl_3$ in dichloromethane at room temperature for 1.5 h led to isolation of the butenolide **78** in 96% yield (Eq. (12.42)). The protocol tolerated a range of substitution at the allenyl carbon atoms.

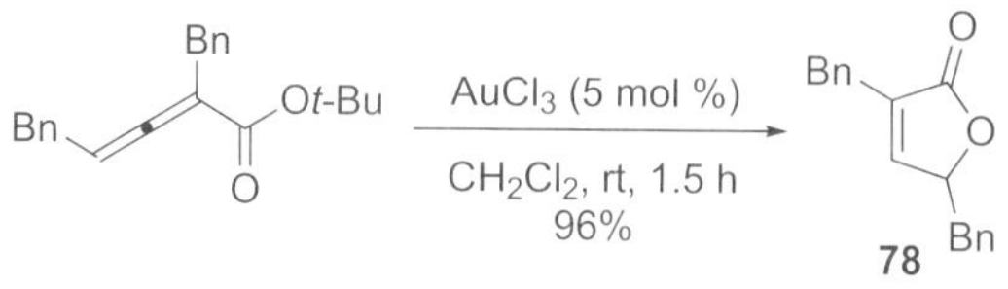

(12.42)

Gagosz has reported the stereoselective gold(I)-catalyzed isomerization of α-allenyl esters to form 1,3-butadien-2-ol esters [117]. As an example, treatment of α-allenyl benzoate ester **79** with a catalytic amount of the gold(I) triflamide complex (**80**) $AuNTf_2$ [**80** = 2-(dicyclohexylphosphino)-2′,4′,6′-triisopropyl-1,1′-biphenyl] in dichloromethane at room temperature for 9 h formed the diene **81** in quantitative yield as a single stereoisomer (Scheme 12.14). The transformation presumably occurs via attack of the pendant carbonyl oxygen atom on the center carbon of gold(I) allene complex **82** followed by ring-opening/deauration to form **81**. The transformation tolerated both alkyl and aryl substitution at the α-carbon atom of the allene and was effective for both secondary and tertiary carbinol esters.

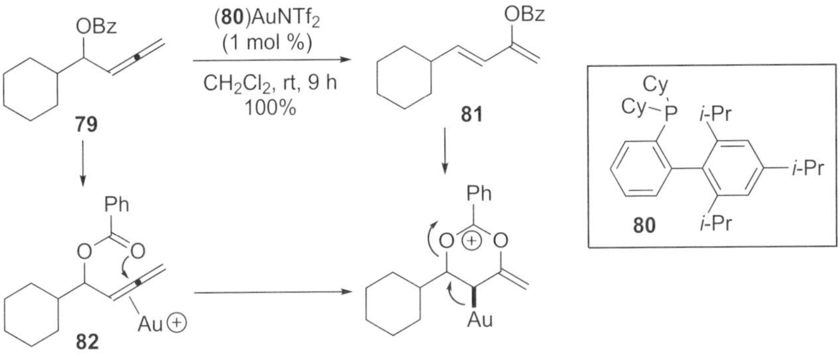

(**80**)$AuNTf_2$ (1 mol %)
CH_2Cl_2, rt, 9 h
100%

Scheme 12.14

Toste has demonstrated the hydroalkoxycarbonylation of β-allenyl carboxylic acid derivatives catalyzed by a mixture of a chiral, enantiomerically pure bis(gold) complex and a chiral, enantiomerically enriched silver phosphonate salt [107]. As an example, treatment of 6-methyl-4,5-heptadienoic acid with a catalytic mixture of [(*S*)-binap] $(AuCl)_2$ and Ag-(*R*)-**67** formed the γ-lactone **83** in 88% yield with 82% *ee* (Eq. (12.43)) [107].

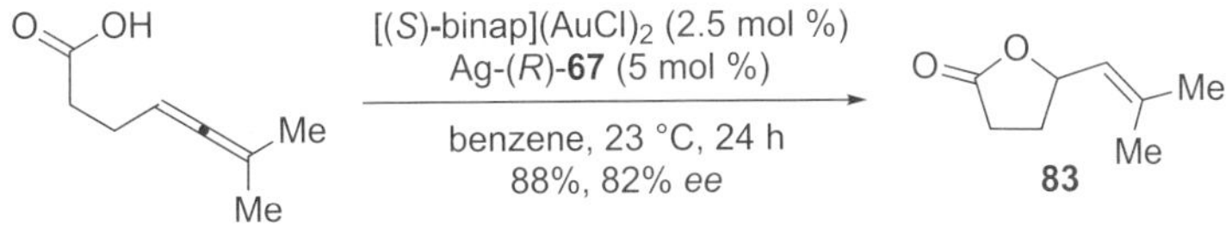

[(*S*)-binap]$(AuCl)_2$ (2.5 mol %)
Ag-(*R*)-**67** (5 mol %)
benzene, 23 °C, 24 h
88%, 82% *ee*

(12.43)

12.4 Addition to Alkenes

12.4.1 Alkenes and Dienes

He has reported a gold(I)-catalyzed protocol for the intermolecular hydroalkoxylation and hydroalkoxycarbonylation of alkenes with phenols and carboxylic acids, respectively [118]. For example, reaction of *p*-methoxyphenol with cyclohexene catalyzed by a 1 : 1 mixture of $(PPh_3)AuCl$ and AgOTf gave aryl cyclohexyl ether **84** in 78% isolated yield (Eq. (12.44)). In comparison, gold(I)-catalyzed hydroalkoxycarbonylation of 1-heptene with benzoic acid formed a 5.5 : 1 mixture of 2-heptyl and 3-heptyl benzoate in 75% combined yield, which suggests the intermediacy of carbonium ions. Corma has described an $AuCl_3$–$CuCl_2$-catalyzed protocol for the intermolecular hydroalkoxylation of vinyl arenes with alcohols at ≥120 °C [119].

MeO–C6H4–OH + cyclohexene → [(PPh3)AuCl (2 mol %), AgOTf (2 mol %), toluene, 85 °C, 78%] → MeO–C6H4–O–cyclohexyl **84** (12.44)

Importantly, He has demonstrated that the intermolecular hydroalkoxylation and hydroalkoxycarbonylation of alkenes and vinyl arenes is also catalyzed by AgOTf at 85 °C with rates and yields comparable to those obtained in the presence of catalytic amounts of $(PPh_3)AuCl$ and AgOTf [120]. Furthermore, both He [121] and Harwig [122] have shown that triflic acid catalyzes the intermolecular hydroalkoxylation and hydroalkoxycarbonylation of alkenes and vinyl arenes with rates and yields comparable to those obtained via $(PPh_3)AuCl$/AgOTf catalysis. These observations call into question the role of gold as the active catalytic species in the hydroalkoxylation and hydroalkoxycarbonylation of alkenes and vinyl arenes and point to triflic acid, generated under reaction conditions, as the active catalyst in these transformations.

Functionalized alkenes and dienes have also been employed as substrates for gold-catalyzed C—O bond formation. Aponick has disclosed a gold(I)-catalyzed procedure for the cyclization of monoallylic diols to form tetrahydropyrans in good yields with good diastereoselectivity (Eq. (12.45)) [123]. Nolan has reported the gold(I)-catalyzed rearrangement of allylic acetates [124]. Li has shown that cationic gold(III) complexes catalyze the annulation of phenols with dienes to form benzofuran derivatives [125]. As an example, reaction of 4-hydroxytoluene with 1,3-cyclohexadiene and a catalytic 1 : 3 mixture of $AuCl_3$ and AgOTf at 40 °C led to isolation of benzofuran **85** in 74% yield as a 4 : 1 mixture of diastereomers (Eq. (12.46)). The efficiency of the transformation was diminished by the presence of electron-withdrawing groups on the phenol moiety, but annulation of 2-naphthol with dienes led to regiospecific addition of the diene to the C1 carbon atom of the naphthol.

EtO_2C–CH2–CH(OH)–(CH2)3–CH=CH–CH2OH → [(PPh3)AuCl (1 mol %), AgOTf (1 mol %), CH_2Cl_2, −10 °C, 9 h, 99% (12:1)] → 2-(EtO_2CCH2)-6-vinyltetrahydropyran (12.45)

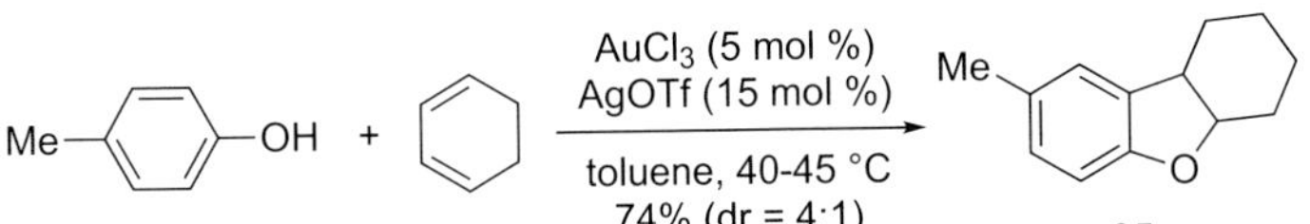

(12.46)

12.4.2
Cyclization/Nucleophile Capture of Enynes

Kozmin has reported the formation of polycyclic ethers via the gold(I) or gold(III)-catalyzed tandem enyne cyclization/alkoxylation of 1,5-enynes that contained a tethered hydroxy group [126]. As an example, treatment of homopropargylic alcohol **86** with a catalytic 1 : 1 mixture of $(PPh_3)AuCl$ and $AgClO_4$ at 20 °C led to isolation of 6-oxabicyclo[3.2.1]octene **87** in 89% yield (Scheme 12.15). Conversion of **86** to **87** presumably occurs via attack of the pendant alkene on a gold-complexed alkyne, which may proceed in concert with attack of the hydroxy group at the disubstituted alkenyl carbon atom. Protodeauration then releases heterocycle **87** (Scheme 12.15).

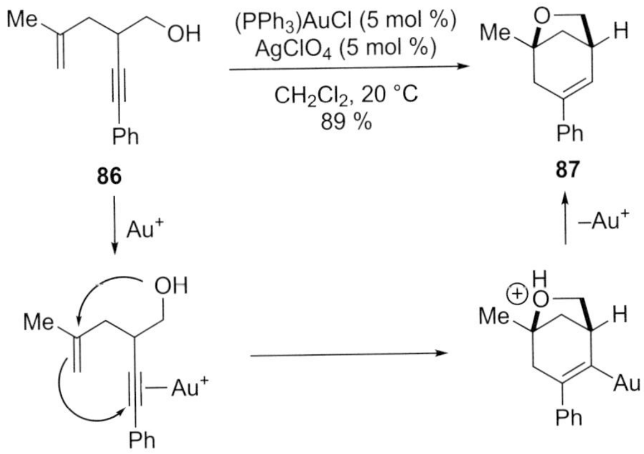

Scheme 12.15

Gagosz has described the gold(I)-catalyzed *endo*-cyclization/alkoxylation of 1,5-enynes in the presence of alcohol to form functionalized cyclopentenes [127]. As an example, reaction of allylic acetate **88** with a catalytic amount of the gold(I) complex (**80**)$AuNTf_2$ (1 mol%) in methanol at room temperature led to isolation of cyclopentene **89** in 95% yield with exclusive formation of the *trans*-4,5-isomer (Eq. (12.47)). The acetoxy group was not required for efficient cyclization, but the protocol was restricted to terminally disubstituted alkenes or conjugated dienes. Echavarren has reported a similar transformation involving the gold(I)-catalyzed cyclization/alkoxylation of 1,6 and 1,7-enynes (Eq. (12.48)) [128]. Genêt and Michelet have reported the gold(III)-catalyzed 5-*exo*-cyclization/alkoxylation of 1,6-enynes to form methylenecyclopentanes employing water, methanol, and other primary alcohols [129]. Echavarren has recently described a closely related protocol involving the

5-*exo*-cyclization/alkoxylation of 1,6-enynes catalyzed by cationic gold(I) complexes [130].

88 (Ar = 4-MeOC$_6$H$_4$) → (**80**)AuNTf$_2$ (1 mol %), MeOH, rt, 95 % → **89** (12.47)

(**80**)AuCl (5 mol %), AgSbF$_6$ (5 mol %), MeOH, rt, 144 h, 88 % (12.48)

Shin has described the gold(I)-catalyzed tandem cyclization/acetoxylation of the *tert*-butyl carbonate derivatives of 1-hexen-5-yn-3-ols to form 4-cyclohexene-1,2-diol derivatives [131]. As a representative example, treatment of **90** with a catalytic 1 : 1 mixture of [P(t-Bu)$_2$*o*-biphenyl]AuCl and AgSbF$_6$ led to isolation of cyclohexene **91** in 87% yield as an 8 : 1 mixture of diastereomers (Eq. (12.49)).

90 → [P(t-Bu)$_2$*o*-biphenyl]AuCl (2 mol %), AgSbF$_6$ (2 mol %), DCE, rt, 30 min, 87% (8:1) → **91** (12.49)

References

1 (a) Li, Z., Brouwer, C., and He, C. (2008) *Chem. Rev.*, **108**, 3239–3265; (b) Widenhoefer, R.A. (2008) *Chem.-Eur. J.*, **14**, 5382–5391; (c) Gorin, D.J., Sherry, B.D., and Toste, F.D. (2008) *Chem. Rev.*, **108**, 3351–3378; (d) Arcadi, A. (2008) *Chem. Rev.*, **108**, 3266–3325; (e) Jiménez-Núñez, E. and Echavarren, A.M. (2008) *Chem. Rev.*, **108**, 3326–3350; (f) Bongers, N. and Krause, N. (2008) *Angew. Chem. Int. Ed.*, **47**, 2178–2181.

2 Norman, R.O.C., Parr, W.J.E., and Thomas, C.B. (1976) *J. Chem. Soc. Perkin Trans. 1*, 1983–1987.

3 (a) Fukuda, Y. and Utimoto, K. (1991) *J. Org. Chem.*, **56**, 3729–3731; (b) Fukuda, Y. and Utimoto, K. (1991) *Bull. Chem. Soc. Jpn.*, **64**, 2013–2015.

4 Deetlefs, M., Raubenheimer, H.G., and Esterhuysen, M.W. (2002) *Catal. Today*, **72**, 29–41.

5 Casado, R., Contel, M., Laguna, M., Romero, P., and Sanz, S. (2003) *J. Am. Chem. Soc.*, **125**, 11925–11935.

6 Teles, J.H., Brode, S., and Chabanas, M. (1998) *Angew. Chem. Int. Ed.*, **37**, 1415–1418.

7 Mizushima, E., Sato, K., Hayashi, T., and Tanaka, M. (2002) *Angew. Chem. Int. Ed.*, **41**, 4563–4565.

8 Santos, L.L., Ruiz, V.R., Sabater, M.J., and Corma, A. (2008) *Tetrahedron*, **64**, 7902–7909.

9 Roembke, P., Schmidbaur, H., Cronje, S., and Raubenheimer, H. (2004) *J. Mol. Catal. A-Chem.*, **212**, 35–42.

10 Schneider, S.K., Herrmann, W.A., and Herdtweck, E. (2003) *Z. Anorg. Allg. Chem.*, **629**, 2363–2370.

11 Wessig, P. and Teubner, J. (2006) *Synlett*, 1543–1546.

12 Antoniotti, S., Genin, E., Michelet, V., and Genêt, J.-P. (2005) *J. Am. Chem. Soc.*, **127**, 9976–9977.

13 Liu, B. and De Brabander, J.K. (2006) *Org. Lett.*, **8**, 4907–4910.

14 Belting, V. and Krause, N. (2006) *Org. Lett.*, **8**, 4489–4492.

15 Hashmi, A.S.K., Schwarz, L., Choi, J.-H., and Frost, T.M. (2000) *Angew. Chem. Int. Ed.*, **39**, 2285–2288.

16 Liu, Y., Song, F., Song, Z., Liu, M., and Yan, B. (2005) *Org. Lett.*, **7**, 5409–5412.

17 Liu, Y., Song, F., and Guo, S. (2006) *J. Am. Chem. Soc.*, **128**, 11332–11333.

18 Zhang, Y., Xue, J., Xin, Z., Xie, Z., and Li, Y. (2008) *Synlett*, 940–944.

19 Schuler, M., Silva, F., Bobbio, C., Tessier, A., and Gouverneur, V. (2008) *Angew. Chem. Int. Ed.*, **47**, 7927–7930.

20 (a) Jung, H.H. and Floreancig, P.E. (2006) *Org. Lett.*, **8**, 1949–1951; (b) Jung, H.H. and Floreancig, P.E. (2007) *J. Org. Chem.*, **72**, 7359–7366.

21 Li, Y., Zhou, F., and Forsyth, C.J. (2007) *Angew. Chem. Int. Edit.*, **46**, 279–282.

22 Lee, S.I., Baek, J.Y., Sim, S.H., and Chung, Y.K. (2007) *Synthesis*, 2107–2114.

23 Skouta, R. and Li, C.-J. (2007) *Angew. Chem. Int. Ed.*, **46**, 1117–1119.

24 Barluenga, J., Diéguez, A., Fernández, A., Rodríguez, F., and Fañanás, F.J. (2006) *Angew. Chem. Int. Ed.*, **45**, 2091–2093.

25 Zhang, C., Cui, D.-M., Yao, L.-Y., Wang, B.-S., Hu, Y.-Z., and Hayashi, T. (2008) *J. Org. Chem.*, **73**, 7811–7813.

26 (a) Das, A., Chang, H.-K., Yang, C.-H., and Liu, R.-S. (2008) *Org. Lett.*, **10**, 4061–4064; (b) Lian, J.-J. and Liu, R.-S. (2007) *Chem. Commun.*, 1337–1339; (c) Tang, J.-M., Liu, T.-A., and Liu, R.-S. (2008) *J. Org. Chem.*, **73**, 8479–8483.

27 Liu, Y., Liu, M., Guo, S., Tu, H., Zhou, Y., and Gao, H. (2006) *Org. Lett.*, **8**, 3445–3448.

28 (a) Yao, T., Zhang, X., and Larock, R.C. (2004) *J. Am. Chem. Soc.*, **126**, 11164–11165; (b) Yao, T., Zhang, X., and Larock, R.C. (2005) *J. Org. Chem.*, **70**, 7679–7685.

29 Liu, X., Pan, Z., Shu, X., Duan, X., and Liang, Y. (2006) *Synlett*, 1962–1964.

30 Zhang, J. and Schmalz, H.-G. (2006) *Angew. Chem. Int. Ed.*, **45**, 6704–6706.

31 (a) Kirsch, S.F., Binder, J.T., Liébert, C., and Menz, H. (2006) *Angew. Chem. Int. Ed.*, **45**, 5878–5880; (b) Binder, J.T., Crone, B., Kirsch, S.F., Liébert, C., and Menz, H. (2007) *Eur. J. Org. Chem.*, 1636–1647.

32 Zhang, G., Huang, X., Li, G., and Zhang, L. (2008) *J. Am. Chem. Soc.*, **130**, 1814–1815.

33 Li, G., Huang, X., and Zhang, L. (2008) *J. Am. Chem. Soc.*, **130**, 6944–6945.

34 Crone, B. and Kirsch, S.F. (2007) *J. Org. Chem.*, **72**, 5435–5438.

35 (a) Asao, N., Takahashi, K., Lee, S., Kasahara, T., and Yamamoto, Y. (2002) *J. Am. Chem. Soc.*, **124**, 12650–12651; (b) Asao, N., Nogami, T., Lee, S., and Yamamoto, Y. (2003) *J. Am. Chem. Soc.*, **125**, 10921–10925.

36 (a) Hildebrandt, D. and Dyker, G. (2006) *J. Org. Chem.*, **71**, 6728–6733; (b)Asao, N., Sato, K., Menggenbateer, and Yamamoto, Y. (2005) *J. Org. Chem*, **70**, 3682–3685; (c) Sato, K., Asao, N., and Yamamoto, Y. (2005) *J. Org. Chem.*, **70**, 8977–8981; (d) Kim, N., Kim, Y., Park, W., Sung, D., Gupta, A.K., and Oh, C.H. (2005) *Org. Lett.*, **7**, 5289–5291.

37 (a) Asao, N., Aikawa, H., and Yamamoto, Y. (2004) *J. Am. Chem. Soc.*, **126**, 7458–7459; (b) Asao, N. and Aikawa, H. (2006) *J. Org. Chem.*, **71**, 5249–5253.

38 Asao, N. and Sato, K. (2006) *Org. Lett.*, **8**, 5361–5363.

39 Dyker, G., Hildebrandt, D., Liu, J., and Merz, K. (2003) *Angew. Chem. Int. Ed.*, **42**, 4399–4402.

40 Zhu, J., Germain, A.R., and Porco, J.A. (2004) *Angew. Chem. Int. Ed.*, **43**, 1239–1243.

41 (a) Yao, X. and Li, C.-J. (2006) *Org. Lett.*, **8**, 1953–1955; (b) Oh, C.H., Lee, S.J., Lee, J.H., and Na, Y.J. (2008) *Chem. Commun.*, 5794–5796.

42 Harkat, H., Weibel, J.-M., and Pale, P. (2006) *Tetrahedron Lett.*, **47**, 6273–6276.

43 (a) Genin, E., Toullec, P.Y., Marie, P., Antoniotti, S., Brancour, C., Genêt, J.-P., and Michelet, V. (2007) *ARKIVOC*, 67–78; (b) Genin, E., Toullec, P.Y., Antoniotti, S., Brancour, C., Genêt, J.-P., and Michelet, V. (2006) *J. Am. Chem. Soc.*, **128**, 3112–3113.

44 Marchal, E., Uriac, P., Legouin, B., Toupet, L., and van de Weghe, P. (2007) *Tetrahedron*, **63**, 9979–9990.

45 (a) Marion, N. and Nolan, S.P. (2007) *Angew. Chem. Int. Ed.*, **46**, 2750–2752; (b) Correa, A., Marion, N., Fensterbank, L., Malacria, M., Nolan, S.P., and Cavallo, L. (2008) *Angew. Chem. Int. Ed.*, **47**, 718–721.

46 Buzas, A., Istrate, F., and Gagosz, F. (2006) *Org. Lett.*, **8**, 1957–1959.

47 Yeom, H.-S., Yoon, S.-J., and Shin, S. (2007) *Tetrahedron Lett.*, **48**, 4817–4820.

48 Schwier, T., Sromek, A.W., Yap, D.M.L., Chernyak, D., and Gevorgyan, V. (2007) *J. Am. Chem. Soc.*, **129**, 9868–9878.

49 De Brabander, J.K., Liu, B., and Qian, M. (2008) *Org. Lett.*, **10**, 2533–2536.

50 Kato, K., Teraguchi, R., Kusakabe, T., Motodate, S., Yamamura, S., Mochida, T., and Akita, H. (2007) *Synlett*, 63–66.

51 Davies, P.W. and Albrecht, S.J.-C. (2008) *Chem. Commun.*, 238–240.

52 Shapiro, N.D. and Toste, F.D. (2008) *J. Am. Chem. Soc.*, **130**, 9244–9245.

53 Amijs, C.H.M., Lopez-Carrillo, V., and Echavarren, A.M. (2007) *Org. Lett.*, **9**, 4021–4024.

54 Buzas, A. and Gagosz, F. (2006) *J. Am. Chem. Soc.*, **128**, 12614–12615.

55 (a) Zhang, L. and Wang, S. (2006) *J. Am. Chem. Soc.*, **128**, 1442–1443; (b) Shi, F.-Q., Li, X., Zhang, L., and Yu, Z.-X. (2007) *J. Am. Chem. Soc.*, **129**, 15503–15512.

56 (a) Mamane, V., Gress, T., Krause, H., and Fürstner, A. (2004) *J. Am. Chem. Soc.*, **126**, 8654–8655; (b) Fürstner, A. and Hannen, P. (2006) *Chem.-Eur. J.*, **12**, 3006–3019.

57 Marion, N., de Frémont, P., Lemière, G., Stevens, E.D., Fensterbank, L., Malacria, M., and Nolan, S.P. (2006) *Chem. Commun.*, 2048–2050.

58 Boyer, F.-D., Goff, X.L., and Hanna, I. (2008) *J. Org. Chem.*, **73**, 5163–5166.

59 Gorin, D.J., Dub, P., and Toste, F.D. (2006) *J. Am. Chem. Soc.*, **128**, 14480–14481.

60 Shi, X., Gorin, D.J., and Toste, F.D. (2005) *J. Am. Chem. Soc.*, **127**, 5802–5803.

61 Zhang, L. (2005) *J. Am. Chem. Soc.*, **127**, 16804–16805.

62 Luo, T. and Schreiber, S.L. (2007) *Angew. Chem. Int. Ed.*, **46**, 8250–8253.

63 Oh, C.H., Kim, A., Park, W., Park, D.I., and Kim, N. (2006) *Synlett*, 2781–2784.

64 Oh, C.H. and Kim, A. (2007) *New J. Chem.*, **10**, 1719–1721.

65 Zhao, J., Hughes, C.O., and Toste, F.D. (2006) *J. Am. Chem. Soc.*, **128**, 7436–7437.

66 Li, G., Zhang, G., and Zhang, L. (2008) *J. Am. Chem. Soc.*, **130**, 3740–3741.

67 Wang, S. and Zhang, L. (2006) *Org. Lett.*, **8**, 4585–4587.

68 Wang, S. and Zhang, L. (2006) *J. Am. Chem. Soc.*, **128**, 8414–8415.

69 Yu, M., Li, G., Wang, S., and Zhang, L. (2007) *Adv. Synth. Catal.*, **349**, 871–875.

70 Yu, M., Zhang, G., and Zhang, L. (2007) *Org. Lett.*, **9**, 2147–2150.

71 Marion, N., Carlqvist, P., Gealageas, R., de Frémont, P., Maseras, F., and Nolan, S.P. (2007) *Chem.-Eur. J.*, **13**, 6437–6451.

72 Kang, J.-E. and Shin, S. (2006) *Synlett*, 717–720.

73 Buzas, A. and Gagosz, F. (2006) *Org. Lett.*, **8**, 515–518.

74 Asao, N., Aikawa, H., Tago, S., and Umetsu, K. (2007) *Org. Lett.*, **9**, 4299–4302.

75 Buzas, A. and Gagosz, F. (2006) *Synlett*, 2727–2730.

76 Robles-Machn, R., Adrio, J., and Carretero, J.C. (2006) *J. Org. Chem.*, **71**, 5023–5026.

77 Lee, E.-S., Yeom, H.-S., Hwang, J.-H., and Shin, S. (2007) *Eur. J. Org. Chem.*, 3503–3507.

78 Hashmi, A.S.K., Salathé, R., and Frey, W. (2007) *Synlett*, 1763–1766.

79 Hashmi, A.S.K., Weyrauch, J.P., Frey, W., and Bats, J.W. (2004) *Org. Lett.*, **6**, 4391–4394.

80 Milton, M.D., Inada, Y., Nishibayashi, Y., and Uemura, S. (2004) *Chem. Commun.*, 2712–2713.

81 Hashmi, A.S.K. and Sinha, P. (2004) *Adv. Synth. Catal.*, **346**, 432–438.

82 Hashmi, A.S.K., Bührle, M., Salathé, R., and Bats, J.W. (2008) *Adv. Synth. Catal.*, **350**, 2059–2064.

83 (a) Dai, L.-Z., Qi, M.-J., Shi, Y.-L., Liu, X.-G., and Shi, M. (2007) *Org. Lett.*, **9**, 3191–3194; (b) Dai, L.-Z. and Shi, M. (2008) *Chem.-Eur. J.*, **14**, 7011–7018.

84 Shu, X.-Z., Liu, X.-Y., Ji, K.-G., Xiao, H.-Q., and Liang, Y.-M. (2008) *Chem.-Eur. J.*, **14**, 5282–5289.

85 Ji, K.-G., Shen, Y.-W., Shu, X.-Z., Xiao, H.-Q., Bian, Y.-J., and Liang, Y.-M. (2008) *Adv. Synth. Catal.*, **350**, 1275–1280.

86 Dai, L.-Z. and Shi, M. (2008) *Tetrahedron Lett.*, **49**, 6437–6439.

87 Dubé, P. and Toste, F.D. (2006) *J. Am. Chem. Soc.*, **128**, 12062–12063.

88 Bae, H.J., Baskar, B., An, S.E., Cheong, J.Y., Thangadurai, D.T., Hwang, I.-C., and Rhee, Y.H. (2008) *Angew. Chem. Int. Ed.*, **47**, 2263–2266.

89 Baskar, B., Bae, H.J., An, S.E., Cheong, J.Y., Rhee, Y.H., Duschek, A., and Kirsch, S.F. (2008) *Org. Lett.*, **10**, 2605–2607.

90 Shapiro, N.D. and Toste, F.D. (2007) *J. Am. Chem. Soc.*, **129**, 4160–4161.

91 Li, G. and Zhang, L. (2007) *Angew. Chem. Int. Ed.*, **46**, 5156–5159.

92 Asao, N., Sato, K., and Yamamoto, Y. (2003) *Tetrahedron Lett.*, **44**, 5675–5677.

93 Yeom, H.-S., Lee, J.-E., and Shin, S. (2008) *Angew. Chem. Int. Ed.*, **47**, 7040–7043.

94 Hoffmann-Röder, A. and Krause, N. (2001) *Org. Lett.*, **3**, 2537–2538.

95 Kim, S. and Lee, P.H. (2008) *Adv. Synth. Catal.*, **350**, 547–551.

96 Aksun, Ö. and Krause, N. (2008) *Adv. Synth. Catal.*, **350**, 1106–1112.

97 Deutsch, C., Gockel, B., Hoffmann-Röder, A., and Krause, N. (2007) *Synlett*, 1790–1794.

98 Hashmi, A.S.K., Blanco, M.C., Fischer, D., and Bats, J.W. (2006) *Eur. J. Org. Chem.*, 1387–1389.

99 Hyland, C.J.T. and Hegedus, L.S. (2006) *J. Org. Chem.*, **71**, 8658–8660.

100 Brasholz, M. and Reissig, H.-U. (2007) *Synlett*, 1294–1298.

101 Volz, F. and Krause, N. (2007) *Org. Biomol. Chem.*, **5**, 1519–1521.

102 Erdsack, J. and Krause, N. (2007) *Synthesis*, 3741–3750.

103 Gockel, B. and Krause, N. (2006) *Org. Lett.*, **8**, 4485–4488.

104 Zhang, Z., Liu, C., Kinder, R.E., Han, X., Qian, H., and Widenhoefer, R.A. (2006) *J. Am. Chem. Soc.*, **128**, 9066–9073.

105 Alcaide, B., Almendros, P., and Martínez del Campo, T. (2007) *Angew. Chem. Int. Ed.*, **46**, 6684–6687.

106 Zhang, Z. and Widenhoefer, R.A. (2007) *Angew. Chem. Int. Ed.*, **46**, 283–285.

107 Hamilton, G.L., Kang, E.J., Mba, M., and Toste, F.D. (2007) *Science*, **317**, 496–499.

108 Zhang, Z. and Widenhoefer, R.A. (2008) *Org. Lett.*, **10**, 2079–2081.

109 Nishina, N. and Yamamoto, Y. (2008) *Tetrahedron Lett.*, **49**, 4908–4911.

110 Horino, Y., Takata, Y., Hashimoto, K., Kuroda, S., Kimura, M., and Tamaru, Y. (2008) *Org. Biomol. Chem.*, **6**, 4105–4107.

111 (a) Marshall, J.A. and Robinson, E.D. (1990) *J. Org. Chem.*, **55**, 3450–3451; (b) Marshall, J.A. and Bartley, G.S. (1994) *J. Org. Chem.*, **59**, 7169–7171.

112 Zhou, C.-Y., Chan, P.W.H., and Che, C.-M. (2006) *Org. Lett.*, **8**, 325–328.

113 (a) Sromek, A.W., Rubina, M., and Gevorgyan, V. (2005) *J. Am. Chem. Soc.*, **127**, 10500–10501; (b) Xia, Y., Dudnik, A.S., Gevorgyan, V., and Li, Y. (2008) *J. Am. Chem. Soc.*, **130**, 6940–6941.

114 Dudnik, A.S. and Gevorgyan, V. (2007) *Angew. Chem. Int. Ed.*, **46**, 5195–5197.

115 Piera, J., Krumlinde, P., Strubing, D., and Bäckvall, J.-E. (2007) *Org. Lett.*, **9**, 2235–2237.

116 Kang, J.-E., Lee, E.-S., Park, S.-I., and Shin, S. (2005) *Tetrahedron Lett.*, **46**, 7431–7433.

117 Buzas, A.K., Istrate, F.M., and Gagosz, F. (2007) *Org. Lett.*, **9**, 985–988.

118 Yang, C.-G. and He, C. (2005) *J. Am. Chem. Soc.*, **127**, 6966–6967.

119 Zhang, X. and Corma, A. (2007) *Chem. Commun.*, 3080–3082.

120 Yang, C.-G., Reich, N.W., Shi, Z., and He, C. (2005) *Org. Lett.*, **7**, 4553–4556.

121 Li, Z., Zhang, J., Brouwer, C., Yang, C.-G., Reich, N.W., and He, C. (2006) *Org. Lett.*, **8**, 4175–4178.

122 Rosenfeld, D.C., Shekhar, S., Takemiya, A., Utsunomiya, M., and Hartwig, J.F. (2006) *Org. Lett.*, **8**, 4179–4182.

123 (a) Aponick, A., Li, C.-Y., and Biannic, B. (2008) *Org. Lett.*, **10**, 669–671; (b) Aponick, A. and Biannic, B. (2008) *Synthesis*, 3356–3359.

124 Marion, N., Gealageas, R., and Nolan, S.P. (2007) *Org. Lett.*, **9**, 2653–2656.

125 Nguyen, R.-V., Yao, X., and Li, C.-J. (2006) *Org. Lett.*, **8**, 2397–2399.

126 Zhang, L. and Kozmin, S.A. (2005) *J. Am. Chem. Soc.*, **127**, 6962–6963.

127 Buzas, A.K., Istrate, F.M., and Gagosz, F. (2007) *Angew. Chem. Int. Ed.*, **46**, 1141–1144.

128 (a) Nieto-Oberhuber, C., Muñoz, M.P., López, S., Jiménez-Núñez, E., Nevado, C., Herrero-Gómez, E., Raducan, M., and Echavarren, A.M. (2006) *Chem.-Eur. J.*, **12**, 1677–1693; (b) Cabello, N., Rodríguez, C., and Echavarren, A.M. (2007) *Synlett*, 1753–1758.

129 Genin, E., Leseurre, L., Toullec, P.Y., Genêt, J.-P., and Michelet, V. (2007) *Synlett*, 1780–1784.

130 Jiménez-Núñez, E., Claverie, C.K., Bour, C., Cárdenas, D.J., and Echavarren, A.M. (2008) *Angew. Chem. Int. Ed.*, **47**, 7892–7895.

131 Lim, C., Kang, J.-E., Lee, J.-E., and Shin, S. (2007) *Org. Lett.*, **9**, 3539–3542.

Index

Catalyzed Carbon-Heteroatom Bond Formation. Edited by Andrei K. Yudin

ISBN: 978-3-527-32428-6

d

q

r